Methods in Cell Biology

VOLUME 96

Electron Microscopy of Model Systems

Series Editors

Leslie Wilson
Department of Molecular, Cellular and Developmental Biology
University of California
Santa Barbara, California

Paul Matsudaira
Department of Biological Sciences
National University of Singapore
Singapore

Methods in Cell Biology

VOLUME 96

Electron Microscopy of Model Systems

Edited by

Thomas Müller-Reichert

Medical Theoretical Center

TU Dresden

Germany

ELSEVIER

AMSTERDAM • BOSTON • HEIDELBERG • LONDON
NEW YORK • OXFORD • PARIS • SAN DIEGO
SAN FRANCISCO • SINGAPORE • SYDNEY • TOKYO
Academic Press is an imprint of Elsevier

Academic Press is an imprint of Elsevier
30 Corporate Drive, Suite 400, Burlington, MA 01803, USA
525 B Street, Suite 1900, San Diego, CA 92101-4495, USA
32, Jamestown Road, London NW1 7BY, UK
Linacre House, Jordan Hill, Oxford OX2 8DP, UK

First edition 2010

ISBN–978-0-12-381007-6
ISSN: 0091-679X

For information on all Academic Press publications
visit our website at elsevierdirect.com

Printed and bound by CPI Group (UK) Ltd, Croydon, CR0 4YY

Transferred to digital print 2012

CONTENTS

CONTRIBUTORS

Numbers in parentheses indicate the pages on which the authors' contributions begin.

Johannes G. Achatz, (307) Center for Molecular Biosciences, Institute of Zoology, University of Innsbruck, Innsbruck, Austria

Richard D. Allen, (143) Pacific Biosciences Research Center, Electron Microscope Laboratory, University of Hawaii at Manoa, Honolulu, Hawaii 96822

Claude Antony, (235) Cell Biology and Biophysics Program, European Molecular Biology Laboratories, Heidelberg 69117, Germany

Beate Beer, (649) Department of Therapeutic Radiology and Oncology, Innsbruck Medical University, Innsbruck A-6020, Austria; Institute of Legal Medicine, Innsbruck Medical University, Innsbruck A-6020, Austria

Annett Bellack, (47) Department of Microbiology and Archaea Centre, University of Regensburg, D-93053 Regensburg, Germany

Jürgen Berger, (395) Max-Planck Institute of Developmental Biology, D-72076 Tübingen, Germany

Cédric Bouchet-Marquis, (565) The Boulder Laboratory for 3-D Microscopy of Cells, University of Colorado at Boulder, Boulder, Colorado 80309-0347

Edward Brown, (619) Department of Biochemistry, School of Medical Sciences, University of Bristol, University Walk, Bristol, BS8 1TD, United Kingdom

Tillmann Burghardt, (47) Centre for Electron Microscopy, University of Regensburg, D-93053 Regensburg, Germany

Christopher Buser, (217, 671) Department of Molecular and Cell Biology, University of California, Berkeley, California 94720

Robert Cerny, (395) Department of Zoology, Charles University Prague, CZ-128 44 Prague, Czech Republic

Benjamin Cooper, (475) Department of Molecular Neurobiology, Max-Planck-Institute of Experimental Medicine, Göttingen D-37075, Germany

Peter J. Cullen, (619) Department of Biochemistry, School of Medical Sciences, University of Bristol, University Walk, Bristol, BS8 1TD, United Kingdom

Hannes L. Ebner, (649) Division of Histology and Embryology, Innsbruck Medical University, Innsbruck A-6020, Austria

Bernhard Egger, (307) Center for Molecular Biosciences, Institute of Zoology, University of Innsbruck, Innsbruck, Austria

Hans H. Epperlein, (395) Department of Anatomy, TU Dresden, D-01307 Dresden, Germany

Giada Frascaroli, (603) Institute of Virology, University Hospital Ulm, Ulm D-89081, Germany

Daniel W. Gerlich, (591) Institute of Biochemistry, Swiss Federal Institute of Technology Zurich (ETHZ), CH-8093 Zurich, Switzerland

Thomas H. Giddings, (117) Jr., Department of Molecular, Cellular and Developmental Biology, University of Colorado, Boulder, Colorado 80309

Eva Gluenz, (175) Sir William Dunn School of Pathology, University of Oxford, Oxford OX1 3RE, United Kingdom

Kenneth N. Goldie, (93) Centre for Cellular Imaging and Nano Analytics (C-CINA), Structural Biology and Biophysics Core Biozentrum, University Basel, CH-4058 Basel, Switzerland

Werner Graber, (513) Institute of Anatomy, University of Berne, CH-3000 Bern 9, Switzerland

Ralph Gräf, (197) Department of Cell Biology, Institute for Biochemistry and Biology, University of Potsdam, D-14476 Potsdam-Golm, Germany

Julien Guizetti, (591) Institute of Biochemistry, Swiss Federal Institute of Technology Zurich (ETHZ), CH-8093 Zurich, Switzerland

Keith Gull, (175) Sir William Dunn School of Pathology, University of Oxford, Oxford OX1 3RE, United Kingdom

Sonja Gürster, (47) Centre for Electron Microscopy, University of Regensburg, D-93053 Regensburg, Germany

Eric Hanssen, (93) Electron Microscopy Unit, Bio21 Molecular Science and Biotechnology Institute, University of Melbourne, Parkville, VIC 3010, Australia

Klaus Hausmann, (143) Institute of Biology/Zoology, Laboratory for Protozoology, Free University of Berlin, 14195 Berlin, Germany

Thomas Heimerl, (47) Centre for Electron Microscopy, University of Regensburg, D-93053 Regensburg, Germany

Daniel Hekl, (649) Department of Therapeutic Radiology and Oncology, Innsbruck Medical University, Innsbruck A-6020, Austria

Michael W. Hess, (285, 307, 649) Division of Histology and Embryology, Innsbruck Medical University, A-6020 Innsbruck, Austria

Andreas Hoenger, (565) The Boulder Laboratory for 3-D Microscopy of Cells, University of Colorado at Boulder, Boulder, Colorado 80309-0347

Thomas W. Holstein, (285) Institute of Zoology, Heidelberg University, D-69120 Heidelberg, Germany

Johanna L. Höög, (175) Sir William Dunn School of Pathology, University of Oxford, Oxford OX1 3RE, United Kingdom; The Boulder Laboratory for 3D Electron Microscopy of Cells, Department of MCD Biology, University of Colorado, Boulder, Colorado 80309

Harald Huber, (47) Department of Microbiology and Archaea Centre, University of Regensburg, D-93053 Regensburg, Germany

Cordelia Imig, (475) Department of Molecular Neurobiology, Max-Planck-Institute of Experimental Medicine, Göttingen D-37075, Germany

Sonja Jacob, (529) Institute of Molecular Biotechnology of the Austrian Academy of Sciences, 1030 Vienna, Austria

Grant J. Jensen, (21) Division of Biology, California Institute of Technology, Pasadena, California 91125; Division of Biology, Howard Hughes Medical Institute, Pasadena, California 91125

Byung-Ho Kang, (259) Microbiology and Cell Science Department, Electron Microscopy and Bioimaging Lab, Interdisciplinary Center for Biotechnology Research, University of Florida, Gainesville, Florida 32611

Walter A. Kaufmann, (475) Department of Pharmacology, Innsbruck Medical University, Innsbruck A-6020, Austria

Douglas R. Keene, (443) Shriners Hospitals for Children, Micro-Imaging Center, Portland, Oregon 97239

Thomas A. Keil, (363) Department Molecular Structural Biology, Max-Planck-Institute for Biochemistry, D-82152 Martinsried, Germany

Robert Kirmse, (565) The Boulder Laboratory for 3-D Microscopy of Cells, University of Colorado at Boulder, Boulder, Colorado 80309-0347

Andreas Klingl, (47) Centre for Electron Microscopy, University of Regensburg, D-93053 Regensburg, Germany

Michael P. Koonce, (197) Division of Translational Medicine, Wadsworth Center, Albany, New York 12201-0509

Susanne Kretschmar, (395) Center for Regenerative Therapies, TU Dresden, D-01307 Dresden, Germany

Ulf Küper, (47) Department of Microbiology and Archaea Centre, University of Regensburg, D-93053 Regensburg, Germany

Thomas Kurth, (395) Center for Regenerative Therapies, TU Dresden, D-01307 Dresden, Germany

Mark S. Ladinsky, (21) Division of Biology, California Institute of Technology, Pasadena, California 91125

Peter Ladurner, (307) Center for Molecular Biosciences, Institute of Zoology, University of Innsbruck, Innsbruck, Austria

Michael Laue, (1) Electron Microscopy Centre, Medical Faculty, University of Rostock, D-18057 Rostock, Germany

Ben Lich, (331) FEI Company, 5651 Eindhoven, The Netherlands

Sandra Ließem, (603) Institute of Pathology, University Hospital Hannover, Hannover D-30625, Germany

Jan Löfberg, (395) Section of Animal Development and Genetics, Uppsala University, S-752 36 Uppsala, Sweden

Joel Mancuso, (331) Gatan, Inc., Pleasanton, California 94588

Judith Mantell, (619) Department of Biochemistry, School of Medical Sciences, University of Bristol, University Walk, Bristol, BS8 1TD, United Kingdom; Wolfson Bioimaging Facility, School of Medical Sciences, University Walk, Bristol, BS8 1TD, United Kingdom

Jana Mäntler, (591) Max Planck Institute of Molecular Cell Biology and Genetics (MPI-CBG), 01307 Dresden, Germany

Kent McDonald, (331, 671) Electron Microscope Laboratory, University of California, Berkeley, California 94720

Alasdair W. McDowall, (21) Division of Biology, California Institute of Technology, Pasadena, California 91125

Janet B. Meehl, (117) Department of Molecular, Cellular and Developmental Biology, University of Colorado, Boulder, Colorado 80309

Carolin Meyer, (47) Centre for Electron Microscopy, University of Regensburg, D-93053 Regensburg, Germany

Wiebke Möbius, (475) Department of Neurogenetics, Max-Planck-Institute of Experimental Medicine, Göttingen D-37075, Germany

Mary Morphew, (671) Boulder Laboratory for 3D Electron Microscopy of Cells, University of Colorado, Boulder, Colorado 80309

Thomas Müller-Reichert, (331, 591, 671) Medical Theoretical Center, TU Dresden, 01307 Dresden, Germany

Susan J. Nixon, (425) Institute for Molecular Bioscience, The University of Queensland, Brisbane, 4072 Queensland, Australia

Viola M. J. Oorschot, (425) Department of Cell Biology, Cell Microscopy Center, University Medical Center Utrecht, 3584 CX Utrecht, The Netherlands

Eileen T. O'Toole, (71) Department of Molecular, Cellular, and Developmental Biology, Boulder Laboratory for 3-D Electron Microscopy of Cells, University of Colorado, Boulder, Colorado 80309-0347

Cynthia Page, (565) The Boulder Laboratory for 3-D Microscopy of Cells, University of Colorado at Boulder, Boulder, Colorado 80309-0347

Robert G. Parton, (425) Centre for Microscopy and Microanalysis, The University of Queensland, Brisbane, 4072 Queensland, Australia; Institute for Molecular Bioscience, The University of Queensland, Brisbane, 4072 Queensland, Australia

Chad G. Pearson, (117) Department of Molecular, Cellular and Developmental Biology, University of Colorado, Boulder, Colorado 80309

Kristian Pfaller, (649) Division of Histology and Embryology, Innsbruck Medical University, Innsbruck A-6020, Austria

Thomas Piendl, (395) NET-Network for Educational Technology, ETH Zürich, CH-8092 Zürich, Switzerland

Martin Pilhofer, (21) Division of Biology, California Institute of Technology, Pasadena, California 91125; Division of Biology, Howard Hughes Medical Institute, Pasadena, California 91125

Reinhard Rachel, (47) Centre for Electron Microscopy, University of Regensburg, D-93053 Regensburg, Germany

Guenter P. Resch, (529) IMP-IMBA-GMI Electron Microscopy Facility, Institute of Molecular Biotechnology of the Austrian Academy of Sciences, 1030 Vienna, Austria

Hélio Roque, (235) Cell Biology and Biophysics Program, European Molecular Biology Laboratories, Heidelberg 69117, Germany

Torben Ruhwedel, (475) Department of Neurogenetics, Max-Planck-Institute of Experimental Medicine, Göttingen D-37075, Germany

Aiman S. Saab, (475) Department of Neurogenetics, Max-Planck-Institute of Experimental Medicine, Göttingen D-37075, Germany

Willi Salvenmoser, (285, 307) Center for Molecular Biosciences, Institute of Zoology, University of Innsbruck, A-6020 Innsbruck, Austria

Nicole L. Schieber, (425) Centre for Microscopy and Microanalysis, The University of Queensland, Brisbane, 4072 Queensland, Australia; Institute for Molecular Bioscience, The University of Queensland, Brisbane, 4072 Queensland, Australia

Simone Schopf, (47) Department of Microbiology and Archaea Centre, University of Regensburg, D-93053 Regensburg, Germany

Heinz Schwarz, (395, 671) Electron Microscopy Unit, Max Planck Institute of Developmental Biology, D-72076 Tübingen, Germany

Thomas Seppi, (649) Department of Therapeutic Radiology and Oncology, Innsbruck Medical University, Innsbruck A-6020, Austria

Thomas H. Sharp, (619) Department of Biochemistry, School of Medical Sciences, University of Bristol, University Walk, Bristol, BS8 1TD, United Kingdom; School of Chemistry, University of Bristol, Cantock's Close, Bristol, BS8 1TS, United Kingdom

Nicolas Snaidero, (475) Department of Neurogenetics, Max-Planck-Institute of Experimental Medicine, Göttingen D-37075, Germany

R. Alexander Steinbrecht, (363) Max Planck Institute for Ornithology, D-82319 Seewiesen, Germany

Daniel Studer, (513) Institute of Anatomy, University of Berne, CH-3000 Bern 9, Switzerland

Leann Tilley, (93) Department of Biochemistry and Center of Excellence for Coherent X-ray Science, La Trobe University, Melbourne, VIC 3086, Australia

Sara F. Tufa, (443) Shriners Hospitals for Children, Micro-Imaging Center, Portland, Oregon 97239

Edit Urban, (529) Institute of Molecular Biotechnology of the Austrian Academy of Sciences, 1030 Vienna, Austria

Jan R.T. van Weering, (619) Department of Biochemistry, School of Medical Sciences, University of Bristol, University Walk, Bristol, BS8 1TD, United Kingdom

Dimitri Vanhecke, (513) Institute of Anatomy, University of Berne, CH-3000 Bern 9, Switzerland

Frédérique Varoqueaux, (475) Department of Molecular Neurobiology, Max-Planck-Institute of Experimental Medicine, Göttingen D-37075, Germany

Sue Vaughan, (175) Sir William Dunn School of Pathology, University of Oxford, Oxford OX1 3RE, United Kingdom; Oxford Brookes University, Headington Hill site, Headington, Oxford OX3 0BP, United Kingdom

Paul Verkade, (619) Department of Biochemistry, School of Medical Sciences, University of Bristol, University Walk, Bristol, BS8 1TD, United Kingdom; Wolfson Bioimaging Facility, School of Medical Sciences, University Walk, Bristol, BS8 1TD, United Kingdom; Department of Physiology and Pharmacology, School of Medical Sciences, University Walk, Bristol, BS8 1TD, United Kingdom

Paul Walther, (603) Central Electron Microscopy Facility, Ulm University, Ulm D-89081, Germany

Li Wang, (603) Institute of Virology, University Hospital Ulm, Ulm D-89081, Germany

Gerhard Wanner, (47) Biocentre, University of Munich, D-82152 Planegg-Martinsried, Germany

Nadine Wasserburger, (47) Centre for Electron Microscopy, University of Regensburg, D-93053 Regensburg, Germany

Richard I. Webb, (425) Centre for Microscopy and Microanalysis, The University of Queensland, Brisbane, 4072 Queensland, Australia

Rick Webb, (671) Centre for Microscopy and Microanalysis, University of Queensland, 4072 Queensland, Australia

Michaela Wilsch-Bräuninger, (395) Max-Planck Institute of Molecular Cell Biology and Genetics, D-01307 Dresden, Germany

Mark Winey, (117) Department of Molecular, Cellular and Developmental Biology, University of Colorado, Boulder, Colorado 80309

Reinhard Wirth, (47) Department of Microbiology and Archaea Centre, University of Regensburg, D-93053 Regensburg, Germany

PREFACE

The familiar leaves us tranquil, but the unexpected makes us productive.

J. W. v. Geothe

Recently, two volumes of "Methods in Cell Biology" were devoted to biological electron microscopy. "Cellular Electron Microscopy" (Volume 79, edited by J. Richard McIntosh) covers specimen preparation, 3-D imaging, labeling of macromolecules, and technologies for data acquisition and analysis. "Introduction to Electron Microscopy for Biologists" (Volume 88, edited by Terry D. Allen) intends to give information to scientists who may be considering electron microscopy as a tool to extend molecular, biochemical, or light microscopical findings to the ultrastructural level. The reader might ask: Why is there a third MCB volume about electron microscopy within a rather short period of time? The reason is simple: "Electron Microscopy of Model Systems" approaches the subject from a different "angle". The primary goal of this volume is not to give an in-depth, systematic description, or introduction to techniques related to biological electron microscopy. In this book the methods are not the main focus. Here, the intent is to cover EM methods in the context of specific specimen preparation requirements for a particular model organism or system. Specific requirements differ a lot from model system to model system, and this is the challenge. "Electron Microscopy of Model Systems" is the first compendium covering the various aspects of sample preparation for very diverse biological systems. Comparing the different model systems, it is the differences in specimen preparation that make us "productive".

This book covers the preparation of unicellular organisms, invertebrate and vertebrate model systems, tissue samples, and cultured cells for electron microscopy. The list includes the most "traditional" or popular systems, such as budding (Buser) or fission yeast (Roque and Antony), the roundworm *Caenorhabditis elegans* (Müller-Reichert *et al.*), the fruitfly *Drosophila* (and other insects, Keil and Steinbrecht), the zebrafish (Schieber *et al.*), and the plant *Arabidopsis* (Kang). This volume, however, also covers the preparation of single cells and organisms that are less frequently used, such as archaea (Rachel *et al.*), *Chlamydomonas* (O'Toole), *Tetrahymena* (Giddings *et al.*), *Paramecium* (Hausmann and Allen), *Dictyostelium* (Koonce and Gräf), *Hydra* (Holstein *et al.*), flatworms (Salvenmoser *et al.*), and the Axolotl (Kurth *et al.*). Model systems of medical importance are also included, such as viruses (Laue), *Trypanosoma* (Höög and Gull), and *Plasmodium* (Hanssen *et al.*).

A set of chapters is devoted to the preparation of specific tissues such as cartilage and bone (Keene and Tufa), mouse (Moebius *et al.*), and rat tissues (Vanhecke *et al.*). Two chapters deal with the specifics of sample preparation for the cytoskeleton.

The different fixation requirements for actin (Resch *et al.*) and intermediate filaments (Kirmse *et al.*) nicely illustrate that the biological question "dictates" the selection of a specific method out of a repertoire of several techniques. Some systems are also included here because they are used for exemplary cell biology studies on abscission (Guizetti *et al.*), viral infection (Walther *et al.*), and intracellular transport (van Weering *et al.*). Last but not least, the advantages of 2-D versus 3-D cell culture (Hess *et al.*) and "Tips and Tricks" for high-pressure freezing are presented (McDonald *et al.*).

Finally, this book is also a reflection of an ongoing discussion in the field of biological electron microscopy. What is the best method of fixation? Some authors argue for the exclusive application of cryopreparation and imaging, while others emphasize the need for initial chemical fixation, or the requirement for inactivation of infectious material. Again, a decision for any of the methods presented here strongly depends on the biological question asked, the size of the biological system, and the practicality of the approach. The list of model systems presented here is by no means complete, but it is hoped that the models and techniques that are represented will help the reader to find appropriate methods for the preparation of her/his favorite system for electron microscopy.

ACKNOWLEDGMENTS

I would like to thank all the contributors to "Electron Microscopy of Model Systems" for their enthusiasm in putting this volume together. Special thanks go to Kent McDonald (Berkeley) for his recommendations on chapter authors and topics, and for many stimulating discussions over the years. I am grateful to Tony Hyman and Ivan Baines for continued support and to my colleagues of the EM facility at MPI-CBG (Dresden). Finally, I wish to thank Judith Nicholls for proofreading, and Tara Hoey and Zoe Kruse of Elsevier for their professional help in bringing this project to completion.

Dresden, April 2010
Thomas Müller-Reichert

CHAPTER 1

Electron Microscopy of Viruses

Michael Laue

Electron Microscopy Centre, Medical Faculty, University of Rostock, D-18057 Rostock, Germany

Abstract

Electron microscopy is widely used in virology because viruses are generally too small for a direct inspection by light microscopy. Analysis of virus morphology is necessary in many circumstances, e.g., for the diagnosis of a virus in particular clinical situations or the analysis of virus entry and assembly. Moreover, quality control of virus particle integrity is required if a virus is propagated in cell culture, particularly if the virus genome has changed.

978-0-12-381007-6
DOI: 10.1016/S0091-679X(10)96001-9

In most cases already the basic methodology for transmission electron microscopy, i.e., negative staining and ultrathin sectioning, is sufficient to give relevant information on virus ultrastructure. This chapter gives detailed information on the principles of these basic methodologies and provides simple but reliable protocols for a quick start. Moreover, the description of standard protocols for negative staining and ultrathin sectioning are supplemented by protocols on immuno-negative staining and rapid ultrathin sectioning. Finally, principles of methods for an extended ultrastructural research using more elaborate techniques, such as cryotechniques or methods to reveal the three-dimensional virus architecture, are briefly reviewed.

I. Introduction

Each virus is unique, because of its host-specific evolution. Therefore, no virus model of general relevance exists, but several models that represent the general features of particular virus groups are available (e.g., vaccinia virus for poxviruses; cytomegalovirus for herpesviruses). The preparation methods for studying different viruses, however, are more or less the same. Two major fields of interest can be distinguished in virology: basic research and diagnostics.

In basic research, virus structure and virus/host interactions, such as virus entry, propagation, and egress, are subjects of interest. For an initial characterization of a new virus or virus strain, usually routine methods, such as negative staining and ultrathin sectioning, are used as the first line before more sophisticated methods, such as single particle analysis, electron tomography, and cryopreparation methods, are employed (see Section V).

Diagnostic electron microscopy of viruses is used in plant and veterinary pathology and in particular infectious diseases of humans, often after other methods have failed or if an independent control is necessary (Biel and Gelderblom, 1999a). It provides an open view on the sample without the need for specific probes, like antibodies or nucleic acids, and helps in the search for the infectious pathogen in host organisms or in assessing the risk of a potential bioterrorist attack. For obvious reasons, simple and quick preparation methods are preferred.

In virology, biosafety is an important issue. Many virus strains can only be cultivated in particular containments at a higher biosafety class. They must be inactivated, usually by fixation with aldehydes, before they can be transferred to lower biosafety levels where preparation for electron microscopy can be performed. Thus, in those cases, resolution and quality of the preparation are restricted to chemically based preparation methods.

This chapter concentrates on the basic methodologies, i.e., negative staining and thin section electron microscopy, for studying virus structure and cell biology, because these methods are of pivotal importance in virology and are generally used at the very beginning of a research project. Advanced preparation methods, however, are listed and references for further reading are provided.

II. Rationale

Negative-staining electron microscopy is easy to perform and may give a result within a few minutes. It is therefore the most productive approach in electron microscopy in terms of sample numbers. Particles of a suspension are adsorbed onto the surface of a specimen support, stabilized, and contrasted usually by heavy metal stains. By this approach, particles can be visualized down to subnanometer size and categorized based on their morphology. The original term "negative staining" was coined by Brenner and Horne (1959). However, even the early users of electron microscopy have adsorbed small particles (e.g., poxvirus) on a support and stained them with osmium tetroxide or other chemicals (e.g., von Borries *et al.*, 1938).

Because of the ease of use and the comparatively high throughput, negative staining is frequently used for quality assurance, e.g., for testing virus cultures. Many samples can be easily transferred into a suspension without destroying the structure of the virus particles (e.g., by freezing and thawing cycles or by grinding in a potter). Efficiency regarding preparation speed is also important in diagnostic electron microscopy. Therefore, negative-staining electron microscopy is a front-line method in this field (Curry *et al.*, 2006). Moreover, the open view of electron microscopy provides direct information on all nanoparticles present in a sample. Virus particles are identified by morphological parameters, such as size, shape, surface structure, and peculiarities (e.g., appendages). The diagnosis often only leads to a systematic group rather than to a specific virus. However, since morphology of a virus is rather stable during evolution, a diagnosis is still possible, even if nucleic acids have been considerably changed by mutation, rendering identification by other methods more difficult. Therefore, diagnostic electron microscopy is valuable for the identification of viruses in emerging infectious diseases or in cases of presumed bioterrorism (Biel and Gelderblom, 1999a; Miller, 2003). In veterinary and plant pathology, diagnostic electron microscopy plays an even more important role because other diagnostic tools are often not available.

To combine structural information with molecular information, negative staining can be combined with immunolabeling. This immuno-negative staining can increase the specificity in diagnostic electron microscopy or give insights into the molecular topology of viruses (Biel and Gelderblom, 1999b).

Negative staining is a whole mount preparation method. Thus, all structural details are viewed as a projection in two dimensions. To study virus structure in more detail, especially in the cellular context, ultrathin sectioning must be performed. Three-dimensional reconstructions of negatively stained virus preparations give additional information but are restricted to samples of 200–300 nm thickness, depending on the density (Mast and Demeestre, 2009). Thicker samples must be prepared by ultrathin sectioning.

The preparation steps for ultrathin sectioning usually involve several fixation steps, dehydration and embedding into a resin to give a hard block which can be sectioned. Ultrathin sectioning is done with a diamond knife using an ultramicrotome. If chemical modification during preparation is reduced and particular resins are used, localization of molecules by using antibodies is possible for all antigens exposed by the sectioning process. Usually the whole preparation process takes some days. However, for diagnostic purposes rapid preparation protocols have been developed, which allow a "same-day diagnosis," or in specific cases, a diagnosis within 2 h (Laue *et al.*, 2007; Schröder *et al.*, 2006; Zechmann and Zellnig, 2009).

III. Methods

A. Biosafety

Experiments using pathogen viruses must be conducted according to the legal rules of each country. Usually permission is needed to work with pathogenic or genetically modified viruses. Viruses are classified in different biosafety classes according to their pathogenecity (American Biological Safety Association; http://www.absa.org/riskgroups/index.html), and laboratory safety is related to these classes.

In general, most viruses are inactivated by a treatment with at least 2–4% paraformaldehyde in buffer (Gelderblom *et al.*, 2007). However, testing of virus infectivity after inactivation provides further safety. In most cases inactivation does not interfere with the negative staining, even if small amounts of glutaraldehyde (e.g., 0.05%), which is a potent inactivator (Rubo *et al.*, 1967), are added (Gelderblom *et al.*, 2007).

B. Negative Staining

Negative-staining electron microscopy allows the visualization of particles in suspensions. The biology of viruses usually facilitates sampling of particle suspensions, e.g., by simply collecting the released virus together with the cell culture medium of a virus cell culture. Even if cells do not release virus, cells and tissues may be destroyed by repeated cycles of freezing and thawing or by mechanical grinding in a potter, thereby releasing usually intact virus particles or premature stages of the infectious virus. If the primary suspension contains too many unwanted materials, which may cover the structures of interest, cleaning procedures may be performed, e.g., ultracentrifugation through a sucrose cushion (compare e.g., Bartolomé *et al.*, 2007). However, since negative staining is a rapid technique, testing of the raw suspension should be performed in any case.

A detailed description of the negative-staining technique is found in the excellent book of Harris (1997). Specific information on negative staining in virology and diagnostic electron microscopy is given by Biel and Gelderblom (1999b).

The negative-staining procedure can be divided into the following preparation steps:

1. Charging or conditioning of the specimen support
2. Adsorption of the particles at the specimen support
3. Staining with heavy metals

The specimen support should be a metal grid with a small mesh size (e.g., 400 mesh) to give suitable mechanical stability and thermal conductivity, which is covered by a stable but transparent plastic film (e.g., Webster and Webster, 2007). To increase the stability of plastic films, a carbon layer can be evaporated onto the film, which increases thermal conductivity of the support under the electron beam. This procedure needs particular instrumentation (i.e., a carbon evaporator) of a suitable quality. However, negative staining can be performed on the naked plastic film. If particular high resolution is needed, pure carbon film with a low granularity may be used as a support (Harris, 1997).

To increase the stickiness of the plastic or carbon film, several pretreatment procedures can be used. The main principle of the procedure is to add charge to the surface of the carbon or plastic film, which makes the surface hydrophilic and allows binding of charged biomolecules. The easiest way to increase charging is to add charged molecules, such as poly-L-lysine, Alcian blue, or bacitracin. A more elegant method is charging by exposing grids to a plasmon. The original method is called "glow discharge" and needs a particular device (e.g., stand-alone instrument or carbon evaporator with a glow discharge unit). However, a plasma cleaner or sputter coater (without metal target in place) may work as well since both use a plasmon for operation.

Particle adsorption onto the charged film surface may be reinforced by gravity, if particles are dense enough to significantly sediment in a reasonable time (Laue and Bannert, 2010). In this case, small drops (10 μl) of the suspension are placed directly onto the grid. Since small and light particles tend to increase concentration at the air–liquid interface of a droplet (Johnson and Gregory, 1993), grids can be put on droplets (e.g., 30 μl) with the charged side down to increase the number of particles adsorbed.

After adsorption of particles, washing of the grids helps to remove solutes or unbound particles which may interfere with the final staining step (e.g., phosphate salts interfere with uranyl acetate). Washed grids are brought in contact with a drop of the staining solution for a few seconds and dried with a filter paper, thereby producing a thin layer of amorphous stain in which the particles are embedded. The list of negative stains is long (Harris, 1997, 2007), but good results, in most cases, are achieved by using uranyl acetate and phosphotungstic acid (PTA) in parallel (on different grids), because staining results in a different appearance of particles giving additional information (Biel and Gelderblom, 1999b) (Fig. 1).

If the particle number on a grid is too low, particle concentration of the suspension can be increased by ultracentrifugation or ultrafiltration. Enhancement of the particle number on the grid can also be achieved by direct ultracentrifugation of a small volume of the suspension onto the grid. However, a particular air-driven centrifuge (i.e., the Beckman Airfuge) must be used for this purpose (Gelderblom, 2006; Laue and Bannert, 2010).

The following protocol is a simple negative-staining protocol for general use (for a discussion of variables affecting particle recovery on the grid, see also Laue and Bannert (2010)):

1. All steps are conducted on a desk that is covered with a strip of Parafilm. Droplets (30 μl unless otherwise stated) of the different solutions are administered onto the clean surface and filmed grids are incubated on top of them.

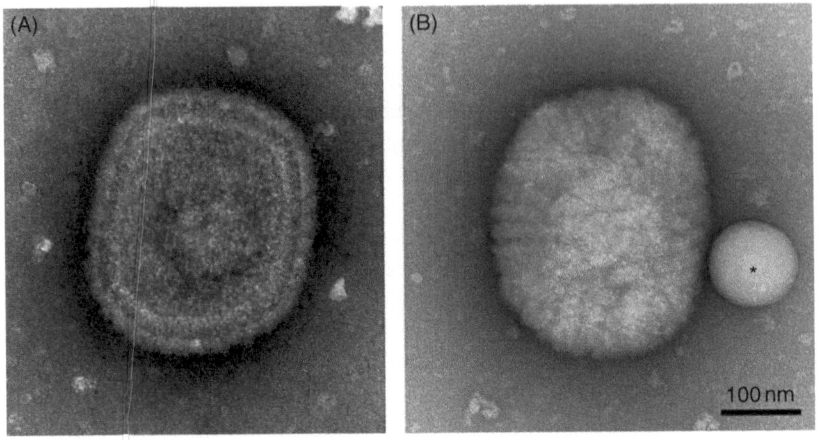

Fig. 1 Negative staining of orthopoxviruses with 1% uranyl acetate. Long staining times (**A**) reveal internal structures of the virus, like the inner membrane, while brief staining (**B**) restricts binding of heavy metals to the surface emphasizing the representation of surface structures. Note: In (**B**), a latex bead (*), with a diameter of approximately 100 nm, is localized close to the virus particle.

2. Incubation of the filmed surface of the grid on Alcian blue solution (1% in 1% acetic acid) for 10 min.
3. Washing of the filmed surface by successively touching four droplets of double-distilled water.
4. Removing of surplus water using a filter paper (do not dry the surface completely).
5. Incubation of the filmed surface with the sample suspension for 10 min. If dense particles (e.g., poxvirus and bacteria) should be adsorbed onto the grids, 10 µl of the suspension may be directly dispensed onto the filmed surface.
6. Washing of the filmed surface by successively touching four droplets of double-distilled water.
7. Removing of the residual water on the grid surface using a filter paper (do not dry the surface completely).
8. Touching of a droplet of the negative stain (1% uranyl acetate or PTA, both in double-distilled water) for a few seconds.
9. Carefully removing the staining solution using a filter paper to generate a thin homogeneous film of stain for optimal preservation of surface details.
10. After a brief drying period (a few minutes are sufficient), samples can be analyzed in the microscope.

Evaluation of grids with the microscope is simply a pattern recognition procedure, which is based on experience (Miller, 1986) (Fig. 2). A diagnostic reference is built up by practicing and is supported by substantial literature (e.g., Doane and Anderson, 1987; Madeley and Field, 1988; Palmer and Martin, 1988). In addition to pattern recognition, particle size is an important diagnostic feature. Therefore, calibration of

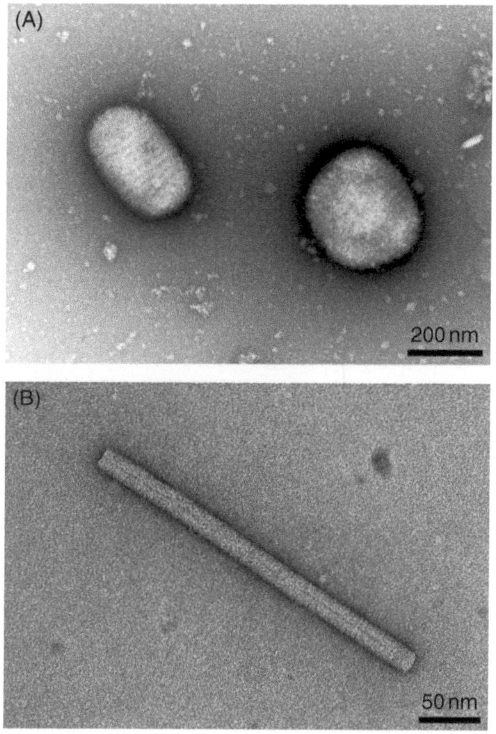

Fig. 2 Negative staining of morphologically different viruses. (**A**) Two different viruses of the poxvirus family. On the left side the more oval particle shows a ring-like surface sculpture that is typical for the subgroup of parapoxviruses. The orthopoxvirus particle on the right side has a rather rectangular shape and a more floccus surface without a closed ring-like sculpture. (**B**) The tobacco mosaic virus (TMV) is a thin (18–20 nm width) filamentous virus with regular surface stripes (pitch = 2.3 nm) and an internal central channel.

the microscope even at high magnification is important or should at least be checked using size standards (e.g., plastic or gold beads; catalase crystals) to guarantee the accurate measurement of particle size. Counting of virus particles may give an idea of the total number of virus particles in the original suspension. A simple reference-based method, using plastic beads at a known concentration, as a standard, is sufficient in most cases to determine the particle concentration with a suitable accuracy (Geister and Peters, 1963; Laue and Bannert, 2010; Miller, 1982).

C. Immuno-Negative Staining

Negative-staining electron microscopy provides high-resolution structural information of virus particles. By combination with specific antibodies and small gold markers, the ultrastructure can be linked to molecular topology at a high spatial resolution (Fig. 3). The negative-staining protocol is simply extended by the antibody

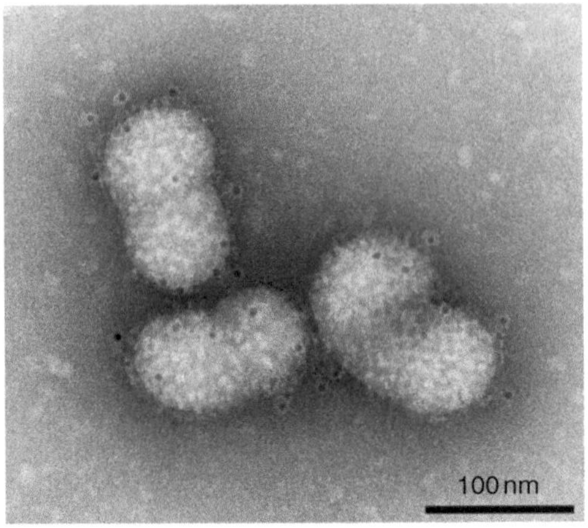

Fig. 3 Immuno-negative staining of influenza virus. Antibodies were directed against the hemagglutinin (HA) of the virus. Binding of the primary antibody was visualized by a secondary antibody coupled to 5 nm of colloidal gold that appears as a dark dot close to the surface or on top of the virus particles.

incubation steps. Colloidal gold serves as an electron dense marker, usually applied with a secondary detecting antibody. The indirect approach reduces somewhat the spatial resolution of the labeling, but is more convenient and provides maximum sensitivity, because modification of the primary antibody is not needed. A gold size of 5 or 10 nm is appropriate. Smaller gold has a broader and more irregular size distribution and is difficult to localize in conjunction with the heavy metal staining. As in any immunological experiment, a set of controls must be performed (positive, negative, and labeling controls).

The following protocol may be a good starting point for labeling studies:

1. Steps 1–5 of the negative-staining protocol (see above). All incubations are carried out on droplets of 30 μl.
2. Incubation on glycine (50 mM in phosphate-buffered saline (PBS)) for 5 min to block free aldehyde groups.
3. Incubation on blocking buffer (0.1% BSA in PBS) for 10 min.
4. Incubation on primary antibody solution (diluted in blocking buffer) for 15 min.
5. Washing on blocking buffer (three droplets).
6. Incubation on secondary antibody solution (diluted 1:20 with blocking buffer) for 10 min.
7. Washing on PBS (two droplets).
8. Washing on double-distilled water (five droplets).
9. Removal of residual water on the grid by using a filter paper.
10. Touching of a droplet of the negative stain (0.5% uranyl acetate in double-distilled water) for a few seconds.
11. Follow steps 9 and 10 of the negative-staining protocol.

Note that the protocol applies a lower concentration (i.e., 0.5%) of the staining solution than the standard negative-staining protocol to reduce the risk of masking gold particles by the heavy metal stain. Particles are lost during the whole procedure. Particle loss may be reduced by reducing the incubation times and the number of washing steps. Another possibility to cope with this problem is to use a more concentrated sample suspension and/or to increase the adsorption time (Laue and Bannert, 2010).

D. Ultrathin Section Electron Microscopy

The production of ultrathin sections reduces a sample to a thin layer, which is transparent for electrons. The method involves several preparation steps, an ultramicrotome, and needs manual expertise. The most common procedure is to embed the sample into a plastic to make it hard enough for ultrathin sectioning (for an overview compare Bozzola and Russel, 1998). Embedding in frozen sucrose is used as an alternative and is basically used for immunolabeling studies (see Section V). The description in this section is focused on plastic embedding since it is the general approach for solving many questions of interest in structural virology. The procedure consists of the following individual preparation steps:

1. Chemical fixation
2. Preparation of virus or cell pellets
3. Dehydration
4. Infiltration and embedding in a plastic
5. Ultrathin sectioning
6. Staining of sections

Fixation is not only important because of biosafety issues (see above), but it is also needed to stabilize the sample prior to following the preparation steps, which otherwise would destroy the sample. Organic solvents used for dehydration and the plastic monomers are efficient in extracting tissue components that are not tightly linked with each other (Weibull *et al.*, 1983). The use of heavy metal-containing fixatives (e.g., osmium tetroxide or uranyl acetate) not only adds more stability to the sample, but it also generates an *en bloc* contrast by selectively binding to certain structures (e.g., biomembranes).

Dehydration is necessary, because plastic monomers are not usually soluble in water. Therefore, water has to be exchanged for an organic solvent (acetone or ethanol) and may be an additional intermedium (e.g., propylene oxide) that can be mixed more easily with the plastic monomer.

For embedding, different plastic formulations are used: (1) epoxy resin (e.g., Epon, Araldite, Spurrs); (2) acrylate resin (e.g., LR White, Lowicryl). While the first group generates highly cross-linked and stable samples, the latter possesses only a low tendency to bind to structural components and rather surrounds them thereby producing "soft" samples that are more suitable for post-embedding immunolabeling (Causton, 1986). However, the epoxy resins give a better resolution than the acrylate resins

(Kellenberger, 1998) and are therefore preferred in pure structural studies. Resins are introduced into the samples as monomers, usually by mixing them with an appropriate organic solvent. The concentration of the resin is then gradually increased. Finally the resin is hardened by polymerization. To increase the speed of polymerization usually the temperature is increased (Doane *et al.*, 1974). With acrylates, polymerization can be performed at low temperature using UV light, thereby reducing the risk of extracting or destroying structures and molecules (Acetarin *et al.*, 1986; Carlemalm *et al.*, 1982).

Ultrathin sectioning is done using an ultramicrotome. The operation needs some experience and is best learned by getting advice from somebody who is operating the individual machine regularly (e.g., in a core facility). It is not only the manipulation of the small samples and sections, but also adjusting the correct settings of parameters at the ultramicrotome that needs experience, because setting of parameters is dependent on sample quality and environmental conditions (compare Hagler, 2007).

Sections can be collected and stored for quite a long time. In many cases, it is necessary to add further contrast to the sample. This can be done by placing the grids, sections facing down, on a drop of heavy metal solutions, usually uranyl acetate followed by lead citrate (compare Bozzola and Russell, 1998).

The following protocol may give reasonable results for many samples and applications (for immunolabeling studies using sections see Section V):

1. Primary fixation of cell cultures: Removal of cell culture medium. Adding of 2.5% glutaraldehyde in 0.05 M HEPES buffer (pH 7.4). Incubation for at least 1 h at room temperature. A slight movement of the fixative helps to promote fixation. Primary fixation of tissue: Pieces of tissue should be as small as possible to provide quick infiltration of the fixative (otherwise consider perfusion fixation). Immerse tissue in 4% paraformaldehyde with 2.5% glutaraldehyde in 0.05 M HEPES (pH 7.4). Incubation should last for at least 2 h at room temperature for a tissue block of ~1 mm^3.
2. Cell culture: washing of the cells with HEPES buffer (twice). Scraping of cells from the substrate using a cell scraper. Collection of the suspension. Mixing with low melting point agarose (3% in distilled water at 37–40°C). Centrifugation in a swing-out rotor for a few minutes to form a stable pellet. Cooling on ice for a few minutes. Postfixation of the agarose block in 2.5% glutaraldehyde gives additional stabilization.
3. Postfixation: Remove primary fixative by washing twice with HEPES buffer. Immerse sample block into 1–2% osmium tetroxide in distilled water. Incubation for 1 h at room temperature under agitation in a dark container. Removal of the fixative by washing in distilled water (twice).
4. Dehydration in a series of ethanol with increasing concentrations: 50, 70, 90, 96, 100, and 100%, each for 15 min.
5. Infiltration with Epon resin by using propylene oxide: pure propylene oxide (twice, for 5 min each); mixtures of propylene oxide and Epon (1:1, 1:2 for 30 min each, and 1:3 overnight for slow evaporation of the propylene oxide).
6. Incubation with fresh Epon for 4 h.

7. Embedding in Epon using particular embedding molds.
8. Polymerization for at least 1 day at 60°C.

With many samples (e.g., cell cultures) infiltration with the Epon monomer can also be done using acetone instead of propylene oxide.

Additional steps may be included to increase membrane contrast. Especially glycoproteins of membranes are contrasted (and increased in thickness) using the following procedure:

1. Complete steps 1–3 of the standard embedding protocol.
2. Incubation in 0.1% tannic acid (in HEPES buffer) for 30 min.
3. Washing with 1% sodium sulfate (in HEPES buffer) twice, 10 min each.
4. Washing with distilled water twice, 10 min each.
5. Incubation with 2% uranyl acetate in distilled water for 2 h.
6. Follow steps 4–8 of the standard embedding protocol.

Ultrathin sectioning of sample blocks should produce sections of about 60–80 nm (silver to a slightly golden color). Sections may be collected on grids with or without plastic film. Sections on bare grids usually give a better appearance and resolution than sections collected on plastic film, basically because of fewer inelastic electron scattering, but are more prone to mechanical damage during exposure in the beam. To increase beam stability, a thin layer of carbon may be evaporated on the sections using a high vacuum carbon evaporator. To increase contrast, sections may be stained with uranyl acetate and/or lead citrate. For an example of viruses in an ultrathin section of a cell, see Fig. 4A.

E. Rapid Ultrathin Section Electron Microscopy

Routine embedding procedures usually take 1–2 days before samples can be sectioned. This is definitely too slow for diagnostic purposes or quality assurance. To reduce overall preparation time, different rapid embedding procedures have been developed. The most direct approach is to speed up the polymerization by increasing the temperature to 90°C. Together with reduced incubation times during dehydration and infiltration, the whole procedure can be reduced to a few hours (Doane *et al.*, 1974). Microwave processing can be used to transfer heat more quickly, reducing efficiently the incubation times of many steps by increased diffusion (Leong and Sormunen, 1998). However, particular microwave ovens are necessary to control the heat transfer properly.

A simple and quick approach is to use LR White resin, which has a low viscosity to speed up diffusion and which can be quickly polymerized using a chemical accelerator (Hobot and Newman, 1990). Together with a small sample size, overall processing time can be reduced to 1–2 h (Laue *et al.*, 2007). The embedding procedure was originally developed for rapid diagnosis of bacterial endospores using thin-section transmission electron microscopy. However, the procedure is also helpful for the quick

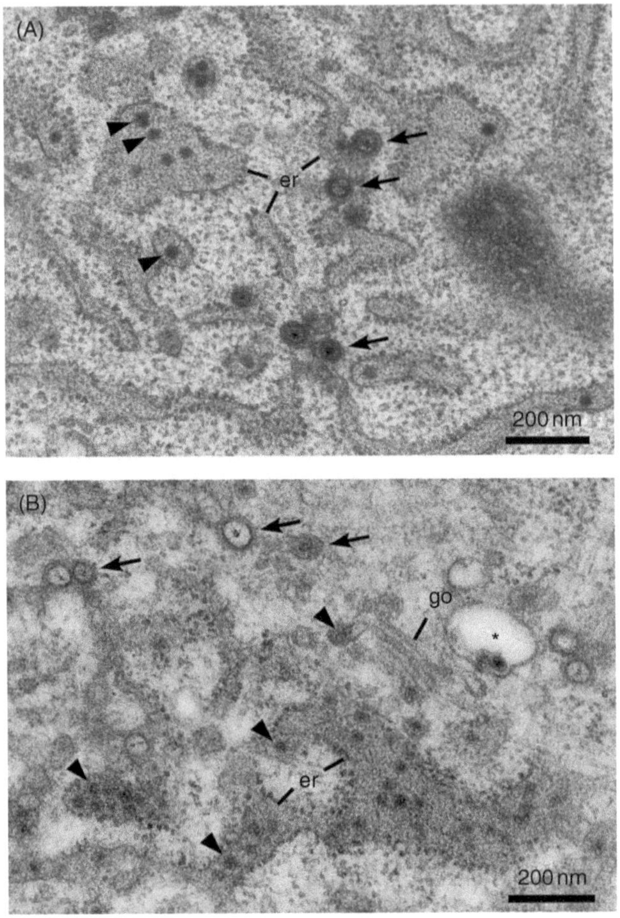

Fig. 4 Ultrathin sections of cells infected with a flavivirus. **(A)** Conventional Epon embedding reveals virus particles (arrowheads) in the endoplasmic reticulum (er) and different vesicles budding at the reticulum membrane (arrows). **(B)** Rapid embedding in LR White using 2.5 μl of accelerator and a brief osmium tetroxide postfixation (3 h processing time including sectioning). The structural appearance is sufficient to detect all relevant ultrastructural aspects of the flavivirus genesis that are also visible in sections of a conventional Epon embedding (compare with A). (g) Golgi apparatus; (*) large vesicle or vacuole where virus particles are concentrated for particle release from the cell.

analysis of various other samples, e.g., eukaryotic cell culture cells and even tissue samples, like biopsies. The structural appearance is sufficient for most questions (compare Fig. 4B with Fig. 4A). In addition, on-section immuno-cytochemistry is possible, but most probably restricted to antigens present at a high concentration (Laue *et al.*, 2007). The following protocol summarizes essential steps of the embedding method:

1. All manipulations using LR White should be performed in a fume hood. Please consult material data safety sheet or local safety officer in case of any questions regarding lab safety.
2. Fixed samples (strong aldehyde fixation is recommended, e.g., 2.5% glutaraldehyde in 0.05 M HEPES, pH 7.4). Sample thickness should be not more than 1 mm (0.15–0.3 mm are optimal). For preparing suspensions in an agarose gel at a defined thickness, see Laue *et al.* (2007).
3. Washing out of primary fixative with buffer (twice for 1 min each).
4. Optional: Postfixation with 1–2% osmium tetroxide for 30 min increases membrane contrast. On-section contrasting may not be necessary with these samples.
5. Dehydration, infiltration, embedding, and polymerization are done on ice. All solutions are used precooled at ice temperature.
6. Incubation in 70 and 100% ethanol for 10 min each.
7. Incubation in 100% ethanol for 5 min.
8. Incubation in a mixture of ethanol and LR White (1:1) for 5 min, in pure LR White for 5 min and for 10 min, respectively.
9. Mixing of LR White with the accelerator is done in portions of 1 ml using a magnetic stirrer. Precooled LR White is filled in a cylindrical glass vial containing a stirring bar. The accelerator (5 μl) is quickly added to the monomer using a microliter pipette while stirring. Finally, the mixture is filled in precooled reaction tubes (0.5 ml). The mixing of the accelerator with the resin monomer must be done very quickly otherwise polymerization starts only locally. With 2.5 μl accelerator per milliliter of monomer, polymerization is completed within 30–40 min on ice; 5 μl needs only 15 min. Accelerator concentrations above 5 μl/ml monomer are not useful because the resin gels too quickly for handling.
10. Transfer of samples into the reaction tubes containing resin mixtures. Any airtight vial of a volume equal or smaller than 0.5 ml will allow polymerization if it is not dissolved by the monomer (polyethylene and polypropylene will work, polystyrene will not). Vials with thinner walls than reaction tubes may be removed easily from the polymerized blocks than from the reaction tubes.
11. Polymerization on ice for 15 min. Polymerization in a small volume and on ice (or even better, ice water) is necessary because the reaction generates heat in a very short time which must be extracted from the resin otherwise bubbles will appear in the polymerized block around the sample rendering ultrathin sectioning difficult. Residual liquid LR White may be soaked with a tissue before final polymerization is conducted.
12. Final polymerization in a preheated (60°C) oven for 5 min.
13. Sections may be observed without poststaining or after quick on-section staining according to Roth *et al.* (1990) using uranyl/methyl cellulose.
14. The protocol can be easily tailored to the needs of the sample (e.g., thickness) to the desired speed and the final quality of structural preservation, e.g., by changing dehydration, infiltration, and polymerization times gradually.

IV. Instrumentation and Materials

A. Biosafety

Instrumentation: Laboratory facility according to the biosafety class of the virus of interest.

Materials: No specific materials are needed if the biosafety laboratory is equipped with standard tools (pipette, table desk centrifuges, reaction and centrifuge tubes, waste container).

Reagents: For inactivation and fixation a concentrated stock solution of paraformaldehyde (provided as powder) must be prepared. Paraformaldehyde (20% final concentration) is dissolved in 0.5 M HEPES buffer (pH 7.4) by heating to 60°C for at least 1 h. The solution must be cleared by adding a few droplets of 10 N NaOH and can be stored at −20°C or below. Before use, it must be heated again to 60°C to shift equilibrium from formaldehyde polymers to monomers (Griffiths, 1993).

For mixtures of paraformaldehyde with glutaraldehyde (e.g., provided as a solution of 25% in water in ampoules; e.g., by Electron Microscopy Services, No. 16000) a stock solution of paraformaldehyde (20%) is prepared in double-distilled water and diluted with water, glutaraldehyde, and concentrated buffer solution (e.g., 1 M HEPES buffer, pH 7.4) to the final concentration.

B. Negative Staining

Instrumentation: High-vacuum carbon evaporator (e.g., Emitech K950X; Leica SCD500, Cressington 208HR). Glow discharge device (e.g., Emitech K100X).

Materials: Grids (e.g., Agar Scientific, Stansted, UK, 400 mesh, G2400C). Grids with a plastic film (e.g., Electron Microscopy Services, No. FF400-Cu).

Reagents: Double-distilled water or equivalent quality is needed for the preparation of all solutions and washing steps. Alcian blue (Sigma-Aldrich, No. 5500) is prepared as a solution of 1% in 1% acetic acid. Solution must be centrifuged to remove crystals. Cleared solution can be stored in the refrigerator for at least 6 months (Laue and Bannert, 2010). Uranyl acetate (e.g., Electron Microscopy Services, No. 22400) and PTA (e.g., Electron Microscopy Services, No. 19500) are prepared as a 1% solution in double-distilled water. PTA should be adjusted to pH 8.5 or 7 by using 1 N NaOH. The alkaline PTA breaks the biomembranes of enveloped viruses thereby revealing internal structures.

C. Immuno-Negative Staining

Instrumentation and Materials (same as in Section IV.B)

Reagents: Same as in Section IV.B. In addition, secondary antibodies, coupled to colloidal gold (e.g., 5 nm size, British Biocell). PBS, e.g., 8 g NaCl, 0.2 g KCl, 1.5 g Na_2HPO_4 dihydrate, 0.2 g KH_2PO_4 in 1 l of distilled water (pH 7.4). Bovine serum albumin (BSA; e.g., Sigma Aldrich, No. A3294, low protease activity) as a 0.1% solution in PBS. Glycine, 50 mM in PBS.

D. Ultrathin Section Electron Microscopy

Instrumentation: Ultramicrotome (different models by Leica/Reichert or RMC). Laboratory oven (capable of heating up to 60°C). Table desk centrifuge (swing-out rotor facilitates collection of cell pellets in low melting point agarose).

Materials: Grids (e.g., Agar Scientific, 300 × 75 mesh, G2375C). Grids with a plastic film (e.g., Electron Microscopy Services, 2 × 1 slots, No. FF2010-Cu). Embedding molds (different sizes and designs are available; e.g., Electron Microscopy Sciences, No. 70907).

Reagents: Double-distilled water or equivalent quality is needed for the preparation of all solutions and washing steps. Low melting point agarose (3% in distilled water; heating once to 60°C is necessary to achieve proper gelling), HEPES buffer (0.05 M, pH 7.4), glutaraldehyde (2.5% in 0.05 M HEPES), osmium tetroxide (2–4% stock solution in distilled water; dissolve overnight at 4°C), tannic acid (e.g., Electron Microscopy Services, No. 21710), sodium sulfate, uranyl acetate (see above), Epon (23.52 g glycidyl ether 100, 12.35 g dodenyl succinic anhydride, 14.13 g nadic methyl anhydride, 0.65 g dimethyl phthalate; Serva, Electrophoresis, Heidelberg, Germany), propylene oxide, acetone, ethanol, and lead citrate.

E. Rapid Ultrathin Section Electron Microscopy

Instrumentation (compare Section IV.D)

Materials (in addition to materials listed in Section IV.D): tightly closing reaction vials for embedding (e.g., Eppendorf, Hamburg, Germany, Safelock, 0.5 ml).

Reagents (in addition to materials listed in Section IV.D): LR White (hard grade, Electron Microscopy Services, No. 14383-UC), LR White accelerator (Electron Microscopy Services, No. 14385), methyl cellulose (25 cps; Sigma Aldrich, M6385).

V. Discussion

Negative staining and classical ultrathin section electron microscopy are the basic techniques to study virus morphology and virus/host interactions. In most cases, relevant results can be achieved by using these techniques. Studies may be extended by using more elaborated methods to gain a further quality of results. One of these methods is on-section immunolabeling which allows the combination of structural data provided by ultrathin sections with molecular topology. Starting from chemically fixed samples the embedding procedure must be changed in comparison to the routine protocol. A resin with a low tendency to co-polymerize with the sample structure, like for instance the LR or Lowicryl resins, should be used together with dehydration and embedding at low temperature (Schwarz and Humbel, 1989). However, modification of antigens by the different preparation steps may impair on-section localization of antigens. A useful variant of ultrathin sectioning is the so-called Tokuyasu technique, where the chemically fixed sample is embedded in sucrose, frozen, and sectioned at low temperature (compare e.g., Webster and Webster, 2007). However, the procedure needs particular equipment and experience. Preembedding immunolabeling is another variant, which may be easily combined with standard embedding procedures, if the antigen is accessible for the antibodies (e.g., virus proteins at the cell surface) or if permeabilization or even extraction methods (e.g. Bohn *et al.*, 1986; Kolesnikova *et al.*, 2007) are used. An excellent comprehensive overview on most aspects of immunocytochemistry is given in a book by Griffiths (1993) and in a review by Skepper (2000).

A more realistic representation of virus ultrastructure is definitely achieved by using cryopreparation methods. The first step involves the so-called vitrification of samples by cryofixation, which simply means freezing without the formation of ice crystals. The size of the virus particles facilitates vitrification, which is pivotal for the preservation of ultrastructural details. Details of the various cryopreparation methods are given in a book by Steinbrecht and Zierold (1987) and in a review by Quintana (1994). In virology, basically two strategies are used: (1) bare-grid method (Lepault *et al.*, 1983) and (2) high-pressure freezing, followed by either direct cryo-electron microscopy or freeze-substitution to transfer samples back to room temperature without generation of ice crystals.

The bare-grid method is basically used for studying virus structure in suspension (e.g., Adrian *et al.*, 1984; Yamaguchi *et al.*, 2008). The technique needs highly concentrated and purified virus suspensions and particular equipment, such as a plunge-freezer, a cryo-transfer system, and a cryo-electron microscope. Suitable protocols can be found in the book by Cavalier *et al.* (2009; see also Chapter 2 by Pilhofer *et al.*, this volume).

High-pressure freezing allows suitable cryofixation of thicker samples than individual virus particles (up to 200–300 μm, depending on variables such as water content and thermal conductivity) (Moor, 1987). This method allows study of virus/cell interaction using cell culture (Chapter 25 by Walther *et al.,* 2010, this volume). Cryo-microscopy can be carried out directly at the thin periphery of the cells (Schwartz *et al.*, 2007) or by using cryo-sections (CEMOVIS; e.g., Grünewald and Cyrklaff, 2006; Vanhecke *et al.*, 2007). However, cryo-electron microscopy is not very useful if the region of interest is not known precisely or an overview on the various virus/cell interactions is needed, because imaging must be done at a low dose of electrons (low magnification during searching and very short illumination times at higher magnification) otherwise the sample would be immediately destroyed (Koster and Bárcena, 2006; Vanhecke *et al.*, 2007). Because of these reasons, high-pressure freezing is frequently combined with freeze-substitution to replace ice at low temperature (Schwarz *et al.*, 1993; Steinbrecht and Müller, 1987) thereby eliminating the risk of ice crystallization during rewarming at higher temperatures. Freeze-substitution usually ends up in plastic embedding and ultrathin sectioning. The procedures and application of the technique in studying virus/host interactions are presented in more detail by Chapter 25 by Walther *et al.*, 2010 (this volume).

Conventional bright-field transmission-electron microscopy reveals a two-dimensional projection of all structures illuminated parallel to the electron beam. Thus, three-dimensional structure is not accessible by this imaging approach. However, tilting the specimen reveals three-dimensional information of the structures. Even two images taken at two different illumination angles (usually by tilting the sample) may be sufficient to generate stereo images for analyzing structures included in the three-dimensional volume (Heuser, 2000). An almost complete reconstruction of the three-dimensional structure is achieved by electron tomography (Baumeister *et al.*, 1999). A series of two-dimensional images taken with a small increment of the tilt angle is used as a basis for an *in silico* backprojection of the images into a three-dimensional volume. Image processing, like segmentation and color-coding of individual structures, allows the generation of models providing information on virus structure. A review on different virus structures resolved by this method is given by Grünewald and Cyrklaff (2006). Detailed information on electron tomography is available in a comprehensive book by Frank (2006a). The paper by Geerts *et al.* (2009) gives an overview of the different processing steps involved. An interesting approach to use electron tomography for diagnostic purposes was recently presented by Mast and Demeestre (2009).

Another method to study virus structure in three dimensions is the so-called single particle analysis. It can be performed with images acquired by negative-staining electron microscopy or by the bare-grid method using cryo-electron microscopy. The principle behind the method is to record images from different randomly orientated particles and to combine those views in a three-dimensional reconstruction. The method may achieve high resolution, but needs optimal conditions, like highest purity of the original sample and some knowledge about the particles to be studied. Moreover, computational procedures are sophisticated and need experience. An overview on single-particle analysis is given by Frank (2006b). A particular focus on studying virus morphology by using single-particle analysis and X-ray diffraction may be found in Baker and Johnson (1997). Protocols for the preparation and imaging of protein aggregates (including viruses) are provided by Grassucci *et al.* (2008a, 2008b) and Harris (2007).

The spectrum of imaging methods to study virus structure and virus/host interactions is still increasing. For a long time, transmission electron microscopy was the only method to reveal virus ultrastructure at high resolution. Today, other methods such as scanning (transmission) electron microscopy, scanning force microscopy, and high-resolution light microscopy provide information at a high spatial resolution. Scanning electron microscopy, especially if field-emission systems are used, can give a rapid overview about the events taking place at the surface of cells (e.g., budding or infection of virus particles). With a resolution of about 1 nm or better, even small viruses can be visualized at a suitable quality (Ng *et al.*, 2004; Watanabe *et al.*, 2004).

The combination of electron microscopy and light microscopy is helpful in many research fields. In virology, especially the dynamics of virus infection, replication and egress can be studied by high-resolution life-cell imaging (e.g., Brandenburg and Zhuang, 2007). Life-cell imaging of dynamic events may be correlatively combined with an end-point study of the subcellular structure and molecular arrangement by employing high-resolution electron microscopy (Brown *et al.*, 2009; Larson *et al.*, 2005) leading to a more comprehensive understanding of virus cell biology. However, even the basic methodology provided within this chapter is still valuable and helpful for virus detection and investigations of virus/host interactions.

Acknowledgments

I thank Dr. Muhsin Özel, Gudrun Holland, Freya Kaulbars, and Andrea Männel for helping me in my "early days" at the Robert Koch Institute (RKI) with the preparation of viruses for electron microscopy. I am grateful for the support of Hans Gelderblom and Norbert Bannert (both RKI). Moreover, I thank the numerous colleagues at the Robert Koch Institute for providing interesting samples of which some are depicted in this chapter.

References

Acetarin, J. -D., Carlemalm, E., and Villiger, W. (1986). Developments of new lowicryl resins for embedding biological specimens at even lower temperatures. *J. Microsc. (Oxford)* **143**, 81–88.

Adrian, M., Dubouchet, J., Lepault, J., and McDowall, A. W. (1984). Cryo-electron microscopy of viruses. *Nature* **308**, 32–36.

Baker, T. S. and Johnson, J. E. (1997). Principles of virus structure determination. *In* "Structural Biology of Viruses" (W. Chiu, R. Burnett, R. L. Garcea, eds.) pp. 38–78. Oxford Press, NY, Oxford.

Bartolomé, J., López-Alcorocho, J. M., Castillo, I., Rodriguez-Iñigo, E., Quiroga, J. A., Palacios, R., and Carreño, V. (2007). Ultracentrifugation of serum samples allows detection of hepatitis C virus RNA in patients with occult hepatitis C. *J Virol.* **81**, 7710–7715.

Baumeister, W., Grimm, R., and Walz, J. (1999). Electron tomography of molecules and cells. *Trends Cell Biol.* **9**, 81–85.

Biel, S. S. and Gelderblom, H. R. (1999a). Diagnostic electron microscopy is still a timely and rewarding method. *J. Clin. Virol.* **13**, 105–119.

Biel, S. S. and Gelderblom, H. R. (1999b). Electron microscopy of viruses. *In* "Virus Cell Culture" (A. Cann, ed. pp. 111–147. Oxford University Press, Oxford.

Bohn, W., Rutter, G., Hohenberg, H., Mannweiler, K., and Nobis, P. (1986). Involvement of actin filaments in budding of measles virus: Studies on cytoskeletons of infected cells. *Virology* **149**, 91–106.

Bozzola, J. J. and Russell, L. D. (1998). "Electron Microscopy: Principles and Techniques for Biologists". Jones and Bartlett Publishers, Boston.

Brandenburg, B. and Zhuang, X. (2007). Virus trafficing—learning from single-virus tracking. *Nat. Rev. Microbiol.* **5**, 197–208.

Brenner, S. and Horne, R. W. (1959). A negative staining method for high resolution electron microscopy of viruses. *Biochim. Biophys. Acta* **34**, 103–110.

Brown, E., Mantell, J., Carter, D., Tilly, G., and Verkade, P. (2009). Studying intracellular transport using high-pressure freezing and correlative light and electron microscopy. *Semin. Cell Dev. Biol.* **20**, 910–919.

Carlemalm, E., Garavito, R. M., and Villiger, W. (1982). Resin development for electron microscopy and an analysis of embedding at low temperature. *J. Microsc. (Oxford)* **1(26)**, 123–143.

Causton, B. E. (1986). Does the embedding chemistry interact with tissue? *In* "The Science of Biological Specimen Preparation. Proceedings of the 4th Pfefferkorn Conference" (M. Müller, R. P. Becker, A. Boyde, J. J. Wolosewick, eds.) pp. 209–292. SEM Inc, AMF O'Hare, Chicago.

Cavalier, A., Sphener, D., and Humbel, B. (2009). "Handbook of Cryo-Preparation Methods for Electron Microscopy". CRC Press, Boca Raton.

Curry, A., Appleton, H., and Dowsett, B. (2006). Application of transmission electron microscopy to the clinical study of viral and bacterial infections: Present and future. *Micron* **37**, 91–106.

Doane, F. W. and Anderson, N. (1987). "Electron Microscopy in Diagnostic Virology". Cambridge University Press, Cambridge.

Doane, F. W., Anderson, N., Chao, J., and Noonan, A. (1974). Two-hour embedding procedure for intracellular detection of viruses by electron microscopy. *Appl. Microbiol.* **27**, 407–410.

Frank, J. (2006a). "Electron Tomography. Methods for Three-Dimensional Visualization of Structures in the Cell. Springer, NY.

Frank, J. (2006b). "Three-Dimensional Electron Microscopy of Macromolecular Assemblies". Oxford University Press, Oxford, NY.

Geerts, W. J.C., Humbel, B. M., and Verkleij, A. J. (2009). 3D-electron tomography of cells and organelles. *In* "Handbook of Cryo-Preparation Methods for Electron Microscopy" (A. Cavalier, D. Sphener, B. Humbel, eds.) pp. 618–649. CRC Press, Boca Raton.

Geister, R. and Peters, D. (1963). Ein vereinfachtes direktes Zählverfahren für Virussupensionen ab 10E5 Partikel/ml. *Zeitschr. Naturf.* **18b**, 266–267.

Gelderblom, H. R. (2006). Virus enrichment using the airfuge for rapid diagnostic EM in infectious diseases. *Rotor* **4**, 4–5.

Gelderblom, H., Bannert, N., and Pauli, G. (2007). Arguments pro disinfection in diagnostic electron microscopy: A response to Madeley and Biel. *J. Infect.* **54**, 307–308.

Grassucci, R. A., Taylor, D. J., and Frank, J. (2008a). Preparation of macromolecular complexes for cryo-electron microscopy. *Nat. Protoc.* **2**, 3239–3246.

Grassucci, R. A., Taylor, D. J., and Frank, J. (2008b). Visualization of macromolecular complexes using cryo-electron microscopy with FEI tecnai transmission electron microscope. *Nat. Protoc.* **3**, 330–339.

Griffiths, G. (1993). "Fine Structure Immunocytochemistry". Springer, Heidelberg, Berlin.

Grünewald, K. and Cyrklaff, M. (2006). Structure of complex viruses and virus-infected cells by electron cryo tomography. *Curr. Opin. Microbiol.* **9**, 437–442.

Hagler, H. K. (2007). Ultramicrotomy for biological electron microscopy. *In* "Methods in Molecular Biology" (J. Kuo, ed.) **vol. 369**, pp. 67–96. Humana Press, Totowa.

Harris, J. R. (1997). Negative staining and cryoelectron microscopy: Thin film techniques. "Royal Microscopy Society Handbooks" **Vol. 35**. BIOS Scientific Publishers, Oxford [the book has different publishers in different years 1996/1997. My copy is ISBN 1 85996 120 7].

Harris, J. R. (2007). Negative staining of thinly spread biological samples. *In* "Methods in Molecular Biology" (J. Kuo, ed.) **vol. 369**, pp. 107–142. Humana Press, Totowa.

Heuser, J. E. (2000). Membrane traffic in anaglyph stereo. *Traffic (Oxford, U. K.)* **1**, 35–37.

Hobot, J. A. and Newman, G. R. (1990). Electron microscopy for bacterial cells. *In* "Molecular Biological Methods for Bacillus" (C. R. Harwood, S. M. Cutting, eds.) pp. 533–566. Wiley & Sons, London.

Johnson, R. P.C. and Gregory, D. W. (1993). Viruses accumulate spontaneously near droplet surfaces: A method to concentrate viruses for electron microscopy. *J. Micros.* **171**, 125–136.

Kellenberger, E. (1998). Learning about truth and biases through experience—section surface corrugation, protein denaturation, and staining. *Micros. Res. Tech.* **42**, 33–42.

Kolesnikova, L., Bohil, A. B., Cheney, R. E., and Becker, S. (2007). Budding of marburgvirus is associated with filopodia. *Cell. Microbiol.* **9**, 939–951.

Koster, A. J. and Bárcena, M. (2006). Cryotomography: Low-dose automated tomography of frozen-hydrated specimens. *In* "Electron Tomography. Methods for Three-Dimensional Visualization of Structures in the Cell" (L. Frank, ed.) pp. 113–161. Springer, NY.

Larson, D. R., Johnson, M. C., Webb, W. W., and Vogt, M. V. (2005). Visualization of retrovirus budding with correlated light and electron microscopy. *Proc. Natl. Acad. Sci.* **102**, 15453–15458.

Laue, M. and Bannert, N. (2010). Detection limit of negative staining electron microscopy for the diagnosis of bioterrorism-related microorganisms. *J. Appl. Microbiol.* (in press) doi: 10.1111/j.1365-2672.2010.04737.x.

Laue, M., Niederwöhrmeier, B., and Bannert, N. (2007). Rapid diagnostic thin section electron microscopy of bacterial endospores. *J. Mod. Ital. Stud.* **70**, 45–54.

Leong, A. D. -Y. and Sormunen, R. T. (1998). Microwave procedures for electron microscopy and resin-embedded sections. *Micron* **29**, 397–409.

Lepault, J., Booy, F. P., and Dubochet, J. (1983). Electron microscopy of frozen biological suspensions. *J. Microsc. (Oxford)* **129**, 89–102.

Madeley, C. R. and Field, A. M. (1988). "Virus Morphology". Churchill Livingstone, Edinbourough, London, Melbourne and New York.

Mast, J. and Demeestre, L. (2009). Electron tomography of negatively stained complex viruses: Application in their diagnosis. *Diagn. Pathol.* **4**, 5.

Miller, M. F. (1982). Virus particle counting by electron microscopy. *In* "Electron Microscopy in Biology" (J. D. Griffith, ed.) pp. 306–339. Wiley, NY.

Miller, S. E. (1986). Detection and identification of viruses by electron microscopy. *J. Electron Micros. Techn.* **4**, 265–301.

Miller, S. E. (2003). Bioterrorism and electron microscopy differentiation of poxviruses from herpesviruses: Dos and don'ts. *Ultrastruct. Pathol.* **27**, 133–140.

Moor, H. (1987). Theory and practice of high pressure freezing. *In* "Cryotechniques in Biological Electron Microscopy" (R. A. Steinbrecht, K. Zierold, eds.) pp. 175–191. Springer, Berlin, Heidelberg.

Ng, M. L., Lee, J. W.M., Leong, M. L.N., Ling, A. -L., Tan, H. -C., and Ooi, E. E. (2004). Topographic changes in SARS coronavirus-infected cells during late stages of infection. *Emerging Infect. Dis.* **10**, 1907–1914.

Palmer, E. L. and Martin, M. L. (1988). "Electron Microscopy in Viral Diagnosis". CRC Press, Boca Raton.

Quintana, C. (1994). Cryofixation, cryosubstitution, cryoembedding for ultrastructural, immunocytochemical and microanalytical studies. *Micron* **25**, 63–99.

Roth, J., Taatjes, D. J., and Tokuyasu, K. T. (1990). Contrasting of lowicryl K4M thin sections. *histochemica* **95**, 123–136.

Rubo, S. D., Gardner, J. F., and Webb, R. L. (1967). Biocidal activities of glutaraldehyde and related compounds. *J. Appl. Bacteriol.* **30**, 78–87.

Schröder, J. A., Gelderblom, H. R., Hauröder, B., Schmetz, C., Milios, J., and Hofstädter, F. (2006). Microwave-assisted tissue processing for same-day EM-diagnosis of potential bioterrorism and clinical samples. *Micron* **37**, 577–590.

Schwartz, C. L., Sarbash, V. I., Ataullakhanov, F. I., McIntosh, J. R., and Nicastro, D. (2007). Cryo-fluorescence microscopy facilitates correlations between light and cryo-electron microscopy and reduces the rate of photobleaching. *J. Microsc. (Oxford)* **227**, 98–109.

Schwarz, H., Hohenberg, H., and Humbel, B. (1993). Freeze- substitution in virus research: A preview. *In* "Immuno-Gold Electron Microscopy in Virus Diagnosis and Research" (A. D. Hyatt, B. T. Eaton, eds.) pp. 349–376. CRC Press, Boca Raton, Ann Arbor.

Schwarz, H. and Humbel, B. (1989). Influence of fixatives and embedding media on immunolabelling of freeze-substituted cells. *In* "Scanning Microscopy. Supplement 3. The Science of Biological Specimen Preparation" (R. M. Albrecht, ed.) pp. 57–64. AFM O'Hare, Scanning Microscopy International, Chicago.

Skepper, J. N. (2000). Immunocytochemical strategies for electron microscopy: Choice or compromise. *J. Microsc. (Oxford)* **199**, 1–36.

Steinbrecht, R. A. and Müller, M. (1987). Freeze-substitution and freeze-drying. *In* "Cryotechniques in Biological Electron Microscopy" (R. A. Steinbrecht, K. Zierold, eds.) pp. 149–172. Springer, Berlin, Heidelberg.

Steinbrecht, R. A. and Zierold, K. (1987). "Cryotechniques in Biological Electron Microscopy". Springer, Berlin, Heidelberg.

Vanhecke, D., Studer, L., and Studer, D. (2007). Cryoultramicrotomy. *In* "Methods in Molecular Biology" (J. Kuo, ed.) **vol. 369**, pp. 175–197. Humana Press, Totowa.

Watanabe, S., Watanabe, T., Noda, T., Takada, A., Feldmann, H., Jasenosky, L. D., and Kawaoka, Y. (2004). Production of novel ebola virus-like particles from cDNAs: An alternative generation of ebola virus generation by reverse genetics. *J Virol.* **78**, 999–1005.

Webster, P. and Webster, A. (2007). Cryosectioning fixed and cryoprotected biological material form immunocytochemistry. *In* "Methods in Molecular Biology" (J. Kuo, ed.) **vol. 369**, pp. 257–288. Humana Press, Totowa.

Weibull, C., Christiansson, A., and Carlemalm, E. (1983). Extraction of membrane lipids during fixation, dehydration and embedding of acholeplasma laidlawii-cells for electron microscopy. *J. Microsc. (Oxford)* **129**, 201–207.

Yamaguchi, M., Danev, R., Nishiyama, K., Sugawara, K., and Nagayama, K. (2008). Zernike phase contrast electron microscopy of ice-embedded influenza A virus. *J. Struct. Biol.* **162**, 271–276.

Zechmann, B. and Zellnig, G. (2009). Rapid diagnosis of plant virus diseases by transmission electron microscopy. *J. Virol. Meth.* **162**, 163–169.

von Borries, B., Ruska, E., and Ruska, H. (1938). Bakterien und Virus in übermikroskopischer Aufnahme. *Klin. Wochenschr.* **17**, 921–925.

CHAPTER 2

Bacterial TEM: New Insights from Cryo-Microscopy

Martin Pilhofer[*,†], **Mark S. Ladinsky**[*], **Alasdair W. McDowall**[*], *and* **Grant J. Jensen**[*,†]

[*]Division of Biology, California Institute of Technology, Pasadena, California 91125

[†]Division of Biology, Howard Hughes Medical Institute, Pasadena, California 91125

Abstract

Some bacteria are amongst the most important model organisms for biology and medicine. Here we review how electron microscopes have been used to image bacterial cells, summarizing the technical details of the various methods, the advantages and

978-0-12-381007-6
DOI: 10.1016/S0091-679X(10)96002-0

disadvantages of each, and the major biological insights that have been obtained. Three specific example structures, "mesosomes," "cytoskeletal filaments," and "nucleoid," are used to illustrate how methodological advances have shaped our understanding of bacterial ultrastructure. Methods that involve dehydration and metal stains are widely practiced and have revealed many ultrastructural features, but they can generate misleading artifacts and have failed to preserve important structures such as the bacterial cytoskeleton. The invention of cryo-electron microscopy, which allows bacterial cells to be imaged in a frozen-hydrated, near-native state without the need for dehydration and stains, has now led to important new insights. Efforts to identify structures and localize specific proteins in cryo-EM images are summarized.

I. Introduction

Bacteria are among the most widely studied biological model systems for many reasons: they are the most abundant organisms on earth, they play vital roles in health and disease, and they are increasingly important industrial tools. They are also easy to culture, easy to manipulate genetically, and have quick life cycles. As a result, model bacteria are the focus of many current genomic, transcriptomic, proteomic, and metabolomic projects. These studies are generating exhaustive cellular "parts lists," raising hopes that we might soon understand these cells in complete molecular detail. A key piece of missing information is, however, how all the molecules within bacterial cells are arranged and how their arrangement supports function. Providing this information is the role of microscopy.

Early light microscopy (LM) revealed the amazing diversity of bacterial cell shapes, interesting developmental processes such as sporulation and binary fission, and the variety of reactions bacteria have with different stains. The invention of the electron microscope (EM) opened the possibility of imaging cells at much higher resolution, but methods had to be developed first to preserve cells within the high vacuum of the microscope column. The first methods to be developed used combinations of fixatives, solvents, and/or metal stains to prepare dehydrated samples for viewing in standard EMs at room temperature. These methods include negative staining, traditional "thin-section EM," freeze-fracture/freeze-etch, rapid freezing/freeze-substitution, and immunoEM. These methods have contributed considerably to our knowledge of bacterial ultrastructure, revealing membrane layers, stalks, flagella, pili, fimbriae, phages, and other similar features. Unfortunately, the nature of these preparative techniques often preclude them from providing reliable insight into molecular structures within the bacterial cell, and so for several decades bacteria were thought to be simple "bags of enzymes." As an example, it was long thought that bacteria lacked cytoskeletons, and moreover that this absence was one of the key characteristics that distinguished prokaryotes from eukaryotes.

Fluorescent light microscopy (fLM) then basically changed our view of the bacterial cell, showing that a multitude of proteins and also certain genetic loci are spatially and temporally localized (Lewis, 2004; Margolin, 2000; Thanbichler *et al.*, 2005). This was done either by immunofluorescence (using fluorescently labeled antibodies) or, more recently, by genetically fusing the green-fluorescent protein (GFP) or its derivatives to target proteins. Unfortunately, tagging proteins with GFP can interfere with the function or localization of proteins (Werner *et al.*, 2009), and immunofluorescence requires cells to be fixed and permeabilized, which obviously both also perturb structure.

The development of "cryo-EM" techniques led to additional breakthroughs. Cryo-EM techniques allow cells to be observed in a fully hydrated state, without the need for chemical fixatives or contrast-enhancing metal stains. In particular, when cryo-EM is combined with electron tomography, intact bacterial cells can be imaged in a nearly native state, in three dimensions, with "molecular" (~4 nm) resolution. This approach (cryo-electron tomography, or CET) has revealed a wealth of new details, including an extensive and complex bacterial cytoskeleton and the architecture of various large macromolecular complexes (Li and Jensen, 2009; Milne and Subramaniam, 2009; Tocheva *et al.*, 2010). Identifying structures and localizing specific proteins remains a challenge, as does imaging thick cells and further increasing resolution.

In this review, we present a technical overview of the different EM methods that are used to image bacteria and discuss their advantages and disadvantages (Fig. 1).

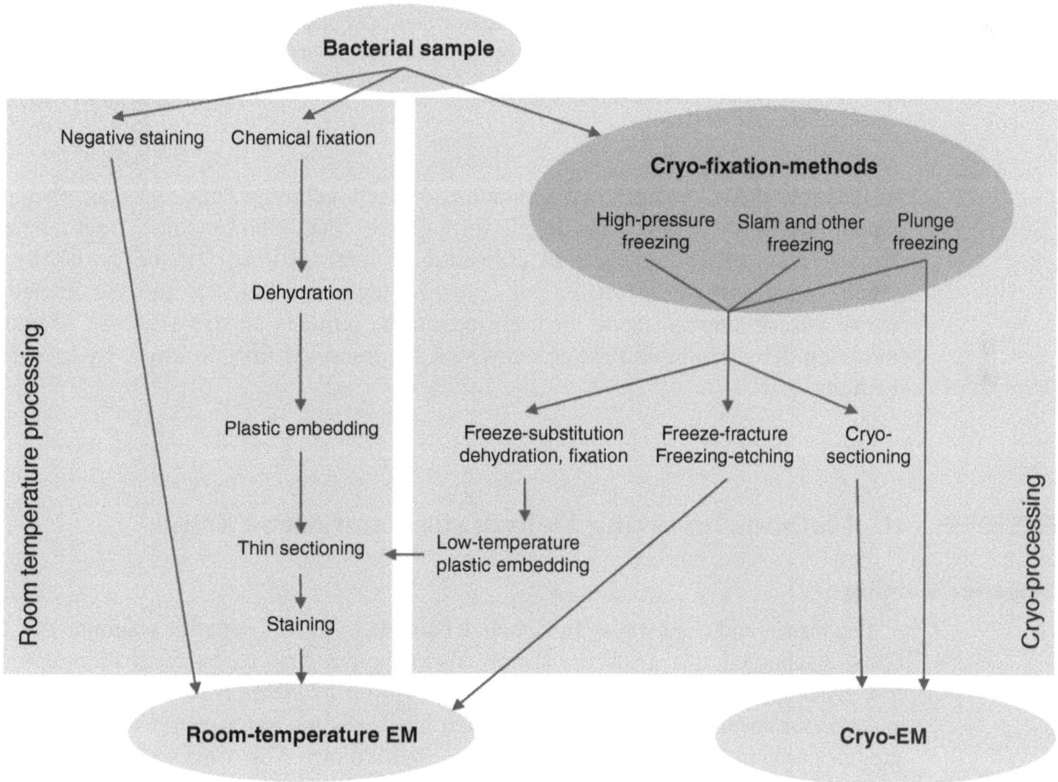

Fig. 1 EM methods to image bacteria. (See Plate no. 1 in the Color Plate Section.)

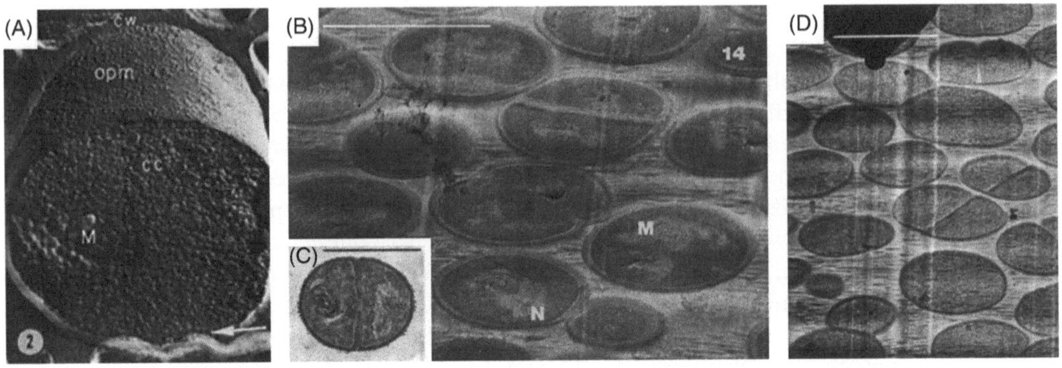

Fig. 2 Example 1: Mesosomes: Mesosomes are convoluted cytoplasmic membranous structures that were seen by methods such as freeze-fracture (A) and traditional thin-section EM (C). These were later proven by cryo-EM methods to be artifacts, since they were absent in straight-forward cryo-sections of frozen-hydrated cells (D), but were present if the cells were first fixed with osmium tetroxide and then frozen and cryo-sectioned (B). (A) Freeze-etch image of a *Bacillus subtilis* cell. Cell content (cc); outer surface of plasma membrane (opm); cross-fractured cell wall (cw); mesosome (M) composed of numerous vesicles. (B) Cryo-section through *Staphylococcus aureus* cells. Cells were fixed with osmium tetroxide before cryo-fixation, producing mesosomes (M). Nucleoid (N): (C) Conventional EM preparation of *S. aureus*. A membranous mesosome is present in the cytoplasm. (D) Cryo-section through *S. aureus* cells. No mesosomes are seen. A adapted from Nanninga (1968). B–D adapted from Dubochet *et al.* (1983). B–D, bar 1 μm.

To illustrate the advantages and limitations of each technique, three specific example ultrastructures will be considered throughout: the membranous "mesosomes" that turned out to be artifacts of conventional methods (Fig. 2); the cytoskeleton, which is curiously only visible by cryo-EM methods (Fig. 3); and the nucleoid, whose ultrastructure, despite all these advances, remains elusive (Fig. 4). Methods involving dehydration and metal stains will be described first, followed by cryo-EM methods.

II. Methods Involving Dehydration and Metal Stains

A. Negative Staining

The fastest and easiest way to visualize bacteria by EM is negative staining (Fig. 1). Glow-discharged EM grids are simply floated on a drop of bacterial culture for a few seconds, partially drained, moved onto a drop of staining solution (e.g., uranyl acetate or phosphotungstic acid), and then air-dried. As water evaporates, stain congregates around the cell, revealing its basic morphology. This works well for visualizing appendages such as flagella, pili, and fimbriae as well as associated phages (Fig. 3B).

B. Traditional Thin-Section EM

In order to obtain insights into intracellular and finer structures, bacteria can be prepared through a typical "thin-section" preparation procedure. In summary, this involves chemical fixation, dehydration into transitional solvents, infiltration with resin, embedding and curing of resin, sectioning, and finally heavy metal staining (see also Bozzola and Russell (1998) and Fig. 1). The goal of fixation is to immobilize the sample in as native state as possible. Most chemical fixation protocols are based on a method consisting of a primary fixation with glutaraldehyde and/or formaldehyde, followed by secondary exposure to osmium tetroxide (Sabatini *et al.*, 1963). The aldehydes cross-link proteins and, to a lesser degree, lipids, carbohydrates and nucleic acids. The result is that the constituents of the cell are linked into a continuous mesh. Secondary fixation with osmium tetroxide oxidizes unsaturated bonds of fatty acids and stabilizes other cell components. In addition, the reduced osmium molecules add density and contrast. Next, the sample is dehydrated by immersion in an organic solvent such as ethanol, acetone, or both, usually in a series of graded steps. The sample is then infiltrated with plastic resins, which are subsequently cured (polymerized) by either exposure to ultraviolet light or heating, producing a hard, solid block. The now "embedded" sample is placed in an ultramicrotome, trimmed of excess plastic, and sectioned. Depending on the experiment, 40- to 400-nm thick sections are then collected onto EM grids and often "post-stained" with uranium (as uranyl acetate) and/or lead (as lead citrate) to enhance the contrast of certain cellular features (Reynolds, 1963).

The advantages of this traditional thin-section EM technique are that it employs well-known and commonly practiced methods and it produces a high-contrast, radiation-resistant sample that can be imaged in standard EMs at room temperature, either in 2-D by simple projection or in 3-D through tomography. Thin-section EM reveals membranes, flagella, pili, fimbriae, microcompartments, and large macromolecular complexes such as chemoreceptor arrays, for example (Fig. 2C, 3C and 4A–C). It was by thin-section EM that the fundamental differences in the cell wall structures of gram-positive and gram-negative organisms were first visualized (Bayer, 1974; Chapman and Hillier, 1953). The major disadvantage is, however, that the chemical cross-linking, dehydration, and staining seriously perturb the fine structure of the cells. In addition to destroying some macromolecular structures (such as cytoskeletal filaments, as will be seen later) and essentially all high-resolution details, this can even create misleading artifacts. As an example, in the 1950s and 1960s prominent invaginations of the cytoplasmic membrane were detected in a multitude of bacterial species with this technique and called "mesosomes," "chondrioids," or "peripheral bodies" (Fig. 2C). These were assumed to be authentic bacterial features with a variety of functions (Greenawalt and Whiteside, 1975). More advanced preservation methods have since shown that mesosomes are simply artifacts of chemical fixation and dehydration (Dubochet *et al.*, 1983).

Concerning the nucleoid, traditional thin-section techniques have produced confusingly inconsistent results (reviewed in Eltsov and Zuber, 2006). Sometimes "vacuoles"

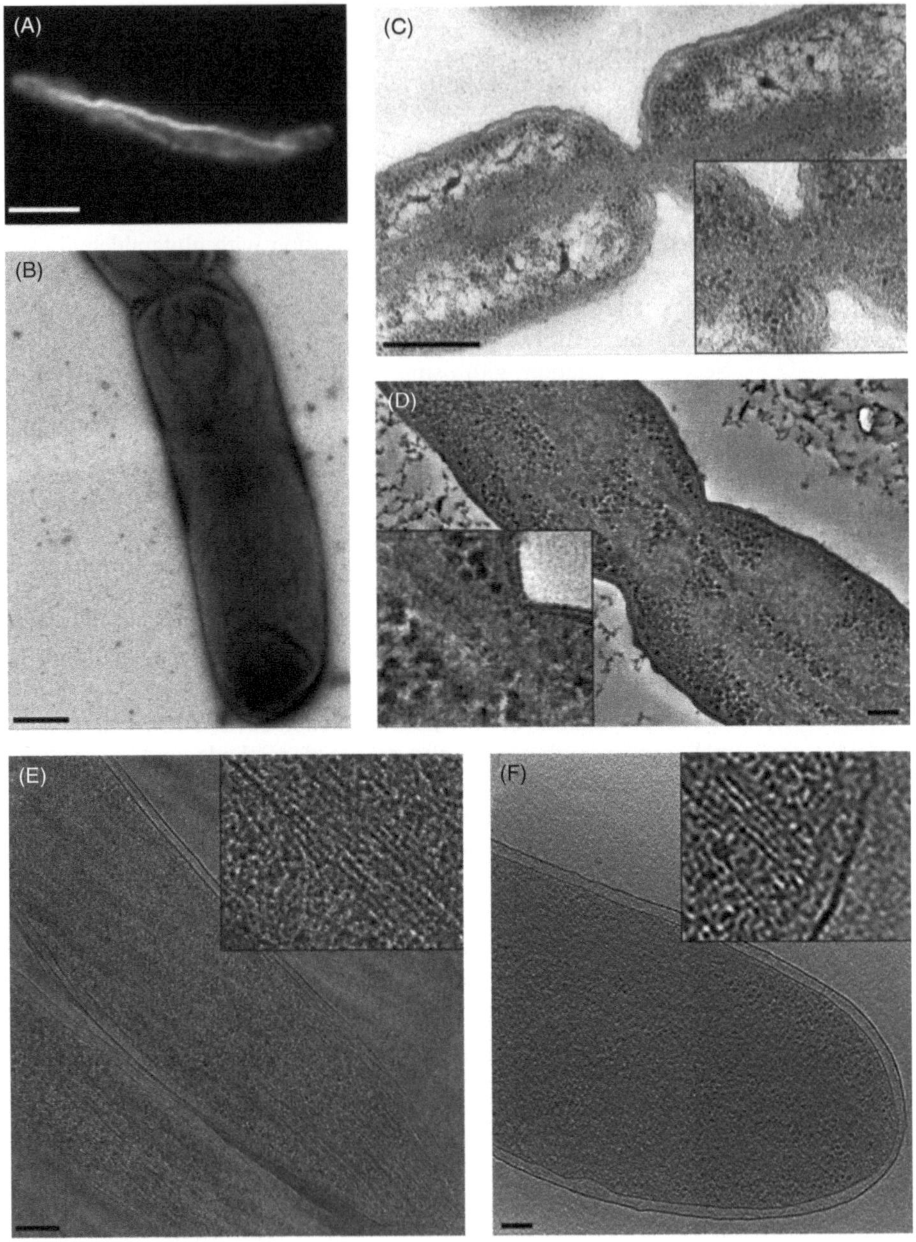

Fig. 3 *(Continued)*

of low electron density are seen which in some regions contain aggregated genomic material (arrows in Fig. 4A). Such aggregation has not been observed in most eukaryotes and archaea however, probably because of differences in chromatin composition (Kellenberger and Arnold-Schulz-Gahmen, 1992). Variations of the chemical fixation procedures were explored to reduce the aggregation of chromatin. The so-called Ryter-Kellenberger procedure was based on fixation with osmium tetroxide under special conditions, whereas another method used glutaraldehyde fixation and post-fixation with uranyl acetate. Fibrillar structures were detected using both methods (Fig. 4B and C); however, the fibril structures themselves and the shape of the nucleoid differed substantially. This led to the hypothesis that the results are biased by the fixation and dehydration protocol and that chromatin aggregation is an artifact (Eltsov and Zuber, 2006).

C. Cryo-Fixation

One of the problems of room-temperature chemical fixation is that the fixative cannot reach every molecule simultaneously and cross-link it in its natural position. Instead, certain structures become cross-linked while others continue diffusing, which can lead to artifactual meshworks and aggregations. Cryo-fixation methods (methods that stop molecular motion through cooling) can act more instantaneously. If samples are cooled gradually at ambient pressures, however, water will crystallize and denature dissolved macromolecules. One way to overcome this is to cool the sample so quickly that the water molecules stop moving before they have time to form the hydrogen bond network of crystalline ice. The resulting amorphous ice, called "vitreous ice," remains in a disordered, liquid-water-like arrangement that does not perturb the native structure of the cell (Angell, 2004; Dubochet, 2009). In fact, it has been shown that vitrified cells are preserved so well that many continue living if later thawed (Erk *et al.*, 1998).

The easiest way to increase cooling speed is to minimize the size of the specimen. This of course makes the method especially favorable to bacterial samples. Indeed, the most straight-forward way to vitrify bacteria is to spread them as a thin layer on an EM grid and

---

Fig. 3 Example 2: Cytoskeletal filaments: In contrast to mesosomes, cytoskeletal filaments are seen by cryo-EM but not plastic-embedding methods. This is why it was thought for decades that bacterial cells lacked cytoskeletons. To illustrate, images of *E. coli* cells expressing two bacterial tubulin genes, *btubA* and *btubB*, are shown, prepared by different methods. Immunofluorescence light microscopy shows that in *E. coli*, the Btub proteins localize in long, rod-like patterns suggestive of filaments (Sontag *et al.*, 2005) (A). Almost no internal ultrastructure is visible by negative staining (B). "Channels" of a different texture but no actual filaments are seen in traditional thin-section EM (C) or in high-pressure freezing/freeze-substitution preparations (D). Cryo-EM of frozen sections (E) and whole cell cryo-electron tomography (F) reveal Btub-filament bundles with single protofilament-subunit resolution. (A) Immunofluorescence of *btubA-btubB* expressing *E. coli* using Btub-specific fluorescently labeled antibodies. Bar, 2 μm. Adapted from (Sontag *et al.*, 2005). (B) Negative Staining with uranyl acetate. Bar, 500 nm. (C) Traditional thin-section EM. Bar, 500 nm. (D) XY-slice (13 nm) through tomogram of a section of high-pressure-frozen and freeze-substituted/low-temperature embedded cells. Bar, 100 nm. (E) Cryo-EM 2-D image of vitreous section. Bar, 100 nm. (F) XY-slice (10 nm) through whole cell cryo-electron tomogram. Bar, 100 nm.

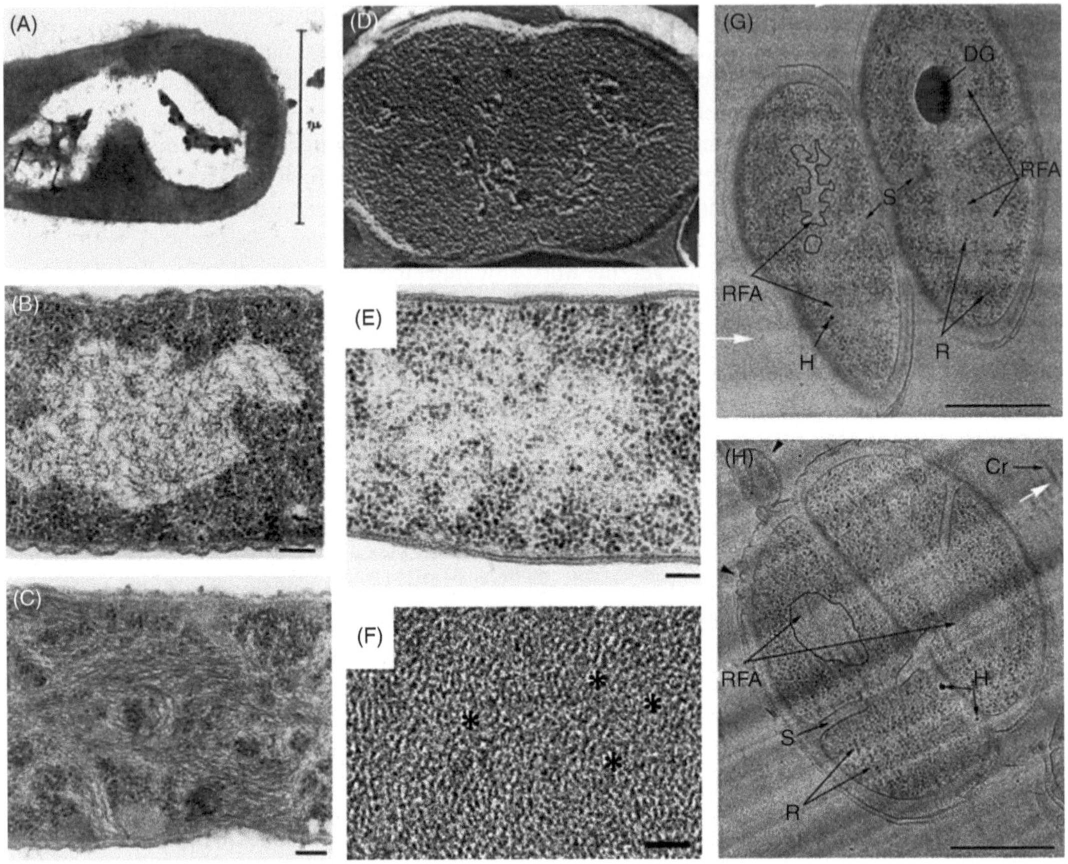

Fig. 4 Example 3: Nucleoid: While each of the established EM methods contributes some information about the organization of chromatin in bacteria, none have revealed it clearly, highlighting our continued need for additional methodological advances. Traditional "thin-section EM" methods show inconsistent results: coarse aggregates of chromatin (arrows) in electrolucent cavities (A), and different shapes of the nucleoid and different patterns of fibrils, which are likely cross-linked aggregates of chromatin (B, C). Only glutaraldehyde-prefixed cells show a nucleoid structure using freeze-fracture (D). Freeze-substitution reveals a coralline-shaped ribosome-free area with grains and fibers (E). Cryo-EM of frozen section shows a dispersed, coralline-shaped ribosome-free area in growing cells (G) and a more confined ribosome-free area in stationary phase cells (H). A characteristic fine-granularity of DNA structure is seen in the confined nucleoid of stationary phase cells, but it is difficult to interpret (F). (A) Osmium tetroxide fixation of *Bacillus megaterium*. Bar, 1 μm. (B) Osmium tetroxide fixation of *E. coli* under R-K conditions. Bar, 100 nm. (C) Glutaraldehyde fixation/uranyl acetate post-fixation of *E. coli*. Bar, 100 nm. (D) Freeze-fracture of glutaraldehyde-fixed *Streptococcus feacalis*. (E) Freeze-substitution and low-temperature embedding of *E. coli*. Bar, 100 nm. (F) Nucleoid fine-structure in cryo-section of stationary phase *Deinococcus radiodurans*. Dotted and stripy patterns are spaced by regions with indefinable structure (asterisks). Bar, 20 nm. (G) Cryo-section of exponentially growing *D. radiodurans*. Dispersed corraline ribosome-free area (RFA; outlined in one cell of the tetrad). Ribosome (R). Septum (S). Hexagonal ice (H). Electron-dense granule (DG). Bar, 500 nm. (H) Cryo-section of stationary phase *D. radiodurans*. Confined, roundish ribosome-free area (RFA; outlined in one cell of the tetrad). Ribosome (R). Septum (S). Hexagonal ice (H). Crevasses (Cr). Bar, 500 nm. (A) adapted from Eltsov and Zuber (2006); Giesbrecht and Piekarski (1958). (B–C), (E) adapted from Eltsov and Zuber (2006); Kellenberger and Arnold-Schulz-Gahmen (1992). (D) adapted from Edelstein *et al.* (1981). (F–H) adapted from Eltsov and Dubochet (2005).

plunge freeze them into a cryogen (De Carlo, 2009), as will be described in more detail below. Plunge freezing vitrifies samples of up to a micron thick (Dubochet and McDowall, 1981). The surface of a thicker sample can also be vitrified by thrusting it against a cold metal block, a technique called "slam freezing" (Spehner and Edelmann, 2009). Bacteria can be embedded into agarose before slamming (Shimizu and Miyata, 2002).

Alternatively, since crystalline ice has a lower density than liquid water, the formation of ice crystals can be inhibited by applying pressure during freezing. In high-pressure freezing (HPF), the specimen is first enclosed within a small metal (aluminum or brass) carrier. High-pressure (~2100 bars) is applied as quickly as possible to avoid alteration of the specimen structure, and then the sample is cooled with a jet of liquid nitrogen. Before the freezing step, bacterial cells are typically pelleted by centrifugation. Because growth media are relatively dilute compared with cell cytoplasm, they have a tendency to crystallize even when the contents of the cell are vitrified. The cell pellet is therefore either resuspended in medium plus a cryo-protectant (an anti-freeze agent such as dextran, bovine serum albumin (BSA), 1-hexadecene, or other high molecular weight branched sugars such as Ficoll or Mannitol), pelleted again, and transferred to the specimen carrier; or the pellet is added to a sealed pipette-tip, spun through a media-cryo-protectant solution, and transferred into the metal specimen carrier. Care is taken to avoid drying by maintaining humidity while the specimen carrier is assembled and frozen in the HPF machine. Two technically different approaches to HPF have been described, which led to the development of different types of commercially available freezing-machines including the BAL-TEC HPM 010, the LEICA EMPACT and the Wohlwend machine (their underlying principles and usage are described in Kaech (2009); McDonald *et al.* (2007); Vanhecke and Studer (2009); and Verkade (2008)). The freezing protocol (i.e., especially pelleting, percentage and kind of cryo-protectant, the way of adding the cells to the specimen holder) has to be adapted to each bacterial species because the shape of the cells, their different sedimentation coefficients, and their media all influence the freezing quality and subsequent processing of the sample. Artifacts can arise from the pelleting of the cells, mixing them with the cryo-protectant, or the impact of high pressure.

Besides the major cryo-fixation methods plunge freezing, HPF, and slam freezing, a variety of other freezing protocols have been tried through the years with mixed success. Historically, these methods have typically been followed by some form of chemical fixation, dehydration, and metal staining/replication, so that the sample can ultimately be imaged at room temperature in a conventional EM. These procedures will therefore be discussed next. The advent of cryo-EMs, which can keep cryo-fixed samples frozen during observation, however, now also allows cells to be observed immediately after cryo-fixation in near-native, "frozen-hydrated" states. These approaches are the subject of the later "cryo-EM" section.

D. Freeze-Fracture/Freeze-Etching

Once cells are frozen, they can be fractured and/or etched. For this procedure, pellets of bacterial cells are typically resuspended in 20% glycerol, sandwiched between two aluminum or copper plates, and frozen either by dropping this sandwich into liquid propane or, for better structural preservation, by HPF. The sandwich is then introduced into the high-vacuum chamber (-160°C) of the freeze-fracture/freeze-etching device and split into two pieces with a cold knife. The fracturing process typically happens along natural planes of weakness, often through the hydrophobic part of membranes. For freeze-etching, the sample is then "freeze-dried" slightly by raising the temperature to −100°C and at the same time positioning a cold knife above the sample to trap sublimated water. As water molecules leave the surface, macromolecular structures like membranes remain and become exposed. A coat of metal (typically 0.5–3 nm of Pt/C or Ta/W) is then deposited on the exposed surfaces by unidirectional shadowing. Subsequently, 20–25 nm carbon is evaporated by rotary shadowing on top of the sample to provide structural support. The sample is then thawed in distilled water and biological material is removed by sulfuric or chromic acid. After rinsing in water and/or bleach, the metal replicas are collected on copper grids and imaged in a room-temperature EM.

In addition to surface details (Studer *et al.*, 1981; bacteriorhodopsin), freeze-fracture and freeze-etching can reveal internal membrane structures (Radolf *et al.*, 1989; outer membrane) and cross-sectional views of the cell (Nanninga, 1970; cell envelope). Disadvantages are the rather low resolution (based on the grain size of the metal), the unpredictability of the fracturing process, shrinkage due to freeze-drying, stress during preparation, insufficient vitrification (i.e., non-HPF), or devitrification during the freeze-etching. Depending on the sample preparation and fixation method, even artifacts such as mesosomes can be introduced (Fig. 2A) (Higgins *et al.*, 1976; Nanninga, 1968; Remsen, 1968). To our knowledge, cytoskeletal filaments were only observed once within bacteria using freeze-fracture/freeze-etching (Martins *et al.*, 2007). Nucleoid fibers were only observed in freeze-fractured/freeze-etched cells when prefixed with glutaraldehyde (Fig. 4D) (Edelstein *et al.*, 1981), suggesting that the fibers were artifacts of chemical fixation.

E. Freeze-Substitution

Dehydration is one of the most damaging steps in conventional EM. "Freeze-substitution" (FS) minimizes damage by simultaneously fixing and dehydrating the vitrified sample at temperatures low enough to maintain cellular water frozen. Typically, the cryo-fixed sample is placed on a block of frozen organic "substitution solvent" under liquid nitrogen. There are numerous recipes for the substitution medium: acetone or ethanol is most often used with glutaraldehyde, formaldehyde, or osmium tetroxide added as fixatives. Contrast-enhancing agents such as uranyl acetate or tannic acid may also be added. Samples are immersed in such substitution media at −90°C for 12–72 h or longer, depending on the type and thickness of the sample. At this temperature the solvent melts but the cellular water remains frozen (Giddings, Jr. *et al.*, 2001). Over time the cellular water is dissolved and replaced by the fixative-containing solvent. Pellets of bacterial cells are well-preserved by a two-step substitution method: cells are first substituted at −90°C with acetone containing glutaraldehyde for 48 h, warmed to approximately −60°C, rinsed with acetone, then further substituted with acetone containing osmium tetroxide for an additional 24 h. After substitution, samples are slowly warmed over ~12 h or up to 24 h to the embedding temperature, where they are infiltrated with resins. If aldehydes are present in the mixture, they begin to cross-link at approximately −70°C. Efficient fixation with osmium tetroxide requires slightly warmer temperatures, −70 to −60°C. Infiltration with low-viscosity methacrylate resins can be done at −50°C to −30°C, and these can be polymerized with UV light while still at that low temperature. For embedding in standard plastic resins, such as Epon or Epon/Araldite, the samples are warmed to room temperature, infiltrated, and then heat-cured at +60°C. The choice of an appropriate substitution medium and resin is based on the structures intended to be analyzed and on whether the sections will be used for morphological analysis or immunolabeling.

Compared to room-temperature chemical fixation and embedding, freeze-substitution and low-temperature embedding of cryo-fixed cells has been shown to better preserve a multitude of different bacterial structures including membranes, cell walls, and various ultrastructures such as the gliding apparatus of Mycoplasmas (Graham, 1992; McCarren *et al.*, 2005; Paul and Beveridge, 1992; Shimizu and Miyata, 2002). The individual leaflets of membrane bilayers are often distinguishable, for instance. Mesosomes were not seen in freeze-substituted bacteria, which raised the first serious doubts about their authenticity. There are still artifacts associated with freeze-substitution, however, some authors speculate, for example, that when the samples are first warmed for freeze-substitution, the amorphous ice crystallizes into damaging cubic ice at approximately −135°C, before fixation (Eltsov and Zuber, 2006; Humbel, 2009; Humbel *et al.*, 2001). The organic solvents used for substitution can still extract or rearrange lipids, though substantially less than at room temperature (Paul and Beveridge, 1994). As a result of these and perhaps other issues, HPF/FS does not routinely preserve cytoskeletal filaments in bacteria (Fig. 3D). Following freeze-substitution, genomic DNA seems to localize in a ribosome-free, coralline-shaped area with a fine structure comprised of a mixture of grains and fibers (Hobot *et al.*, 1985) (Fig. 4E). Based on cryo-EM studies, however, Eltsov and Zuber (2006) later speculated that while the shape of the nucleoid may have been reliably preserved, the fine structure was probably still unreliable.

F. Immunolabeling

One of the advantages of room-temperature EM techniques is that they allow the localization of specific antigens through immunolabeling. Immunolabeling approaches can be classified by whether the antibodies are applied before or after the sample is embedded or if the sample is embedded in plastic at all. Pre-embedding methods are difficult, if not impossible, to apply to bacterial cells, since most bacteria have rigid cells walls that are impervious to all but the most destructive detergents. In order to restrict the damage, typically cells are first lightly fixed, but they are then permeabilized enzymatically (e.g., lysozyme) and/or by detergents. After the primary and secondary antibodies are added, the cells are prepared as for traditional thin-section EM (fixation, dehydration, embedment, sectioning, and staining).

Post-embedding methods are more effective for bacterial cells. The sample is first processed and thin-sectioned as described above, but then the primary and secondary antibodies are applied to the sections. This approach has the advantage of exposing

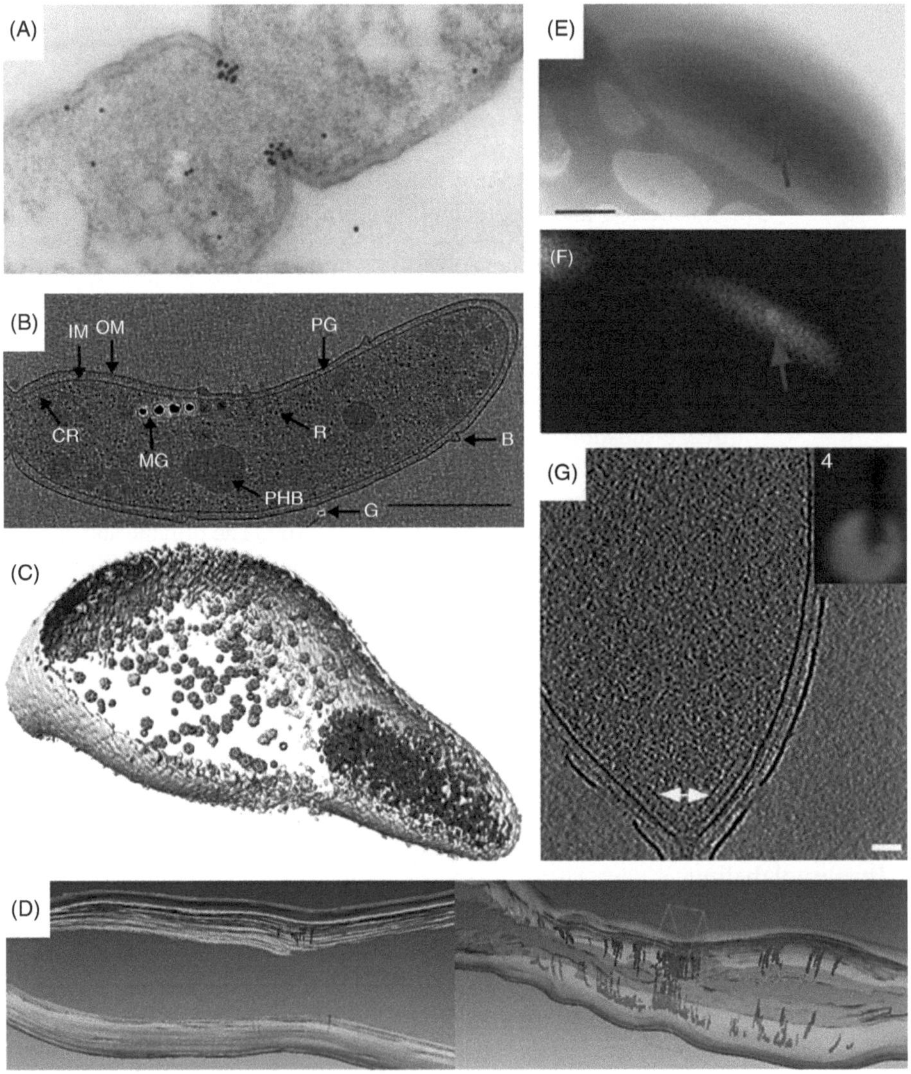

Fig. 5 *(Continued)*

intracellular epitopes to the antibodies without the need for permeabilization, but their antigenicity can be compromised by fixation and embedment. Special methacrylate-embedding resins (e.g., LR White, LR-Gold, and Lowicryl) have been developed that better preserve antigenicity; however, these resins are highly extractive and some ultrastructures such as membranes can be badly damaged. Nevertheless as an example of the success of these techniques, one of the first bacterial proteins localized by immunoEM was FtsZ, the major building block of the bacterial cell division ring (Bi and Lutkenhaus, 1991). While the location of FtsZ was seen, exemplifying the power of immunoEM, the filaments FtsZ forms were not, presumably because of fixation, dehydration, and embedment (Fig. 5A).

A third approach, which avoids plastic embedment altogether, is the method of Tokuyasu (Tokuyasu, 1973) and its variants (Ladinsky and Howell, 2007; Slot *et al.*, 1989), which has become one of the standards for eukaryotic cells (Liou *et al.*, 1996). Cells are first pelleted, mildly fixed with aldehyde, infiltrated with concentrated sucrose (i.e., >2 M), frozen in liquid nitrogen, and then cryo-sectioned (sectioned while frozen, at $\sim-110°C$). The frozen sections are transferred onto EM grids, warmed to room temperature, immunolabeled with the primary and secondary antibody solutions, and then both negative stained and infused with methylcellulose. While this method has only rarely been applied to bacteria, it has accurately defined the presence and locations of certain bacterial antigens (Brusca *et al.*, 1991; van Niftrik L. *et al.*, 2009; M. S. Ladinsky unpublished). In a new variant of this

Fig. 5 Methods to localize proteins and to identify structures. In conventional EM, immunogold labeling can be used to localize proteins, but structures are often not preserved and can be biased by artifacts (A). In cryo-EM, large ultrastructures and protein complexes can be identified by their shapes, since macromolecules are preserved in their near-native states (B–C), but in other cases, less-direct methods must be used such as manipulating the abundance or stability of a certain gene product (D), electron spectroscopic imaging (E), or correlated light and electron microscopy (F). (A) Immunogold labeling of the cell-division protein FtsZ. Note that the filamentous structure of the Z-ring is not preserved. Adapted from Bi and Lutkenhaus (1991). (B) Visual identification of large ultrastructures in whole-cell CET of *Magnetospirillum magneticum.* Outer membrane, OM; inner membrane, IM; peptidoglycan layer, PG; ribosomes, R; outer membrane bleb, B; chemoreceptor bundle, CR; polyb-hydroxybutyrate granule, PHB; gold fiduciary marker, G; magnetosome chain, MG. Bar, 500 nm. Adapted from Komeili *et al.* (2006). (C) Visual atlas of the positions of several proteins in a *Mycoplasma pneumoniae* cell produced by CET followed by pattern recognition/template matching. Structural core of pyruvate dehydrogenase shown in blue, ribosome in yellow, RNA polymerase in purple, and GroEL in red. The cell membrane is shown in light blue and the rod, a prominent structure filling the space of the tip region, is depicted in green. Adapted from Kuhner *et al.* (2009). (D) Segmented tomograms of a *Caulobacter* wild-type cell (left) and an FtsZ-overexpressing cell (right). Membranes shown in cyan and yellow, FtsZ in red. Adapted from Li *et al.* (2007). (E) Spectroscopic difference imaging of a *Caulobacter crescentus* cell. 2-D projection image (upper) and phosphorus map (lower) of the same cell. A phosphorus-rich body is indicated by the red arrow. Bar, 500 nm. Adapted from Comolli *et al.* (2006). (F) Correlated fluorescence LM and CET. Tomographic slice through a *C. crescentus* cell with mCherry fused to the chemoreceptor McpA, showing the putative chemoreceptor array structure (arrows). fLM image of same cell (inset), showing that the chemoreceptors are in the same location as the array seen by CET. The correlation was reproduced in many cells, even after the position of the array was perturbed with mutations. Adapted from Briegel *et al.* (2008). (See Plate no. 2 in the Color Plate Section.)

technique, cryo-fixed cells are freeze-substituted, rehydrated, cryo-sectioned according to Tokuyasu, and then immunolabeled. This approach has been shown to give better structural preservation with the same labeling efficiency on eukaryotic cells, and may work well for difficult-to-fix materials such as bacteria (Stierhof and Schwarz, 1989; van Donselaar E. *et al.*, 2007).

III. Cryo-Electron Microscopy

A. Plunge Freezing Thin Films

As mentioned above, because bacterial cells are small, they can be efficiently vitrified simply by spreading them into a thin film across an EM grid and plunging them into a cryogen. The grids are usually first coated with a thin, perforated/holey layer of carbon and glow-discharged just before use. A few microliters of a bacterial culture is simply applied directly to the grid, or, when working with adherent cells, EM grids can be incubated within the bacterial culture so that the cells can naturally attach to them (Seybert *et al.*, 2006). If the sample is to be used for CET, 5 or 10 nm colloidal gold particles are either mixed with the sample, dried onto the grid prior to freezing, or both to serve as fiducial makers for subsequent image alignment. In order to produce the thin film required for vitrification, the grid is blotted briefly (1–2 s) with filter paper. The grid is then rapidly plunged into the liquid cryogen, which is in turn cooled through thermal contact with liquid nitrogen. As a result, the sample cools faster than 10^6 K/s (Dubochet *et al.*, 1988), vitrifying the water without damaging macromolecules. Most laboratories use liquid ethane as a cryogen. Recent advances have shown that a mixture of propane and ethane freezes samples just as effectively as pure ethane, but the mixture does not solidify and is therefore more experimentally convenient (Tivol *et al.*, 2008). Importantly, plunging non-cryo-protected samples directly into liquid nitrogen is not usually effective because of its lower thermal conductivity. If the freezing process is not rapid enough, ice crystals will form, which can denature biomolecules. Many simple plunge freezing machines have been constructed in individual laboratories, but commercial devices are also available (Frederik and Hubert, 2005). Automatic plunge-freezers allow careful control of parameters such as temperature, humidity, and the extent of blotting prior to freezing. This dramatically improves the efficiency and reproducibility of the method. Plunge freezing is fast, simple, and widely applicable to a broad range of different bacteria, but the most important advantage is that it preserves intact cells in an essentially native state in physiological buffers without any cryo-protectants and without chemical fixation, dehydration, or staining. Plunge-frozen samples are immediately ready for imaging by cryo-EM or they can be stored in liquid nitrogen essentially indefinitely. A detailed plunge-freezing description using the FEI Vitrobot can be found in Iancu *et al.* (2006a).

B. Cryo-Electron Tomography

While tomography is not strictly a cryo-EM method (it is routinely used with plastic-embedded samples as well), its application to cryo-fixed samples (CET) is a powerful combination that is opening new windows into bacterial ultrastructure. Frozen grids are loaded at liquid nitrogen temperature into the cryo-stage of a cryo-EM. After a target cell is located, a series of low-dose projection images is acquired as the grid (and therefore the cell) is rotated around an axis, producing a "tilt-series" of 2-D images. The cell is typically imaged from around $-65°$ to $+65°$ with an increment of 1–2° depending on the thickness of the cell. The tilt range is limited by either the increasing sample thickness (as the cell is rotated) or by parts of the grid or grid-holder blocking the beam. For a movie showing a tilt-series of a vitrified bacterium see Movie 1 at http://www.elsevierdirect.com/companions/9780123810076. Since no goniometer is mechanically perfect, the target cell moves laterally and vertically within the column as the sample is tilted. To correct this, the beam and the image have to be shifted and the focus has to be adjusted before each image is acquired. Several software packages are now available which perform these corrections and acquire tilt-series images automatically (Mastronarde, 2005; Nickell *et al.*, 2005; Suloway *et al.*, 2009; Zheng *et al.*, 2007).

The single images of a tilt-series are merged computationally into a 3-D reconstruction, or "tomogram" (Movie 2 at http://www.elsevierdirect.com/companions/9780123810076). First, the images must be precisely aligned. Changes in specimen height, for instance, cause image rotation, changes in magnification, and changes in focus. Lateral movement of the sample causes shifts between images, and the tilt-increment varies slightly. The translation, rotation, magnification, tilt-axis, and tilt-angle of each image is therefore determined by tracking the gold fiducial markers (which were added to the sample before freezing) throughout the tilt-series. 3-D reconstructions can then be calculated using a variety of software packages (Amat *et al.*, 2008; Mastronarde, 2008; Nickell *et al.*, 2005). Automatic EM data collection packages (e.g., Leginon (Suloway *et al.*, 2009), SerialEM (Mastronarde, 2005), TOM (Nickell *et al.*, 2005), etc.) and methods for automatic tomogram reconstruction (e.g., Raptor (Amat *et al.*, 2008)) now allow tens of tomograms to be produced in a single day. This makes it possible to screen cells for structural features present in very low abundance, for instance, or screen cells from many different strains, with different mutations, and/or different growth stages. Large numbers of tomograms also make it possible to obtain more reliable reconstructions of regular objects through averaging (Liu *et al.*, 2009).

Tomograms can be denoised to improve image contrast and enhance interpretability (Frangakis and Hegerl, 2001; Narasimha *et al.*, 2008). Real-space digital filters such as Gaussian smoothing, as well as more advanced filters such as nonlinear anisotropic diffusion have been used. Segmentation and surface rendering can be used to decompose a complex, 3-D object into its individual components (Pruggnaller *et al.*, 2008). These methods help with the perception of the 3-D data and can be used to interrogate the continuity of membranes or spatial relationships of macromolecular complexes (Movie 3 at http://www.elsevierdirect.com/companions/9780123810076). Some structures can now be segmented automatically (by computer) (Sandberg and Brega, 2007), but others must still be done manually.

C. Cryo-Sectioning and EM of Frozen Sections

While tomography allows samples to be imaged in 3-D, a major limitation is that even high-voltage electrons (300 KeV) are usually inelastically scattered after passing through just ~0.5 μm of biological material, and are thereby lost to the image inducing "noise." Bacteria that are thicker than that must therefore first be sectioned (Fig. 1). In order to preserve the native molecular structure, sections must be cut in a cryo-ultramicrotome cooled by liquid nitrogen (cryo-fixed materials devitrify above −135°C). Under these conditions, samples remain vitreous throughout sectioning (McDowall *et al.*, 1983). Sections are typically 80–150 nm thick.

Cutting plastic sections is facilitated by floating them on water. A liquid which performs the same functions at cryo-temperatures and which is also not toxic to the operator has not yet been found, so cryo-sections have to be cut "dry." A low-angle diamond knife that the sections can glide off smoothly is typically used. Vitreous water is a liquid with very high viscosity, which means that it flows under the force applied by the knife. Vitreous samples are therefore considerably compressed along the cutting direction (generally between 15 and 60%), and thicken proportionally. Nonhomogeneous material and discontinuities in the cutting process lead to irregular deformations and fractures (crevasses) perpendicular to the cutting direction. The thickness of the section along the cutting direction also varies due to friction of the knife against the sample. This is referred to as "chatter." Damage along the edge of the knife or particles adhering on its surface can cause knife marks along the cutting direction (Al-Amoudi *et al.*, 2005). Some of these cutting artifacts can be reduced by trimming the block into a square-based pyramid to reduce tension before sectioning.

Because cryo-sections are sensitive to electrostatic charges (which causes them to fly or to stick to any surface), a low relative humidity and an adjustable ion shower in the vicinity of the knife makes them easier to control. Under good cutting conditions, the sections come off the knife as regular ribbons. Sections or ribbons are caught with an eyelash or other suitable fiber and transferred to a prepositioned grid. Ladinsky *et al.* (2006) employed a micromanipulator to hold and control the cryo-sections. This allows slower cutting speeds to be used and more reproducible placement of the ribbon onto the grid. Once cut, the sections are flattened onto the grid by pressing with a flat surface (metal, ceramic, or glass) or by applying an electrostatic charge (Pierson *et al.*, 2010). The grids can be stored in liquid nitrogen before examination. More detailed technical descriptions can be found in Al-Amoudi *et al.* (2004) and Bouchet-Marquis and Fakan (2009).

The major drawbacks of cryo-sectioning are that it is tedious, it is technically very demanding (material and skills), and it is hampered by cutting artifacts. The cutting artifacts are quite predictable, however, and can therefore be recognized easily in the image or tomogram and discounted. Vitrified sections can also be harder to image since they are mechanically less stable. Nevertheless as the technology of cryomicrotomes, grids and substrates improves, the quality and quantity of data from cryosections will increase. Serial cryo-sectioning may even be possible to image a larger volume of the same cell, though it is unclear how much material will be lost and how much distortion will exist above and below each section (Leis *et al.*, 2008).

D. Limitations of Cryo-EM

Cryo-EM methods are fundamentally different than the previously discussed techniques because contrast from biological macromolecules in their native state dominates the image rather than metal stains. The disadvantages are, however, that since biomolecules comprise mainly light atoms, the contrast is low, and it cannot be increased by applying higher doses because unfixed biological samples are destroyed by electron irradiation. As a result, during image acquisition the originally sharp edges of macromolecular structures gradually degrade and eventually "bubbles" of (presumably) radiolytic fragments appear and catastrophically disrupt the structure (Comolli and Downing, 2005; Iancu *et al.*, 2006b; Wright *et al.*, 2006). Thus the most fundamentally limiting factor in cryo-EM is the total number of electrons that can be used to record images before the sample is destroyed. Depending on how much damage is tolerated, in practice total doses of ~50–200 electrons/$Å^2$ are typically used for a CET tilt-series. As a result, individual atoms or even protein secondary structures are not resolved, but instead only the position of domains and the rough morphology of complexes.

As the sample is tilted, the distance the electrons must traverse through the sample increases as the secant of the tilt-angle, and more and more electrons are lost to inelastic scattering. Thus another fundamental limitation in CET is that the interpretability (clarity and resolution) of reconstructions degrades with sample thickness, with 0.5 μm being a useful practical limit for intermediate accelerating voltages (~300 kV). As mentioned earlier, thicker cells must be cryo-sectioned, or at least gently lysed just before freezing, for instance, with cell wall-degrading enzymes (Briegel *et al.*, 2009). The locations of many macromolecules are obviously perturbed by lysis, but some structures remain intact.

As a further consequence of inelastic scattering, it is usually not possible to collect data past about ±70°, resulting in a "missing wedge" of nonsampled information. As a result, the resolution of the 3-D reconstruction in the direction parallel to the electron beam is significantly worse than the resolution perpendicular. In simple visual terms, this causes spherical objects to appear somewhat ellipsoidal (smeared in the direction of the beam) and continuous objects such as filaments and membranes to be more visible in some orientations than in others. Unfortunately, plunge-frozen rod-shaped cells almost always lie flat across the grid, so the missing wedge can therefore always obscure the same features. While the missing wedge may be reduced to a missing pyramid by rotating the grid 90° and collecting a second, orthogonal tilt-series, this procedure is more than twice as time consuming, the dose that can be used per image is halved, and alignment errors between the tilt-series can erode the benefit (Iancu *et al.*, 2005; Nickell *et al.*, 2003). HPF and cryo-sectioning can overcome this problem, however, since essentially random cross-sections of cells can be cut.

E. Identification of Structures in Cryo-EM

Identifying structures of interest in cryo-EM images and tomograms remains a major challenge. Large ultrastructures such as membranes, S-layers, peptidoglycan, flagella, pili, storage granules, carboxysomes, and gas vesicles can be easily identified by their characteristic appearance and structure (Fig. 5B). Because molecules are imaged in their native state, however, other smaller ultrastructures and even individual macromolecules can also be recognized directly by their shape. The power spectrum of ParM filaments in cryo-sectioned cells, for instance, revealed a subunit repeat and interfilament distance matching the crystal structure of ParM filaments (Salje *et al.*, 2008). Pattern recognition algorithms have been developed that identify target macromolecular complexes based on the correspondence of their shape with a known template. The number and orientation of 70S ribosomes in a *Spiroplasma* cell were identified by such an approach (Ortiz *et al.*, 2006). Very recently, "visual atlases" of the positions of several different macromolecules in a single cell have been produced (Beck *et al.*, 2009; Kuhner *et al.*, 2009) (Fig. 5C).

Comparing images of wild-type and mutant (deletion, depletion, or overexpression) strains can also reveal the identity of certain structures by perturbing their abundance, position, frequency, or size (Fig. 5D) (Komeili *et al.*, 2006; Li *et al.*, 2007). While this approach has become practical by the automation of CET, it can still only be applied to genetically manipulable model organisms. Electron energy loss spectroscopy (EELS) and electron spectroscopic imaging (ESI) in combination with CET can be used to map elemental compositions in the bacterial cell. These methods are based on the fact that electrons transfer specific amounts of energy to different elements when they scatter, and this energy loss can be measured using the energy filter. Bacterial storage granules have been distinguished as enriched in carbon, nitrogen, and/or phosphorous in this way (Fig. 5E) (Comolli *et al.*, 2006; Iancu *et al.*, 2010).

Other methods to identify objects involve tagging targets of interest with either fluorescent or electron-dense markers. If a fluorescent protein can be fused to a target, for instance, its approximate location within a set of cells can be determined by fLM. If the same cells can then be imaged by CET before any molecular rearrangements occur, and if a particular ultrastructure is always found in the same location as the fluorescence signal and nowhere else, it likely contains the target protein. To facilitate correlated LM/EM, cryo-stages have been constructed for light microscopes which allow frozen samples to be imaged (Sartori *et al.*, 2007; Schwartz *et al.*, 2007). Because the sample must remain frozen, however, oil-immersion lenses cannot be used and the resolution is limited. For this reason, Briegel *et al.* (2008) developed a method to immobilize bacteria on an EM grid and image them with a high-resolution, 100×, oil-immersion lens at room temperature before plunge freezing. Chemoreceptor arrays were identified in *Caulobacter crescentus* in this way (Fig. 5F). One drawback of this method is that it requires mild fixation, which is known to perturb some structures and introduce artifacts. Excitingly, genetically encodable tags which nucleate metal clusters have very recently been described (Diestra *et al.*, 2008; Mercogliano and DeRosier, 2007; Nishino *et al.*, 2007), which may allow specific proteins to be localized directly through cryo-tomography. One of the problems with any method that requires labels is, however, that labels sometimes disrupt their target's ability to form complexes or localize to their native position. This is in fact frequently the case with GFP (Werner *et al.*, 2009).

F. Contributions of Cryo-EM

CET, and cryo-EM more generally, has now provided dramatic advances in our understanding of bacterial ultrastructure. In this review we have followed three specific features (mesosomes, cytoskeleton, and nucleoid) as examples. For more comprehensive reviews, see Li and Jensen (2009), Tocheva *et al.* (2009), and Milne and Subramaniam (2009).

After mesosomes were detected using traditional EM and freeze-fracture methods, and after the freeze-substitution method raised serious doubts about their authenticity, cryo-EM showed unambiguously that mesosomes were artifacts introduced by sample preparation (Fig. 2D). It was even shown that mesosomes could be artificially introduced (Fig. 2B) by fixing the sample with osmium tetroxide before cryo-fixation and cryo-EM (Dubochet *et al.*, 1983). This raises the important question about how fixation produces mesosomes. One possibility is that during chemical fixation, osmotic pressure gradients cause water to leave the cell rapidly, which causes the membrane to invaginate. Dehydration steps could obviously aggravate the problem.

Concerning the bacterial cytoskeleton, filaments were only rarely seen in chemically fixed, plastic-embedded, or metal-stained samples (Bermudes *et al.*, 1994; Martins *et al.*, 2007; van Iterson *et al.*, 1967). In contrast, CET has now revealed numerous distinct bundles of filaments in almost every species that has been imaged (Briegel *et al.*, 2006; Tocheva *et al.*, 2009). As a specific example, in Fig. 3 images of *Escherichia coli* cells over-expressing two particular bacterial cytoskeletal proteins are shown, prepared by the various techniques described in this review. The two genes, *btubA* and *btubB*, are organized within a single operon (Pilhofer *et al.*, 2007a) and were first noticed in different species of the bacterial genus *Prosthecobacter* because of their high similarity to eukaryotic tubulin (Jenkins *et al.*, 2002; Pilhofer *et al.*, 2007b). We recombinantly overexpressed this *Prosthecobacter* bacterial tubulin operon in *E. coli* and prepared the cells in different ways. In contrast to all other EM methods (negative-staining, traditional thin-section EM, and HPF/FS), only cryo-EM of either whole cells (CET) or cryo-sections (2D-imaging) showed filament bundles running the length of the cells (Figure 3E and F). This is consistent with previous immunofluorescence observations showing fluorescent signals resembling rods running the length of the cells (Figure 3A) (Sontag *et al.*, 2005). Using cryo-EM methods, the resolution was good enough to distinguish even single bacterial tubulin subunits within the filaments. The comparison of Btub filaments in cryo-sections and whole cell tomograms also shows that, at least in this particular case, the two main cryo-fixation methods (i.e., HPF and plunge freezing) deliver roughly comparable structural preservation. Knowing what the filament bundles look like and where they localize, looking back at the traditional thin-section EM and HPF/FS images, "channels" of different texture can be seen that are likely remnants of the filaments. It is possible that the filaments depolymerized during chemical fixation or dehydration, but that the monomers still remained roughly in their original position, producing the appearance of a channel. A second possibility is that the stain failed to resolve individual filaments. The latter seems improbable since cytoskeletal filaments (and especially microtubules) are very well resolved in eukaryotic cells using traditional EM and especially freeze-substitution. Why then are cytoskeletal filaments so much less likely to be preserved in bacteria than in eukaryotes? It seems unlikely that all bacterial filaments are biochemically different than eukaryotic filaments in some EM-critical way, since by sequence they are highly diverse, and in fact cluster more closely to their eukaryotic counterparts than to other bacterial cytoskeletal protein families. The bacterial tubulins shown here, for instance, are more closely related to eukaryotic tubulins then to all other known bacterial proteins (Jenkins *et al.*, 2002; Pilhofer *et al.*, 2007b). Instead, one might speculate that differences in cell size, cell wall, composition/number of membranes, and the composition of the cytoplasm result in different fixation/dehydration kinetics, causing the filaments to depolymerize before they are cross-linked.

In the case of the nucleoid, no EM technique has produced clear results yet. The different EM techniques involving dehyration and stains produced conflicting results, showing many different patterns of fibers. Since these patterns must be derivatives of the native structure (probably local aggregates), however, these patterns combined with additional data might be helpful clues that at least rule out certain models of chromatin organization. Likewise, the structure of bacterial chromatin has not been made clear by cryo-EM yet either. Ribosome-free regions are typically seen by cryo-EM in actively growing cultures (Fig. 4G) corresponding to the corraline shape seen by freeze-substitution (Fig. 4E). While it is not normally possible to discern any ordered fine-structure in such regions of growing cells (suggesting a random organization of chromatin), characteristic textures have been found there in stationary phase cells, and their "ribosome-free" nucleoids seem to be more confined (Fig. 4H). Dotted and stripy patterns, for instance, have been interpreted as single, parallel DNA filaments in different orientations (Fig. 4F), inspiring at least one model for DNA packing (Eltsov and Dubochet, 2005, 2006; Eltsov and Zuber, 2006). It is likely that in order to confirm the present models for stationary phase chromatin structure (Eltsov and Zuber, 2006) and in order to get insights into the fine-structure of chromatin in exponentially growing cells, further improvements in cryo-EM instrumentation (like the advent of direct detectors, phase plates, and aberration correctors) will be needed. The continual development of new and better EM techniques has clearly played a major role in the history of microbiology, and will continue for years to come.

Acknowledgments

Giulio Petroni and Karl-Heinz Schleifer are acknowledged for providing their laboratories and assistance for traditional thin-section EM of recombinant *E. coli* cells. Dylan M. Morris, H. Jane Ding, and Sarah Cheng are acknowledged for providing movies. MP and GJJ are supported by the Howard Hughes Medical Institute. AWM is supported in part by NIH grants R01 AI067548, R01 GM081520, R01 GM086200, R01 AI049194, and by the Beckman Institute at Caltech. MSL is supported by NIH grant 2R37-A1041239-06A1 to Pamela S. Björkman. EM studies of bacterial ultrastructure in the Jensen laboratory at Caltech are supported in part by gifts to Caltech from the Gordon and Betty Moore Foundation and Agouron Institute.

References

Al-Amoudi, A., Chang, J. J., Leforestier, A., McDowall, A., Salamin, L. M., Norlen, L. P., Richter, K., Blanc, N. S., Studer, D., and Dubochet, J. (2004). Cryo-electron microscopy of vitreous sections. *EMBO J.* **23**, 3583–3588.

Al-Amoudi, A., Studer, D., and Dubochet, J. (2005). Cutting artefacts and cutting process in vitreous sections for cryo-electron microscopy. *J. Struct. Biol.* **150**, 109–121.

Amat, F., Moussavi, F., Comolli, L.R., Elidan, G., Downing, K.H., and Horowitz, M. (2008). Markov random field based automatic image alignment for electron tomography. *J. Struct. Biol.* **161**, 260–275.

Angell, C.A. (2004). Amorphous water. *Ann. Rev. Phys. Chem.* **55**, 559–583.

Bayer, M. E. (1974). Ultrastructure and organization of the bacterial envelope. *Ann. NY. Acad. Sci.* **235**, 6–28.

Beck, M., Malmstrom, J. A., Lange, V., Schmidt, A., Deutsch, E. W., and Aebersold, R. (2009). Visual proteomics of the human pathogen leptospira interrogans. *Nat. Methods* **6**, 817–823.

Bermudes, D., Hinkle, G., and Margulis, L. (1994). Do prokaryotes contain microtubules. *Microbiol. Rev.* **58**, 387–400.

Bi, E., and Lutkenhaus, J. (1991). FtsZ ring structure associated with division in *Escherichia coli*. *Nature* **354**, 161–164.

Bouchet-Marquis, C., and Fakan, S. (2009). Cryoelectron microscopy of vitreous sections: A step further towards the native state. *Methods Mol. Biol.* **464**, 425–439.

Bozzola, J. J., and Russell, L. D. (1998). "Electron Microscopy: Principles and Techniques for Biologists". Jones and Bartlett Publishers, Sudbury, MA.

Briegel, A. <, et al., (2009). Universal architecture of bacterial chemoreceptor arrays. *Proc Natl. Acad. Sci. U. S.A.* **106**, 17181–17186.

Briegel, A., Dias, D. P., Li, Z., Jensen, R. B., Frangakis, A. S., and Jensen, G. J. (2006). Multiple large filament bundles observed in *Caulobacter crescentus* by electron cryotomography. *Mol. Microbiol.* **62**, 5–14.

Briegel, A., Ding, H. J., Li, Z., Werner, J., Gitai, Z., Dias, D. P., Jensen, R. B., and Jensen, G. J. (2008). Location and architecture of the *Caulobacter crescentus* chemoreceptor array. *Mol. Microbiol.* **69**, 30–41.

Brusca, J. S., McDowall, A. W., Norgard, M. V., and Radolf, J. D. (1991). Localization of outer surface proteins A and B in both the outer membrane and intracellular compartments of *Borrelia burgdorferi*. *J. Bacteriol.* **173**, 8004–8008.

Chapman, G. B., and Hillier, J. (1953). Electron microscopy of ultra-thin sections of bacteria I. Cellular division in *bacillus cereus*. *J. Bacteriol.* **66**, 362–373.

Comolli, L. R., and Downing, K. H. (2005). Dose tolerance at helium and nitrogen temperatures for whole cell electron tomography. *J. Struct. Biol.* **152**, 149–156.

Comolli, L. R., Kundmann, M., and Downing, K. H. (2006). Characterization of intact subcellular bodies in whole bacteria by cryo-electron tomography and spectroscopic imaging. *J. Microsc.* **223**, 40–52.

De Carlo, S. (2009). Plunge freezing. *In* "Handbook of Cryo-Preparation Methods for Electron Microscopy" (A. Cavalier, D. Spehner, and B. M. Humbel, eds.) pp. 48–68. CRC Press, Boca Raton.

Diestra, E., Fontana, J., Guichard, P., Marco, S., and Risco, C. (2008). Visualization of proteins in intact cells with a clonable tag for electron microscopy. *J. Struct. Biol.* **165**, 157–168.

Dubochet, J. (2009). Vitreous water. In: "Handbook of Cryo-Preparation Methods for Electron Microscopy" (A.Cavalier, D. Spehner, and B. M. Humbel, eds.) pp. 1–17. CRC Press, Boca Raton.

Dubochet, J., Adrian, M., Chang, J. J., Homo, J. C., Lepault, J., McDowall, A. W., and Schultz, P. (1988). Cryo-electron microscopy of vitrified specimens. *Q. Rev. Biophys.* **21**, 129–228.

Dubochet, J., and McDowall, A. W. (1981). Vitrification of pure water for electron microscopy. *J. Microsc.* **124**, RP.

Dubochet, J., McDowall, A. W., Menge, B., Schmid, E. N., and Lickfeld, K. G. (1983). Electron microscopy of frozen-hydrated bacteria. *J. Bacteriol.* **155**, 381–390.

Edelstein, E., Parks, L., Tsien, H. C., Daneo-Moore, L., and Higgins, M. L. (1981). Nucleoid structure in freeze fractures of *streptococcus faecalis*: Effects of filtration and chilling. *J. Bacteriol.* **146**, 798–803.

Eltsov, M., and Dubochet, J. (2005). Fine structure of the *Deinococcus radiodurans* nucleoid revealed by cryoelectron microscopy of vitreous sections. *J. Bacteriol.* **187**, 8047–8054.

Eltsov, M., and Dubochet, J. (2006). Study of the *Deinococcus radiodurans* nucleoid by cryoelectron microscopy of vitreous sections: Supplementary comments. *J. Bacteriol.* **188**, 6053–6058.

Eltsov, M., and Zuber, B. (2006). Transmission electron microscopy of the bacterial nucleoid. *J. Struct. Biol.* **156**, 246–254.

Erk, I., Nicolas, G., Caroff, A., and Lepault, J. (1998). Electron microscopy of frozen biological objects: A study using cryosectioning and cryosubstitution. *J. Microsc.* **189**, 236–248.

Frangakis, A. S., and Hegerl, R. (2001). Noise reduction in electron tomographic reconstructions using nonlinear anisotropic diffusion. *J. Struct. Biol.* **135**, 239–250.

Frederik, P. M., and Hubert, D. H. (2005). Cryoelectron microscopy of liposomes. *Methods Enzymol.* **391**, 431–448.

Giddings, T. H., Jr., O'Toole, E. T., Morphew, M., Mastronarde, D. N., McIntosh, J. R., and Winey, M. (2001). Using rapid freeze and freeze-substitution for the preparation of yeast cells for electron microscopy and three-dimensional analysis. *Methods Cell Biol.* **67**, 27–42.

Giesbrecht, P., and Piekarski, G. (1958). Zur organisation des zellkerns von *Bacillus megaterium*. *Arch. Mikrobiol.* **31**, 68–81.

Graham, L. L. (1992). Freeze-substitution studies of bacteria. *Electron Microsc. Rev.* **5**, 77–103.

Greenawalt, J. W., and Whiteside, T. L. (1975). Mesosomes: membranous bacterial organelles. *Bacteriol. Rev.* **39**, 405–463.

Higgins, M. L., Tsien, H. C., and Daneo-Moore, L. (1976). Organization of mesosomes in fixed and unfixed cells. *J. Bacteriol.* **127**, 1519–1523.

Hobot, J. A., Villiger, W., Escaig, J., Maeder, M., Ryter, A., and Kellenberger, E. (1985). Shape and fine structure of nucleoids observed on sections of ultrarapidly frozen and cryosubstituted bacteria. *J. Bacteriol.* **162**, 960–971.

Humbel, B. M. (2009). Freeze-substitution. *In* "Handbook of Cryo-Preparation Methods for Electron Microscopy" (A.Cavalier, D. Spehner, and B. M. Humbel, eds.) pp. 319–341. CRC Press, Boca Raton.

Humbel, B. M., Konomi, M., Takagi, T., Kamasawa, N., Ishijima, S. A., and Osumi, M. (2001). In situ localization of beta-glucans in the cell wall of *Schizosaccharomyces pombe*. *Yeast* **18**, 433–444.

Iancu, C. V. <, et al., (2006a). Electron cryotomography sample preparation using the vitrobot. *Nat. Protoc.* **1**, 2813–2819.

Iancu, C. V., Morris, D. M., Dou, Z., Heinhorst, S., Cannon, G. C., and Jensen, G. J. (2010). Organization, structure, and assembly of alpha-carboxysomes determined by electron cryotomography of intact cells. *J. Mol. Biol.* **396**, 105–117.

Iancu, C. V., Wright, E. R., Benjamin, J., Tivol, W. F., Dias, D. P., Murphy, G. E., Morrison, R. C., Heymann, J. B., and Jensen, G. J. (2005). A "flip-flop" rotation stage for routine dual-axis electron cryotomography. *J. Struct. Biol.* **151**, 288–297.

Iancu, C. V., Wright, E. R., Heymann, J. B., and Jensen, G. J. (2006b). A comparison of liquid nitrogen and liquid helium as cryogens for electron cryotomography. *J. Struct. Biol.* **153**, 231–240.

Jenkins, C., Samudrala, R., Anderson, I., Hedlund, B. P., Petroni, G., Michailova, N., Pinel, N., Overbeek, R., Rosati, G., and Staley, J. T. (2002). Genes for the cytoskeletal protein tubulin in the bacterial genus *prosthecobacter*. *Proc. Natl. Acad. Sci. U. S.A.* **99**, 17049–17054.

Kaech, A. (2009). BAL-TEC HPM 010 high-pressure freezing machine. *In* "Handbook of Cryo-Preparation Methods for Electron Microscopy" (A.Cavalier, D. Spehner, and B. M. Humbel, eds.) pp. 101–128. CRC Press, Boca Raton.

Kellenberger, E., and Arnold-Schulz-Gahmen, B. (1992). Chromatins of low-protein content: Special features of their compaction and condensation. *FEMS Microbiol. Lett.* **79**, 361–370.

Komeili, A., Li, Z., Newman, D. K., and Jensen, G. J. (2006). Magnetosomes are cell membrane invaginations organized by the actin-like protein MamK. *Science* **311**, 242–245.

Kuhner, S. <, et al., (2009). Proteome organization in a genome-reduced bacterium. *Science* **326**, 1235–1240.

Ladinsky, M. S., and Howell, K. E. (2007). Electron tomography of immunolabeled cryosections. *Methods Cell Biol.* **79**, 543–558.

Ladinsky, M. S., Pierson, J. M., and McIntosh, J. R. (2006). Vitreous cryo-sectioning of cells facilitated by a micromanipulator. *J. Microsc.* **224**, 129–134.

Leis, A., Rockel, B., Andrees, L., and Baumeister, W. (2008). Visualizing cells at the nanoscale. *Trends Biochem. Sci.* **34**, 60–70.

Lewis, P. J. (2004). Bacterial subcellular architecture: Recent advances and future prospects. *Mol. Microbiol.* **54**, 1135–1150.

Li, Z., and Jensen, G. J. (2009). Electron cryotomography: A new view into microbial ultrastructure. *Curr. Opin. Microbiol.* **12**, 333–340.

Li, Z., Trimble, M. J., Brun, Y. V., and Jensen, G. J. (2007). The structure of FtsZ filaments *in vivo* suggests a force-generating role in cell division. *EMBO J.* **26**, 4694–4708.

Liou, W., Geuze, H. J., and Slot, J. W. (1996). Improving structural integrity of cryosections for immunogold labeling. *Histochem. Cell. Biol.* **106**, 41–58.

Liu, J., Lin, T., Botkin, D. J., McCrum, E., Winkler, H., and Norris, S. J. (2009). Intact flagellar motor of *Borrelia burgdorferi* revealed by cryo-electron tomography: Evidence for stator ring curvature and rotor/C ring assembly flexion. *J. Bacteriol.* **191**, 5026–5036.

Margolin, W. (2000). Green fluorescent protein as a reporter for macromolecular localization in bacterial cells. *Methods* **20**, 62–72.

Martins, J. L., Keim, C. N., Farina, M., Kachar, B., and Lins, U. (2007). Deep-etching electron microscopy of cells of *Magnetospirillum magnetotacticum*: Evidence for filamentous structures connecting the magnetosome chain to the cell surface. *Curr. Microbiol.* **54**, 1–4.

Mastronarde, D. N. (2005). Automated electron microscope tomography using robust prediction of specimen movements. *J. Struct. Biol.* **152**, 36–51.

Mastronarde, D. N. (2008). Correction for non-perpendicularity of beam and tilt axis in tomographic reconstructions with the IMOD package. *J. Microsc.* **230**, 212–217.

McCarren, J., Heuser, J., Roth, R., Yamada, N., Martone, M., and Brahamsha, B. (2005). Inactivation of swmA results in the loss of an outer cell layer in a swimming synechococcus strain. *J. Bacteriol.* **187**, 224–230.

McDonald, K. L., Morphew, M., Verkade, P., and Muller-Reichert, T. (2007). Recent advances in high-pressure freezing: Equipment- and specimen-loading methods. *Methods Mol. Biol.* **369**, 143–173.

McDowall, A. W., Chang, J. J., Freeman, R., Lepault, J., Walter, C. A., and Dubochet, J. (1983). Electron microscopy of frozen hydrated sections of vitreous ice and vitrified biological samples. *J. Microsc.* **131**, 1–9.

Mercogliano, C., and DeRosier, D. (2007). Concatenated metallothionein as a clonable gold label for electron microscopy. *J. Struct. Biol.* **160**, 70–82.

Milne, J. L., and Subramaniam, S. (2009). Cryo-electron tomography of bacteria: Progress, challenges and future prospects. *Nat. Rev. Microbiol.* **7**, 666–675.

Nanninga, N. (1968). Structural features of mesosomes (chondrioids) of *Bacillus subtilis* after freeze-etching. *J. Cell Biol.* **39**, 251–263.

Nanninga, N. (1970). Ultrastructure of the cell envelope of *Escherichia coli* B after freeze-etching. *J. Bacteriol.* **101**, 297–303.

Narasimha, R., Aganj, I., Bennett, A. E., Borgnia, M. J., Zabransky, D., Sapiro, G., McLaughlin, S. W., Milne, J. L., and Subramaniam, S. (2008). Evaluation of denoising algorithms for biological electron tomography. *J. Struct. Biol.* **164**, 7–17.

Nickell, S., Forster, F., Linaroudis, A., Net, W. D., Beck, F., Hegerl, R., Baumeister, W., and Plitzko, J. M. (2005). TOM software toolbox: Acquisition and analysis for electron tomography. *J. Struct. Biol.* **149**, 227–234.

Nickell, S., Hegerl, R., Baumeister, W., and Rachel, R. (2003). *Pyrodictium cannulae* enter the periplasmic space but do not enter the cytoplasm, as revealed by cryo-electron tomography. *J. Struct. Biol.* **141**, 34–42.

Nishino, Y., Yasunaga, T., and Miyazawa, A. (2007). A genetically encoded metallothionein tag enabling efficient protein detection by electron microscopy. *J. Electron Microsc.* **56**, 93–101.

Ortiz, J. O., Förster, F., Kürner, J., Linaroudis, A. A., and Baumeister, W. (2006). Mapping 70s ribosomes in intact cells by cryoelectron tomography and pattern recognition. *J. Struct. Biol.* **156**, 334–341.

Paul, T. R., and Beveridge, T. J. (1992). Reevaluation of envelope profiles and cytoplasmic ultrastructure of mycobacteria processed by conventional embedding and freeze-substitution protocols. *J. Bacteriol.* **174**, 6508–6517.

Paul, T. R., and Beveridge, T. J. (1994). Preservation of surface lipids and determination of ultrastructure of *Mycobacterium kansasii* by freeze-substitution. *Infect. Immun.* **62**, 1542–1550.

Pierson, J., Fernandez, J. J., Bos, E., Amini, S., Gnaegi, H., Vos, M., Bel, B., Adolfsen, F., Carrascosa, J. L., and Peters, P. J. (2010). Improving the technique of vitreous cryo-sectioning for cryo-electron tomography: Electrostatic charging for section attachment and implementation of an anti-contamination glove box. *J. Struct. Biol.* **169**, 219–225.

Pilhofer, M., Bauer, A. P., Schrallhammer, M., Richter, L., Ludwig, W., Schleifer, K. H., and Petroni, G. (2007a). Characterization of bacterial operons consisting of two tubulins and a kinesin-like gene by the novel two-step gene walking method. *Nucleic Acids Res.* **35**, e135.

Pilhofer, M., Rosati, G., Ludwig, W., Schleifer, K. H., and Petroni, G. (2007b). Coexistence of tubulins and *ftsZ* in different *prosthecobacter* species. *Mol. Biol. Evol.* **24**, 1439–1442.

Pruggnaller, S., Mayr, M., and Frangakis, A. (2008). A visualization and segmentation toolbox for electron microscopy. *J. Struct. Biol.* **164**, 161–165.

Radolf, J. D., Norgard, M. V., and Schulz, W. W. (1989). Outer membrane ultrastructure explains the limited antigenicity of virulent *Treponema pallidum*. *Proc. Natl. Acad. Sci. U. S.A.* **86**, 2051–2055.

Remsen, C. C. (1968). Fine structure of the mesosome and nucleoid in frozen-etched *bacillus subtilis*. *Arch. Mikrobiol.* **61**, 40–47.

Reynolds, E. S. (1963). The use of lead citrate at high pH as an electron-opaque stain in electron microscopy. *J. Cell Biol.* **17**, 208–212.

Sabatini, D. D., Bensch, K., and Barrnett, R. J. (1963). Cytochemistry and electron microscopy. The preservation of cellular ultrastructure and enzymatic activity by aldehyde fixation. *J. Cell Biol.* **17**, 19–58.

Salje, J., Zuber, B., and Löwe, J. (2008). Electron cryomicroscopy of *E. Coli* reveals filament bundles involved in plasmid DNA segregation. *Science* **323**, 509–512.

Sandberg, K., and Brega, M. (2007). Segmentation of thin structures in electron micrographs using orientation fields. *J. Struct. Biol.* **157**, 403–415.

Sartori, A., Gatz, R., Beck, F., Rigort, A., Baumeister, W., and Plitzko, J. M. (2007). Correlative microscopy: Bridging the gap between fluorescence light microscopy and cryo-electron tomography. *J. Struct. Biol.* **160**, 135–145.

Schwartz, C. L., Sarbash, V. I., Ataullakhanov, F. I., McIntosh, J. R., and Nicastro, D. (2007). Cryo-fluorescence microscopy facilitates correlations between light and cryo-electron microscopy and reduces the rate of photobleaching. *J. Microsc.* **227**, 98–109.

Seybert, A., Herrmann, R., and Frangakis, A. (2006). Structural analysis of *Mycoplasma pneumoniae* by cryo-electron tomography. *J. Struct. Biol.* **156**, 342–354.

Shimizu, T., and Miyata, M. (2002). Electron microscopic studies of three gliding mycoplasmas, *Mycoplasma mobile, M. Pneumoniae*, and *M. Gallisepticum*, by using the freeze-substitution technique. *Curr. Microbiol.* **44**, 431–434.

Slot, J. W., Posthuma, G., Chang, L. Y., Crapo, J. D., and Geuze, H. J. (1989). Quantitative aspects of immunogold labeling in embedded and in nonembedded sections. *Am. J. Anat.* **185**, 271–281.

Sontag, C. A., Staley, J. T., and Erickson, H. P. (2005). *In vitro* assembly and GTP hydrolysis by bacterial tubulins BtubA and BtubB. *J. Cell Biol.* **169**, 233–238.

Spehner, D., and Edelmann, L. (2009). Slam-freezing, metal-mirror freezing. *In* "Handbook of Cryo-Preparation Methods for Electron Microscopy" (A.Cavalier, D. Spehner, and B. M. Humbel, eds.) pp. 18–47. CRC Press, Boca Raton.

Stierhof, Y. D., and Schwarz, H. (1989). Labeling properties of sucrose-infiltrated cryosections. *Scanning Microsc. Suppl.* **3**, 35–46.

Studer, D., Moor, H., and Gross, H. (1981). Single bacteriorhodopsin molecules revealed on both surfaces of freeze-dried and heavy metal-decorated purple membranes. *J. Cell Biol.* **90**, 153–159.

Suloway, C., Shi, J., Cheng, A., Pulokas, J., Carragher, B., Potter, C. S., Zheng, S. Q., Agard, D. A., and Jensen, G. J. (2009). Fully automated, sequential tilt-series acquisition with leginon. *J. Struct. Biol.* **167**, 11–18.

Thanbichler, M., Viollier, P. H., and Shapiro, L. (2005). The structure and function of the bacterial chromosome. *Curr. Opin. Genet. Dev.* **15**, 153–162.

Tivol, W. F., Briegel, A., and Jensen, G. J. (2008). An improved cryogen for plunge freezing. *Microsc. Microanal.* **14**, 375–379.

Tocheva, E. I., Li, Z., and Jensen, G. J. (2010). "Electron Cryotomography". *Cold Spring Harb. Perspect. Biol.* 2, a003442.

Tokuyasu, K. T. (1973). A technique for ultracryotomy of cell suspensions and tissues. *J. Cell Biol.* **57**, 551–565.

van Donselaar, E., Posthuma, G., Zeuschner, D., Humbel, B. M., and Slot, J. W. (2007). Immunogold labeling of cryosections from high-pressure frozen cells. *Traffic* **8**, 471–485.

Vanhecke, D., and Studer, D. (2009). High-pressure freezing LEICA EMPACT. *In* "Handbook of Cryo-Preparation Methods for Electron Microscopy" (A. Cavalier, D. Spehner, and B. M. Humbel, eds.) pp. 129–156. CRC Press, Boca Raton.

van Iterson, W., Hoeniger, J. F., and Nijman van, Z. E. (1967). A "microtubule" in a bacterium. *J. Cell Biol.* **32**, 1–10.

van Niftrik, L., et al. (2009). Cell division ring, a new cell division protein and vertical inheritance of a bacterial organelle in anammox planctomycetes. *Mol. Microbiol.* **73**, 1009–1019.

Verkade, P. (2008). Moving EM: The rapid transfer system as a new tool for correlative light and electron microscopy and high throughput for high-pressure freezing. *J. Microsc.* **230**, 317–328.

Werner, J. N., Chen, E. Y., Guberman, J. M., Zippilli, A. R., Irgon, J. J., and Gitai, Z. (2009). Quantitative genome-scale analysis of protein localization in an asymmetric bacterium. *Proc. Natl. Acad. Sci. U. S.A.* **106**, 7858–7863.

Wright, E. R., Iancu, C. V., Tivol, W. F., and Jensen, G. J. (2006). Observations on the behavior of vitreous ice at ~ 82 and ~ 12 K. *J. Struct. Biol.* **153**, 241–252.

Zheng, S. Q., Keszthelyi, B., Branlund, E., Lyle, J. M., Braunfeld, M. B., Sedat, J. W., and Agard, D. A. (2007). UCSF tomography: an integrated software suite for real-time electron microscopic tomographic data collection, alignment, and reconstruction. *J. Struct. Biol.* **157**, 138–147.

CHAPTER 3

Analysis of the Ultrastructure of Archaea by Electron Microscopy

Reinhard Rachel*, Carolin Meyer*, Andreas Klingl*, Sonja Gürster*, Thomas Heimerl*, Nadine Wasserburger*, Tillmann Burghardt*, Ulf Küper†, Annett Bellack†, Simone Schopf†, Reinhard Wirth†, Harald Huber†, *and* Gerhard Wanner‡

*Centre for Electron Microscopy, University of Regensburg, D-93053 Regensburg, Germany

†Department of Microbiology and Archaea Centre, University of Regensburg, D-93053 Regensburg, Germany

‡Biocentre, University of Munich, D-82152 Planegg-Martinsried, Germany

METHODS IN CELL BIOLOGY, VOL. 96

978-0-12-381007-6
DOI: 10.1016/S0091-679X(10)96003-2

Abstract

The ultrastructural characterization of archaeal cells is done with both types of electron microscopy, transmission electron microscopy, and scanning electron microscopy. Depending on the scientific question, different preparation methods have to be employed and need to be optimized, according to the special cultivation conditions of these—in many cases extreme—microorganisms. Recent results using various electron microscopy techniques show that archaeal cells have a variety of cell appendages, used for motility as well as for establishing cell–cell and cell–surface contacts. Cryo-preparation methods, in particular high-pressure freezing and freeze-substitution, are crucial for obtaining results: (1) showing the cells in ultrathin sections in a good structural preservation, often with unusual shapes and subcellular complexity, and (2) enabling us to perform immunolocalization studies. This is an important tool to make a link between biochemical and ultrastructural studies.

I. Introduction

Archaea were defined as the third kingdom in the universal phylogenetic tree of life initially by the determination of oligonucleotide maps of ribosomal RNAs and, later, by the full sequence of the 16S rRNA gene (Ludwig and Schleifer, 1994; Woese, 2000; Woese and Fox, 1977; Woese *et al.*, 1990). In the meantime, the sequencing of full genomes and also biochemical analyses have confirmed this hypothesis (Wolf *et al.*, 2002). It is now well documented that many molecular components of archaeal cells are distinct to those of the other prokaryotic kingdom, the bacteria. Selected examples are cell wall components such as pseudomurein (reviewed in Kandler and König, 1998) or S-layers (König *et al.*, 2007), membrane components such as the lipids (Langworthy and Pond, 1986; Rosa and Gambacorta, 1986; Boucher, 2007), and proteins such as the elongation factors (Cammarano *et al.*, 1992). The molecular architecture of several macromolecular complexes such as the DNA-dependent RNA polymerase (reviewed by Werner, 2007), the chaperonins (Archibald *et al.*, 1999; Phipps *et al.*, 1991, 1993), and again the elongation factors, suggested a distinct relationship of Archaea to eukaryotic cells. Of particular interest and value was the analysis of the ATP synthase complex (Gogarten *et al.*, 1989; Iwabe *et al.*, 1989); the sequence comparison of the A and B subunits provided the first experimental evidence to set up a root of the universal phylogenetic tree.

The current status of knowledge about the Archaea was recently summarized in two books, which cover almost all aspects of the cell and molecular biology of these unique microorganisms (Cavicchioli, 2007; Garrett and Klenk, 2007). Ultrastructural transmission electron microscopy (TEM) investigations using electron tomography with *Sulfolobus* and *Pyrobaculum* cells were performed by Grimm *et al.* (1998), and another study on *Sulfolobus* was recently published in Methods Cell Biol. (Robertson, 2007).

The analysis of Archaea at the University of Regensburg started in the early 1980s (Stetter, 2006). Numerous new species, genera, even orders, and phyla were described. Among those are *Pyrodictium* (Stetter, 1982), *Methanopyrus* (Kurr *et al.*, 1991), *Pyrolobus* (Blöchl *et al.*, 1997), *Ignicoccus* (Huber *et al.*, 2000), and *Nanoarchaeum* (Huber *et al.*, 2002). Very early ultrastructural investigations by electron microscopy played an important role in the description of these species, and led to the discovery of several unusual structural features and molecules of Archaea, like the cannulae of *Pyrodictium* (König *et al.*, 1988; Nickell *et al.*, 2003; Rieger *et al.*, 1995), the chaperonin complex, present in *Pyrodictium* (Phipps *et al.*, 1991, 1993), and in most other Archaea (Archibald *et al.*, 1999), and the two membranes of *Ignicoccus* cells (Huber *et al.*, 2000; Rachel *et al.*, 2002).

Since 1995, cryo-preparation methods were introduced in our laboratories in the course of these studies and used stringently in order to improve the preservation of the fine structure of the archaeal cells. The methods included cryofixation by plunge-freezing in liquid nitrogen, propane jet freezing, and high-pressure freezing (HPF), followed by either freeze-substitution (FS) and resin embedding or freeze-etching. For the principles and theory of these methods, the reader is referred to textbooks (e.g., Steinbrecht and Zierold, 1987) and to recent reviews (e.g., McDonald, 2007; Severs and Shotton, 2006).

II. Growth of Cells for Electron Microscopy

Many of the archaeal cells under investigation in the Archaea Centre in Regensburg, Germany, thrive exclusively under extreme conditions, mostly at high temperature (hence the terms "thermophiles" and "hyperthermophiles") and also under the exclusion of oxygen, i.e., they are strict anaerobes. There is considerable interest to study and understand the molecular biology and biochemistry of hyperthermophiles, because they cluster around the root in rRNA-based phylogenetic trees, and therefore represent cells which might be close relatives of early forms of life on Earth (Stetter, 1992). The preparation of culture media and cultivation of these cells are highly demanding. When cells are prepared for electron microscopy using classical methods, the first two steps are chemical fixation (usually by addition of glutardialdehyde) and centrifugation to obtain a high cell density, which is a prerequisite for efficient work with the electron microscope. Both steps are counterproductive for preserving the native state of the cells; they will ultimately lead to alterations of the ultrastructure, as directly shown in the case of *Pyrodictium* (Rieger *et al.*, 1995, 1997) and *Ignicoccus* (Müller *et al.*, 2009; Rachel *et al.*, 2002). In order to avoid these two steps, we developed several ways to keep the cells in an "undisrupted status" before cryofixation is achieved. In particular, the anoxic growth media are supplemented with "surfaces," which are compatible with the follow-up procedures for electron microscopy (Näther *et al.*, 2006; Schopf *et al.*, 2008): pieces of freshly cleaved mica; glass, such as fragments of microscope slides or cover slips; carbon-coated gold grids; cellulose capillary tubes (Hohenberg *et al.*, 1994; Rieger *et al.*,

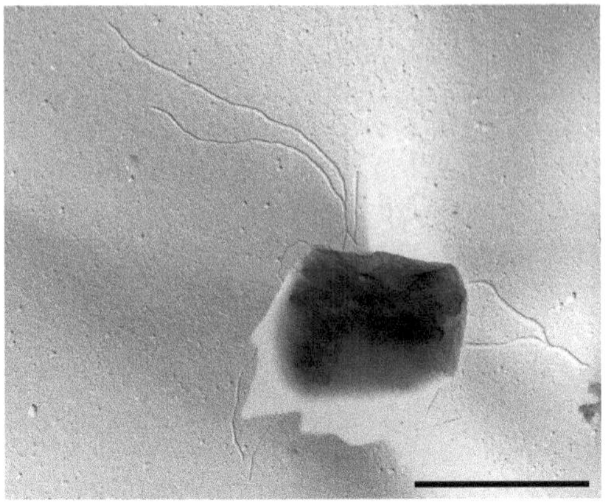

Fig. 1 Transmission electron micrograph of a cell of the genus *Archaeoglobus*, exhibiting several flagella. Specimen shadowed unidirectionally with platinum/carbon (1 nm; angle: 15°). Shadowing from top right. Bar: 1 μm.

1997)—just to mention a few. The medium is either sterilized after addition of these surfaces or they are added in a sterile way. An important "by-product" of these experiments was the result that many (if not most) of the Archaea, which were described to grow as planctonic cells, were now observed to also live and thrive attached to surfaces. This considerably changed our view of these microorganisms, and also, in part, the direction of our research.

Procedure 1, for TEM or SEM analysis of cells, whole mounts, and cell appendages:

1. Prepare growth medium as previously described (for anoxic media of hyperthermophilic sulfur reducers, see e.g., Huber *et al.*, 2000; Paper *et al.*, 2007). Routinely, 120 ml serum bottles are used, containing 20 ml liquid medium.
2. Before pressurizing and sterilizing the medium, add several pieces of glass, mica, sand grains, or carbon-coated gold grids (see Näther *et al.*, 2006). Grids are placed into Teflon holders (~9 × 22 mm) with four holes, each 3.8 mm in diameter.
3. Inoculate 20 ml of fresh medium with 0.1–1 ml of a fresh pre-culture using a sterile syringe. Incubate the serum bottles with Teflon holders horizontally. Usually, we incubate samples at least in triplicate. Shaking may need to be adjusted or reduced.
4. Check growth of cells by counting the planctonic cells (e.g., using a Thoma counting chamber); this gives a rough estimate of the number of cells on the surfaces.
5. Start chemical fixation by adding glutardialdehyde (anoxic; 25 or 50% stock solution; final concentration: 1–2.5%); leave for 10–50 min at room temperature. This step might be omitted, depending on the cell type and experiment.
6. Remove pieces of glass, mica, grids, and investigate some (still wet) samples by phase contrast light microscopy (see Fig. 1 in Schopf *et al.*, 2008).
7. Continue with the EM preparation as given below.

Procedure 2, for TEM analysis of (ultra-)thin sections of cells:

1. Prepare growth medium as described above (for anoxic media of hyperthermophilic sulfur reducers, see e.g., Huber *et al.*, 2000; Paper *et al.*, 2007).
2. Pressurize and sterilize growth medium. Cool down.
3. Depressurize medium bottles, open in a hood with CO_2 or a sink filled with argon. Place a 0.2- to 0.5-ml drop of a fresh pre-culture of the Archaea onto a Petri dish. By using capillary forces, take up a minute amount (about 1 μl) of this fresh pre-culture into pieces of cellulose capillaries (each about 10–30 mm in length). Capillaries can be sealed with a drop of superglue and briefly washed in a few milliliters of fresh medium, in order to ensure that cells only grow inside the capillary lumen. Several capillaries can be added to one bottle with 20 ml medium (for mesophilic Archaea, this step is best done in a sterile bench).
4. Close medium bottles and apply pressurized gas phase again via a sterile gas filter; incubate as usual, with or without shaking. Usually, we incubate samples at least in triplicate.
5. Check growth of cells every 4 h (depends on the species) by counting the planctonic cells (e.g., using a Thoma counting chamber) or by measuring the concentration of metabolic end products, such as H_2S; this gives a rough estimate of the growth of cells in the capillaries. Remove one or several capillaries and observe by phase contrast light microscopy for growth of cells inside the capillary lumen (see, e.g., Figs. 1 and 2 in Rieger *et al.* (1997)). Not all capillaries will be tightly filled with cells; this depends on the strain of Archaea and also the culture conditions. The success rate in our hands is between 30 and 80%, at this stage.
6. Before cryofixation, a chemical fixation under anoxic conditions by adding glutardialdehyde to the culture medium can be done; we prefer to omit this in order to carry out the cryofixation with chemically unperturbed "native" cells. For this, put capillaries into a Petri dish with either growth medium, or *n*-hexadecene, or 20% bovine serum albumin (BSA) (this depends on the microorganisms). Cut the capillaries into small pieces, about 2 mm in length; one or two of these small pieces will fit into the specimen carrier of the high-pressure freezer. Capillaries longer than 2 mm can directly be frozen in the copper tubes of the Leica (Wetzlar, Germany) EM-PACT 2.
7. Continue with HPF and follow-up procedures (see below).

III. Preparation for Transmission Electron Microscopy

A. Uranyl Acetate Staining, Platinum Shadowing

Procedure for TEM analysis of cells and their appendages, with uranyl acetate staining or Pt/C shadowing:

1. Remove 1–20 ml from the freshly grown culture and add glutardialdehyde (final concentration: 1–2.5%).
2. Centrifuge for 5–10 min in a table-top centrifuge at about 10,000 × *g* (if necessary, rpm can be adjusted).

3. Resuspend the pellet in a minute amount of the medium (10–30 μl) and apply onto glow-discharged carbon-coated copper grids. Leave for 30 s and blot with filter paper.
4. Wash briefly (1–2 s) on, or in, a drop of H_2O (or suitable buffer). Air-dry if Pt/C shadowing is done. The necessity of this step—washing once with H_2O—depends on the salt concentration of the medium. It might be omitted for uranyl acetate staining because the one-step-uranyl-acetate staining also dilutes the salt in the sample.

5a. Uranyl acetate: Add a drop of uranyl acetate or place grid with sample side on a drop of uranyl acetate (2%, in H_2O) for 20–60 s. Blot with filter paper and air-dry. Do not wash with H_2O.

5b. Platinum/carbon: Air-dry the grids. Place onto a carrier and shadow with Pt/C (15°, 1–2 nm), preferably by electron-gun evaporation.

For carbon-coated gold grids, which were incubated in the culture medium, perform steps 1, 4 (if necessary; depending on salt concentration of the medium, and whether staining or shadowing is done), and 5.

In the course of describing unknown Archaea, we perform as many electron microscopy experiments as possible. For TEM this usually involves a first quick check of planctonic cells, following glutardialdehyde fixation, centrifugation, adsorption onto carbon-coated copper grids, a brief wash with H_2O or a suitable buffer, and either uranyl acetate staining or air-drying in combination with platinum/carbon shadowing (angle: 15° /unidirectional).

Results of experiments done by this procedure give an overview of the cell shape and, in particular, of the number, abundance, and diameter/type of cell appendages, as far as they can be observed with this (simple and crude) method (Fig. 1). For many archaeal species, the presence, amount, and, e.g., insertion at one pole or at single/multiple sites is used as one characteristic feature in their classification (e.g., *Thermococcales; Methanococcales; Archaeoglobales*). In addition, the observed diameter correlates with their putative function(s): 10-nm appendages of *Thermococcales* and *Methanococcales*, and 13-nm appendages of *Sulfolobales* are supposed to be used for both motility and adhesion (Bellack *et al.*, submitted; Näther *et al.*, 2006; Schopf *et al.*, 2008; Szabó *et al.*, 2007); 13-nm fibers of *Ignicoccus hospitalis* (Müller *et al.*, 2009), and 5-nm fibers, observed on cells of some members of the *Methanobacteriales* and the *Sulfolobales* (Zolghadr *et al.*, 2010) are discussed to be involved in adhesion. In this respect, we observe a significant higher number of cells adhering to the carbon film (using carbon-coated gold grids), and often an improved preservation of the structure, which is achieved by adding grids directly to the growth medium, as described above. The follow-up procedure is similar, but not identical.

B. Freeze-Etching

Freeze-etching of a suspension of archaeal cells has been routinely performed for more than 18 years in our laboratory. Here, a protocol is presented, which was briefly described in the course of the characterization of the genus *Ignicoccus* (Huber *et al.*, 2000; Rachel *et al.*, 2002). After cultivation, the cells are concentrated by (preferably a single step) a brief centrifugation for 3 min, e.g., in a table-top centrifuge in 1.5-ml cups (5000–13,000 Upm). The resulting pellet is resuspended in a minimum volume (about 2–5 μl, up to 10 μl) depending on the cell density reached. In many early experiments, the sample was cryo-immobilized between two copper plates by propane jet freezing. Later, we found this unnecessary. Preservation of ultrastructure, as judged by the appearance of the cytoplasm, membranes, S-layer, and other cell wall components, was not better than after freezing in liquid nitrogen. Therefore, the sample is routinely applied onto a gold carrier, frozen by plunging it manually into liquid nitrogen, and then transferred in the ultra-high vacuum ($p < 10^{-6}$ mbar) of the freeze-etch unit. Several aspects of freeze-etching of cell suspensions were discussed recently in a review (Severs and Shotton, 2006).

Procedure for TEM analysis of archaeal cells by freeze-etching:

1. Remove 1–20 ml from the freshly grown culture. Do not fix chemically by glutardialdehyde or formaldehyde, unless absolutely necessary.
2. Centrifuge for 5–10 min in a table-top centrifuge at about 10,000 rpm (rpm can be adjusted).
3. Resuspend the pellet in a minute amount of the medium (2–5, up to 10 μl) and apply 1.5 μl onto a gold specimen carrier (before use, clean in 70% sulfuric acid for > 60 min, water, and finally acetone; e.g., from Leica; or from BALTIC Präparation, Niesgrau, Germany). Freeze sample immediately by plunging it into liquid nitrogen (alternatively, use liquid ethane or propane cooled down by liquid nitrogen, if available). Under liquid nitrogen mount carrier onto a precooled specimen transfer holder (our machine can hold three specimens at once). Transfer the holder with the samples into a freeze-etch unit (e.g., CFE 50, Cressington, Watford, UK).
4. Place a cryo-shield above the specimen (<−190°C) to cryo-protect the sample against contamination.
5. Adjust the temperature to about −97°C (about −100°C; depending on the freeze-etch unit). Leave the specimen at this temperature for about 7 min for temperature stabilization.
6. Freeze-fracture the specimen with a cold (<−190°C) knife. Again, place a cryo-shield above the specimen. Leave specimen at −97°C for 4 min. This is now the "freeze-etching" process. In our freeze-etch device, we have an etch-rate of about 100 nm min^{-1}. Therefore, when etching for 4 min, about half of an archaeal cell (average diameter of 1 μm) is freed from the surrounding ice/medium.
7. After this process, immediately shadow with Pt/C, preferably by electron-beam evaporation (fast and easy to control). Usually, we apply 1–1.5 nm Pt/C (45° unidirectional) within 15 s, and immediately after this, 10–15 nm C (90°) is applied, within another 15 s.
8. Remove the holder with the samples from the freeze-etch unit.

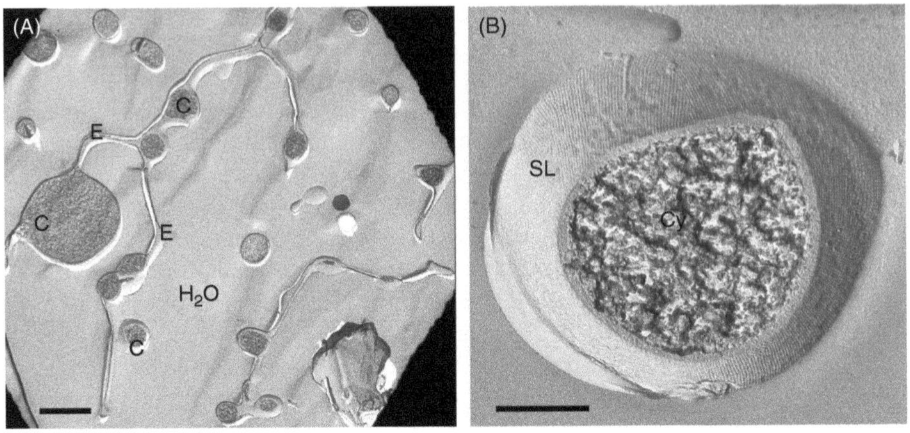

Fig. 2 Transmission electron micrographs of freeze-etched archaeal cells of a yet uncharacterized species of the Crenarchaeota. (A) Low-magnification overview. (B) Higher resolution micrograph of a single cell. Labels: E: eutecticum; C: cell; H_2O: water/ice; Cy: cytoplasm; SL: S-layer. Bars: (A) 3 μm, (B) 0.5 μm.

9. Remove the replicas from the biological sample by immersing the gold carrier directly into freshly diluted H_2SO_4 (70%). Leave overnight (although 2–3 h might be sufficient).
10. Wash the replicas twice on bi-distilled H_2O by transferring the replicas with a platinum loop. Collect the replicas on 600 or 700 mesh EM grids (preferably hexagonal, 600 or 700 mesh copper grids, hydrophilized by glow-discharging). Let the samples air-dry.

A typical result of freeze-etching is shown in Fig. 2. A low magnification view (Fig. 2A) shows the result of slow freezing of aqueous suspensions in liquid nitrogen. The contents of the culture medium (seawater, i.e., basically a 3% NaCl solution) become separated into a frozen pure water phase (about 5–20 μm in diameter) and a eutecticum (containing the salt at high concentration). Although this segregation process is highly undesirable, most cells appear structurally unaltered when compared with cells frozen in a propane jet. A possible explanation is that the disintegration of the cells is slower than the segregation process. Another explanation is that almost all archaeal cells, which we have investigated so far, can tolerate pure H_2O for 1 or 2 s before the cell starts to disintegrate. Most archaeal cells are highly "rewarding" when investigated by electron microscopy following freeze-etching. They display a beautiful regular pattern on their cell surface, an S-layer (Fig. 2B), which is a two-dimensional crystal of 2-, 3-, 4-, or 6-fold symmetry (König *et al.*, 2007). Another unique feature of hyperthermophilic Archaea is the fact that their membranes do not split during the freeze–fracture process. Because they contain both diether and tetra-ether lipids (Boucher, 2007; Langworthy and Pond, 1986), a separation of two membrane leaflets is not possible (Huber *et al.*, 2000; Rachel *et al.*, 2002). The cytoplasm is tightly packed, with no internal features visible. This result correlates with the view of the cytoplasm in ultrathin sections (see below).

C. High-Pressure Freezing, Freeze-Substitution, and Embedding

For a detailed introduction to HPF/FS, many papers and reviews are available. We refer the reader to two recent, comprehensive reviews about HPF applications (McDonald, 2007; Müller-Reichert *et al.*, 2007), covering many technical aspects.

Procedure for TEM analysis of ultrathin sections of archaeal cells:

1a. Grow cells in cellulose capillaries as described above (II. Growth of Cells for Electron Microscopy: Procedure 2).

1b. Cells which are robust during centrifugation, or cells which only grow planctonically in liquid medium, can be collected by a brief centrifugation (at the lowest tolerable speed). Resuspend the pellet and directly transfer 1–2 μl of this concentrated suspension to a specimen carrier for HPF. These steps should be performed quickly to reduce the time of exposure to oxygen (for anaerobic microorganisms) and to avoid drying out of the pellet.

2. Freeze specimens and store frozen samples in liquid nitrogen. We have used the high-pressure freezers of BAL-TEC (HPF 010), Wohlwend (HPF Compact 01), and Leica (EM PACT2), all with good success.

3. Freeze-substitute the samples as described (Hohenberg *et al.*, 1994; Rieger *et al.*, 1995, 1997). Our standard procedure is: −90°C for ≥8 h (up to 24 or 36 h), −60°C for 8 h, and −30°C for 6–8 h. As substitution media we have successfully used acetone pure, containing: 2% OsO_4; 2% OsO_4 plus 5% H_2O; 2% OsO_4, 0.25% uranyl acetate, 5% H_2O; and, as alternative solvent, ethanol 99%, containing 0.5% uranyl acetate, 1% glutardialdehyde (Rachel *et al.*, 2002); 0.5% uranyl acetate, 1% formaldehyde, 0.5% glutardialdehyde, 5% H_2O (Burghardt *et al.*, 2007). Raise the temperature of the samples to 0°C. Rinse twice with pure acetone. Infiltrate in Epon and polymerize at 60°C for 2 days. Prepare ultrathin sections with a diamond knife, and stain the sections with uranyl acetate for 15–30 min and lead citrate for 1–2 min (if desired and necessary; be aware: adding contrast at this step might be useful, but not always: it may also obscure details, and will reduce the visibility of small gold labels on a heavily contrasted cytoplasm). Recommendations and comments on the media for FS: Various recipes for freeze substitution media were suggested, tested, and discussed in the literature (see e.g., Buser and Walther, 2008; Humbel and Müller, 1986; McDonald, 2007). Our own experience is that the OsO_4-containing media are suitable for ultrastructural investigations, while those without OsO_4 are now routinely and successfully used for immunolocalization studies. We have to state, however, that testing all possible variations with ones own sample is recommended—although very costly and time consuming. In addition, we observe an improved preservation of ultrastructure, in particular, visibility of membranes, by adding 5% water to the FS medium (this is now routinely added), and that cells which were substituted without OsO_4 exhibit lower contrast, but show some features (like bundles resembling a cytoskeleton; Junglas *et al.*, 2008) which were not observed after substitution in acetone plus 2% OsO_4. Two questions need to be addressed, for each sample: How much or how little OsO_4 is needed for preservation of ultrastructure and for sufficient contrast? How little, if any, OsO_4 is tolerated if immunolocalization is to be performed?

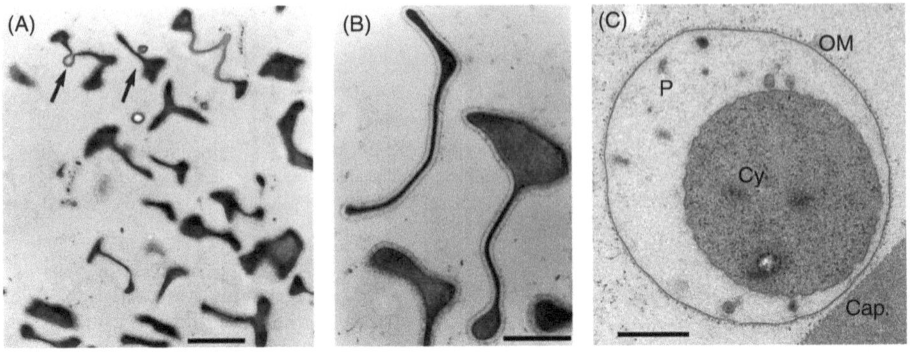

Fig. 3 Transmission electron micrographs of ultrathin sections of *Pyrodictium abyssi* and *Ignicoccus hospitalis*. Cells were grown in cellulose capillary tubes, high-pressure frozen, freeze-substituted, and embedded in resin. (A) *P. abyssi*, low-magnification overview. (B) *P. abyssi*, higher resolution micrograph of few highly lobed cells, which exhibit ultraflat areas. Arrows point to cells that display evaginations of the ultraflat areas. (A) and (B): from Rieger *et al.*, 1997, with permission. (C) *I. hospitalis*. Cap: cellulose capillary; Cy: cytoplasm; P: periplasm with vesicles; OM: outer membrane. Bars: (A) 1 μm, (B) and (C) 0.5 μm.

During our earlier studies on hyperthermophilic Archaea, we noted several results in ultrathin sections, which could not be adequately explained. First, within a single sample, cells were seen to be either smoothly round or lobed in shape; some cells exhibited a tightly packed cytoplasm, in other cells, it was loosely packed. Similar conflicting results had been published in the older literature. At the same time, results on some bacteria provided strong evidence that cryopreparation was able to much better preserve the ultrastructure of microorganisms than "conventional" processing at room temperature (Graham and Beveridge, 1990; Graham *et al.*, 1991; Hohenberg *et al.*, 1994). Our first rigorous application of cryo-preparation was on *Pyrodictium* cells (Rieger *et al.*, 1995). These cells form an unusual and complex multicellular network in suspension, and have two different types of appendages, cannulae (diameter: 25 nm) and fibers (diameter: 10 nm). Ultrathin sections were obtained of cells that had been cryo-immobilized by HPF in the native state (after growth in cellulose capillaries; Rieger *et al.*, 1997), FS in acetone containing 2% OsO_4, and resin embedding. The micrographs confirmed that these cells have a highly irregular shape, including parts of the cells which are "ultraflat," a feature which had not been observed or not been correctly interpreted before (Fig. 3A and B, from Rieger *et al.* (1997)). In other micrographs, the cytoplasm was tightly packed and contained (para-)crystalline areas, consisting of yet unknown macromolecules; the cell wall with the membrane and the S-layer was perfectly preserved, and in the resin surrounding the cells cross sections of cannulae were observed (Fig. 4 in Rieger *et al.* (1995)). We were aware that this route of preparation was and is "sub-optimal" compared with the study of fully hydrated cells using cryo-electron tomography (see, e.g., Nickell *et al.*, 2003). Resin-embedded samples

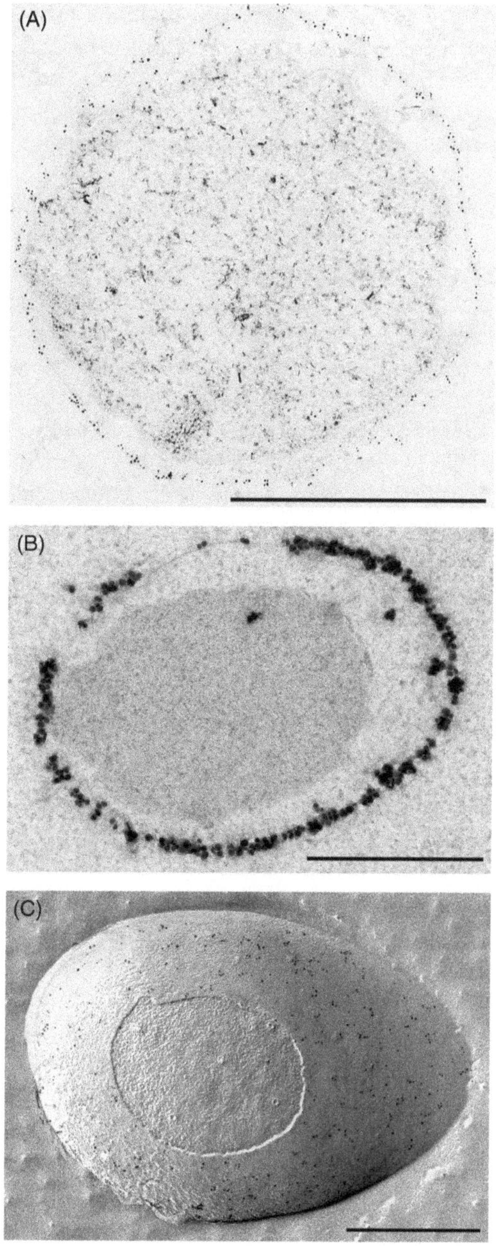

Fig. 4 Transmission electron micrographs showing various ways to immuno-localize a protein of *Ignicoccus* cells. (A) Immuno-localization by post-embed labeling on sections; secondary antibody with 6-nm gold. (B) Same procedure, but secondary antibody with ultrasmall gold, visualized by silver enhancement. (C) Immuno-localization by fracture-labeling. Bars: 1 μm. From: Burghardt *et al.*, 2007, with permission.

are, however, advantageous for several reasons: centrifugation and chemical fixation are avoided and, therefore, a "close-to-native" state of the cells is preserved; cells were finally embedded in resin, with the advantage that the imaging was possible with a standard 120 kV transmission electron microscope (compared with cryo-electron tomography: for imaging an intact *Pyrodictium* or *Sulfolobus* cell—or other archaea—with an average diameter/in an ice film of about 1 μm, a 200 or better 300 kV-cryo-TEM with energy filter is highly recommended; see Grimm *et al.*, 1998). Cells embedded in resin via this route are amenable to immunolabeling, i.e., the same samples can be used in later studies to address questions such as the subcellular distribution of specific proteins.

Another example in which HPF, FS, and resin embedding led to unexpected results was the investigation of cells of the genus *Ignicoccus*: *I. islandicus* (Huber *et al.*, 2000; Rachel *et al.*, 2002) and later, *I. hospitalis* (Paper *et al.*, 2007). Following chemical fixation and centrifugation, the cells are badly damaged. Their ultrastructure can only be properly visualized by avoiding both steps in the course of preparation for ultrathin sections (Rachel *et al.*, 2002). These cells (Fig. 3C) have no rigid cell wall and no cell wall polymer, but are surrounded by two membranes, an inner and an outer one, which are remarkably different in biochemical composition, with regard to both the lipids and the proteins. The outer membrane is the only lipid membrane of a hyperthermophilic archaeon known today that fractures in freeze-etch experiments, i.e., it splits into two halves. This correlates with the fact that it does not contain tetra-ether lipids, in contrast to the inner one (Rachel *et al.*, 2002; Jahn *et al.*, 2004). The two membranes enclose a huge intermembrane space, quasi a periplasm, whose volume is larger than that of the inner compartment, the cytoplasm (Rachel *et al.*, 2002). The main constituent of the outer membrane is about 10^6 copies of an oligomeric membrane protein complex, forming pores (Burghardt *et al.*, 2007). Unexpectedly, this membrane also contains the hydrogen : sulfur oxidoreductase, and the A_1A_O ATP synthase (Küper *et al.*, 2010).

D. Immunolabeling for Electron Microscopy

In recent projects we have performed several experiments to identify the subcellular distribution of proteins. In particular, we were successful not only in the case of *I. hospitalis* (Burghardt *et al.*, 2007; Küper *et al.*, 2010), but also with *Pyrodictium abyssi, Sulfolobus solfataricus*, and *Methanopyrus kandleri* (unpublished results). Essentially, we follow protocols as established for many years in the laboratories of Heinz Schwarz, York-Dieter Stierhof (Tübingen, Germany), and Bruno Humbel (EPFL Lausanne). Immunolabeling was done in two different ways: first, on ultrathin sections of Epon-embedded samples as described and discussed above, and second, by fracture-labeling (Burghardt *et al.*, 2007), essentially following the protocol developed and described earlier (Fujimoto, 1995).

Procedure for labeling ultrathin sections of Archaea for TEM:

Quite a number of variations of the protocol given below exist in the literature. The following route was successfully employed many times, but might need to be adjusted for specific needs:

1. After FS in ethanol containing 0.5% uranyl acetate, 1% formaldehyde, 0.5% glutardialdehyde, and 5% H_2O (instead of 1% formaldehyde and 0.5% glutardialdehyde, we also used 0.5 and 1% glutardialdehyde only, with success), the samples are embedded in Epon. Prepare ultrathin sections on the same day or 1 day before labeling.
2. Block sections by incubation with phosphate-buffered saline (PBS) + 0.1% glycin for 5 min.
3. Rinse sections two times in PBS + blocking agent (e.g., 1% BSA), 5 min each.
4. Incubate sections on a drop with primary antibody diluted in PBS + 0.1% BSA, for 45–60 min.
5. Rinse sections (up to) five times in PBS + 0.1% BSA, 2 min each.
6. Incubate sections on a drop with secondary antibody, coupled preferably to ultrasmall or 6-nm gold, diluted in PBS + 0.1% BSA, for 45 up to 120 min.
7. Rinse sections (up to) five times in PBS + 0.1% BSA, 2 min each.
8. Rinse section two times in PBS, 2 min each.
9. Incubate sections on a drop of PBS + 2% glutardialdehyde, for 5 min.
10. Rinse sections two times on PBS, 2 min each.
11. Rinse sections two times on H_2O, 2 min each.
12. Incubate sections on a drop of uranyl acetate, for 15 min.
13. Rinse sections two times on H_2O, 1 min each, and air-dry.

Lead citrate counter-staining is usually omitted or done only briefly (<1 min).

In the case of using a secondary antibody with ultrasmall gold, the final steps need to be modified and replaced by silver enhancement (Stierhof *et al*., 1992).

Procedure for labeling freeze–fracture replicas of Archaea for TEM:

1. Prepare freeze–fracture replicas as described above, but do not clean on H_2SO_4. Instead, float off the replicas on a porcelain spot plate on PBS + 2.5% sodium dodecyl sulfate (SDS) for 120 min. In the following, the replicas are not transferred from one well to the next. Instead, the solution underneath the replica is replaced with fresh solution using a Pasteur pipette.
2. Rinse sections four times with PBS, 5 min each.
3. Rinse sections twice in PBS + blocking agent (0.5% BSA, 0.2% gelatin) for 1 h.
4. Incubate replicas with the diluted primary antibody in PBS/BSA/gelatin, for 1 h.
5. Rinse replicas three times with PBS + BSA + gelatin, 5 min each.
6. Incubate replicas with the diluted secondary antibody (with 6-nm gold) in PBS + BSA + gelatin, for 30 min.
7. Rinse replicas three times with PBS + BSA + gelatin, 5 min each.
8. Rinse replicas three times with PBS, 5 min each.
9. Incubate replicas with PBS + 0.5% glutardialdehyde, 5 min.

10. Rinse replica twice on H_2O.
11. Collect replicas with hydrophilized (glow discharge), hexagonal 600 or 700 mesh copper grids and air-dry[1].

In our recent studies with *I. hospitalis*, postembedding labeling on ultrathin Epon sections enabled us to demonstrate that a novel membrane protein (Ihomp1) is exclusively located in the outer membrane (Burghardt *et al.*, 2007). Technically, this was done using several variations of immunolabeling, first with 6-nm gold coupled to the secondary antibody, and, even more convincingly, with ultrasmall-gold/secondary antibody complexes, which resulted in an almost even and exclusive distribution of the label on the outer membrane (Fig. 4A and B). In our studies, detection with ultrasmall-gold/secondary antibody complexes (see Fig. 6 in Schroeder-Reiter *et al.*, 2006) always results in higher density of labeling, and with a somewhat higher background. The nice result with the ultrasmall-gold/secondary antibody has two important consequences: first, the image is so easy to visualize that it can be identified in the electron microscope even at low magnification; second, because the important message is visible at low magnification, the final figure can be reduced in the printed manuscript to a ridiculously small size, resulting in "stamps" (which is often done in prestigious journals, against the wish of the authors). For Ihomp1, the unique location of this protein in a membrane allowed us to complement these results by fracture-labeling (Fig. 4C). With this method, a membrane surface is exposed to the antibodies, and not only a cross section of the membrane, resulting in a higher probability that the antibodies can bind to their target. More recently, sections of the same embedded sample of *I. hospitalis* enabled us to successfully perform immunolocalization studies with antibodies directed against >10 other proteins (Meyer, Huber, Rachel, unpublished). One unexpected result was obtained using antibodies directed against several protein subunits, showing that the membrane-bound A_1A_0 ATPase and the hydrogen : sulfur oxidoreductase complex are both located in the outer membrane (Küper *et al.*, 2010).

E. Imaging

In our laboratory, we use a standard 120 kV transmission electron microscope, equipped with a LaB_6 cathode. Any other TEM can be used, with 80–120 or even 200 keV. The samples are usually scanned at low magnification (100–1000 fold), with

[1] For the incubation steps with the primary and secondary antibodies, several dilutions need to be tested in order to find the one which is "best" for the epitopes and antibodies in use, and for the section/sample under investigation. Usually, we use a dilution series with a factor of 5 or 10, e.g., 1:5; 1:25; 1:125; or 1:10; 1:100; 1:1000. In general, we find that a much lower dilution (i.e., higher absolute concentration, 10- or even 100-fold) of the primary antibody has to be used for the detection of a given protein by section-labeling, in comparison to a western blot following SDS-PAGE. This is due to the fact that the number of protein epitopes, which are exposed, i.e., freely accessible, is by far lower on a section surface than in a band of a protein blot. In addition, the expression level of the protein in the cell must be taken into account: It might be low compared with its high concentration in a protein band following purification and SDS-PAGE.

the aid of a TV-rate side-entry CCD camera (Gatan model 673; Munich, Germany), which is also used for focusing and adjusting the astigmatism. Images are digitally recorded at a final magnification between 4500 and 33,000 or even higher using a slow-scan CCD camera (1k $\times$ 1k; TVIPS; Gauting, Germany). If the specimen area is larger than the size of the detector, montages are produced, with 2 $\times$ 2 or up to 9 $\times$ 9 images; stitching is done using a built-in tool in EM-MENU4 (TVIPS); other freeware or commercial tools can also be used.

IV. Preparation for Scanning Electron Microscopy

A. Room-Temperature Preparation

For a quick and easy preparation of cells for SEM analysis with a more qualitative rather than a quantitative view, it is sufficient to perform the fixation and dehydration of adhering microorganisms at room temperature.

Procedure for fixation, dehydration, and contrast enhancement (Fig. 5A; Zolghadr *et al.*, 2010):

1. The various steps which have to be performed before sample preparation are given above, (II. Growth of Cells for Electron Microscopy: Procedure 2, steps 1–4).
2. Chemical fixation with 1% glutardialdehyde (final conc.).
3. Dehydration with a series of ethanols: 20, 50, 70, 96, and 100%. Materials with adhering cells are incubated in each dilution step for 30 min and air-dried after the final step.
4. Electron beam gun shadowing with platinum/carbon (1–2 nm; 45°, rotary shadowing) or magnetron sputtering (platinum) for contrast enhancement.

B. Advanced Preparation

Several of our more recent studies include observations on the adherence of Archaea onto surfaces and on the formation of archaeal biofilms by either single species (*Pyrococcus furiosus*: Näther *et al.*, 2006; *S. solfataricus*: Zolghadr *et al.*, 2010) or two defined species (*P. furiosus* with *M. kandleri*: Schopf *et al.*, 2008). These samples were prepared by an approach, including growth on surfaces, chemical fixation with glutardialdehyde, and subsequent processing for critical-point drying and sputtering with platinum.

Procedure 1 for SEM analysis of cells, whole mounts, and cell appendages (see Figs. 4 and 5 in Näther *et al.*, 2006):

1. Prepare growth medium as described above (II. Growth of Cells for Electron Microscopy: Procedure 1).
2. Collect the materials with the adhering cells and fix by immersion into a solution containing 2.5% glutardialdehyde in a suitable buffer, e.g., 140 mM cacodylate buffer pH 7.0 (dimethylarsinic acid: handle with care, contains arsenic), 2 mM

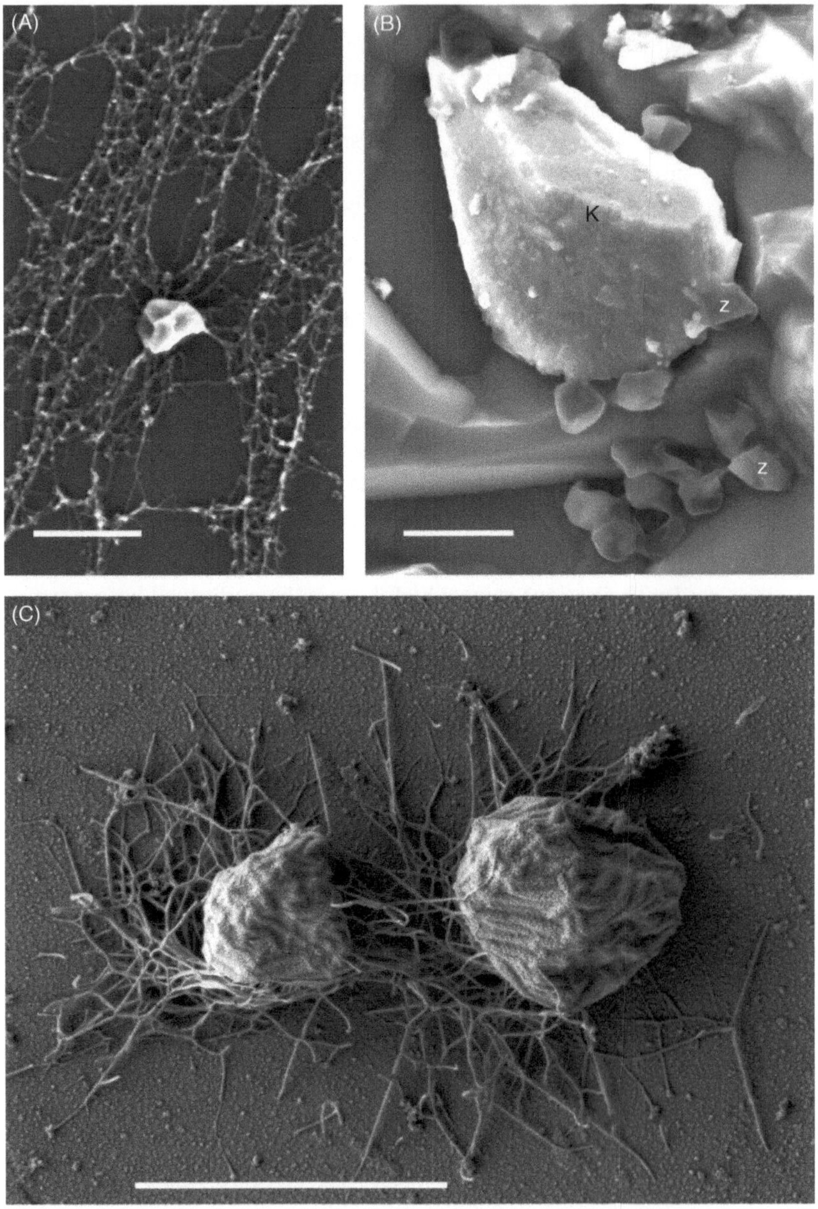

Fig. 5 Scanning electron micrographs of archaeal cells. (A) Cells of *Sulfolobus solfataricus* P2, attached to mica via a network of flagella/pili. (B) Cells of *Sulfolobus metallicus* Kra23 (Z), adherently growing on a Pyrite crystal (K). (C) Two cells of *Methanocaldococcus villosus* Kin24T80, with numerous flagella. Bars: 2 μm.

$CaCl_2$, 2.5% NaCl or artificial seawater (SME; Huber *et al.*, 2000), buffered with 0.1 M HEPES, pH 7.0.

3. Continue as previously described (Näther *et al.*, 2006): wash samples in bi-distilled water, dehydrate in a graded series of acetone, critical-point dry from liquid CO_2, mount on stubs with conductive tabs, and sputter-coat with 3–5 nm platinum (magnetron sputter coater: SCD 050, formerly BAL-TEC, now Leica).

Procedure 2 for SEM analysis of adherent cells on various surfaces (Fig. 5B; Klingl, 2007):

1. Prepare growth medium as described in Procedure 1, steps 1 and 2.
2. Fix the adhering cells with 1% glutardialdehyde.
3. Freeze-dry for 120 min at −80°C (CFE 50; Cressington Ltd.).
4. Perform electron beam gun rotary shadowing with platinum/carbon (1–2 nm as described above).

Procedure 3 for SEM analysis of adherent cells by cryo-preparation (Fig. 5C; Wanner *et al.*, 2008):

1. Grow cells as described and concentrate with a brief centrifugation.
2. Place drops of the sample onto a glass slide, cover with a coverslip, and freeze rapidly by plunging the sample into liquid nitrogen.
3. Remove coverslip with a razor blade, then immerse the glass slide into a buffer with chemical fixative (75 mM cacodylate buffer pH 7.0; 2.5% glutardialdehyde), and postfix in buffer with 1% OsO_4 (1 h each).
4. Degrade in a graded series of acetone solutions.
5. Critical-point dry after transfer to liquid CO_2.
6. Mount on stubs and coat with 2–3 nm platinum (magnetron sputtering).

The samples for SEM were prepared using the different methods, depending on the laboratory, in which the preparation was performed, the availability of instruments, and on the number of samples, e.g., method A was mainly used in a study comparing a huge number of samples. The methods 1–3 (under B) were employed depending on the type of sample and the scientific question to be addressed. For example, freeze-drying at −80°C combined with electron beam gun Pt/C shadowing may result in a finer metal film on the surface of the cells and appendages, compared with critical-point drying and magnetron sputtering. Effectively, the differences between the results obtained with Procedures 1–3 are comparatively small (Fig. 5C, vs SEM micrographs published in Näther *et al.* (2006) and Schopf *et al.* (2008)). The samples are imaged in modern scanning electron microscopes with field emitters, which results in fast acquisition times for a single image and in improved contrast compared with basic SEMs, in particular, due to improved detector technology: Hitachi S-4100 (image recording using Digiscan hardware and Digital micrograph; Gatan, München, Germany); FEI Quanta 400F, often in the low-vacuum mode; Zeiss Auriga. The high tension was varied, between 1 and 1.5 kV, up to 5 or even 25 kV, depending on the specimen, the SEM, and the detector used.

V. Instrumentation and Methods

A. Growth of microorganisms

Instrumentation: Growth of archaeal cells: as described, e.g., by Rieger *et al.* (1997), and Paper *et al.* (2007). Phase contrast light microscopy: e.g., Olympus BX 60, equipped with two objective lenses, UPlanFl 40/0.75 and UPlanFl 100/1.3 oil-immersion, and a digital camera (PCO, Kelheim, Germany; 1600 × 1200 pixels). For documentation: PCO software.

Materials: Archaeal cells: as stock cultures from either the DSMZ (Braunschweig, Germany), our own culture collection (Regensburg Culture Collection; Archaea Centre, University of Regensburg, Regensburg, Germany), or from other culture collections. Cellulose capillary tubes: e.g., from Leica, or removed from suitable hollow-fiber dialysis devices containing cellulose fibers (check carefully); cellulose capillary tubes might also be obtained from manufacturers (Cuprophan: e.g., from Membrana GmbH, Wuppertal, Germany). Micropipettes (e.g., 0.5–10 μl), 1-ml syringes (e.g., ERSTA, from CODAN Medical ApS, Rødby, Denmark), and needles (e.g., Fine-Ject, Henke Sass Wolf, Tuttlingen; or Braun, Melsungen, Germany).

Reagents: Fast glue

B. Uranyl Acetate Staining, Platinum Shadowing

Instrumentation: For platinum shadowing, we use a CFE-50 freeze-etch unit (Cressington), equipped with three electron-beam guns, two for platinum shadowing (45°; 5–20°) and one for carbon shadowing (90°).

Materials: Parafilm; fine-tipped tweezers, filter paper.

Reagents: Uranyl acetate, aqueous stock solution (2%).

C. Freeze-Etching and Fracture-Labeling

Instrumentation: For freezing: propane jet freezer (model JFD 030; BAL-TEC); for freeze-etching, CFE-50 freeze-etch unit (see above; Cressington). This unit is equipped with three electron-beam guns, two for platinum shadowing (45°; 5–20°), and one for carbon shadowing (90°). Alternatively, the BAF060 (now marketed as EM BAF060 by Leica) can also be used.

Materials: Gold carriers (e.g., from Leica, formerly from BAL-TEC; or from BALTIC Präparation); copper plates, for the propane jet freezing, from Leica (formerly from BAL-TEC); hexagonal 600 or 700 mesh copper grids; liquid nitrogen; Parafilm; fine-tipped tweezers.

Reagents: 70% sulfuric acid, freshly diluted from a stock solution of concentrated sulfuric acid (handle with care!); 2.5% SDS; bi-distilled water; acetone; PBS, glycine, BSA, or gelatin as blocking agents.

D. High-Pressure Freezing, Freeze Substitution, and Embedding

Instrumentation: Stereomicroscope with LED light source (Leica); high-pressure freezer (Leica EMPACT2+RTS; Wohlwend Compact 01); FS unit (EM-AFS2; Leica).

Materials: Membrane carriers (copper, plated with gold; 100 μm deep; Leica). Aluminum carrier (Wohlwend Compact 01). Other specimen carriers, e.g., made from brass, might be suitable, too, but have not been tested.

Reagents: Acetone (EM grade), ethanol, formaldehyde (freshly made 20% stock solution, from paraformaldehyde). Glutardialdehyde (25 or 50%; Fluka): we prefer to refill the 100 ml quantity of glutaraldehyde in smaller aliquotes (5 ml) into serum bottles, which are tightly sealed with rubber stoppers (butylseptum or natural rubber; Glasgerätebau Ochs, Bovenden, Germany), and then apply nitrogen gas; aliquotes are removed using a syringe with a needle. This ensures anoxic handling of this oxygen-sensitive reagent, and prolonged shelf life. Osmium tetroxide (crystalline, e.g., from Plano, Wetzlar, Germany). Uranyl acetate, Epon or Epon/Araldite, Formvar, copper, and nickel grids, uranyl acetate (5% stock solution, for preparation of the media used for FS; 2%: for staining), lead citrate.

E. Immunolabeling for Electron Microscopy

Materials: Parafilm; filter paper; fine-tipped tweezers. Copper and Nickel slot or 100 mesh grids, uncoated or coated with formvar or pioloform. Uranyl acetate, lead citrate.

Reagents: PBS. Glutardialdehyde (25% or 50%; Fluka). Antisera. Secondary antibodies, coupled with 10 nm, 6 nm, or ultrasmall gold. Glycin, BSA. Uranyl acetate. Formvar, Pioloform. Copper and nickel grids, uranyl acetate, lead citrate. Further reagents for fracture labeling are given above, under "Freeze-etching and fracture-labeling."

F. Imaging

Instrumentation: Transmission electron microscope CM12 (FEI, Eindhoven, the Netherlands) equipped with a LaB_6 cathode, which is highly recommended for standard applications, due to its high brightness and good coherence compared with a W cathode; in our EM, the lifetime of a LaB_6 is at about 3 years, at least. CCD-camera with fast readout, in the wide angle, side-entry port of the TEM, for searching (e.g., from Gatan; or from other manufacturers). CCD-camera with 1k × 1k pixel, better 2k × 2k, or larger; e.g., from TVIPS (Tietz, Gauting, Germany) or other manufacturers.

Materials: Fine-tipped tweezers.

G. Preparation for SEM

Instrumentation: Stereomicroscope with LED light source (Leica, Wetzlar); unit for controlled freeze-drying, e.g., freeze-etch unit CFE50 (Cressington); alternatively, a freeze-drying unit (e.g., LMC-1; Christ, Osterode, Germany), in case of a huge quantity of samples and if the requirements regarding sample quality are not too high. Unit for critical-point-drying (e.g., from Polaron, GaLa, Bad Schwalbach, Germany; or Leica, Wetzlar). Magnetron sputter unit: e.g., Leica EM-SCD 050 (formerly from BAL-TEC). Scanning electron microscopy with field emitter (for high coherence), and modern SE and BSE detectors; e.g., from Hitachi, Krefeld, Germany; Zeiss, Oberkochen, Germany; FEI, the Netherlands; JEOL, München, Germany.

Materials: Fine-tipped tweezers; aluminum stubs; Leit-C Plast (from EMS, München, Germany; or Plano).

Reagents: Acetone (EM grade), ethanol, OsO_4, glutardialdehyde (25 or 50%; Fluka); cacodylate buffer, 140 mM, pH 7.0. Carbon dioxide gas.

VI. Conclusions

The methods and results presented in this overview give a framework for similar studies on other Archaea. They can also be used for ultrastructural studies of bacteria with equal success, as some of our own investigations have shown (Huber *et al.*, 1998). In particular, the samples obtained so far can now be used for three-dimensional analyses of cells and subcellular structures (e.g., periplasmic vesicles in *Ignicoccus* cells), either by serial sectioning and combining these images to 3D datasets at low resolution (Junglas *et al.*, 2008) or by electron tomography of (serial) semi-thin sections. Both investigations have already yielded first and very promising results (Heimerl, 2009). In addition, the feasibility of immunolabeling on sections and on replicas is highly promising. For two of the archaeal species under investigation, *I. hospitalis* and *Nanoarchaeum equitans*, the genomes are sequenced and the proteomes are under investigation. Proteins with possible roles in the interaction between these two Archaea can now be investigated by biochemical and molecular methods. In parallel, antibody-based methods can help to identify interaction partners in the test tube. The subcellular localization can then be performed on ultrathin sections and on freeze–fracture replicas. In future experiments, we plan to investigate, in a complementary way, the same specimens by light and electron microscopy, i.e., by correlative light and electron microscopy (Müller-Reichert *et al.*, 2007). The combination of all these methods will lead to a better understanding of archaea and of the kind of interactions taking place between these two unique archaeal cells.

Acknowledgments

The authors thank Cordula Neuner, Silvia Dobler, and Elke Pabst for excellent technical assistance; Prof. Dr. Michael Thomm, Prof. Dr. Karl O. Stetter, and Prof. Dr. Ralph Witzgall for continuous support; and the Faculty of Natural Sciences III for financial support to maintain the transmission electron microscope and other equipment. The work presented here was supported by several grants from the DFG.

References

Archibald, J. M., Logsdon, J. M.Jr., and Doolittle, W. F. (1999). Recurrent paralogy in the evolution of archaeal chaperonins. *Curr. Biol.* **9**, 1053–1056.

Blöchl, E., Rachel, R., Burggraf, S., Hafenbradl, D., Jannasch, H. W., and Stetter, K. O. (1997). *Pyrolobus fumarii*, gen. And sp. Nov., Represents a novel group of archaea, extending the upper temperature limit for life to 113°C. *Extremophiles* **1**, 14–21.

Boucher, Y. (2007). Lipids: biosynthesis, function, and evolution. *In*: "Archaea: Molecular and Cellular Biology" (R. Cavicchioli, ed.) pp. 341–353. ASM Press, Washington DC.

Burghardt, T., Näther, D. J., Junglas, B., Huber, H., and Rachel, R. (2007). The dominating outer membrane protein of the hyperthermophilic archaeum *Ignicoccus hospitalis*: A novel pore-forming complex. *Mol. Microbiol.* **63**, 166–176.

Buser, C., and Walther, P. (2008). Freeze-substitution: The addition of water to polar solvents enhances the retention of structure and acts at temperatures around −60°C. *J. Microsc.* **230**, 268–277.

Cammarano, P., Palm, P., Creti, R., Ceccarelli, E., and Tiboni, O. (1992). Early evolutionary relationships among known life forms inferred from elongation factor EF-2/EF-G sequences: Phylogenetic coherence and structure of the archaeal domain. *J. Mol. Evol.* **34**, 396–405.

Cavicchioli, R., ed. (2007). "Archaea: Molecular and Cellular Biology." ASM Press, Washington DC.

Fujimoto, K. (1995). Freeze-fracture replica electron microscopy combined with SDS digestion for cytochemical labeling of integral membrane proteins. *J. Cell. Sci.* **108**, 3443–3449.

Garrett, R. A., and Klenk, H. -P., eds. (2007). "Archaea: Evolution, Physiology, and Molecular Biology." Wiley Blackwell Publishing, Malden, MA, USA.

Gogarten, J. P., Kibak, H., Dittrich, P., Taiz, L., Bowman, E. J., Bowman, B. J., Manolson, M. F., Poole, R. J., Date, T., Oshima, T., Konishi, J., Denda, K., et al., (1989). Evolution of the vacuolar H^+-ATPase: implications for the origin of eukaryotes. *Proc. Natl. Acad. Sci. USA.* **86**, 6661–6665.

Graham, L. L., and Beveridge, T. J. (1990). Evaluation of freeze-substitution and conventional embedding protocols for routine electron microscopic processing of eubacteria. *J. Bacteriol.* **172**, 2141–2149.

Graham, L. L., Harris, R., Villiger, W., and Beveridge, T. J. (1991). Freeze-substitution of Gram-negative eubacteria: General cell morphology and envelope profiles. *J. Bacteriol.* **173**, 1623–1633.

Grimm, R., Singh, H., Rachel, R., Typke, D., Zillig, W., and Baumeister, W. (1998). Electron tomography of ice-embedded prokaryotic cells. *Biophys. J.* **74**, 1031–1042.

Heimerl, T. (2009). Ultrastruktur der Cokultur KIN4/M: Serienschnitte, 3D-Modelle und Immunmarkierungsversuche. Diploma thesis, University of Regensburg, Germany.

Hohenberg, H., Mannweiler, K., and Müller, M. (1994). High-pressure freezing of cell suspensions in cellulose capillary tubes. *J. Microsc.* **175**, 34–43.

Huber, H., Burggraf, S., Mayer, T., Wyschkony, I., Rachel, R., and Stetter, K. O. (2000). *Ignicoccus* gen. Nov., A novel genus of hyperthermophilic, chemolithoautotrophic archaea, represented by two new species, *ignicoccus islandicus* sp. Nov. and *ignicoccus pacificus* sp. Nov. *Int. J. Syst. Evol. Microbiol.* **50**, 2093–2100.

Huber, R., Eder, W., Heldwein, S., Wanner, G., Huber, H., Rachel, R., and Stetter, K. O. (1998). *Thermocrinis ruber* gen. Nov., Sp. Nov., A pink-filament-forming hyperthermophilic bacterium isolated from Yellowstone national park. *Appl. Envir. Microbiol.* **64**, 3576–3583.

Huber, H., Hohn, M. J., Rachel, R., Fuchs, T., Wimmer, V. C., and Stetter, K. O. (2002). A new phylum of archaea represented by a nanosized hyperthermophilic symbiont. *Nature* **417**, 63–67.

Humbel, B., and Müller, M. (1986). Freeze substitution and low temperature embedding. *In* "The Science of Biological Specimen Preparation 1985" (M. Müller, R. P. Becker, A. Boyde, and J. Wolosewick, eds.) pp. 175–183. SEM, Inc., AMF O'Hare, IL.

Iwabe, N., Kuma, K.-I., Hasegawa, M., Osawa, S., and Miyata, T. (1989). Evolutionary relationship of archaebacteria, eubacteria, and eukaryotes inferred from phylogenetic trees of duplicated genes. *Proc. Natl. Acad. Sci. USA.* **86**, 9355–9359.

Jahn, U., Summons, R., Sturt, H., Grosjean, E., and Huber, H. (2004). Composition of the lipids of *Nanoarchaeum equitans* and their origin from its host *Ignicoccus* sp. strain KIN4/I. *Arch. Microbiol.* **182**, 404–413.

Junglas, B., Briegel, A., Burghardt, T., Walther, P., Huber, H., Rachel, R. (2008). *Ignicoccus hospitalis* and *Nanoarchaeum equitans*: Ultrastructure, cell-cell interaction, and 3D reconstruction from serial sections of freeze-substituted cells and by electron cryotomography. *Arch. Microbiol.* **190**, 395–408.

Kandler, O., and König, H. (1998). Cell wall polymers in archaea (archaebacteria). *Cell Mol. Life Sci.* **54**, 305–308.

Klingl, A. (2007). Ultrastrukturelle Aspekte Pyrit-oxidierender Mikroorganismen. Diploma thesis, University of Regensburg, Germany.

König, H., Messner, P., and Stetter, K. O. (1988). The fine structure of the fibres of *pyrodictium occultum*. *FEMS Microbiol. Lett.* **49**, 207–212.

König, H., Rachel, R., and Claus, H. (2007). Proteinaceous surface layers of archaea: Ultrastructure and biochemistry. *In* "Archaea: Molecular and Cellular Biology" (R. Cavicchioli, ed.) pp. 315–340. ASM Press, Washington DC.

Kurr, M., Huber, R., König, H., Jannasch, H. W., Fricke, H., Trincone, A., Kristjansson, J. K., and Stetter, K. O. (1991). *Methanopyrus kandleri*, gen. And sp. Nov. Represents a novel group of hyperthermophilic methanogens, growing at 110°C. *Arch. Microbiol.* **156**, 239–247.

Küper, U., Meyer, C., Müller, V., Rachel, R., and Huber, H. (2010). Energized outer membrane and spatial separation of metabolic processes in the hyperthermophilic archaeon *ignicoccus hospitalis*. *Proc. Natl. Acad. Sci. USA* **107**, 3152–3156.

Langworthy, T. A., and Pond, J. L. (1986). Archaebacterial ether lipids and chemotaxonomy. *System. Appl. Microbiol.* **7**, 253–257.

Ludwig, W., and Schleifer, K. H. (1994). Bacterial phylogeny based on 16s and 23s rRNA sequence analysis. *FEMS microbiol. Rev.* **15**, 155–173.

McDonald, K. (2007). Cryopreparation methods for electron microscopy of selected model systems. *Methods Cell Biol.* **79**, 23–56.

Müller, D. W., Meyer, C., Gürster, S., Küper, U., Huber, H., Rachel, R., Wanner., G., Wirth, R., and Bellack, A. (2009). The iho670 fibers of *ignicoccus hospitalis*: A new type of archaeal cell surface appendage. *J. Bacteriol.* **191**, 6465–6468.

Müller-Reichert, T., Srayko, M., Hyman, A., O'Toole, E. T., and McDonald, K. (2007). Correlative light and electron microscopy of early *Caenorhabditis elegans* embryos in mitosis. *Methods Cell Biol.* **79**, 101–119.

Näther, D. J., Rachel, R., Wanner, G., and Wirth, R. (2006). Flagella of *Pyrococcus furiosus*: Multifunctional organelles, made for swimming, adhesion to various surfaces, and cell–cell contacts. *J. Bacteriol.* **188**, 6915–6923.

Nickell, S., Hegerl, R., Baumeister, W., and Rachel, R. (2003). *Pyrodictium* cannulae enter the periplasmic space but do not enter the cytoplasm, as revealed by cryo-electron tomography. *J. Struct. Biol.* **141**, 34–42.

Paper, W., Jahn, U., Hohn, M. J., Kronner, M., Näther, D. J., Burghardt, T., Rachel, R., Stetter, K. O., and Huber, H. (2007). *Ignicoccus hospitalis* sp. Nov., The host of "*nanoarchaeum equitans*". *Int. J. Syst. Evol. Microbiol.* **57**, 803–808.

Phipps, B. M., Hoffmann, A., Stetter, K. O., and Baumeister, W. (1991). A novel ATPase complex selectively accumulated upon heat shock is a major cellular component of thermophilic archaebacteria. *EMBO J.* **10**, 1711–1722.

Phipps, B. M., Typke, D., Hegerl, R., Volker, S., Hoffmann, A., Stetter, K. O., and Baumeister, W. (1993). Structure of a molecular chaperone from a thermophilic archaebacterium. *Nature* **361**, 475–477.

Rachel, R., Wyschkony, I., Riehl, S., and Huber, H. (2002). The ultrastructure of *Ignicoccus*: Evidence for a novel outer membrane and for intracellular vesicle budding in an archaeon. *Archaea* **1**, 9–18.

Rieger, G., Müller, K., Hermann, R., Stetter, K. O., and Rachel, R. (1997). Cultivation of hyperthermophilic archaea in capillary tubes resulting in improved preservation of fine structure. *Arch. Microbiol.* **168**, 373–379.

Rieger, G., Rachel, R., Hermann, R., and Stetter, K. O. (1995). Ultrastructure of the hyperthermophilic archaeon *pyrodictium abyssi*. *J. Struct. Biol.* **115**, 78–87.

Robertson, C. E. (2007). Electron microscopy of archaea. *Methods Cell Biol.* **79**, 169–191.

Rose, Md. e, and Gambacorta, A. (1986). Lipid biogenesis in archaebacteria. *System. Appl. Microbiol.* **7**, 278–285.

Schopf, S., Wanner, G., Rachel, R., and Wirth, R. (2008). An archaeal bi-species biofilm formed by *Pyrococcus furiosus* and *Methanopyrus kandleri*. *Arch. Microbiol.* **190**, 371–377.

Schroeder-Reiter, E., Houben, A., Grau, J., and Wanner, G. (2006). Characterization of a peg-like terminal NOR structure with light microscopy and high-resolution scanning electron microscopy. Chromosoma **115**, 50–59.

Severs, N. J., and Shotton, D. M. (2006). Rapid freezing of biological specimens for freeze fracture and deep etching. *In* "Cell Biology" (J. E. Celis, ed.), 3rd edn., **Vol. 3**, pp. 249–263. Elsevier Academic Press, Amsterdam.

Steinbrecht, R. A., and Zierold, K. (1987). "Cryotechniques in Biological Electron Microscopy." Springer, Berlin.

Stetter, K. O. (1982). Ultrathin mycelia-forming organisms from submarine volcanic areas having an optimum growth temperature of 105°C. *Nature* **300**, 258–260.

Stetter, K. O. (1992). Life at the upper temperature border. *In* "Frontiers of Life" (J. Tran Than Van, K. Tran Than Van, J. C. Mounolou, J. Schneider and C. McKay, eds.) pp. 195–219. Editions Frontieres, Gif-sur-Yvette.

Stetter, K. O. (2006). History of discovery of the first hyperthermophiles. *Extremophiles* **10**, 357–362.

Stierhof, Y. D., Humbel, B. M., Hermann, R., Otten, M. T., and Schwarz, H. (1992). Direct visualization and silver enhancement of ultrasmall antibody-bound gold particles on immunolabeled ultrathin resin sections. *Scanning Microsc.* **6**, 1009–1012.

Szabó, Z., Sani, M., Groeneveld, M., Zolghadr, B., Schelert, J., Albers, S. -V., Blum, P., Boekema, E. J., and Driessen, A. J.M. (2007). Flagellar motility and structure in the hyperthermophilic archaeon *Sulfolobus solfataricus*. *J. Bacteriol.* **189**, 4305–4309.

Wanner, G., Vogl, K., and Overmann, J. (2008). Ultrastructural characterization of the prokaryotic symbiosis in "*chlorochromatium aggregatum*". *J. Bacteriol.* **190**, 3721–3730.

Werner, F. (2007). Structure and function of archaeal RNA polymerases. *Mol. Microbiol.* **65**, 1395–1404.

Woese, C. R. (2000). Interpreting the universal phylogenetic tree. *Proc. Natl. Acad. Sci. U. S.A.* **97**, 8392–8396.

Woese, C. R., and Fox, G. E. (1977). Phylogenetic structure of the prokaryotic domain: The primary kingdoms. *Proc.Natl.Acad.Sci. U. S.A.* **74**, 5088–5090.

Woese, C. R., Kandler, O., and Wheelis, M. L. (1990). Towards a natural system of organisms: Proposal for the domains archaea, bacteria and eucarya. *Proc.Natl.Acad.Sci. U. S.A.* **87**, 4576–4579.

Wolf, Y. I., Rogozin, I. B., Grishin, N. V., and Koonin, E. V. (2002). Genome trees and the tree of life. *Trends Genet.* **18**, 472–479.

Zolghadr, B., Klingl, A., Koerdt, A., Driessen, A. J.M., Rachel, R., and Albers, S. -V. (2010). Appendage-mediated surface adherence of *Sulfolobus solfataricus*. *J. Bacteriol.* **192**, 104–110.

CHAPTER 4

Chlamydomonas: Cryopreparation Methods for the 3-D Analysis of Cellular Organelles

Eileen T. O'Toole

Department of Molecular, Cellular, and Developmental Biology, Boulder Laboratory for 3-D Electron Microscopy of Cells, University of Colorado, Boulder, Colorado 80309-0347

Abstract

Chlamydomonas reinhardtii is a popular model organism in modern cell biology. Historically, methods for preparing this cell for transmission electron microscopy have used conventional chemical fixation that can result in artifacts that affect the 3-D organization of the cell. We have developed improved methods of specimen preparation that involve high-pressure freezing followed by freeze-substitution that are

978-0-12-381007-6
DOI: 10.1016/S0091-679X(10)96004-4

particularly well suited for 3-D studies (O'Toole *et al.*, 2003, 2007). In this chapter, we describe the details of our cryopreparation methods for the optimal preservation of whole cells for immunocytochemistry and electron tomography. Examples are presented that show the utility of this approach for studying the 3-D architecture of membrane systems and cytoskeletal arrays in intact cells.

I. Introduction and Historical Perspectives

The unicellular alga, *Chlamydomonas*, is an excellent model organism for many areas of research including proteomics, cell biology, and molecular genetics (reviewed in Harris, 2001). This small eukaryote shares organelles common to both animal and plant cells, which makes it a popular model system in which to study photosynthesis, phototaxis, motility and centriole/basal body function. Importantly, powerful genetic and molecular analyses are possible due to the availability of the genome sequence, and it is now possible to use these techniques in combination with electron microscopy to analyze genes and characterize the gene products that are important in specific cellular processes (Merchant *et al.* 2007).

Landmark papers using transmission electron microscopy to describe the ultrastructure of *Chlamydomonas* were first published over half a century ago (Cavalier-Smith, 1974; Johnson and Porter, 1968; Ringo, 1967; Sager and Palade, 1954, 1957). These studies used conventional chemical fixation, followed by room temperature dehydration and embedding. However, this method can lead to artifacts, particularly in organisms that have natural barriers to diffusion. Indeed, the cell wall of *Chlamydomonas* presents a formidable barrier to diffusion of chemical fixatives due to the fact that it is composed of at least seven layers (Woessner and Goodenough, 1994). Methods to remove the cell wall using trypsin or shaking have also been used to improve the diffusion of the chemical fixative into the cell, yet these treatments lead to osmotic damage and substantial artifacts (Brand and Arnold, 1986). Despite these drawbacks, the early papers describing the ultrastructure of *Chlamydomonas* remain the landmark publications for the structural overview of this model organism.

Cryofixation and freeze-substitution methods were first proposed by Brand and Arnold (1986) to improve the structural preservation of the *Chlamydomonas* zygote. Dramatic improvements in cell ultrastructure were shown in the plunge frozen, freeze substituted samples as compared to cells prepared using conventional chemical fixation; the nucleus maintained its spherical shape and mitochondria were well preserved. However, the chloroplasts were particularly sensitive to freeze damage, and the overall yield of well-frozen cells was very low (Brand and Arnold, 1986).

Recently, methods using high-pressure freezing followed by freeze-substitution have been used for the optimum preservation of algal cell structure, and result in a greater yield of well-frozen cells compared to plunge freezing. The high-pressure (~2000 bar) used in this method lowers the freezing point of water, thus suppressing ice crystal formation (Moor, 1987). In practice samples or cell suspensions 100–200 μm thick are frozen well, which makes high-pressure freezing an ideal method for preserving algal cells. Cellular structures such as the basal body complex in *Chlamydomonas reinhardtii* (O'Toole *et al.*, 2003; Preble *et al.*, 2001) and the Golgi apparatus in the scale-producing alga, *Scherffelia dubia* (Donohoe *et al.*, 2006) have been well preserved with this method. Together, these studies illustrate the great potential of cryopreparation methods for the optimum preservation of algal cell ultrastructure.

Electron tomography (ET) has been shown to be a powerful method to study complex 3-D structures, such as cell organelles and cytoskeletal arrays (reviewed in Baumeister, 2002; Frank, 1992; Hoenger and McIntosh, 2009; McIntosh *et al.*, 2005). This 3-D imaging technique results in computer-generated reconstructions that can be imaged in any orientation at ~6 nm resolution. Further, serial section tomography can be used to enable fine scale resolution over a larger cellular volume (McIntosh *et al.*, 2005). Because of the increased z-resolution of this method, the optimum preservation of cell structure is an inherent prerequisite. In *Chlamydomonas*, cryo-ET of isolated axonemes has been used to study the 3-D organization of flagellar dynein (Bui *et al.*, 2008, 2009; Ishikawa *et al.*, 2007; Nicastro *et al.*, 2006) and axonemal doublet microtubules (Sui and Downing, 2006). Recent studies using ET of plastic-embedded material have revealed the 3-D organization of intraflagellar transport particles (Pigino *et al.*, 2009) and the basal body complex (O'Toole *et al.*, 2003, 2007) within their cellular context. It is clear that important insights into the 3-D architecture of cellular organelles are now possible at unprecedented resolution using ET.

Here we describe the steps necessary to prepare *Chlamydomonas* for the localization of antigens for immunocytochemistry and for 3-D studies using ET. This chapter illustrates use with *Chlamydomonas reinhardtii*, although the same methods can be applied to other species. We provide examples that illustrate how these methods can be used to study the 3-D morphology of membrane systems as well as cytoskeletal arrays in intact cells. The figures we present are selected, single frames, extracted from complete tomographic volumes or models. Movies of complete tomographic reconstructions and models are available as video supplements.

II. Rationale

The rationale for this chapter is to describe a detailed method of specimen preparation that results in the excellent preservation of cellular fine structure. These methods use high-pressure freezing, followed by freeze-substitution fixation to optimally preserve organelles for 3-D studies and allow immunolocalization of antigens within their cellular context. The steps required for tomographic analysis to image and model these structures in 3-D are also outlined. Examples are shown that illustrate the utility of these methods for studies of membrane organization in the cell as well as microtubule arrays that form the basal body complex and mitotic spindle.

III. Methods

A. Cryopreparation of Cells Using High-Pressure Freezing and Freeze-Substitution

1. Cell Culture

Healthy cells grown under physiologic conditions are used for high-pressure freezing. An excellent and comprehensive resource of protocols and methods for cell culture can be found in *The Chlamydomonas Sourcebook* (Harris, 2008). Briefly, cells are inoculated into 150 ml sterile flasks containing 50 ml of Sager and Granick medium I supplemented with 8 mM sodium acetate (Sager and Granick, 1953). The liquid cultures are then grown at 25°C under constant illumination until the cultures are in log-phase growth (typically 2–3 days).

2. High-Pressure Freezing

The work described in this chapter has been performed with a BAL-TEC HPM-010 high-pressure freezer. This model is now manufactured by ABRA (ABRA HPM010) and is available through RMC (Boeckler Instruments, Tucson, AZ). Other models of high-pressure freezers are available commercially that differ in their loading design and size (see Section IV).

In preparation for high-pressure freezing, the cells are incubated with a cryoprotectant. Aliquots of the liquid cultures are transferred to 50 ml conical centrifuge tubes, spun at 500 × g for 5 min, and the resulting pellets are then gently resuspended in medium containing 150 mM mannitol. We found that other external cryoprotectants, such as dextran or ficol had a negative effect on the cell and often resulted in release of the flagella. Mannitol was chosen as it has very minor effects on cell viability and the cells remained intact during incubation in this media. After the cells have incubated for 1 h, the samples are spun at 500 × g for 5 min and a loose pellet is obtained.

When loading specimens for high-pressure freezing it is important to work quickly, but carefully. We typically use aluminum hats with a bottom hat containing a 50 or 100 μm well and a top hat with a flat side. The flat side of the hat is rubbed on a piece of filter paper that is saturated with 1-Hexadecene, and then allowed to dry. This allows for easy separation of the hats after freezing. For sample loading, the well of the bottom aluminum hat is filled with the loose cell pellet and the top aluminum hat is then placed flat-side down on top of the sample. The sample is then transferred to the specimen holder and high-pressure frozen. After freezing, the samples are quickly transferred to liquid nitrogen where they are stored until further use (O'Toole *et al.*, 2007; Preble *et al.*, 2001).

3. Freeze-Substitution and Embedding for Electron Tomography

In preparation for electron microscopy, the frozen cells must first be freeze-substituted and embedded in resin (Table I). Our standard method of freeze-substitution involves transferring the frozen samples under liquid nitrogen to cryovials

Table I
Freeze-Substitution Protocols for Immunocytochemistry and EM

Tomography Immunoctytochemistry		EM Tomography	
Reagents	Time/Temperature	Reagents	Time/Temperature
0.2% glutaraldehyde and 0.1% uranyl acetate in acetone	3 days at −90°C then warm to −50°C for 24 h	1% OsO_4 and 0.1% uranyl acetate in acetone	3 days at −90°C then warm to room temperature
Rinse in acetone	3 × 15 min at −50°C	Rinse in acetone	3 × 15 minutes, room temperature
1:1 mix HM20: acetone	24 h at −50°C	1:1 mix Epon: acetone	24 h at room temperature rocking platform
2:1 mix HM20: acetone	24 h at −50°C	2:1 mix Epon: acetone	24 h at room temperature rocking platform
100% HM20	3 changes over 2 days at −50°C	100% Epon	3 changes over 2 days at room temperature on rocking platform
100% HM20 plus accelerator	4 h at −50°C	100% Epon plus accelerator (DMP30)	4 h at room temperature on rocking platform
Polymerize with UV light	−50°C 2days then while warming to 0°C for 1 day	Polymerize in oven	2 days at 60°C

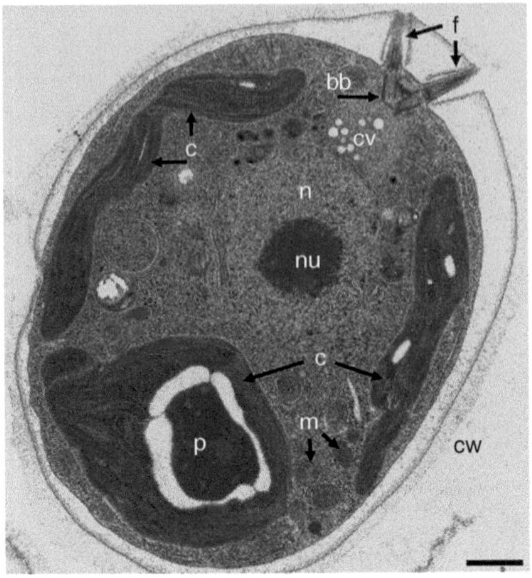

Fig. 1 *Chlamydomonas reinhardtii* cell prepared by high-pressure freezing followed by freeze-substitution in 1% OsO_4 and 0.1% uranyl acetate, and embedded in epon/araldite resin. The cell is surrounded by a cell wall (cw) and two flagella (f) can be seen emerging from the basal bodies (bb) in the apical region of the cell. The nucleus (n) appears round and contains a prominent nucleolus (nu). Organelles such as chloroplasts (c), mitochondria (m), and the contractile vacuole (cv) can be seen distributed throughout the cell. The pyrenoid (p) does not embed well in these preparations, as indicated by the white halos surrounding this organelle. Bar = 1 μm.

containing 1% OsO_4 and 0.1% uranyl acetate in anhydrous acetone. The samples are freeze-substituted for 3 days at −90°C, and then gradually warmed to room temperature. Alternatively, samples can be freeze-substituted in 0.5% glutaraldehyde and 0.1% tannic acid in acetone for 3 days at −90°C, rinsed in acetone, followed by freeze-substitution in 2% OsO_4 in acetone at −20°C for 1 day, then warmed to 4°C (Preble *et al.*, 2001). After freeze-substitution, the samples are gently removed from the high-pressure freezer hats using the tip of a #5 forceps and then rinsed several times in EM grade acetone.

The freeze-substituted cells are embedded in resin by infiltrating the samples in a graded series of resin and acetone over the course of several days. We typically use Epon resin for 3-D tomography studies. Infiltration is done at room temperature on a rocking platform beginning with 3:1 (acetone: resin) for 24 h, followed by 1:1 (acetone: resin), a 1:3 (acetone: resin) and two changes of 100% resin for 24 h each resin change. The samples are then transferred to BEEM capsules containing 100% resin with an accelerator (DMP30 for Epon/Araldite) and polymerized in a 60°C oven.

An example of a *Chlamydomonas* cell prepared by high-pressure freezing followed by freeze-substitution is shown in Fig. 1. The typical body plan of the cell is illustrated with two flagella extending from the apical region, a central nucleus with prominent nucleolus, and a basal chloroplast that extends laterally. The overall structure of the cell is well preserved, as indicated by the round nucleus, numerous organelles embedded in a dense and uniform cytoplasm and a contractile vacuole that has not been extracted or collapsed.

4. Freeze-Substitution and Embedding for Immunocytochemistry

For immunocytochemistry, the frozen samples are transferred under liquid nitrogen to cryovials (Nalge Nunc, Intl. Rochester, NY) containing 0.2% glutaraldehyde and 0.1% uranyl acetate in anhydrous acetone (Table I). The samples are freeze-substituted for 3 days at −90°C and are then gradually warmed to −50°C. The samples are then gently removed from the high-pressure freezer hats using the pointed end of precooled (-50°C) #5 forceps and rinsed several times in EM grade acetone.

We generally embed our samples in Lowicryl HM20 resin, although LR White resin can also be used (see details of infiltration and embedding by McDonald and Muller-Reichert, 2002). Lowicryl HM20 resin is prepared fresh according to the manufacturer's instructions (Electron Microscopy Sciences; Hatfield, PA). Samples are infiltrated at −50°C over a period of several days with a graded series of acetone and Lowicryl HM20 resin. Infiltration begins with 1:1 (acetone: resin) followed by 1:2 (acetone: resin) for 24 h each resin change. The samples are then infiltrated three times in 100% resin over the course of 2 days. The samples are then transferred to precooled (-50°C) BEEM capsules and polymerized under a UV light for 2 days and then warmed to 0°C for 1 day.

B. Immunocytochemistry

The method described below works well for material embedded in Lowicryl HM20 or LR White resin, and has been previously described (Morphew, 2007). It is useful to cut serial, thin (70 nm) sections in order to confirm specific label through multiple sections. Ribbons of serial sections are placed onto formvar-coated nickel slot grids. For antibody labeling, a humid chamber is assembled using a glass Petri dish (with top and bottom) containing a moistened kimwipe. A schematic of the humid chamber setup for immunolabeling can be seen in Fig. 6 of Chapter 12. Our standard blocking buffer contains 0.8% bovine serum albumin (BSA), 0.1% fish gelatin (Amersham), and 0.01% Tween-20 in phosphate buffered saline (PBS). We have also used 2% nonfat dry milk in PBS, or 10% normal goat serum in PBS as a blocking buffer for some antibodies. The grids are first floated section-side down on drops of blocking buffer. The top Petri dish is then placed over the bottom and sealed with parafilm to form a humid chamber. The samples are then incubated at room temperature for 30 min. After this time, the blocking buffer is blotted carefully from the side of the grid using filter paper. The grids are then placed section side down onto drops of primary antibody (diluted in blocking buffer) for 2 h at room temperature or overnight at 4°C. The dilution to use is highly dependent on the antibody and it is therefore necessary to carry out a dilution series to determine the optimum concentration. The grids are then rinsed three times with PBS containing 0.1% Tween-20 (PBST). The secondary

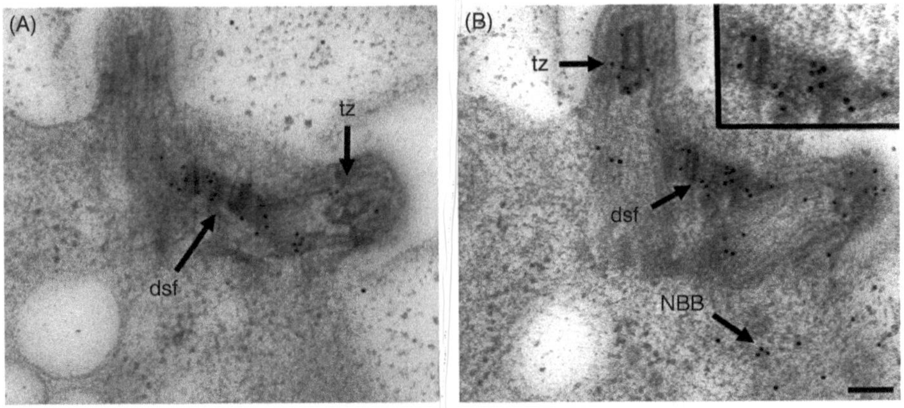

Fig. 2 Localization of centrin using post-embed immunoEM labeling. Cells were high-pressure frozen, freeze substituted in 0.2% glutaradehyde and 0.01% uranyl acetate, and embedded in Lowicryl HM20. Two serial sections through a wild-type basal body complex labeled with a monoclonal antibody to centrin and a secondary antibody conjugated to 10 nm gold. Two serial sections confirm the localization of centrin as a component of the distal striated fiber (dsf) that connects the two flagella, the transition zone (tz), and the nucleus basal body connector (NBB). Bar = 100 nm.

gold-conjugated antibody is diluted in blocking buffer (typically 1: 20) and the grids are incubated for 2 h at room temperature, rinsed twice with PBST and then twice with PBS. We found it was important to then post-fix the grids in 1% paraformaldehyde for 10 min. This step helps to retain bound gold and maintain cell structure during post-staining steps. Finally, the grids are rinsed briefly in distilled water, and post stained for 5 min in 2% aqueous uranyl acetate followed by 3 min in lead citrate.

An example of serial thin sections containing a portion of the basal body complex labeled with an antibody to centrin and a secondary antibody conjugated with 10 nm colloidal gold is shown in Fig. 2. Numerous gold particles can be seen associated with the distal striated fiber, the transition zone and a portion of the nucleus-basal body connector. This confirms results by others using post-embed labeling techniques to localize centrin to isolated basal body complexes (Geimer and Melkonian, 2005).

C. Electron Tomography

1. Preparation of Sections

Because cellular organelles are usually much larger than a typical thick (200–300 nm) section, it is useful to collect serial sections both to find the organelle of interest and to track it completely through its volume. For tomography, the serial, thick (200–300 nm) sections are collected onto Formvar-coated (0.7% w/v) copper slot grids and post stained using 2% uranyl acetate in 70% methanol for 8 min, rinsed and stained with Reynold's lead citrate for 5 min.

An important step in preparing grids for tomography is the addition of alignment markers on the top and bottom surfaces of the sections. Grids are floated on ~20 μl drop of 15 or 10 nm colloidal gold particles for 5 min, and the grid is then blotted from the side with filter paper. The grid is then inverted, and the process is repeated for the opposite side. The gold particles will serve as fiducial markers for the subsequent alignment of the tilted views (Ladinsky *et al.*, 1994; O'Toole *et al.*, 2007).

2. Tilt Series Acquisition

Imaging a tilt series of thick (> 200 nm) sections requires microscopes with intermediate (200–300 kV) or high (400–1000 kV) accelerating voltages. In our lab we currently use a Tecnai F30 EM (operating at 300 kV) equipped with a eucentric tilting stage although other tomography-capable microscopes are available from JEOL, Zeiss, and Hitachi. To facilitate the acquisition of the serial images, we use the SerialEM image acquisition software package developed in our lab (Mastronarde, 2005), although other software packages are available (see Section IV). To begin, the grid is placed in a high tilt holder (see Section IV), and the region containing the particular organelle of interest is located. Serial, tilted views of the section are then collected every degree over a ± 60° or 70° range, at a pixel size of 1 nm. After the first tilt series has been acquired, the grid is rotated 90°, and a second tilt series is acquired.

Examples of selected views from a tilt series of a 200 nm thick section containing a portion of the eyespot is shown in Fig. 3A. Not much detail is evident in the thick section because there is so much material projected in any given view. Likewise, the profile of this organelle changes dramatically in the different tilted views. This is especially true at high tilt (–60° or +60°), where the section thickness doubles (Fig. 3A). The tilt series is the raw data from which a tomogram is computed and a movie of the complete tilt series (121 views) can be seen in Video Supplement 1 at http://www.elsevierdirect.com/companions/9780123810076.

3. Tomographic Reconstruction and Image Analysis

Our lab has developed a software package, IMOD, which contains all of the programs needed to compute dual-axis tomographic reconstructions (Kremer *et al.*, 1996; Mastronarde, 1997). The IMOD suite of programs is managed by a user-friendly graphical user interface, Etomo, which allows the user to step through the series of programs used to compute a tomographic reconstruction from the original tilt series. The process is described in detail in O'Toole *et al.* (2007) and there is also a tutorial with a sample data available on our website (http://bio3d.colorado.edu/IMOD) to help people learn how to use the software. An overview of this process is outlined below.

To begin, a pre-processing step is run to eliminate pixels with extreme intensity values, which limits the dynamic range of the data set. A coarse alignment step is then done that uses cross correlation to create a pre-aligned stack. The next step in the process is to generate a fiducial model. Modeling software is used to mark the positions

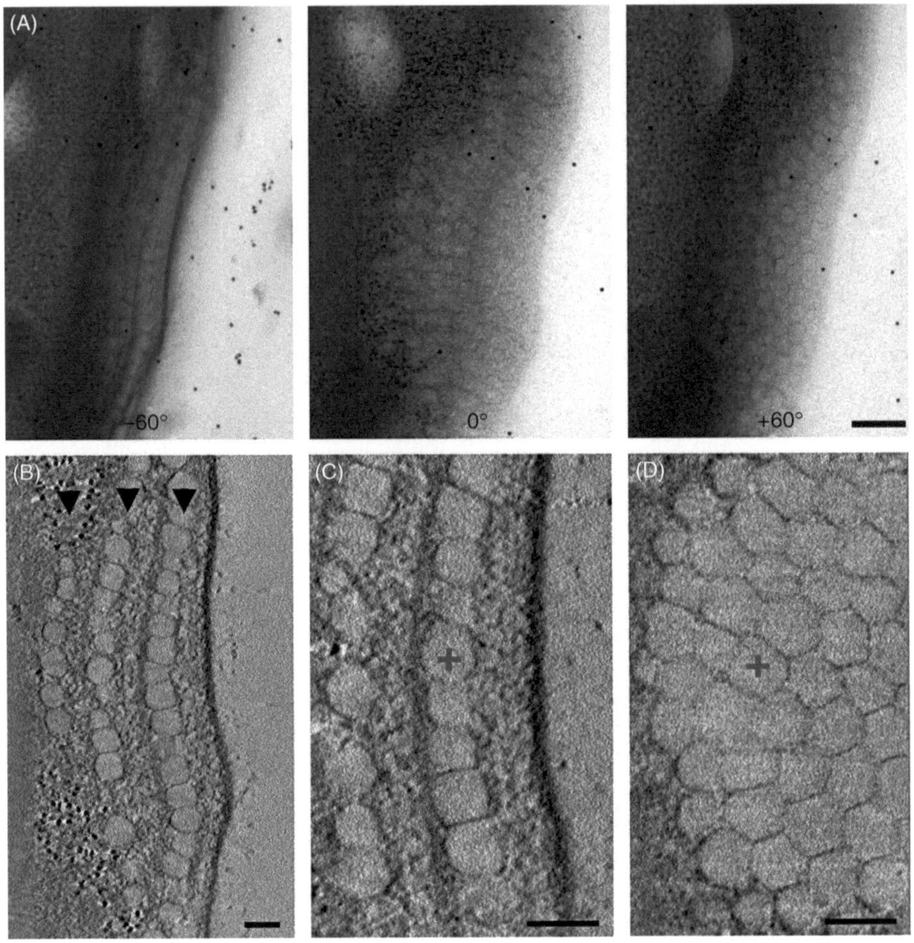

Fig. 3 Electron tomography reveals the 3-D architecture of the cell. (A) The raw data for a tomographic reconstruction are a series of tilted views of a single, thick section. Images are collected every degree over a ± 60 or 70° range about two axes. Here, tilted views at −60°, 0°, and +60° of a portion of the cell containing the eye spot illustrate how the image changes at different tilted views. Colloidal gold particles on the section surface are used as alignment markers. Bar = 200 nm. A movie of the complete tilt series can be viewed in Video Supplement 1 at http://www.elsevierdirect.com/companions/9780123810076. (B) A selected 1.6 nm slice through the tomographic reconstruction. Here, a tremendous amount of cellular fine structure can be seen including the 3 lipid-granule layers of the eyespot, shown here in cross section (arrowheads). Bar = 100 nm. A movie of the complete tomogram can be viewed in Video Supplement 2 at http://www.elsevierdirect.com/companions/9780123810076. (C, D) An image slicer tool is used to image the individual layers of the eyespot. The region of the tomographic slice marked (+) was rotated about a *y*-axis to view the outer eyespot layer in a single plane. Bars = 100 nm. A movie that simulates the use of the slicer tool can be viewed in Video Supplement 3 at http://www.elsevierdirect.com/companions/9780123810076. (See Plate no. 3 in the Color Plate Section.)

of gold particles (generally 20–40 per 2k × 2k image) on a single, low-tilt view, and a program is then run to automatically track the gold in all of the other views.

The most interactive and important step in the process is the fine alignment step. In this iterative step, an alignment program uses the positions of the gold particles and a least-squares approach to solve for the alignment transforms that will best align the fiducials on the images from all the tilted views (Luther *et al.*, 1988). The operator adjusts the positions of model points with large errors and the alignment program is run again. After the best solution is obtained, the final transforms are then applied to the images of the tilted views to create an aligned image stack. A tomographic reconstruction is then computed using the density information in the aligned, tilted views using an R*-weighted back projection algorithm (Gilbert, 1972).

The tomogram can then be viewed using the 3dmod program that is part of the IMOD software package. Fig. 3B shows a selected, 1.4 nm-thick tomographic slice computed from the tilt series. Here, a tremendous amount of cellular detail can now be seen, including details of the three rows of hexagonal lipid granules of the eyespot, bounded by an electron-dense matrix. The surrounding membranes of the chloroplast and plasma membrane can also be seen. A movie of serial, tomographic slices through the complete volume can be seen in Video Supplement 2 at http://www.elsevierdirect.com/companions/9780123810076.

Because the number of tilted views is limited in ET (±60° or 70°), there is missing information in the computed tomogram, referred to as the missing wedge. This missing information can lead to artifacts that lower the resolution of the reconstruction and is one of the biggest limitations of ET. We routinely perform dual-axis ET to minimize this limitation, and reduce the missing wedge to a smaller, missing pyramid of information (Mastronarde, 1997). After the two tomograms have been computed independently from the two tilt series, they are aligned to each other and their information is combined into a single reconstruction.

One of the strengths of ET is the fact that a tomographic slice is a sample through a 3-D volume, and as such its orientation can be rotated to obtain the best view of a particular organelle. Our viewing software, 3dmod, contains an image slicer window that displays a slice extracted from the 3-D volume in any position or orientation (Fig. 3C and D). In this example, the three rows of the eyespot are seen in cross section (Fig. 3C, arrowheads). The morphology of one of the eyespot layers (Fig. 3C, +) can be viewed by extracting a slice of image data one voxel thick, and rotating its orientation about the *y*-axis to view the layer *en face* (Fig. 3D). A simulation of the associated rotations applied to view the eyespot can be seen in the Video Supplement 3 at http://www.elsevierdirect.com/companions/9780123810076.

Because many cellular organelles are much larger than a typical 200–300 nm thick section, it is necessary to perform serial section tomography to reconstruct the entire volume of large organelles. An interface is available in Etomo to align and join tomograms computed from serial sections, and a tutorial on its use is available on our website (http://bio3d.colorado.edu). This approach has been used to reconstruct the entire *Chlamydomonas* basal body complex within a ~600 nm volume (O'Toole *et al.*, 2003).

D. Three-Dimensional Modeling of Cellular Organelles

1. Modeling membrane components

The tomographic data presented in this chapter were modeled using the 3dmod program from the IMOD software package (Kremer *et al.*, 1996; see Section IV for alternative modeling software). Figure 4 shows an example of how this modeling software can be

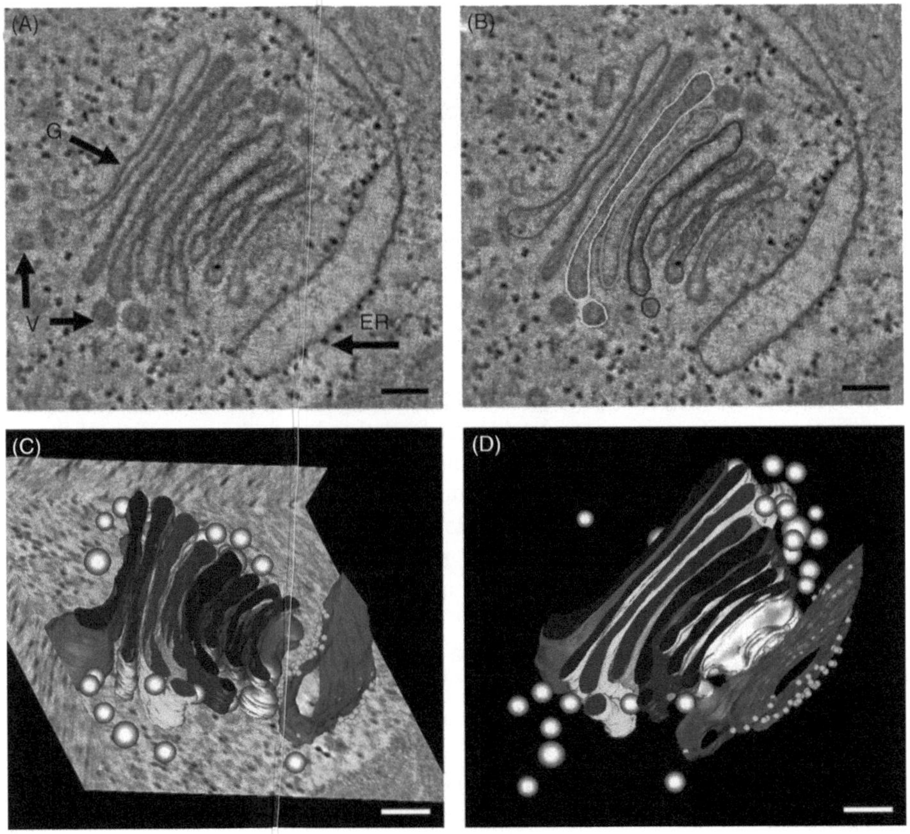

Fig. 4 Modeling membrane organization in three dimensions. (A) Selected tomographic slice showing the multiple cisternae of a Golgi complex (G), surrounding vesicles (V), and a portion of the endoplasmic reticulum (ER). (B) Contours are drawn around the eight, closely apposed cisternal membranes (red, light orange, yellow, green, blue, light blue, orange, violet) and associated vesicles (tan) on a selected tomographic slice. The contours of an associated piece of rough ER are shown (pink) with associated ribosomes (small blue spheres). The contours are drawn on each tomographic slice to build the model of the organelle through the whole volume. A movie showing serial, tomographic slices through the complete volume and with model contours drawn is shown in Video Supplements 4 and 5 at http://www.elsevierdirect.com/companions/9780123810076, respectively. (C, D) The 3-D model can be projected onto planes of tomographic data (C) and the complete model can be rotated to get a better view of objects of interest (D). A movie showing the 3-D model of the *Chlamydomonas* Golgi complex can be seen in Video supplement 6 at http://www.elsevierdirect.com/companions/9780123810076. Bar = 100 nm (See Plate no. 4 in the Color Plate Section.)

used to study Golgi membrane organization in three dimensions. Individual slices from the tomogram show detail of a portion of the cell volume containing the Golgi complex, surrounding vesicles, ribosomes, and endoplasmic reticulum (Fig. 4A; a movie showing serial, tomographic slices through the complete volume can be seen in Video Sequence 4 at http://www.elsevierdirect.com/companions/9780123810076). Organelles of interest are modeled separately and represented as different objects with a unique color. Membrane compartments are typically modeled as objects consisting of closed contours, where the boundaries of each membrane compartment are traced by hand around the object of interest (Fig. 4B). A new contour is created on each successive tomographic slice to build a model of the organelle through the whole volume. Vesicles and ribosomes are modeled as scattered points where the software builds a sphere of a designated diameter from a single model point. Video Sequence 5 is a movie of the serial, 1.4 nm tomographic slices with model contours marking the positions of each Golgi cisternae (red, orange, yellow, green, light and dark blue, violet), surrounding vesicles (tan), and a portion of the ER (purple) with associated ribosomes (light blue spheres).

The resulting model can then be viewed as a 3-D projection in the model view window and rotated to study the relationships of the organelles in the volume (Fig. 4C and D). Objects can be "meshed" to create a skin over the contours, and transparencies and lighting can also be adjusted to best visualize the objects of interest. Planes of tomographic slices from the image data can also be displayed in the model view window to better relate the position of the modeled organelles within the context of the cell (Fig. 4C). The full model can be viewed (Fig. 4D) or subsets of objects can be turned on or off to visualize relationships of particular organelles to one another. Finally, quantitative measurements of surface areas and volumes of the membrane compartment can be extracted from the model data using the companion program, imodinfo. A movie of the rotating model projected on planes of image data is shown in Video Sequence 6 at http://www.elsevierdirect.com/companions/9780123810076.

2. Modeling Cytoskeletal Arrays

Cytoskeletal arrays such as microtubules (MTs), form complex 3-D arrays within the cell and are often present at oblique angles within the volume of the tomogram. To facilitate tracking and modeling of cytoskeletal elements it is necessary to use the slicer window. Figure 5A shows an example of how the slicer window can be used to best visualize MTs for accurate modeling. A tomographic slice from the tomogram containing a segment of a MT is shown in Fig. 5A (left, arrowhead). The slice of image data is rotated about an x-axis to view the MT along its length (Fig. 5A, middle). MTs are typically modeled as objects with open contours in which model points are deposited by hand along the length of the MT (Fig. 5A, right).

An example showing a model of the MTs forming the basal body complex and mitotic spindle is shown in Fig. 5B. In this example, it was necessary to perform serial section tomography to reconstruct the entire basal body complex and most of the mitotic spindle pole within a ~750 nm thick volume. In this cell the basal body complex remained closely associated with the plasma membrane. Nascent daughter

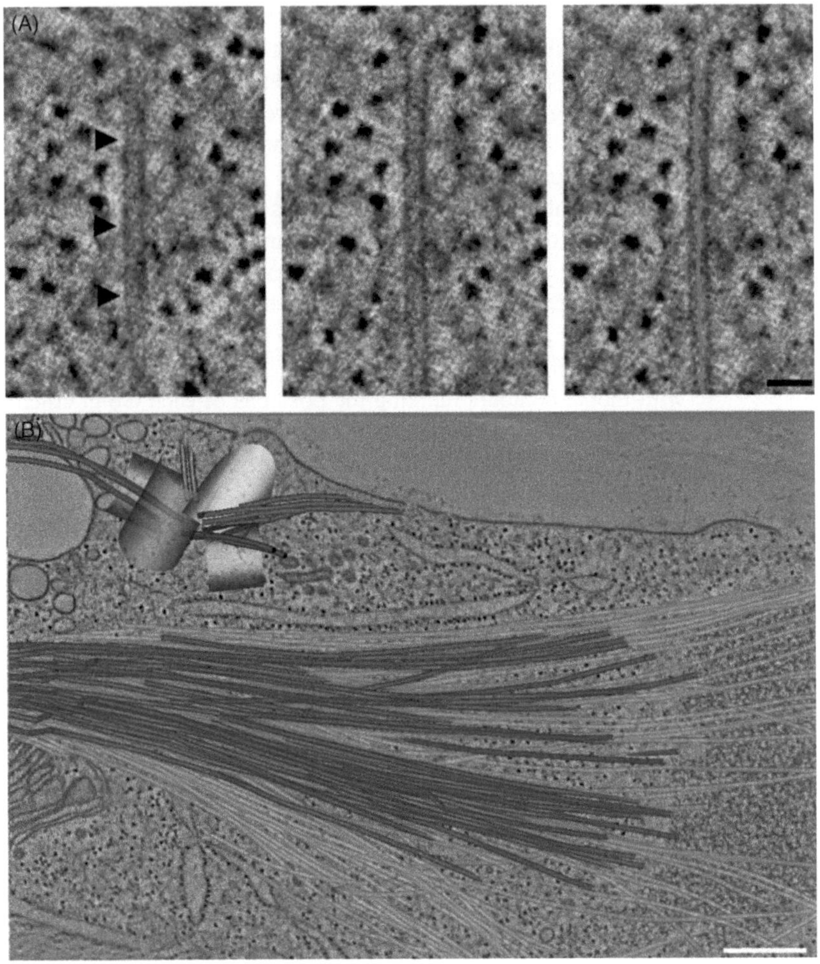

Fig. 5 Modeling cytoskeleton arrays in three dimensions. (A) Use of the slicer tool to model MTs. Mts often run at oblique angles within the volume of the reconstruction (left; arrowheads). The slicer tool was used to rotate the slice of image data about an *x*-axis to view the MT in a single view (middle). Model points are then deposited by hand along the length of the MT (right; green line). Bar = 50 nm. (B) 3-D model of the basal body complex and a portion of the mitotic spindle projected on a single, tomographic slice. The basal body complex (blue cylinders) remains associated with the plasma membrane during mitosis. Bundles of rootlet MTs (red lines) can be seen emanating from the basal body complex. Kinetochore and nonkinetochore MTs were modeled (purple and green lines, respectively) and were focused in large polar fenestrae that formed the pole of the mitotic spindle. Bar = 100 nm. A movie of the 3D model projected onto serial, tomographic slices is shown in Video Supplement 7 at http://www.elsevierdirect.com/companions/9780123810076. (See Plate no. 5 in the Color Plate Section.)

basal bodies could be seen orthogonal to the mature basal bodies (Fig. 5B, blue cylinders). Cytoplasmic MTs forming the rootlet bundles were also seen associated with the basal body complex (Fig. 5B, red). During mitosis, the nuclear envelope remains partially intact and large, polar fenestrae form the pole from which spindle

MTs assemble. Numerous kinetochore and nonkinetochore microtubules were modeled and were organized much like a bouquet (Fig. 5B, purple, green lines). A movie of the 3-D model projected on serial, tomographic slices is shown in Video Sequence 7 at http://www.elsevierdirect.com/companions/9780123810076.

IV. Instrumentation and Materials

Unless otherwise noted, all electron microscopy supplies are supplied by Electron Microscopy Sciences, (Hatfield, PA).

A. High-Pressure Freezing

Instrumentation: There are currently four commercially available high-pressure freezing instruments: the Wohlwend Compact HPF01 from Wohlwend Engineering, Sennwald, Switzerland (distributed and serviced by Technotrade International (Manchester, NH), the Leica EM PACT2 and the Leica EM HPM 100 (Leica Microsystems in Vienna, Austria), and the ABRA HPM010 (available through RMC; Boeckler Instruments, Tucson, AZ).

Materials and Reagents: Clinical centrifuge, 50 ml conical centrifuge tubes, 1.5 ml Eppendorf tubes, pipets, medium with 150 mM D-Mannitol (Sigma, St. Louis, MO) prior to freezing. Aluminum high-pressure freezer hats (Technotrade, Manchester, NH). 1-Hexadecene (Sigma, St. Louis, MO) saturated filter paper.

B. Freeze-Substitution

Instrumentation: Automated freeze-substitution devices are available such as the Leica AFS where one can accurately set and control temperature. However, homemade devices can also be assembled using dry ice and a Styrofoam box (details of their use can be found in McDonald, 2007).

Materials and reagents: Nunc cryovials, 1% OsO4, 0.1% uranyl acetate in acetone, EM grade acetone, Epon resin, and BEEM capsules.

C. Immunocytochemistry

Materials and Reagents: A humid chamber assembled from two glass Petri dishes with moistened kimwipe and square of parafilm (see Fig. 6 of Chapter 12.). Nickel slot grids, blocking buffer containing 0.8% BSA, 0.1% fish gelatin (Amersham), 0.01% Tween-20 in phosphate buffered saline (PBS), secondary gold/antibody (Ted Pella, Redding, CA), and 1% paraformaldehyde.

D. Electron Tomography

Instrumentation: Electron tomography of thick (200–400 nm) sections requires specialized microscopes operating at higher accelerating voltages (200–400 kV). A high tilt specimen holder is also necessary and several are commercially available:

Fishione model 2020 (Export, PA) and several models made by Gatan (Pleasanton, CA). There are currently four NIH-supported National Resource Research Facilities in the United States that accept tomography service projects and collaborations. A list of these facilities can be found at: http://www.ncrr.nih.gov/biomedical_technology/biomedical_technology_research_centers.

Automated image-acquisition programs are available commercially: FEI Company (Eindhoven,The Netherlands); Tietz Video and Image Processing Systems (TVIPS) (Gauting, Germany), and Gatan, Inc. (Pleasanton, CA). In addition, several packages are freely available and include TOM (http://www.biochem.mpg.de/tom), UCSF Tomography (http:// msg.ucsf.edu/tomography), and Serial EM (http://bio3d.colorado.edu/SerialEM).

Materials and Reagents: Copper/rhodium slot grids, 0.7% (w/v) formvar in ethylene dichloride, 2% uranyl acetate in 70% methanol, rinse solutions of 70, 50, 20% methanol, Reynold's lead citrate, 10 or 15 nm colloidal gold (BBI International).

E. 3-D Modeling

The examples presented in this chapter used the IMOD software package developed in our lab. This package is freely available, runs on multiple platforms, and there are guides and tutorials for its use (http://bio3d.colorado.edu). Within the IMOD package is a user-friendly interface, Etomo, that manages all of the programs needed for calculating tomograms, joining serial section tomograms, and additional image processing interfaces. Image display and modeling are carried out with the 3dmod viewing program within the IMOD software package.

Other freely available software packages used for computing and/or modeling tomograms include EM3D (http://em3d.stanford.edu), TOM (http://www.biochem.mpg.de/tom), SPIDER (http://www.wadsworth.org), XMIPP (http://xmipp.cnb.csic.es/), and Chimera (http://www.cgl.ucsf.edu/chimera). Commercially available software packages include Xplore3D from FEI (Hillsboro, Oregon, US; Eindhoven, The Netherlands) an integrated acquisition and tomography software package from Gatan (Pleasanton, CA), and Amira (http://www.amiravis.com).

F. Discussion

Chlamydomonas is a powerful model organism because numerous genetic, biochemical, and molecular biology techniques can be combined with light and electron microscopy to elucidate important cellular processes. However, 3-D structure work on this organism has been limited due to the difficulty of preserving cell structure using conventional methods. The cell wall of *Chlamydomonas* presents a formidable barrier to chemical diffusion and thus conventional methods of fixation often lead to artifacts that affect the 3-D organization of the cell. We have developed an improved method of specimen preparation that involves high-pressure freezing followed by freeze-substitution that preserves antigenicity and results in excellent preservation of cell structure. The strengths of this method are that samples prepared in this manner can be used for immunocytochemistry and also for ET, where the 3-D arrangements of cellular organelles can be studied in unprecedented detail.

The use of pre-embed or post-embed immunocytochemistry has been an important technique to localize specific antigens at the EM level. For pre-embed labeling, the antibody is applied prior to embedding in resin. Advantages of pre-embed labeling is the increased accessibility of the antibody to numerous epitopes. This method, however, suffers from the fact that the organelles need to be isolated from their cellular context or the cell needs to be permeablized to provide antibody access into the cell, which can compromise cell ultrastructure. Post-embed labeling techniques, like the one described in this chapter, require that the cell be fixed, embedded, and sectioned prior to labeling. Our method results in excellent preservation of the antigens of interest and overall cell structure, but is limited in that only the surface of the section is available for labeling.

Studies coupling serial section immunocytochemistry and 3-D modeling have been used to document the spatial distribution of antigens within the cell. Indeed, this approach has been used to study the localization of specific proteins within the yeast spindle pole body and throughout the yeast mitotic spindle (Bridge *et al.*, 1998; Mueller-Reichert *et al.*, 2003). Recently, immunoelectron tomography has been used to gather more detail about the structures directly below the surface label to confirm the identity of proteins associated with the Golgi membranes in *Arabidopsis* root cells (Staehelin and Kang, 2008). A major goal for future 3-D studies involves the development of electron dense labels that can be used for the localization of specific proteins in well-preserved samples.

ET has been an important tool to study the 3-D organization of cellular organelles. Numerous studies have used ET to understand Golgi membrane organization in mammalian tissue culture cells (Ladinsky *et al.*, 1994, 1999; Marsh *et al.*, 2001), the yeast *Pichia pastoris* (Mogelsvang *et al.*, 2003), *Arabidopsis,* and the scale-producing algae, *S. dubia* (Donohoe *et al.*, 2006). In these studies, important structural insights about membrane trafficking through the Golgi were made possible due to the increased resolving power of ET to visualize vesicles and their relationship to the Golgi cisternea in 3-D.

Likewise, ET has been key to understanding the complex arrays that the cytoskeleton can form in a wide variety of cell systems. For example, ET of frozen hydrated or plastic-embedded specimens has been used to study the 3-D organization of actin in *Dictyostelium* and insect flight muscle (Medalia *et al.*, 2002; Taylor *et al.*, 1999), cytoplasmic MTs in fission yeast, cultured mammalian cells, and trypanosomes (Höög *et al.*, 2007; Koning *et al.*, 2007; Lacomble *et al.*, 2009), the mitotic spindle in budding yeast (O'Toole *et al.*, 1999), the kinetochore MTs of mammalian cells (Dong *et al.*, 2007; McIntosh *et al.*, 2008), and mitotic and meiotic spindle organization in *Caenorhabditis elegans* (O'Toole *et al.*, 2003; Srayko *et al.*, 2006). The time-intensive nature of ET and its expense are limitations, yet this method is becoming more popular in laboratories around the world.

Recently, genetic analyses have been combined with ET to compare wild-type and mutant structure in model organisms. For example, the mitotic spindle of

Saccharomyces cerevisiae and mitotic and meiotic spindles in *C. elegans* have been studied in wild-type and mutant strains to analyze defects in spindle MT organization (Gardner *et al.*, 2008; Ozlu *et al.*, 2005; Srayko *et al.*, 2006; Yoder *et al.*, 2005). Similar studies have also been performed with strains of *Chlamydomonas* with mutations that affect intraflagellar transport (Pigino *et al.*, 2009), flagellar dynein organization (Bui *et al.*, 2008, 2009; Ishikawa *et al.*, 2007), and basal body complex structure (O'Toole *et al.*, 2003, 2007). In the future, coupling genetics with 3-D structure studies of optimally preserved cells will continue to enhance our understanding of structure–function relationships in a number of model organisms.

V. Summary

ET is becoming a mainstream approach to study the 3-D architecture of cells in a number of model systems. In this chapter we described a method of cryopreservation involving high-pressure freezing followed by freeze-substitution that results in excellent preservation of *Chlamydomonas* cells. Material prepared in this way is amendable for both immunocytochemistry and 3-D studies using ET. The increased resolving power of ET, coupled with genetic analysis in model systems such as *Chlamydomonas*, should further enhance our understanding of important processes in cell biology.

Acknowledgments

We thank Thomas H. Giddings, Jr. for the initial development and continued advice for high-pressure freezing and freeze-substitution of *Chlamydomonas*, and David Mastronarde for developing the automated image capture software and for helpful advice on tomographic reconstruction. We thank Andrew Staehelin for use of the high-pressure freezer and Mary Morphew for useful suggestions for specimen preparation. This work was supported by Grant RR-000592 from the National Center for Research Resources of the National Institutes of Health to Andreas Hoenger.

References

Baumeister, W. (2002). Electron tomography: Towards visualizing the molecular organization of the cytoplasm. *Curr. Opin. Struct. Biol.* **12**, 697–684.

Brand, N., and Arnold, C. G. (1986). Improved structural preservation of the mature zygote of *Chlamydomonas reinhardtii* by freeze-substitution fixation compared with chemical fixation with special attention to the mitochondria. *Endocyt. Cell. Res.* **3**, 79–95.

Bridge, A., Morphew, M., Bartlett, R., and Hagan, I. (1998). The fission yeast SPB component cut12 links bipolar spindle formation to mitotic control. *Genes Dev.* **12**, 927–942.

Bui, K. H., Sakakibara, H., Movassagh, T., Oiwa, K., and Ishikawa, T. (2008). Molecular architecture of inner dynein arms in situ in *Chlamydomonas reinhardtii* flagella. *J. Cell Biol.* **183**, 923–932.

Bui, K. H., Sakakibara, H., Movassagh, T., Oiwa, K., and Ishikawa, T. (2009). Asymmetry of inner dynein arms and inter-doublet links in *chlamydomonas* flagella. *J. Cell Biol.* **186**, 437–446.

Cavalier-Smith, T. (1974). Basal bodies and flagellar development during the vegetative cell cycle and the sexual cycle of *chlamydomonas reinhardtii*. *J. Cell. Sci.* **16**, 529–556.

Dong, Y., Vanden Beldt, K. J., Meng, X., Khodjakov, A., and McEwen, B. F. (2007). The outer plate in vertebrate kinetochores is a flexible network with multiple microtubule interactions. *Nat. Cell Biol.* **9**, 516–522.

Donohoe, B. S., Kang, B. -H., and Staehelin, A. (2006). Identification and characterization of COPIa- and COPIb-type vesicle classes associated with plant and algal golgi. *Proc. Natl. Acad. Sci. U. S.A.* **104**, 163–168.

Frank, J. (1992). Introduction: Principles of electron tomography. *In* "Electron Tomography: Three Dimensional Imaging with the Transmission Electron Microscope" (J. Frank, ed.) pp. 1–13. Plenum Press, NewYork.

Gardner, M. K., Bouck, D. C., Paliulis, L. V., Meehl, J., O'Toole, E. T., Haase, J., Soubry, A., Joglekar, A. P., Winey, M., Salmon, E. D., Bloom, K., and Odde, D. J. (2008). Chromosome congression by kinesin-5 motor-mediated disassembly of longer kinetochore microtubules. *Cell* **135**, 894–906.

Gardner, M. K., Haase, J., Mythreye, K., Molk, J. N., Anderson, M., Joglekar, A. P., O'Toole, E. T., Winey, M., Salmon, E. D., Odde, D. J., and Bloom, K. (2008). The microtubule-based motor kar3 and plus end-binding protein bim1 provide structural support for the anaphase spindle. *J. Cell Biol.* **180**, 91–100.

Geimer, S., and Melkonian, M. (2005). Centrin scaffold in *chlamydomonas reinhardtii* revealed by immunoelectron microscopy. *Eurkaryot. Cell* **4**, 1253–1263.

Gilbert, P. F.C. (1972). The reconstruction of a three-dimensional structure from projections and its application to electron microscopy. II. Direct methods. *Proc. R. Soc. Lond. B Biol. Sci.* **182**, 89–102.

Harris, E. H. (2001). *Chlamydomonas* as a model organism. *Annu. Rev. Plant Mol. Biol.* **52**, 363–406.

Harris, E. H. (2008). "The *Chlamydomonas* Sourcebook", 2nd ed., San Diego Academic, San Diego. (780 pp.)

Hoenger, A., and McIntosh, J. R. (2009). Probing the macromolecular organization of cells by electron tomography. *Curr. Opin. Cell Biol.* **21**, 89–96.

Höög, J. L., Schwartz, C., Noon, A. T., O'Toole, E. T., Mastronarde, D. N., McIntosh, J. R., and Antony, C. (2007). Organization of interphase microtubules in fission yeast analyzed by electron tomography. *Dev. Cell.* **12**, 349–361.

Ishikawa, T., Sakakibara, H., and Oiwa, K. (2007). The architecture of outer dynein arms in situ. *J. Mol. Biol.* **368**, 1249–1258.

Johnson, U. G., and Porter, K. R. (1968). Fine structure of cell division in *chlamydomonas reinhardtii*. Basal bodies and MTs. *J. Cell Biol.* **38**, 403–425.

Koning, R. I., Zovko, S., Bárcena, M., Oostergetel, G. T., Koerten, H. K., Galjart, N., Koster, A. J., and Mieke Mommaas, A. (2007). Cryo electron tomography of vitrified fibroblasts: Microtubule plus ends in situ. *J. Struct. Biol.* **161**, 459–468.

Kremer, J. R., Mastronarde, D. N., and McIntosh, J. R. (1996). Computer visualization of three-dimensional image data using IMOD. *J. Struct. Biol.* **116**, 71–76.

Lacomble, S., Vaughan, S., Gadelha, C., Morphew, M. K., Shaw, M. K., McIntosh, J. R., and Gull, K. (2009). Three-dimensional cellular architecture of the flagellar pocket and associated cytoskeleton in trypanosomes revealed by electron microscope tomography. *J. Cell Sci.* **122**, 1081–1090.

Ladinsky, M. S., Kremer, J. R., Furcinitti, P. S., McIntosh, J. R., and Howell, K. E. (1994). HVEM tomography of the trans-golgi network: Structural insights and identification of a lace-like vesicle coat. *J. Cell Biol.* **127**, 29–38.

Ladinsky, M. S., Mastronarde, D. N., McIntosh, J. R., Howell, K. E., and Staehelin, L. A. (1999). Golgi structure in three dimensions: Functional insights from the normal rat kidney cell. *J. Cell Biol.* **144**, 1135–1149.

Luther, P. K., Lawrence, M. C., and Crowther, R. A. (1988). A method for monitoring the collapse of plastic sections as a function of electron dose. *Ultramicroscopy* **24**, 7–18.

Marsh, B. J., Mastronarde, D. N., Buttle, K. F., Howell, K. E., and McIntosh, J. R. (2001). Organellar relationships in the golgi region of the pancreatic beta cell line, HIT-T15, visualized by high resolution electron tomography. *Proc. Natl. Acad. Sci.* **98**, 2399–2406.

Mastronarde, D. N. (1997). Dual-axis tomography: an approach with alignment methods that preserve resolution. *J. Struct. Biol.* **120**, 343–352.

Mastronarde, D. N. (2005). Automated electron microscope tomography using robust prediction of specimen movements. *J. Struct. Biol.* **152**, 36–51.

McDonald, K. (2007). Cryopreparation methods for electron microscopy of selected model systems. *Meth. Cell Biol.* **79**, 23–56.

McDonald, K. L., Morphew, M. K., Verkade, P., and Mueller-Reichert, T. (2007). Recent advances in high-pressure freezing: Equipment and specimen loading methods. *Methods Mol. Biol.* **369**, 143–173.

McDonald, K., and Muller-Reichert, T. (2002). Cryomethods for thin section electron microscopy. *In* "Guide to Yeast Genetics and Molecular and Cell Biology, Parts B and C" (C. Guthrie, and G. Fink, eds.) Methods in Enzymology. **Vol. 351**, pp. 96–123. Academic Press, San Diego.

McIntosh, J. R., Grishchuk, E. L., Morphew, M. K., Efremov, A. K., Zhudenkov, K., Volkov, V. A., Cheeseman, I. M., Desai, A., Mastronarde, D. N., and Ataullakhanov, F. I. (2008). Fibrils connect microtubule tips with kinetochores: A mechanism to couple tubulin dynamics to chromosome motion. *Cell* **135**, 322–333.

McIntosh, J. R., Nicastro, D., and Mastronarde, D. (2005). New views of cells in 3d: An introduction to electron tomography. *Trends Cell Biol.* **15**, 43–51.

Medalia, O., Weber, I., Frangakis, A. S., Nicastro, D., Gerisch, G., and Baumeister, W. (2002). Macromolecular architecture in eukaryotic cells visualized by cryoelectron tomography. *Science* **298**, 1209–1213.

Merchant, S. S., et al., (2007). The *chlamydomonas* genome reveals the evolution of key animal and plant functions. *Science* **318**, 245–250.

Mogelsvang, S., Gomez-Ospina, N., Soderholm, J., Glick, B. S., and Staehelin, L. A. (2003). Tomographic evidence for continuous turnover of golgi cisternae in *Pichia pastoris.*. *Mol. Biol. Cell* **14**, 2277–2291.

Moor, H. (1987). Theory and practice of high-pressure freezing. *In* "Cryotechniques in Biological Electron microscopy" (R. A. Steinbrecht and K. Zierold, eds.). Springer-Verlag, Berlin Heidelberg.

Morphew, M. K. (2007). 3D immunolocalization with plastic sections. *Meth. Cell Biol.* **79**, 493–513.

Muller-Reichert, T., Sassoon, I., O'Toole, E. T., Romao, M., Ashford, A. J., Hyman, A. A., and Antony, C. (2003). Analysis of the distribution of the kinetochore protein ndc10p in *saccharomyces cerevisiae* using 3-D modeling of mitotic spindles. *Chromosoma* **111**, 417–428.

Nicastro, D., Schwartz, C., Pierson, J., Gaudette, R., Porter, M. E., and McIntosh, J. R. (2006). The molecular architecture of axonemes revealed by cryoelectron tomography. *Science* **313**, 944–948.

O'Toole, E. T., Giddings, T. H. Jr., and Dutcher, S. K. (2007). Understanding microubule organizing centers by comparing mutant and wild-type structures with electron tomography. *Meth. Cell Biol.* **79**, 125–143.

O'Toole, E. T., Giddings, T. H. Jr., McIntosh, J. R., and Dutcher, S. K. (2003). Three-dimensional organization of basal bodies from wild-type and δ-tubulin deletion stains of *chlamydomonas reinhardtii. Mol. Biol. Cell* **14**, 2999–3012.

O'Toole, E. T., McDonald, K. L., Mantler, J., McIntosh, J. R., Hyman, A. A., and Mueller-Reichert, T. (2003). Morphologically distinct microtubule ends in the mitotic centrosome of *Caenorhabditis elegans. J. Cell Biol.* **163**, 451–456.

O'Toole, E. T., Winey, M., and McIntosh, J. R. (1999). High voltage electron tomography of spindle pole bodies and early mitotic spindles in the yeast *saccharomyces cerevisiae. Mol. Biol. Cell* **10**, 2017–2031.

Ozlu, N., Srayko, M., Kinoshita, K., Habermann, B., O'Toole, E. T., Muller-Reichert, T., Schmaltz, N., Desai, A., and Hyman, A. A. (2005). An essential function of the *C. elegans* ortholog of TPX2 is to localize activated aurora A kinase to mitotic spindles. *Dev. Cell* **9**, 237–248.

Pigino, G., Geimer, S., Lanzavecchia, S., Paccagnini, E., Cantele, F., Diener, D. R., Rosenbaum, J. L., and Lupetti, P. (2009). Electron-tomographic analysis of intraflagellar transport particle trains in situ. *J. Cell Biol.* **187**, 135–148.

Preble, A. M., Giddings, T. H., and Dutcher, S. K. (2001). Extragenic bypass suppressors of mutations in the essential gene *BLD2* promote assembly of basal bodies with abnormal microtubules in *chlamydomonas reinhardtii. Genetics* **157**, 163–181.

Ringo, D. L. (1967). Flagellar motion and fine structure of the flagellar apparatus in *chlamydomonas. J. Cell Biol.* **33**, 543–571.

Sager, R., and Granick, S. (1953). Nutritional studies with *chlamydomonas reinhardtii. Ann. NY Acad. Sci.* **56**, 831–838.

Sager, R., and Palade, G. E. (1954). Chloroplast structure in green and yellow strains of *chlamydomonas*. *Exp. Cell Res.* **7**, 584–548.

Sager, R., and Palade, G. E. (1957). Structure and development of the chloroplast in Chlamydomonas. I. The normal green cell. *J. Biophys. Biochem. Cytol.* **3**, 463–488.

Srayko, M., O'Toole, E. T., Hyman, A. A., and Müller-Reichert, T. (2006). Katanin disrupts the microtubule lattice and increases polymer number in *C. elegans* meiosis. *Curr. Biol.* **16**, 1944–1949.

Staehelin, L. A., and Kang, B. -H. (2008). Nanoscale architecture of endoplasmic reticulum export sites and of Golgi membranes as determined by electron tomography. *Plant Physiol.* **147**, 1454–1468.

Sui, H., and Downing, K. H. (2006). Molecular architecture of axonemal microtubule doublets revealed by cryoelectron tomography. *Nature* **442**, 475–478.

Taylor, K. A., Schmitz, H., Reedy, M. C., Goldman, Y. E., Franzini-Armstrong, C., Sasaki, H., Tregear, R. T., Poole, K., Lucaveche, C., Edwards, R. J., Chen, L. F., Winkler, H., et al., (1999). Tomographic 3D reconstruction of quick-frozen, Ca2+-activated contracting insect flight muscle. *Cell* **99**, 421–431.

Woessner, J. P., and Goodenough, U. W. (1994). Volvocine cell walls and their constituent glycoproteins: An evolutionary perspective. *Protoplasma* **181**, 245–258.

Yoder, T. J., McElwain, M. A., Francis, S. E., Bagley, J., Muller, E. G., Pak, B., O'Toole, E. T., Winey, M., and Davis, T. N. (2005). Analysis of a spindle pole body mutant reveals a defect in biorientation and illuminates spindle forces. *Mol. Biol. Cell* **16**, 141–152.

CHAPTER 5

Ultrastructure of the Asexual Blood Stages of *Plasmodium falciparum*

Eric Hanssen[*], **Kenneth N. Goldie**[†], *and* **Leann Tilley**[‡]

[*]Electron Microscopy Unit, Bio21 Molecular Science and Biotechnology Institute, University of Melbourne, Parkville, VIC 3010, Australia

[†]Centre for Cellular Imaging and Nano Analytics (C-CINA), Structural Biology and Biophysics Core Biozentrum, University Basel, CH-4058 Basel, Switzerland

[‡]Department of Biochemistry and Center of Excellence for Coherent X-ray Science, La Trobe University, Melbourne, VIC 3086, Australia

METHODS IN CELL BIOLOGY, VOL. 96

978-0-12-381007-6
DOI: 10.1016/S0091-679X(10)96005-6

Abstract

Plasmodium falciparum is the most deadly of the human malaria parasites. The particular virulence of this species derives from its ability to subvert the physiology of its host during the blood stages of its development. The parasite grows and divides within erythrocytes, feeding on the hemoglobin, and remodeling its host cells so they adhere to blood vessel walls. The advent of molecular transfection technology, coupled with optical microscopy of fluorescent protein reporters, has greatly improved our understanding of the ways in which the malaria parasite alters its host cell. However, a full interpretation of the information from these studies requires similar advances in our knowledge of the ultrastructure of the parasite. Here we give an overview of different electron microscopy techniques that have revealed the fine structure of the parasite at different stages of development. We present data on some of the unusual organelles of *P. falciparum*, in particular, the membrane structures that are elaborated in the erythrocyte cytoplasm and are thought to play an important role in trafficking of virulence proteins. We present and discuss some of the exciting whole cell imaging techniques that represent a new frontier in the studies of parasite ultrastructure.

I. Introduction

A. The Importance of Studying the Ultrastructure of *P. falciparum*

Plasmodium falciparum is a protozoan parasite that is responsible for the most virulent form of human malaria. Transmission from *Anopheles* mosquitoes to humans involves a remarkable series of morphological transformations. Following injection of motile sporozoites into the host's blood stream, the parasite enters the liver where it multiplies, then re-differentiates to generate many thousands of merozoites. These merozoites invade the host's red blood cells (RBCs) to initiate the blood stage of the infection. The intraerythrocytic parasite morphs through the ring, trophozoite, and schizont stages, eventually bursting to release 16–32 daughter merozoites (Bannister *et al.*, 2000a; Garcia *et al.*, 2008) (Fig. 1A). Each asexual cycle takes ~48 h and cell rupture induces periodic waves of fever in the patient as the disease progresses (Miller *et al.*, 2002). The intraerythrocytic cycle can continue until the patient dies or is cured by drugs or by the development of immunity. After a couple of weeks of asexual cycling, some of the merozoites develop into highly specialized gametocytes that are capable of transfer to a mosquito. Fertilization ensues in the mosquito's stomach and the parasite forms an ookinete that migrates through the wall of the mosquito's stomach, then re-differentiates into a sporozoite and migrates to the salivary glands ready to begin a new infection.

Cycling of the asexual forms of the parasite within RBCs (see Fig. 1A) is responsible for the clinical symptoms of the disease, which range from uncomplicated fevers to life-threatening cerebral and placental malaria (Miller *et al.*, 2002). The complications are due to the adhesion of infected RBCs to receptors on brain venule endothelial

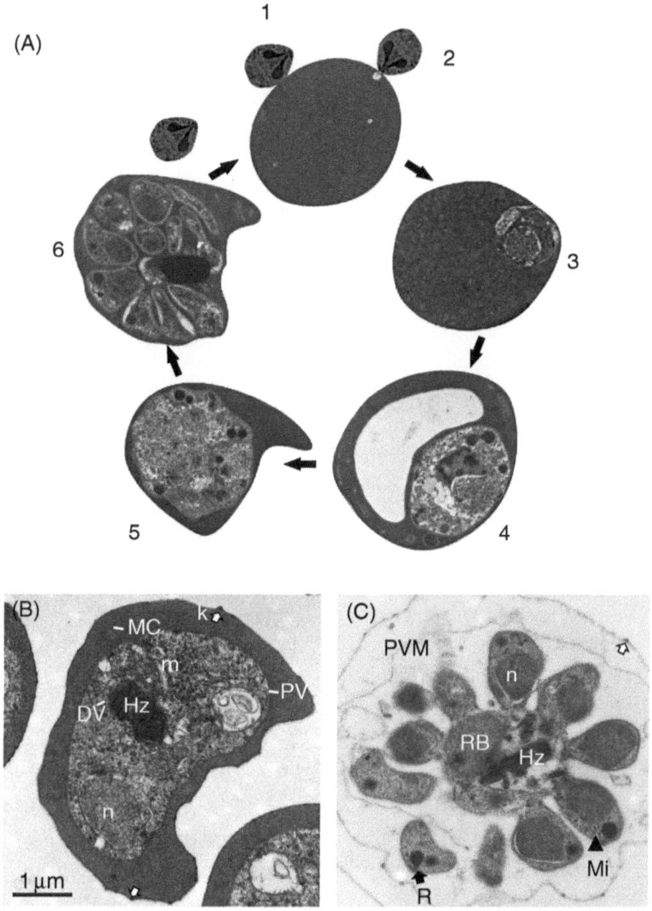

Fig. 1 Thin section TEM images of *P. falciparum*-infected RBCs. (A) Individual sections representing the different stages of development of the parasite in the intraerythrocytic cycle. Merozoite attachment (1) and reorientation (2) leads to invasion and the development of the ring stage (3), the mid and late trophozoite stages (4, 5), followed by cell division in the schizont stage (6). (B and C) Sections through a late trophozoite (B) and a permeabilized schizont. The parasite develops within a PV surrounded by a PV membrane (PVM). The following features are visible: nucleus (n), digestive vacuole (DV), hemozoin crystals (Hz), mitochondrion (m), Maurer's cleft (MC), knob (k), rhoptry (R), remnant body (RB), microneme (Mi). Scale bar, 1 µm.

cells or to placental syncytiotrophoblasts in pregnant women. The adhesion process prevents phagocytic clearance in the spleen, which contributes to virulence, while an inappropriate host immune response to the sequestered parasites can induce coma and death (Duffy and Fried, 2003; Haldar and Mohandas, 2007; Scherf *et al.*, 2008).

The intraerythrocytic parasite develops within a parasitophorous vacuole (PV) (Figs. 1 and 2A) that is initially formed during the invasion step (Bannister *et al.*,

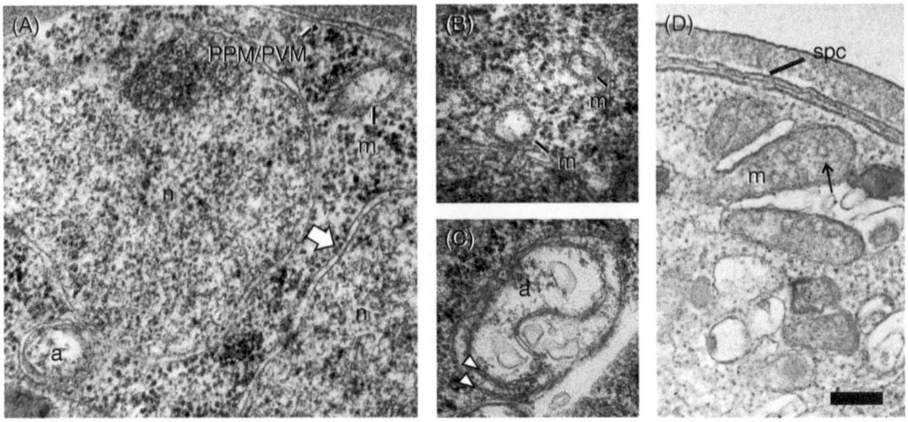

Fig. 2 TEM analysis of some of the core complement of organelles in the parasite cytoplasm. Sections through trophozoites (A–C) and gametocytes (D) showing different membrane features. The parasite plasma membrane (PPM) and PV membrane (PVM) are tightly apposed (A) as are the two membranes around the nucleus (n) (A, thick white arrow). The apicoplast (a) is surrounded by four membranes with the inner and outer double membrane features evident in C (white arrowheads). The mitochondrion is surrounded by two membranes (A and B) and is acrystate. By contrast, in the gametocyte crystae are observed in the mitochondria (D, arrow). The gametocyte is surrounded by a tri-laminar subpellicular complex (spc). Scale bar, 200 nm.

2000a). As the intraerythrocytic parasite develops, it needs to create space for expansion and division and to obtain a source of amino acid building blocks (Goldberg, 2005; Lew *et al.*, 2003). To achieve this, the parasite digests its host cell from the inside out, consuming ~75% of the host hemoglobin (Bakar *et al.*, 2010; Loria *et al.*, 1999). Early biochemical and ultrastructural studies of the feeding process in *P. falciparum*-infected RBCs showed that uptake of host cytoplasm involves morphologically distinct endocytic structures at the surface of the parasite (Garnham *et al.*, 1961; Slomianny, 1990; Yayon *et al.*, 1984). A number of recent studies have further examined this process (Elliott *et al.*, 2008; Lazarus *et al.*, 2008).

There are some logistical challenges associated with life in an enclosed niche in the highly differentiated environment of the human RBC. The parasite needs to generate the machinery required for its own growth and replication; however, it also needs to modify the properties of the host cell membrane to facilitate uptake of nutrients and to promote adhesion to endothelial cells. To do this, the parasite exports a range of proteins into the RBC cytoplasm and establishes an extracellular secretory apparatus to transfer virulence proteins to the RBC membrane (Maier *et al.*, 2009; Tilley *et al.*, 2008). In this review we discuss some of the recent studies using electron microscopy that provide new insights into the morphology and function of this unusual exomembrane system and point to a novel and sophisticated pathway for trafficking of virulence proteins. We also refer the reader to several earlier excellent reviews of parasite ultrastructure (Atkinson and Aikawa, 1990; Blackman and Bannister, 2001; Bannister and Dluzewski, 1990; Bannister and Mitchell, 2009; Bannister *et al.*, 2000a).

B. *P. falciparum*-Specific Advantages and Challenges

The intraerythrocytic stages of *P. falciparum* can be maintained indefinitely in culture (Trager and Jensen, 1976). Mature-stage-infected RBCs can be enriched by flotation on gelatin (Pasvol *et al.*, 1978) or Percoll (Aley *et al.*, 1986), or using the paramagnetic properties of the malaria pigment (Trang *et al.*, 2004), routinely achieving parasitemias (ratio of infected to uninfected RBCs) of $>95\%$. Ring-stage parasites can be enriched by selectively lysing uninfected RBCs with streptolysin-O (Jackson *et al.*, 2007). The formation of a subpopulation of gametocytes can be induced by maintaining the parasites for lengthy periods without the addition of fresh RBCs (Day *et al.*, 1993; Dixon *et al.*, 2008).

Infected RBCs have a diameter of ~8 μm, which permits ready fixation or freezing of suspensions of intact cells; however, artifacts can be induced during sample preparation. The intraerythrocytic parasite is surrounded by three membrane layers, the RBC membrane, the PV membrane, and its own plasma membrane. Moreover, several of the organelles within the parasite are surrounded by multiple membranes. These membrane-bound features are susceptible to dehydration artifacts when using room temperature methods, often resulting in separation of adjacent membranes. In Section III subsection C, we describe the sample preparation protocols for electron microscopy that permit the most reliable results in our hands.

There are some health and safety issues associated with handling human malaria parasites, and culturing them in the presence of blood and serum, especially if the parasites have been genetically modified and are known to be drug resistant. A culture facility approved for Biosafety Level 2 containment is required and the blood and serum should be tested and shown to be free of pathogens. If a ready source of certified pathogen-free serum is not available, fatty-acid-loaded bovine serum albumin (BSA; Albumax®) can be used as a substitute but this can compromise the delivery of cytoadhesion proteins to the RBC surface (Frankland *et al.*, 2007). In the absence of the mosquito vector, laboratory infections with *Plasmodium* can only occur by needle-stick injury, thus it is important to avoid the use of sharps where possible.

C. Transmission Electron Microscopy (TEM) of Thin Sections

1. Membrane Systems Within the Parasite

The growing parasite establishes a core complement of organelles needed for metabolism. These include the nucleus (Fig. 2A), the ER, and a rudimentary Golgi (Bannister *et al.*, 2004; Struck *et al.*, 2008). The single mitochondrion (Fig. 2A and B) is characterized by its inner and outer membranes (Slomianny and Prensier, 1986).

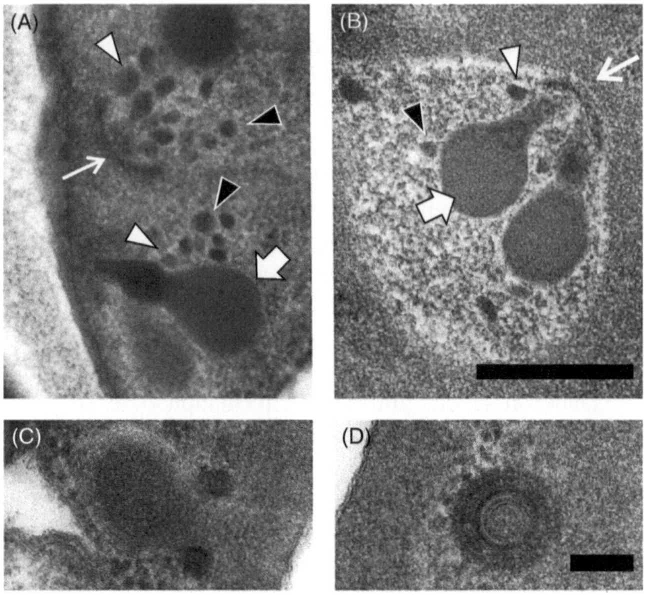

Fig. 3 TEM analysis of some of the unusual organelles of *P. falciparum* (A and B). Sections through schizont-infected RBCs showing daughter merozoites developing within the parent cell and featuring some of the apical organelles. The rhoptries (thick white arrows) are a pair of pear-shaped organelles with a more electron-dense neck, consistent with a different protein composition is this region of the organelle. The elongated micronemes (white arrowheads) and round dense granules (black arrowheads) accumulate in the apical region of the merozoite and the nascent polar ring region (white arrow) is evident at the apex of the daughter cells. Scale bar, 500 nm. (C and D) Longitudinal and transverse sections through a cytostome at the parasite surface. Double membraned endocytic flasks comprising invaginations of the parasite plasma membrane and PV membrane are formed at the region of an electron-dense cytostomal ring (internal diameter ~90 nm). Scale bar, 100 nm.

This organelle is acrystate in asexual-stage parasites but regains the crystae in the gametocyte stage (Fig. 2D). The parasite also develops a suite of novel organelles, including a plastid-like organelle (apicoplast) that is surrounded by four membranes (Fig. 2A and C) (Foth and McFadden, 2003; Hopkins *et al.*, 1999). A modified lysosome (referred to as the digestive vacuole) is the site of hemoglobin degradation and accumulation of hemozoin crystals (Figs. 1B and C). The parasite-feeding apparatus involves a characteristic structure referred to as the cytostome (Fig. 3C and D). Cytostome-mediated invaginations of the PV and parasite plasma membranes enable the uptake of host cytoplasm for transfer to the digestive vacuole. The daughter merozoites assemble a series of apical organelles that are involved in invasion. The rhoptries are characterized by a flask-shaped appearance (Fig. 3A and B); the dense granules, micronemes, and mononemes (Fig. 3A and B) store material for release onto the interface between the parasite and the host cell during or shortly after invasion (Tonkin *et al.*, 2006).

2. The Exomembrane System

Early TEM studies revealed unusual membrane structures in the RBC cytoplasm upon *P. falciparum* infection; these are referred to as the Maurer's clefts and the tubulovesicular network (TVN; (Aikawa, 1988; Langreth *et al.*, 1978)). The TVN consists of extensions and whorls emanating from the PV membrane (Haldar and Mohandas, 2007). Maurer's clefts (Fig. 4) appear at early ring stage (Kriek *et al.*, 2003) and are thought to originate from the PV membrane, then mature to form functionally independent structures that are tethered to the RBC plasma membrane (Atkinson *et al.*, 1988; Elford *et al.*, 1997; Hanssen *et al.*, 2008b; Lanzer *et al.*, 2006; Spycher *et al.*, 2006). Maurer's clefts are believed to be an important intermediate compartment involved in the sorting and trafficking of *P. falciparum* erythrocyte membrane protein-1 (PfEMP1) and other virulence determinants *en route* to the RBC membrane (Marti *et al.*, 2005; Tilley *et al.*, 2008). In thin sections, the Maurer's cleft bodies usually appear as slit-like structures (Fig. 4A and B), although in transverse sections (Fig. 4C) or in 3D reconstructions (Hanssen *et al.*, 2008b) their disk shape is revealed. The morphology of the Maurer's clefts varies between different strains. In the commonly used laboratory strain, 3D7, the Maurer's clefts usually comprise a single lamella (Fig. 4A); however, in the cytoadherence-defective D10 parasite strain, the Maurer's clefts have stacked and whorled lamellae (Fig. 4B) (Hanssen *et al.*, 2008b).

Tubular connecting elements (Fig. 4C) appear to tether the Maurer's cleft bodies to other membrane structures (Hanssen *et al.*, 2008b; Hanssen *et al.*, 2010). Indeed, the exomembrane system comprises a series of modular units, consisting of the flattened cisternae of the Maurer's cleft bodies and tubular connecting elements. The membrane network is not continuous, pointing to an important role for vesicle-mediated transport in the delivery of cargo to different destinations in the host cell. The parasite surface is decorated with knob structures that appear in transmission electron micrographs as electron-dense cups with a height of 30–40 nm and a diameter of about 90 nm (Fig. 4D) (Aikawa *et al.*, 1985). These "knobs" are formed by self-association of the knob-associated histidine-rich protein (KAHRP) underneath the RBC membrane and serve as platforms for the presentation of the membrane-embedded cytoadherence protein, PfEMP1 (Maier *et al.*, 2009). Effective presentation of PfEMP1 is needed for parasite virulence, and knobs are found in all field strains of the parasite. However, parasites that are maintained for lengthy periods in culture may undergo deletion of a chromosomal region containing the KAHRP gene, with consequent loss of knobs (Kilejian, 1979).

Recently, some novel membrane-bound structures in the RBC cytoplasm have been described, including two different vesicle populations. Uncoated vesicles with a diameter of 25 nm and coated vesicles with a diameter of 80 nm are observed in electron micrographs of intact and permeabilized infected RBCs (Hanssen *et al.*, 2008b; Hanssen *et al.*, 2010; Kriek *et al.*, 2003; Wickert *et al.*, 2003) (Fig. 4C and F). In some sections, electron-dense structures (possibly derived from the 80-nm vesicles) appear to have fused with the RBC membrane (Fig. 4E). This suggests that the electron-dense vesicles could be involved in the delivery of integral membrane

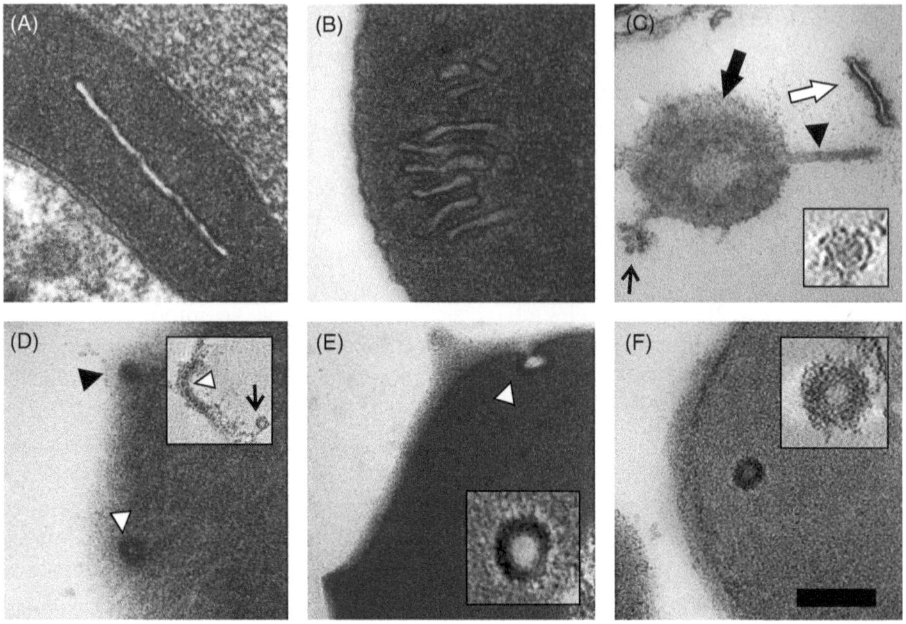

Fig. 4 TEM analysis of features elaborated in the host cell cytoplasm by the intracellular parasite (A and B). Sections through RBCs infected with parasites of the 3D7 (A) and D10 (B) strains showing single lamellar (A) and stacked lamellar (B) Maurer's clefts with characteristic electron-lucent lumens and electron-dense coats. The Maurer's cleft bodies appear tubular in cross section. (C) Region in the host cell cytoplasm of an Eqt II-permeabilized-infected RBC (K1 strain) showing a transverse view (thick black arrow) and a cross section (thick white arrow) through adjacent Maurer's cleft bodies. The organelle is roughly disk shaped. A tether-like structure (arrowhead) with a tubular profile is attached to the Maurer's cleft body. A cross section from a 1.8-nm virtual slice from a tomogram (inset) shows that the tether-like structure has an electron-dense core. A number of 25-nm uncoated vesicles are visible in the section (thin arrow). (D) Electron-dense knobs (arrowheads) with a diameter of ~90 nm are observed at the surface of the infected RBC (3D7 strain). A transverse section through the region of one of the knobs (white arrowhead) reveals the cup-shape of the structure. A 1.8-nm virtual slice from a tomogram (inset) shows that the cup shape is generated by deposition of protein material beneath the RBC membrane. A 25-nm vesicle is observed in the same tomogram. (E–F) Recently described novel features of infected RBCs. Electron-dense vesicles (80 nm diameter) are observed in the RBC cytoplasm in knob-positive and knob-negative parasites. The example shown in F is a K1 strain knob-minus parasite. Eqt II permeabilization increases the contrast revealing the fuzzy coat of these 80-nm vesicles (inset). (E) Caveolae-like structures are observed at the RBC membrane that may represent electron-dense vesicles that have fused with the RBC membrane. The inset shows a 1.8-nm tomogram cross section. Scale bar, 200 nm. Inset panel widths: C, 50 nm; D, 150 nm; E, 200 nm; F, 150 nm.

protein cargo to the RBC membrane. The electron-dense structures at the RBC membrane have an appearance that is very similar to that of caveolae (Atkinson and Aikawa, 1990).

A recent study evaluated different techniques for preparing samples for electron tomography, including whole cell cryo-preparations, vitreous sectioning, high pressure freezing/freeze-substitution, and chemical fixation (Henrich *et al.*, 2009). The authors concluded that each of these different approaches has some merits. In the vitrified, frozen-hydrated samples the Maurer's cleft lumen showed a similar density to the RBC cytoplasm indicating that the clefts are filled with more material than has been appreciated from studies using conventional sample preparation protocols that use room temperature fixation and dehydration. The frozen-hydrated samples also showed a subtle increase in the apparent thickness of the membrane at the centre of the Maurer's cleft disks; the authors suggested that this may be due to a different protein composition in the central region of each cisterna.

D. Electron Tomography

Most ultrastructural studies deduce the 3D organization of morphological features of plasmodium-infected RBCs from extensive analyses of multiple randomly cut sections of different cells. A more accurate view of the parasite's 3D ultrastructure can be provided by reconstructing data from serial thin sections. This approach has been used in a number of studies of *P. falciparum* organelles, including the mitochondrion (Slomianny and Prensier, 1986), the digestive vacuole (Elliott *et al.*, 2008; Slomianny, 1990), the apicoplast (Hopkins *et al.*, 1999), the developing rhoptries (Bannister *et al.*, 2000b), and membrane features in the RBC cytoplasm (Wickert *et al.*, 2003; Wickert *et al.*, 2004). However, the reconstructions that have been published are usually restricted to stacks of ~500 nm thickness. Moreover, the resolution in the *z*-direction is limited by the section thickness (i.e., ~70 nm for routine sections), which makes it difficult to resolve fine details of connections between different compartments.

Electron tomography is increasingly popular as a method that can be used to obtain 3D ultrastructural views of thicker sections (up to 400 nm) of *P. falciparum*-infected RBCs (Hanssen *et al.*, 2008b; Henrich *et al.*, 2009). Electron absorption and scattering, radiation damage, and signal-to-noise issues hamper the analysis of thicker sections at intermediate voltage (Baumeister *et al.*, 1999; Grunewald *et al.*, 2003; Lucic *et al.*, 2005; Steven and Aebi, 2003); however, methods have been developed that permit tiling and stitching of serial tomograms to generate whole cell images (Hanssen *et al.*, 2008b; Hanssen *et al.*, 2010; Noske *et al.*, 2008). Thus, it is possible to generate global 3D views of the parasite ultrastructure using multi-section electron tomography. While technically more challenging, these methods permit interrogation of the organization and connectivity of whole cells at increased resolution. We routinely use 250-nm semi-thin sections for datasets acquired within the protein- and membrane-rich regions in intact cells (Fig. 5A) and 300-nm sections to examine features in the infected RBC cytoplasm after permeabilization of the cells to release the hemoglobin.

The parasite surface and selected features within the parasite can be rendered to reveal details of the 3D organization. In the mature-stage parasite shown in Fig. 5B, the RBC surface is depicted in translucent cream and the four nuclei are in gold. A large digestive vacuole (blue) is evident along with a series of cytostomal invaginations arising from cytostomal rings (yellow) located at the parasite surface. The mitochondria and apicoplast are rendered in pink and aqua, whereas a series of acidocalcisomes are depicted in green. The 3D organization of the cells is best appreciated by examining a translation

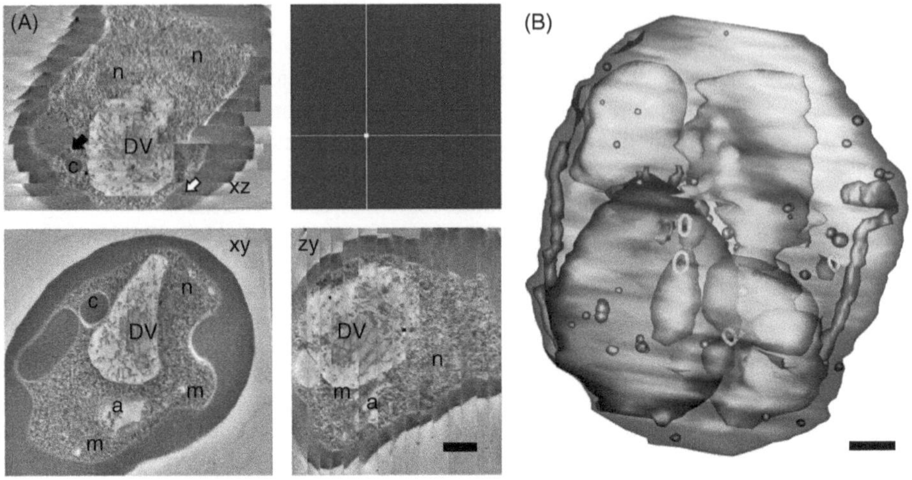

Fig. 5 Whole cell electron tomography. Tomograms are acquired for each of 15 serial sections across an entire cell, using dual-axis tilting. (A) The stitched tomogram is represented along the three axes (*xy*, *xz*, and *zy*, with *xy* being the plane perpendicular to the electron beam). The following features are evident: DV, digestive vacuole; c, cytostomal invagination; cytostomal ring, indicated with a black arrow; n, nucleus; a, apicoplast; m, mitochondrion. An 80-nm electron-dense vesicle in the RBC cytoplasm is indicated with a white arrow. (B) The cell is segmented and modeled. The rendered featured are depicted as follows: Nuclei, gold; cytostomal invaginations, translucent yellow; digestive vacuole, dark blue; mitochondrion, pink; apicoplast, aqua; acidocalsisomes, green. Sections contract up to 30% in the direction of the beam; the model is corrected for the shrinkage by increasing the *z* by a factor of 1.5. Scale bars, 500 nm. (See Plate no. 6 in the Color Plate Section.)

through the tomogram and a rotation of the rendered model (see also Supplement Movies 1 and 2 at http://www.elsevierdirect.com/companions/9780123810076).

Recently, scanning transmission electron tomography (STEM) of 1-μm thick samples of *P. falciparum* was achieved at a resolution approaching that of conventional electron tomography using axial detection of scattered electrons (Hohmann-Marriott *et al.*, 2009). This STEM-based technique employed a tightly focused electron probe and collected and analyzed only those electrons that were scattered to low angles, using an axial bright field detector. This technique has enormous potential for rapidly imaging whole cells (see also Chapter 25).

E. Immuno-Labeling Techniques for TEM

Protocols that enhance structural preservation invariably reduce antigenicity. This is exacerbated by the low abundance of many of the proteins of interest, which makes labeling of resin-embedded thin sections difficult. Moreover, the epitopes of many antigens of interest appear to be susceptible to disruption during preparation for electron microscopy. However, several methods have been employed to determine the locations of proteins in infected RBCs using antibody reagents.

For antigens exposed at the surface of infected RBCs (or free merozoites, sporozoites, etc.) pre-embedding labeling techniques can be used (Bannister and Kent,

1993). In this strategy, labeling of cells in suspension is carried out prior to fixation and preparation for TEM or SEM. For example, the variant surface antigen exposed at the RBC surface can be detected using human hyperimmune polyclonal antisera or reagents that recognize specific variants (Horrocks *et al.*, 2005; Tilley and Hanssen, 2008).

An alternative approach is pre-embedding labeling of epitopes that are located within the RBC cytoplasm following selective permeabilization of the RBC membrane. The release of hemoglobin also facilitates analysis of details of the membrane features in the RBC cytoplasm. A range of pore-forming toxins (e.g., streptolysin O, tetanolysin, equinatoxin II (Eqt II)) and detergents (e.g., digitonin and saponin) have been used to permeabilize *P. falciparum*-infected RBCs (Baumeister *et al.*, 2001; Jackson *et al.*, 2007; Kriek *et al.*, 2003; Spycher *et al.*, 2006). We have found that Eqt II performs particularly well for this purpose permitting good preservation of the parasite exomembrane system. Eqt II binds preferentially to sphingomyelin-containing membranes and lyses the limiting membrane of infected and uninfected red blood cells with similar efficiency. By contrast, streptolysin O binds to cholesterol and permeabilizes uninfected RBCs more efficiently than infected RBCs (Jackson *et al.*, 2007).

To stabilize the infected RBCs and to prevent loss of small organelles, the samples are pre-fixed with 2% paraformaldehyde. The samples are treated with 2 units of Eqt II (Anderluh *et al.*, 1999), then re-fixed to further stabilize the samples and to deactivate any excess Eqt II. The Eqt II pores are large enough to allow exit of hemoglobin and entry of the primary antibody and of protein A conjugated to 6-nm gold (Hanssen *et al.*, 2008a; Hanssen *et al.*, 2008b; Spycher *et al.*, 2008). Gold-conjugated antibodies (or larger gold-conjugated protein A) have more limited access to the intracellular targets. Cells are fixed with glutaraldehyde, postfixed in osmium tetroxide, and stained with uranyl acetate prior to embedding in LR White resin (Hanssen *et al.*, 2008b, Fig. 6A).

To access structures within the parasite itself, post-embedding immunolabeling is required. In an attempt to preserve antigenicity, low concentrations of aldehydes are generally used (Bannister and Kent, 1993; Proellocks *et al.*, 2009); however, this usually results in less well defined cellular features. The Tokuyasu method with labeling before the dehydration step, or cryofixation by rapid freezing combined with cryo-substitution, are alternative procedures that can give improved results (Bannister and Kent, 1993; Bannister *et al.*, 2000b; Henrich *et al.*, 2009; Howard *et al.*, 1986; Trager *et al.*, 1992).

Some laboratories routinely use high-pressure freezing and freeze-substitution to perform immunolabeling (Hodder *et al.*, 2009; Waller *et al.*, 2000). The formation of crystals during freezing can lead to a dramatic loss of structural information of biological samples. Freezing specimens at high pressure speeds up the freezing rate and largely prevents the formation of ice crystals. High pressure (~2000 bar) is applied to the sample (~3 μl) while a double jet of liquid nitrogen is blown over it. After about 1 s, the nitrogen flow stops, the pressure is released, and the specimen is transferred to liquid nitrogen for further processing. The frozen material is dehydrated by serial treatment with organic solvents at liquid nitrogen temperature followed by substitution

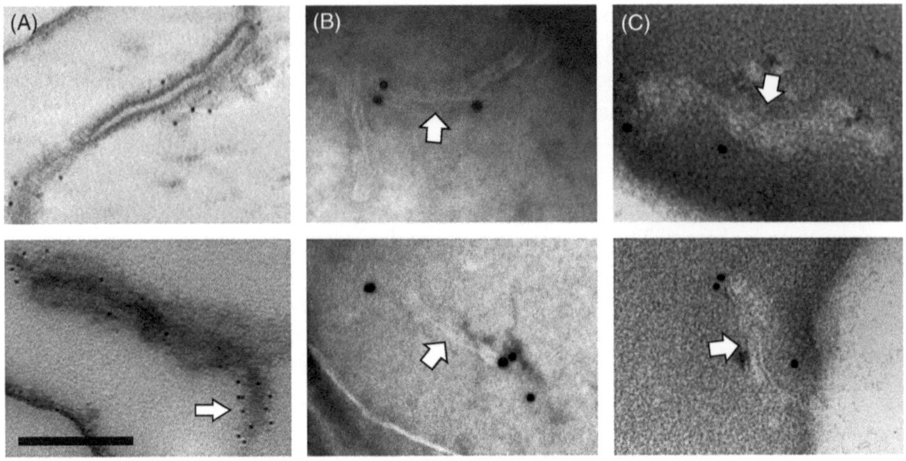

Fig. 6 Comparison of three immunolabeling techniques for determining the location of REX1. A rabbit anti-REX1 polyclonal antiserum (Hawthorne *et al.*, 2004) was used (A) in a pre-embedding method to label Eqt II-permeabilized-infected RBCs, and in post-embedding methods in which (B) cryo-sectioned samples were prepared using the Tokuyasu method or (C) samples were fixed with a low concentration of aldehyde. All three methods permit visualization of the REX1 labeling at the Maurer's clefts (white arrowheads); however, the level of labeling and preservation of the membrane structures is enhanced in the pre-embedding method.

with resin. The instrumentation for high-pressure freezing is expensive and the real benefits and advantages of this technique are still a matter of some discussion (Dubochet, 1995). If access to such instrumentation is limited, other methods can be used with some success.

We have compared three different methods for labeling a Maurer's cleft resident antigen, the ring-exported protein-1 (REX1), using a rabbit polyclonal antibody (Fig. 6). We immunolabeled chemically fixed resin-embedded samples following the protocol of Bannister and Kent (Fig. 6B) (Bannister and Kent, 1993). It should be remembered that this only allows access to the surface of a section thereby limiting the amount of accessible antigen. We also used cryo-sectioning and immunolabeling using the Tokuyasu method (Tokuyasu, 1980) following the protocol published by Slot and Geuze (Slot and Geuze, 2007) (Fig. 6C). We compared these methods with Eqt II-permeabilized cells that were labeled with anti-REX1 prior to embedding (Fig. 6A). In all cases labeling of the Maurer's clefts was observed. In the Eqt II-permeabilized samples, an accumulation of label in the subdomains of the Maurer's clefts is observed, which is consistent with the distribution of REX1 observed by super-resolution optical microscopy analysis of GFP-labeled transfectants (Hanssen *et al.*, 2010). We have found that the post-embedding labeling techniques are more challenging and require extensive optimization to obtain an acceptable and reliable signal. These methods are best employed when high-avidity polyclonal antisera are available and when the epitope to be labeled is abundant and relatively stable to fixation.

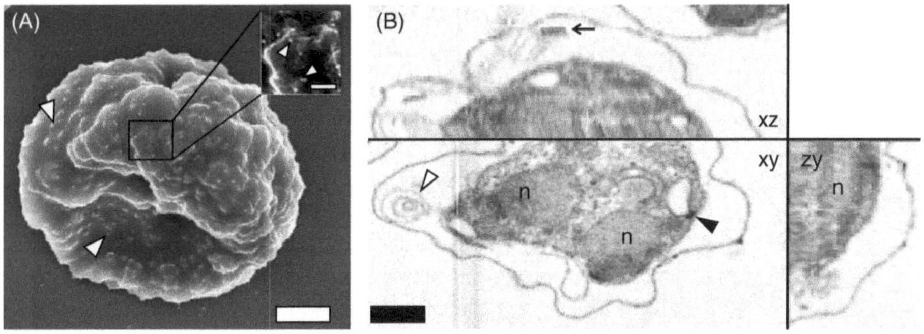

Fig. 7 SEM techniques. (A) Scanning electron microscopy of an infected RBC. Cultures of mature stage parasite-infected RBCs (3D7 strain) were surface immunolabeled with an antibody recognizing the RBC protein, CD59, then fixed, dehydrated and coated with platinum. The RBC surface is distorted and marked by protruding knobs (arrowhead). Silver-enhanced gold particles on the RBC surface are evident in the enlarged inset. (B) Dual beam microscopy. Infected RBCs were permeabilized with Eqt II and embedded as for TEM. The surface of a sample block was repeatedly carved at 50 nm increments using a gallium ion beam. One hundred serial block faces were imaged with the electron beam (back-scatter detector). Even at this low resolution internal structures of the parasites and features in the RBC cytoplasm are revealed: n, nucleus; black arrowhead, cytostome neck in *xy* plane; black arrow, Maurer's cleft in *xz* plane; white arrowhead, tubulovesicular network. Scale bars, 1 μm. Inset scale bar, 250 nm.

F. Scanning Electron Microscopy (SEM)

SEM generates a surface image by raster-scanning a tightly focused high-energy beam of electrons across the sample. The electrons interact with atoms within the sample producing secondary electrons and back-scattered electrons that provide information about the sample's surface topography and composition.

1. Surface Topography of Gold-Coated Samples

Infected RBCs can be sputter-coated with a thin layer of gold so they are electrically conductive and show high topographic contrast. SEM reveals the distortions of RBCs infected with mature-stage *P. falciparum* (Fig. 7) and has been used in a number of studies of knob morphology (Gruenberg *et al.*, 1983; Horrocks *et al.*, 2005; Maier *et al.*, 2008; Rug *et al.*, 2006). The knobs are formed by the deposition of KAHRP at the RBC cytoskeleton, resulting in distortions of the host membrane that are observed in SEM as conical protrusions. Figure 7A (inset) shows immunogold labeling of an RBC surface protein, CD59 (Frankland *et al.*, 2006), observed by SEM after silver enhancement.

2. Block-Face SEM

Three-dimensional volume information equivalent to that available from reconstructing TEM images of serial thin sections or by electron tomography can be obtained by sectioning a sample block and repeatedly imaging the block face by

SEM (Leighton, 1981). Automated block-face imaging has been combined with serial sectioning with a microtome integrated into a variable-pressure, field emission gun SEM (FEGSEM). Low vacuum (20–60 Pa H_2O) conditions are used to prevent charging of the uncoated block face. Stacks of several hundred sections (50–70 nm thickness) have been imaged at remarkably fine depth resolution (Armer *et al.*, 2009; Denk and Horstmann, 2004; Macke *et al.*, 2008; Rouquette *et al.*, 2009; West *et al.*). This method has not yet been applied to the study of malaria parasites.

A related technique is the combination of focused ion beam (FIB) milling and SEM observation in a dual-beam apparatus (Drobne *et al.*, 2005; Leser *et al.*, 2009; Marko *et al.*, 2006, 2007; Yonehara *et al.*, 1989). A thin layer of the block surface is ablated with an ion beam and an image of the newly formed surface is obtained with the scanning electron beam. The process has been automated and, while the level of resolution is less than can be achieved by serial section TEM or electron tomography, large volumes can be readily analyzed (Yonehara *et al.*, 1989). Since heavy elements (high atomic number) back-scatter electrons more strongly than lighter elements (low atomic number), osmium/uranium-labeled features appear brighter in the image. Thus, back-scattered electrons can be used to detect contrast between areas with different chemical compositions (z-contrast).

This technique is evolving rapidly with the recent release of FIB microscopes with smaller ion beams. This permits very thin layers of material to be removed from the block face, thereby improving the resolution in the *Z*-direction. The technique will soon be limited not by the size of the ion beam that carves the sample but by the depth from which back-scattered electron are generated. It is of no advantage to have a beam less than few nanometers as the back-scattered electrons are generated from a depth of up to 30 nm depending on the material (Schroeder-Reiter *et al.*, 2009). The automated routine permits collection of data through a depth of ~20 μm in an overnight run (18 h). To maximize the chance of having a whole cell in a sample volume of this thickness, we suggest packing the cells by centrifugation between each step of the embedding process. Reconstruction of the image data is performed as for other serial section techniques, with each image being registered using a suitable alignment program. Coupled with immunolabeling protocols, this technique should prove invaluable for mapping protein locations in whole cells at high resolution.

II. Rationale

The aim of this chapter is to provide an overview of the electron microscopy techniques than can be applied to *P. falciparum*-infected erythrocytes. We present an ultrastructural survey of the unusual organelles of *P. falciparum* and present a description of different immunolabeling protocols. Finally, we describe the preparation of samples for tomography and more specifically for whole cell tomography using a serial section approach.

III. Methods

A. Culturing *P. falciparum*

Parasite-infected RBCs are maintained in continuous culture in medium comprising HEPES-buffered RPMI 1640, supplemented with L-glutamine, glucose, and hypoxanthine (Frankland *et al.*, 2006). The medium is further supplemented by adding 8% pooled human serum or 1% fatty-acid-loaded albumin, or a mixture of the two (Frankland *et al.*, 2006). Cultures are maintained at 37°C under a controlled atmosphere of 5% CO_2, 1% O_2, and 94% N_2. The medium is replaced daily and the level of parasitemia is kept below 10% (usually 1–2%) by subculturing with fresh aliquots of RBCs. Growth synchronization at the ring stage is achieved by sorbitol lysis (Lambros and Vanderberg, 1979).

B. Immunolabeling

1. Pre-embedding Immunolabeling

The cells are harvested and fixed in 2% paraformaldehyde in RPMI medium for 15 min. The sample is twice pelleted at 300 g for 1 min and rinsed with PBS. The cells are permeabilized with 5 μl of EqtII (500 μg/ml) per 10 μl of pelleted cells in a total volume of 95 μl for 6 min at room temperature. The cells are pelleted at 300 g. The cells are re-fixed with 2% paraformaldehyde for 5 min, washed twice in PBS, and the excess aldehyde groups blocked with 3% BSA in PBS for 30 min at room temperature. The sample is incubated with the primary antibody at the required concentration (typically a 10-fold increase in concentration compared with immunofluorescence working concentrations) for 2 h at room temperature. The cells are washed twice with PBS/BSA and incubated with protein A conjugated with 6-nm gold particles (Aurion, NL), diluted 1:10 with PBS/BSA for 1 h at room temperature.

2. Post-embedding Immunolabeling

Parasitized RBCs are suspended in 0.1 M cacodylate buffer pH 7.4 containing 2% paraformaldehyde and 0.0075% glutaraldehyde for 20 min on ice. The cells are rinsed three times with PBS and partially dehydrated by serial treatment with 50, 70, and 80% ethanol, then embedded and polymerized in LR White resin (medium grade). Thin sections (~70 nm) are cut and collected on formvar-coated nickel grids. The sections are rehydrated in PBS for 5 min, and then blocked with PBS containing 5% skim milk and 0.01% Tween 20 for 30 min. After rinsing in PBS containing 1% BSA and 0.01% Tween 20, the grids are incubated with antiserum for 2 h, then rinsed, and incubated with secondary antibody conjugated to gold particles (usually 15 nm, Aurion) for 1 h. The grids are washed three times with PBS /BSA/Tween 20 and twice with PBS, fixed with 1% glutaraldehyde in water for 5 min and stained with uranyl acetate.

3. "Tokuyasu" Immunolabeling

The cells are harvested and fixed in 4% paraformaldehyde/0.1% glutaraldehyde in PBS for 1 h on ice, rinsed twice in PBS, and embedded in 10% gelatin (w/v) solution (37°C, 10 min). The sample is then cooled to 4°C to polymerize the gelatin and then fixed in 1% paraformaldehyde, and stored until needed. The pellet is cut into small blocks (1 mm^3) and transferred to 20 mM disodium carbonate containing 0.8 M sucrose and 10% (w/v) polyvinlylpyrolydone, at 4°C, then mounted onto 2 mm flat head pins and frozen by dipping into liquid nitrogen. Sections of 50 nm thickness are prepared and recovered onto carbon-coated grids layered with a 1:1 mixture of 2% methyl cellulose and 2.3 M sucrose. The sections can be stored at 4°C. The grids are placed, sections down on a Petri dish layered with 2% gelatin (w/v), and the gelatin is melted at 37°C for 20 min. The sections are rinsed with PBS containing 0.1% glycine for 2 min, incubated with 1% BSA in PBS, then with the primary antibody in the same buffer. (The concentration and time are optimized for each antibody.) The sections are rinsed in 1% BSA in PBS, and incubated with the secondary antibody. The grids are rinsed twice in 1% BSA in PBS, four times in PBS, fixed for 5 min in 1% glutaraldehyde in PBS, and rinsed twice in PBS for 5 min. The grids are floated on 2% aqueous uranyl acetate, pH 7.0, for 5 min, then on 3 drops of 0.4% uranyl acetate in 2% methyl cellulose for 10 min each. The excess methyl cellulose is absorbed with a piece of filter paper and the sections are air-dried and observed at 80 kV.

C. Embedding for TEM

Freshly harvested infected RBCs are fixed in 75 mM NaCl, 5 mM sodium phosphate, pH 7.0 (0.5 × PBS), containing 2.5% glutaraldehyde for 1 h, pelleted at 300 g for 2 min at room temperature, rinsed in 0.2 M Sorensen phosphate buffer for 10 min, pelleted and resuspended in 1 vol of 4% (w/v) low-melting agarose in a 200-μl pipette tip. When the agarose has set (1 min) the block is removed from the pipette tip, cut into 1 mm^3 blocks, and washed in Sorensen buffer four times over a period of 1 h. The agarose blocks are post-fixed in 1% osmium tetroxide in 0.1 M phosphate buffer for 1 h, extensively rinsed in water, *en bloc* stained with 1% aqueous uranyl acetate for 1 h, and rinsed in water. The blocks are serially dehydrated in ethanol (10 min each in 50, 70, 85, 90, 95%, and three times 10 min in 100%). The blocks are incubated in acetone/ethanol (1:1), then in pure acetone, for 10 min each. The samples are substituted in 1:1 resin/acetone for 1 h at 4°C, then impregnated for 36 h in Epoxy resin with three changes of resin. The blocks are polymerized in an open 0.5-ml microfuge tube for 48 h at 60°C. Alternatively samples can be prepared in LR White resin. Following dehydration in ethanol, the blocks are incubated in ethanol/LR White (1:1) for 30 min then in three changes of LR White resin over a period of 36 h. During the last change the blocks are transferred into 0.5-ml microfuge tubes. The tubes are overfilled with resin and closed to expel air. The tubes are sealed under nitrogen in a 10-ml Falcon tube and placed at 60°C for 24 h to polymerize.

D. Grid Preparation

The sections are collected on carbon/formvar-coated copper grids. We find that nickel grids, especially single slotted grids, are unsuitable for tomography, due to their electromagnetic properties; they distort the electron beam when the sample is tilted, resulting in degradation of the image data at tilts higher than 45°. Carbon-coated grids are pre-treated with poly-L-lysine for 10 s, quickly rinsed with two drops of water, then coated with a mixture of 6, 15, and 50 nm colloidal gold particles (used as fiducials to align the tilt series images).

E. Preparing Serial Sections

In order to cut the serial sections we use a glass knife to trim the cutting face and the opposite face. Glass knives provide an efficient and cheap (US$0.25 per knife) way of trimming blocks. To make the glass knife we shift the scoring of the glass square to one side (Fig. 8A, arrow), break the square to produce two unequal triangles one with two cutting faces and one with two blunt edges. The two-sided knife (Fig. 8B) had two corners that can be used for trimming the block (Fig. 8B, black arrow heads). One of the corners is used to trim the side of the block (Fig. 8C), which will become the cutting face. This is typically done to a depth of ~50 μm. The knife is then rotated to expose the second trimming corner (Fig. 8D), which is used to trim the opposite side of the block. The block is then rotated by 90° and cut using standard procedures (Fig. 8E). This technique gives two perfectly parallel edges allowing the formation of a straight ribbon. Moreover, the faces formed are perpendicular to the surface of the block

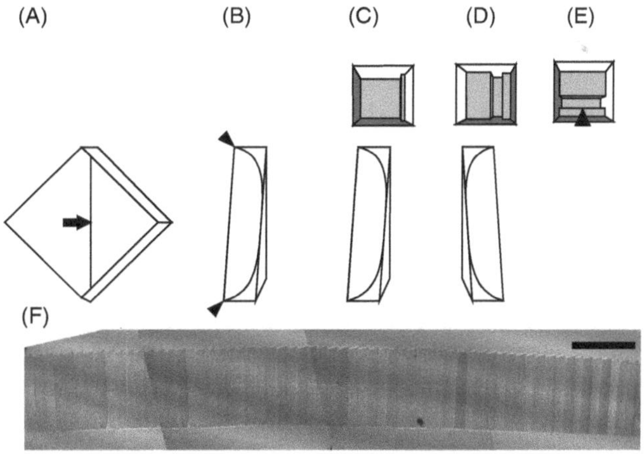

Fig. 8 Glass knife preparation and block trimming for the generation of serial sections. We score the knife off-center from the diagonal (A, arrow). This creates two cutting sides and two hard corners that can be used for trimming (B, arrowheads). The block is trimmed on one side with the first corner to a depth of about 50 μm (C). The knife is rotated and the second corner is used to trim the other side of the block (D). The block is rotated 90° (E) and serial sections are cut. (F) 65 serial sections forming a ribbon. Scale bar, 200 μm.

delivering sections that do not increase in width during the cutting process. Typically the dimensions of our sections are 200–500 μm by 20–50 μm and 50–300 nm thick. The sections are stained with saturated aqueous uranyl acetate for 5 min then post-stained with Reynold's lead citrate for a further 5 min.

F. Tomogram Acquisition and Reconstruction

The full depth of an infected RBC (6–8 μm) can be surveyed using 20–25 serial sections (each ~300 nm thickness) prepared for electron tomography. We have acquired data for whole cell imaging at moderate resolution with a pixel size between 2.33 nm and 0.93 nm (i.e., between 4700 and 12,000 magnification). This allows us to collect a whole cell section in one image. Each section is also observed at higher magnification (0.2 nm/pixel) to obtain higher-resolution information about particular features within the sample. The specimen is tilted around two orthogonal axes to improve the fidelity of the 3D reconstruction by increasing the angular views (i.e., by decreasing the so-called missing wedges) (Mastronarde, 1997). The two separate tomograms for the dual-axis imaging are acquired on the same day to minimize sample handling outside the microscope. Some sections are bent due to wrinkling during the recovery process or to folding of the plastic/carbon coating on the grid. As illustrated in Fig. 7, bent sections are computationally flattened before the reconstruction process to limit empty regions in the joined tomograms (Hanssen *et al.*, 2010).

For our studies the samples are imaged using a Tecnai G2 TF30 (FEI Company) operating at 200 kV, equipped with a 2k × 2k CCD camera (Gatan). Tilt series for tomographic reconstruction are acquired using the Xplore 3-D tomography software (FEI Company). The sections are tilted routinely between −69° and +69° at 1.5° intervals for the first axis. The grid is then removed from the microscope, rotated by 90° and reloaded. Second-axis images of the same region are acquired between −69° and +69° at every 3°, which speeds up the process and improves the resolution compared with single-axis acquisition due to decreased effects of the missing wedge (Mastronarde, 1997). Tomograms are generated using the IMOD software package (Kremer *et al.*, 1996; Mastronarde, 1997). Segmentation models are generated using IMOD-auto routine (3dmod (http://bio3d.colorado.edu/). Contours are assigned manually. The models are computationally stretched by a factor of 1.5 in the *z*-direction to account for shrinkage of the resin during exposure to the electron beam.

G. Sample Preparation for SEM

Harvested infected RBCs (or a mixed cell culture) are washed twice with PBS, fixed in 0.5 × PBS, pH 7.0, containing 2.5% glutaraldehyde, for 1 h, pelleted at 300 g for 2 min at room temperature, and rinsed in 0.2 M Sorensen phosphate buffer for 2 min. A drop of the cell suspension is allowed to settle on a glass coverslip coated with poly-L-lysine. After 10 min the unattached cells are removed and the sample is serially dehydrated. The cells can be critical point dried in order to preserve cell morphology; however, the high protein content of the RBC cytoplasm (20 mM hemoglobin) helps stabilize the cells and they usually survive drying with a blow-dryer. The coverslip is sputter coated with a thin layer of gold and observed under the microscope at 5 kV.

H. Sample Preparation for FIB/SEM

We have examined samples of *P. falciparum*-infected RBCs by FIB-SEM (Fig. 7B). The cells were harvested, fixed, stained, and embedded as for TEM. After polymerization the block was trimmed to remove regions outside the cell pellet, planed with a diamond knife to create a flat surface at the top of the block, carbon-coated to improve conductivity, and glued to a standard SEM disk. After loading the sample into the microscope (Nova 200 Nanolab, FEI, Eindhoven, NL), a thin layer of platinum was evaporated onto a selected region of the block and a "U"-shaped cavity was carved into the block using the ion beam. The middle face was imaged with the SEM beam and analyzed using a 1:1 mixed signal from secondary and back-scattered electrons at 2 kV. The major internal structures of the parasite and features in the RBC cytoplasm are revealed (Fig. 7B) and the technique could be used to obtain images of large numbers of cells for morphological analyses.

IV. Materials

A. Culturing *P. falciparum*

Instrumentation: 37°C incubator with chambers (or flasks) with a controlled atmosphere of 5% CO_2, 1% O_2, and 94% N_2. Biosafety centrifuges. Biosafety Level 2 Culture Hood.

Materials: *P. falciparum* parasites (biological hazard), packed human RBCs (biological hazard), sterile 75 mm^2 culture flasks, and other cell culture materials.

Reagents: RPMI medium (Invitrogen), human serum (biological hazard) or Albumax®, sorbitol, L-glutamine, glucose, hypoxanthine, Percoll, and PBS.

B. Electron Microscopy

Instrumentation: TEM: FEI TF30 with a 2k × 2k bottom-mounted camera (Gatan), JEOL 2010HC with a 2k × 2k side-mounted camera (Soft Imaging Systems), Ultramicrotomes with and without cryo-chamber (Leica); SEM: JEOL JSM 6340F, FEI Nova 200 Nanolab; Polaron SC7640 sputter coater.

Materials: Glass knifes, single slot copper grids, and 200 mesh nickel grids.

Reagents: Sodium cacodylate (hazardous), uranyl acetate (hazardous), osmium tetroxide (hazardous), lead citrate (hazardous), glutaraldehyde (hazardous), formaldehyde (hazardous), ethanol, London Resin White, hard and medium grades, Formvar, antibodies, protein A coupled with 6–15 nm colloidal gold, and BSA.

C. Perspectives

The first definitive ultrastructural description of *P. falciparum*-infected RBCs was provided in 1978 (Langreth *et al.*, 1978) and the continued use of a range of electron microscopy techniques has underpinned our view of the cellular organization of the intraerythrocytic parasite; however much remains to be elucidated. Efforts to map the locations of proteins in cells could further be enhanced by the development of improved immunolabeling techniques; this would be enormously aided by the development of a genetically encoded marker that can be visualized in the electron microscope. Recent developments in electron microscopy emerging from leading research laboratories and microscope manufacturers appear to revolutionize our ability to image whole cells quickly and at unprecedented resolution. The possibility of determining the proteome of an entire infected RBC by electron tomography is within reach and the ability to rapidly analyze the morphology of significant numbers of whole cells makes it feasible to use these techniques for screening purposes. We now have available an ultrastructural tool box that will allow us to ask critical new question about the cell biology of this important pathogen.

Acknowledgments

The authors would like to acknowledge Ms Samantha Deed, La Trobe University, for technical assistance with the sample preparation; Dr David Elliott, University of Arizona, for advice on serial sectioning; Dr Sergey Rubanov, Bio21 Institute, Melbourne; and Dr Eugenieu Belaur, La Trobe University, for technical assistance with the focused ion beam microscopy; Prof David Ferguson, Nuffield Department of Pathology, Oxford, for helpful discussions regarding sample preparation; and Dr Jason Mackenzie and Ms Jennifer Hyde, La Trobe University, for sharing their knowledge about the Tokayasu method.

References

Aikawa, M., Udeinya, I. J., Rabbege, J., Dayan, M., Leech, J. H., Howard, R. J., and Miller, L. H. (1985). Structural alteration of the membrane of erythrocytes infected with *Plasmodium falciparum*. *J Protozool.* **32**, 424–429.

Aikawa, M. (1988). Morphological changes in erythrocytes induced by malarial parasites. *Biol. Cell.* **64**, 173–181.

Aley, S. B., Sherwood, J. A., Marsh, K., Eidelman, O., and Howard, R. J. (1986). Identification of isolate-specific proteins on sorbitol-enriched *Plasmodium falciparum*-infected erythrocytes from Gambian patients. *Parasitology* **92**, 511–525.

Anderluh, G., Krizaj, I., Strukelj, B., Gubensek, F., Macek, P., and Pungercar, J. (1999). Equinatoxins, pore-forming proteins from the sea anemone, *Actinia equina*, belong to a multigene family. *Toxicon* **37**, 1391–1401.

Armer, H. E., Mariggi, G., Png, K. M., Genoud, C., Monteith, A. G., Bushby, A. J., Gerhardt, H., and Collinson, L. M. (2009). Imaging transient blood vessel fusion events in zebrafish by correlative volume electron microscopy. *PLoS ONE* **4**, e7716.

Atkinson, C. T., Aikawa, M., Perry, G., Fujino,T., Bennett, V., Davidson, E. A., and Howard, R. J. (1988). Ultrastructural localization of erythrocyte cytoskeletal and integral membrane proteins in *Plasmodium falciparum*-infected erythrocytes. *Eur. J. Cell. Biol.* **45**, 192–199.

Atkinson, C. T., and Aikawa, M. (1990). Ultrastructure of malaria-infected erythrocytes. *Blood Cells* **16**, 351–368.

Bakar, N. A., Klonis, N., Hanssen, E., Chan, C., and Tilley, L. (2010). Digestive-vacuole genesis and endocytic processes in the early intraerythrocytic stages of *Plasmodium falciparum. J. Cell. Sci.* **123**, 441–450.

Bannister, L. H., and Dluzewski, A. R. (1990). The ultrastructure of red cell invasion in malaria infections: a review. *Blood Cells.* **16**, 257–292; discussion 293–297.

Bannister, L. H., and Kent, A. P. (1993). Immunoelectron microscopic localization of antigens in malaria parasites. *Methods Mol. Biol.* **21**, 415–429.

Bannister, L. H., Hopkins, J. M., Fowler, R. E., Krishna, S., and Mitchell, G. H. (2000a). A brief illustrated guide to the ultrastructure of *Plasmodium falciparum* asexual blood stages. *Parasitol. Today* **16**, 427–433.

Bannister, L. H., Hopkins, J. M., Fowler, R. E., Krishna, S., and Mitchell, G. H. (2000b). Ultrastructure of rhoptry development in *Plasmodium falciparum* erythrocytic schizonts. *Parasitology* **121**(Pt 3), 273–287.

Bannister, L. H., Hopkins, J. M., Margos, G., Dluzewski, A. R., and Mitchell, G. H. (2004). Three-dimensional ultrastructure of the ring stage of *Plasmodium falciparum*: evidence for export pathways. *Microsc. Microanal.* **10**, 551–562.

Bannister, L. H., and Mitchell, G. H. (2009). The malaria merozoite, forty years on. *Parasitology* **136**, 1435–1444.

Baumeister, W., Grimm, R., and Walz, J. (1999). Electron tomography of molecules and cells. *Trends Cell Biol.* **9**, 81–85.

Baumeister, S., Paprotka, K., Bhakdi, S., and Lingelbach, K. (2001). Selective permeabilization of infected host cells with pore-forming proteins provides a novel tool to study protein synthesis and viability of the intracellular apicomplexan parasites *Plasmodium falciparum* and Toxoplasma gondii. *Mol. Biochem. Parasitol.* **112**, 133–137.

Blackman, M. J., and Bannister, L. H. (2001). Apical organelles of Apicomplexa: biology and isolation by subcellular fractionation. *Mol. Biochem. Parasitol.* **117**, 11–25.

Day, K. P., Karamalis, F., Thompson, J., Barnes, D. A., Peterson, C., Brown, H., Brown, G. V., and Kemp, D. J. (1993). Genes Necessary for Expression of a Virulence Determinant and for Transmission of *Plasmodium falciparum* Are Located on a 0.3 Megabase Region of Chromosome 9. *Proceedings of the National Academy of Sciences of the United States of America.* **90**, 8292–8296

Denk, W., and Horstmann, H. (2004). Serial block-face scanning electron microscopy to reconstruct three-dimensional tissue nanostructure. *PLoS Biol.* **2**, e329.

Dixon, M. W., Thompson, J., Gardiner, D. L., and Trenholme, K. R. (2008). Sex in Plasmodium: a sign of commitment. *Trends Parasitol.* **24**, 168–175.

Drobne, D., Milani, M., Zrimec, A., Leser, V., and Berden Zrimec, M. (2005). Electron and ion imaging of gland cells using the FIB/SEM system. *J. Microsc.* **219**, 29–35.

Dubochet, J. (1995). High-pressure freezing for cryoelectron microscopy. *Trends Cell Biol.* **5**, 366–368.

Duffy, P. E., and Fried, M. (2003). *Plasmodium falciparum* adhesion in the placenta. *Curr. Opin. Microbiol.* **6**, 371–376.

Elford, B. C., Cowan, G. M., and Ferguson, D. J. (1997). Transport and trafficking in malaria-infected erythrocytes. *Trends. Microbiol.* **5**, 463–465; discussion 465–466.

Elliott, D. A., McIntosh, M. T., Hosgood, H. D., 3rd, Chen, S., Zhang, G., Baevova, P., and Joiner K. A. (2008). Four distinct pathways of hemoglobin uptake in the malaria parasite *Plasmodium falciparum. Proc. Natl. Acad. Sci. U.S.A.* **105**, 2463–2468.

Foth, B. J., and McFadden, G. I. (2003). The apicoplast: a plastid in *Plasmodium falciparum* and other Apicomplexan parasites. *Int. Rev. Cytol.* **224**, 57–110.

Frankland, S., Adisa, A., Horrocks, P., Taraschi, T. F., Schneider, T., Elliott, S. R., Rogerson, S. J., Knuepfer, E., Cowman, A. F., Newbold, C. I., and Tilley, L. (2006). Delivery of the malaria virulence protein PfEMP1 to the erythrocyte surface requires cholesterol-rich domains. *Eukaryot. Cell* **5**, 849–860.

Frankland, S., Elliott, S. R., Yosaatmadja, F., Beeson, J. G., Rogerson, S. J., Adisa, A., and Tilley, L. (2007). Serum lipoproteins promote efficient presentation of the malaria virulence protein PfEMP1 at the erythrocyte surface. *Eukaryot. Cell* **6**, 1584–1594.

Garnham, P. C., Bird, R. G., Baker, J. R., and Bray, R. S. (1961). Electron microsope studies of motile stages of malaria parasites. II. The fine structure of the sporozoite of Laverania (Plasmodium) falcipara. *Trans. R. Soc. Trop. Med. Hyg.* **55**, 98–102.

Garcia, C. R., de Azevedo, M. F., Wunderlich, G., Budu, A., Young, J. A., and Bannister, L. (2008). Plasmodium in the postgenomic era: new insights into the molecular cell biology of malaria parasites. *Int. Rev. Cell. Mol. Biol.* **266**, 85–156.

Goldberg, D. E. (2005). Hemoglobin degradation. *Curr. Top. Microbiol. Immunol.* **295**, 275–291.

Gruenberg, J., Allred, D. R., and Sherman, I. W. (1983). Scanning electron microscope-analysis of the protrusions (knobs) present on the surface of *Plasmodium falciparum*-infected erythrocytes. *J. Cell. Biol.* **97**, 795–802.

Grunewald, K., Medalia, O., Gross, A., Steven, A. C., and Baumeister, W. (2003). Prospects of electron cryotomography to visualize macromolecular complexes inside cellular compartments: Implications of crowding. *Biophys. Chem.* **100**, 577–591.

Haldar, K., and Mohandas, N. (2007). Erythrocyte remodeling by malaria parasites. *Curr. Opin. Hematol.* **14**, 203–209.

Hanssen, E., Carlton, P., Deed, S., Klonis, N., Sedat, J., DeRisi, J., and Tilley, L. (2010). Whole cell imaging reveals novel modular features of the exomembrane system of the malaria parasite, *Plasmodium falciparum*. *Int. J. Parasitol.* **40**, 123–134.

Hanssen, E., Hawthorne, P., Dixon, M. W., Trenholme, K. R., McMillan, P. J., Spielmann, T., Gardiner, D. L., and Tilley, L. (2008a). Targeted mutagenesis of the ring-exported protein-1 of *Plasmodium falciparum* disrupts the architecture of Maurer's cleft organelles. *Mol. Microbiol.* **69**, 938–953.

Hanssen, E., Sougrat, R., Frankland, S., Deed, S., Klonis, N., Lippincott-Schwartz, J., and Tilley, L. (2008b). Electron tomography of the Maurer's cleft organelles of *Plasmodium falciparum*-infected erythrocytes reveals novel structural features. *Mol. Microbiol.* **67**, 703–718.

Hawthorne, P. L., Trenholme, K. R., Skinner-Adams, T. S., Spielmann, T., Fischer, K., Dixon, M. W., Ortega, M. R., Anderson, K. L., Kemp, D. J., and Gardiner, D. L. (2004). A novel *Plasmodium falciparum* ring stage protein, REX, is located in Maurer's clefts. *Mol. Biochem. Parasitol.* **136**, 181–189.

Henrich, P., Kilian, N., Lanzer, M., and Cyrklaff, M. (2009). 3-D analysis of the *Plasmodium falciparum* Maurer's clefts using different electron tomographic approaches. *Biotechnol. J.* **4**, 888–894.

Hodder, A. N., Maier, A. G., Rug, M., Brown, M., Hommel, M., Pantic, I., Puig-de-Morales-Marinkovic, M., Smith, B., Triglia, T., Beeson, J., and Cowman, A. F. (2009). Analysis of structure and function of the giant protein Pf332 in *Plasmodium falciparum*. *Mol. Microbiol.* **71**, 48–65.

Hohmann-Marriott, M. F., Sousa, A. A., Azari, A. A., Glushakova, S., Zhang, G., Zimmerberg, J., and Leapman, R. D. (2009). Nanoscale 3D cellular imaging by axial scanning transmission electron tomography. *Nat. Methods* **6**, 729–731.

Hopkins, J., Fowler, R., Krishna, S., Wilson, I., Mitchell, G., and Bannister, L. (1999). The plastid in *Plasmodium falciparum* asexual blood stages: A three-dimensional ultrastructural analysis. *Protist* **150**, 283–295.

Horrocks, P., Pinches, R. A., Chakravorty, S. J., Papakrivos, J., Christodoulou, Z., Kyes, S. A., Urban, B. C., Ferguson, D. J., and Newbold, C. I. (2005). PfEMP1 expression is reduced on the surface of knobless *Plasmodium falciparum* infected erythrocytes. *J. Cell. Sci.* **118**, 2507–2518.

Howard, R. J., Uni, S., Aikawa, M., Aley, S. B., Leech, J. H., Lew, A. M., Wellems, T. E., Rener, J., and Taylor, D. W. (1986). Secretion of a malarial histidine-rich protein (Pf HRP II) from *Plasmodium falciparum*-infected erythrocytes. *J. Cell. Biol.* **103**, 1269–1277.

Jackson, K. E., Spielmann, T., Hanssen, E., Adisa, A., Separovic, F., Dixon, M. W., Trenholme, K. R., Hawthorne, P. L., Gardiner, D. L., Gilberger, T., and Tilley, L. (2007). Selective permeabilization of the host cell membrane of *Plasmodium falciparum*-infected red blood cells with streptolysin O and equinatoxin II. *Biochem. J.* **403**, 167–175.

Kilejian, A. (1979). Characterization of a protein correlated with the production of knob-like protrusions on membranes of erythrocytes infected with *Plasmodium falciparum*. *Proc. Natl. Acad. Sci. U.S.A.* **76**, 4650–4653.

Kremer, J. R., Mastronarde, D. N., and McIntosh, J. R. (1996). Computer visualization of three-dimensional image data using IMOD. *J. Struct. Biol.* **116**, 71–76.

Kriek, N., Tilley, L., Horrocks, P., Pinches, R., Elford, B. C., Ferguson, D. J., Lingelbach, K., and Newbold, C. I. (2003). Characterization of the pathway for transport of the cytoadherence-mediating protein, PfEMP1, to the host cell surface in malaria parasite-infected erythrocytes. *Mol. Microbiol.* **50**, 1215–1227.

Lambros, C., and Vanderberg, J. P. (1979). Synchronization of *Plasmodium falciparum* erythrocytic stages in culture. *J. Parasitol.* **65**, 418–420.

Langreth, S. G., Jensen, J. B., Reese, R. T., and Trager, W. (1978). Fine structure of human malaria *in vitro*. *J Protozool.* **25**, 443–452.

Lanzer, M., Wickert, H., Krohne, G., Vincensini, L., and Braun Breton, C. (2006). Maurer's clefts: A novel multi-functional organelle in the cytoplasm of *Plasmodium falciparum*-infected erythrocytes. *Int. J. Parasitol.* **36**, 23–36.

Lazarus, M. D., Schneider ,T. G., and Taraschi, T. F. (2008). A new model for hemoglobin ingestion and transport by the human malaria parasite *Plasmodium falciparum*. *J. Cell. Sci.* **121**, 1937–1949.

Leighton, S. B. (1981). SEM images of block faces, cut by a miniature microtome within the SEM—a technical note. *Scan. Electron Microsc.* **233(2)**, 73–76.

Leser, V., Drobne, D., Pipan, Z., Milani, M., and Tatti, F. (2009). Comparison of different preparation methods of biological samples for FIB milling and SEM investigation. *J. Microsc.* **233**(2), 309–319.

Lew, V. L., Tiffert, T., and Ginsburg, H. (2003). Excess hemoglobin digestion and the osmotic stability of *Plasmodium falciparum*-infected red blood cells. *Blood* **101**, 4189–4194.

Loria, P., Miller, S., Foley, M., and Tilley, L. (1999). Inhibition of the peroxidative degradation of haem as the basis of action of chloroquine and other quinoline antimalarials. *Biochem. J.* **339**, 363–370.

Lucic, V., Forster, F., and Baumeister, W. (2005). Structural studies by electron tomography: From cells to molecules. *Ann. Rev. Biochem.* **74**, 833–865.

Macke, J. H., Maack, N., Gupta, R., Denk, W., Scholkopf, B., and Borst, A. (2008). Contour-propagation algorithms for semi-automated reconstruction of neural processes. *J. Neurosci. Methods* **167**, 349–357.

Maier, A. G., Rug, M., O'Neill, M. T., Brown, M., Chakravorty, S., Szestak, T., Chesson, J., Wu, Y., Hughes, K., Coppel, R. L., Newbold, C., Beeson, *et al.* (2008). Exported proteins required for virulence and rigidity of *Plasmodium falciparum*-infected human erythrocytes. *Cell.* **134**, 48–61.

Maier, A. G., Cooke, B. M., Cowman, A. F., and Tilley, L. (2009). Malaria parasite proteins that remodel the host erythrocyte. *Nat. Rev. Microbiol.* **7**, 341–354.

Marko, M., Hsieh, C., Moberlychan, W., Mannella, C. A., and Frank, J. (2006). Focused ion beam milling of vitreous water: Prospects for an alternative to cryo-ultramicrotomy of frozen-hydrated biological samples. *J. Microsc.* **222**, 42–47.

Marko, M., Hsieh, C., Schalek, R., Frank, J., and Mannella, C. (2007). Focused-ion-beam thinning of frozen-hydrated biological specimens for cryo-electron microscopy. *Nat. Methods* **4**, 215–217.

Marti, M., Baum, J., Rug, M., Tilley, L., and Cowman, A. F. (2005). Signal-mediated export of proteins from the malaria parasite to the host erythrocyte. *J. Cell. Biol.* **171**, 587–592.

Mastronarde, D. N. (1997). Dual-axis tomography: An approach with alignment methods that preserve resolution. *J. Struct. Biol.* **120**, 343–352.

Miller, L. H., Baruch, D. I., Marsh, K., and Doumbo, O. K. (2002). The pathogenic basis of malaria. *Nature* **415**, 673–679.

Noske, A. B., Costin, A. J., Morgan, G. P., and Marsh, B. J. (2008). Expedited approaches to whole cell electron tomography and organelle mark-up in situ in high-pressure frozen pancreatic islets. *J. Struct. Biol.* **161**, 298–313.

Pasvol, G., Wilson, R. J., Smalley, M. E., and Brown, J. (1978). Separation of viable schizont-infected red cells of *plasmodium falciparum* from human blood. *Ann. Trop. Med. Parasitol.* **72**, 87–8.

Proellocks, N. I., Kats, L. M., Sheffield, D. A., Hanssen, E., Black, C. G., Waller, K. L., and Coppel, R. L. (2009). Characterisation of PfRON6, a *Plasmodium falciparum* rhoptry neck protein with a novel cysteine-rich domain. *Int. J. Parasitol.* **39**, 683–692.

Rouquette, J., Genoud, C., Vazquez-Nin, G. H., Kraus, B., Cremer, T., and Fakan, S. (2009). Revealing the high-resolution three-dimensional network of chromatin and interchromatin space: A novel electron-microscopic approach to reconstructing nuclear architecture. *Chromosome Res.* **17**, 801–810.

Rug, M., Prescott, S. W., Fernandez, K. M., Cooke, B. M., and Cowman, A. F. (2006). The role of KAHRP domains in knob formation and cytoadherence of *P. falciparum*-infected human erythrocytes. *Blood.* **108**, 370–378.

Scherf, A., Lopez-Rubio, J. J., and Riviere, L. (2008). Antigenic variation in *Plasmodium falciparum*. *Annu. Rev. Microbiol.* **62**, 445–470.

Schroeder-Reiter, E., Perez-Willard, F., Zeile, U., and Wanner, G. (2009). Focused ion beam (FIB) combined with high resolution scanning electron microscopy: A promising tool for 3D analysis of chromosome architecture. *J. Struct. Biol.* **165**, 97–106.

Slomianny, C., and Prensier, G. (1986). Application of the serial sectioning and tridimensional reconstruction techniques to the morphological study of the *Plasmodium falciparum* mitochondrion. *J. Parasitol.* **72**, 595–598.

Slomianny, C. 1990. Three-dimensional reconstruction of the feeding process of the malaria parasite. *Blood Cells.* **16**, 369–378.

Slot, J. W., and Geuze, H. J. (2007). Cryosectioning and immunolabeling. *Nat. Protoc.* **2**, 2480–2491.

Steven, A. C., and Aebi, U. (2003). The next ice age: Cryo-electron tomography of intact cells. *Trends Cell Biol.* **13**, 107–110.

Spycher, C., Rug, M., Klonis, N., Ferguson, D. J., Cowman, A. F., Beck, H. P., and Tilley, L. (2006). Genesis of and trafficking to the Maurer's clefts of *Plasmodium falciparum*-infected erythrocytes. *Mol. Cell. Biol.* **26**, 4074–4085.

Spycher, C., Rug, M., Pachlatko, E., Hanssen, E., Ferguson D., Cowman A. F., Tilley, L., and Beck, H. P. (2008). The Maurer's cleft protein MAHRP1 is essential for trafficking of PfEMP1 to the surface of *Plasmodium falciparum*-infected erythrocytes. *Mol. Microbiol.* **68**, 1300–1314.

Struck, N. S., Herrmann, S., Schmuck-Barkmann, I., de Souza Dias, S., Haase, S., Cabrera, A. L., Treeck, M., Bruns, C., Langer, C., Cowman, A. F., Marti, M., Spielmann, T., *et al.* (2008). Spatial dissection of the cis- and trans-Golgi compartments in the malaria parasite *Plasmodium falciparum*. *Mol. Microbiol.* **67**, 1320–1330.

Tilley, L., and Hanssen, E. (2008). A 3d view of the host cell compartment in *P. Falciparum*-Infected erythrocytes. *Transfus. Clin. Biol.* **15**, 72–81.

Tilley, L., Sougrat, R., Lithgow, T., and Hanssen, E. (2008). The twists and turns of Maurer's cleft trafficking in *P. falciparum*-infected erythrocytes. *Traffic* **9**, 187–197.

Tokuyasu, K. T. (1980). Immunochemistry on ultrathin frozen sections. *Histochem. J.* **12**, 381–403.

Tonkin, C. J., Pearce, J. A., McFadden, G. I., and Cowman, A. F. (2006). Protein targeting to destinations of the secretory pathway in the malaria parasite *Plasmodium falciparum*. *Curr. Opin. Microbiol.* **9**, 381–387.

Trager, W., and Jensen, B. J. (1976). Human malaria parasites in continuous culture. *Science*, **193**, 673–675.

Trang, D. T., Huy, N. T., Kariu, T., Tajima, K., and Kamei, K. (2004). One-step concentration of malarial parasite-infected red blood cells and removal of contaminating white blood cells. *Malar J.* **3**, 7.

Waller, R. F., Reed, M. B., Cowman, A. F., and McFadden, G. I. (2000). Protein trafficking to the plastid of *Plasmodium falciparum* is via the secretory pathway. *Embo. J.* **19**, 1794–1802.

West, J.B., Fu, Z., Deerinck, T.J., Mackey, M.R., Obayashi, J.T., and Ellisman, M.H. (2010). Structure–function studies of blood and air capillaries in chicken lung using 3D electron microscopy. *Respir. Physiol. Neurobiol.* **170**, 202–209.

Wickert, H., Wissing, F., Andrews, K. T., Stich, A., Krohne, G., and Lanzer, M. (2003). Evidence for trafficking of PfEMP1 to the surface of *P. falciparum*-infected erythrocytes via a complex membrane network. *Eur. J. Cell. Biol.* **82**, 271–284.

Wickert, H., Gottler, W., Krohne, G., and Lanzer, M. (2004). Maurer's cleft organization in the cytoplasm of *Plasmodium falciparum*-infected erythrocytes: new insights from three-dimensional reconstruction of serial ultrathin sections. *Eur. J. Cell. Biol.* **83**, 567–582.

Yayon, A., Timberg, R., Friedman, S., and Ginsburg, H. (1984). Effects of chloroquine on the feeding mechanism of the intraerythrocytic human malarial parasite *Plasmodium falciparum*. *J. Protozool.* **31**, 367–372.

Yonehara, K., Baba, N., and Kanaya, K. (1989). Application of ion-beam etching techniques to the fine structure of biological specimens as examined with a field emission SEM at low voltage. *J. Electron Microsc. Tech.* **12**, 71–77.

CHAPTER 6

Electron Tomography and Immuno-labeling of *Tetrahymena thermophila* Basal Bodies

Thomas H. Giddings Jr., Janet B. Meehl, Chad G. Pearson, *and* Mark Winey

Department of Molecular, Cellular and Developmental Biology, University of Colorado, Boulder, Colorado 80309

978-0-12-381007-6
DOI: 10.1016/S0091-679X(10)96006-8

Abstract

Basal bodies and centrioles are highly ordered, microtubule-based organelles involved in the organization of the mitotic spindle and the formation of cilia and flagella. The ciliate *Tetrahymena thermophila* has more than 700 basal bodies per cell, making it an excellent choice for the study of the structure, function, and assembly of basal bodies. Here, we describe methods for cryofixation of *Tetrahymena* by high-pressure freezing and freeze-substitution (HPF/FS) for the analysis of basal body structure with advanced electron microscopy techniques. Electron tomography of semi-thick HPF/FS sections was used to generate high-resolution three-dimensional images and models that reveal the intricate structure of basal bodies and associated structures. Immuno-labeling of thin sections from the same HPF/FS samples was used to localize proteins to specific domains within the basal body. To further optimize this model system, we used cell cycle synchronization to increase the abundance of assembling basal bodies. The *Tetrahymena* genome has been sequenced and techniques for genetic manipulations, such as construction of gene deletion strains, inducible expression and epitope tagging of proteins are now available. These advances have helped to make *Tetrahymena* a tractable experimental model system. Collectively, these methods facilitate studies of the mechanism of basal body assembly, the functions of basal body constituents and the cytological role of the basal body as a whole.

I. Introduction

Centrioles primarily serve two functions in cells. They comprise the core of centrosomes and mitotic spindle poles, and they act as basal bodies (BB) to template the formation of cilia. Regardless of their cytological role or cell type, the basic structure, consisting of a cylinder of microtubules arranged in 9-fold symmetry, is highly conserved (Beisson and Wright, 2003). A number of ciliated or flagellated cell types have been developed as model systems to investigate centriole and BB structure, function and mechanism of assembly (e.g. Marshall and Rosenbaum, 2000; O'Toole *et al.*, 2007; Pearson and Winey, 2009). In particular, *Tetrahymena thermophila* are single-cell, motile, ciliated protists containing approximately 750 basal bodies per cell. Here, we present methods to analyze the structure of assembling and mature *T. thermophila* BBs at high resolution and in three dimensions. These methods can be combined with molecular techniques to investigate the structure and function of components by gene disruption and localization of tagged proteins.

Approximately 600 basal bodies are aligned in rows in the cortical cytoplasm of *T. thermophila* to form the cilia responsible for motility (Fig. 1). About 150 more are tightly packed in the oral apparatus, a cavity involved in nutrient uptake. In *Tetrahymena* and other ciliates, cortical basal bodies anchor a complex network of cytoskeletal elements arrayed in a reiterated pattern. Each basal body nucleates the microtubules (MTs) of the ciliary axoneme and is associated with at least two other

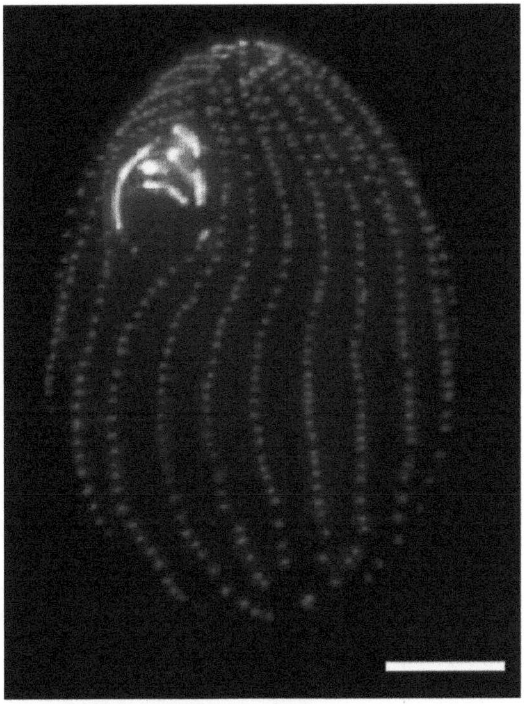

Fig. 1 Immuno-fluorescence micrograph of a *Tetrahymena* cell labeled with an antibody to centrin, a pan specific marker for BBs (Stemm-Wolf *et al.*, 2005). Cortical rows of BBs run the length of the cell. The oral apparatus is located near the anterior of the cell (top). Scale bar = 10 μm.

MT arrays, the transverse MTs and the post-ciliary MTs (Allen, 1969; Dippell, 1968; Iftode and Fleury-Aubusson, 2003). Mature *Tetrahymena* BBs are approximately 200 nm in diameter and 600 nm in length (Fig. 2). The proximal end is defined as the base, located in the cytoplasm facing the cell center, and the distal end is linked with the cilium (Allen, 1969). The wall of the basal body is comprised of nine MT triplets arranged in a cylinder (Fig. 3). Near the proximal end, the triplets are noticeably angled toward the center. The innermost MT is designated the "a" tubule, the middle MT as "b" and the outermost as the "c" tubule (Fig. 3B). The a and b MTs of the basal body triplets are continuous with the outer doublet MTs of the cilia. A cartwheel-shaped structure occupies the center of the BB cylinder at the proximal end (Fig. 2). A long filamentous structure, the kinetodesmal (KD) fiber, attaches laterally to the proximal end of the BB (Fig. 3). The formation of new basal bodies begins near the site of attachment of the KD fiber (Fig. 3). An electron dense core occupies the central BB lumen extending from the cartwheel to the transition zone (Fig. 2). Near the cell surface, the transition zone (TZ) marks the distal end of the BB (Fig. 2). The outermost MT of each BB triplet terminates at the TZ. Distal to the TZ is the

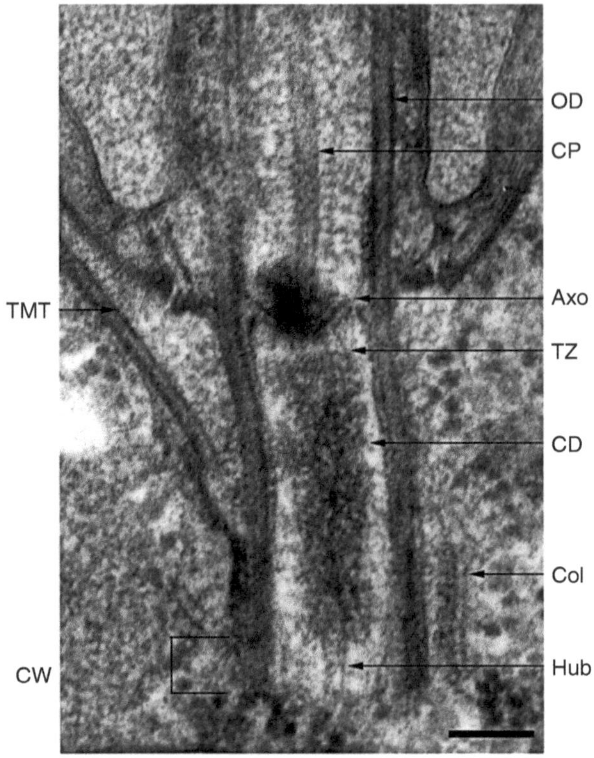

Fig. 2 Electron micrograph of a longitudinal section through a cortical row BB and attached cilium from a HPF/FS-fixed *Tetrahymena* cell. The hub of the cartwheel (CW) is visible at the proximal end of the BB. The core density (CD) extends from the top of the CW nearly to the transition zone (TZ). On the distal surface of the TZ, an electron-dense area known as the axosome (Axo) anchors one of the central pair (CP) MTs of the cilium's axoneme. Outer doublet (OD) MTs of the axoneme are continuous with the triplet MTs that form the BB cylinder. Some of the structures associated with the proximal end of the BBs are visible in this image, including one of the transverse microtubules (TMT) and the collar (Col). The cell anterior is to the left in this view; HPF/FS thin section (Epon). Scale bar = 100 nm.

axoneme of the cilium. The central pair of MTs of the axoneme is anchored in an electron density known as the axosome on the distal side of the TZ.

In addition to making it possible to sample many BBs in a few micrographs, the large number of BBs and repetitive nature of the ciliary row organization make it feasible to find transient structures such as early stages of assembly. The position of new BBs relative to the mother BBs is predictable; they are always assembled on the anterior side at the proximal end. Synchronized cultures where most of the cells (> 95%) have a high frequency of assembling BBs enable straightforward identification of basal body assembly intermediates. The abundance of basal bodies in *Tetrahymena* makes this organism an excellent model system for ultrastructural analysis of basal body and cilia biogenesis. Often multiple BBs can be viewed in cross section at

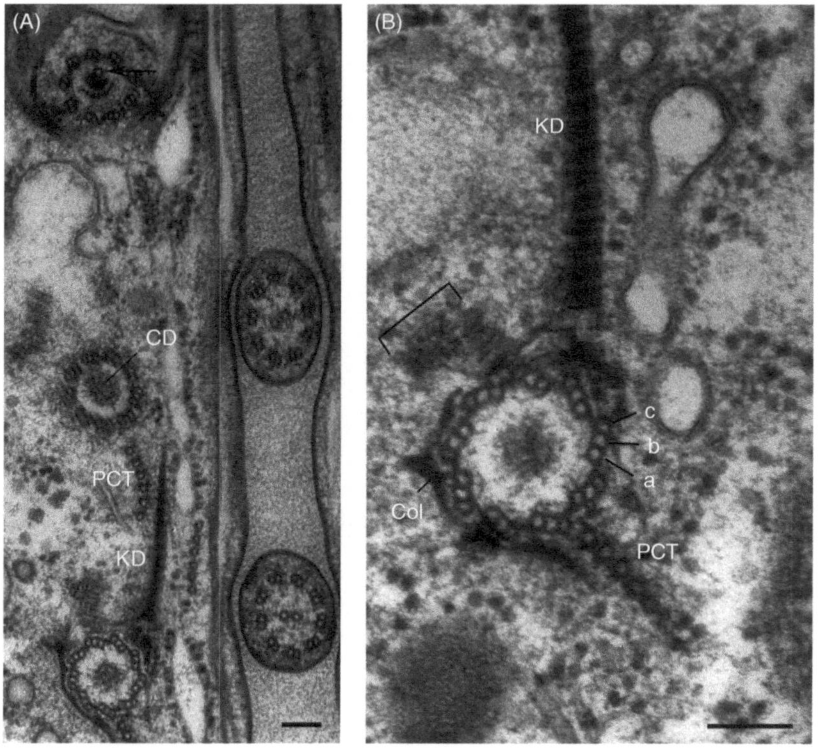

Fig. 3 Cortical row BBs in cross section. (A) Three BBs in a cortical row (left side) and two cilia (right side) are seen in cross section. A band of post-ciliary MTs (PCT) lies adjacent to each BB along the side facing the posterior of the cell. The KD fiber (KD) attaches to the anterior-facing side of the BB at the proximal end. The core density (CD) is visible in the lower two BBs. Part of the axosome and a single central pair MT (arrow) is visible in the upper BB. (B) Higher magnification view of a BB in cross section showing a KD fiber, post-ciliary MTs, some of the collar (Col) and the site of nascent BB assembly (bracket). The a, b, and c MTs of one of the BB triplets are identified; HPF/FS thin sections (Epon). Scale bars = 100 nm.

sequential levels along the long axis of the basal body in a single micrograph (Fig. 4). A number of molecular techniques have been developed in recent years that allow the use of *Tetrahymena* as a versatile model system for experimental cell biology (reviewed by Turkewitz *et al.*, 2002). The complete *Tetrahymena* genome has been sequenced (Eisen *et al.*, 2006) and genetic tools have been developed to construct strains with epitope-tagged gene products, inducible gene expression, and gene deletions (Bruns and Cassidy-Hanley, 2000a,b; Frankel, 2000; Gaertig and Kapler, 2000; Hai *et al.*, 2000; Malone *et al.*, 2008; Pearson and Winey, 2009; Pearson *et al.*, 2009b; Stemm-Wolf *et al.*, 2005; Yu and Gorovsky, 2000). Kilburn *et al.* (2007) have characterized a *Tetrahymena* BB proteome. In addition to the genetic tools, basal body duplication can be temporarily suppressed by cell cycle arrest or synchronized basal body amplification can be induced to maximize the frequency of BB duplication

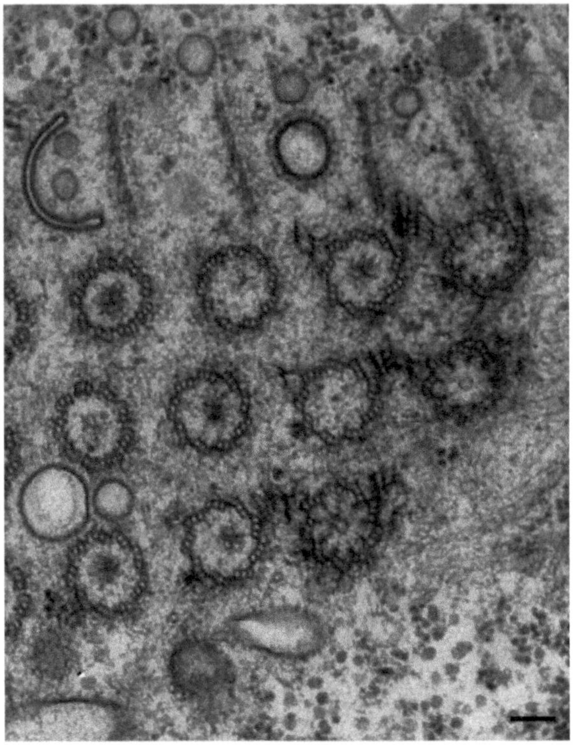

Fig. 4 Thin section through an oral apparatus showing BBs cross sectioned at different levels along their long axis. The basal bodies toward the right side of the image were sectioned through the cartwheel structure located at the proximal end. HPF/FS. Scale bar = 100 nm.

at a known time point (Pearson *et al.*, 2009b). Experiments related to basal body assembly, maintenance and turnover are now possible (Pearson and Winey, 2009).

II. Rationale

The need for precise 3-D EM imaging of BBs in *Tetrahymena* is derived from their intricate internal structure, their dynamic nature, and the complexity of the associated cytoskeletal elements. *Tetrahymena* provides an excellent model system in which these features can be studied in the cellular context of basal body assembly and the defects that arise when specific molecular components are depleted. Here, we describe modified techniques for high-pressure freezing (HPF)/freeze-substitution (FS), combined with either immuno-electron microscopy (IEM) or electron tomography (ET) to study mature and assembling BBs in *Tetrahymena* cells. This combination of methods was used effectively to resolve fine structural detail within basal bodies and associated structures in *Chlamydomonas* (O'Toole *et al.*, 2003, 2007) and to obtain high-resolution 3D models of centrioles in mitotic spindles in *Caenorhabditis elegans* (Müller-Reichert *et al.*, 2007). Although 3D BB models have been generated from reconstructions of serial sections of *Tetrahymena* cells (Allen, 1969), we believe that the cryo-fixation and ET methods described here yield more precise models and convey a better understanding of the complex 3D structure of mature and assembling BBs. These techniques can also be used to assess the ultrastructural phenotype of mutant BBs at high resolution.

III. Methods

A. Asynchronous and Synchronous *Tetrahymena* Culture

Quiescent (G1) *Tetrahymena* cells exhibit very low rates of BB duplication, whereas there is a rapid increase in the number of basal bodies beginning early in cell division (Frankel *et al.*, 1981; Kaczanowski, 1978; Nanney, 1975; Pearson *et al.*, 2009b). Starved cultures were used to obtain a population of cells in G1 with no new BB assembly. Virtually all of the BBs observed in cells fixed directly from depleted or minimal media can be assumed to be mature. Alternatively, synchronized cell growth and division was obtained by releasing starved cells back into the cell cycle by the addition of rich media (Pearson *et al.*, 2009b). Fixation when most cells were in mitosis yielded a population of *Tetrahymena* cells with maximal numbers of basal body assembly events per cell.

1. *Asynchronous Cultures*: *T. thermophila* strains were grown in super proteose peptone (SPP) media to mid-log phase, approximately 2×10^5 cells/ml, in a 30°C incubator without shaking.
2. *Synchronization of cell growth*: To arrest cells in G1, log-phase growing cultures were washed into starvation media (10 mM Tris, pH 7.4) and incubated at 30°C 12–14 h. Cells were fixed directly from starvation media to visualize cells with mature BBs. To visualize cells in which there is significant amplification of basal body assembly, starved cells were washed and cultured in fresh SPP. Assembly of new basal bodies peaks about 2 h after release into the rich medium (Pearson *et al.*, 2009a). A similar increase in the number of assembling basal bodies can be achieved by growing cultures to confluency or stationary phase, followed by re-feeding with the addition of an equal volume of 2× SPP.

B. High-pressure Freezing and Freeze-substitution of Cells

HPF and FS of *Tetrahymena* specimens resulted in good preservation of basal body ultrastructure (Figs. 2–5) as well as overall cell structure (Meehl *et al.*, 2009). Staining of the finest structures, such as the transition zone and cartwheel, is often lighter but

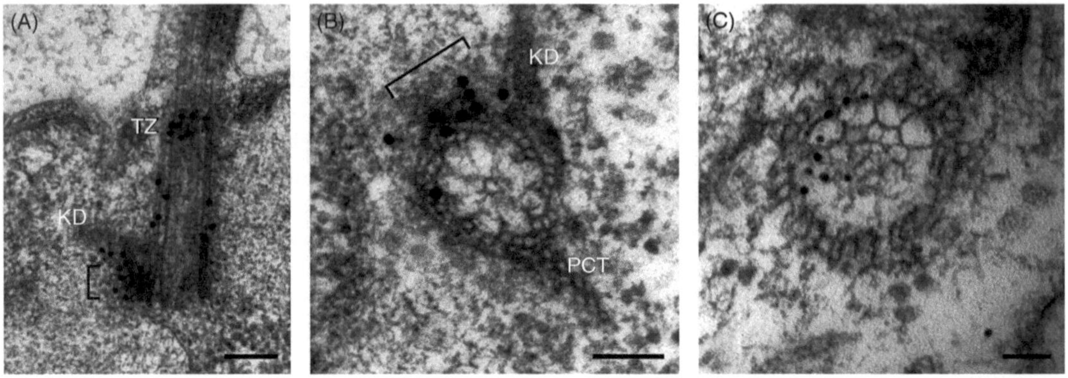

Fig. 5 Immuno-electron microscopy using a polyclonal primary antibody specific to the basal body marker, Centrin (Cen1). Lowicryl HM20 sections prepared by HPF followed by FS in 0.25% glutaraldehyde and 0.1% UA in acetone were used. Cen1 localized primarily to the transition zone (TZ) and the site of nascent BB assembly (brackets) adjacent to the kinetodesmal fiber (KD). (A) Localization of Cen1 along the proximal–distal axis of a cortical basal body. Scale bar = 200 nm. (B) A cross section through the cartwheel at the proximal end of a cortical BB shows the radial distribution of Cen1. Post ciliary MTs (PCT). Scale bar = 100 nm. (C) High magnification view of the transition zone in cross section. Scale bar = 50 nm.

reveals more detail than in comparable views of chemically fixed cells. The difference is most apparent when thin tomographic sections are compared (Fig. 6).

1. *HPF*: 8–10 ml of log phase or synchronized *Tetrahymena* culture was centrifuged in a 15 ml conical centrifuge tube at 500 × g for 2 min. The supernatant was quickly removed from the pellets to prevent cells from swimming out of the pellet. The pellet was gently resuspended in 0.5 ml of a cryoprotectant solution consisting of SPP media supplemented with 15% dextran and 5% bovine serum albumin (BSA). After centrifugation at 800–1000 × g for 4 min, the supernatant was removed, leaving a minimal residue of cryoprotectant media with the pellet. This allows the cells to resuspend slightly, resulting in separation between cells. Cells that are loosely packed retain their normal shape, freeze better, and retain more of their cortical cilia. Two to three microliters of cells were loaded into the 100 μm deep well (shallow side) of an aluminum Type B specimen carrier (Technotrade International) by pipette. After loading the cells under a dissecting microscope and confirming that the cells were still actively swimming, the sample was capped with a Type A specimen carrier. The flat side of the specimen carrier was coated with hexadecene (Sigma-Aldrich, St. Louis, MO, USA) for easier separation of the two pieces after HPF. We find it quicker and easier to have the Type B specimen carrier placed into the HPF's specimen holder before loading the cells. The holder's clamp is closed and tightened gently before inserting into the HPF. Working in an open tray of liquid nitrogen, HPF samples were transferred to cryovials containing 1 ml of FS medium. The samples lie on top of the frozen FS medium and sink down into it once the vial is warmed to initiate FS.

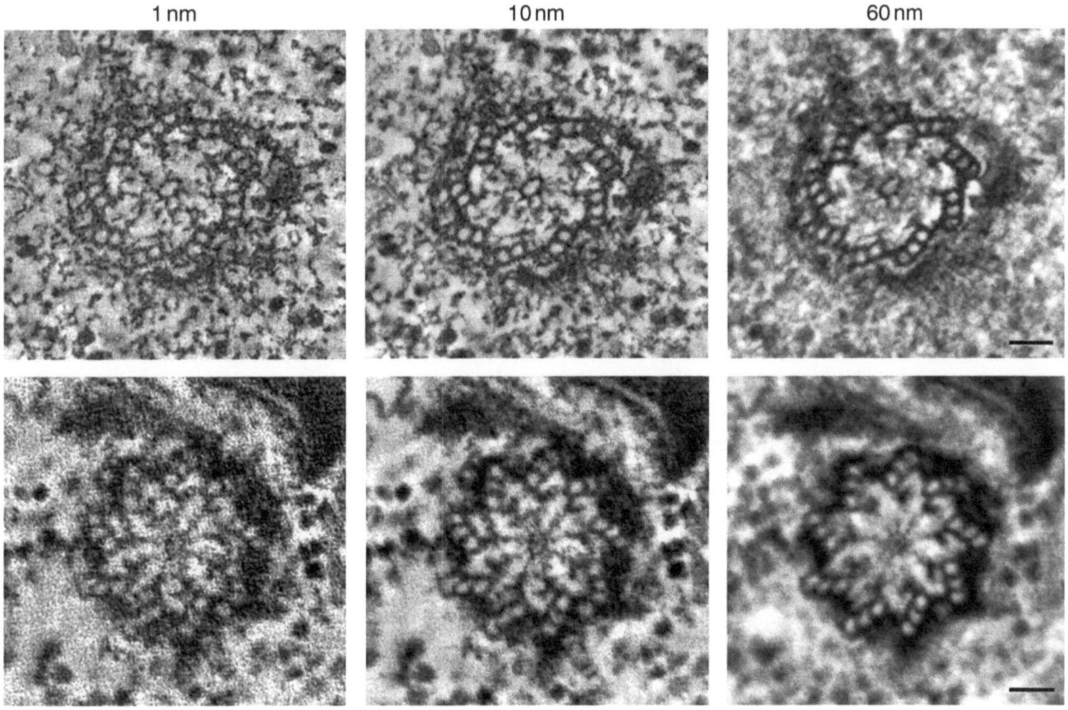

Fig. 6 Comparison of high-pressure frozen cells and conventional chemically fixed cells by electron tomography. Using IMOD Slicer, the thickness, or Z-dimension, of the volume being viewed may be adjusted to simulate sections of variable thickness. Top row. HPF/FS cortical row basal body viewed in cross section at 1, 10, and 60 nm thicknesses. The finest details of the triplet MTs and cartwheel are best resolved in thinner tomographic slices. (HM20). Scale bar = 50 nm. Bottom row. A chemically fixed cortical row basal body in tomographic cross sections representing 1, 10, and 60 nm thick volumes. The fixation is good but fine details are more difficult to resolve due to the grainy or fuzzy appearance of BB structures; (chemical fixation, Epon). Scale bars = 50 nm.

2. *FS*: We currently employ two FS protocols for fixation and embedding of high-pressure frozen *Tetrahymena* cells (Meehl *et al.*, 2009). One uses FS in osmium tetroxide (OsO_4) and uranyl acetate (UA) in acetone followed by embedding in Epon-Araldite to achieve a comprehensive, thorough fixation and stronger staining of both membranous and cytoskeletal organelles. The other is a milder fixation with glutaraldehyde and UA in acetone followed by embedding in Lowicryl HM20. The Lowicryl low-temperature embedding method was initially chosen to optimize the retention of antigenicity for immuno-labeling plastic-embedded sections but we have found that it also yields excellent preservation of cellular ultrastructure for high-resolution EM analysis including tomography. We generally freeze enough samples for both FS protocols.

1. Freeze-Substitution in 2% Osmium Tetroxide and 0.1% Uranyl Acetate in Acetone Followed by Embedding in Epon-Araldite Epoxy Resin

Cryovials containing the FS media and samples were placed in a metal block cooled to –80°C. The block with samples was nestled in a chest of dry ice and placed in a standard –20°C freezer for 3–4 days. Samples were warmed gradually to –20°C overnight by removing the lid from the box, allowing a small amount of the dry ice to evaporate. The samples, still in the metal block, were moved to 4°C for 4–6 h and finally to room temperature for 1 h. Alternatively, automated FS devices or –80°C freezers may be used to maintain the desired temperature throughout FS. The FS media was removed and the samples were rinsed 2× with acetone. The freeze-substituted cells and cryoprotectant solution typically form a cohesive disc that either falls off or can be removed gently from the specimen carriers by means of dissecting needles or other sharp tools. It may be necessary to work under a dissecting microscope to retrieve any small fragments. Samples were rinsed again in fresh acetone, then infiltrated with increasing concentrations of Epon-Araldite resin (without DMP30 accelerator) diluted in acetone: 25% Epon in acetone overnight; 50% Epon 8–10 h; 75% Epon overnight; and two changes of 100% Epon during the next day. The samples were left in Epon with accelerator overnight, transferred to BEEM capsules with fresh embedding resin the next day, and placed in a 60°C oven for polymerization for at least 48 h.

2. Freeze-Substitution in 0.25% Glutaraldehyde and 0.1% Uranyl Acetate in Acetone Followed by Embedding in Lowicryl HM20

As above, samples were freeze-substituted at –80°C for 3–4 days followed by gradual warming to –20°C overnight. Acetone rinses and infiltration with increasing concentrations of Lowicryl HM20 in acetone were all done at –20°C. After rinsing with acetone chilled to –20°C, the FS samples were separated from the specimen carriers. The procedure for separating the specimens from the specimen carriers is the same as above except that chilled acetone is used. It is best to work quickly to minimize sample warming that can lead to extraction and potential morphological changes. Samples were immediately rinsed in fresh –20°C acetone as soon as they were returned to the cryovial and were infiltrated with increasing concentrations of Lowicryl HM20 diluted in acetone: 25% HM20 in acetone overnight; 50% HM20 for 6–8 h; and 75% HM20 overnight. Samples were finally incubated in 100% HM20 for about 1.5 days. During that time, four changes with fresh resin were made to ensure that any residual acetone was removed. Embedding capsules were half filled with fresh HM20 before transferring the samples, then filled to the top and capped. Polymerization under UV illumination was carried out at –45°C in a homemade device (see Section IV.B.1).

We have recently freeze-substituted high-pressure frozen *Tetrahymena* cells using only 0.1% UA in acetone. The rest of the procedure was identical to the glutaraldehyde/UA FS and HM20-embedding protocol. The resulting preparations were nearly indistinguishable in morphology from those generated using the glutaraldehyde/UA FS media. In other systems, this has allowed us to obtain significant labeling of aldehyde-sensitive antigens (Pearson *et al.*, 2009b) whereas little or no label was visible after using the glutaraldehyde/UA FS protocol.

3. Notes on Ultramicrotomy and Staining

Epon or Lowicryl HM20 plastic resin block faces were trimmed to short, wide trapezoids to optimize both the number of cells per section and the number of serial sections per grid. Cells can then be easily tracked from one section to the next. Thin sections (50–70 nm) were picked up on copper slot grids and stained in 2% UA in 70% methanol for 6 min; rinsed in the same solvent and dried; then stained in Reynolds lead citrate for 4 min and thoroughly rinsed with water. Aqueous 2% UA stain yields a less intense and less grainy staining, making it a better choice for thick sections destined for tomography. For IEM, staining times for both UA and lead can be reduced to better visualize colloidal gold particles over electron dense structures.

C. Chemical Fixation

We have used aqueous chemical fixation for relatively quick preliminary assessments of new experimental samples. We modified the method of Orias *et al.* (1983) in which cells are fixed in a mixture of glutaraldehyde and OsO_4. Cells were gently pelleted and resuspended in 0.5 ml glutaraldehyde fixative (2% glutaraldehyde, 1% sucrose in 5 mM $NaPO_4$ buffer at pH 7.0). After 1–2 min, an equal volume (0.5 ml) of 2% OsO_4 in 20 mM $NaPO_4$ buffer was added. Cells were fixed in the mixture of glutaraldehyde and osmium for 10 min. After gentle centrifugation, the blackened fixative solution was removed and the cells were resuspended in 1 ml of 2% OsO_4 in $NaPO_4$ buffer for 30 min. The fixed cells were washed twice with 20 mM $NaPO_4$ buffer, pH 7.0. Storing samples overnight at 4°C in the $NaPO_4$ buffer seemed to reduce the presence of small particulate precipitate (presumably a glutaraldehyde–osmium reaction product). The fixed cells were washed in 50% ethanol for about 2 min, *en bloc* stained in 1% UA/70% ethanol for 10 min, then further dehydrated in 95% ethanol for 5 min. Dehydration was completed with two 5 min washes in 100% ethanol and two rinses (2 and 5 min) in propylene oxide (PO). The cells were infiltrated for 1 h in a 1:1 mixture of Epon-Araldite (without accelerator) and PO, 1 h in 3:1 Epon-Araldite : PO, and 6 h in Epon-Araldite-containing accelerator. Cells were pipetted into BEEM capsules, allowed to settle, and the capsules were filled with resin. The samples were polymerized 60°C for 2 days.

D. Immuno-labeling Thin Sections

For IEM, primary antibodies to selected native proteins or to fused tags such as green fluorescent protein (GFP) were used to label sections of *Tetrahymena* cells prepared by HPF/FS and embedded in Lowicryl HM20 as described above. Longitudinal and cross sections of BBs were used to determine the localization of proteins along the proximal–distal axis and radially (Fig. 5). In Figure 5A and B for example, intense labeling of centrin at the site of future BB assembly can be distinguished from other positions around the proximal end of the BB. For less-abundant proteins, data from many median longitudinal sections and serial cross sections have been combined by placing a dot on a simple schematic map of a basal body for each secondary gold particle observed on a basal body in the TEM, making it possible to discern the localization pattern (Kilburn *et al.*, 2007; Pearson *et al.*, 2009b). Sections of lightly fixed cells embedded in LR White have also been used for immuno-localization studies in *Tetrahymena* (Ueno *et al.*, 2003).

We used established IEM methods for these studies of *Tetrahymena* BB proteins (Meehl *et al.*, 2009). Serial thin sections (50–70 nm) of Lowicryl-embedded cells on Formvar-coated nickel slot grids were placed, section side down, onto 15 µl drops of blocking solution for 30 min, followed by 2 h on primary antibody diluted in blocking solution. Grids were rinsed with a steady stream of phosphate-buffered saline with Tween (PBST) for 20 s then labeled with an appropriate secondary antibody, generally 10 or 15 nm colloidal gold conjugated to goat-anti-rabbit or anti-mouse immunoglobulin for 1 h. Grids were rinsed first with PBST, then with distilled water, and finally carefully blotted and allowed to dry. To improve the visibility of colloidal gold secondary antibody on the basal bodies and associated structures, we use thinner sections and reduced staining times, usually less than 3 min with 2% aqueous UA, and 1 min with Reynolds' lead citrate.

Expression of GFP-fusion proteins is a versatile and valuable technique for the study of *Tetrahymena* BBs. We have used fluorescence microscopy of living *Tetrahymena* cells to identify new basal body components (Kilburn *et al.*, 2007), monitor changes in the number and distribution of BBs under various experimental conditions, and follow the incorporation and turnover of tagged BB components (Pearson *et al.*, 2009b). To date, we have used two GFP rabbit polyclonal antibodies (see Section IV.B.2), both of which give a strong signal with very low background following this IEM protocol.

A fundamentally different approach to mapping the distribution of proteins within BBs is *in situ* labeling of isolated pellicles, cell cortices that retain a high percentage of their BBs. Unfixed or very lightly fixed pellicles can be used, allowing the antibodies to recognize proteins in a near native state. Many more antigens are exposed to antibodies than with plastic-embedded sections. However, there is clearly a potential for loss of BB proteins under these conditions. For the *in situ* localization of BB proteins, pellicles were prepared as described in Kilburn *et al.* (2007), based on the methods of Nozawa and Thompson (1971). Cells were lysed in ice-cold buffer with a Dounce homogenizer and the cell lysate was fractionated on a sucrose step gradient. Alternatively, pellicles were prepared according to Coue *et al.* (1991). Pellicles were incubated in a range of dilutions of primary antibody in PBS, thoroughly rinsed in PBS, and incubated in 5 nm gold anti-rabbit immunoglobulin (or other appropriate secondary antibody). Use of a small size colloidal gold yields a stronger overall signal and may give a more accurate labeling pattern due to better penetration into and around basal bodies and easier removal of unbound secondary antibody. Following antibody incubation, pellicles may be fixed with the same glutaraldehyde and osmium fixation used for whole cells, dehydrated in acetone, and embedded in Epon-Araldite resin (Suppl. Fig. A at http://www.elsevierdirect.com/companions/9780123810076).

E. Electron Tomography

A 300 KV intermediate-voltage electron microscope (IVEM) was used to obtain a series of tilted views collected from semi-thick (200–300 nm) sections that are then compiled into high-resolution tomograms. Sections of 200–250 nm thickness are optimal because the ability of the IVEM to penetrate the sample and resolve detail in three dimensions is retained (Mastronarde, 1997), while the number of sections and corresponding tomograms to be combined is kept to a minimum. The best strategy is to use these thicker sections to reduce distortions and information loss within the reconstructed volume due to physical damage at each section interface. Since basal bodies almost always span more than one 250 nm section, dual axis tilt series were collected from each section in which a part of the basal body was present. Data sets can be collected from both longitudinal sections and cross sections of *Tetrahymena* basal bodies. The whole basal body can be captured in fewer sections in the longitudinal orientation than in cross section. However, the cross sections reveal more detail in the cartwheel structure as well as the transition zone (Figs. 7 and 8). Data from both orientations are useful for optimal visualization of basal body structure in the context of the cortical rows (Fig. 7).

Prior to undertaking ET of a new sample, we assessed the quality of fixation by conventional transmission electron microscopy (TEM) of thin sections. To prepare samples for tomography in the IVEM, serial semi-thick sections (200–250 nm) were cut and placed onto copper/rhodium slot grids (Electron Microscopy Sciences, Hatfield, PA, USA) coated with 0.7% Formvar. It is important to collect serial sections of consistent, known thickness to accurately calculate the sample volume in the final tomogram. The semi-thick sections require longer post-stain times than thin sections, 8 min in 2% aqueous UA and 5 min in Reynolds lead stain. Fifteen nanometer gold particles serve as fiducial markers for alignment of the tilt series images collected for tomography (O'Toole *et al.*, 2007). Droplets (5 μl) of 15 nm colloidal gold (BBI, Itnl., Cardiff, UK) were placed onto each side of the grid for 5 min. The droplets were gently wicked away with a Kimwipe and the grids were allowed to dry for 30 s. Excess salt was rinsed from the grids by applying 5 μl of water to each side of the grid and the droplet was wicked away immediately using a Kimwipe. If there is a hole in the Formvar, droplets will not remain confined to one side of the grid and the sample as a whole will be very unstable under the electron beam. Damaged grids were "repaired" by floating a new Formvar film on water and placing the grids section side down, onto the film. Grids were then retrieved and freed from the surrounding Formvar film.

The plane of section through a basal body has a significant impact on the resolution of fine structural detail within or around the basal body, making it desirable to

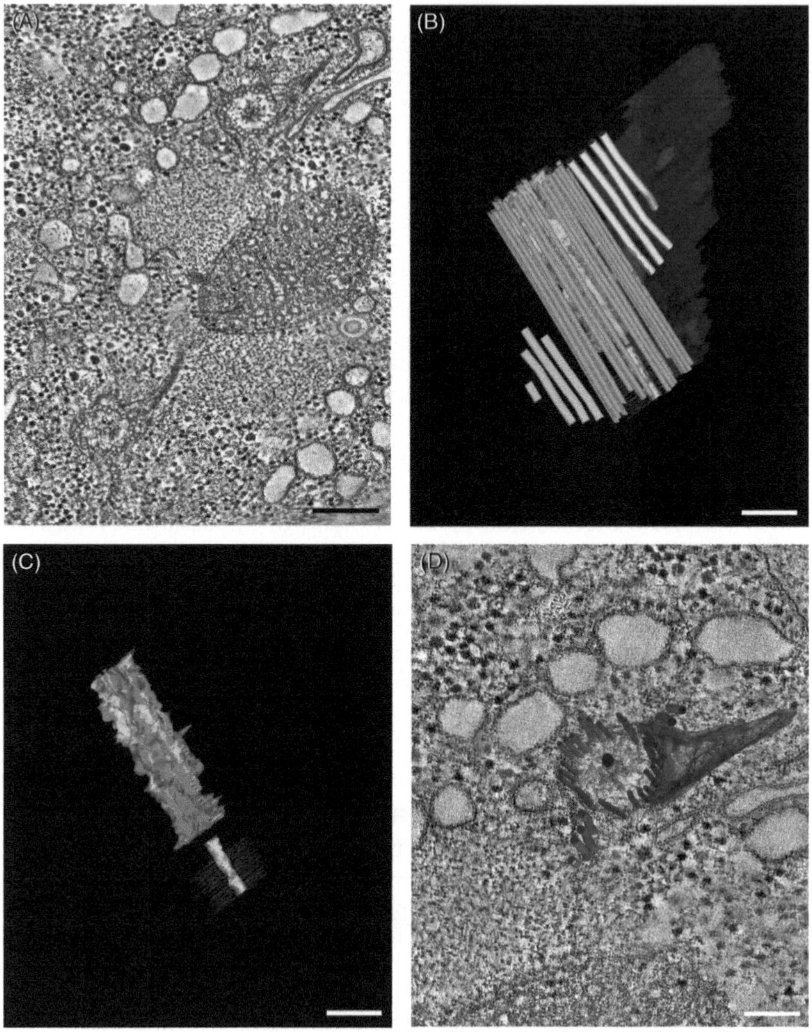

Fig. 7 A tomogram and 3D projections of modeled cortical row basal bodies. The basal body itself is composed of many structures. The cylinder walls are constructed of nine sets of triplet microtubules (green), a hub (orange) and spokes (periwinkle) radiating from the hub form the cartwheel located at the proximal end of the basal body. Distal to the cartwheel is the core density (yellow). Other cytoskeletal structures are associated with the BBs in the cortical rows. The kinetodesmal fiber (red) attaches to the proximal end of the BB. The post-ciliary MTs (blue) and the transverse MTs (tan) are located on opposite sides of the BB. The spatial relationships among these structures are more easily visualized by rotating the models (supplemental movie—Fig. 7 model.mov at http://www.elsevierdirect.com/companions/9780123810076, scale bar = 100 nm). (A) Tomographic slice displaying two BBs in a cortical row; HPF/FS (Os, Epon) tomogram. Scale bar = 200 nm. (B) A 3D projection of modeled features of one of the cortical row BBs. Scale bar = 100 nm. (C) Certain contours may be removed to reveal internal structures of the BB such as the core density and cartwheel. Scale bar = 100 nm. (D) The 3D projection model superimposed on the tomographic slice ties together the modeled structures with the actual structures. Scale bar = 100 nm. (See Plate no. 7 in the Color Plate Section.)

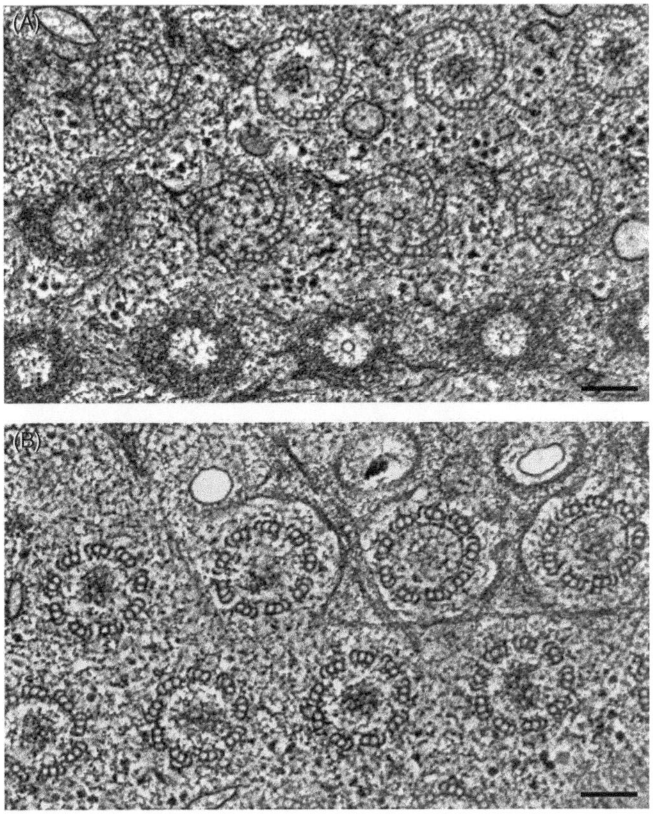

Fig. 8 Computed image slices from a WT *Tetrahymena* oral apparatus tomogram. Within one tomographic slice, many cross-sectional views of the BB are represented. (A) The hub and spokes of the cartwheel are visible in the BBs of the bottom row. The triplet microtubules and the cartwheel are well defined in the middle row and the BBs in the top row are transected through the mid-region where the core density resides. (B) In a more distal tomographic slice of the same oral apparatus, the fine lacey structure of the transition zone (TZ) is visible in the center of the two BBs on the right side of the top row; HPF/FS (HM20) tomogram. Scale bars = 100 nm. A movie of the whole tomogram is available for viewing (supplemental movie—Fig.8 tomo.mov at http://www.elsevierdirect.com/companions/9780123810076, scale bar = 200 nm).

first search for basal bodies of interest that happen to be favorably oriented within the section. This can be achieved by previewing the semi-thick sections in a standard TEM at 100 KV. Low magnification images are used to map the chosen basal bodies of specific cells. Errors in tracking a selected basal body from section to section can easily occur due to the reiterated nature of the basal bodies and associated cytoskeletal elements. Non-cytoskeletal features of the cortical cytoplasm, such as mitochondria and vacuoles can be used as large-scale fiducial landmarks. Time spent previewing a given grid on a 100 KV TEM can save a significant amount of time on the IVEM.

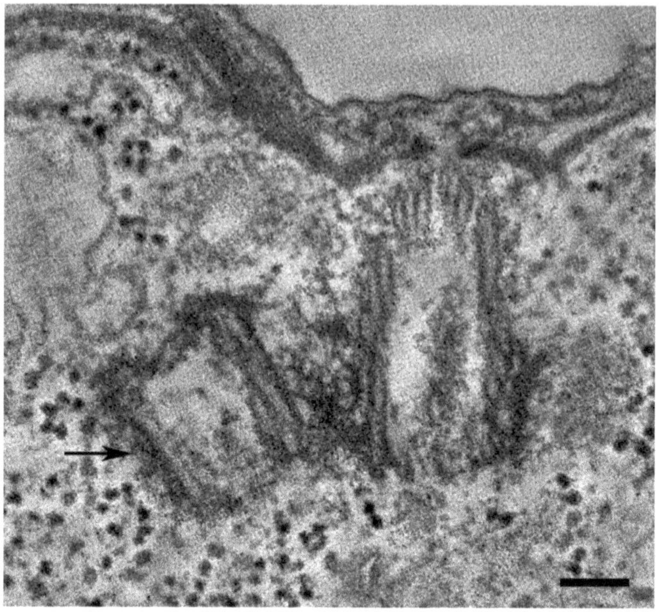

Fig. 9 Longitudinal view of new basal body assembly. The cells were starved and released and then chemically fixed at selected time points to generate samples enriched for stages of new basal body assembly. The new basal body (arrow) arises on the anterior side of the mature basal body; Chemical fixation, tomographic slice. Scale bar = 100 nm.

The methods of image acquisition and tomogram generation are described in O'Toole *et al.* (2007) and Chapter 4 by O'Toole (this volume). Images were collected at 1° increments from +60° to –60° using a Technai F30 IVEM equipped with a eucentric stage, SerialEM software (available from the Boulder 3D EM Lab website: http://bio3d.colorado.edu) and Digital Micrograph software (Gatan, Inc., Pleasanton, CA). SerialEM automates the stage tilt, sample tracking, and sample autofocusing (Mastronarde, 2005). After collection of the first tilt series around one axis, referred to as the "A stack," the grid is rotated 90° and images of the same area are collected for the second tilt series around the orthogonal axis, called the "B stack." Since the grid cannot be rotated through ±90° range, dual axis data sets are collected to minimize the effect of the "missing wedge" of information when the images are processed to generate tomograms (Mastronarde, 1997).

A set of programs collectively known as the IMOD software package (Kremer *et al.*, 1996; Mastronarde, 1997) is used to generate a tomogram from the dual axis tilt series data. A graphical user interface, eTomo, manages the set of IMOD programs used to align the tilt series data and generate the tomograms. First, a pre-processing step eliminates pixels of extreme intensity caused by X-rays. Next, the images in the data set are coarsely aligned based on X and Y translational cross correlation and a temporary coarsely aligned stack is generated. Following the course alignment of the stack, a seed model is generated using the image display and modeling program, 3dmod. Fiducial markers are chosen by adding contours over 20–40 gold particles at the 0° tilt view. These fiducial markers should be distributed equally over the viewing area as well as on both sides of the section. The gold particles are then automatically tracked through the entire stack and their model positions are used for the fine alignment. The user corrects fiducial model points that need adjustment to improve the alignment. Other factors such as magnification changes, tilt axis rotation, translation shifts, and distortion are corrected when this program is executed. A fully aligned stack is generated and a tomogram is produced. The same process is repeated to generate a tomogram from the B stack data. Tomograms from both axes are then aligned and combined to generate a final, dual-axis tomogram (Figs. 6–9 and supplemental movies).

Tomograms were generated for each section and then combined using the Join interface in eTomo. Details of this procedure can be found in Höög and Antony (2007). To begin, the tomograms were manually aligned with each other by aligning sample slices from the top of one tomogram with the bottom of the adjacent tomogram. A model was then generated to refine the alignment of the serial tomograms. For example, a microtubule may continue from one section into the next. A contour is traced along the length of the microtubule contained in each section. MTs are more easily visualized by using a tool called Slicer, in which a 3D volume is extracted from the tomogram and rotated in X, Y, or Z to best visualize the MT along its length. Model points are placed along the microtubule fragments from adjacent sections and these contours are joined together; the more contours that are joined across a boundary, the better the alignment. The aligned tomograms were then rejoined into a combined volume and the model was transformed to fit the new tomogram. The 3dmod program was used to model features of interest in the joined tomogram (Fig. 7B). Fine substructures within the BB were modeled (Fig. 7C and supplemental movie Fig. 7 model.mov at http://www.elsevierdirect.com/companions/9780123810076) along with the BB triplet MTs (green), post-ciliary MTs (blue) and transverse MTs (tan), and the KD fiber (red). Overlaying the model on the tomogram is a useful tool to correlate the model with the actual biological sample (Fig. 7D).

IV. Instrumentation and Materials

A. Culture and Synchronization Media

Materials: T. thermophila strain B2086 and CU428 (for construction of knockout strains), Tetrahymena Stock Center, Cornell University.

Media: SPP media [2% proteose peptone, 0.1% yeast extract, 0.2% glucose, 0.003% Ferric EDTA supplemented with antibiotic/antimycotic mix (100 units/ml penicillin,

100 μg/ml streptomycin, 0.25 μg/ml amphotericin B (fungizone)]; Starvation media [10 mM Tris, pH 7.4]; 2× SPP was prepared for adding to cultures of starved cells. The final medium for re-feeding was 1× SPP.

B. High-pressure Freezing and Freeze-Substitution

1. Instrumentation

High-pressure freezers: General techniques for HPF have been described (e.g., Dahl and Staehelin, 1989; Gilkey and Staehelin, 1986; McDonald, 1999, 2007; Moor, 1987). We use a Bal-Tec HPM 010 (currently available from RMC, Tucson, AZ). Other available models include the Wohlwend HPM 01 (available in the USA through Technotrade International, Manchester, NH) and two models from Leica, the Leica EM PACT2 (McDonald *et al.*, 2007) and the Leica EM HPM100.

FS system: A Styrofoam box filled with dry ice was used to maintain the samples at –80°C for initial FS and was placed in a standard refrigerator-freezer unit for intermediate temperatures and low-temperature embedding. We used a metal block with holes drilled in it to hold the cryovials of FS media upright and to provide a slower rate of temperature change during warming from –80 to –20°C.

Our UV polymerization chamber is a homemade insulated box with two 7-watt UV lights mounted into the lid. BEEM capsules with samples in liquid resin are held in a wire rack immersed in a temperature-controlled bowl of isopropyl or methyl alcohol. Dry ice is placed in the bottom of the box and the temperature is maintained at –45°C by means of a thermocouple-based controller and a heating element wrapped around the bowl. Commercially available alternatives include automated freeze substitution devices such as the Leica EM AFS (Leica Microsystems, Vienna, Austria). These are versatile and convenient means of achieving controlled, reproducible FS and UV polymerization of low-temperature embedding resins. These units have the advantage of offering a wide range of temperatures for initial FS, low-temperature fixation, resin infiltration and polymerization, and controlled rates of temperature change throughout the protocol.

2. Materials

Cryoprotectant solutions for HPF: 15% dextran (Avg. MW 9.5 KD, Sigma), 5% BSA in SPP was used. We have evaluated the quality of freezing of *Tetrahymena* cells achieved by using a variety of cryoprotectants. Lower MW dextran (9.5–11 KD) is less viscous at the same concentration than the more typically used 40 KD dextran, permitting easier handling of the *Tetrahymena* cells. The most consistent results were obtained with a mixture of 15% dextran (Avg. MW 9.5 KD; Sigma) and 5% BSA in culture media; 18% dextran in SPP also produced samples free of detectable ice crystal damage in freeze-substituted samples. Factors considered in choosing the cryoprotectant included not only quality of freezing but also ease of embedding, sectioning, and staining; 20% dextran, for example, typically yields very good freezing but apparently surrounds the cells in a hydrophilic shell that retards penetration of fixatives and embedding resin. The results of poor embedding include a block face that attracts water during sectioning and sections that expand or disintegrate on the boat of the microtome knife.

Specimen carriers: Type A and Type B aluminum specimen carriers were purchased from Technotrade International and are available from several sources. These and many other styles of specimen carriers have been reviewed (McDonald *et al.*, 2007). Use of the 100 μm deep well gave consistently better freezing than deeper wells. In addition, the discs of sample exhibited a more uniform and clear color after FS presumably indicating thorough FS and penetration of the fixatives.

FS media for Epon embedding: 2% OsO_4 and 0.1% UA in acetone was prepared by placing 12.25 ml of anhydrous acetone in a vial and using 1 ml of the acetone to dissolve the 0.25 g OsO_4 in a glass ampoule (EMS: Electron Microscopy Sciences, Hatfield, PA). The dissolved osmium was returned to the vial and placed on dry ice. Repeating the process quickly dissolved all of the OsO_4; 0.25 ml of 5% UA (EMS) in methanol (stored at –20°C) was added to the solution. The FS mixture was kept on dry ice until it was aliquoted (1 ml/vial) to 1.8 ml cryovials (Nalge Nunc International, Rochester, NY, USA), which were then stored under liquid nitrogen until needed.

FS media for Lowicryl HM20 embedding: 0.25% glutaraldehyde and 0.1% UA in acetone was prepared by adding 0.25 ml 10% glutaraldehyde in acetone (EMS) and 0.2 ml of a 5% UA/methanol stock solution to 9.55 ml acetone. The FS media was then aliquoted to cryovials as described above.

C. Immuno-labeling Thin Sections

1. Equipment

Immuno-labeling was done in a covered glass Petri dish lined with moist filter paper and Parafilm. The droplets of blocking solution and antibodies were placed on the Parafilm and the dish was set on a magnetic stir plate. The speed of the stirrer is adjusted to provide very slow rotation of the nickel grids. Non-magnetic-self-closing tweezers are useful for handling nickel grids.

2. Reagents

1. PBST: 10 mM sodium phosphate, 150 mM sodium chloride, and 0.1% Tween 20.
2. Blocking solution: 1% nonfat dry milk powder was dissolved in PBST, vortexed, and allowed to stand for 5 min, then centrifuged at 1500 × g to remove undissolved solids.
3. Primary polyclonal antibodies to GFP were generous gifts from M. Rout (Rockefeller University, New York, NY) or P. Silver (Dana Farber Cancer Institute, Boston, MA). Goat-anti-rabbit-15 nm gold or 10 nm gold (Ted Pella, Redding, CA) secondary antibodies were diluted 1:20 in blocking solution.

D. Electron Tomography

IVEM: Semi-thick (200–300 nm) sections were imaged in an IVEM equipped with an automated serial tilt goniometer and a rotating specimen holder. We use a TECNAI F30 FEG at the Boulder Laboratory for 3D Electron Microscopy of Cells (website: http://bio3d.colorado.edu). Tilt series were captured using Serial EM software. 3D reconstruction and modeling were accomplished with the aid of IMOD. These programs were developed by, and may be obtained from, the Boulder 3D EM Lab (website: http://bio3d.colorado.edu).

100 KV TEMs: FEI-Philips CM10 or CM100 operated at 80 KV for thin sections (50–100 nm) or 100 KV for previewing semi-thick sections for ET. These and most other research-grade TEMs are equipped with ± 60° goniometers for tilting specimens. A specimen holder capable of rotating grids can be used to align BBs or other structures on a desired tilt axis, such as for viewing the triplet MTs of the BB in perfect cross section. With the abundance of BBs in *Tetrahymena* cells, we often found it to be more efficient to simply search around for optimal views.

V. Discussion

High-resolution, 3D electron microscopy is needed to generate accurate depictions of the intricate structure of BBs. This serves as the basis for the precise mapping of constituent proteins, the models of assembly, and the elucidation of the functions of those molecules and of the BB as a whole. The application of methods for manipulating *Tetrahymena* cultures to produce cells in predictable stages of the basal body cycle, using HPF/FS methods of specimen preparation to faithfully preserve the structures, and the use of 3D imaging techniques are discussed below. We have employed these methods to study the assembly of new BBs in *Tetrahymena*. This allows for analysis of the phenotypes resulting from mutations in basal body constituent proteins and the localization of those proteins.

A. Synchronization of the Cell Cycle to Control Basal Body Duplication

Basal bodies are dynamic structures. In addition to initial assembly and subsequent maturation, basal body proteins continuously turn over at their binding sites (Pearson *et al.*, 2009a). The simple method of nutrient removal described here was used to ensure that all of the basal bodies being monitored for protein turnover were not newly assembled during the experiment. Conversely, synchronization by starvation followed by re-feeding enabled us to find cells with a high frequency of assembling basal bodies. From fluorescence images and stained cells, it is known that the highest frequency of duplication occurs in the midzone of *Tetrahymena* cells during the early stages of cell division (Frankel *et al.*, 1981; Kaczanowski, 1978; Nanney, 1975). Preparations from starved and released cultures allow visualization of BBs at a range of stages of assembly in both wild-type (Fig. 9) and mutants. In addition, this enables the determination of the timing of incorporation of selected proteins.

B. Fixation

Traditional chemical fixation of *Tetrahymena* and *Paramecium* cells with aqueous glutaraldehyde and OsO_4 generated a wealth of structural information about basal bodies, their duplication, and the associated cortical cytoplasm (Allen, 1969; Dippell, 1968; Iftode and Fleury-Aubusson, 2003; Iftode *et al.*, 1989, 1996; Jerka-Dziadosz *et al.*, 2001; Orias *et al.*, 1983; Sharma *et al.*, 2007). Following the successful application of HPF, FS, and tomography to the study of basal bodies and overall cytology in *Chlamydomonas* (O'Toole *et al.*, 2007; Chapter 4 by O'Toole, this volume), we sought to develop effective HPF/FS methods for *Tetrahymena* to generate similarly well fixed and stained cells for ET. We used chemical fixation in parallel with HPF/FS to monitor potential HPF artifacts and to determine whether any structural element was inadequately preserved or stained. Those results were used to modify cell harvesting and cryoprotectant composition for HPF (Meehl *et al.*, 2009). Mannitol, a monomeric sugar-alcohol used as a penetrating cryoprotectant in HPF protocols for *Chlamydomonas* (O'Toole *et al.*, 2007), caused plasmolysis and disruption of cellular organization in *Tetrahymena*. We replaced it with dextran, a high-MW glucose polymer. A mixture of 15% (w/v) dextran and 5% BSA dissolved in growth medium (SPP) yielded cryofixation that was free of detectable ice crystal damage and served as a stable but permeable encasement around the cells through FS and embedding.

Together with careful handling (minimal centrifugation for harvesting and gentle resuspension), use of this cryoprotectant medium resulted in reduced disruption of the cytoplasm and improved retention of cilia.

FS in the presence of OsO_4 in the FS media followed by embedding in Epon Araldite resulted in fixation that more closely resembles traditional room temperature, aqueous chemical fixation. Staining of both cytoskeletal elements and membranes was strong but frequently grainy, a significant problem when the samples were imaged by tomography. FS with low concentrations of glutaraldehyde and UA yielded high-quality ultrastructure with the added benefit of good preservation of antigenicity for IEM. We believe that the light staining of fine structure is less likely to mask fine detail and permits a higher resolution rendition of the structure.

C. Electron Tomography

The complex cortical cytoskeleton of *Tetrahymena* makes it a desirable candidate for the use of 3D imaging techniques to understand the interrelationships of the various elements as well as the detailed structure of the basal bodies themselves. We know from immunofluorescence observations that mutants in basal body components display disrupted cortical row organization, underscoring the importance of 3D ultrastructural analysis of this system (Culver *et al.*, 2009; Pearson *et al.*, 2009b; Stemm-Wolf *et al.*,

2005). Although imaging serial thin sections plays an important role in viewing and analyzing this system, we found that ET provides superior detail and conveys 3D structure much more efficiently.

Tomographic slices and model projections derived from some of our tomograms are shown in Figs. 7 and 8. The tomograms and associated models are also presented as Quick Time movies (Supplemental Materials). The kinetodesmal fiber, transverse microtubules, and post-ciliary microtubules can be traced and modeled along with the basal body (Fig. 7). Projections of the model can be rotated in three dimensions to get a better idea of the spatial relationship of all the structures (supplemental movie—Fig. 7 model.mov at http://www.elsevierdirect.com/companions/9780123810076). Several components of the basal body itself are incorporated into the model, including the triplet MTs, the hub and spokes of the cartwheel at the proximal end of the basal body, and the core density (Fig. 7D).

Favorable cross sections through the BB-rich oral apparatus contain a wealth of information in a single tomogram. One tomogram can incorporate every cross-sectional view through a basal body. In Fig. 8A, the proximal end of the basal body is identified by the characteristic spoke and hub components of the cartwheel. The core density is clearly visible in several basal bodies, and distal to the core density, the lacey layers of the transition zone can be seen (Fig. 8B).

We believe that the ET methods described here permit the visualization, modeling, and understanding of *Tetrahymena* basal body and cytoskeleton organization in three dimensions at higher resolution and more effectively than traditional serial sectioning approaches. Chemical fixation was occasionally used to generate samples for ET and yielded some useful data. However, the cell ultrastructure had a grainy, fuzzy appearance. When compared to a HPF/FS sample (Fig. 6), the chemically fixed sample cannot be resolved to the same degree as the HPF sample. In addition to revealing the structure of mature, wild-type BBs at improved resolution, ET of HPF/FS *Tetrahymena* cells can now be used to describe the process of basal body assembly, examine basal body mutants, and detect abnormalities in the basal body accessory structure organization.

D. Concluding Remarks

Excellent 3D descriptions and models of mature and developing basal bodies in *Tetrahymena* (Allen, 1969) and *Paramecium* (Dippell, 1968; Iftode and Fleury-Aubusson, 2003; Iftode *et al.*, 1989, 1996) were generated 40 years ago from TEM of serial thin sections of chemically fixed cells. The goal of the cryofixation and ET methods described here is to push the preservation and resolution of *Tetrahymena* basal body structure to the molecular level. Following HPF/FS/ET, it is possible to more precisely visualize detail in many BB substructures such as the cartwheel, transition zone, site of duplication, and peripheral densities. Selected gene products can be localized to specific sub-domains within, or associated with basal bodies. A combination of immuno-localization and the comparison of deletion mutants to wild-type structures can now be used to identify the function of individual gene products and visualize the structures that they form at the molecular level. Following experimental manipulation of the strains and cell culture conditions, the same methods can be used to visualize the assembly process, identify the order of assembly of the various basal body components, and reveal the mechanisms for the maintenance and modification of basal bodies.

Acknowledgments

We thank Eileen O'Toole (Boulder Laboratory for 3D EM of Cells) for helpful advice and assistance with ET. Christina Clarissa provided excellent technical assistance and contributed some of the thin section images. Alex Stemm-Wolf provided the Cen1 antibody and immuno-fluorescence images and initiated many of the experiments. *Tetrahymena* work in the Winey lab is supported by NIH (RO1 GM74746) and by the March of Dimes Birth Defects Foundation (#1-FY07-422).

References

Allen, R. D. (1969). The morphogenesis of basal bodies and accessory structures of the cortex of the ciliated protozoan *Tetrahymena pyriformis*. *J. Cell Biol.* **40**, 716–733.

Beisson, J. and Wright, M. (2003). Basal body/centriole assembly and continuity. *Curr. Opin. Cell Biol.* **15**, 96–104.

Bruns, P. J. and Cassidy-Hanley, D. (2000a). Methods for genetic analysis. *Meth. Cell Biol.* **62**, 229–240.

Bruns, P. J. and Cassidy-Hanley, D. (2000b). Biolistic transformation of macro- and micronuclei. *Meth. Cell Biol.* **62**, 501–512.

Coue, M., Lombillo, V. A., and McIntosh, J. R. (1991). Microtubule depolymerization promotes particle and chromosome movement *in vitro*. *J. Cell Biol.* **112**, 1165–1175.

Culver, B. P., Meehl, J. B., Giddings, T. H.Jr., and Winey, M. (2009). The two SAS-6 homologs in *Tetrahymena thermophila* have distinct functions in basal body assembly. *Mol. Biol. Cell* **20**, 1865–1877.

Dahl, R. and Staehelin, L. A. (1989). High pressure freezing for the preservation of biological structure: Theory and practice. *J. Electron Microsc.* **13**, 165–174.

Dippell, R. V. (1968). The development of basal bodies in paramecium. *Proc. Natl. Acad. Sci. USA* **61**, 461–468.

Eisen, J. A., Coyne, R. S., Wu, M., Wu, D., Thiagarajan, M., Wortman, J. R., Badger, J. H., Ren, Q., Amedeo, P., Jones, K. M., et al., (2006). Macronuclear genome sequence of the ciliate *Tetrahymena thermophila*, a model eukaryote. *PLoS Biol.* **4**, e286.

Frankel, J. (2000). Cell biology of *Tetrahymena thermophila*. *Meth. Cell Biol.* **62**, 27–125.

Frankel, J., Nelsen, E. M., and Martel, E. (1981). Development of the ciliature of *Tetrahymena thermophila* II. Spatial subdivision Prior To cytokinesis. *Dev. Biol.* **88**, 39–54.

Gaertig, J. and Kapler, G. (2000). Transient and stable DNA transformation of *Tetrahymena thermophila* by electroporation. *Meth. Cell Biol.* **62**, 485–500.

Gilkey, J. C. and Staehelin, L. A. (1986). Advances in ultrarapid freezing for the preservation of cellular ultrastructure. *J. Electron Microsc. Tech.* **3**, 177–210.

Hai, B., Gaertig, J., and Gorovsky, M. A. (2000). Knockout heterokaryons enable facile mutagenic analysis of essential genes in *tetrahymena*. *Methods Cell Biol.* **62**, 513–531.

Höög, J. L. and Antony, C. (2007). Whole-cell investigation of microtubule cytoskeleton architecture by electron microscopy. *Meth. Cell Biol.* **79**, 145–167.

Iftode, F., Adoutte, A., and Fleury, A. (1996). The surface pattern of *Paramecium tetraurelia* in interphase: An electron microscopic study of basal body variability, connections with associated ribbons and their epiplasmic environment. *Eur. J. Protistol.* **32**, 46–57.

Iftode, F., Cohen, J., Ruiz, F., Torres-Rueda, A., Chen-Sban, L., Adoutte, A., and Beisson, J. (1989). Development of surface pattern during division in *paramecium*. I. Mapping of duplication and reorganization of cortical cytoskeletal structures in the wild type. *Development* **105**, 191–211.

Iftode, F. and Fleury-Aubusson, A. (2003). Structural inheritance in *paramecium*: Ultrastructural evidence for basal body and associated rootlets polarity transmission through binary fission. *Biol. Cell.* **95**, 39–51.

Jerka-Dziadosz, M., Strzyzewska-Jowko, I., Wojsa-Lugowska, U., Krawczynska, W., and Krzywicka, A. (2001). The dynamics of filamentous structures in the apical band, oral crescent, fission line and the postoral meridional filament in *Tetrahymena thermophila* revealed by the monoclonal antibody 12G9. *Protist* **152**, 53–67.

Kaczanowski, A. (1978). Gradients of proliferation of ciliary basal bodies and the determination of the position of the oral primordium in tetrahymena. *J. Exp. Zool.* **204**, 417–430.

Kilburn, C. L., Pearson, C. G., Romijn, E. P., Meehl, J. B., Giddings, T. H., Jr., Culver, B. P., Yates, J. R.3rd, and Winey, M. (2007). New *tetrahymena* basal body protein components identify basal body domain structure. *J. Cell Biol.* **178**, 905–912.

Kremer, J. R., Mastronarde, D. N., and McIntosh, J. R. (1996). Computer visualization of three-dimensional image data using IMOD. *J. Struct. Biol.* **116**, 71–76.

Malone, C. D., Falkowska, K. A., Li, A. Y., Galanti, S. E., Kanuru, R. C., LaMont, E. G., Mazzarella, K. C., Micev, A. J., Osman, M. M., Piotrowski, N. K., Suszko, J. W., Timm, A. C., et al., (2008). Nucleus-specific importin alpha proteins and nucleoporins regulate protein import and nuclear division in the binucleate *Tetrahymena thermophila*. *Eukaryot. Cell* **7**, 1487–1499.

Marshall, W. F. and Rosenbaum, J. L. (2000). How centrioles work: Lessons from green yeast. *Curr. Opin. Cell Biol.* **12**, 119–125.

Mastronarde, D. N. (1997). Dual-axis tomography: An approach with alignment methods that preserve resolution. *J. Struct. Biol.* **120**, 343–352.

Mastronarde, D. N. (2005). Automated electron microscope tomography using robust prediction of specimen movements. *J. Struct. Biol.* **152**, 36–51.

McDonald, K. L. (1999). High-pressure freezing for preservation of high resolution fine structure and antigenicity for immunolabeling. *Methods Mol. Biol.* **117**, 77–97.

McDonald, K. L. (2007). Cryopreparation methods for electron microscopy of selected model systems. *Meth. Cell Biol.* **79**, 24–52.

McDonald, K. L., Morphew, M., Verkade, P., and Mueller-Reichert, T. (2007). Recent advances in high–pressure freezing: Equipment and specimen loading methods. *Methods Mol. Biol.* **369**, 143–173.

Meehl, J. B., Giddings, T. H., and Winey, M. (2009). High pressure freezing, electron microscopy, and immuno-electron microscopy of *Tetrahymena thermophila* basal bodies. *Methods Mol. Biol.* **586**, 227–241.

Moor, H. (1987). Theory and practice of high pressure freezing. *In* "Cryotechniques in Biological Electron Microscopy" (R. A. Steinbrecht and K. Zierold, eds.) pp. 175–191. Springer, Berlin.

Müller-Reichert, T., Srayko, M., Hyman, A., O'Toole, E. T., and McDonald, K. (2007). Correlative light and electron microscopy of early *Caenorhabditis elegans* embryos in mitosis. *Meth. Cell Biol.* **79**, 101–119.

Nanney, D. L. (1975). Patterns of basal body addition in ciliary rows in tetrahymena. *J. Cell Biol.* **65**, 503–512.

Nozawa, Y. and Thompson, G. A., Jr. (1971). Studies of membrane formation in *tetrahymena pyriformis*. II. Isolation and lipid analysis of cell fractions. *J. Cell Biol.* **49**, 712–721.

Orias, J. D., Hamilton, E. P., and Orias, E. (1983). A microtubule meshwork associated with gametic pronucleus transfer across a cell-cell junction. *Science* **222**, 182–184.

O'Toole, E. T., Giddings, T. H.Jr., and Dutcher, S. K. (2007). Understanding microtubule organizing centers by comparing mutant and wild-type structures with electron tomography. *Meth. Cell Biol.* **79**, 125–143.

O'Toole, E. T., Giddings, T. H., Jr., McIntosh, J. R., and Dutcher, S. K. (2003). Three dimensional organization of basal bodies from wild-type and delta-tubulin deletion strains of *Chlamydomonas reinhardtii*. *Mol. Biol. Cell* **14**, 2999–3012.

Pearson, C. G., Giddings, T. H.Jr., and Winey, M. (2009a). Basal body components exhibit differential protein dynamics during nascent basal body assembly. *Mol. Biol. Cell* **20**, 904–914.

Pearson, C. G., Osborne, D.P.S., Giddings, T. H., Jr., Beales, P. L., and Winey, M. (2009b). Basal body stability and ciliogenesis requires the conserved component poc1. *J. Cell Biol.* **187**, 905–920.

Pearson, C. G., and Winey, M. (2009). Basal body assembly in ciliates: The power of numbers. *Traffic* **10**, 461–471.

Sharma, N., Bryant, J., Wloga, D., Donaldson, R., Davis, R. C., Jerka-Dziadosz, M., and Gaertig, J. (2007). Katanin regulates dynamics of microtubules and biogenesis of motile cilia. *J. Cell Biol.* **178**, 1065–1079.

Stemm-Wolf, A. J., Morgan, G., Giddings, T. H., Jr., White, E. A., Marchione, R., McDonald, H., and Winey, M. (2005). Basal body duplication and maintenance require one member of the *Tetrahymena thermophila* centrin gene family. *Mol. Biol. Cell* **16**, 3606–3619.

Turkewitz, A. P., Orias, E., and Kapler, G. (2002). Functional genomics: The coming of age for *Tetrahymena thermophila. Trends Genet.* **18**, 35–40.

Ueno, H., Gonda, K., Takeda, T., and Numata, O. (2003). Identification of elongation factor-1a as a Ca^{2+}/calmodulin-binding protein in cilia. *Cell Motil. Cytoskeleton* **55**, 51–60.

Yu, L. and Gorovsky, M. A. (2000). Protein tagging in *tetrahymena*. *Methods Cell Biol.* **62**, 549–559.

CHAPTER 7

Electron Microscopy of *Paramecium* (Ciliata)

Klaus Hausmann* ***and*** **Richard D. Allen†**

*Institute of Biology/Zoology, Laboratory for Protozoology, Free University of Berlin, 14195 Berlin, Germany

†Pacific Biosciences Research Center, Electron Microscope Laboratory, University of Hawaii at Manoa, Honolulu, Hawaii 96822

Paramecium may be the best known single-celled organism in existence (Hausmann *et al.*, 2003). Today its image often appears on television programs where the producers use it to illustrate a stereotypic microorganism, be it pathogenic or nonpathogenic, prokaryotic or eukaryotic. *Paramecium* was probably one of the first single-celled organisms observed with a light microscope by the Dutch cloth vendor and amateur lens maker Antoni van Leuwenhoek (1632–1723) (Dobell, 1932), and it is still being investigated in the 21st century in the days of the modern electron microscopes.

I. Introduction

The relatively large size of *Paramecium* (Fig. 1A), for a microorganism, and its complex fine structure have made it the organism of choice for innumerable morphological and

METHODS IN CELL BIOLOGY, VOL. 96

978-0-12-381007-6
DOI: 10.1016/S0091-679X(10)96007-X

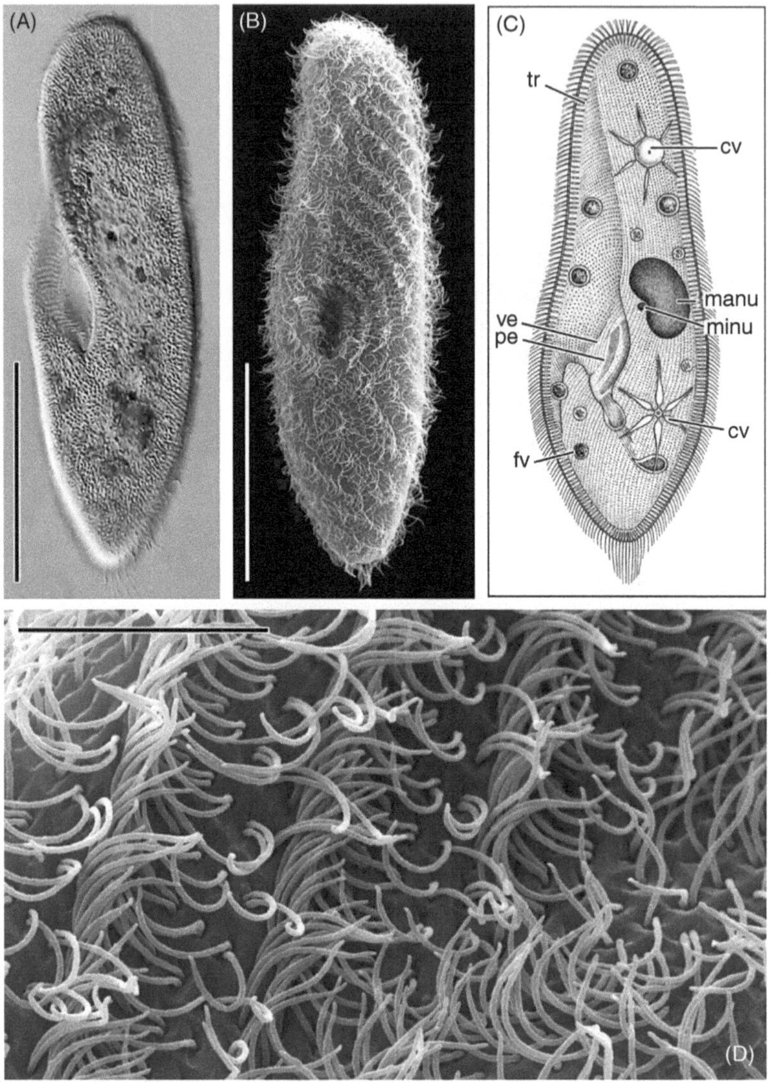

Fig. 1 (A) and (B) Light and scanning electron microscopical appearance of *Paramecium caudatum*. (C) Drawing of *Paramecium* illustrating the light microscopical features. cv, contractile vacuole; fv, food vacuole; manu, macronucleus; minu, micronucleus; pe, peristome; tr, trichocysts; ve, vestibulum. (D) Higher magnification of the metachronal waves of the cilia (A Courtesy of Gerd Günther, Düsseldorf, Germany; C modified from Grell (1973)). Scale bar in A and B = 100 μm, in D = 10 μm).

physiological studies. In addition, as an organism it is required to carry out all of the essential functions of any living organism, including gathering and digesting food, eliminating waste, growth and mitotic cell division, and sexual mating, as well as gene recombination (for reviews, see Görtz, 1988; van Wagtendonk, 1974; Wichterman, 1986).

As a cell living in a hypotonic liquid environment, without a rigid cell wall, it is also required to regulate its internal osmolarity under normal environmental conditions, as water will constantly enter its cytosol by osmosis. Even when exposed to hyper-osmotic conditions, the cell must adjust its osmolarity to keep its cytosol hypertonic to its environment. To maintain its water balance, *Paramecium*, as well as other wall-less single-celled organisms, contains special organelles, contractile vacuole complexes (CVC), to serve this function (Hausmann and Allen, 1977; Patterson, 1980).

Finally, to cope with predators such as haptorid ciliates it extrudes paracrystalline filaments, called trichocysts, i.e., a special type of extrusome (Hausmann, 1978).

After detailing some special culturing techniques we will describe the preparation methods for conventional scanning (SEM) and transmission electron microscopy (TEM). Then we will deal with several other techniques that have been used in our laboratories to enhance our understanding of this ciliate. We will end this chapter by describing the techniques of (1) replica formation following rapid-freezing, fracturing, and deep-etching cells; (2) cryo-immunogold labeling of frozen and thawed sections; and (3) triggering fixation electrophysiologically.

II. Methods

A. Preparing *Paramecium* for Electron Microscopic Studies

For studying *Paramecium* in detail it is essential to be able to culture these cells in large quantities and to be able to identify the ages of some of their organelles, e.g., food vacuoles. The following protocols meet this demand.

1. Growing *Paramecium* in Bacterized Culture Medium

Paramecium can be cultivated successfully at room temperature in a bacterized culture medium prepared from a decoction of fresh or desiccated lettuce as well as on a powder prepared from rye grass, called Cerophyl^{TM}. Nonpathogenic strains of bacteria such as *Klebsiella pneumonia* or *Enterobacter aerogenes* are added as food supply. Detailed culture instructions are given by Vivier (1974), Wichterman (1986), Nerad and Daggett (1992), and Hausmann (2009).

2. Growing *Paramecium* in Axenic Culture Medium

As mentioned above, *Paramecium* is normally maintained in the laboratory at room temperature in dishes containing bacterized lettuce or hay infusion. However, in order to rid the *Paramecium* culture of the contaminating bacteria and fungi, which is most critical for biochemical studies, *Paramecium multimicronucleatum* and *Paramecium tetraurelia*, as well as other species of the genus *Paramecium* (Allen and Nerad, 1978), can be cultured in a partially defined medium or sterile crude medium in which the bacteria and fungi have been eliminated (Fok and Allen, 1979). This insures that *Paramecium* or its isolated organelles are being studied and not contaminating bacteria as has happened in the past, e.g., when the micronucleus of *Paramecium aurelia* was being isolated, at least one published picture showed isolated bacteria (Fig. 9 in Cummings and Tait, 1975) not micronuclei. It is also important to ensure that the species of *Paramecium* is confirmed, as the supply houses do not always supply the species one orders. This happened to one of us; when *Paramecium caudatum* was ordered, we actually received a very similar *P. multimicronucleatum*. Thus the cell we thought we were establishing in axenic culture (Fok and Allen, 1979) turned out to be *P. multimicronuleatum*. Although the protocol reported in Fok and Allen (1979) will work for *P. caudatum*, it actually works better for *P. multimicronucleatum*.

To free cells of bacteria and other contaminates that grow in the Cerophyl infusion medium, the following method is adopted:

1. Transfer 20–30 cells into sterilized infusion medium such as Cerophyl and wash by transferring cells aseptically through 10 or more changes of sterile medium containing 2000 U/ml penicillin in depression plates over a period of 6 h.
2. Transfer cells aseptically to sterile infusion containing 2000 U/ml of penicillin and 100 μg of streptomycin per milliliter of medium.
3. Add axenically cultivated *Chlamydomonas* for food and allow *Paramecium* cells to multiply.
4. Transfer cells to the medium of Soldo *et al.* (1966) modified by adding 10× the folic acid concentration and replace TEM-4T with 0.05–0.5 g/l phosphatidylethanolamine (commercial grade from bovine brain). Stir this lipid into the mixture vigorously (without foaming) over a long time until it is homogeneous.
5. Place cells in the dark at 25°C and test for bacterial contamination using agar plates containing the same medium. Incubate plates at 37°C overnight and check periodically for 2 weeks for bacterial colony growth. *Chlamydomonas* will eventually die in the dark.
6. Aseptically subculture the stock cultures every week and store in the dark in an incubator set at 25°C in slanted culture tubes.

3. Marking the Age of Food Vacuoles for TEM

P. multimicronucleatum cells growing in axenic culture medium will attain concentrations in mid-log phase of 3000–6000 cells per milliliter. These cells can be concentrated in a clinical centrifuge at 95 g for 30 s. To be able to tell the age of a digestive (food) vacuole within the fixed cell we routinely exposed cells to polystyrene latex spheres (PLS) of different sizes according to the following protocol: At 0 time, one or two drops of 0.1-μm-diameter PLS was added from the dropper bottle of undiluted PLS (10% solid) to 5 ml of cell culture in a conical centrifuge tube, and this was quickly mixed by capping and inverting the tube. Capping was usually accomplished with the operator's thumb. At 2 min, the tube was spun for 30 s at 95 g in the clinical centrifuge. The rotation was quickly stopped by placing a finger on the central nut of the arms of the spinning centrifuge (this took 30 s). Just before 3 min, the supernatant was quickly aspirated and fresh axenic medium was added at exactly 3 min to the concentrated cells to return the total volume of the mixture to 5 ml. Mixing was always accomplished by inverting the tube. At 5 min, two drops of 0.3 μm PLS were added and the culture was mixed. At 7 min, the cells were again spun down and the supernatant aspirated and fresh axenic medium was added at exactly 8 min time. In this way, 0.5 μm PLS was added at 10 min, 0.8 μm PLS was added at 15 min, and 1.1 μm PLS was added at 20 min. Spinning and addition of fresh axenic medium was completed at 13 and 18 min and, finally, fixative was added to the cells at 23 min. Digestive vacuoles containing PLS of 0.1 μm were thus 20–23 min old, those containing PLS of 0.3 μm were 15–18 min old, those containing PLS of 0.5 μm were 10–13 min old, those containing 0.8 μm PLS were 5–8 min old, and those with 1.1 μm PLS were 0–3 min old.

The concentration of glutaraldehyde fixative used depended on how the fixed cells were to be handled subsequent to fixation. For conventional TEM, cells were routinely fixed with 1% glutaraldehyde buffered in 0.05 M cacodylate buffer, pH 7.4, for 15–20 min at room temperature. For cryo-immunogold observation, cells were fixed in 0.25% glutaraldehyde in 0.05 M phosphate buffer, pH 7.5, for 20 min at room temperature. For rapid-freezing the cells were not chemically fixed at all.

B. Scanning Electron Microscopy

SEM provides a good overview of the general structure of a *Paramecium* cell (Fig. 1B). The preparation procedure normally involves conventional fixation with buffered glutaraldehyde and osmium tetroxide, dehydration, critical point drying, and sputter coating (Foissner, 1991, 1992).

To document the vivid, metachronal ciliary beat (Fig. 1B and D) it is necessary to use methods beyond conventional fixatives. Párducz developed a protocol in 1952 for light microscopical investigations of the ciliary beat, which was then modified in 1967 and adopted for SEM. Since the Párducz fixation is not widely known, we explain it here in some detail.

1. Solutions

1. *Fixative*: 6 volumes 2% aqueous OsO_4 and 1 volume saturated aqueous $HgCl_2$.
2. *Mordant*: 10 g pale violet iron alum crystal (ferric ammonium sulfate) dissolved in 100 ml distilled water.
3. *Stain*: Heidenhain hematoxylin: 0.5 g iron hematoxylin plus 10 ml 96% ethyl alcohol plus 90 ml distilled water. The solution to be used should be 1–6 weeks old and light red in color.

2. Procedure

1. Place a sample of a rich, clean culture in a standard (12–15 ml) calibrated, clinical centrifuge tube. Concentrate cells and, if possible, wash twice with distilled water, using mild centrifugation.

2. To 1 volume of the concentrated culture add 4 volumes of fixative. Fix for 15 min.
3. Centrifuge gently and decant the supernatant.
4. Wash with distilled water, centrifuge gently, and decant the supernatant.
5. Add 1 volume distilled water, then 4 volumes iron alum solution to fill tube. Mordant for 2 min (including centrifuging time).
6. Centrifuge gently and decant the supernatant.
7. Wash with distilled water, then add 3–4 volumes hematoxylin solution to fill the tube.
8. Remove the sample for microscopic control, watch for staining of cilia: 10–20 min.
9. When staining has progressed to the appropriate point, centrifuge gently and decant the supernatant.
10. Wash quickly with distilled water, centrifuge, and decant the supernatant.
11. Fill the tube with tap water.
12. The preparation is best studied as a temporary mount in tap water. It may be dehydrated and mounted in balsam, but this inevitably causes some distortion, and the hematoxylin stain gradually fades.

3. Comments

1. For SEM preparations, only steps 1–4 have to be applied. After these procedures, dehydration, critical point drying, and sputtering follow, as described by Foissner (1991, 1992).
2. Centrifugation at every step should be the absolute minimum necessary to sediment the majority of cells to avoid ciliary damage. Five to ten revolutions per second of a hand-operated centrifuge suffice for *Paramecium*. After each reagent or washing, water is added; the tube should be stoppered and gently inverted a few times to ensure mixing.
3. If observation of cell movement up to the moment of fixation is desirable, fixation may be carried out in a flat-bottomed container suitable for viewing with a microscope. The fixative should be poured quickly into the container to ensure that osmium vapors do not drift over the culture in advance. Moreover, the vapors are dangerous for humans. One should work fast, but not carelessly.

Grębecki modified this protocol in 1964 for a better staining of the cilia for light microscopical examinations. In 1979 Barlow and Sleigh developed a new freeze-substitution technique for the preservation of ciliated surfaces for scanning electron microscopy, including ciliates. The arresting of rapid movement of living cells and the preservation of their surface fine structure for scanning electron microscopy are reported to be guaranteed by rapid freezing of the organisms in liquid propane. It is said that this technique is more reliable than $OsO_4/HgCl_2$ fixation, both in terms of obtaining a much higher proportion of good results and in the improved preservation of structural details.

C. Transmission Electron Microscopy

The techniques used to prepare *Paramecium* for TEM viewing are essentially identical to those used for all animal cells and tissues. Thus any manual on biological techniques in electron microscopy such as that of Dykstra (1992) or Hayat (2000) can be used as a guide to general protocols to be followed. In our study we found that 0.34 M s-collidine (trimethylpyridine) buffered 1% glutaraldehyde, followed by post-fixation in OsO_4, provided well-defined images of the cell, particularly the fibrous and filamentous systems in the cell's cortex (Fig. 2A, B) (Allen, 1971). However, this fixation procedure did result in a more extracted appearance of the cytoplasm and to swelling of endomembrane compartments, such as the endoplasmic reticulum, the Golgi complex, and the alveoli. On the one hand, leaching of cytoplasmic proteins did have the beneficial effect of making filamentous cell components and the ribosomes much more visible but, on the other hand, trichocysts were frequently seen to "fire" internally before fixation had been completed.

Thus, for most of our later TEM studies we used 0.05 M cacodylate-buffered 1% glutaraldehyde as the fixative of choice. This protocol preserved more of the cytoplasmic proteins as well as the food-storage, glycogen bodies. Such bodies were particularly evident in *P. caudatum*.

The following protocols are intended to point out how a standard procedure is adjusted for *Paramecium*.

1. Fixing and Embedding

1. Concentrate cells by mild centrifugation in a clinical centrifuge at ~95 g for 30 s.
2. To 1 volume of concentrated culture, add an equal volume of 2% glutaraldehyde in 0.1 M cacodylate buffer at pH 7.4. Fix for 25 min at room temperature.
3. Concentrate again, aspirate fixative, and wash by adding fresh buffer solution.
4. Spin, aspirate, and add 1% OsO_4 in 0.05 M cacodylate buffer for 45 min.
5. Decant OsO_4 fixative in the hood into a bottle used for OsO_4 waste, and add distilled water.
6. Dehydrate cells in a graded series of ethanol (from 30 to 100%) and end with 100% propylene oxide.
7. Embed cells in resin and pour mixture into a shallow flat-bottomed aluminum dish.
8. Place in desiccator overnight to remove gas bubbles from mixture.
9. Transfer to 60°C oven to polymerize resin for 1–2 days. The cells are now ready for sectioning but can be stored for years, preferably in a desiccator so that the resin does not soften with time.

2. Sectioning

1. Select one cell lying at the bottom of the dish and remount for sectioning.
2. Trim around the cell, leaving it in a tiny trapezoid. Ensure that the top and bottom of the trapezoid are perfectly parallel in order to get straight ribbons of serial sections that do not curve.

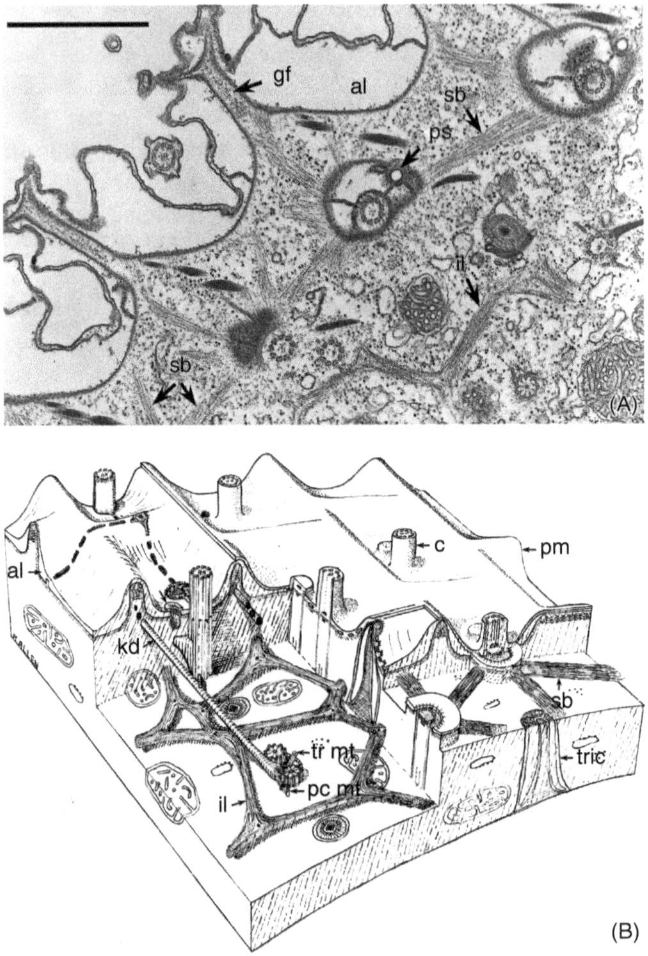

Fig. 2 (A) Fibrous systems in the cortex of *P. multimicronucleatum* as seen by TEM of chemically fixed samples. A granulo-fibrillar (gf) system fills the tops of the polygonally shaped ridges. Below this network are the striated bands (sb). Below this is the continuous network of the infraciliary lattice (il), which lies at the level of the proximal ends of the basal bodies. al, alveolus; ps, parasomal sac. Scale bar = 1 μm. (B) 3D reconstruction of the cortical architecture of *P. multimicronucleatum* based on ultrathin sections. al, alveolus; c, cilium; il, infraciliary lattice; kd, kinetodesmal fiber; pm, plasma membrane; pc mt, postciliary microtubules; ps, parasomal sac; sb, striated band; tric, trichocyst; tr mt, transverse microtubules. Reproduced from *J. Cell Biol.* **49,** 1–20, 1971.

3. Mount in a specimen holder of ultramicrotome and section with a diamond knife.
4. Separate the ribbon of sections into shorter segments with an eyelash glued to a short stick and tease the segments of the ribbon so they lie side by side. Pick up the ribbons onto a Formvar^{TM}-supported one-hole grid. Wick away excess water with a wedge of filter paper.

3. Positive Staining

Sections are routinely stained with aqueous uranyl acetate (UA) followed by lead citrate (Dykstra, 1992; Hayat, 2000).

4. Negative Staining

The negative staining technique is mainly used for the visualization of small structures such as virus particles or isolated cell constituents, e.g., macromolecules (enzymes), surface components of cells (bacterial pili), mastigonemes of eukaryotic flagella, small organelles such as dictyosomes, or structures such as nuclear pores and components of the cytoskeleton.

Suspensions of the structures to be analyzed are transferred in a drop of the suspension fluid to a Formvar-coated grid and surrounded by a solution of heavy-metal stain. This covers the grid but is excluded by the biological specimens which do not interact with the heavy-metal stain. The structures stand out as light on an electron-dense background. The four most common aqueous staining solutions of heavy metals in use, in a concentration of 0.2–4%, are UA, phosphotungstic acid, ammonium molybdate, and methylamine tungstate.

a. Methods The staining procedure is relatively simple and fast. The basic staining procedure includes the following steps (Hausmann *et al.*, 1972):

1. A droplet of the suspension to be examined is placed on a Formvar-coated grid.
2. After a few minutes, during which the particles attach (primarily by electrostatic forces) to the Formvar film, the grid is partially, but not completely, dried with a piece of filter paper.
3. A drop of the negative stain solution is then added to the grid.
4. After 30 s to several minutes, the grid is completely dried with fresh filter paper.
5. The grid is then deposited on a piece of filter paper in a Petri dish until its examination.
6. A brief distilled water rinse might be used after the negative staining to achieve a lighter staining for better resolution. However, this procedure may result in too little contrast to be of any use.

For further modifications, preparation guides like those of Dykstra (1992) or Hayat (2000) should be consulted.

In a typical preparation, some of the structures to be analyzed will completely exclude the stain, which might form a slightly gray background at the edge of the specimen. Other structures will exclude the stain but will have large amounts of stain around their edges, which result in a completely electron-opaque zone that covers many particles of interest. Some areas of the grid may appear to be unstained with no gray- or black-stained background and thus with no detectable structures, while another part of the grid contains a deep layer of negative stain, completely impenetrable to the electron beam.

Consequently, the success of this method is unpredictable and a good deal of luck is needed to get good results.

b. Research Applications In *Paramecium* the structure and mode of function of the trichocysts, a type of extrusome used for defence, are of special interest (Fig. 3). Every single *Paramecium* possesses several thousand trichocysts. Extrusomes are vesicles containing an organized substance that usually changes its form when the vesicle undergoes exocytosis and releases its contents to the exterior of the cell, often within milliseconds (Hausmann, 1978).

The spindle trichocysts of *Paramecium* originate in the cytoplasm, and, in their resting states, they are docked at predictable positions within the cortex (Fig. 1C) (Allen and Hausmann, 1976; Hausmann and Allen, 1976). The interior of mature trichocysts has a cross-striated periodicity of 7 nm. The extrusomal material is delivered to the outside of the cell via exocytosis. Until the introduction of rapid-freezing and deep etching (Section II.E.), the ultrastructure of both the resting and the ejected states was best analyzed using the negative staining technique, which guaranteed the highest resolution of these delicate structures in TEM.

The ejected trichocyst is about eight times longer than the resting trichocysts (Fig. 3A and B). In the extruded state, there are also periodic cross striations, but their periodicity is about 56 nm (Fig. 3C and D). This periodicity results from a three-dimensional arrangement of thin filaments, which are connected in an orderly manner (Fig. 3E). A three-dimensional model of the extruded shaft of the spindle trichocysts has been produced at the macromolecular level (Fig. 3F). Based on this model it can be demonstrated that the shafts of trichocysts always have the same structure, even though the arrangement of the filamentous connections seems to change (Hausmann *et al.*, 1972).

5. Cytochemistry

In order to study the functions of cellular membrane compartments and organelles, it is often necessary to determine what enzymes are associated with a particular compartment. Knowing the by-product of the enzyme activity can help the researcher design methods for localizing the enzyme inside the cell.

a. Methods

Acid Phosphatases

1. An axenic culture of *Paramecium* was fed PLS of different sizes in the protocol described in Section II.A.c. in order to be able to determine the age of each labeled digestive vacuole.
2. Cells were fixed in 1% glutaraldehyde in 0.05 M cacodylate buffer (pH 7.2) for 15 min.
3. Cells were washed 3× in 0.05 M cacodylate buffer and left in buffer overnight at 4°C.
4. The following day, cells were washed 3× in 0.05 M acetate buffer (pH 4.0), and incubated in Gomori medium (Gomori, 1952) for 30 min. Gomori medium consists

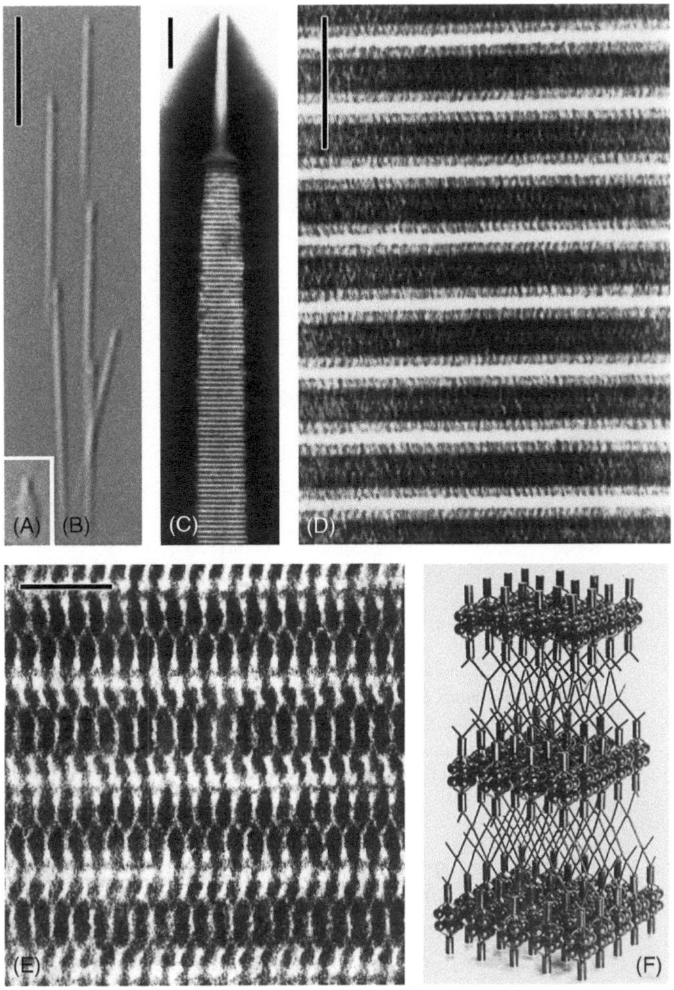

Fig. 3 Trichocysts of *Paramecium* in light (A and B) and electron micrographs (C–E). (A) Spindle-like resting state. (B) Thread-like extruded state. (C–E) Negatively stained trichocysts with progressively increasing magnification. (F) 3D reconstruction of the network of protein filaments visible in E. Reproduced from *Int. Rev. Cytol.* **53,** 197–276, 1978. Scale bar in A and B = 20 μm, in C = 1 μm, in D = 100 nm, in E = 60 nm.

of 10 ml of 0.05 M acetate buffer (pH 4.0), in which 0.12 g of lead nitrate (~3 mM) are dissolved. To this solution, 1 ml of 3% (~0.1 M) β-glycerophosphate is added. This mixture is incubated for 24 h at 37°C and the resulting turbid mixture is filtered just before use.

5. After incubation, the cells were washed 3× in acetate buffer, pH 4.0, and then 3× in cacodylate buffer (pH 7.2).

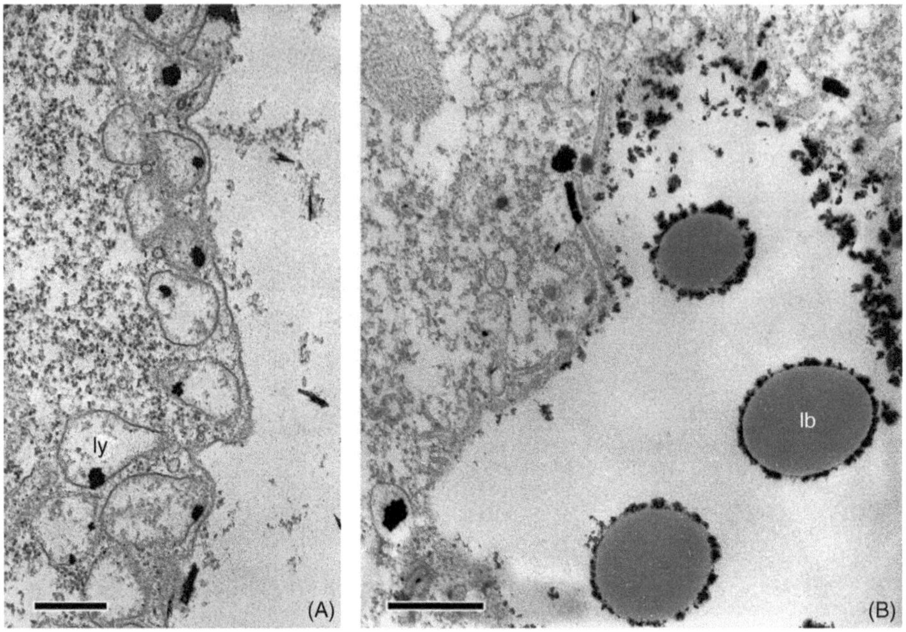

Fig. 4 (A) Secondary lysosomes (ly) have foci of acid phosphatase electron-opaque reaction product after applying the Gomori technique. Reproduced from *J. Protozool.* **31,** 216–220, 1984. Scale bar = 0.2 μm. (B) The tubules and rounded expansions that arise from the membrane of a late phagolysosome and the periphery of marker latex beads (lb) are labeled with acid phosphatase reaction product. Reproduced from *J. Cell Biol.* **99,** 1955–1959, 1984. Scale bar = 0.5 μm.

6. Following the wash, 2% OsO_4 in 0.05 M cacodylate (pH 7.2) was added as a second fixative for 40 min. The cells were then processed for TEM as in Section II.C.a., except that propylene oxide was not used as a final dehydration step as this tended to dissolve the latex beads.
7. The localization of acid phosphatases is visible by an electron-opaque reaction product (Fig. 4A, B).

Cellular Peroxidases Endogenous peroxidase can be demonstrated by the Graham and Karnovsky (1966) technique in which glutaraldehyde-fixed cells are thoroughly washed in phosphate buffer for 30 min to remove all unreacted glutaraldehyde and then washed in several changes (20 min total) of 0.05 M Tris buffer (pH 9.0) and stained in a freshly prepared solution of 0.05% 3,3′-diaminobenzidine and 0.01% hydrogen peroxide in Tris buffer (pH 9.0) at 37°C for 1–4 h (Fok and Allen, 1975).

Tracer Studies with Horseradish Peroxidase

1. *Paramecium* was exposed to horseradish peroxidase (HRP) (2 mg/ml of type II HRP, Sigma Chem. Co., St. Louis, MO) in axenic medium for predetermined times

with or without a chase in fresh axenic medium. If desired, PLS (Section II.A.c.) can be combined with the HRP.

2. Cells were fixed in 1% glutaraldehyde in 0.05 M cacodylate buffer at pH 7.2 as described in the TEM protocol (Section II.C.3.) and washed in fresh buffer.
3. Fixed and washed cells were then incubated for 20 min in a solution containing 0.01% H_2O_2 and 0.05% 3,3′-diaminobenzidine (DAB, Sigma Chem. Co.) according to the procedure of Graham and Karnovsky (1966).
4. After incubation the cells were washed and postfixed in OsO_4 for 30 min at room temperature and processed for TEM as in Section II.C.3.

b. Research Applications One group of naturally occurring enzymes that has been studied extensively in many organisms, as well as in *Paramecium*, are the acid phosphatases. The acid phosphatases (phosphomonoesterases II) are a group of non-specific hydrolases that liberate phosphate ions from organic esters at a pH of ~4.5–6.0 (Essner, 1973). The acid phosphatase reaction product has been localized to lysosomes and phagolysosomes of *Paramecium* (Fok *et al.*, 1984; Müller and Törö, 1962; Müller *et al.*, 1963) as well as to the region of the endoplasmic reticulum next to the Golgi complex and to some Golgi stacks and to the trans-Golgi reticulum.

While acid phosphatase reaction product is used to localize the phagocytic pathway compartments, as well as some biosynthetic compartments, a technique used to demonstrate the presence of peroxidases, such as catalase, has been used to localize peroxisomes in ciliates such as *Paramecium* and *Tetrahymena* (Fok and Allen, 1975; Stelly *et al.*, 1975). A naturally occurring peroxidase was also found in the mitochondria of the cell.

It is sometimes useful to expose *Paramecium* to external enzymes that can function as tracers when taken into the cell. Thus we have made extensive use of the enzyme HRP to follow membrane trafficking in the endosomal pathway (Allen *et al.*, 1992) and in the phagocytic pathway of *Paramecium* (Allen and Fok, 1980). This technique was also used to see what happens to trichocyst membranes (Allen and Fok, 1984) upon the release of these extrusomes.

D. Standard Freeze-Fracture and Replica Formation

Freeze-fracturing glutaraldehyde-fixed, glycerinated (to prevent ice crystal formation), and frozen *Paramecium* followed by replica formation of subsequently fractured surfaces provide for a more three-dimensional view of intracellular structure than is provided by thin sectioning. This technique proved invaluable for our understanding of the complexities of the oral region of this cell (Allen, 1974, 1978) as well as in our working out details of the membranes of the digestive vacuole cycle (Allen and Staehelin, 1981).

a. Methods

1. If it is important to know the age of a vacuole, prelabel the digestive vacuoles (DVs) by adding 1 drop of PLS to 2 ml of concentrated cells followed with no chase or with an appropriate chase time as in Section II.A.3.

2. Fix 0.5 ml of concentrated cells for 15 min in 0.5% glutaraldehyde in 0.05 M cacodylate buffer at pH 7.2.
3. Wash in fresh 0.05 M cacodylate buffer.
4. Glycerinate cells for 2 h in 30% glycerol (not buffered) at room temperature and concentrate cells into a pellet.
5. Place cells on a copper support appropriate for the freeze-etch unit to be used.
6. Freeze cells by plunging them into monochlorodifluoromethane (Ucon 22 of Union Carbide Corp., New York, NY) bathed in liquid nitrogen.
7. Fracture cells at –100 or –106°C with the blade of the microtome while in the vacuum of a freeze-etch unit, such as a Balzers 400, while the gun of the evaporator is heating to keep any etching to a minimum.
8. Evaporate carbon–platinum at a 45° angle, without rotation, to the fractured surface and then evaporate carbon at 90° for 2 min.
9. Clean the replicas of all biological debris with bleach for 1 h and with 20% chromic acid overnight. Finally, replace 20% chromic acid with 40% chromic acid for 1 h. If latex beads were used the replicas will need to be washed in tetrahydrofuran to dissolve the beads for 1–2 h.
10. Wash replicas with 2 changes of distilled water over 1–2 h.
11. Pick up replicas on Formvar-supported single-hole grids.
12. Observe replica in a TEM.

b. Research Applications The advantage of viewing replicas of freeze-fractured cells lies in the fact that fracture planes in frozen cells tend to follow the lines of least resistance, which often happens along the middle hydrophobic core of membranes. Thus one will see large expanses of membranes in the replicas such as that of the somatic plasma membranes in *Paramecium* as well as of the membranes of the buccal cavity (Fig. 5A, B) (Allen, 1974, 1978) and of the digestive vacuoles (Allen and Staehelin, 1981). Not only can one view the general topography of these membranes at normal TEM magnifications but, at higher magnifications, one can also view the intramembrane particles (IMPs), which are mostly proteins that penetrate into or through the middle of those membranes. Thus a fractured membrane has both fracture faces corresponding to its inner membrane leaflet and its outer membrane leaflet exposed and these two faces, from the same membrane, frequently display very different appearances as the IMPs tend to stay on only one face of the membrane.

Freeze-fracture images have demonstrated that the alveolar sacs are "welded" to the plasma membrane by bands of IMPs that are extensions of intermembrane links between the two membranes. Such bands are found at the edge of the cytopharynx, around the tips of the trichocysts, and along the two lips of the cytoproct. Microtubules and filaments at the cytopharynx and the CV pore are also linked to the cytopharynx and plasma membranes, respectively (Allen, 1978).

Other specialized modifications of the plasma membrane of *Paramecium* that have been reported include plaques or plates of IMPs that are found mostly on the anterior ventral surface of the cell (Fig. 6). The function of these plaques, which may or may

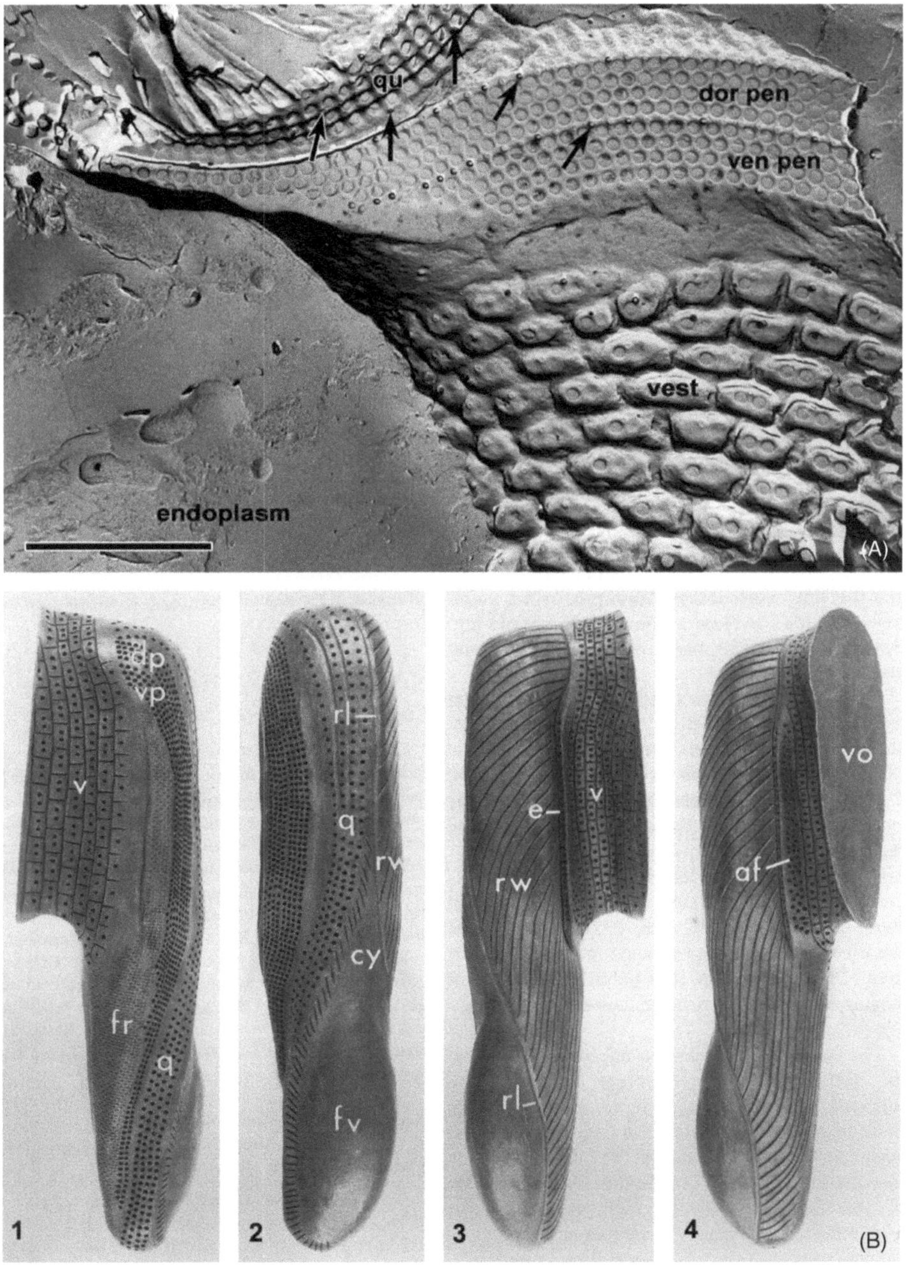

Fig. 5 (A) The oral region of *Paramecium*, illustrated in a freeze-fracture image that passed through its left side and opens at the mid-ventral part of the cell. Arrows point to rows of parasomal sacs. dor pen, dorsal peniculus; qu, quadrulus; ven pen, ventral peniculus; vest, vestibulum. Scale bar = 1 μm. (B) A clay model of the space occupied by the buccal cavity, its connection to the vestibulum (v), and the nascent food vacuole (fv). Dots indicate sites of basal body/cilium complexes of the somatic ciliature on the vestibulum and the membranelles of the ventral peniculus (vp), the dorsal peniculus (dp), and the quadrulus (q). af, anarchic field of basal bodies; cy, cytostome–cytoharynx complex; fr, filamentous reticulum; rl, right lip of cytopharynx; rw, ribbed wall; vo, vestibular opening. Reproduced from *J. Cell Biol.* **63,** 904–922, 1974.

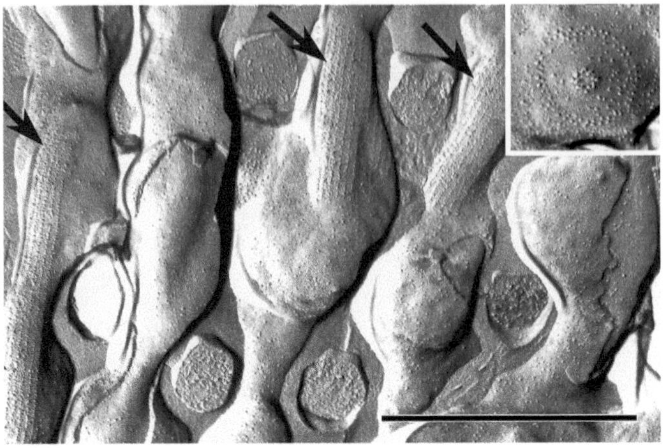

Fig. 6 Freeze-fracture image of the surface ridges at the anterior end of *Paramecium* with numerous plates of regularly arranged IMPs (arrows) in the P-face of the plasma membrane. Inset above right: Ring and rosette IMPs at the trichocyst exocytic site. Reproduced from *J. Ultrastruct. Res.* **63,** 64–78, 1978. Scale bar = 1 μm.

not be linked to the alveoli, has not yet been determined with certainty. However, IMPs in the plasma membrane that are located at the tips of trichocysts (Allen and Hausmann, 1976; Plattner *et al.*, 1973) (Fig. 6, inset above right) and at the base of cilia (Bardele, 1981) may be involved in signal transduction or calcium transport across the membrane at these sites.

Thus, this technique along with the next technique to be discussed, which provides similar images of cryo-immobilized and deep-etched cells, can provide views of a cell impossible to obtain with thin-section TEM.

E. Rapid-Freeze, Fracture, Deep-Etch, Rotary-Shadow Technique

Next, we describe the production of rapid-frozen, deep-etched replicas that allowed us to view fractured cells, which had never been exposed to chemical fixatives. This technique provides a near-native view of the cell in which membrane shrinking and swelling artifacts are avoided that are often introduced during the steps of chemical fixation and dehydration in conventional TEM preparations. This technique also provides a more three-dimensional view of cellular structures than is seen in thin-section TEM and also allows one to see and study the internal intramembrane molecular protein structure and distribution in cellular membranes.

*a. **Methods*** Rapid-freezing of cells (to obtain vitreous ice formation) was achieved in our laboratory using a slam freezer (KF 80) produced by Reichert and Jung (Vienna, Austria). *P. multimicronucleatum* cells (not fed PLS) growing in axenic medium (free of any other organisms) were briefly spun in a clinical centrifuge to concentrate the cells into a loose pellet. An aliquot of the concentrated living cells swimming in the axenic medium was pipetted onto a thin copper hat designed for use in the Balzers 400 freeze-etch apparatus. The hat was affixed to the plunger of the Reichert KF-80 rapid-freezing device assembled with its MM-80 attachment which is provided for impact freezing. Cells were slammed onto a highly polished copper disk, which was cooled to the temperature of liquid nitrogen, –195°C. The MM-80 attachment has a bounce-free variable-force control, which, when optimally adjusted, prevents the suspended cells from bursting on impact. The frozen cells, attached to the hats, were then stored in liquid nitrogen.

When needed, the frozen cells were transferred from liquid nitrogen to the cooled stage of a Balzers 400 apparatus. Air was pumped out of the Balzers 400 apparatus and the temperature of the stage was adjusted to –150°C for fracturing. Only one pass with the microtome knife was made so that the plane of fracturing would be within 15–20 μm of the surface that had impacted the polished copper plate. After fracturing, the temperature within the Balzers apparatus was raised to –95°C for 5–15 min to permit slow etching of the cellular water from the fractured surface. Platinum, to a thickness of 1.5–2 nm, was then evaporated onto the fractured surface of the specimen at an angle of 24° as the specimen was rotated. Carbon was then evaporated at an angle of 80° during rotation to give support to the platinum replica. The specimens were then retrieved from the Balzers stage and quickly placed in liquid nitrogen until the replicas were ready to be processed. At that time the hat was removed from the specimen holder and thawed.

For cleaning, the thawed replica was first floated onto a drop of double-distilled water (DH_2O) and then transferred by a platinum wire loop to the surface of a drop of bleach for 3–5 h and then onto a drop of chromic acid overnight to digest away all biological material. The replica was then floated onto three to five changes of DH_2O and finally picked up onto a Formvar-coated single-hole grid by touching the Formvar film to the replica surface.

b. Research Applications Although we have used the rapid-freezing, deep-etched, and rotary shadowing technique to make replicas of *Paramecium,* this technique has not been extensively used by other laboratories to study protozoa. In our hands, this technique has revealed the three-dimensional fine structure at the molecular level as no other technique is capable of doing. For example, the tubular cristae of the mitochondria were shown to bear highly ordered arrays of F_1F_0-ATP synthases along their outer helical turns, which allowed us to estimate the number of F_1F_0-ATPases present in a unit of length of the helically wound tubular cristae (Allen *et al.*, 1989). This technique has also revealed for the first time large membrane-associated particles, thought to be complex I of the respiratory chain, that are nonrandomly located along the inside curve of the tubular cristae as a single helically curved row (Fig. 7A). Estimation of the ratio of the number of these complexes to the F_1F_0-ATP synthases was then possible, and this figure could then be compared to the ratio that had been obtained biochemically for beef heart mitochondria (Capaldi, 1982).

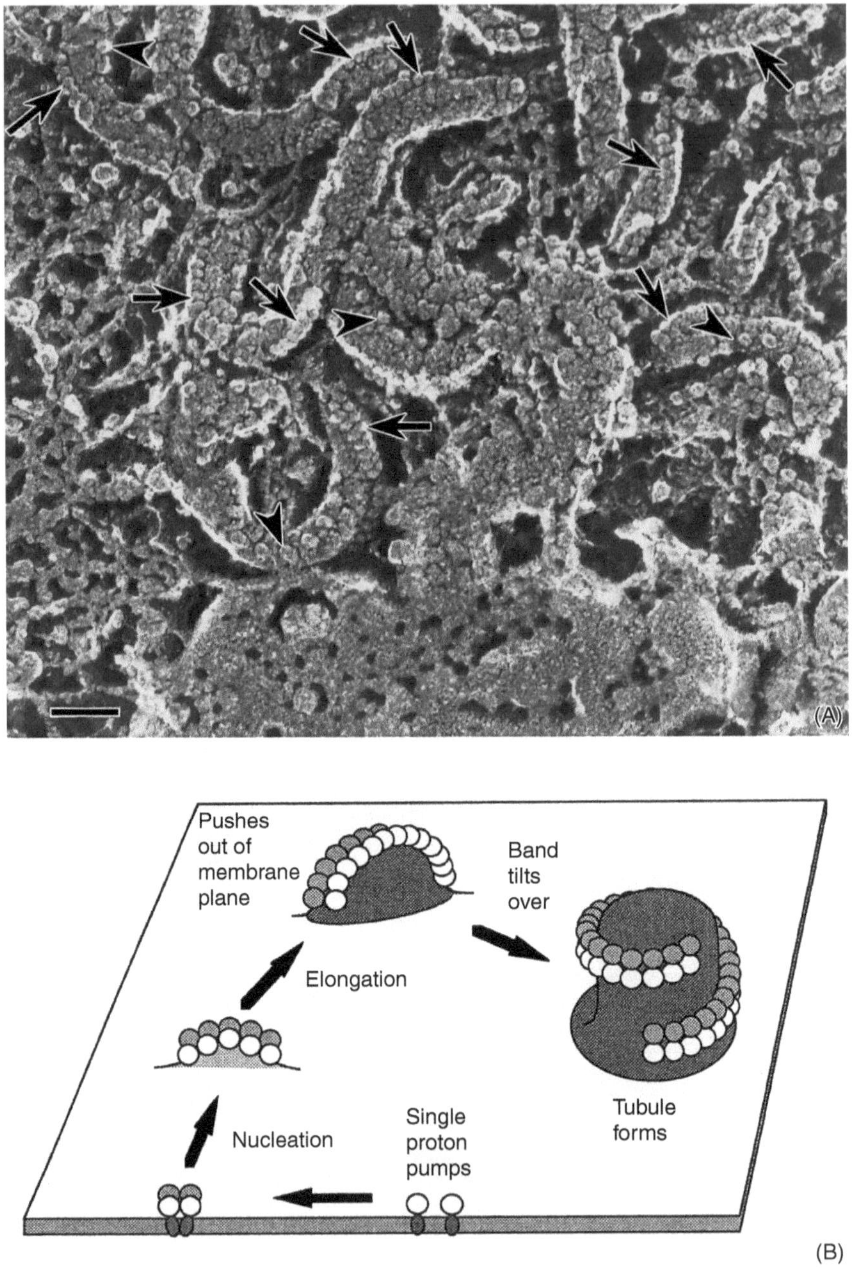

Fig. 7 (*Continued*)

In addition, the proton-transporting V-ATPases of the CVC have also been shown to form two helical arrays around the 50-nm tubules of the decorated spongiome of this organelle (Fok *et al.*, 1995). Thus, in *Paramecium*, it appears that lateral associations of proton pumps within a planar membrane that then rearrange into helical arrays may lead to tubularization of membranes both in mitochondria and in the CVC (Fig. 7B) (Allen, 1995).

The study of replicas of cells rapidly frozen, fractured, and deep-etched also provided us with a more reliable view of the internal structure of cellular membranes than was provided by conventional freeze-fracture replicas where cells had been chemically fixed. This was instrumental in documenting some of the changes that occur in the phagosomal membrane as it passes from the phagosome to the phagoacidosome stage, then to the phagolysosome stage, and finally to the defecation-competent stage (Allen and Fok, 2000). This technique also allowed us to detect the fusion of one kind of vesicle with another vesicle (Fig. 8A) when the fracture faces of the two vesicle populations bear significantly different concentrations of IMPs (Allen *et al.*, 1993) or even when one vesicle leaflet is smooth and the other has pits such as those occurring on the E faces of carrier vesicles fusing with acidosomes in *Paramecium*.

Finally, segments of the true surfaces of membranes, as compared to their fractured internal surfaces, can also be seen in the deep-etched replicas. From these areas we can detect the attachment of peripherally bound proteins such as motors, e.g., cytoplasmic dyneins, which are linked to vesicle membranes, and also observe their attachments to nearby microtubules. Thus, snapshots of the native structure of dynein can be viewed while this motor is actually in the process of moving vesicles along microtubules (Fig. 8B).

F. Cryo-Immunogold Studies

The next technique to be described is the use of lightly fixed cells that are then infiltrated with 2.3 M sucrose before being frozen in liquid nitrogen and thin-sectioned at a temperature of −100°C (Tokuyasu method). Frozen sections from these preparations are then thawed, washed, and exposed to a primary antibody that was raised in

---

Fig. 7 (A) Quick-freeze deep-etched replica of unfixed and crushed mitochondrion from *Paramecium*. Each helically curved tubular crista has a tightly packed zipper-like double row of 9-nm diameter F_1 subunits of F_1F_0-ATP synthases located on the outer curve of the crista (arrows). A single row of larger particles (arrowheads) is located on the inner curve, which most closely corresponds in frequency and size to complex 1 of the respiratory electron transport chain. Reproduced from *J. Cell Biol.* **108,** 2233–2240, 1989. Scale bar = 0.1 μm. (B) A model for one type of membrane tubulation has been proposed for *Paramecium*. Single F_1F_0 or V_1V_0 (V-type proton pumps) holoenzymes may bind identical proton pumps within the plane of the fluid membrane at a slight tilt to each other to form a double row. Additional proton pumps binding to this nucleus form the beginning of an arch that deforms the planar membrane. Continued elongation of this double row completes a helical turn and pushes the membrane plane into a short tube. Adding additional proton pumps to the chain will elongate the tube. Reproduced from *Protoplasma* **189,** 1–8, 1995.

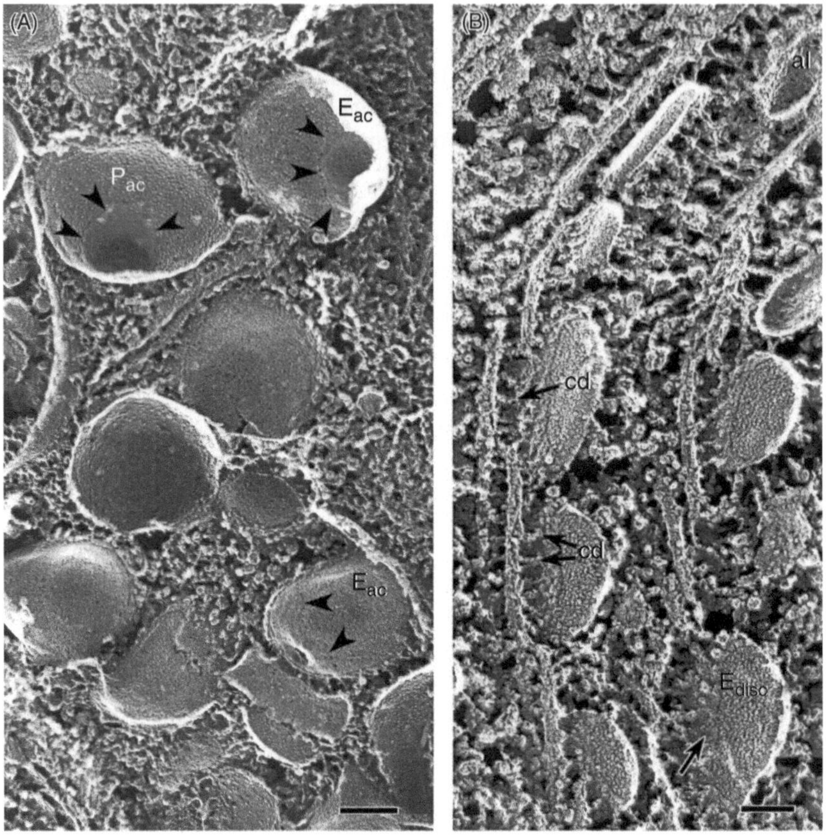

Fig. 8 (A) Quick-frozen deep-etched replica of acidosomes of *Paramecium* showing evidence for vesicle fusion. Acidosomes have a relatively particle-free E-fracture face (E_{ac}) and a P-fracture face that contains many intramembrane particles (P_{ac}). Where acidosome membranes exhibit two distinct areas on their fracture faces (arrowheads on P_{ac} and E_{ac}) it appears that carrier vesicles have recently fused with the larger acidosomes. Thus carrier vesicles coming from the endocytic pathway fuse with acidosomes of the phagocytic pathway in *Paramecium*. Reproduced from *J. Cell Sci.* **106,** 411–422, 1993. Scale bar = 0.2 μm. (B) Quick-frozen deep-etched replicas in unfixed cells provide images of molecular motors in action. The bridges that link discoidal vesicles to the microtubules of the cytopharyngeal microtubular ribbons represent active cytoplasmic dynein (cd) motors. In favorable regions, two bridges (legs) appear to arise from the true vesicle surface, cross the 40-nm gap, and contact an adjacent microtubule (lowest arrow). Such views of cytoplasmic dynein in action are quite different from the axonemal dynein in action in this same cell. Currently, there is a need to reconcile images such as that presented here with the published models for cytoplasmic dynein. Scale bar = 0.1 μm.

mice or rabbit/goat and to colloidal gold-complexed goat-anti-mouse/rabbit secondary antibodies. In this way, one can visualize the location of specific protein antigens within the cell's structure.

In our laboratory, we preferred to work with monoclonal antibodies (Mabs) as they are each specific for an epitope of a particular protein and it is not necessary to affinity purify the antibodies as is required to make polyclonal antibodies specific. We characterized the antibodies by first determining what structures were being labeled in immunofluorescence microscopy of permeabilized whole fixed cells. When proven to be useful we then ran gels to determine the molecular mass of the antigen and then we determined the immunoglobulin species of the Mab.

a. Methods The methods we used were modified from Tokuyasu (1984).

1. Cells were concentrated from axenic medium and fixed lightly in 0.25% glutaraldehyde in 0.05 M phosphate buffer, pH 7.5, for 20 min at room temperature.
2. Cells were then washed 2× in 0.05 M phosphate buffer and 1× in DH_2O, 10 min in each wash.
3. An aliquot of these cells was added to the top of an equal volume of 8% liquefied gelatin in a microfuge tube and briefly spun through the gelatin to concentrate the cells at the bottom of the tube. The tubes were held at 4°C overnight to allow the gelatin to set.
4. The gelled preparations were then sliced into thin wafers (by slicing the tube and its contents) and transferred into 2.3 M sucrose overnight.
5. The wafers from the bottom of the tube were pressed out of their tube enclosures and cut into triangular wedges. A triangle was mounted such that one edge is flat against the microtome pin. These pins were dropped into liquid nitrogen for freezing and storage.
6. A pin with a frozen wedge of cells was mounted in the ultramicrotome of a low-temperature sectioning system (Ultracut E/FC 4D, Reichert-Jung, Vienna, Austria). Sectioning was at –100°C. No trimming was necessary if the wedge was sufficiently thin and the specimen was correctly aligned such that the wedge is parallel to the knife edge. With proper alignment, a ribbon of frozen sections would soon be attained as sectioning progressed.
7. A ribbon of frozen sections can be picked up from the back of the knife when a 1.5- to 2-mm diameter platinum loop dipped in 2.3M sucrose is touched on the frozen sections just before the sucrose freezes.
8. The sections are then allowed to thaw and are transferred to a one-hole Formvar-coated grid. The sections on the grid are placed face down on 2% gelatin for 10 min or more in a small dish cooled on ice to allow the sucrose to diffuse out of the sections into the gelatin.
9. Washing and incubation of the thin sections are done by floating the grid face down onto a series of drops on a sheet of Parafilm™ laid on a bench. This series includes (1) Tris buffer mixed with 0.1% bovine serum albumin (BSA) 2× for 15 min each, (2) normal goat serum (15 min), (3) Tris mixed with 1% BSA (30 min), (4) Mab (as little as 5 μl) for 2 h, (5) Tris buffer mixed with 0.1% BSA 5× for 5 min each, (6) Tris buffer mixed with 1%

BSA 1× for 5 min, (7) secondary antibody–colloidal gold complex, 30 min, (8) Tris buffer mixed with 0.1%BSA 6× for 5 min each, (9) 3 drops of DH_2O for 5 min total, (10) 2% UA mixed with 0.5 M oxalic acid adjusted to pH 7 with NH_4OH, for 10 min, (11) 1.5% methyl cellulose mixed with 2% UA, 3× for 30 s total.

10. Pick up and wick off excess methyl cellulose. The final interference color should be light straw to gold.
11. Examine in the TEM.

b. Research Applications Examples of the use of immunogold labeling to study proteins as well as the fine structure of *Paramecium* include (1) studies of the V-ATPase proton pumps in the cell, (2) studies of membrane replacement, (3) development of trichocysts, (4) antigens in the endocytic pathway, and (5) double-labeling experiments to distinguish between fibrous systems.

1. One of the first Mabs we raised to *P. multimicronucleatum* was found to be specific for antigens of the CVC (Fig. 9A). Immunofluorescence demonstrated its specificity for the radial arms of the CVC (Fig. 9B) rather than the CV itself or the collecting canals and ampulli (Allen *et al.*, 1990). Immunogold labeling of frozen thin sections demonstrated its specificity for the 50-nm decorated tubules and epitopes of the V-ATPase proton pumps (Fig. 9C). This was confirmed (Fok *et al.*, 1995) by showing identical labeling with the affinity-purified polyclonal antibodies raised to the V-ATPases of *Dictyostelium discoideum* (Nolta *et al.*, 1993) and another to the 57-kDa B-subunit of the V-ATPase of chromaffin granules (Nelson, 1992). Thus cryo-sectioning combined with the immunogold technique provides a means of pin-pointing the molecular complexes in the cell that carry out a particular physiological role.

   Other techniques may allow the same type of localization, e.g., high-pressure freezing (Murray, 2008; Vanhecke *et al.*, 2008), but we have had little experience with this technique and to our knowledge it has yet to be applied to *Paramecium*, so we cannot evaluate its applicability for *Paramecium*.
2. In another series of studies, we made extensive use of the immunogold technique to confirm what our previous freeze-fracturing (Allen *et al.*, 1993) and regular TEM had suggested regarding membrane replacement and the dramatic changes occurring in the digestive vacuole system of *Paramecium*. Mabs were raised to at least four different antigens and each occurs only at a specific stage of the digestive vacuole's cycle. The ages of the digestive vacuoles were labeled as described in Section II.A.3. As previously shown by the freeze-fracture studies, the DV-I are formed out of discoidal vesicle membrane and these vesicles bear two antigens of 140 and 150 kDa, respectively, localized by our Mabs. As the vacuole forms a second group of vesicles, the acidosomes that bear antigens of 29 and 130 kDa, respectively, move to and dock with the forming DV-I (Allen *et al.*, 1993). Following the release of the DV-1 from the cytopharynx the acidosomes quickly fuse with the DV-I at the same time that the 140- and 150-kDa antigens are removed

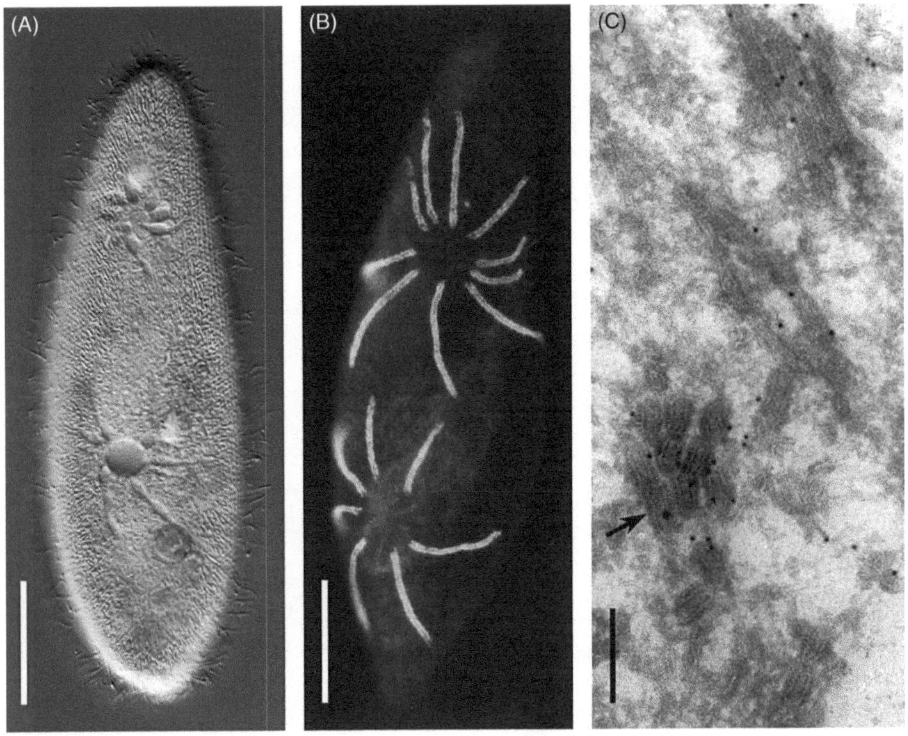

Fig. 9 (A) Light microscopical aspects of the two contractile vacuole complexes in *Paramecium*. (B) Monoclonal antibody DS-1 is highly specific for the decorated tubules of the contractile vacuole complex of *Paramecium multimicronucleatum* only. (C) Higher resolution using the same antibody as in B but on lightly fixed cells, frozen in liquid nitrogen, and then thin sectioned at –100°C before thawing and reacting the sections with immunogold. DS-1 proved to be specific for the V-type proton pumps that extend from the cytosolic surface of the decorated membranes of the contractile vacuole complex (arrow). A courtesy of Gerd Günther, Düsseldorf, Germany, C reproduced from *J. Cell Sci.* **108,** 3163–3170, 1995. Scale bar in A and B = 100 μm, in D = 2 μm.

from the DV-I. Thus the DV-I is converted into a phagoacidosome (DV-II), which now bears only the 29- and 130-kDa antigens (Allen *et al.*, 1995). Lysosomes, which have none of the above antigens, can then dock at the DV-II and fuse to give rise to the phagolysosome (DV-III). Digestion occurs only in this stage of the digestive vacuole. Finally, lysosome-specific membrane is removed from the DV-III and this membrane, in the form of tubules, expands into secondary lysosomes. The spent vacuole (DV-IV) is now defecation competent. In this study we have used Mabs not to study proteins per se, but as a means of studying changes in membrane composition over time.

3. The structure of *Paramecium* trichocysts has been analyzed using a trichocyst core-specific Mab and indirect immunogold labeling on LR-Gold-embedded as well as on fixed, frozen, and cryosectioned material (Fig. 10) (Hausmann *et al.*, 1988).

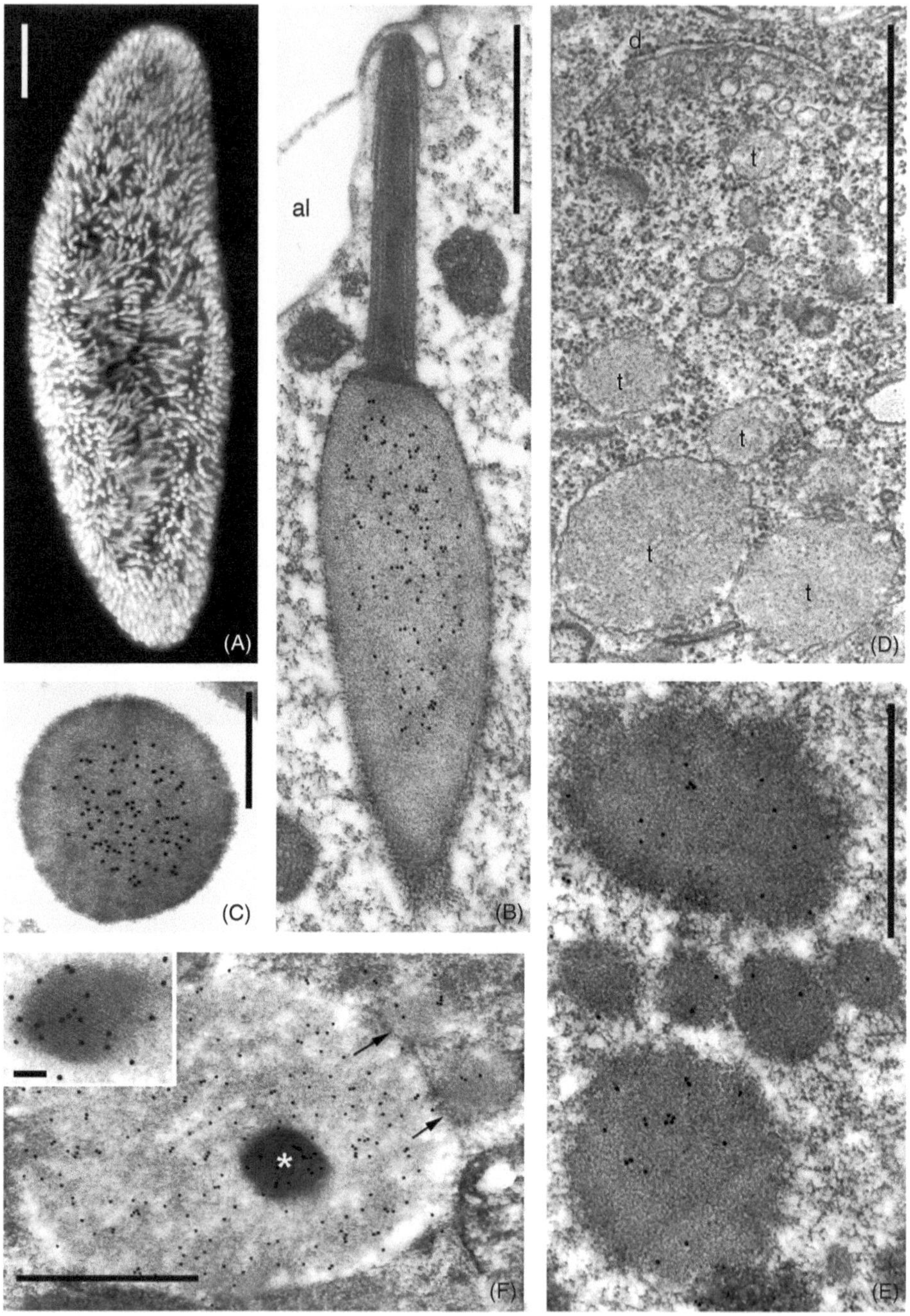

Fig. 10 (*Continued*)

During the morphogenesis of trichocysts, small immunogold-positive vesicles, originating in all likelihood from dictyosomes (Golgi stacks), fuse to form larger vacuoles. A paracrystalline core forms within the vacuole matrix. The label becomes progressively concentrated within this core and, ultimately, is restricted entirely to the core. The other constituents of the trichocyst, e.g., the cortex of the body, the tip, and the cap covering the tip, could be labeled with other specific Mabs (Fok *et al.*, 1988).

4. Immunogold studies have also been used to follow the movements of antigens from the cell surface into the early endosome system (Allen *et al.*, 1992)
5. Finally, immunogold can also be used to label two antigens on the same section provided the Mabs are of two different immunoglobulin species such as IgG and IgM. We were able to confirm that the two microfilament arrays in the cortex of *Paramecium* are not two views of the same system by using this double-labeling technique (Allen *et al.*, 1998).

G. Membrane Voltage-Triggered Fixation

Last, but not least, we describe the use of a specific electrophysiological event occurring within the living cell to trigger the nearly instantaneous fixation of the cell at the moment this event occurs. This technique can be used when the membrane voltage of a specific cellular event is known to fluctuate and is measurable in the live cell using microelectrodes. For *Paramecium* we used this technique to trigger cellular fixation at the moment that the CV entered systole and the cell begins to expel the CV contents. At this moment the voltage across the CV membrane relative to the cytosol drops precipitously from +80 mV to +10 mV. Fixation of the cell at this precise time (fixation was triggered at +50 mV) allowed us to see what was happening to the membranes of the CV as the CV is emptied.

a. Methods A single *P. multimicronucleatum* cell equilibrated in a saline solution of 1.0 mM KCl, 1.0 mM $CaCl_2$, and 1.0 mM MOPS-KOH buffer, pH 7.0, was pipetted in a small droplet of saline solution into a somewhat larger droplet of silicone oil on a glass slide. Excess saline was removed by a suction micropipette until the oil droplet pressed against the cell and immobilized the cell (Fig. 11A) (Tominaga *et al.*, 1999).

Fig. 10 Immunogold labeling of trichocysts. (A) Light micrograph of *Paramecium* in fluorescence illumination treated with a trichocyst core-specific Mab and stained with FITC. (B and C) Longitudinal and cross-section of resting trichocysts. The cortical region of the shaft is free of label. (D–F) Steps of trichocyst morphogenesis. (D) Dictyosome (d) with numerous vesicles of different sizes, some of them resembling trichocyst vesicles (t). (E) Vesicles similar to those marked with t in Fig. (D) The immunogold label indicates that these vesicles represent developmental stages of trichocysts. (F) Large vacuole with paracrystalline core (*) showing a periodically striated texture (inset). Small vesicles lie close to the vacuole (arrows). Reproduced from *J. Ultrastruct. Mol. Struct. Res.* **99**, 213–225, 1988. Scale bar in A = 10 μm, in B, D, E, F = 1 μm, in C = 0.5 μm, inset = 0.1 μm.

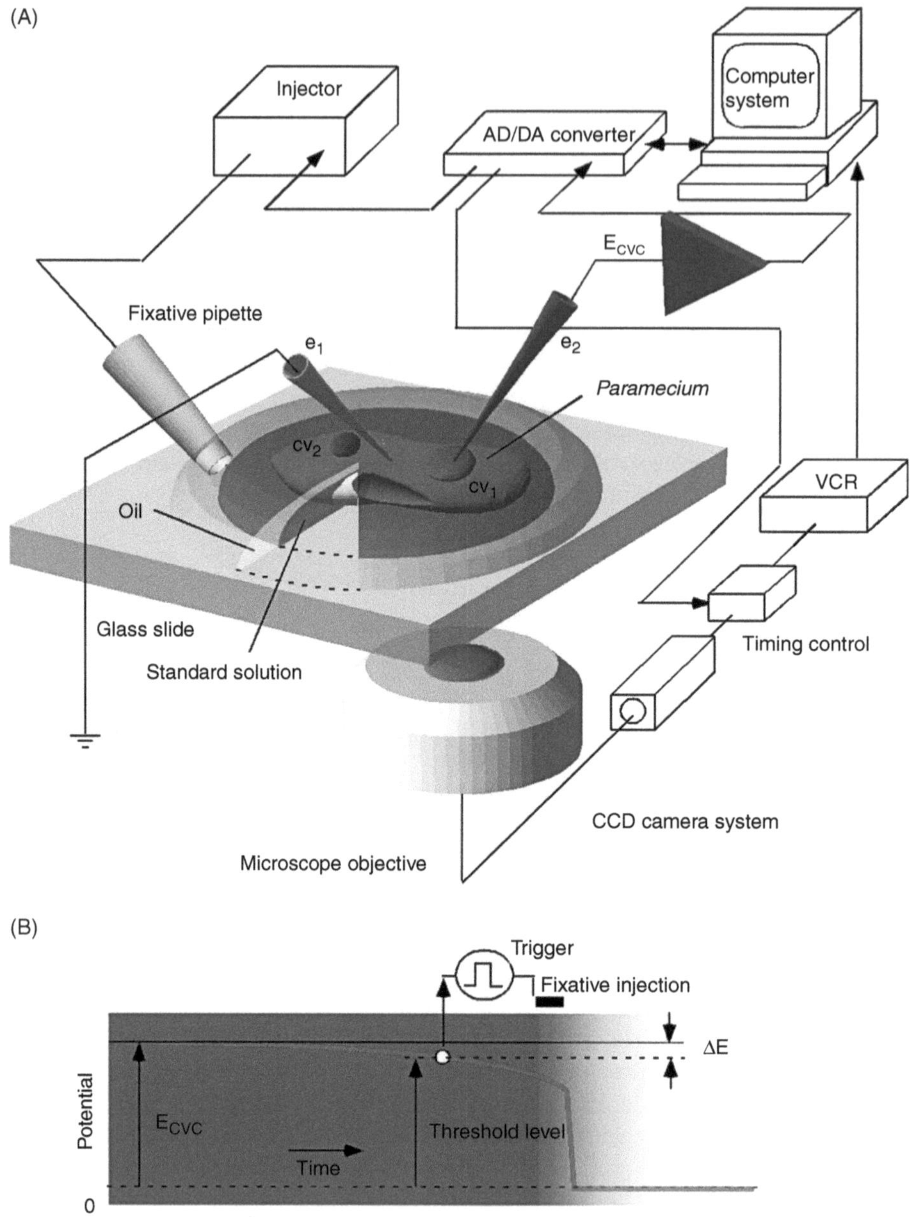

Fig. 11 Electrophysiologically controlled fixation of *Paramecium* is accomplished with the following setup. (A) A living *Paramecium* cell in standard saline solution is injected into the bottom of a small drop of silicone oil on a glass slide on the stage of an inverted microscope. A grounded microcapillary electrode (e1) is inserted into the cytosol of the cell. A fine-tipped microcapillary electrode (e2) is inserted through the cytosol into one of the CVs (CV_1) in the cell to record the potential difference between the CV membrane (E_{CVC}) and the cytosol. See text for other instrumentation. (B) E_{CVC} gradually falls as the CV enters the rounding phase and separates from its ampulli. At a preset threshold value for E_{CVC} the fixative injector is activated and squirts fixative through the oil onto the cell. Within 90 ms the fixative hits the cell and E_{CVC} drops to 0. Reproduced from *J. Cell Sci.* **112,** 3733–3745, 1999.

A microcapillary electrode filled with 3 M KCL (with an electrical resistance of ~50 MΩ) was inserted into the cytosol of the living cell and grounded. A fine-tipped (~1 μm diameter) microcapillary electrode, also filled with 3 M KCL (~100 MΩ), was inserted into one CV of the cell. This electrode also served to inject 0.3 nA square current pulses into the cell that were used to monitor the input resistance of the cell cavities so one could tell when the electrode had successfully entered the CV (Tominaga *et al.*, 1998). When this happened, the input resistance increased ~5 times over that of the cytosol resistance.

A microcapillary of ~10 μm inner diameter was filled with 2% glutaraldehyde in 50 mM cacodylate buffer and inserted into the silicone oil at a distance of ~50 μm from the saline droplet that contained the cell, being careful not to allow the fixative to come into contact with the saline solution. This pipette was linked to the pressure outlet of an electronic microinjector (IM-200, Narishiga USA Inc., Greenvale, NY), which was used to squirt the fixative against the cell when the injector received a signal from the A/D–D/A converter (ITC 16, Instrutech Corp., Great Neck, NY). The temperature of the stage area of the inverted microscope was kept at 17°C.

The potential difference across the CV membrane (E_{CVC}) was constantly monitored with a computer (Power Macintosh 7600/136, Apple Computer Inc., Cupertino, CA) through the A/D–D/A converter and was found to remain, depending on the CV, between +60 and +80 mV during the CVC-filling phase (Tominaga *et al.*, 1998). During rounding of the CV prior to emptying (i.e., systole), the CV breaks away from the collecting canals and the E_{CVC} begins a rapid decline as the CV is no longer connected to the proton pumps that generate the voltage. At a predetermined E_{CVC} value (+50 mV), the injector is activated by the computer and, with a 90-ms delay, the injector squirts the fixative against the cell and the cell is very rapidly fixed (Tominaga *et al.*, 1999). To keep the fixative from leaking out of the injector and unintentionally fixing the cell, the hydrostatic pressure inside the injector is kept slightly lower than the atmospheric pressure. About 0.2 μl of fixative was squirted against the cell, which was about 10 times the volume of the saline solution surrounding the cell. Software for feeding the electrical signals into the computer and for generating electric signals were based on IgorPro (WaveMetrics, Inc., Oswego, OR) and pulseControl XOP (Herrington *et al.*, 1995).

b. Research Applications In order to view the structure of the cell at a precise moment in time, it is sometimes important to be able to fix the cell at the precise moment that the event occurs. This requires working with a single cell and in making use of a parameter of the cell's physiology that is known to vary over time, and which can be constantly monitored.

In the case of single protistan cells, events related to the CV were poorly understood until recently. For years it was a matter of debate as to what happened to the CV membrane during systole when the vacuole empties (Patterson, 1980). At the light microscope level, the CV could no longer be visualized in the cell so it was not possible to determine whether the CV flattened into an empty container or whether it flowed into and became part of the plasma membrane, with which it fuses in order to expel its content to the cell's exterior. This question was not completely resolved until methods were developed that allowed us to trigger cell fixation at the precise time at

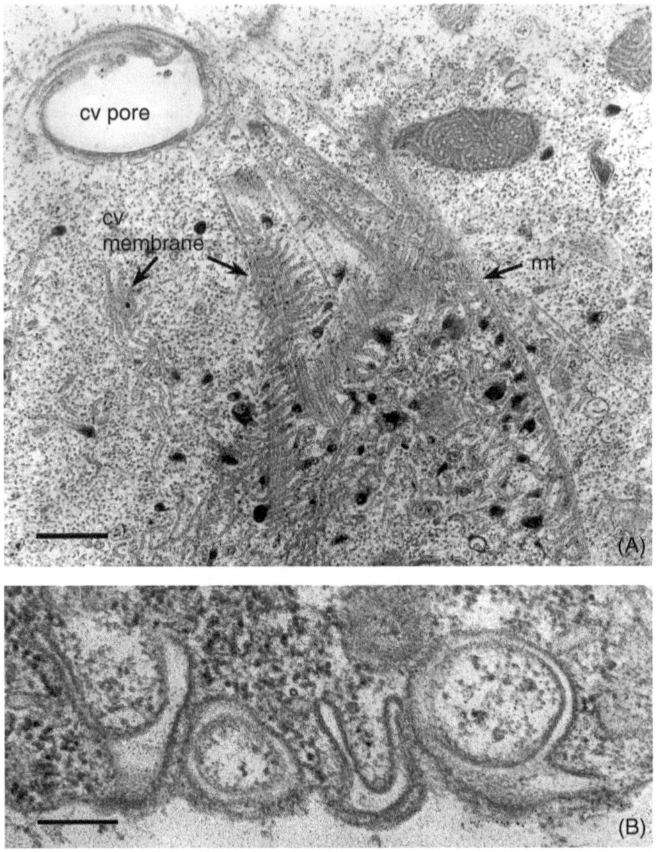

Fig. 12 (A) Using the electrophysiological trigger technique demonstrates that the CV membrane instantly collapses into a network of 40-nm tubules that remain in close contact with the microtubular ribbons (mt) that radiate out from the CV pore. Thus the planar CV membrane returns to a tubular form when the CV's internal osmotic pressure is released, and this shows that high curvature is the relaxed configuration of the CV membrane. Scale bar = 0.5 μm. (B) With love from *Paramecium*! A cipher of acidosomes docked at a phagosome. Scale bar = 0.5 μm.

which the CV was in the process of expelling its content. This was done not by relying on visual monitoring but on electrophysiologically monitoring the voltage across the CVC membrane system relative to the cell's cytosol (Tominaga *et al.*, 1999). Electrophysiological studies had determined that the voltage across the membrane of the CVC during the CV filling remained fairly constant at about +60 to +80 mV. However, this value rapidly dropped to +10 mV when the CV rounds up ("contracts") as it separates from the bulk of the CVC membranes and just before the CV fuses with the plasma membrane. These events are completed within 0.5–1 s. In our experiments, fixation was set to be triggered automatically when the voltage dropped from +80 to +50 mV. TEM then showed that the CV collapsed into a network of 40 nm tubules (Fig. 12A) that remained connected to the CVC microtubular ribbons. These membrane tubules are too small to be resolved by light microscopy.

Though this technique requires a detailed knowledge of the cell membrane's potential voltages as they vary during cellular activity, it offers a level of precision to follow changes in cellular structure that is not possible with any other currently known technique. This technique also showed that fixation was very rapid under these conditions, and cell activity had stopped within 90–100 ms of activating the electronic trigger (Fig. 11B). Fixation was so rapid that the CV expulsion was stopped before the CV had time to completely empty its content, which is usually completed in about 1 s.

III. Final Remarks

In this chapter, which reflects our lifelong scientific engagement with *Paramecium*, we summarized our different attempts and approaches to elicit some of the structural and functional mysteries of this ciliate by applying diverse strategies of electron microscopy. We believe that we finally found one shy indication that our beloved ciliated friend wanted to communicate with us its own sensations via an ultrastructural cipher. We hope that we did not overinterpret this micrograph (Fig. 12B).

References

Allen, R. D. (1971). Fine structure of membranous and microfibrillar systems in the cortex of *Paramecium caudatum*. *J. Cell Biol.* **49**, 1–20.

Allen, R. D. (1974). Food vacuole membrane growth with microtubule-associated membrane transport in *Paramecium*. *J. Cell Biol.* **63**, 904–922.

Allen, R. D. (1978). Particle arrays in the surface membrane of *Paramecium*: Junctional and possible sensory sites. *J. Ultrastruct. Res.* **63**, 64–78.

Allen, R. D. (1995). Membrane tubulation and proton pumps. *Protoplasma* **189**, 1–8.

Allen, R. D., Aihara, M. S., and Fok, A. K. (1998). The striated bands of *Paramecium* are immunologically distinct from the centrin-specific infraciliary lattice and cytostomal cord. *J. Eur. Microbiol.* **45**, 202–209.

Allen, R. D., Bala, N. P., Ali, R. F., Nishida, D. M., Aihara, M. S., Ishida, M., and Fok, A. K. (1995). Rapid bulk replacement of acceptor membrane by donor membrane during phagosome to phagoacidosome transformation in *Paramecium*. *J. Cell. Sci.* **108**, 1263–1274.

Allen, R. D., and Fok, A. F. (1980). Membrane recycling and endocytosis in *Paramecium* confirmed by horseradish peroxidase pulse-chase studies. *J. Cell. Sci.* **45**, 131–145.

Allen, R. D., and Fok, A. K. (1984). Membrane behaviour of exocytic vesicles. III. Flow of horseradish peroxidase labeled trichocyst membrane remnants in *Paramecium*. *Eur. J. Cell Biol.* **35**, 27–34.

Allen, R. D., and Fok, A. K. (2000). Membrane trafficking and processing in *Paramecium*. *Int. Rev. Cytol.* **198**, 277–318.

Allen, R. D., and Hausmann, K. (1976). Membrane behaviour of exocytic vesicles. I. The ultrastructure of *Paramecium* trichocysts in freeze-fracture preparations. *J. Ultrastruct. Res.* **54**, 224–234.

Allen, R. D., Ma, L., and Fok, A. K. (1993). Acidosomes: Recipients of multiple sources of membrane and cargo during development and maturation. *J. Cell. Sci.* **106**, 411–422.

Allen, R. D., Schroeder, C. C., and Fok, A. K. (1989). An investigation of mitochondrial inner membranes by rapid-freeze and deep-etch techniques. *J. Cell Biol.* **108**, 2233–2240.

Allen, R. D., Schroeder, C. C., and Fok, A. K. (1992). Endosomal system of *Paramecium*: Coated pits to early endosomes. *J. Cell. Sci.* **101**, 449–461.

Allen, R. D., and Staehelin, L. A. (1981). Digestive system membranes: Freeze-fracture evidence for differentiation and flow in *Paramecium*. *J. Cell Biol.* **89**, 9–20.

Allen, R. D., Ueno, M. S., Pollard, L. W., and Fok, A. K. (1990). Monoclonal antibody study of the decorated spongiome of contractile vacuole complexes of *Paramecium*. *J. Cell. Sci.* **96**, 469–475.

Allen, S. L., and Nerad, T. A. (1978). Method for the simultaneous establishment of many axenic cultures of *Paramecium*. *J. Protozool.* **25**, 134–139.

Bardele, C. F. (1981). Functional and phylogenetic aspects of the ciliary membrane: A comparative freeze-fracture study. BioSystems **14**, 403–421.

Barlow, D. I., and Sleigh, M. A. (1979). Freeze substitution for preservation of ciliated surfaces for scanning electron microscopy. *J. Microsc.* **115**, 81–95.

Capaldi, R. A. (1982). Arrangement of proteins in the mitochondrial inner membrane. *Biochim. Biophys. Acta* **694**, 291–306.

Cummings, D. J., and Tait, A. (1975). The isolation of nuclei from *Paramecium aurelia*. *Methods Cell Biol.* **9**, 281–309.

Dobell, C. (1932). "Antony van Leeuwenhoek and his Little Animals." Swets & Zeitlinger, Amsterdam (reprinted in 1960 by Dover, New York).

Dykstra, M. J. (1992). "Biological Electron Microscopy—Theory, Techniques, and Troubleshooting" Plenum Press, New York and London.

Essner, E. (1973). Phosphatases. *In* "Electron Microscopy of Enzymes: Principles and Methods" (M. A. Hayat, ed.) **Vol. 1**, pp. 44–76. Van Nostrand Reinhold Co., New York.

Foissner, W. (1991). Basic light and scanning electron microscopic methods for taxonomic studies of ciliated protozoa. *Eur. J. Protistol.* **27**, 313–330.

Foissner, W. (1992). Preparation of samples for scanning electron microscopy. *In* "Protocols in Protozoology" (J. J. Lee and A. T. Soldo, eds.) pp. C-20.1–C-20.5. Society of Protozoologists, Lawrence.

Fok, A. K., and Allen, R. D. (1975). Cytochemical localization of peroxisomes in *Tetrahymena pyriformis*. *J. Histochem. Cytochem.* **23**, 599–606.

Fok, A. K., and Allen, R. D. (1979). Axenic *Paramecium caudatum*. I. Mass culture and structure. *J. Protozool.* **26**, 463–470.

Fok, A. K., Muraoka, J. H., and Allen, R. D. (1984). Acid phosphatase in the digestive vacuoles and lysosomes of *Paramecium caudatum*: A timed study. J. Protozool. **31**, 216–220.

Fok, A. K., Olegario, R. P., Ueno, M. S., and Allen, R. D. (1988). Characterization of monoclonal antibodies to trichocyst antigens in *Paramecium*. *J. Cell. Sci.* **91**, 191–199.

Fok, A. K., Aihara, M. S., Ishida, M., Nolta, K. V., Steck, T. L., and Allen, R. D. (1995). The pegs on the decorated tubules of the contractile vacuole complex of *Paramecium* are proton pumps. *J. Cell. Sci.* **108**, 3163–3170.

Gomori, G. (1952). "Microscopic Histochemistry. Principles and Practice." University of Chicago Press, Chicago.

Görtz, H.-D. (ed.) (1988). "*Paramecium*." Springer Verlag, Heidelberg.

Graham, R. C., Jr., and Karnovsky, M. J. (1966). The early stages of absorption of injected horseradish peroxidase in the proximal tubules of mouse kidney: Ultrastructural cytochemistry by a new technique. *J. Histochem. Cytochem.* **14**, 291–302.

Grębecki, A. (1964). Calcium substitution in staining the cilia. *Acta Protozool.* **2**, 375–377.

Grell, K. G. (1973). "Protozoology" 3rd edn., Springer Verlag, Heidelberg.

Hausmann, K. (1978). Extrusive organelles in protists. *Int. Rev. Cytol.* **52**, 197–276.

Hausmann, K. (2009). *Paramecium caudatum*. *In* "Protozoological Monographs, Vol. 4: A Course in Protozoology" (R. Röttger, R. Knight and W. Foissner, eds.) pp. 194–201. Shaker Verlag, Aachen.

Hausmann, K., and Allen, R. D. (1976). Membrane behaviour of exocytic vesicles. II. Fate of the trichocysts membranes in *Paramecium* after induced trichocysts discharge. *J. Cell Biol.* **69**, 313–326.

Hausmann, K., and Allen, R. D. (1977). Membranes and microtubules of the excretory apparatus of *Paramecium caudatum*. *Eur. J. Cell Biol.* **15**, 303–320.

Hausmann, K., Fok, A. K., and Allen, R. D. (1988). Immunocytochemical analysis of trichocyst structure and development in *Paramecium*. *J. Ultrastruct. Mol. Struct. Res.* **99**, 213–225.

Hausmann, K., Hülsmann, N., and Radek, R. (2003). "Protistology" 3rd edn. Schweizerbart'sche Verlagsbuchhandlung, Stuttgart.

Hausmann, K., Stockem, W., and Wohlfarth-Bottermann, K.-E. (1972). Cytologische Studien an Trichocysten. I. Die Feinstruktur der gestreckten Spindeltrichocyste von *Paramecium caudatum*. *Cytobiologie* **5**, 208–227.

Hayat, M. A. (2000). "Principles and Techniques of Electron Microscopy—Biological applications" 4th edn. Cambridge University Press, Cambridge.

Herrington, J., Newton, K. R., and Bookman, R. J. (1995). "Pulse Control V4. 6: Igor XOPs for Patch Clamp Data Acquisition and Capacitance Measurements" University of Miami, Miami, FL.

Müller, M., Röhlich, P., Tóth, J., and Törö, I. (1963). Fine structure and enzymic activity of protozoan food vacuoles. *In* "Ciba Foundation Symposium on Lysosomes" (A.V.S. de Reuck and M. P. Cameron, eds.) pp. 201–216. J. & A. Churchill, London.

Müller, M., and Törö, I. (1962). Studies on feeding and digestion in protozoa. III. Acid phosphatase activity in food vacuoles of *Paramecium multimicronucleatum*. *J. Protozool.* **9**, 98–102.

Murray, S. (2008). High-pressure freezing and freeze substitution of *Schizosaccharomyces pombe* and *Saccharomyces cerevisiae* for TEM. *Methods Cell Biol.* **88**, 3–17.

Nelson, N. (1992). The vacuolar H^+-ATPase—one of the most fundamental ion pumps in nature. *J. Exp. Biol.* **172**, 19–27.

Nerad, T. A., and Daggett, P.-M. (1992). Cultivation of *Paramecium*. *In* "Protocols in Protozoology" (J. J. Lee and A. T. Soldo, eds.) pp. A-46.1–A-46.4. Society of Protozoologists, Allen Press, Lawrence.

Nolta, K. V., Padh, H., and Steck, T. L. (1993). An immunocytochemical analysis of the vacuolar proton pump in *Dictyostelium discoideum*. *J. Cell. Sci.* **105**, 849–859.

Párducz, B. (1952). úJ gyorsfestö eljárás a véglénykutatás és oktatás szolgálatában (Eine neue Schnellfärbemethode im Dienste der Protistenforschung und des Unterrichtes) (with German and Russian summary). *Ann. Mus. Nat. Hung. N. S.* **2**, 5–12.

Párducz, B. (1967). Ciliary movement and coordination in ciliates. *Int. Rev. Cytol.* **21**, 91–128.

Patterson, D. J. (1980). Contractile vacuoles and associated structures: Their organization and function. *Biol. Rev.* **55**, 1–46.

Plattner, H., Miller, F., and Bachmann, L. (1973). Membrane specializations in the form of regular membrane-to-membrane attachment sites in *Paramecium*. A correlated freeze-etching and ultrathin-sectioning analysis. *J. Cell. Sci.* **13**, 687–719.

Soldo, A. T., Godoy, G. A., and van Wagtendonk, W. J. (1966). Growth of particle-bearing and particle-free *Paramecium aurelia* in axenic culture. *J. Protozool.* **13**, 492–497.

Stelly, N., Balmefrezol, M., and Adoutte, A. (1975). Diaminobenzidine reactivity of mitochondria and peroxisomes in *tetrahymena* and in wild-type and cytochrome oxidase-deficient *Paramecium*. *J. Histochem. Cytochem.* **23**, 686–696.

Tokuyasu, K.T.. (1984) Immuno-cryoultramicrotomy. *In* "Immunolabeling for Electron Microscopy" (J. M. Polak and I. M. Varndell, Eds.) pp. 71–82. Elsevier, New York.

Tominaga, T., Allen, R. D., and Naitoh, Y. (1998). Electrophysiology of the *in situ* contractile vacuole complex of *Paramecium* reveals its membrane dynamics and electrogenic site during osmoregulatory activity. *J. Exp. Biol.* **201**, 451–460.

Tominaga, T., Naitoh, Y., and Allen, R. D. (1999). A key function of non-planar membranes and their associated microtubular ribbons in contractile vacuole membrane dynamics is revealed by electrophysiologically controlled fixation of *Paramecium*. *J. Cell. Sci.* **112**, 3733–3745.

Vanhecke, D., Graber, W., and Studer, D. (2008). Close-to-native ultrastructural preservation by high-pressure freezing. *Methods Cell Biol.* **88**, 151–164.

van Wagtendonk, W. J. (ed.) (1974). "*Paramecium*—A current survey" Elsevier Science Publisher, Amsterdam.

Vivier, E. (1974). Morphology, taxonomy and general biology of the genus *Paramecium*. *In* "*Paramecium*—A current survey" (W. J. van Wagtendonk, ed.) pp. 1–89. Elsevier Science Publisher, Amsterdam.

Wichterman, R. (1986). "The Biology of *Paramecium*" 2nd edn., Plenum Press, New York.

CHAPTER 8

Ultrastructural Investigation Methods for *Trypanosoma brucei*

Johanna L. Höög[*,†], **Eva Gluenz**[*], **Sue Vaughan**[*,‡], *and* **Keith Gull**[*]

[*]Sir William Dunn School of Pathology, University of Oxford, Oxford OX1 3RE, United Kingdom

[†]The Boulder Laboratory for 3D Electron Microscopy of Cells, Department of MCD Biology, University of Colorado, Boulder, Colorado 80309

[‡]Oxford Brookes University, Headington Hill site, Headington, Oxford OX3 0BP, United Kingdom

978-0-12-381007-6
DOI: 10.1016/S0091-679X(10)96008-1

Abstract

Trypanosoma brucei is a unicellular parasite causing African sleeping sickness in cattle and humans. Due to the ease with which these cells can be cultured and genetically manipulated, it has emerged as a model organism for the kinetoplastids. In this chapter we describe the preparation of *T. brucei* for transmission electron microscopy. A thorough explanation of conventional sample preparation through chemical fixation of whole cells and detergent extracted cytoskeletons followed by dehydration and Epon embedding is given. We also introduce a novel high-pressure freezing protocol, which followed by rapid freeze substitution and HM20 embedding generates *T. brucei* samples displaying good cell morphology, which are suitable for immunocytochemistry.

I. Introduction

Trypanosoma brucei is a eukaryotic single-celled, flagellated protozoan parasite, which is the causative agent of African sleeping sickness. It is spread through the bite of the tsetse fly (*Glossina sp.*), and has a complex lifecycle. Each of the different life cycle stages of *T. brucei* have different cell shapes and internal organization (Gull, 1999; Sharma *et al.*, 2009; Vickerman, 1985). Laboratory studies tend to focus on two easily cultured forms: the procyclic form (PCF), normally found in the fly midgut, and the form found in the bloodstream of humans and cattle (BSF).

T. brucei has emerged as the model organism of the Kinetoplastida, a protozoan order named after a DNA-containing structure found within their single mitochondrion. Other kinetoplastids include *Leishmania spp.* (causing cutaneous and/or visceral leishmaniasis) and *T. cruzi* (cause of Chagas disease). All three pose major public health problems in their respective endemic areas. Kinetoplastids have a large evolutionary distance to yeast, worms, and other common model systems, making the study of their cellular mechanisms also a study of conservation and divergence in biological systems. Studies of trypanosomes have brought insight into many areas of biology including mitochondrial DNA, RNA editing, GPI anchors, antigenic variation and mono-allelic exclusion, flagellum and cytoskeleton, glycolytic metabolism and gene expression systems.

African sleeping sickness is lethal if untreated in humans and the drugs available have severe side effects, and drug resistance is emerging. The need for new, cheap drugs that can be easily administered is large but only one such drug has been brought into use since the 1950s (Bacchi, 2009; Delespaux and de Koning, 2007). Therefore, it is increasingly important to better understand this parasite's cell biology.

The subspecies *T. brucei brucei* is not human infective (Pays *et al.*, 2006) and the procyclic and bloodstream forms are easily kept cultured. Molecular genetic approaches are very tractable since homologous recombination is a dominant phenomenon. Gene knockouts, inducible RNAi for gene silencing, and inducible expression of tagged (GFP/epitope) or mutant proteins are well established (Beverley, 2003; LaCount *et al.*, 2000; Shi *et al.*, 2000). The *T. brucei* genome sequencing has been completed, as well as the genomes of the related *L. major* and *T. cruzi* (Berriman *et al.*, 2005; El-Sayed *et al.*, 2005; Ivens *et al.*, 2005). It was found that these three species have similar genomic architecture and their genomes are now gathered in a common database named tritrypDB (Aslett *et al.*, 2009).

For electron microscopic (EM) purposes, these cells are ideal, because of their small size (~4 μm wide and ~20 μm long). Their internal architecture is defined by a subpellicular microtubule array that maintains the cell shape. The other organelles are often in single copies and occupy distinct subcellular locations (Gull, 2003). This invariant architecture and reproducibility in position and division facilitates the description of mutant phenotypes.

Excellent EM methods have been developed including those for chemically fixed samples and, more recently, results were obtained using high pressure freezing (Gadelha *et al.*, 2009; Grunfelder *et al.*, 2002; Lacomble *et al.*, 2009; Overath and Engstler, 2004; Sherwin and Gull, 1989; Weise *et al.*, 2000, 2003). In this chapter we detail some existent methods and their uses and present a new high-pressure freezing protocol that has yielded very good cell morphology, and is useable for immunolocalization of proteins.

II. Rationale

This chapter seeks to provide a toolbox for performing electron microscopy of *T. brucei*. The established methods of chemical fixation, positive staining of cytoskeletons, and detection of nucleic acids using EDTA is reiterated, and a new protocol using high-pressure freezing is introduced. The improvements made when using high-pressure freezing, followed by freeze substitution (FS) and plastic embedding for studying *T. brucei* have already provided novel views of these parasites. The largest structural preservation improvements of high-pressure freezing are to be found when studying membranous cellular compartments such as the Golgi.

III. Methods

A. Cell Culture

The procyclic form of *T. brucei* is grown in SDM-79 (Brun, 1979) medium, supplemented with 10% (v/v) heat-inactivated fetal calf serum (HIFCS) at 28°C in closed top tissue culture flasks. The procyclic culture is kept between 10^5 and 10^7 cells/ml, and best harvested at ~8×10^6 cells/ml. Bloodstream form *T. brucei* is grown in HMI-9 medium supplemented with 15% (v/v) HIFCS at 37°C in vented top culture flasks in a 5% CO_2 environment (Hirumi and Hirumi, 1989), and best harvested at ~8×10^5 cells/ml. Cell line stocks are kept in liquid nitrogen in a medium containing 10% glycerol.

B. Chemical Fixation for Morphology Studies

Chemical fixations provide a means for rapid crosslinking and immobilization of cellular structures. The most widely used fixative for ultra structural studies is glutaraldehyde. Primary fixation with an aldehyde is typically followed by postfixation in osmium tetroxide, en bloc staining with uranyl acetate, dehydration in ethanol or acetone, and embedding in Epon resin. The following is a standard protocol for the processing of cultured trypanosomes. This protocol has been optimized for *T. brucei* procyclic forms, but will give good results for a number of trypanosomatid species and life cycle stages. Where the processing of *T. brucei* bloodstream forms requires modifications, this is specifically indicated. Procyclic *T. brucei* prepared with this method are shown in Fig. 1.

1. Primary Fixation of Cells in Suspension Culture

To minimize manipulation of cells prior to fixation, the initial fixation occurs in the culture flask by adding one volume of 25% EM-grade glutaraldehyde to nine volumes of cell culture (2.5% final glutaraldehyde concentration). Cells are fixed for 5–10 min at room temperature. The fixed cells are then harvested by gentle centrifugation (800 g for 5–10 min) and the supernatant decanted. The cells are resuspended in 1 ml buffered fixative (2.5% glutaraldehyde, 2% formaldehyde in 100 mM phosphate buffer pH 7–7.4) and centrifuged again to form a pellet and left to continue fixation for at least 2 h at room temperature (the samples can be left for longer fixation at 4°C). The buffer should maintain the physiological pH and osmolarity of the sample and the choice of buffer will therefore vary depending on the type of cell used. To maintain the osmolarity of the fixative, sucrose, glucose or NaCl can be added to the buffer. For *T. brucei* bloodstream forms, the addition of 50 mM sucrose to the buffered fixative enhances the ultrastructural preservation.

To avoid losing cells during the subsequent steps they are processed in a pellet, rather than in suspension. It is most convenient to continue sample processing in a 1.5 ml Eppendorf tube, using ~1 ml volumes for each of the following steps. After each centrifugation step the pellet can be dislodged from the wall of the tube by gentle tapping. The fixative is removed and the sample washed extensively in buffer (200 mM phosphate buffer pH 7–7.4). It is important to remove any remaining free aldehyde, which otherwise react with osmium and cause formation of small electron dense particles. This requires a minimum of three buffer changes over a period of 2–3 h at room temperature (samples may be left at 4°C overnight).

2. Preparation of Trypanosomes Isolated from the Tsetse Fly

Dissect the trypanosome-infected tsetse flies (Sharma *et al.*, 2008, 2009), isolate the gut, proventriculus, or salivary glands as required and transfer the organs directly into buffered fixative (Section III.B.1). Cut the organs into small pieces (<1 mm) to facilitate penetration of the fixative and then process in the same way as the cell pellets described in Section III.B.1.

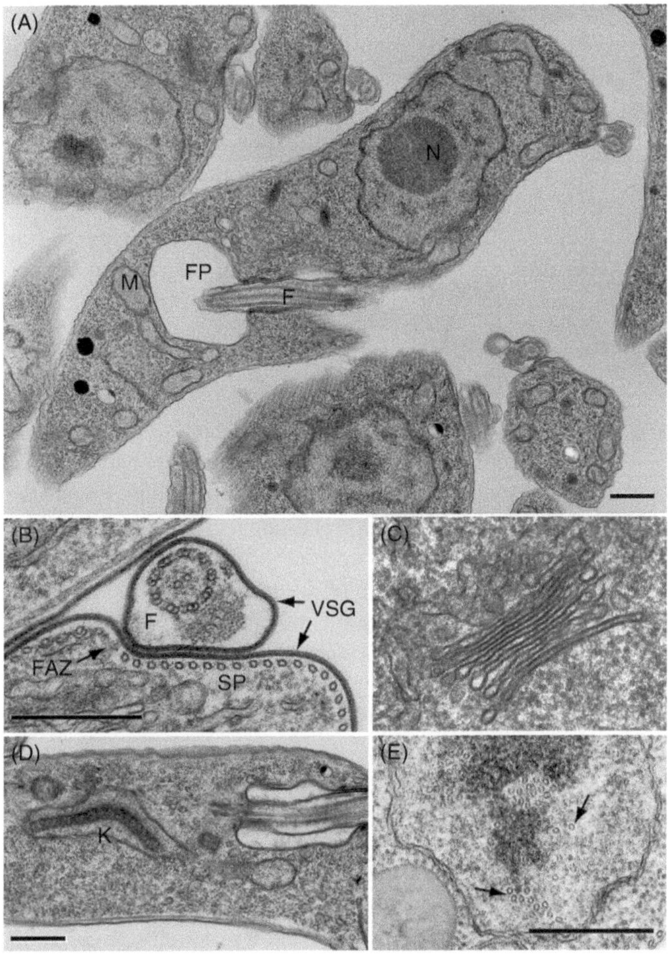

Fig. 1 Chemical fixation of *T. brucei* (A) Low magnification view of procyclic form (PCF) showing nucleus (N), flagellar pocket (FP), flagellum (F), and mitochondrion (M). (B) Bloodstream form (BSF), cross-section of flagellum, subpellicular microtubules (SP), flagellum attachment zone (FAZ), VSG surface coat (arrows). (C) BSF Golgi and (D) BSF dividing kinetoplast (K), flagellar pocket. (E) PCF nucleus showing spindle microtubules (arrows) in cross-section fixed in 2.5% glutaraldehyde and embedded in Epon. Scale bars: 500 nm.

3. Postfixation, Staining, and Dehydration

Glutaraldehyde-fixed samples are postfixed in 1% osmium (in 100 mM phosphate buffer or distilled water) for 1–2 h in the dark. Osmium tetroxide acts as a fixative as well as a stain. It reacts with lipids and oxidizes unsaturated bonds of fatty acids, adds electron density, which gives contrast, and it permeabilizes the cells instantly. It is important not to leave the samples in osmium too long because this will result in the extraction of cellular material. Osmium fixation will cause the pellet to go dark brown. After osmium fixation, the samples are washed multiple times in distilled water. It is important to remove excess phosphate ions to prevent the precipitation of uranyl acetate in the subsequent step. To provide en-bloc stain, the cell pellet is immersed in 2% magnesium uranyl acetate in water for 2 h at room temperature or 4°C overnight. Uranyl acetate compounds are photo reactive and will precipitate when exposed to light, so samples need to be kept in the dark. Samples are then washed with water. Dehydration enables penetration of resin, and is typically performed in a series of washes with increasing concentration of dehydrated ethanol or acetone: 30, 50, 70, 90% and three times 100% for 10 min each.

4. Resin Infiltration and Embedding

In the final step, the solvent is gradually replaced by resin monomers, which are then polymerized to form a solid resin block in which the sample is embedded. A typical infiltration consists of submersion of the cell pellet two times into 100% propylene oxide for 15 min each, then several hours in 1:3, 1:1, and 1:0 resin : propylene oxide baths. The choice of resin will depend on individual preference and the nature of the experiment. For routine preparation of trypanosomes, Epon or Spurr's resin gives good results. Propylene oxide is highly miscible with the plastic embedding medium, and thus facilitates infiltration. However, it is very volatile and dehydration of the sample should be prevented by keeping tube lids closed and the sample immersed. The pellets are then transferred to silicone embedding moulds containing fresh resin. These are placed at 60°C for 24–48 h during which the epoxy monomers that permeate the sample polymerize until a solid block is formed that can be sectioned.

C. High-Pressure Freezing for Morphology Studies

High-pressure freezing is a physical means to cryoimmobilize samples very rapidly by exposure to liquid nitrogen under a pressure of over 2000 bar. The pressure is applied to prevent the formation of cubic and hexagonal ice (Richter, 1994) (see also chapters by O'Toole, Giddings, Müller-Reichert, this volume). Ice crystals are known to damage the cell's fine structure. Using high-pressure freezing, "large" samples with a thickness of up to 200 μm can be routinely vitrified, i.e., frozen without such ice-crystals).

A sample transfer follows the cryoimmobilization from the high-pressure freezer into a freeze substitution (FS) machine, where the water is extracted using a solvent, such as ethanol or acetone and a simultaneous slow fixation occurs at very low temperatures through addition of fixatives to the FS "cocktail". The FS is then followed by resin infiltration and the sample preparation is completed with resin embedding/polymerization.

The benefits of high-pressure freezing are usually most apparent in membrane morphology, which appear wrinkled after dehydration but are smooth when using cryoimmobilization before fixation and dehydration at low temperatures. Although some aspects of cell morphology are greatly improved in high-pressure frozen

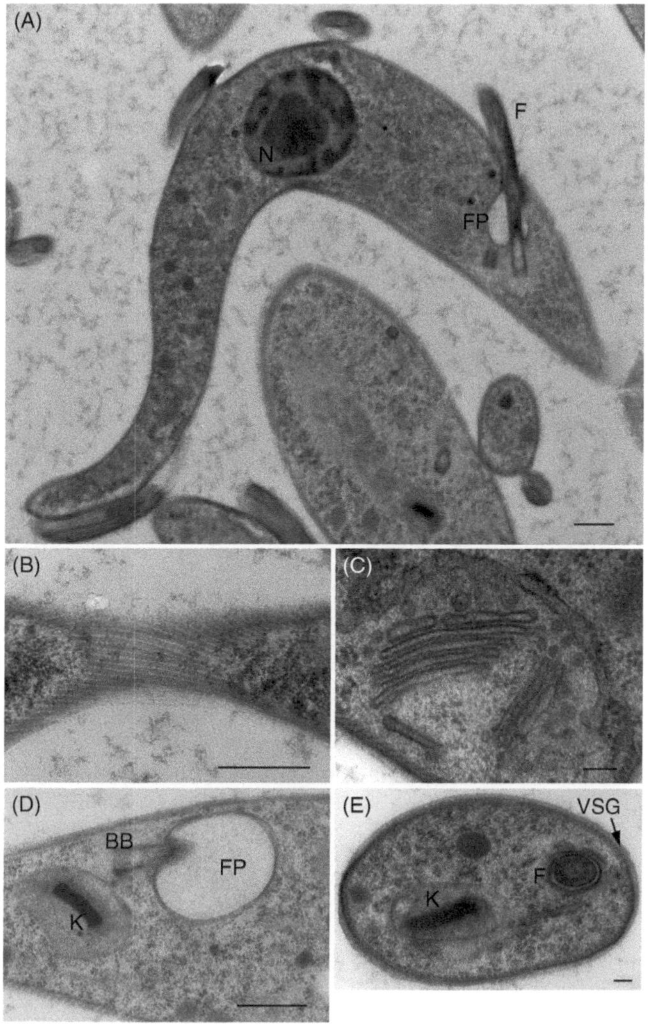

Fig. 2 High-pressure frozen *T. brucei* embedded in HM20 resin. (A) Low magnification of a PCF, note the round nucleus (N), well-defined chromatin, and smooth membranes around the cell and in the flagellar pocket (FP), where the flagellum (F) originates. (B) PCF subpellicular microtubule array. (C) Golgi apparatus in a BSF. (D) The FP is plump and the kinetoplast (K) DNA is an electron dense disc without clear filamentous circles as seen in chemical fixation. BB, basal body. (E) BSF in cross-section showing a VSG-coat, a flagellum in cross section, and a kinetoplast. Scale bars: 500 nm, in (C) 100 nm.

T. brucei (Fig. 2), these cells are not straight forward to freeze (Fig. 3). This is probably due to a cytoplasm relatively "empty" of solutes that would act as internal cryoprotectants. Freezing is variable inside the pellet, so taking time to search for a well-frozen area is recommended.

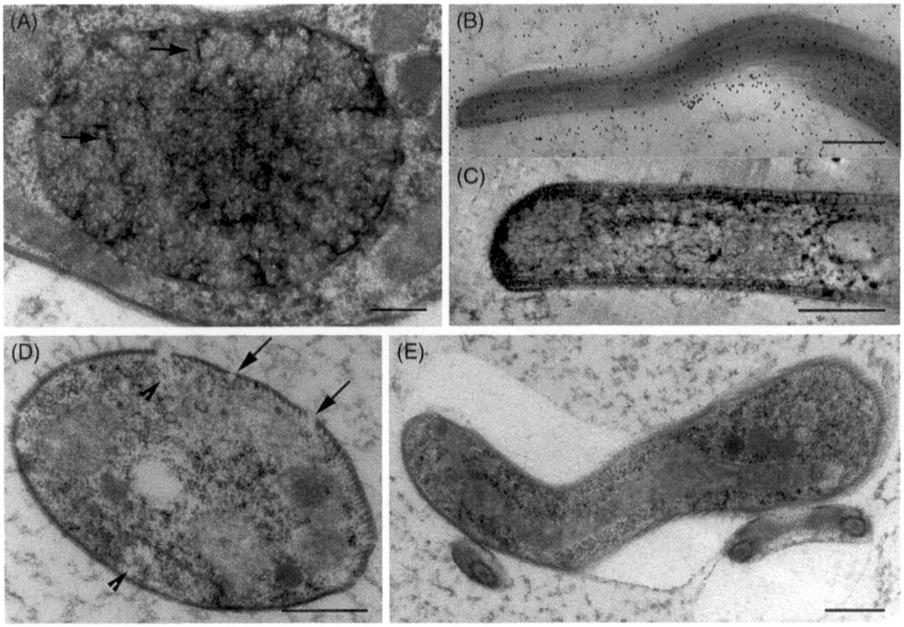

Fig. 3 Sample preparation problems (A) A severely ice-damaged *T. brucei* nucleus. Note the chicken-wire appearance of the nucleoplasm (arrows). (B) A zero-tilt low-magnification image of an anterior end in a semi-thick section shows no traces of ice damage. The section displays colloidal gold particles to be used for image alignment when calculating a tomogram. (C) The tomographic reconstruction of the same anterior end shows extensive ice damage by smaller ice crystals. (D) A cell with cracks in the membrane and underlying sub-pellicular microtubule cytoskeleton (arrows). These fractures could have been caused by centrifugation prior to freezing or by ice-crystal formation since ice damage is also apparent (arrowheads). (E) Poor infiltration of resin is often seen around narrow structures, such as the flagellum. Here the whole cell has "fallen out" of the plastic. Scale bar: 200 nm in (A) and (C), 500 nm (B, D) and (E).

1. Procyclic Form (PCF)

To increase the amount of well-frozen cells, the amount of solutes in the medium is increased by using 20 or 50% heat-inactivated FCS instead of the regular 10%. The increased FCS medium does not inhibit growth. This is likely to work as an external cryoprotectant (McDonald *et al.*, 2007). Other cryoprotectants tested but with less successful results were 20% bovine serum albumin (McDonald *et al.*, 2007) and 2% gum arabic.

For high-pressure freezing, cells need to be concentrated in a pellet. It is convenient to use a ~50 ml culture of procyclic forms at a density of $0.8–1.0 \times 10^7$ cells/ml (seed cultures at 1.0×10^6 cells/ml 24 h prior freezing) and prepare pellets by gently spinning down 10–15 ml of culture at a time (2–3 min at 600 g; $\sim 1.0 \times 10^8$ cells/pellet).

Prepare the high-pressure freezing and FS machines and have everything ready for freezing before harvesting the cells.

2. Bloodstream Form (BSF)

Since the bloodstream form must be kept at low density to ensure high viability the culture needs to be expanded to > 200 ml to get enough cells for freezing. Use healthy looking cells just below 1×10^6 cells/ml. Spin down a full 50 ml falcon tube for 2–3 min at 600 g. Proceed immediately to freezing. Since the bloodstream form *T. brucei* is more sensitive to temperature changes than the procyclic form, they may not benefit from centrifugation before freezing and an alternative method harvesting cells from colonies growing on agar plates may prove superior (Carruthers and Cross, 1992; Gadelha, 2008). However, we have found good cell preservation using centrifugation as a means to concentrate the cells.

3. High-Pressure Freezing

Without disturbing the pellet, gently remove the supernatant and pipette up a few μl of sample from inside the pellet. White lumps of cell pellet should be visible in the pipette tip (but this might not be possible for the bloodstream form). Fill the smallest freezing carriers available with cells; smaller volumes increase the efficiency of the subsequent freezing. For a thorough description of high-pressure freezers and carriers, see McDonald *et al.* (2007).

Once the cells have been frozen, be extremely careful to precool all the tools in liquid nitrogen before using them anywhere near the sample. When moving the sample, if there is any doubt whether it has been raised above the nitrogen surface, discard and do another. This might appear wasteful but it saves a lot of sectioning and microscopy time. Freeze as fast as possible after centrifugation to ensure cells remain healthy, no longer than 5 min after centrifugation. Then spin down a new pellet. Try to make at least 2–3 samples of each condition. Once all samples have been frozen, transfer them under liquid nitrogen into the precooled (–90°C) FS container.

4. Freeze Substitution and Embedding into HM20 Resin

This protocol is an adaptation of a protocol worked out for mammalian cells (Hawes *et al.*, 2007), but we have seen it work well on *Schizosaccharomyces pombe* as well as *T. brucei* (Fig. 2). It has a much shortened FS and excludes many of the most toxic chemicals, such as glutaraldehyde and osmium tetroxide, in the FS solution. The contrast is good, the HM20 resin is more electron transparent than Epon, which is especially beneficial for electron tomography when thick sections are used. In addition, it can also be used for immunolocalization of proteins.

We find increased electron beam induced shrinkage in x- and y-axis when cells were imaged in HM20 compared to Epon. This might not be a disadvantage, since the shrinkage is more uniform at around 25–30% in all directions (as measured from MT shrinkage), compared to Epon that has only 12% shrinkage in x and y but up to 40–60% in the z-axis (Luther *et al.*, 1988). These considerations are especially important if you prepare the sample for electron tomography. If the increased shrinkage in x and y would pose a problem, use the alternative protocol using Epon (see below).

The FS solution is made up of 2% uranyl acetate (UA; from a 20% stock in dehydrated methanol kept at –20°C) in dehydrated acetone. The solution is precooled and immediately applied to the samples after the transfer. After 1 h of FS at –90°C, the UA is washed out with two changes of precooled dehydrated acetone. Whilst the acetone washes occur, the temperature is slowly increased (15°C/h) to –50°C where HM20 infiltration starts. Infiltration occurs in steps with increasing resin to acetone ratio; start with 3:1 acetone to resin, increase to 2:1, 1:1, 1:3 ratios. Then do at least three 100% HM20 exchanges, with at least one overnight. The whole infiltration protocol should take 2–4 days, but the longer the better. Start UV polymerization at –50°C for 48 h, heat to room temperature (20°C/h), and continue polymerization for a further 48 h.

5. Freeze Substitution and Embedding into Epoxy Resin

We have better experience with using HM20 resin than Epoxy resins, probably due to the low infiltration temperatures. However, cells can also be embedded into Epon which is a standard resin for morphology studies. Since Epon infiltration occurs at higher temperatures, an FS solution containing more fixatives is used to preserve cell morphology.

FS takes place for 50 h at –90°C by adding precooled 0.01% OsO_4, 0.25% uranyl acetate, 0.1% glutaraldehyde in acetone containing 1% H_2O (Muller-Reichert *et al.*, 2003). The temperature is then raised by 10°C/h to –30°C where the samples are left for 6–8 h and then washed twice with precooled dehydrated acetone. The infiltration starts at –30°C with a 1:1 ratio of acetone : Epon for 4 h. Change solution to a 1:2 acetone : Epon mixture and leave for another 3 h. Warm up to room temperature and do several pure Epon changes before starting to polymerize at 60°C for 50 h.

D. Sectioning and On-Section Stain

It is very practical to use serial thin sections to gain 3D understanding of the structures studied.

To gain ultra thin or semi-thick serial sections, trim the block surface into a trapezoid shape (Fig. 4A). The section area can be changed depending on how many sections one has to follow. We have found that it is much easier to shrink the area of each section, by reducing the height of the trapezoid, and fitting many sections onto the same grid, than it is to keep the large area and have a series of sections over many grids. In cases where many serial sections are needed, the block surface can start as a “line”, with a pyramid below (Fig. 4B). With this geometry, a ribbon of ~50 sections can fit onto a single slot grid. Two or even three ribbons may be picked up onto the same grid, so one can follow over a hundred serial sections without changing the grid.

If serial sections do not stick together in a ribbon, a drop of rubber cement diluted 1:20 in xylene can be applied onto the corner of the block and allowed to dry. This will even allow poorly trimmed sections to form nice ribbons.

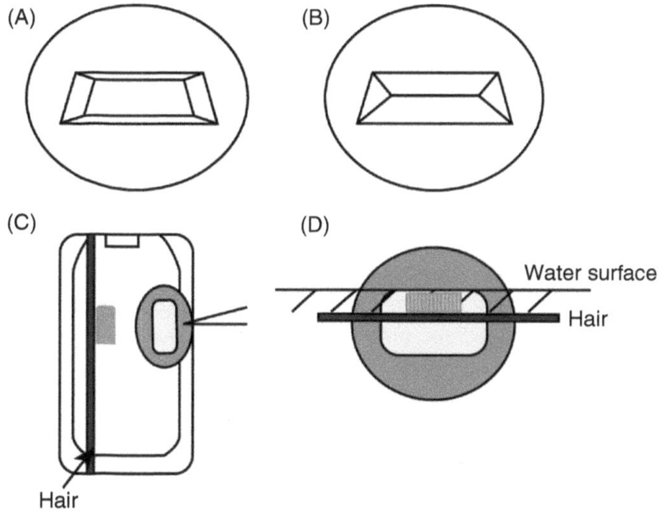

Fig. 4 Preparing serial sections. (A) The trapezoid block surface trimmed to fit few to medium amount of sections on a grid. (B) This "long pyramid" trimming of a block will yield increasing size of sections, starting with extremely thin ones. This enables fitting 30–50 sections in a single ribbon on the slot. (C) To pick up the sections we use a hair that has been spanned across the water surface, next to which the ribbon can be parked. (D) The slot grid is lowered to the top bar of the slot and pushed in close to the sections, angled to catch the sections and pulled out of the water.

Picking up the sections is made easier by "parking" them next to a hair that has been put in place over the knife's boat with a piece of tape on each side of the knife (Fig. 4C–D). The sections are then picked up using the hair to trap them on one side, and the grid on the other.

Finally, the sections are stained for 8 min in 2% aqueous UA (this time may be increased up to 15 min if contrast is bad), followed by 3 min in Reynold's lead citrate (Reynolds, 1963).

E. Immunolocalization and Cytochemistry for the Study of Cellular Compartments or Proteins

There are a large number of cytochemical and immunocytochemical methods to visualize specific cellular structures or proteins within cells.

In *T. brucei* cytochemical techniques have been particularly informative in the study of the functional domain organization of the cell nucleus (Fakan, 2004; Moyne, 1980), and have recently been used to provide novel insights into the ultrastructural organization of the kinetoplast (Gluenz *et al.*, 2007).

For immunolocalization of proteins on sections to work, the antigen needs to survive the fixation and embedding process, and be exposed on the section surface where the antibody has free access. Alternatively, antigens can be detected before embedding (i.e., the so-called preembed labeling), using plastic-embedded detergent-extracted cytoskeletons (Stephan *et al.*, 2007), or in whole-mounted cytoskeletons (Sherwin and Gull, 1989). The antibody ideally needs to be highly specific and have a high affinity for the antigen. Here, we present a method where we used antibodies on the HM20 embedded high-pressure frozen samples with great success.

1. Immunolocalization of Proteins on High-Pressure Frozen Samples

To test the immunogenicity of the high-pressure frozen samples embedded in HM20 we detected *T. brucei* β-tubulin using 3B15.9 and 4A1, two monoclonal antibodies raised against sea-urchin and *Drosophila melanogaster* β-tubulin (Piperno and Fuller, 1985; Scholey *et al.*, 1984) (Fig. 5A–C).

Thin sections (80 nm) on nickel mesh grids are postfixed for 10 min using 1% paraformaldehyde in phosphate-buffered saline (PBS) buffer. The grids are then washed three times in PBS. To avoid unspecific binding the grids are then floated for 1 h on drops of blocking buffer (0.1% fish skin gelatin, 8% bovine serum albumin in PBS). The excess blocking buffer is blotted off with a filter paper, and the grids are floated on drops with the primary antibody (in blocking buffer, dilution dependent on the antibody used) in a wet chamber overnight at 4°C. The grids are then washed three times with PBS for 20 min each before they are floated on drops of blocking buffer containing the secondary antibody (dilution 1:20) conjugated to gold particles for 1 h in room temperature. Wash the grids three times 20 min in PBS again. The labeled sections are then postfixed in 1% aqueous glutaraldehyde for 10 min before the final three washes in water. The sections are then stained in UA and Reynold's lead citrate as in Section III.D.

2. EDTA Regressive Staining—To Visualize DNA Containing Structures

Uranyl acetate, which is used as a general stain, can be specific to nucleic acids when used under certain conditions. Uranyl ions react with phosphate groups, including those in the backbone of nucleic acids, and with amino groups. UA staining therefore contributes to the stabilization of nucleic acids and confers electron density, staining nucleic acids and certain proteins. Bernhard's EDTA regressive staining method differentiates between ribonucleoprotein and deoxyribonucleoprotein particles (Bernhard, 1969). The *T. cruzi* kinetoplast was among the original test materials used by Bernhard to demonstrate the specificity of this method to a wide range of DNA-containing structures (Bernhard, 1969). Using Bernhard's EDTA regressive staining has recently enabled us to define a new domain in the kinetoplast, the inner unilateral filaments of the tripartite attachment complex (TAC) of *T. brucei* and the insect trypanosome *Crithidia* (Gluenz *et al.*, 2007).

Samples are fixed in glutaraldehyde (as described in Section III.B), without osmium postfixation and en-bloc stain using UA. Samples are dehydrated using ethanol and embedded into Epon resin. Thin sections are then stained with 5% aqueous uranyl acetate for 5 min, followed by three washes in filtered distilled water (1 min each). The

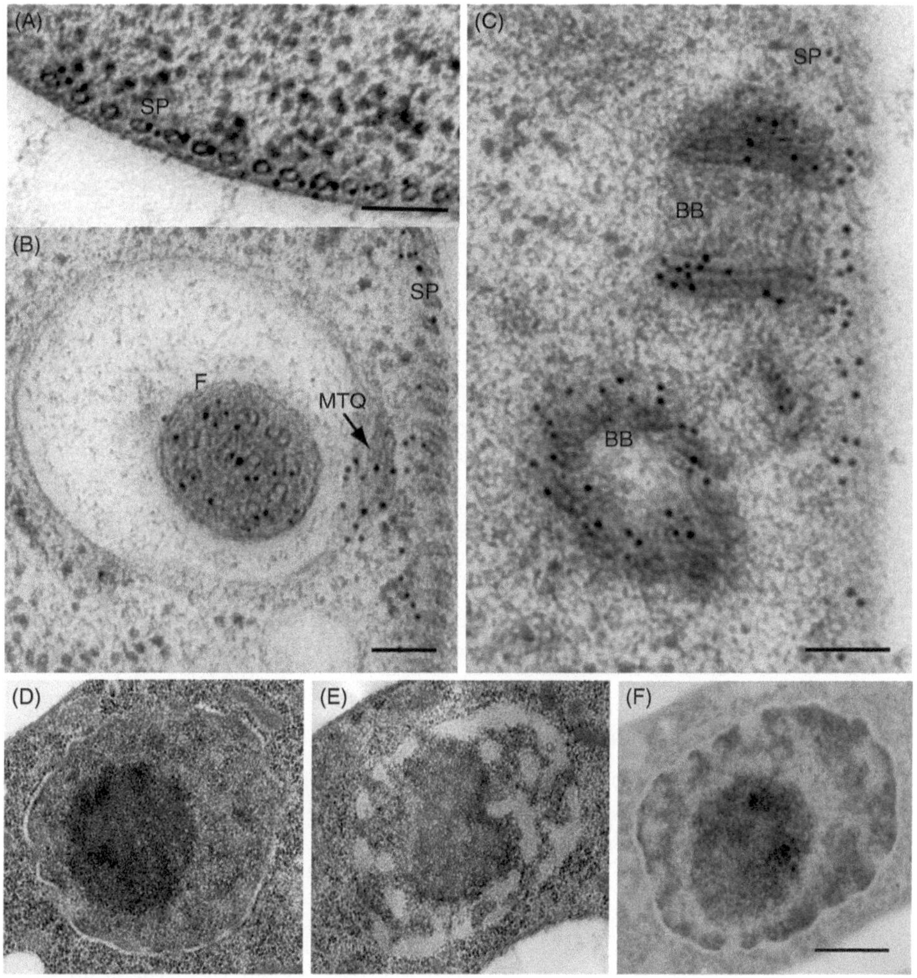

Fig. 5 Immunolocalisation and cytochemistry to localize proteins or nucleotides. Anti-β-tubulin antibody detecting (A) subpellicular microtubule array (SP), (B) flagellum (F), microtubule quartet (MTQ) and SP, (C) basal bodies (BB) and SP in high-pressure frozen cells embedded in HM20, (D) uranyl acetate-stained nucleus in chemically fixed cells, (E) nucleus stained with uranyl acetate and then treated with EDTA (30 min) to bleach areas containing DNA, and (F) nucleus treated with E-PTA to reveal location of basic proteins. All images are of procyclic cells. Scale bars: 100 nm in (A–C), 500 nm in (D–F).

sections are then treated with EDTA (0.2 M EDTA in distilled water; pH 7) for 15 min to 1 h (timing is critical and must be determined experimentally), and grids washed three times in filtered distilled water (1 min each). This results in preferential removal of uranyl stain from DNA, giving the DNA-containing structures a bleached appearance while RNA-containing particles remain strongly stained. The EDTA bleaching effect is reversible; restaining with UA will again contrast the DNA-containing structures. Add more contrast with a lead citrate stain (1 min), followed by three filtered water washes. This technique is applied to a *T. brucei* nucleus (Fig. 5D–E).

3. Ethanolic Phosphotungstic Acid Staining of Basic Proteins

Phosphotungstic acid (PTA) selectively binds to the basic groups (lysine and arginine residues) of proteins (Sheridan and Barrnett, 1969). This property has been used in electron microscopy to study the ultrastructure of synapses and chromosomes (Bloom and Aghajanian, 1966; Sheridan and Barrnett, 1969), and the flagellar attachment zone of *Trypanosoma cruzi* (Rocha *et al.*, 2006). Trypanosome nuclear DNA is associated with histones, which stain strongly with ethanolic PTA (E-PTA) and the kinetoplast DNA is also associated with basic proteins. E-PTA staining is therefore a useful reagent to reveal ultrastructural details in the nucleus and kinetoplast that are not seen in samples stained with osmium and uranyl acetate (Gluenz *et al.*, 2007; Souto-Padron and De Souza, 1979).

Cells are fixed in 2.5% glutaraldehyde, as above (Section III.B), without postfixation in osmium and without uranyl acetate staining. Samples are dehydrated in ethanol. After the first change of 100% ethanol, pellets are immersed in 2% PTA in 100% ethanol at 4°C overnight. Samples are then rinsed twice more in 100% ethanol and embedded in Epon, as above (Section III.B). No on-section stain is necessary (Fig. 5F).

F. Detergent Extraction of Cells

The following methods are used to visualize the detergent insoluble cytoskeleton of trypanosomes, which includes both the subpellicular and flagellar microtubules and associated structures.

1. Whole-Mount Cytoskeletons

Detergent extraction of whole cells on an EM grid is an excellent protocol that preserves the cell shape and form as a whole and allows visualization of individual cytoskeletal elements including microtubules and their connections (Fig. 6A).

To enable cells to attach to the formvar, coated copper mesh grids were carbon coated and plasma etched. Grids should ideally be used immediately after plasma etching, but can be used for up to 1 h afterwards. Grids can be reetched if needed.

Centrifuge approximately 5×10^7 cells for 5 min at 800 g. Carefully remove most of the supernatant, leaving 500 μl of cells and carefully resuspend. Hold grid (coated side up) in forceps and place a drop of cell suspension on it so that the grid is completely covered. Allow cells to adhere for 5 min. Mutant cell lines that produce very small, aflagellate, or very big cells have more difficulty adhering, so extend the time allowed for the cells to adhere accordingly.

Transfer grid (cell side down) to a 100 μl drop 1% (v/v) NP-40 (Sigma) in PEME (100 mM PIPES, 2mM EGTA, 1 mM $MgSO_4$, 0.1 mM EDTA, pH 6.9) for 2–3 min.

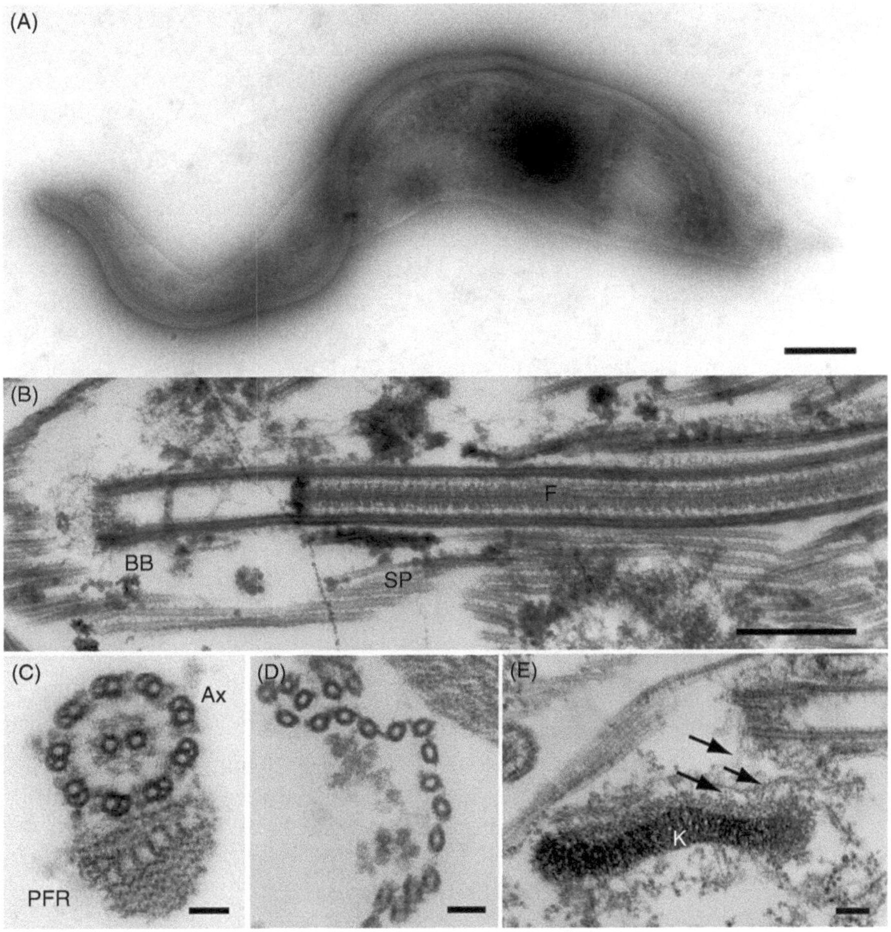

Fig. 6 Cytoskeletal preparations of *T. brucei*. (A) Whole mount cytoskeleton negatively stained using aurothioglucose. Plastic-embedded cytoskeleton showing (B) a longitudinal view of flagellum (F) with basal body (BB) and subpellicular microtubules (SP), (C) axoneme (Ax) and the paraflagellar rod (PFR), and (D) subpellicular microtubules. (E) Simultaneous detergent extraction and glutaraldehyde fixation preserves some cellular features, such as the kinetoplast (K) and tripartite attachment complex (arrows). (E, Courtesy of Emmanuel Ogbadoyi). Scale bars: 1 μm in (A), 500 nm in (B), 100 nm in (E), 50 nm in (C–D).

Repeat with a fresh drop. Gently transfer the grid using forceps onto the surface of the detergent droplet. NP-40 is a nonionic detergent that solubilizes both the plasma membrane and the internal membranous organelles, leaving the cytoskeleton intact (Sherwin and Gull, 1989). Keep the back of the grid dry to avoid sinking the grid – this is important to extract the cells fully. Larger or denser cells may need to be extracted for longer. To avoid problems in picking up and releasing grids, clean and dry forceps with filter paper regularly.

Fix cells by transferring to a drop of 2.5% (w/v) EM grade glutaraldehyde in PEME for 3 min. Transfer into double distilled water (ddH_2O) for 1 min to wash, repeat on fresh water drop. Pick up grids and blot off excess water, but do not allow drying out.

Apply negative stain with a 15 μl drop of 0.5% aurothioglucose in ddH_2O and immediately remove with a plastic pipette tip attached to a vacuum pump or, alternatively by blotting the side of the grid with filter paper. The staining reaction is nonspecific and occurs immediately. Aurothioglucose is supplied in powder form, but is best made up as a stock solution (2.5–3% w/v) in ddH_2O and stored in the fridge where it will last at least a year.

This protocol can be used for immunolocalization of antigens as it is useful to see labeling in a whole cell environment (Sherwin and Gull, 1989). Prepare cells on grids and fix as above, then instead of applying the negative stain transfer the grid using forceps to a 100 μl drop of 100 mM glycine in PBS for 5 min to neutralize free aldehyde groups. Incubate the grids in 1% (w/v) BSA, 0.1% (v/v) Tween 20 in PBS for 30 min to block nonspecific binding of antibody, then incubate with primary antibody diluted in 1% (w/v) BSA + 0.1% (v/v) Tween 20 in PBS for 30 min. Wash grids at least 5 times in 1% (w/v) BSA in PBS for 5 min each. Incubate secondary antibody (gold conjugated) 1% (w/v) BSA + 0.1% (v/v) Tween 20 in PBS for 30 min. Wash grids at least 5 times in 0.1% (w/v) BSA in PBS. Finally fix in 2.5% (w/v) glutaraldehyde in PBS and then apply a negative stain as above.

2. Plastic-Embedded Cytoskeletons

Centrifuge trypanosomes (800 g for 5–10 min) and resuspend in 1% NP-40 in PEME buffer pH 6.9 for 5 min. Wash the cells twice with PEME buffer and fix with 2.5% glutaraldehyde and 2% formaldehyde in 100 mM phosphate buffer pH 7–7.4 and follow the description for chemically fixed cells, starting with the second fixation step in buffer (Section III.B; Fig. 6B–D).

3. Simultaneous Fixation and Detergent Extraction for Plastic-Embedded Cytoskeletons

This method to prepare cytoskeletons has been developed to visualize the set of filaments linking the kinetoplast DNA to the basal body of the flagellum, a structure termed tripartite attachment complex (TAC; Fig. 6E) (Ogbadoyi *et al.*, 2003). Harvest procyclic cells from 5–10 ml of a log phase trypanosome culture by centrifugation (800 g for 5–10 min), and discard the supernatant. Resuspend cells in the residual small volume of media before adding 5 ml fixative (1% glutaraldehyde, 1% formaldehyde, 0.1% NP-40 in 1× PEM [0.05 M PIPES, 1 mM EGTA, 0.5 mM $MgSO_4$, pH 6.9]), which is mixed gently with the cells by inverting the tube a few times. Cells are then left to fix for 1 h at room temperature, before they are collected by centrifugation and resuspended in 1ml PBS. Continue washing cells in PBS three times and collect cells by centrifugation (1800 g for 3 min), without resuspending the cells (just tap on the tube gently to dislodge the pellet from the tube wall). Let it stand for 3 min.

To postfix the cytoskeletons, add 1% OsO_4 in PBS to pellets and leave for 30 min in the dark, at room temperature. The pellet should turn dark brown. Remove the osmium and wash the pellet thoroughly in distilled water (add 1 ml of distilled water to tube, leave for a few minutes, remove; repeat at least seven times). Stain the pellet by adding 1 ml 2% UA in water, leave overnight at 4°C in the dark. Wash with distilled water, dehydrate in ethanol, and embed in Epon resin as for standard fixation, described above (Section III.B).

IV. Materials

A. Cells, Media, and Buffers

Lister 427 *T. brucei* cells were used throughout, see Section III.A.
PEME (100 mM PIPES, 2 mM EGTA, 1 mM $MgSO_4$, 0.1 mM EDTA, pH 6.9)

B. High-Pressure Freezing and Freeze Substitution

Cells were frozen using 100 μl deep membrane carriers (McDonald *et al.*, 2007) in a Leica EM PACT2 (Leica Microsystems GmbH, Vienna, Austria). The small volume of these cell carriers is essential for good freezing. Carriers were placed in flow-through rings inside reagent bath holders (both by Leica Microsystems GmbH, Vienna, Austria), in which they were freeze substituted, infiltrated, and embedded. FS was performed in a Leica EM AFS device.

C. Chemicals/Reagents

Paraformaldehyde (Sigma Aldrich Company Ltd., Gillingham, UK)
Formaldehyde (EM-grade, TAAB Laboratory Equipment Ltd.)
Glutaraldehyde (Science Services GmbH, Berlin, Germany)
Uranyl acetate (Amsbio, Abingdon, UK)
Dehydrated methanol (Sigma Aldrich Company Ltd., Gillingham, UK)
Dehydrated acetone (Science Services GmbH, Berlin, Germany)
Osmium tetroxide (Amsbio, Abingdon, UK)
General chemicals (i.e., PBS and BSA; Sigma-Aldrich Company Ltd, Dorset, England)
Goat α-mouse F(ab′) conjugated to 10 nm gold (Ted Pella, Inc., Redding, CA, USA)

D. Resins

We embed samples in Agar 100 Resin (equivalent to Epon 812), Spurr's, or HM20 (all from Agar Scientific, Standsted, UK).

E. Microtomy

Sections were cut using a Leica Ultracut E Microtome or a Leica Ultracut UCT (Leica Microsystems GmbH, Vienna, Austria), and collected on Cu–Pd slot grids (Agar Scientific, Standsted, UK). For cytoskeletons, copper mesh grids from TAAB Laboratories Equipment Ltd, Berks, UK, were used.

F. Electron Microscopy

Micrographs were acquired using the AMTV600 software operating an Advantage HS-B camera (2000×2000 pixels; AMT, Danvers, MA, USA) on a FEI CM100 microscope or DigitalMicrograph software operating a Gatan Ultrascan 1000 CCD camera on a FEI Tecnai 12 microscope (FEI Company, Eindhoven, the Netherlands). Alternatively, micrographs were acquired on Kodak electron microscopy film 4489 (Agar Scientific Ltd).

V. Discussion

A fixative should preserve the morphology and structure of the cell as close to the living state as possible without altering the volume and spatial relationships of subcellular structures. Moreover, the cellular ultrastructure must remain intact during the harsh conditions encountered during postfixation treatment and withstand the electron beam in the microscope.

In this chapter we have described both chemical fixation and cryoimmobilization using high-pressure freezing. The most widely used fixation method for ultrastructural studies is chemical fixation using glutaraldehyde. Glutaraldehyde has two terminal aldehyde groups that react with amino groups in proteins, some carbohydrates, and other molecules. This results in irreversible crosslinking of proteins, creating a dense meshwork of immobilized intracellular structures (Bozzola and Russell, 1998).

Variants of this method have been used to study trypanosome fine structure since the late 1950s and provided the basis for discovery of many fundamentally important aspects of trypanosome biology and virulence. Among these are the discovery of the variant surface glycoprotein (VSG) coat in bloodstream form *T. brucei* and its role in antigenic variation in the mammalian host (Vickerman and Luckins, 1969), the structure of the single mitochondrion and changes that occur during metabolic adaptation in different life cycle stages (Brown *et al.*, 1973), the structure of the kinetoplast (Meyer *et al.*, 1958), the flagellar pocket as the site of endocytosis (Brown *et al.*, 1965), the cell cycle cytoskeletal changes (Sherwin and Gull, 1989), the flagella connector (Moreira-Leite *et al.*, 2001), and this method is still used today (Lacomble *et al.*, 2009).

Glutaraldehyde can be added directly to the culture medium of trypanosomes, where fixation can be seen as immobilization of cells within seconds. Handling of cells before fixation is thus kept to an absolute minimum, without subjecting live cells to centrifugation or temperature shifts. This facilitates experiments such as time series, where the intervals between samples are within the second to minute range, since high-pressure freezing individual samples cannot be frozen faster than ~1 min one after the other. Furthermore, fixation of cells is achieved without the need for specialized equipment, thus providing a routine method for preservation of trypanosomes collected in field studies.

However, the process of dehydration and crosslinking of proteins using chemical fixation causes aggregation artifacts, where molecules in solution will cluster and form matrices not representative of the undisturbed cellular environment (Dubochet and Sartori Blanc, 2001). In micrographs, such dehydration artifacts often show as extracted cells with wrinkled membranes (Vanhecke *et al.*, 2008). Therefore, striving toward visualizing the most native state of the cell is important especially with a view to building faithful 3D reconstructions of cellular structures, and ultimately whole cells, using electron tomography.

High-pressure freezing followed by FS provides a step toward better cellular preservation since the cells are dehydrated and fixed slowly at very low temperatures (i.e., at –90°C in the FS "cocktail"). The cells may still have some extraction, but membranes are smooth and the cytoplasm is more homogenous than in chemical fixation. In budding yeast (*Saccharomyces cerevisiae*) it has been shown that the volume of organelles such as endoplasmatic reticulum, nucleus, and vacuole, and their interactions changes significantly between chemical fixation and high-pressure freezing (Perktold *et al.*, 2007). We have shown that in *Chlamydomonas reinhardii* and *T. brucei*, axonemal microtubules at the distal end of the flagellum were depolymerized in chemically fixed samples but were well preserved in high-pressure frozen cells (Höög *et al.* manuscript in preparation). In general, trypanosome cell morphology benefits greatly from using cryoimmobilization.

The next step toward a more native cell preservation is to freeze the cells and then image them in ice directly, completely avoiding dehydration, resin infiltration, and embedding. However, the cells are too thick to be imaged without sectioning (3–4 μm), therefore cryomicroscopy of whole cells will prove challenging. Methods for cryomicrotomy of hydrated bacteria (Matias *et al.*, 2003), human skin cells (Al-Amoudi *et al.*, 2005, 2007), and frozen rat hepatoma (HTC) cells (Bouchet-Marquis and Fakan, 2009) have been developed. In the future, cryomicroscopy of such frozen hydrated sections may provide new insights into trypanosome cell biology.

VI. Summary

T. brucei is an excellent model system to study cell motility, morphology, and gene expression because of the many tools available for its genetic manipulation, ease of growing cells in culture, and suitability for electron microscopy studies. We summarize a range of well-tested methods to prepare cells for electron microscopy using chemical fixation, and introduce a protocol using rapid freezing followed by FS, which yields much improved cellular morphology. The ease with which genes can be silenced in this organism combined with good structural preservation makes *T. brucei* a powerful tool when examining structure–function relationships in cells.

Acknowledgments

We would like to thank Mike Shaw, University of Oxford, for help with sample preparation and protocol development. This work was funded by The Wellcome Trust and the EP Abraham Trust and an EMBO long-term fellowship (to JLH). JHL is a Sir Henry Wellcome Fellow, and K.G. is a Wellcome Trust Principal Research Fellow. The Boulder Laboratory for 3D Electron Microscopy of Cells is supported by National Institutes of Health Biotechnology grant RR00592 to A. Hoenger.

References

Al-Amoudi, A., Diez, D. C., Betts, M. J., and Frangakis, A. S. (2007). The molecular architecture of cadherins in native epidermal desmosomes. *Nature* **450**, 832–837.

Al-Amoudi, A., Dubochet, J., and Norlen, L. (2005). Nanostructure of the epidermal extracellular space as observed by cryo-electron microscopy of vitreous sections of human skin. *J. Invest. Dermatol.* **124**, 764–777.

Aslett, M., Aurrecoechea, C., Berriman, M., Brestelli, J., Brunk, B. P., Carrington, M., Depledge, D. P., Fischer, S., Gajria, B., Gao, X., *et al.* (2009). TriTrypDB: A functional genomic resource for the trypanosomatidae. *Nucleic Acids Res.* D457–D462.

Bacchi, C. J. (2009). Chemotherapy of human African trypanosomiasis. *Interdiscip. Perspect. Infect. Dis.* **2009**, 1–5.

Bernhard, W. (1969). A new staining procedure for electron microscopical cytology. *J. Ultrastruct. Res.* **27**, 250–265.

Berriman, M., Ghedin, E., Hertz-Fowler, C., Blandin, G., Renauld, H., Bartholomeu, D. C., Lennard, N. J., Caler, E., Hamlin, N. E., Haas, B., *et al.* (2005). The genome of the African trypanosome trypanosoma brucei. *Science* **309**, 416–422.

Beverley, S. M. (2003). Protozomics: Trypanosomatid parasite genetics comes of age. *Nat. Rev. Genet.* **4**, 11–19.

Bloom, F. E., and Aghajanian, G. K. (1966). Cytochemistry of synapses: Selective staining for electron microscopy. *Science* **154**, 1575–1577.

Bouchet-Marquis, C., and Fakan, S. (2009). Cryoelectron microscopy of vitreous sections: A step further towards the native state. *Methods Mol. Biol.* **464**, 425–439.

Bozzola, J., and Russell, L. (1998). "Electron Microscopy: Principles and Techniques for Biologists", 2nd edn. Jones & Bartlett Publishers, Inc, Boston.

Brown, K. N., Armstrong, J. A., and Valentine, R. C. (1965). The ingestion of protein molecules by blood forms of *Trypanosoma rhodesiense*. *Exp. Cell Res.* **39**, 129–135.

Brown, R. C., Evans, D. A., and Vickerman, K. (1973). Changes in oxidative metabolism and ultrastructure accompanying differentiation of the mitochondrion in trypanosoma brucei. *Int. J. Parasitol.* **3**, 691–704.

Brun, R. (1979). Schonenberger: Cultivation and *in vitro* cloning or procyclic culture forms of trypanosoma brucei in a semi-defined medium. Short communication. *Acta Trop.* **36**, 289–292.

Carruthers, V. B., and Cross, G. A. (1992). High-efficiency clonal growth of bloodstream- and insect-form trypanosoma brucei on agarose plates. *Proc. Natl. Acad. Sci. U. S.A* **89**, 8818–8821.

Delespaux, V., and de Koning, H. P. (2007). Drugs and drug resistance in African trypanosomiasis. *Drug Resist. Updat.* **10**, 30–50.

Dubochet, J., and Sartori Blanc, N. (2001). The cell in absence of aggregation artifacts. *Micron* **32**, 91–99.

El-Sayed, N. M., Myler, P. J., Bartholomeu, D. C., Nilsson, D., Aggarwal, G., Tran, A. N., Ghedin, E., Worthey, E. A., Delcher, A. L., Blandin, G., *et al.* (2005). The genome sequence of trypanosoma cruzi, etiologic agent of chagas disease. *Science* **309**, 409–415.

Fakan, S. (2004). The functional architecture of the nucleus as analysed by ultrastructural cytochemistry. *Histochem. Cell Biol.* **122**, 83–93.

Gadelha, (2008). Using colonies of T. brucei as a mean to concentrate cells for high pressure freezing.

Gadelha, C., Rothery, S., Morphew, M., McIntosh, J. M., Severs, N. J., and Gull, K. (2009). Membrane domains and boundaries: Cytoskeletal influences on the flagellar pocket of bloodstream trypanosomes. *PNAS.* **106**(41), 17425–17430.

Gluenz, E., Shaw, M. K., and Gull, K. (2007). Structural asymmetry and discrete nucleic acid subdomains in the trypanosoma brucei kinetoplast. *Mol. Microbiol.* **64**, 1529–1539.

Grunfelder, C. G., Engstler, M., Weise, F., Schwarz, H., Stierhof, Y. D., Boshart, M., and Overath, P. (2002). Accumulation of a GPI-anchored protein at the cell surface requires sorting at multiple intracellular levels. *Traffic* **3**, 547–559.

Gull, K. (1999). The cytoskeleton of trypanosomatid parasites. *Annu. Rev. Microbiol.* **53**, 629–655.

Gull, K. (2003). Host-parasite interactions and trypanosome morphogenesis: A flagellar pocketful of goodies. *Curr. Opin. Microbiol.* **6**, 365–370.

Hawes, P., Netherton, C. L., Mueller, M., Wileman, T., and Monaghan, P. (2007). Rapid freeze-substitution preserves membranes in high-pressure frozen tissue culture cells. *J. Microsc.* **226**, 182–189.

Hirumi, H., and Hirumi, K. (1989). Continuous cultivation of trypanosoma brucei blood stream forms in a medium containing a low concentration of serum protein without feeder cell layers. *J. Parasitol.* **75**, 985–989.

Höög, J. L., Lacomble, S., O'Toole, E. T., Gull, K. (Manuscript in preparation), How to Grow a Flagellum.

Ivens, A. C., Peacock, C. S., Worthey, E. A., Murphy, L., Aggarwal, G., Berriman, M., Sisk, E., Rajandream, M. A., Adlem, E., Aert, R., *et al.* (2005). The genome of the kinetoplastid parasite, leishmania major. *Science* **309**, 436–442.

LaCount, D. J., Bruse, S., Hill, K. L., and Donelson, J. E. (2000). Double-stranded RNA interference in trypanosoma brucei using head-to-head promoters. *Mol. Biochem. Parasitol.* **111**, 67–76.

Lacomble, S., Vaughan, S., Gadelha, C., Morphew, M. K., Shaw, M. K., McIntosh, J. R., and Gull, K. (2009). Three-dimensional cellular architecture of the flagellar pocket and associated cytoskeleton in trypanosomes revealed by electron microscope tomography. *J. Cell. Sci.* **122**, 1081–1090.

Luther, P. K., Lawrence, M. C., and Crowther, R. A. (1988). A method for monitoring the collapse of plastic sections as a function of electron dose. *Ultramicroscopy* **24**, 7–18.

Matias, V. R., Al-Amoudi, A., Dubochet, J., and Beveridge, T. J. (2003). Cryo-transmission electron microscopy of frozen-hydrated sections of *Escherichia coli* and *Pseudomonas aeruginosa. J. Bacteriol.* **185**, 6112–6118.

McDonald, K. L., Morphew, M., Verkade, P., and Muller-Reichert, T. (2007). Recent advances in high-pressure freezing: Equipment- and specimen-loading methods. *Methods Mol. Biol.* **369**, 143–173.

Meyer, H., De Oliveira Musacchio, M., and De andrade Mendonca, I. (1958). Electron microscopic study of trypanosoma cruzi in thin sections of infected tissue cultures and of blood-agar forms. *Parasitology* **48**, 1–8.

Moreira-Leite, F. F., Sherwin, T., Kohl, L., and Gull, K. (2001). A trypanosome structure involved in transmitting cytoplasmic information during cell division. *Science* **294**, 610–612.

Moyne, G. (1980). Methods in ultrastructural cytochemistry of the cell nucleus. *Prog. Histochem. Cytochem.* **13**, 1–72.

Muller-Reichert, T., Sassoon, I., O'Toole, E., Romao, M., Ashford, A. J., Hyman, A. A., and Antony, C. (2003). Analysis of the distribution of the kinetochore protein ndc10p in *Saccharomyces cerevisiae* using 3-D modeling of mitotic spindles. *Chromosoma* **111**, 417–428.

Ogbadoyi, E. O., Robinson, D. R., and Gull, K. (2003). A high-order trans-membrane structural linkage is responsible for mitochondrial genome positioning and segregation by flagellar basal bodies in trypanosomes. *Mol. Biol. Cell* **14**, 1769–1779.

Overath, P., and Engstler, M. (2004). Endocytosis, membrane recycling and sorting of GPI-anchored proteins: Trypanosoma brucei as a model system. *Mol. Microbiol.* **53**, 735–744.

Pays, E., Vanhollebeke, B., Vanhamme, L., Paturiaux-Hanocq, F., Nolan, D. P., and Perez-Morga, D. (2006). The trypanolytic factor of human serum. *Nat. Rev. Microbiol.* **4**, 477–486.

Perktold, A., Zechmann, B., Daum, G., Zellnig, G. (2007). Organelle association visualized by three-dimensional ultrastructural imaging of the yeast cell. *FEMS Yeast Res.* **7**(4), 629–638.

Piperno, G., and Fuller, M. T. (1985). Monoclonal antibodies specific for an acetylated form of alpha-tubulin recognize the antigen in cilia and flagella from a variety of organisms. *J. Cell Biol.* **101**, 2085–2094.

Reynolds, E. S. (1963). The use of lead citrate at high pH as an electron-opaque stain in electron microscopy. *J. Cell Biol.* **17**, 208–212.

Richter, K. (1994). High-density morphologies of ice in high-pressure frozen biological specimens. *Ultramicroscopy* **53**, 237–249.

Rocha, G. M., Brandao, B. A., Mortara, R. A., Attias, M., de Souza, W., and Carvalho, T. M. (2006). The flagellar attachment zone of trypanosoma cruzi epimastigote forms. *J. Struct. Biol.* **154**, 89–99.

Scholey, J. M., Neighbors, B., McIntosh, J. R., and Salmon, E. D. (1984). Isolation of microtubules and a dynein-like MgATPase from unfertilized sea urchin eggs. *J. Biol. Chem.* **259**, 6516–6525.

Sharma, R., Gluenz, E., Peacock, L., Gibson, W., Gull, K., and Carrington, M. (2009). The heart of darkness: Growth and form of trypanosoma brucei in the tsetse fly. *Trends Parasitol.* **25**(11), 517–524.

Sharma, R., Peacock, L., Gluenz, E., Gull, K., Gibson, W., and Carrington, M. (2008). Asymmetric cell division as a route to reduction in cell length and change in cell morphology in trypanosomes. *Protist* **159**, 137–151.

Sheridan, W. F., and Barrnett, R. J. (1969). Cytochemical studies on chromosome ultrastructure. *J. Ultrastruct. Res.* **27**, 216–229.

Sherwin, T., and Gull, K. (1989). The cell division cycle of *Trypanosoma brucei brucei*: Timing of event markers and cytoskeletal modulations. *Philos. Trans. R. Soc. Lond. B Biol. Sci.* **323**, 573–588.

Sherwin, T., and Gull, K. (1989). Visualization of detyrosination along single microtubules reveals novel mechanisms of assembly during cytoskeletal duplication in trypanosomes. *Cell* **57**, 211–221.

Shi, H., Djikeng, A., Mark, T., Wirtz, E., Tschudi, C., and Ullu, E. (2000). Genetic interference in trypanosoma brucei by heritable and inducible double-stranded RNA. *RNA* **6**, 1069–1076.

Souto-Padron, T., and De Souza, W. (1979). Cytochemical analysis at the fine-structural level of trypanosomatids stained with phosphotungstic acid. *J. Protozool.* **26**, 551–557.

Stephan, A., Vaughan, S., Shaw, M. K., Gull, K., and McKean, P. G. (2007). An essential quality control mechanism at the eukaryotic basal body Prior To intraflagellar transport. *Traffic* **8**, 1323–1330.

Vanhecke, D., Graber, W., and Studer, D. (2008). Close-to-native ultrastructural preservation by high pressure freezing. *Methods Cell Biol.* **88**, 151–164.

Vickerman, K. (1985). Developmental cycles and biology of pathogenic trypanosomes. *Br. Med. Bull.* **41**, 105–114.

Vickerman, K., and Luckins, A. G. (1969). Localization of variable antigens in the surface coat of trypanosoma brucei using ferritin conjugated antibody. *Nature* **224**, 1125–1126.

Weise, F., Stierhof, Y. D., Kuhn, C., Wiese, M., and Overath, P. (2000). Distribution of GPI-anchored proteins in the protozoan parasite leishmania, based on an improved ultrastructural description using high-pressure frozen cells. *J. Cell. Sci.* **113**(Pt 24), 4587–4603.

Wiese, M., Kuhn, D., and Grunfelder, C. G. (2003). Protein kinase involved in flagellar-length control. *Eukaryotic Cell* **2**, 769–777.

CHAPTER 9

Dictyostelium discoideum: A Model System for Ultrastructural Analyses of Cell Motility and Development

Michael P. Koonce* *and* **Ralph Gräf†**

*Division of Translational Medicine, Wadsworth Center, Albany, New York 12201-0509

†Department of Cell Biology, Institute for Biochemistry and Biology, University of Potsdam, D-14476 Potsdam-Golm, Germany

Abstract

Dictyostelium occupies an interesting niche in the grand scheme of model organisms. On the one hand, it is a compact, highly motile single cell that presents numerous opportunities to investigate the fundamental mechanisms of signal transduction, cell movement, and pathogen infection. However, upon starvation, individual cells enter a developmental pathway that involves cell aggregation, cell–cell

978-0-12-381007-6
DOI: 10.1016/S0091-679X(10)96009-3

adhesion, pattern formation, and differentiation. Thus, *Dictyostelium* is also well known as a basic model for studying developmental processes. Electron microscopy (EM) has played a large role in both the unicellular and the multicellular life stages, for example, providing image detail for structure/function relationships of cytoskeletal proteins, the deposition of cellulose fibrils in maturing spores, and the identification of intercellular junctional complexes. Powerful combinations of robust molecular genetic tools, high-resolution light microscopy, and EM methods make this organism an attractive model for imaging dynamic cell processes. This chapter serves to highlight the past and current EM approaches that have advanced our understanding of how cells and proteins function.

I. Introduction

Dictyostelium is a relatively simple, compact amoeba that spends much of its vegetative life crawling through the soil searching for food (e.g., bacteria). Under adverse environmental conditions such as starvation, individual amoebae signal one another, follow chemotatic cues to aggregate en masse, develop cell–cell adhesions to form a motile metazoan slug, and then undergo a relatively simple developmental program that produces spore and stalk cells. Spores are encapsulated with a trilaminar coat of proteins, cellulose, and polysaccharides; these coatings resist environmental stresses and permit regeneration of viable cells. The entire developmental process is elegantly covered in the review by Kessin (2001), and a wide variety of images and movies can be found at www.dictybase.org. In the laboratory, *Dictyostelium* is straightforward to grow, easy to work with, and, through the dedicated efforts of a robust experimental community, has been developed into a fully featured model organism. Because of its rapid cell movements, *Dictyostelium* has proven to be an outstanding system in which to study nonmuscle cell motility, from identification and characterization of cytoskeletal proteins to dissection of the signal transduction cascades that regulate cell polarity and chemotaxis. Because *Dictyostelium* transitions, on cue, from solitary individuals into a metazoan of approximately 100,000 cells, it has also been exceptionally useful in the study of developmental mechanisms, in particular, differential gene expression, cell–cell interactions, cell fate determination, and spore formation. More recently, *Dictyostelium* has been used as an experimental factory to synthesize difficult-to-make polypeptide fragments (e.g., myosin and dynein, Manstein *et al.*, 1989; Koonce and Samsó, 1996). Furthermore, *Dictyostelium* has proven to be a versatile model for biomedical research (Williams *et al.*, 2006), e.g., in the study of immune-cell disease, centrosomal abnormalities in cancer, nuclear movement during brain development (Rehberg *et al.*, 2005), cytotoxicity of novel transplantation materials (Shkilnyy *et al.*, 2009), and as a useful host to study pathogenic infection (e.g., *Legionella* and *Mycobacterium*, Annesley and Fisher, 2009; Hagedorn *et al.*, 2009; Hägele *et al.*, 2000; Steinert and Heuner, 2005). Nearly every biological EM method has been applied to examine ultrastructural details of the above processes, from scanning EM (SEM) overviews of developmental stages to single-particle reconstructions of in situ nuclear pores. This review seeks to highlight approaches in which *Dictyostelium* has been particularly useful in developing a basic understanding of eukaryotic cell biology.

A. Strengths of *Dictyostelium* as a Model Organism

The most obvious advantages of *Dictyostelium* as an experimental model are the organism's simplicity, the fact that it performs activities common to most eukaryotic cells, and the fact that a robust combination of approaches can be applied in this model to address a range of important questions. *Dictyostelium* is well suited to large-scale biochemical analyses and can quite literally be grown by the pound (Griffith *et al.*, 1987). The model includes a fully sequenced genome and a molecular toolbox filled with powerful homologous recombination, plasmid, and RNA-mediated methods to modify or induce gene expression, as well as readily available DNA arrays for genome-wide expression analyses (Eichinger and Rivero, 2006; Sillo *et al.*, 2008). Importantly, *Dictyostelium* is readily amenable to live-cell imaging and has been used to study dynamic processes, including phagocytosis and cytokinesis, and in particular, chemotatic events such as rapid whole cell response to 3′-5′-cyclic adenosine monophosphate (cAMP), transient cortical enrichment of effector molecules, and even spatial/temporal movements of individual cAMP receptors on the cell surface.

B. Historical Application of EM Methods

Transmission EM (TEM) and SEM approaches have long provided detail and discovery in *Dictyostelium*. The first PubMed citation of EM use in *Dictyostelium* dates to 1957 (Gezelius and Ranby, 1957); early works in the 1960s to 1970s were dominated by fine structural analyses of membranes and spore cell walls; of cell–cell interactions during development; investigation of cytoskeletal components; and even chromatin, nuclei, and transcription-oriented studies. These works provided basic descriptions of cytoplasmic organization in growing cells and revealed details of the morphological changes that occur during development. In particular, EM revealed the ultrastructural changes that accompany sporulation and germination. Two notable achievements from this period were the initial ultrastructural analyses of actin and myosin (Clarke and Spudich, 1974; Spudich, 1974; Woolley, 1972) and the structural characterizations of mitotic spindle assembly and cell division (Moens, 1976; Roos, 1975).

The actomyosin work demonstrated the existence of muscle contractile proteins in *Dictyostelium* and provided early analyses of cortical actin filament arrangements (Clarke and Spudich, 1974; Clarke *et al.*, 1975; Spudich, 1974; Woolley, 1972). These studies marked the start of a long history of contractile protein investigations in *Dictyostelium*, resulting, so far, in the identification of numerous actin-binding proteins, the characterization of large multi-gene families of related proteins (e.g., there are 17 actin and 13 myosin genes in this simple organism), and domain-mapping/atomic-level analyses of myosin motors (Claviez *et al.*, 1982; Flicker *et al.*,

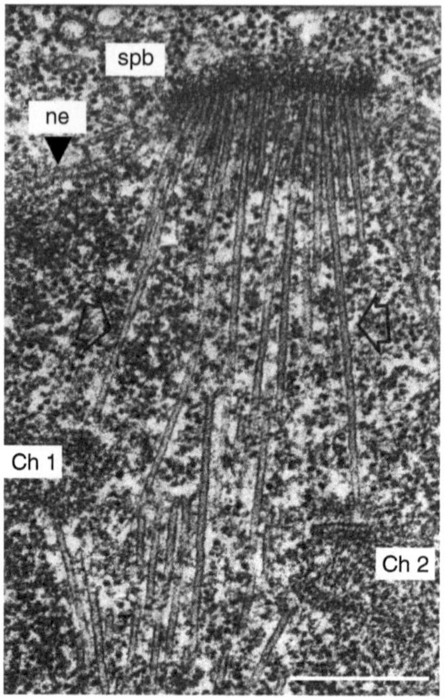

Fig. 1 Conventional chemical preservation of a mitotic spindle. Cells were fixed with 1.25% glutaraldehyde in Sørenson's phosphate buffer (Table 1), followed by postfixation in 1% OsO_4 (McIntosh *et al.*, 1985). Microtubules clearly emanate from an elongate spindle pole body (spb) and extend into the nuclear volume. The second pole of this metaphase spindle is located past the bottom of this micrograph, and in another section. Two kinetochore pairs are visible in this field (ch 1, 2), as well as a portion of the surrounding nuclear envelope (ne). Two microtubules (open arrows) can also be seen to run continuously, connecting the spb with the kinetochores. Micrograph adapted with permission from McIntosh *et al.*, *J. Cell Sci.* **75**, 93–129, 1985. Scale bar = 0.5 μm.

1985). Recent EM studies continue these works by illustrating the in situ arrangement of actin filaments in extending filopodia via cryo-electron tomography (cryo-ET) (Medalia *et al.*, 2007). Optimistically, we can anticipate filling in the details of how intracellular signaling cascades are translated into processes such as actin filament polymerization, pseudopod protrusion, and directional cell movement.

The second notable achievement (from the perspective of the present authors) was the initial analyses of the microtubule (MT) system in *Dictyostelium*. These studies utilized conventional methods of glutaraldehyde/osmium fixation, plastic embedment, and thin-section analyses (Fig. 1; McIntosh *et al.*, 1985; Moens, 1976; Roos, 1975). Key insights gained from this work outlined the organization of the centrosome, its linkage to the nucleus, and the dynamics of mitotic spindle assembly. Although *Dictyostelium* performs many of the whole-cell crawling movements found in vertebrate organisms, it divides via an intranuclear spindle in a manner similar to that seen in yeast and fungi.

A cytoplasmic cuboidal centrosome is tightly tethered to and remains within 1 μm of the interphase nucleus, and functions to organize a radial array of ~30–70 MTs (Omura and Fukui, 1985; Schulz *et al.*, 2009). The MTs support a striking amount of rapid, membrane-bound organelle movement. During prophase, the centrosome sheds the cytoplasmic MT array, duplicates, migrates into an opening in the nuclear envelope, and then directs MT/spindle assembly within the nuclear volume (Moens, 1976; Roos, 1975; Ueda *et al.*, 1999). The detailed elucidation of these processes required fine structural preservation and EM analysis. The cylindrically shaped mitotic spindle is similar to structures that are assembled in the many so-called lower eukaryotes, yet the kinetochores in chemically fixed *Dictyostelium* have the distinctive tri-laminar appearance typically found in cells from higher animals. In general, the MT system has been less well studied than the actomyosin system; nevertheless, the features and functions of the centrosomes and associated MT-based motors are sufficiently robust that further analysis is strongly warranted.

II. Current Approaches

A. Conventional Methods and Applications

Routine negative staining of purified molecules and cell isolates, thin-section TEM, and SEM of whole cells continue to provide valuable structural information about cellular processes in *Dictyostelium*, especially when combined with molecular tools that modify key proteins. Some recent examples include the imaging of cell–cell adhesion during late developmental stages (Grimson *et al.*, 2000), myosin filament assembly (McMahon *et al.*, 2008), dynein motor domain structure (Roberts *et al.*, 2009; Samsó and Koonce, 2004), and pathogen invasion (Hagedorn *et al.*, 2009; Hägele *et al.*, 2000; Steinert and Heuner, 2005). The preparative methods for such studies follow relatively standard protocols, with only a few modifications required by *Dictyostelium* biology. Table I presents a compilation of preparative steps that have been successfully used for conventional fine-structure preservation; several reviews of these methods have also been previously published (Condeelis *et al.*, 1987; De Priester, 1991; Grimson and Blanton, 2006; Hagedorn *et al.*, 2006; Sameshima, 1984).

1. Attachment of Cells and Isolated Organelles

One of the few drawbacks of *Dictyostelium* is that these highly motile cells do not adhere as tightly to a substrate as do metazoan tissue culture cells. As a consequence, buffer turbulence during fixation and subsequent wash steps can cause an appreciable loss of cells, which may impact experiments where individual cells are followed while alive. Clean glass coverslips (e.g., soaked in 1N HCl for 1 h, followed by rinsing in dH_2O and 100% EtOH or acetone before drying) should be used as the substrate for cell attachment. Surface treatment with poly-L-lysine has been shown to be effective in enhancing adhesion, particularly for cells that have been prewashed in

Table I
Conventional Chemical Fixation Methods

Preparative step	Method detail	Key References
Cells attached to coverslips	For optimum adhesion, clean, acid-washed glass coverslips need to be used. Polylysine coating or silanization will also enhance binding	Mazia *et al.*, 1975
Cell suspensions (cells, isolated nuclei, membranes, cytoskeletons, etc.)	Suspensions are typically diluted 1:1 in 2× fixative or resuspended in 1× fix. Cell pellets can be processed directly or mixed 1:1 after fixation with 4% molten agar at 60°C. After cooling, the agar mixture solidifies. Small blocks can then be cut out for dehydration and embedding	Moens, 1976
Fixative	0.5–3% Glutaraldehyde (final concentration), 15–60 min	
Buffers	A number of buffers have been used with excellent results and a few are listed below. One common feature is that they tend to be lower in ionic strength than those buffers used for vertebrate cell lines. Phosphate: 14.7 mM KH_2PO_4, 2.0 mM Na_2HP0_4, pH 6.0 (Sørensen) PIPES: 10 mM, pH 6.8, 10 mM NaCl, 2 mM $MgSO_4$, 10 mM EGTA pH 6.8 PHEM: 12 mM Pipes, 5 mM Hepes, 5 mM EGTA, 0.5 mM $MgCl_2$ pH 6.9 PBS, HL-5 culture medium, or 25–100 mM Na cacodylate, pH 7.2	Hagedorn *et al.*, 2006; Luna *et al.*, 1981; 1984; Marchetti *et al.*, 2004; McIntosh *et al.*, 1985; Neuhaus *et al.*, 1998; Sameshima *et al.*, 2002; Ueda *et al.*, 1999
Detergent	0.01–0.1% Triton X-100 in buffer with fixative. 30 s to 2 min, then continue fixation without detergent	Euteneuer *et al.*, 1998; Ueda *et al.*, 1999
Additives	4% PEG 6000, 10% DMSO, 0.1 M sucrose. These agents, in addition to low concentrations of Triton X-100, reportedly help to reduce cell swelling 2% tannic acid. Frequently useful to stabilize and contrast actin filaments	Condeelis *et al.*, 1987; McDonald, 1984; Roos, 1987
Washes	2–3 × 5 min	
Postfix	0.5–2% OsO_4, 20–60 min on ice 0.8% potassium ferricyanide ($K_3Fe(CN)_6$). Also useful to stabilize and contrast actin filaments and membranes	McDonald, 1984
En bloc staining	1% uranyl acetate, $K_3Fe(CN)_6$ can be added here as well	
Washes	2–3 × 5 min	
Dehydration and embedding	25, 50, 75, 100% EtOH, 50–50% ETOH–propylene oxide, 100% Propylene oxide	
Sectioning and staining	1% uranyl acetate, Reynolds lead citrate are commonly used	

phosphate buffer (Mazia *et al.*, 1975). A washing step removes soluble medium components that can act as substrate-blocking agents. Coverslips are coated with 1 mg/ml poly-L-lysine (70–100 kDa) for 2–5 min and rinsed in dH_2O before cells are applied. Silane derivatization of glass can also be used to tack cells onto coverslips. For example, a fresh 1–3% solution of 3-aminopropyltriethoxysilane (Sigma A3648,

St. Louis, MO, USA) is prepared in 100% EtOH. Clean coverslips are incubated for 30 min to 1 h, washed with 100% EtOH, and dried well before addition of cells. In general, cells become attached in as little as 15 min; prolonging the incubation time longer than 1 h offers little adhesive benefit. Some investigators have also used a brief (2-min) treatment with 5 mM $MgCl_2$ to induce further cell flattening (Ueda *et al.*, 1999).

For many applications, it can be more useful to perform microscopy on isolated organelles. Even crude purification steps often concentrate organelles, thus the structures of interest will likely occur at a much higher frequency than in intact cells. Moreover, the cytoplasmic background that complicates imaging will be greatly reduced and polypeptides of interest will often be more accessible to labeling techniques. This strategy has proven useful for nuclei and centrosomes (Beck *et al.*, 2004; Gräf, 2001; Schulz *et al.*, 2009; Xiong *et al.*, 2008). Intact nuclei can readily be isolated by simple mechanical lysis and centrifugation methods. After the forceful filtration of cells through nuclepore polycarbonate filters (5 μm pore size, Whatman, Piscataway, NJ, USA), nuclei are sedimented by centrifugation at 500 × g for 5 min (Beck *et al.*, 2004). If required to be of higher purity, they can be collected in a specific band after sucrose density gradient centrifugation (Schulz *et al.*, 2009).

Isolated nuclei are also the starting point for the separation of centrosomes, which are tightly attached to the nuclei. After disruption of the nuclei by treatment with a pyrophosphate buffer and mechanical forces, intact centrosomes can be isolated by density gradient centrifugation (Gräf, 2001). Prior to fixation, isolated nuclei or centrosomes are diluted in the desired buffer and then either pipetted directly onto EM grids or spun onto glass coverslips at 2800 × *g* for 10 min at 4°C. Here, there is no need to pretreat coverslips since both nuclei and centrosomes stick readily to glass surfaces. Note that when such methods are used, it is important that activity assays are performed to ensure that structures of interest retain their functional properties. A nuclear import assay was used to demonstrate that nuclear pores retained functional activity (Beck *et al.*, 2004); MT polymerization was assayed to demonstrate that centrosomes retained critical nucleation capacities (Gräf, 2001; Kuriyama *et al.*, 1982).

2. Fixation

Whereas acrolein has sometimes been utilized as a fixative for *Dictyostelium* (Sameshima, 1984), glutaraldehyde is by far the most commonly used chemical preservative. About 1–3% glutaraldehyde in buffer has provided a fine structural preservation that appears consistent across a number of laboratories. Some investigators report that addition of a low concentration of detergent or other additive (DMSO, PEG, sucrose, etc.) helps to reduce cell swelling during the initial stages of chemical fixation, and can thus enhance preservation (Roos, 1987). A lower concentration of glutaraldehyde (e.g., 0.1–0.5%), coupled with moderate concentrations of Triton X-100 (0.1–0.5%), has been used for immunolabeling of cells (Euteneuer *et al.*, 1998). This combination lyses cells, provides good structural preservation, and enables antibody penetration, yet imposes few problems with reduced antigenicity than has been observed with higher glutaraldehyde concentrations (see also Neuhaus *et al.*, 1998). A fluorescent secondary antibody can be used to image and preselect cells of interest prior to embedding; alternatively, a gold-coupled secondary antibody can be used for higher-resolution EM localization. However, some caution is necessary since the detergent eliminates most of the soluble cytosolic compartment and disrupts membranes. Thus this method should only be applied to robust or firmly anchored cell features that can withstand detergent extraction.

Cryopreservation via a number of methods (e.g., spray, plunge, slam, and high pressure), followed by freeze-substitution, embedding, and sectioning, has also been successfully applied to *Dictyostelium* cells. Although these methods typically require specialized equipment, they offer notable improvements in structural preservation, particularly in imaging of multicellular developmental stages or spores (Grimson *et al.*, 2000; Neuhaus *et al.*, 1998; Sameshima *et al.*, 2002). For example, careful analyses of plunge-frozen fruiting bodies reveal new details in cell–cell contacts and in the underlying actin filament arrangements. Importantly, these studies produced a clear identification of adherens junctions in fruiting bodies. This work, coupled with molecular characterization of a β-catenin homolog in *Dictyostelium*, establishes that the β-catenin signaling pathway, well known in metazoan organisms, likely evolved prior to the establishment of multicellular development (Grimson *et al.*, 2000). For greater detail and discussion of these methods, we refer readers to a recent chapter on the cryopreservation of *Dictyostelium* (Grimson and Blanton, 2006).

3. Buffer

Several buffers have been used with excellent results. The K/Na-phosphate buffer (Table I) is particularly useful for washing cells prior to fixation. In general, *Dictyostelium* cells require a lower ionic-strength buffer (~15–25 mM) than do vertebrate cells (~100 mM) for optimal structural preservation.

4. Post-Fixation

Staining, dehydrating, and embedding methods follow routine EM protocols, with no essential steps unique to specific features of *Dictyostelium* biology. A variety of reagents and stains that have been used to enhance fixation and/or contrast (e.g., osmium, tannic acid, and potassium ferricyanide) in other systems provide the same advantages and/or drawbacks in *Dictyostelium*.

5. Scanning EM

Most of the published works that include SEM images have utilized chemically fixed or cryo-fixed samples as described above, followed by dehydration, critical-point drying, and metal (i.e., gold, platinum, and palladium)-coating procedures. Cells can be prepared on solid supports such as glass or plastic coverslips/dishes or on filter paper. Again, there does not seem to be any organism-dependent steps required for *Dictyostelium*, though fully developed stalk/spore structures should be handled with care to prevent breakage. SEM approaches have been important in describing surface topologies during directed movement, interactions with bacteria (as food) and pathogens (during infection), morphologies of the various developmental stages, and details of spore coat formation.

B. Cryo-Electron Tomography

More recent analyses of *Dictyostelium* have utilized direct cryo-fixation methods for rapid preservation and tomographic image reconstruction to generate structural information in three dimensions. The strength of this combination of approaches lies in the potential to capture near atomic-level detail (4–5 nm resolution) of dynamic processes or complexes, with minimal structural perturbation. Because of beam energy and penetration constraints, sample (or vitreous ice) thickness is generally limited to 0.5 μm or less, and thus this methodology is currently applied to very small cells or to exceptionally thin cell processes. Moreover, cytoplasmic complexity represents a significant challenge in the discrimination of structures within the cell. *Dictyostelium* has proven to be an excellent eukaryotic cell model in which subcellular structures can be examined via cryo-ET. Four landmark studies using these methods have revealed details of the cytoplasmic organization, the assembly of actin filaments in filopodia, and the functional mechanisms by which nuclear pore complexes operate (Beck *et al.*, 2004, 2007; Medalia *et al.*, 2002, 2007).

Cells or isolated nuclei were applied to carbon-coated mesh grids, allowed to attach, and then plunge-frozen in liquid ethane (Dubochet *et al.*, 1988). Although these cells require only a short period of time (30–60 min) to adhere, Medalia *et al.* point out that copper grids still present a level of cytotoxicity that minimally requires a carbon coat on both sides to maintain cell survival (Medalia *et al.*, 2002). Subsequent studies utilized Pt grids to avoid this issue (Medalia *et al.*, 2007). Upon transfer to a cryo-holder, the grids were viewed in the EM under low-dose cryo conditions. The potential variables inherent in plunge freezing are numerous, involving apparatus type, plunging distance and speed, and blotting time (Grassucci *et al.*, 2007; Iancu *et al.*, 2007). By extension, the microscopy and image-processing steps that follow are also time consuming, not to mention the identification of structures in a functional context (e.g., Hoenger and McIntosh, 2009; Lucic *et al.*, 2008), and are thus well beyond consideration in this short review (see also Chapter 22 by Resch *et al.*, this volume).

The tomographic reconstructions of actin filament networks in *Dictyostelium* were the first published of cells that were frozen while alive and then imaged via cryo-ET (Fig. 2 and Medalia *et al.*, 2002). The reconstructions depicted not only parallel arrangements of actin filaments, but also isotropic networks of filaments interconnected at varying angles and with different types of linkages. That study revealed structural arrangements in as native an environment as possible, a significant achievement since filament organizations are known to be sensitive to fixation and preparative conditions (Small *et al.*, 2008: see also Chapter 22 by Resch *et al.*, this volume). Potentially, a

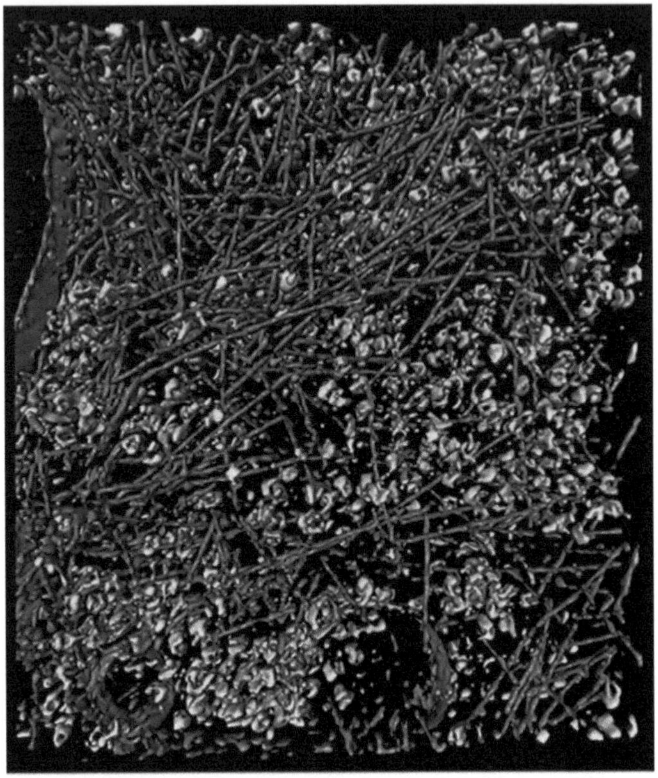

Fig. 2 Cryo-tomographic reconstruction of *Dictyostelium* cytoplasm. Linear elements, predominantly actin filaments, are presented in dark orange, membranes in blue, and macromolecular complexes (ribosomes, etc.) are in gold. This image comprises a cytoplasmic volume of 815 × 870 × 97 nm. From Medalia *et al.* (2002), reprinted with permission from AAAS. (See Plate no. 8 in the Color Plate Section.)

number of physical parameters of the filament networks can be quantitated in cells imaged by cryo-ET (branching angles, filament length, orientation to membrane, etc.) and can be compared with organizations in mutant cells that lack specific actin-binding or regulatory proteins. If one can correlate the function of that particular portion of the cell with its detailed structure, these types of analyses will significantly advance our understanding of how filament networks are modified within cells to produce directional cell movement. Medalia and coworkers took a step forward toward this goal by examining filament organization in filopodia (Medalia *et al.*, 2007). The authors took advantage of a Rac1A-overexpressing strain to increase the number of filopodia per cell and then generated tomographic reconstructions showing actin filament organization. In filopodia that appeared consistent with actively growing structures, membrane protrusion seems to occur in a novel conical region populated by short actin filaments, many with both ends anchored to the plasma membrane. Moreover, there were distinct structural

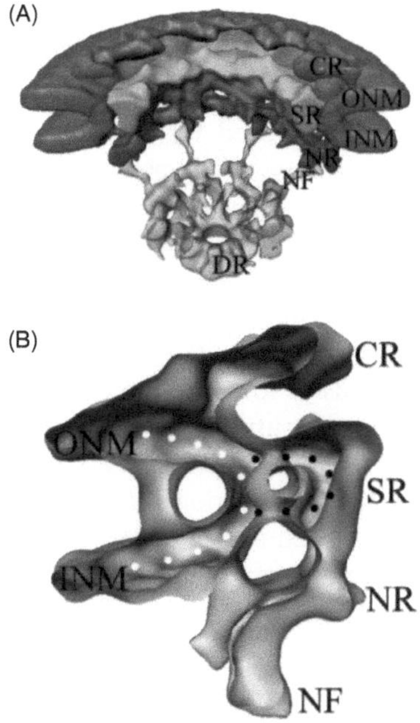

Fig. 3 Reconstructed in situ nuclear pore complex from *Dictyostelium*. A. Cut-away view of the pore complex embedded in the nuclear envelope. ONM, INM: outer and inner nuclear membrane, respectively, CR, SR, NR: cytoplasmic, spoke, and nuclear ring features; NF: nuclear filament; DR: distal ring. The central plug/transporter feature has been omitted for clarity. B. One view of the individual protomer unit that assembles into the intact 8-fold pore complex. The white dots indicate the nuclear envelope; the black dots outline the clamp-shaped spoke structure important for membrane interaction. Adapted by permission from Macmillan Publishers Ltd. (Beck *et al.*, *Nature* **449**, 611–615, 2007). (See Plate no. 9 in the Color Plate Section.)

differences in filament architecture between what were likely growing and retracting filopodia that suggested force redistributions and altered mechanical stability.

The *Dictyostelium* model has also significantly advanced our understanding of nuclear pore structure and function. Since the nuclei in *Dictyostelium* are relatively small (~2 μm diameter), can be readily isolated by a gentle lysis and sedimentation procedure, and have been shown to retain transport activity, their pores are well suited to cryo-ET analysis (Beck *et al.*, 2004). Reconstructions of nuclear pores computed from tilt series of frozen hydrated nuclei were used to map their overall architecture; closer inspection revealed at least two distinct structural states, between which the positions of the nuclear basket and luminal spoke features differed (Beck *et al.*, 2004). A later study from the same group developed a novel image-processing strategy to accommodate variations in the eight-fold unit structure, allowing the authors to average asymmetric units and develop structures for the repeated subassemblies (protomers) (Fig. 3, Beck *et al.*, 2007). The averaging process substantially increased resolution of the reconstructed pores from 8.3 to 5.8 nm, revealing new structural features and permitting imaging/simulations of the molecular import process.

C. Single-Particle Methods

The development of single-particle image averaging procedures has revolutionized 3D EM structure determination of nonsymmetric objects (e.g., Frank, 2006). While not one of the systems used for methods development, *Dictyostelium* does provide an excellent example of single-particle averaging, one that highlights its strength as a model organism. Dynein is a MT-based motor that couples ATP hydrolysis to force-producing conformational changes; such motors drive intracellular transport toward MT minus ends at rates close to 1–2 μm/s. Major challenges lie in understanding how dynein operates and how it is regulated in cells. A primary difficulty in working with dynein lies in managing its exceptionally large size, a single heavy chain of ~4500 amino acid folds to form the motor and cargo-binding domains. Although MT-based kinesin motors are sufficiently small to be functionally expressed in *Escherichia coli*, the large polypeptide fragment that comprises the dynein motor cannot. *Dictyostelium* has proven to be exceptionally useful in this regard, as it was straightforward to assemble coding sequences and promoter on a plasmid, then transform *Dictyostelium* and isolate cell lines producing robust expression levels of dynein polypeptides (Koonce and Samsó, 1996, Nishiura *et al.*, 2004). Moreover, dynein fragments can be readily purified for biochemical, motility, and structural assays (Fig. 4). Early dynein mapping work, combined with negative-staining and EM imaging, identified the polypeptide sequence responsible for forming the ~380-kDa motor domain (Koonce and Samsó, 1996). Single-particle averaging of the isolated motor further demonstrated that dynein is fundamentally different in structure from a kinesin or myosin, in that the motor domain is composed of an annular structure instead of a globular one (Fig. 5; Roberts *et al.*, 2009; Samsó and Koonce, 2004; Samsó *et al.*, 1998). Motor-decorated MTs have also been imaged and subjected to particle averaging, demonstrating the orientations of the ring-shaped motor and MT-binding domains relative to the MT (Mizuno *et al.*, 2007; Narita *et al.*, 2007). More detailed studies have placed a number of site-specific tags into the motor, for use as FRET sensors, as selective crosslinkers, and as fiduciary markers to deduce conformational changes during force production and to correlate sequence with EM structure (Kon *et al.*, 2005, 2009; Meng *et al.*, 2006; Nishiura *et al.*, 2004; Roberts *et al.*, 2009). The most recent single-particle analysis of *Dictyostelium* dynein effectively used GFP insertions to map the positions of the motor's AAA-subunit domains (Roberts *et al.*, 2009). This elegant combination of EM and molecular genetic manipulations demonstrates the functional utility of this organism and has produced the most detailed structural views obtained thus far of the dynein motor domain.

III. Future Challenges

A. Correlative Live Cell Dynamics and Cell Ultrastructure

A long-term approach in cell biology has been to follow events in live cells by light microscopy and then use EM to resolve the structural details that underlie such events.

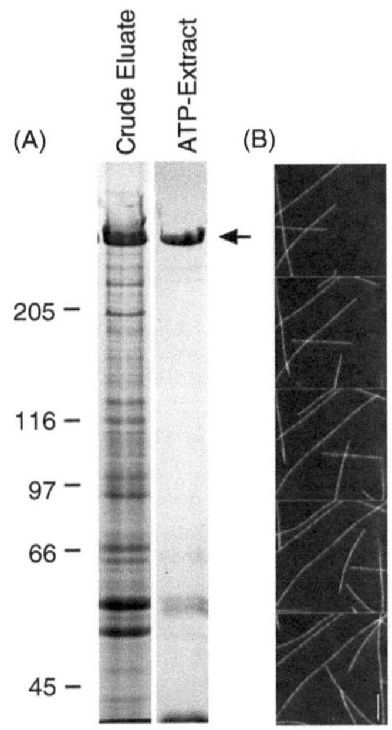

Fig. 4 Purification of recombinant dynein from *Dictyostelium*. His-tagged 380-kDa dynein motor expressed in *Dictyostelium* can be readily bound to nickel-affinity resin and eluted with buffer-containing imidazole. A number of native *Dictyostelium* polypeptides also bind His-affinity resins so this first step only crudely purifies the motor. Two secondary affinity procedures have been successfully used to produce purified protein. One involves binding dynein to taxol-stabilized microtubules, centrifugation, and extraction with buffer-containing ATP, as shown in this figure (M. Koonce, unpublished) and in Nishiura *et al.*, 2004. The second method incorporates binding to a FLAG-tag affinity column and elution with the FLAG peptide (Kon *et al.*, 2004). Either method generates dynein with sufficient purity and activity for motility and/or EM analyses. Panel A shows Coomassie blue-stained gel lanes following crude elution from Talon resin and after ATP extraction of the microtubule pellet. The sequence in panel B shows fluorescently labeled MTs gliding on a coverslip that has been coated with the dynein motor fragment purified in (A). All of the microtubules in this field of view are moving at about 1 μm/s. The five panels were recorded at 4-s intervals. Scale bar = 5 μm.

While correlative light microscopy (LM) and EM (CLEM) studies have been performed for decades, methodological improvements now include steps for high-pressure freezing (Brown *et al.*, 2009; McDonald, 2009; Verkade, 2008), and post-freezing identification of areas and structures of interest (Sartori *et al.*, 2007; Schwartz *et al.*, 2007; van Driel *et al.*, 2009). Tomographic reconstructions of such preparations have conferred improvements in resolution over traditional chemical fixations and sectioning, such that tomography appears to have the potential to dramatically broaden/increase our understanding of dynamic cell processes. *Dictyostelium* offers multiple advantages as a model for these types of studies. Two such strengths lie in investigating the extension of cell membranes during directed cell motion and the structural reorganizations that accompany phagocytosis.

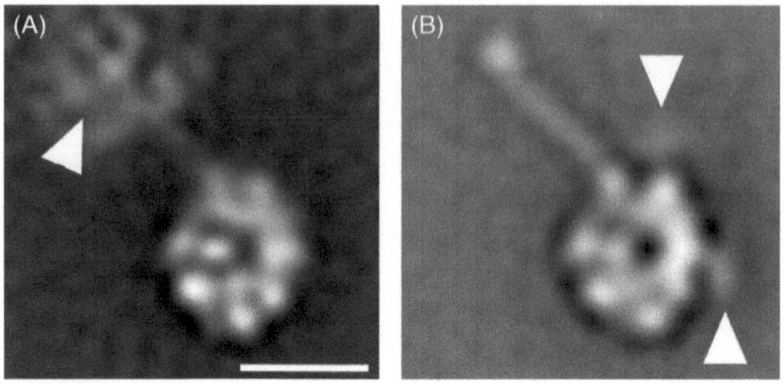

Fig. 5 Single-particle class averages of the dynein motor. Examples of the negatively stained motor domain from two independent studies showing approximately the same orientation. In panel A, the microtubule-binding domain is decorated with a polyclonal pool of Fabs, accounting for the wispy cloud-like appearance in the upper left of the image (arrowhead). Panel B shows a motor that has been decorated with GFP and BFP molecules, visible as diffuse stain-excluding areas just off the motor ring, marked by arrowheads. Though independently prepared, imaged, and image processed, the two studies illustrate a nearly identical molecule. This panel illustrates the robustness of the image-processing procedures and demonstrates two methods used to map sequence position of the dynein motor. Panels A and B are adapted with permission from Elsevier (Samsó and Koonce, *J. Mol. Biol.* **340**, 1059–1072, 2004; Roberts *et al., Cell* **136**, 485–495, 2009). Scale bar = 10 nm.

Dictyostelium has long been known to respond to cAMP as a chemoattractant, and detailed biochemical pathways have been mapped with the goal of understanding how this signal is received, interpreted, and relayed (Chen *et al.*, 2007b; Franca-Koh *et al.*, 2006; Sasaki and Firtel, 2009). Moreover, large numbers of mutants have been generated in which individual or multiple components in these signaling processes have been deleted. *Dictyostelium* cells will follow a micropipette tip filled with cAMP as it is moved across a coverslip. Improvements on the micropipette assay include the use of microfabricated flow chambers and photoactivatable signaling molecules (Beta *et al.*, 2007; Song *et al.*, 2006); thus, it is now possible to generate very precise spatial and temporal signals to direct cell protrusions. The combination of such flow chamber assays with cryo-ET (a challenging prospect, to be sure!) will enable the examination of the very earliest stages of actin filament reorganization at the underlying cell cortex. Strength of the *Dictyostelium* model lies in the capability to integrate these analyses with the use of signaling pathway mutants, thus providing structural data to support robust modeling efforts, as well as to differentiate the contributions of multiple signaling pathways.

A second advantage of *Dictyostelium* is that it readily engulfs bacteria, yeast, latex beads, and several pathogenic organisms and thus has widely been used to examine the molecular mechanisms of phagocytosis (Annesley and Fisher, 2009; Bozzaro *et al.*, 2008; Cosson and Soldati, 2008; Hagedorn *et al.*, 2009; Hägele *et al.*, 2000; Isik *et al.*, 2008; Sasaki and Firtel, 2009; Steinert and Heuner, 2005). With current TEM and SEM methods, it is not difficult to show membrane rearrangements and internalized particles, but extension of these analyses to cryo-ET will enable a much deeper understanding of the actin filament reorganization steps that underlie membrane invagination. Once again, the strength of the *Dictyostelium* model is the existence of an arsenal of mutations that perturb these processes (e.g., Chen *et al.*, 2007a; Isik *et al.*, 2008; Vlahou *et al.*, 2009); thus, combining mutant analysis with cryo-ET methods will enhance the detail of filament dynamics as well as further characterize the key roles of the mutated proteins. Of particular biomedical importance are the mechanisms by which certain pathogens can avoid the normal digestive pathway of endosomal trafficking, then propagate and escape the cellular confines. The detail provided by EM is required to understand these events.

B. Protein Localization

Recently, a novel fusion tag consisting of three metallothionein heavy metal-binding domains followed by a FLAG-tag sequence was used to construct a C-terminal fusion plasmid vector for expression of tagged proteins (Mercogliano and DeRosier, 2007); it has been applied in bacteria (Diestra *et al.*, 2009; Nishino *et al.*, 2007) and in hippocampal neurons (Fukunaga *et al.*, 2007). Treatment of cells with cadmium or gold salts results in metal deposition onto the metallothionein tag. After fast-freezing and freeze-substitution, electron-dense particles were visualized by EM to demonstrate localization of the fusion proteins. Given the relative ease by which *Dictyostelium* can be transformed to generate stable cell lines expressing fusion proteins, it may serve as a particularly good model system for routine EM localizations by this method. Preliminary experiments in *Dictyostelium* have already shown that the nuclear envelope protein, Sun1, and the centrosomal protein, CP39, can each be fused to a triple metallothionein–FLAG-tag and that both fusion proteins are well expressed. Furthermore, immunofluorescence microscopy demonstrates that the fusion proteins localize correctly to the nuclear envelope and centrosomes, respectively (R. Gräf and I. Meyer, unpublished). Although the incorporation of this labeling technique in EM preparations of *Dictyostelium* is only in progress, the method holds much promise for future studies. Metal tagging of fusion proteins provides a means by which to localize proteins in cells at EM resolution without the use of a specific antibody. Its use may increase the fidelity of localization since the electron-dense label is situated much closer to the folded polypeptide chain than is the case when primary and gold-labeled secondary antibodies are used. Deposition can be accomplished by the incubation of live cells with metal salts dissolved in the culture medium, without the need for detergent permeabilization or sectioning to gain access to the target proteins. Finally, it is possible to perform partial gene replacements in *Dictyostelium*, such that tagged proteins are expressed under endogenous promoter control (Daunderer and Gräf, 2002). Thus, native levels of polypeptide can also be analyzed.

IV. Summary Comments

Dictyostelium is a versatile model organism that has been the subject of numerous and varied analyses of cell structure and dynamics. Perhaps its greatest strength lies in the relative ease with which it can be manipulated to examine mechanisms of cell motility, development, signaling, and trafficking. It imposes no unique technical demands; rather, this single motile cell grows in simple cell cultures and is amenable to analysis by virtually every EM method available. Cryofixation techniques have been adapted to study this organism's multicellular developmental stages with excellent ultrastructure preservation. Lastly, the available mutants and well-established presence in the literature make *Dictyostelium* highly suited for ultrastructural analyses that address the fundamental questions of eukaryotic cell biology.

Acknowledgments

Work in the authors' laboratories is supported by the National Science Foundation (MCB-0542731 to M.P.K.) and the Deutsche Forschungsgemeinschaft (DFG GR1642/3-1 to R.G). We thank Drs Stan Burgess, Richard McIntosh, Ohad Medalia, and Kazuo Sutoh for kindly supplying figures.

References

Annesley, S., and Fisher, P. (2009). *Dictyostelium discoideum*—a model for many reasons. *Mol. Cell. Biochem.* **329**, 73–91.

Beck, M., Forster, F., Ecke, M., Plitzko, J. M., Melchior, F., Gerisch, G., Baumeister, W., and Medalia, O. (2004). Nuclear pore complex structure and dynamics revealed by cryoelectron tomography. *Science* **306**, 1387–1390.

Beck, M., Lucic, V., Forster, F., Baumeister, W., and Medalia, O. (2007). Snapshots of nuclear pore complexes in action captured by cryo-electron tomography. *Nature* **449**, 611–615.

Beta, C., Wyatt, D., Rappel, W.-J., and Bodenschatz, E. (2007). Flow photolysis for spatiotemporal stimulation of single cells. *Anal. Chem.* **79**, 3940–3944.

Bozzaro, S., Bucci, C., Steinert, M., and Kwang, W. J. (2008). Phagocytosis and host–pathogen interactions in *Dictyostelium* with a look at macrophages. *Int. Rev. Cell Mol. Biol.* **271**, 253–300.

Brown, E., Mantell, J., Carter, D., Tilly, G., and Verkade, P. (2009). Studying intracellular transport using high-pressure freezing and correlative light electron microscopy. *Semin Cell Dev.. Biol.* **20**, 910–919.

Chen, L., Iijima, M., Tang, M., Landree, M. A., Huang, Y. E., Xiong, Y., Iglesias, P. A., and Devreotes, P. N. (2007b). PLA2 and PI3K/PTEN pathways act in parallel to mediate chemotaxis. *Dev. Cell* **12**, 603–614.

Chen, G., Zhuchenko, O., and Kuspa, A. (2007a). Immune-like phagocyte activity in the social amoeba. *Science* **317**, 678–681.

Clarke, M., Schatten, G., Mazia, D., and Spudich, J. A. (1975). Visualization of actin fibers associated with the cell membrane in amoebae of *Dictyostelium discoideum*. *Proc. Natl. Acad. Sci. U.S.A.* **72**, 1758–1762.

Clarke, M., and Spudich, J. A. (1974). Biochemical and structural studies of actomyosin-like proteins from non-muscle cells. Isolation and characterization of myosin from amoebae of *Dictyostelium discoideum*. *J. Mol. Biol.* **86**, 209–222.

Claviez, M., Pagh, K., Maruta, H., Baltes, W., Fisher, P., and Gerisch, G. (1982). Electron microscopic mapping of monoclonal antibodies on the tail region of *Dictyostelium* myosin. *EMBO J.* **1**, 1017–1022.

Condeelis, J., Ogihara, S., Bennett, H., Carboni, J., and Hall, A. (1987). Ultrastructural localization of cytoskeletal proteins in *Dictyostelium* amoebae. *Methods Cell Biol.* **28**, 191–207.

Cosson, P., and Soldati, T. (2008). Eat, kill or die: When amoeba meets bacteria. *Curr. Opin. Microbiol.* **11**, 271–276.

Daunderer, C., and Gräf, R. O. (2002). Molecular analysis of the cytosolic *Dictyostelium* gamma–tubulin complex. *Eur. J. Cell Biol.* **81**, 175–184.

De Priester, W. (1991). Techniques for the visualization of cytoskeletal components in *Dictyostelium discoideum*. *Electron Micros. Rev.* **4**, 343–376.

Diestra, E., Fontana, J., Guichard, P., Marco, S., and Risco, C. (2009). Visualization of proteins in intact cells with a clonable tag for electron microscopy. *J. Struct. Biol.* **165**, 157–168.

Dubochet, J., Adrian, M., Chang, J. J., Homo, J. C., Lepault, J., McDowall, A. W., and Schultz, P. (1988). Cryo-electron microscopy of vitrified specimens. *Q. Rev. Biophys.* **21**, 129–228.

Eichinger, L., and Rivero, F. (2006). *Dictyostelium discoideum* protocols. *Methods Mol. Biol.* **346**, 564.

Euteneuer, U., Gräf, R., Kube-Granderath, E., and Schliwa, M. (1998). *Dictyostelium* gamma-tubulin: Molecular characterization and ultrastructural localization. *J. Cell. Sci.* **111**, 405–412.

Flicker, P. F., Peltz, G., Sheetz, M. P., Parham, P., and Spudich, J. A. (1985). Site-specific inhibition of myosin-mediated motility in vitro by monoclonal antibodies. *J. Cell Biol.* **100**, 1024–1030.

Franca-Koh, J., Kamimura, Y., and Devreotes, P. (2006). Navigating signaling networks: Chemotaxis in *Dictyostelium discoideum*. *Curr. Opin. Gen. Dev.* **16**, 333–338.

Frank, J. (2006). "Three-Dimensional Electron Microscopy of Macromolecular Assemblies." Oxford University Press, Oxford, UK. (pp. 410)

Fukunaga, Y., Hirase, A., Kim, H., Wada, N., Nishino, Y., and Miyazawa, A. (2007). Electron microscopic analysis of a fusion protein of postsynaptic density-95 and metallothionein in cultured hippocampal neurons. *J. Electron Microsc. (Tokyo)* **56**, 119–129.

Gezelius, K., and Ranby, B. G. (1957). Morphology and fine structure of the slime mold *Dictyostelium discoideum*. *Exp. Cell Res.* **12**, 265–289.

Gräf, R. (2001). Isolation of centrosomes from *Dictyostelium*. *Methods Cell Biol.* **67**, 337–357.

Grassucci, R. A., Taylor, D. J., and Frank, J. (2007). Preparation of macromolecular complexes for cryo-electron microscopy. *Nat. Protoc.* **2**, 3239–3246.

Griffith, L. M., Downs, S. M., and Spudich, J. A. (1987). Myosin light chain kinase and myosin light chain phosphatase from *Dictyostelium*: Effects of reversible phosphorylation on myosin structure and function. *J. Cell Biol.* **104**, 1309–1323.

Grimson, M. J., and Blanton, R. L. (2006). Cryofixation methods for ultrastructural studies of *Dictyostelium discoideum*. *Methods Mol. Biol.* **346**, 339–365.

Grimson, M. J., Coates, J. C., Reynolds, J. P., Shipman, M., Blanton, R. L., and Harwood, A. J. (2000). Adherens junctions and [beta]-catenin-mediated cell signaling in a non-metazoan organism. *Nature* **408**, 727–731.

Hagedorn, M., Neuhaus, E. M., and Soldati, T. (2006). Optimized fixation and immunofluorescence staining methods for *Dictyostelium* cells. *Methods Mol. Biol.* **346**, 327–338.

Hagedorn, M., Rohde, K. H., Russell, D. G., and Soldati, T. (2009). Infection by tubercular mycobacteria is spread by nonlytic ejection from their amoeba hosts. *Science* **323**, 1729–1733.

Hägele, S., Köhler, R., Merkert, H., Schleicher, M., Hacker, J., and Steinert, M. (2000). *Dictyostelium discoideum*: A new host model system for intracellular pathogens of the genus *Legionella*. *Cell Microbiol.* **2**, 165–171.

Hoenger, A., and McIntosh, J. R. (2009). Probing the macromolecular organization of cells by electron tomography. *Curr. Opin. Cell Biol.* **21**, 89–96.

Iancu, C. V., Tivol, W. F., Schooler, J. B., Dias, D. P., Henderson, G. P., Murphy, G. E., Wright, E. R., Li, Z., Yu, Z., Briegel, A., Gan, L., He, Y., *et al.* (2007). Electron cryotomography sample preparation using the vitrobot. *Nat. Protoc.* **1**, 2813–2819.

Isik, N., Brzostowski, J. A., and Jin, T. (2008). An elmo-like protein associated with myosin II restricts spurious F-actin events to coordinate phagocytosis and chemotaxis. *Dev. Cell* **15**, 590–602.

Kessin, R. H. (2001). "*Dictyostelium*: Evolution, Cell Biology, and the Development of Multicellularity." Cambridge University Press, Cambridge, UK, (pp. 294).

Kon, T., Imamula, K., Roberts, A. J., Ohkura, R., Knight, P. J., Gibbons, I. R., Burgess, S. A., and Sutoh, K. (2009). Helix sliding in the stalk coiled coil of dynein couples ATPase and microtubule binding. *Nat. Struct. Mol. Biol.* **16**, 325–333.

Kon, T., Mogami, T., Ohkura, R., Nishiura, M., and Sutoh, K. (2005). ATP hydrolysis cycle-dependent tail motions in cytoplasmic dynein. *Nat. Struct. Mol. Biol.* **12**, 513–519.

Kon, T., Nishiura, M., Ohkura, R., Toyoshima, Y. Y., and Sutoh, K. (2004). Distinct functions of nucleotide-binding/hydrolysis sites in the four AAA modules of cytoplasmic dynein. *Biochemistry* **43**, 11266–11274.

Koonce, M. P., and Samsó, M. (1996). Overexpression of cytoplasmic dynein's globular head causes a collapse of the interphase microtubule network in *Dictyostelium*. *Mol. Biol. Cell* **7**, 935–948.

Kuriyama, R., Sato, C., Fukui, Y., and Nishibayashi, S. (1982). *In vitro* nucleation of microtubules from microtubule-organizing center prepared from cellular slime mold. *Cell Motil.* **2**, 257–272.

Lucic, V., Leis, A., and Baumeister, W. (2008). Cryo-electron tomography of cells: Connecting structure and function. *Histochem. Cell Biol.* **130**, 185–196.

Luna, E. J., Fowler, V. M., Swanson, J., Branton, D., and Taylor, D. L. (1981). A membrane cytoskeleton from *Dictyostelium discoideum*. I. Identification and partial characterization of an actin-binding activity. *J. Cell Biol.* **88**, 396–409.

Luna, E. J., Goodloe-Holland, C. M., and Ingalls, H. M. (1984). A membrane cytoskeleton from *Dictyostelium discoideum*. II. Integral proteins mediate the binding of plasma membranes to F-actin affinity beads. *J. Cell Biol.* **99**, 58–70.

Manstein, D. J., Ruppel, K. M., and Spudich, J. A. (1989). Expression and characterization of a functional myosin head fragment in *Dictyostelium discoideum*. *Science* **246**, 656–658.

Marchetti, A., Mercanti, V., Cornillon, S., Alibaud, L., Charette, S. J., and Cosson, P. (2004). Formation of multivesicular endosomes in *Dictyostelium*. *J. Cell Sci.* **117**, 6053–6059.

Mazia, D., Schatten, G., and Sale, W. (1975). Adhesion of cells to surfaces coated with polylysine. Applications to electron microscopy. *J. Cell Biol.* **66**, 198–200.

McDonald, K. (1984). Osmium ferricyanide fixation improves microfilament preservation and membrane visualization in a variety of animal cell types. *J. Ultrastruct. Res.* **86**, 107–118.

McDonald, K. L. (2009). A review of high-pressure freezing preparation techniques for correlative light and electron microscopy of the same cells and tissues. *J. Microsc.* **235**, 273–281.

McIntosh, J. R., Roos, U. P., Neighbors, B., and McDonald, K. L. (1985). Architecture of the microtubule component of mitotic spindles from *Dictyostelium discoideum*. *J. Cell Sci.* **75**, 93–129.

McMahon, P., Hostetter, D., and Rice, S. (2008). Temperature dependence of myosin-II tail fragment assembly. *J. Mus. Res. Cell Motil.* **29**, 109–118.

Medalia, O., Beck, M., Ecke, M., Weber, I., Neujahr, R., Baumeister, W., and Gerisch, G. (2007). Organization of actin networks in intact filopodia. *Curr. Biol.* **17**, 79–84.

Medalia, O., Weber, I., Frangakis, A. S., Nicastro, D., Gerisch, G., and Baumeister, W. (2002). Macromolecular architecture in eukaryotic cells visualized by cryoelectron tomography. *Science* **298**, 1209–1213.

Meng, X., Samsó, M., and Koonce, M. P. (2006). A flexible linkage between the dynein motor and its cargo. *J. Mol. Biol.* **357**, 701–706.

Mercogliano, C. P., and DeRosier, D. J. (2007). Concatenated metallothionein as a clonable gold label for electron microscopy. *J. Struct. Biol.* **160**, 70–82.

Mizuno, N., Narita, A., Kon, T., Sutoh, K., and Kikkawa, M. (2007). Three-dimensional structure of cytoplasmic dynein bound to microtubules. *Proc. Natl. Acad. Sci. U.S.A.* **104**, 20832–20837.

Moens, P. B. (1976). Spindle and kinetochore morphology of *Dictyostelium discoideum*. *J. Cell Biol.* **68**, 113–122.

Narita, A., Mizuno, N., Kikkawa, M., and Maeda, Y. (2007). Molecular determination by electron microscopy of the dynein–microtubule complex structure. *J. Mol. Biol.* **372**, 1320–1336.

Neuhaus, E. M., Horstmann, H., Almers, W., Maniak, M., and Soldati, T. (1998). Ethane-freezing/methanol-fixation of cell monolayers: A procedure for improved preservation of structure and antigenicity for light and electron microscopies. *J. Struct. Biol.* **121**, 326–342.

Nishino, Y., Yasunaga, T., and Miyazawa, A. (2007). A genetically encoded metallothionein tag enabling efficient protein detection by electron microscopy. *J. Electron Microsc. (Tokyo)* **56**, 93–101.

Nishiura, M., Kon, T., Shiroguchi, K., Ohkura, R., Shima, T., Toyoshima, T. T., and Sutoh, K. (2004). A single-headed recombinant fragment of *Dictyostelium* cytoplasmic dynein can drive the robust sliding of microtubules. *J. Biol. Chem.* **279**, 22799–22802.

Omura, F., and Fukui, Y. (1985). *Dictyostelium* MTOC: Structure and linkage to the nucleus. *Protoplasma* **127**, 212–221.

Rehberg, M., Kleylein-Sohn, J., Faix, J., Ho, T. H., Schulz, I., and Gräf, R. (2005). *Dictyostelium* LIS1 is a centrosomal protein required for microtubule/cell cortex interactions, nucleus/centrosome linkage, and actin dynamics. *Mol. Biol. Cell* **16**, 2759–2771.

Roberts, A. J., Numata, N., Walker, M. L., Kato, Y. S., Malkova, B., Kon, T., Ohkura, R., Arisaka, F., Knight, P. J., Sutoh, K., and Burgess, S. A. (2009). AAA+ ring and linker swing mechanism in the dynein motor. *Cell* **136**, 485–495.

Roos, U. P. (1975). Mitosis in the cellular slime mold *Polysphondylium violaceum*. *J. Cell Biol.* **64**, 480–491.

Roos, U. P. (1987). Probing the mechanisms of mitosis with *Dictyostelium discoideum*. *Methods Cell Biol.* **28**, 261–279.

Sameshima, M. (1984). Preparation of *Dictyostelium discoideum* for electron microscopy. *Cell Struct. Funct.* **9**, 187–192.

Sameshima, M., Kishi, Y., Osumi, M., Mahadeo, D., and Cotter, D. A. (2002). Electron microscopy of actin rods and bundles in *Dictyostelium discoideum* by high-pressure freezing. *J. Electron Microsc. (Tokyo)* **51**, 337–340.

Samsó, M., and Koonce, M. P. (2004). 25 Angstrom resolution structure of a cytoplasmic dynein motor reveals a seven-member planar ring. *J. Mol. Biol.* **340**, 1059–1072.

Samsó, M., Radermacher, M., Frank, J., and Koonce, M. P. (1998). Structural characterization of a dynein motor domain. *J. Mol. Biol.* **276**, 927–937.

Sartori, A., Gatz, R., Beck, F., Rigort, A., Baumeister, W., and Plitzko, J. M. (2007). Correlative microscopy: Bridging the gap between fluorescence light microscopy and cryo-electron tomography. *J. Struct. Biol.* **160**, 135–145.

Sasaki, A. T., and Firtel, R. A. (2009). Spatiotemporal regulation of ras-GTPases during chemotaxis. *Methods Mol. Biol.* **571**, 333–348.

Schulz, I., Baumann, O., Samereier, M., Zoglmeier, C., and Gräf, R. (2009). *Dictyostelium* sun1 is a dynamic membrane protein of both nuclear membranes and required for centrosomal association with clustered centromeres. *Eur. J. Cell Biol.* **88**, 621–638.

Schwartz, C. L., Sarbash, V. I., Ataullakhanov, F. I., McIntosh, J. R., and Nicastro, D. (2007). Cryo-fluorescence microscopy facilitates correlations between light and cryo-electron microscopy and reduces the rate of photobleaching. *J. Micros.* **227**, 98–109.

Shkilnyy, A., Gräf, R., Hiebl, B., Neffe, A. T., Friedrich, A., Hartmann, J., and Taubert, A. (2009). Unprecedented, low cytotoxicity of spongelike calcium phosphate/poly(ethylene imine) hydrogel composites. *Macromol. Biosci.* **9**, 179–186.

Sillo, A., Bloomfield, G., Balest, A., Balbo, A., Pergolizzi, B., Peracino, B., Skelton, J., Ivens, A., and Bozzaro, S. (2008). Genome-wide transcriptional changes induced by phagocytosis or growth on bacteria in *Dictyostelium*. *BMC Genomics* **9**, 291.

Small, J. V., Auinger, S., Nemethova, M., Koestler, S., Goldie, K. N., Hoenger, A., and Resch, G. P. (2008). Unravelling the structure of the lamellipodium. *J. Microsc.* **231**, 479–485.

Song, L., Nadkarni, S. M., Bodeker, H. U., Beta, C., Bae, A., Franck, C., Rappel, W.-J., Loomis, W. F., and Bodenschatz, E. (2006). *Dictyostelium discoideum* chemotaxis: Threshold for directed motion. *Eur. J. Cell Biol.* **85**, 981–989.

Spudich, J. A. (1974). Biochemical and structural studies of actomyosin-like proteins from non-muscle cells. II. Purification, properties, and membrane association of actin from amoebae of *Dictyostelium discoideum*. *J. Biol. Chem.* **249**, 6013–6020.

Steinert, M., and Heuner, K. (2005). *Dictyostelium* as host model for pathogenesis. *Cell Microbiol.* **7**, 307–314.

Ueda, M., Schliwa, M., and Euteneuer, U. (1999). Unusual centrosome cycle in *Dictyostelium*: Correlation of dynamic behavior and structural changes. *Mol. Biol. Cell.* **10**, 151–160.

van Driel, L. F., Valentijn, J. A., Valentijn, K. M., Koning, R. I., and Koster, A. J. (2009). Tools for correlative cryo-fluorescence microscopy and cryo-electron tomography applied to whole mitochondria in human endothelial cells. *Eur. J. Cell Biol.* **88**, 669–684.

Verkade, P. (2008). Moving EM: The rapid transfer system as a new tool for correlative light and electron microscopy and high throughput for high-pressure freezing. *J. Micros.* **230**, 317–328.

Vlahou, G., Schmidt, O., Wagner, B., Uenlue, H., Dersch, P., Rivero, F., and Weissenmayer, B. (2009). *Yersinia* outer protein YopE affects the actin cytoskeleton in *Dictyostelium discoideum* through targeting of multiple rho family GTPases. *BMC Microbiol.* **9**, 138.

Williams, R. S. B., Boeckeler, K., Gräf, R., Muller-Taubenberger, A., Li, Z., Isberg, R. R., Wessels, D., Soll, D. R., Alexander, H., and Alexander, S. (2006). Towards a molecular understanding of human diseases using *Dictyostelium discoideum*. *Trend. Mol. Med.* **12**, 415–424.

Woolley, D. E. (1972). An actin-like protein from amoebae of *Dictyostelium discoideum*. *Arch. Biochem. Biophys.* **150**, 519–530.

Xiong, H., Rivero, F., Euteneuer, U., Mondal, S., Mana-Capelli, S., Larochelle, D., Vogel, A., Gassen, B., and Noegel, A. A. (2008). *Dictyostelium* Sun-1 connects the centrosome to chromatin and ensures genome stability. *Traffic* **9**, 708–724.

CHAPTER 10

Toward Sub-second Correlative Light and Electron Microscopy of *Saccharomyces cerevisiae*

Christopher Buser
Department of Molecular and Cell Biology, University of California, Berkeley, California 94720

Abstract

The yeast *Saccharomyces cerevisiae* is a model organism widely used to study cell biological processes because of its easy genomic manipulation and its close relatedness to higher eukaryotes. For electron microscopy, the good freezing properties and

978-0-12-381007-6
DOI: 10.1016/S0091-679X(10)96010-X

the small size of yeast cells make it a nearly ideal specimen for the development of cryopreparation techniques. Here we report on the development of a method to correlate yeast cells by live-fluorescence and electron microscopy with the potential to achieve sub-second correlation times. This is possible by plunge-freezing of an optically transparent sample sandwich, so that the temporal resolution is only determined by the transfer speed from the fluorescence microscope to the freezing device. While direct correlation was not yet achieved, the system already offers the possibility to verify the state of the identical population of cells by fluorescence microscopy immediately before freezing and processing for transmission electron microscopy.

I. Introduction

The yeast *Saccharomyces cerevisiae* has been a longtime favorite organism in genetics and cell biology research. Extensive research has shown that, while being a rather simple organism, many cell biological processes are highly conserved from yeast up to mammalian cells. A key advantage of budding yeast is the possibility to genomically integrate tags and mutations while maintaining normal expression levels of the protein of interest, whereas mammalian systems rely on protein overexpression from a transfected plasmid. With respect to electron microscopy, yeast has served well as a test specimen in the development of cryopreparation techniques. *S. cerevisiae* has been imaged numerous times by both scanning and transmission electron microscopy (SEM and TEM), either at ambient temperatures or in the frozen-hydrated state. For an in depth discussion of cryo-methods we refer to Echlin (1992). As early as 1963, Moor and Müehlethaler have used plunge freezing in liquid propane followed by freeze-etching, platinum-carbon coating, and imaging of the replica by TEM to reveal the fine structure of *S. cerevisiae* (Moor and Mühlethaler, 1963). The article describes various fine structures with previously unattained detail, among those nuclear pore baskets, hexagonal protein arrays and rod-like invaginations in the plasma membrane, and endocytic sites. The authors already made two key observations important in relation to this chapter: (1) Cryoprotection with 20% glycerol drastically increased the success rate of the plunge freezing preparation and (2) cryofixed samples were structurally superior to chemically fixed cells. Later, the freezing rates and thus the success rates of cryopreparation were improved by the development of propane jet freezing (Müller *et al.*, 1980). This technique allowed cryofixation of *Schizosaccharomyces pombe* without cryoprotection followed by high-resolution TEM and SEM imaging of the plasma membrane, including the cell wall and the characteristic furrows (Walther *et al.*, 1984). The function of these striking furrows in the plasma membrane has been elusive until recently. During their investigation of the membrane compartment of Can1 (MCC), Stradalova *et al.* (2009) have found evidence that the MCC coincides with yeast furrows, thus connecting the furrows with an extensive signaling network. This again underlines the importance of these early structural studies made possible by the advancement of cryopreparation techniques. A further beautiful example of plunge freezing of

S. cerevisiae followed by freeze-substitution (FS) for SEM and TEM is the work by Baba and Osumi (Baba and Osumi, 1987). The authors showed detailed views of nuclear pores, multivesicular bodies, Golgi-apparatus, secretory vesicles, and more, and synthesized their findings into a diagram of the yeast ultrastructure. A notable point of this work is the excellent structural preservation of the samples, even by today's standards. In recent years, cryopreparation of yeast has been used to study a variety of cell biological processes, including mitosis (McDonald *et al.*, 1996), autophagy (Roberts *et al.*, 2003), and interphase microtubule organization in *S. pombe* (Hoog *et al.*, 2007). With the recent "spreading" of Cryo-Electron Microscopy of Vitreous Sections (CEMOVIS) (Al-Amoudi *et al.*, 2004), yeast is again a popular sample.

Sadly, many cell-biology groups still prepare their yeast samples for TEM by chemical fixation, irrespective of the relative ease and vast superiority of cryo-methods. It is clear that a multitude of artifacts are induced by both the aldehyde fixative and the dehydration at room temperature. Of special interest with respect to this chapter are the striking structural changes induced in the endocytic pathway (Murk *et al.*, 2003). To make things worse many protocols first remove the cell wall, which most probably causes changes in the underlying structures before fixation can arrest the native state. While cryopreparation methods avoid the chemical stress by immobilizing cellular structures by non-invasive physical means on a timescale of milliseconds, they usually require concentrating the cells to achieve a sufficiently high density. In the case of high-pressure freezing (HPF) the cells also have to be transferred into the cavity of the freezing hats. These concentration and transfer steps do their part in stressing the cells and can cause large-scale morphological changes or more subtle changes in the cellular process to be studied. In respect to endocytosis, brief centrifugation of yeast cells causes depolarization of actin patches (Soto *et al.*, 2007) and also appears to disrupt endocytosis for several minutes (Buser and McDonald, 2010), while collection by filtration (McDonald and Mueller-Reichert, 2002) is more gentle but can cause osmotic effects by partial drying of the sample during loading into the freezing hats. To avoid both centrifugation and osmotic stress to the yeast cells, we recently proposed a filtration technique using syringe filters (Buser and McDonald, 2010) followed by a transfer into capillary tubes (Hohenberg *et al.*, 1994). While this method is a good compromise, it still requires a resuspension step and immersion in 1-hexadecane (Studer *et al.*, 1995), with which the cells are in contact for up to a minute. Ideally, the yeast cells should be immobilized in their native state without any manipulation whatsoever. This is an important advantage of the correlative light and electron microscopy (CLEM) approach presented below, which does not require collection of the cells. In addition, this approach allows us to verify by fluorescence microscopy if the process of interest is occurring normally in the same population of cells which are immediately plunge-frozen and prepared for TEM.

In this chapter we use *S. cerevisiae* as a model system to study the process of endocytosis by CLEM. Yeast actin patches, which were shown to be sites of endocytosis, have been studied extensively by fluorescence microscopy (Kaksonen *et al.*, 2003; Waddle *et al.*, 1996) and also by immuno-TEM (Idrissi *et al.*, 2008; Mulholland *et al.*, 1994). The emerging picture is that of a temporally and spatially highly regulated assembly of protein modules, consisting of a total of > 60 proteins, which drive the processes of coat recruitment, invagination of the plasma membrane coinciding with a burst of actin polymerization, and ultimately vesicle scission (Kaksonen *et al.*, 2005, 2006). While recruitment of the coat proteins takes place over a period of 1–2 min, membrane invagination and scission occur very rapidly within 15 s. This process is very well defined by fluorescence microscopy, but the exact geometry of the endocytic site and the dynamic localization of the individual proteins at the ultrastructural level can currently only be inferred based on the fluorescence data or alternatively imaged by TEM without the temporal information. True correlation of these datasets requires a technique able to image endocytic sites by live-fluorescence microscopy and to correlate this information with a TEM-image of the same site.

The current cryopreparation method of choice is HPF, which allows the freezing of non-cryoprotected samples up to a thickness of 200 μm without significant ice-crystal damage (Studer *et al.*, 1989). HPF is a very reproducible technique, but requires the sample to be positioned between two metal planchettes or sapphire discs (Hawes *et al.*, 2007). Unfortunately, manual assembly of the sample "sandwich" usually takes > 20 s in both the Abra (formerly BAL-TEC) and the Wohlwend-type HPF machines. The HPF with the shortest loading time is the Leica EM PACT2 equipped with the rapid transfer system which was specifically developed for CLEM (McDonald *et al.*, 2007; Muller-Reichert *et al.*, 2007; Pelletier *et al.*, 2006; Verkade, 2008). The system offers a temporal resolution of approximately 4 s combined with fluorescent imaging with low-numerical aperture (NA) 60× air objectives. This limitation is caused by the relatively thick sample carrier which prevents the use of high-NA oil immersion lenses due to their short working distance. Recently, Brown and colleagues (2009) have reduced the possible working distance by thinning the sample carrier. As outlined above, the final steps of endocytosis occur in a small volume within a few seconds. The imaging of fluorescently tagged endocytic proteins in yeast thus requires the use of high-NA objectives and a high temporal resolution with freezing to be able to the correlate rapid processes of membrane invagination and scission.

To fulfill these criteria, we propose a CLEM approach based on plunge freezing of a sample sandwich consisting of two glass coverslips separated by a thin formvar spacer. This system is ideal for fluorescence microscopy in that it allows imaging with a standard microscope using high-NA oil immersion objectives and can yield temporal correlation times of less than 1 s even by manual plunge freezing. Using a small sample volume, the inherent cryoprotection of yeast cells and the added cryoprotection by sugars in the growth medium, the sample can be fixed by plunge freezing in liquid propane with only minor crystal damage. Unfortunately, we still did not achieve same-cell imaging because of a lack of reproducibility in manually plunging the samples. Nevertheless, we believe that the ability to verify that the kinetics of endocytosis before cryo-immobilization and the potential to achieve sub-second correlation times are very valuable. At the end of this chapter, we will propose several experimental pathways that we intend to pursue to increase the rate of success in the plunge-freezing process.

II. Rationale

S. cerevisiae is frequently used for cryopreparation methods because of its natural cryoprotection by the dense cytoplasm. In addition, yeast can be grown in media with cryoprotectants as a carbon source (e.g., glucose or glycerol). Accordingly, yeast suspensions are easily frozen by HPF with little ice-crystal damage and vitrified for CEMOVIS, if grown in media containing 20–30% dextran. However, working with yeast also has its complications. Yeast cells are highly sensitive to changes in their environment, such as shifts in temperature, osmotic stress, chemicals, and the depletion of carbon sources. In addition, the sugars added to the medium in order to further reduce ice-crystal damage can cause infiltration problems during resin embedding since they are only poorly soluble in acetone during FS and need to be washed out before infiltration. Furthermore, the dense cytoplasm is a natural cryoprotectant, but also frequently obscures fine structures in sections.

Considering all aspects related to cell homeostasis, fluorescence microscopy, plunge freezing, and the preparation for TEM, the general requirements to be fulfilled by a CLEM holder system for the work proposed here are the following:

1. Imaging by fluorescence microscopy should be simple to perform and allow the use of oil immersion objectives.
2. The yeast cells need to be maintained in a stable and non-perturbing environment during assembly and imaging of the sample.
3. Sufficiently high cooling rates have to be achievable by plunge freezing to prevent excessive ice-crystal growth.
4. The sample design needs to be simple to allow sufficiently high throughput.
5. Disassembly of the sample for FS and resin embedding should be feasible without damaging or losing the cells.
6. A coordinate system has to be in place to identify the area of interest for both fluorescence and electron microscopy.

The approach presented here uses two 5 mm diameter glass coverslips separated by a $<10\,\mu m$ thin formvar spacer, which is bonded to the lower coverslip (Fig. 1A). A cavity is cut in the spacer and a carbon image of a finder grid is evaporated into the cavity. The yeast cells are attached to the carbon with concanavalin A (Kaksonen *et al.*, 2003) and embedded in 2% low-melting agarose (Sims and Hardin, 2007). This sandwich is sealed at the edges with vacuum grease (Kaksonen *et al.*, 2003) and attached onto a wire loop with Scotch tape. The sample rests on the microscope stage by attaching the wire loop to a support ring. After recording a time-lapse movie of the area of interest, the holder is picked up manually by the support ring and the sample on the central wire loop is plunged into liquid propane and prepared for TEM by FS and embedding in Epon.

The design presented here was optimized for fluorescence microscopy by using standard glass coverslips to sandwich the yeast, instead of using sapphire coverslips. Sapphire has superior thermal properties but is far more expensive and also has a

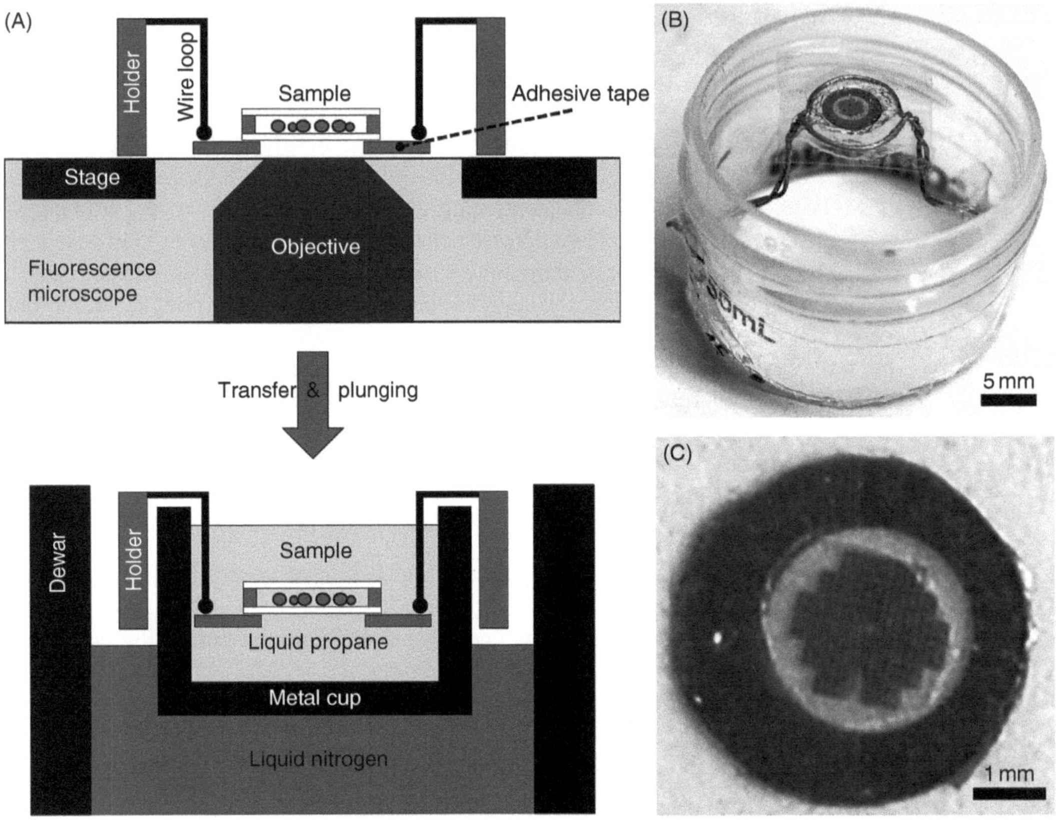

Fig. 1 Sample preparation for CLEM. (A) Schematic representation of the sample setup during fluorescence microscopy and after plunge freezing (not drawn to scale). The yeast cells are sandwiched between two glass coverslips separated by the formvar spacer and supported by a piece of Scotch tape with a central hole. The Scotch tape is sticking to the wire loop and the holder ring. During fluorescence microscopy the holder ring rests on the stage of the microscope, bringing the sample into the focal plane of the oil-immersion objective. At the desired moment, the holder is transferred and the sample plunged into the liquid propane bath. (B) Image of a fully assembled sample attached to the wire loop and the plastic holder ring. The holder is then inverted and positioned on the stage of the fluorescence microscope for imaging. Note the excess vacuum grease surrounding and sealing of the coverslip sandwich. (C) Image of a pre-assembled bottom coverslip with formvar spacer and the evaporated carbon finder grid in the center. The yeast cells are later attached in the central cavity by concanavalin A.

higher refractive index of 1.71, which would have to be matched with a specialized immersion fluid for optimal imaging. It is also crucial for light microscopy to bring the coverslip parallel to and into working distance of the oil immersion objective (common working distances are between 0.1 and 0.2 mm), and thus a thin but stable support is required. Here we used a piece of Scotch tape with a central viewing hole to attach the coverslip sandwich to the holder. This setup keeps the sample in focus of the immersion objective and perfectly level on the stage, thus allowing stage movements without the objective touching the sample at its edges. Furthermore, the tape is soft and prevents accidental scratching of the objective lens.

The next problem we addressed was how to maintain a stable environment. As discussed above, yeast cells respond very quickly to changing conditions. Proper sealing of the sample sandwich is exceedingly important to prevent drying of the very small volume of medium. To achieve a good seal the outer rim of the formvar spacer ring was coated with very little vacuum grease. When pressing the top coverslip in place the excess grease is expelled from the sandwich thus creating a tight seal. In addition, the outer edge of the sandwich was sealed off with more vacuum grease.

Optimizing the thermal properties to ensure adequate freezing was a key issue. The holder ring and the attached loop possess a low heat mass, position the sample parallel to the microscope stage and fit into the cavity of the liquid propane bath for plunge freezing. The design ensures minimal influence on the cooling rates, a robust way of manually transferring the sample for plunge freezing and also a sufficient stability during fluorescent imaging. For optimal freezing by immersion in liquid propane, the sample volume needs to be as small as possible, i.e. the thickness of the aqueous layer should be less than 10 μm and additional cryoprotection is usually required (Echlin, 1992). A low heat mass of the holder can be achieved by preferentially using parts made of metal with small dimensions. Formvar films were chosen as spacers because they can be cast reproducibly and the thickness is easily adjusted by varying the concentration of formvar in the casting solvent. For external cryoprotection, yeast can be grown in high concentrations of sugars without obvious adverse effects (see below). Here we supplemented our medium with 10% glucose, since 2% glucose is routinely present in the media used in our lab. While cryoprotectants like dextran or glycerol are more frequently used, dextran is known to cause infiltration problems while glycerol affects the morphology of mitochondria (Egner *et al.*, 2002) and thus might affect endocytosis as well.

More restrictions are imposed by practical aspects of assembly and post-processing of the frozen sample. The lower coverslip covered with the formvar spacer and the carbon finder pattern can be pre-assembled in large quantities and is stable over months. If the assembled sample is sealed well, the cells remain unperturbed for 30 min or more (see below). This currently allows the full assembly of 3–5 samples in parallel, while fluorescence microscopy, propane plunging, and sample storage together require approximately 10 min per sample under optimal conditions. The most challenging and time-consuming step remains the deposition of the cells and the sealing and mounting of the sample.

Post-processing of the frozen samples is possible with standard protocols for FS, embedding, trimming, and sectioning. Care has to be taken to remove the excess sugar after FS by repeated washing in ethanol, since remaining sugar causes incomplete infiltration with the resin. An important and often underestimated function of the carbon layers (one thin over the entire surface and one thicker for the finder pattern) is to act as a separator when breaking the coverslip off the polymerized resin. Cells directly adhering to uncoated glass slides are frequently ripped out of the resin during this step.

An additional advantage of the setup we used here is that it does not require expensive or highly specialized EM equipment, except for a FS device, a good microtome and diamond knife for serial sectioning, and a standard 80–120 kV TEM. In addition, the parts used are cheap and widely available. Accordingly, this system is also useful for facilities which do not have access to HPF equipment but are interested in using cryopreparation to study the yeast ultrastructure by TEM, with or without correlation with fluorescence microscopy. The ultrastructure of yeast prepared by HPF and FS is well documented (for examples see the following references: Giddings, 2003; Heiman *et al.*, 2007; McDonald and Mueller-Reichert, 2002; Murray, 2008; Walther and Ziegler, 2002). For an inexpensive alternative to HPF, the technique of self-pressurized rapid freezing has been recently introduced (Leunissen and Yi, 2009).

III. Methods

A. Holder Assembly and Light Microscopy

A wire loop with a diameter of approximately 10 mm was made from steel wire, the extensions bent into a "U"-shape and attached to a 2 cm high section of a 50 ml Falcon tube (Fig. 1B). Formvar films were manually cast on glass slides from a 6% formvar solution in ethylene dichloride, attached to a cardboard frame with Scotch tape at the edges and carefully detached while breathing onto the film (Chowdhury *et al.*, 1999). Glass coverslips (diameter 5 mm) were then placed onto the suspended films and baked in an oven at 120°C for 45 min, which causes the formvar to adhere firmly to the coverslip. Under a binocular, the coverslips were excised from the film and circular holes were cut out in the center of the coverslip with a scalpel to create a cavity with an approximate diameter of 3 mm and 10 μm depth (Fig. 1C) and received a first carbon-coat of approximately 5 nm thickness. A copper finder grid was then placed in the cavity and a second carbon-coat (20 nm thickness) was evaporated, creating an image of the finder grid in the cavity of the coverslips. The carbon was then stabilized by baking at 120°C for 1 h.

Scotch tape was placed adhesive side up in a petri dish with wet paper to create a humid chamber. A 4 mm hole was punched in the center of the adhesive tape and the coverslip was then positioned on top of the hole (cavity up), held in place by the adhesive. The formvar spacer ring was carefully coated with 1-hexadecene using a fine brush to prevent wetting of the spacer and a small layer of vacuum grease was applied on the outer edge of the spacer with a toothpick. A 10 μl drop of a concanavalin A solution (0.5 mg/ml in water) was then placed in the cavity for 5 min and removed. Then, a 10 μl drop yeast cell suspension in log phase was pipetted in the cavity and allowed to settle for 5 min before carefully washing once with medium. The medium was then completely removed, quickly replaced by a drop of 2% agarose (in SD medium, kept liquid at 35°C). The top 5 mm coverslip was immediately placed on top and pressed in place with a forceps. Excess agarose was removed with a scalpel and the sample sandwich was carefully sealed off completely by applying more vacuum grease at its outer edge. The wire loop was then pressed in place on the adhesive tape and secured to the sample with two thin strips of scotch tape from the top. The holder was then placed on the stage of an inverted fluorescence microscope and imaged with a 100×, 1.4 NA oil immersion objective using bright field illumination and epifluorescence with GFP-optimized filters. GFP-images and time-lapse movies of cells expressing Abp1-GFP were taken with collection times of 500 ms.

B. Plunge Freezing

Beforehand, commercial grade propane was cooled in a standard liquefier (see Echlin, 1992) and placed next to the fluorescence microscope. At the desired time point, the holder was manually removed from the stage and the sample plunged as quickly as possible into the propane by horizontal entry (lab safety goggles should be worn to protect eyes from splashes of liquid propane). The frozen samples were then stored in liquid nitrogen and prepared for FS.

C. Freeze-Substitution and Post-Processing

Before FS, the supporting Scotch tape was removed under liquid nitrogen and the excess grease was carefully scratched off with a forceps. The intact sample sandwich was then transferred to a 1.5 ml microcentrifuge tube filled with 1 ml pre-cooled FS medium (at LN_2 temp.), consisting of acetone, 2% glutaraldehyde, 0.1% (w/v) uranyl acetate, and 5% (v/v) water (Buser and Walther, 2008; Walther and Ziegler, 2002). FS was performed overnight from −90 to 0°C and programmed the following: −90°C, 1 h; 3°C/h, 10 h; −60°C, 1 h; 9°C/h, 6.5 h; 0°C; total duration 18.5 h. At 0°C the samples were shifted to room temperature, washed 5× with ethanol and infiltrated with 50% Epon in ethanol for 2 h. The top coverslip was then carefully removed and the lower coverslip with the attached yeast cells was infiltrated with pure Epon for 6 h. The Epon was exchanged once more before polymerization at 60°C for 2 days.

The coverslip was removed by repeated short freeze-thawing cycles in liquid nitrogen. The remaining pieces of glass were loosened by placing a drop of water on the surface, followed by a freeze-thaw cycle in liquid nitrogen. The water is pulled into gaps by capillary action and expands on freezing, thus gently separating the glass from the resin surface. If necessary, fragments of glass were then carefully removed with a scalpel.

For electron microscopy, the surface was trimmed down to the area with the cells of interest (approx. 250 × 400 μm), serially sectioned (100 nm thickness, >25 sections), and picked up on formvar-filmed slot grids. The sections were stained in 2% aqueous uranyl acetate for 10 min and 2% lead citrate for 2 min and imaged.

IV. Materials

A. Instrumentation

Carbon evaporator: DV-502A, DentonVacuum, USA.
Freeze-substitution: Leica AFS, Leica Microsystems, Austria.

Fluorescence microscopy: Olympus IX71, Olympus, USA.
TEM: Tecnai 12 (FEI, USA) equipped with an Ultrascan 1000 CCD camera (Gatan Inc., USA).

B. Materials

Finder grid: Maxtaform "London Finder" 200 mesh; Cat# LF200Cu, Electron microscopy Sciences, USA
Glass coverslips: round, 5 mm diameter; Cat# 72296-05, Electron microscopy Sciences, USA
Punch: Harris Uni-core 4 mm diameter; Cat# 15080, Ted Pella, Inc., USA
Slot grids: PELCO©, 1 × 2 mm slot; Ted Pella, Inc., USA
Wire: steel, 0.41 mm diameter

C. Reagents

Agarose: 2% (w/v) in SD medium with 10% glucose; FMC Bioproducts, Cat#50107.
Concanavalin A: type IV; 0.5% (w/v) in water; Cat# C2010, Sigma-Aldrich, USA.
Formvar: 6% in ethylene dichloride; Cat# 15800 and Cat# 13250, Electron Microscopy Sciences, USA.
Growth medium: SD tryptophan dropout medium supplied with 10% glucose.
1-hexadecene: Cat# H2131, Sigma-Aldrich, USA.
Vacuum grease: Dow Corning high vacuum grease; Dow Corning, USA.

V. Results

We explored the possibility of studying endocytosis in *S. cerevisiae* by correlative light- and electron microscopy using a plunge freezing approach to achieve sub-second time resolution. At first glance the combined restrictions imposed by the biological system, fluorescence microscopy, and the preparation for electron microscopy seem irreconcilable. The following results show that by carefully controlling every individual step in the protocol both a high spatial and temporal correlation between fluorescence and TEM appears feasible, and in the future will allow us to correlatively investigate short-lived endocytic intermediates.

A. Light Microscopy

Drying out of the sample due to imperfect sealing was a major issue. With a total liquid volume of less than 0.1 μl, even small leaks will lead to massive drying and osmotic shock thus causing morphological changes. Sealing with vacuum grease (Kaksonen *et al.*, 2003) effectively prevents drying and keeps the cells in a stable environment. After 30 min in the sample sandwich, yeast cells still maintained

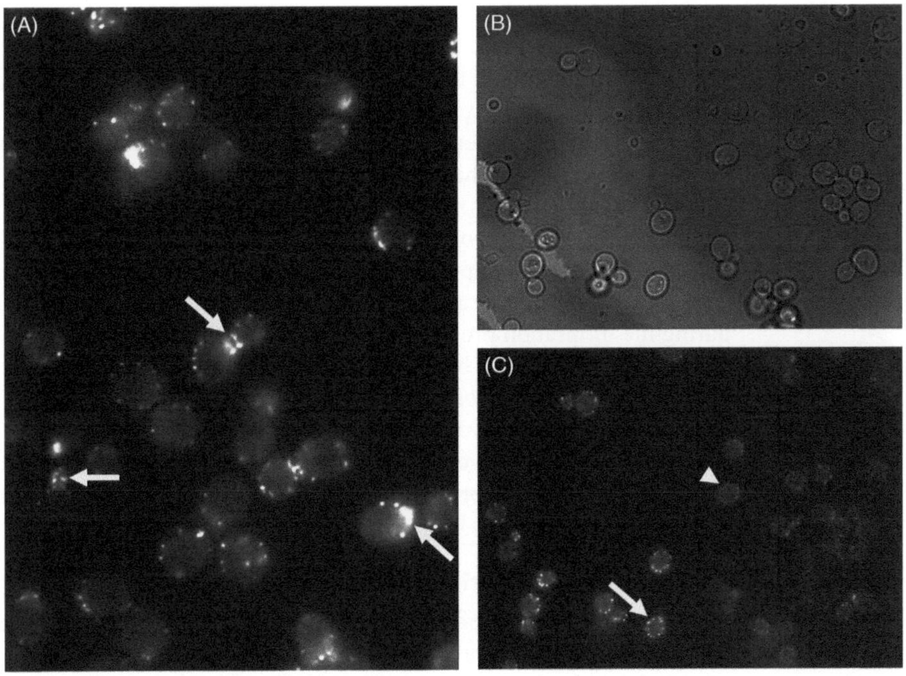

Fig. 2 Light microscopy of *S. cerevisiae* cells imaged within the setup shown in Fig. 1. The cells are expressing the actin-binding protein Abp1-GFP, which is recruited to endocytic sites together with actin during the invagination process. (A) Fluorescence image of yeast cells assembled in the sample sandwich imaged after 30 min show characteristically polarized actin patches (arrows). (B) The carbon finder pattern allows clear identification of the area of interest by bright-field imaging, here the letter "R". (C) Fluorescence image of the same region as in B. Differences in the recorded fluorescence intensity due to absorption of fluorescent light by the carbon film can be seen between cells located only on the 5 nm thin carbon film, i.e. on the letter "R" (arrow) versus cells viewed through both carbon films (total thickness of 25 nm, arrowhead).

polarized and motile actin patches (Fig. 2A), and no signs of drying or oxygen depletion were observed.

The thickness of the carbon films strongly affects the fluorescence intensity and thus should be kept as thin as possible while still being visible by bright field illumination (Fig. 2B and C). We are still experimenting with different thicknesses, but 5 nm for the first layer and 20 nm for the finder pattern was sufficient to image relatively abundant endocytic proteins. Accordingly, live-cell imaging on Abp1-GFP expressing cells was performed with acquisition times of 500 ms per frame.

An important aspect for the imaging with oil-immersion objectives is a perfectly parallel positioning of the sample in the plane of the objective. In our first experiments, a 70 μm thin adhesive copper foil was used instead of Scotch tape as a base support due to its superior thermal properties. Unfortunately, after punching the viewing hole the rigidity of the copper made it difficult to perfectly flatten the foil. This caused parts of the copper foil to contact the objective and thus made it impossible to bring the coverslip in the central hole into working distance. Using the flexible Scotch tape solved this problem and the entire area of the coverslip could now be imaged with oil immersion. Interestingly, the Scotch tape did not affect the quality of freezing.

In our initial experiments the yeast cells were attached to the coverslip by concanavalin A alone. While this method is sufficient with standard media containing 2% glucose (Kaksonen *et al.*, 2003), increasing the concentration to 10% results in incomplete attachment of the cells to the coverslip. This caused the yeast cells to wiggle and made it impossible to track individual endocytic sites by fluorescence microscopy. Furthermore, the cells easily and randomly detach from the coverslip during the infiltration process, which is only noticed in the last step when imaging the sections in the TEM. To better immobilize cells for fluorescence microscopy and at the same time prevent cell loss during infiltration, the yeast cells were first attached to the coverslip with concanavalin A and then embedded in 2% low-melting agarose. This additional encasing prevented both wiggling and the loss of cells and did not disturb the infiltration process or the imaging by fluorescent light microscopy.

B. Freezing Damage and Cryoprotection

Since freezing by manual plunging in liquid propane is not expected to yield high cooling rates, it is important to assess the amount of ice-crystal damage that can be tolerated for the imaging of endocytic sites. In yeast cells, the vacuole is the best indicator for freezing damage, as it has a high water content and is difficult to vitrify routinely without cryoprotection, even by HPF. Intermediate ice-crystal damage is easy to identify in the electron dense lumen of the vacuole, manifesting as brighter patches formed by the exclusion of solutes from growing ice crystals.

We first tried to plunge freeze yeast suspensions sandwiched between two sapphire discs (3 mm diameter, 50 μm thickness) separated by a 20 μm thin copper TEM grid, with 8% glucose present in the medium and a 70 μm thin adhesive copper foil with a central hole as support. This configuration represents a setup in which high-cooling rates are expected, due to the optimal thermal properties of sapphire and copper. As shown, freezing damage was excessive and organelles were barely recognizable (Fig. 3A). The absence of morphologically detectable ice-crystal damage is represented by a high-pressure frozen sample (Fig. 3B).

This comparison argues that even with a low thermal mass of the sample sandwich the aqueous layer was too thick and thus a thinner spacer was needed. Additionally, fluorescence microscopy and sample assembly was very tedious with the small sapphire discs. We thus decided to explore the advantages of 5 mm diameter glass coverslips with a 10 μm thin formvar spacer. During the initial experiments it also became clear that manual plunge freezing in liquid propane would require a certain extent of cryoprotection. Instead of introducing an additional component with reported cryoprotective properties (e.g., dextran, sucrose or glycerol) we decided to increase the concentration of glucose, which is used as carbon source in our yeast growth media. Cryoprotection in 10% glucose already resulted in sufficiently well-preserved

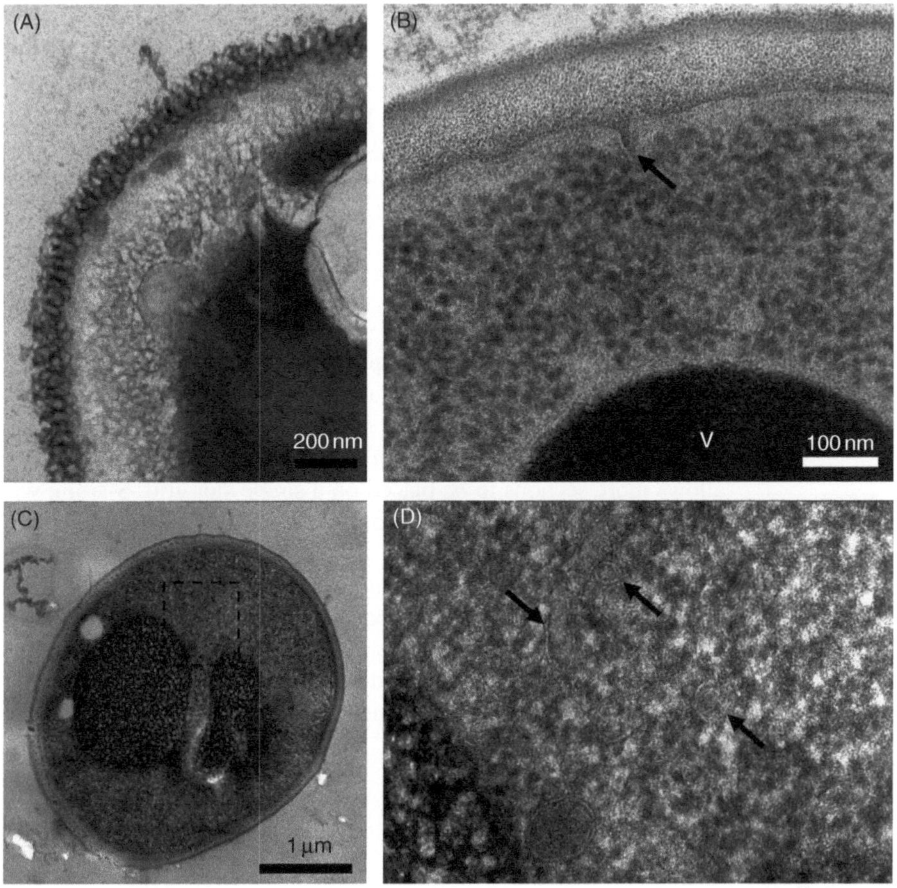

Fig. 3 TEM images of yeast cells prepared with different freezing setups. (A) Cells plunge frozen between two sapphire discs with a 20 μm copper grid as spacer. The cells have suffered from extensive freezing damage completely destroying the fine structure. (B) Cells frozen by HPF showing perfect freezing judged by the absence of segregation patterns in the vacuole (V) and the even density of the cytoplasm. An early endocytic site is visible on the plasma membrane (arrow). (C) Overview of a cell plunge frozen with the proposed setup. Note the patch-like segregation patterns in the vacuole which are typical for intermediate crystal growth. (D) Magnification of the boxed area in C. Surprisingly, many small membranous structures (arrows) can be seen in the vicinity of the vacuole that are hardly visible in well-frozen samples prepared by HPF.

structures with intermediate crystal damage (Fig. 3C). To our surprise, the visibility of membrane compartments was drastically improved compared to high-pressure frozen samples. Small vesicles and cisterns in the vicinity of the vacuole are nearly invisible in well-frozen yeast cells prepared by HPF (Fig. 3B), but easily observed in plunge-frozen samples (Fig. 3D).

The addition of high concentrations of glucose can affect infiltration. Sugars are very poorly soluble in acetone (the standard solvent used for FS), aggregate around the cells during FS, and prevent infiltration of the cell periphery if not removed (Buser and McDonald, 2010). Initially, we washed the samples with acetone before infiltration with 50% Epon in acetone. This procedure resulted in poor infiltration of the cell periphery and cells were frequently ripped out of section. Extended washing in ethanol and infiltration with 50% Epon in ethanol solved this problem due to the increased solubility of sugars in ethanol.

C. Correlating Fluorescence and Electron Microscopy

After removal of the coverslip, the imprinted carbon image of the finder grid was used to determine the position of the observed cells. The block was trimmed down to include the cells in an area as small as possible to facilitate the search for the previously observed cells within the section in the TEM. Furthermore, a small area allows more sections to be placed on a single slot grid. Starting with the first complete sections, up to 40 serial sections of 100 nm thickness were collected, stained and imaged at an acceleration voltage of 120 kV. Median cross-sections were usually obtained in the sections 10–20, corresponding to an approximate Z-depth of 1–2 μm. The same cells imaged by fluorescence microscopy could be identified in thin sections, but unfortunately were extensively damaged by crystallization and thus not usable for correlative microscopy (Fig. 4A). Since good freezing rates were obtained with the identical sample setup without attempting correlation (Fig. 4B), the probable culprit for the poor freezing in correlated samples is a slow entry speed of the sample into the cryogen due to imperfect manual plunging.

VI. Discussion

Endocytosis is a highly conserved process from single-celled eukaryotes to humans and is central to the development of many diseases and a frequent entry point for viral infections. With its easy genomic manipulation and the possibility of preparing yeast for electron microscopy by plunge freezing, *S. cerevisiae* offers unique opportunities to unravel the molecular mechanism of endocytosis. The protocol presented here demonstrates the feasibility of correlative live-cell light and electron microscopy in yeast with oil-immersion fluorescence microscopy, cryopreparation, and the potential to achieve sub-second temporal resolution. Relying on a reduction of the sample volume and cryoprotection by addition of sugars, sufficient cooling rates can be achieved by manual plunge-freezing even when using standard glass coverslips. The approach presented here has three major advantages over HPF: (1) A collection process by filtration is not needed before freezing and the cells can be imaged by fluorescence microscopy immediately before plunging into the cryogen. This ensures that the cells are frozen in their native state. (2) The temporal resolution is only restricted by the time required for transferring the sample from the fluorescence microscope to the plunge

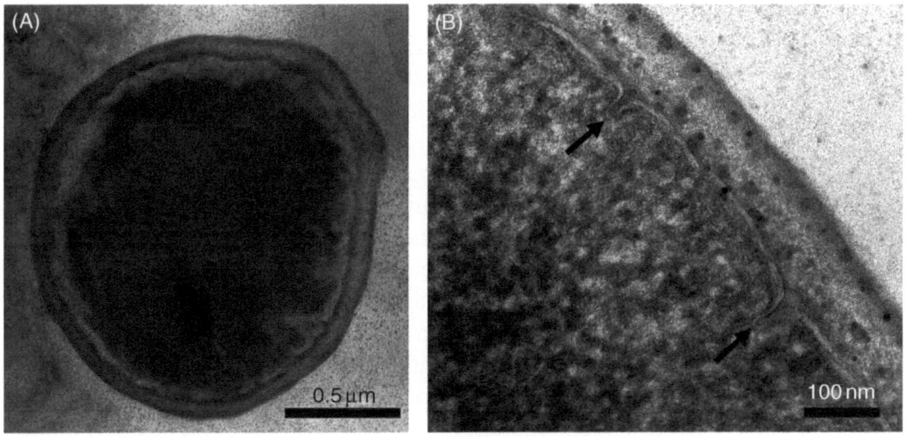

Fig. 4 Attempts at sub-second CLEM. (A) An Abp1-GFP expressing cell previously imaged by time-lapse fluorescence microscopy and plunge-frozen within less than one second of the last frame at the fluorescence microscope. The cells have suffered from extensive crystal damage similar to Fig. 3A. (B) An Abp1-GFP expressing cell prepared with the same setup, but plunge-frozen without attempting correlation by fluorescence microscopy. The freezing quality is sufficient to clearly identify endocytic intermediates during the invagination process of the plasma membrane (arrows), which are labeled with Abp1-GFP in fluorescence microscopy (Fig. 2).

freezing device, thus offering correlation times in the sub-second range especially if automated. (3) The proposed setup does not compromise on fluorescence microscopy. By using Scotch tape as support and glass coverslips, the conditions are optimal for the use of immersion objectives with very short working distances and high-NA values, and the full cavity with the cells is accessible for observation. Furthermore, there is no risk of damaging the expensive objectives. This combination allows a simple imaging of small and highly dynamic GFP-labeled structures. Further optimization in fluorescence microscopy might be required to allow the visualization of very dim fluorophores and dual-color imaging in the usually dimmer red part of the spectrum. Fine-tuning of the thickness of the carbon films is necessary to decrease its absorbance of emitted fluorescent light while still preserving the good visibility of the finder pattern in bright-field mode. The key aspect in allowing proper imaging of the yeast cells was the embedding in 2% low-melting agarose to prevent both the movement of the cells during live-imaging and their loss during infiltration with the resin (Sims and Hardin, 2007). Surprisingly, the agarose did not cause any infiltration problems.

The above results suggest that the quality of freezing by plunging into liquid propane depends on the sample thickness and the entry speed into the cryogen as described previously (Costello, 1980). Reduction of the sample thickness to 10 μm yielded well-frozen samples even with suboptimal materials as glass coverslips and adhesive tape as support, as long as plunging was done quickly. Unfortunately, the freezing failed so far in correlative experiments probably due to suboptimal entry speeds into the cryogen by manual plunging. Thus, the main goal of our further experiments is the improvement of the thermal properties of the sample. The main focus will thus be to further reduce the spacer thickness from approximately 10 down to 5 μm and concurrently increase the glucose content from 10 to 20%. Judging from the data presented above, improving these parameters alone might already suffice to ensure adequate freezing even in a correlative experiment with less than optimal cooling rates at the sample surface. A further compromise would be the use of sapphire discs to utilize the better thermal properties of sapphire compared to quartz glass. The downside of sapphire for routine fluorescence microscopy is the high refractive index of 1.71, which would have to be matched with a special immersion fluid. It is also known that the depth of the cryogen can have an effect on the cooling rate (Robards and Sleytr, 1985). When the entry rate is relatively slow, the samples need to enter deeper into the cryogen to reach the desired freezing rate. We will experiment with a deeper well for the cryogen as a means to achieve a higher percentage of well-frozen cells.

An unsolved problem for both fluorescence and electron microscopy is the lack of a reference marker for the Z-position. Currently, the Z-position of endocytic sites has to be estimated from the diffraction rings around the cells in bright-field mode and usually a medial focal plane is chosen for the cells of interest. The corresponding endocytic site in the section then has to be found by searching the serial sections for the expected invaginated membrane profile. This approach fails in putative cases where no morphological patterns can be expected, e.g., the membrane has not been deformed yet. Furthermore, the depth of field in the fluorescence microscope is approximately 1 μm, which corresponds to ten 100 nm sections. A possible solution would be to add latex or Dynal® beads of a defined size and to use the diameter of the cross-section in the TEM as a measure for the Z-position. Additionally, the FS and infiltration protocol could be modified to allow immunogold labeling (Buser and McDonald, 2010). To further reduce the dependency on specialized equipment, FS could also be carried out in a simple dry-ice system or by using freezers to regulate the temperature Echlin, 1992; McDonald and Mueller-Reichert, 2002).

An interesting aspect would be the automation of the transfer and plunging process, which would allow more reproducible freezing and an even better time resolution of the method. This could offer correlative insights into very rapid processes similar to the original work of Heuser and Reese (Heuser and Reese, 1981; Heuser *et al.*, 1979). Alternatively, the dual coverslip system presented here could be adapted for propane jet or HPF applications for samples thicker and harder to freeze than yeast.

In conclusion, live-cell correlative fluorescence and electron microscopy by plunge freezing is a promising method to investigate rapid cellular processes in *S. cerevisiae*. Further optimization and simplification of the protocol would also allow the technique to be implemented in laboratories without cryopreparation equipment.

Acknowledgments

I thank David Drubin and Kent McDonald for their support and critical discussions, and Yidi Sun and Voytek Okreglak for their help with the fluorescence microscope. I also acknowledge a fellowship for prospective researchers from the Swiss National Science Foundation (SNF).

References

Al-Amoudi, A., Chang, J. J., Leforestier, A., McDowall, A., Salamin, L. M., Norlen, L. P., Richter, K., Blanc, N. S., Studer, D., and Dubochet, J. (2004). Cryo-electron microscopy of vitreous sections. *EMBO J.* **23**, 3583–3588.

Baba, M., and Osumi, M. (1987). Transmission and scanning electron microscopic examination of intracellular organelles in freeze-substituted kloeckera and saccharomyces cerevisiae yeast cells. *J. Electron Microsc. Tech.* **5**, 249–261.

Brown, E., Mantell, J., Carter, D., Tilly, G., and Verkade, P. (2009). Studying intracellular transport using high-pressure freezing and correlative light electron microscopy. *Semin. Cell Dev. Biol.* **20**(8), 910–919.

Buser, C., and McDonald, K. L., (2010). Correlative GFP-immunoelectron microscopy in yeast. *Methods Enzymol.* **470**, 603–618.

Buser, C., and Walther, P. (2008). Freeze-substitution: The addition of water to polar solvents enhances the retention of structure and acts at temperatures around −60 degrees C. *J. Microsc.* **230**, 268–277.

Chowdhury, A., Naik, P. A., and Gupta, P. D. (1999). Development of free-standing submicron formvar films with multiple thickness steps for XUV-soft X-ray applications. *Sadhana* **24**, 551–555.

Costello, M. J. (1980). Ultra-rapid freezing of thin biological samples. *Scan. Electron Microsc.* 361–370.

Echlin, P. (1992). "Low-Temperature Microscopy and Analysis." Plenum Press, NY.

Egner, A., Jakobs, S., and Hell, S. W. (2002). Fast 100-nm resolution three-dimensional microscope reveals structural plasticity of mitochondria in live yeast. *Proc. Natl. Acad. Sci. U. S.A.* **99**, 3370–3375.

Giddings, T. H. (2003). Freeze-substitution protocols for improved visualization of membranes in high-pressure frozen samples. *J. Microsc.* **212**, 53–61.

Hawes, P., Netherton, C. L., Mueller, M., Wileman, T., and Monaghan, P. (2007). Rapid freeze-substitution preserves membranes in high-pressure frozen tissue culture cells. *J. Microsc.* **226**, 182–189.

Heiman, M. G., Engel, A., and Walter, P. (2007). The golgi-resident protease kex2 acts in conjunction with prm1 to facilitate cell fusion during yeast mating. *J. Cell Biol.* **176**, 209–222.

Heuser, J. E., and Reese, T. S. (1981). Structural changes after transmitter release at the frog neuromuscular junction. *J. Cell Biol.* **88**, 564–580.

Heuser, J. E., Reese, T. S., Dennis, M. J., Jan, Y., Jan, L., and Evans, L. (1979). Synaptic vesicle exocytosis captured by quick freezing and correlated with quantal transmitter release. *J. Cell Biol.* **81**, 275–300.

Hohenberg, H., Mannweiler, K., and Muller, M. (1994). High-pressure freezing of cell suspensions in cellulose capillary tubes. *J. Microsc.* **175**(Pt 1), 34–43.

Hoog, J. L., Schwartz, C., Noon, A. T., O'Toole, E. T., Mastronarde, D. N., McIntosh, J. R., and Antony, C. (2007). Organization of interphase microtubules in fission yeast analyzed by electron tomography. *Dev. Cell* **12**, 349–361.

Idrissi, F. Z., Grotsch, H., Fernandez-Golbano, I. M., Presciatto-Baschong, C., Riezman, H., and Geli, M. I. (2008). Distinct acto/myosin-I structures associate with endocytic profiles at the plasma membrane. *J. Cell Biol.* **180**, 1219–1232.

Kaksonen, M., Sun, Y., and Drubin, D. G. (2003). A pathway for association of receptors, adaptors, and actin during endocytic internalization. *Cell* **115**, 475–487.

Kaksonen, M., Toret, C. P., and Drubin, D. G. (2005). A modular design for the clathrin- and actin-mediated endocytosis machinery. *Cell* **123**, 305–320.

Kaksonen, M., Toret, C. P., and Drubin, D. G. (2006). Harnessing actin dynamics for clathrin-mediated endocytosis. *Nat. Rev. Mol. Cell Biol.* **7**, 404–414.

Leunissen, J. L., and Yi, H. (2009). Self-pressurized rapid freezing (SPRF): A novel cryofixation method for specimen preparation in electron microscopy. *J. Microsc.* **235**, 25–35.

McDonald, K. L., Morphew, M., Verkade, P., and Muller-Reichert, T. (2007). Recent advances in high-pressure freezing: Equipment- and specimen-loading methods. *Methods Mol. Biol.* **369**, 143–173.

McDonald, K. L., and Mueller-Reichert, T. (2002). Cryomethods for thin section electron microscopy. *In* "Methods in Enzymology" (C. Guthrie, G. R. Fink, eds.) **Vol. 351**, pp. 96–123. Academic Press, San Diego, CA.

McDonald, K., O'Toole, E. T., Mastronarde, D. N., Winey, M., and Richard McIntosh, J. (1996). Mapping the three-dimensional organization of microtubules in mitotic spindles of yeast. *Trends Cell Biol.* **6**, 235–239.

Moor, H., and Mühlethaler, K. (1963). Fine structure in frozen-etched yeast cells. *J. Cell Biol.* **17**, 609–628.
Mulholland, J., Preuss, D., Moon, A., Wong, A., Drubin, D., and Botstein, D. (1994). Ultrastructure of the yeast actin cytoskeleton and its association with the plasma membrane. *J. Cell Biol.* **125**, 381–391.
Müller, M., Meister, N., and Moor, H. (1980). Freezing in a propane jet and its application in freeze-fracturing. *Mikroskopie (Wien)* **36**, 129–140.
Muller-Reichert, T., Srayko, M., Hyman, A., O'Toole, E. T., and McDonald, K. (2007). Correlative light and electron microscopy of early caenorhabditis elegans embryos in mitosis. *Methods Cell Biol.* **79**, 101–119.
Murk, J. L., Posthuma, G., Koster, A. J., Geuze, H. J., Verkleij, A. J., Kleijmeer, M. J., and Humbel, B. M. (2003). Influence of aldehyde fixation on the morphology of endosomes and lysosomes: Quantitative analysis and electron tomography. *J. Microsc.* **212**, 81–90.
Murray, S. (2008). High pressure freezing and freeze substitution of schizosaccharomyces pombe and saccharomyces cerevisiae for TEM. *Methods Cell Biol.* **88**, 3–17.
Pelletier, L., O'Toole, E., Schwager, A., Hyman, A. A., and Muller-Reichert, T. (2006). Centriole assembly in caenorhabditis elegans. *Nature* **444**, 619–623.
Robards, A. W., and Sleytr, J. B. (1985). "Low Temperature Methods in Biological Electron Microscopy." North-Holland, Amsterdam, NY, Oxford.
Roberts, P., Moshitch-Moshkovitz, S., Kvam, E., O'Toole, E., Winey, M., and Goldfarb, D. S. (2003). Piecemeal microautophagy of nucleus in saccharomyces cerevisiae. *Mol. Biol. Cell.* **14**, 129–141.
Sims, P. A., and Hardin, J. D. (2007). Fluorescence-integrated transmission electron microscopy images: Integrating fluorescence microscopy with transmission electron microscopy. *Methods Mol. Biol.* **369**, 291–308.
Soto, T., Nunez, A., Madrid, M., Vicente, J., Gacto, M., and Cansado, J. (2007). Transduction of centrifugation-induced gravity forces through mitogen-activated protein kinase pathways in the fission yeast schizosaccharomyces pombe. *Microbiology* **153**, 1519–1529.
Stradalova, V., Stahlschmidt, W., Grossmann, G., Blazikova, M., Rachel, R., Tanner, W., and Malinsky, J. (2009). Furrow-like invaginations of the yeast plasma membrane correspond to membrane compartment of can1. *J. Cell Sci.* **122**, 2887–2894.
Studer, D., Michel, M., and Muller, M. (1989). High pressure freezing comes of age. *Scanning Microsc. Suppl.* **3**, 253–268, discussion 268–269.
Studer, D., Michel, M., Wohlwend, M., Hunziker, E. B., and Buschmann, M. D. (1995). Vitrification of articular cartilage by high-pressure freezing. *J. Microsc.* **179**(Pt 3), 321–332.
Verkade, P. (2008). Moving EM: The rapid transfer system as a new tool for correlative light and electron microscopy and high throughput for high-pressure freezing. *J. Microsc.* **230**, 317–328.
Waddle, J. A., Karpova, T. S., Waterston, R. H., and Cooper, J. A. (1996). Movement of cortical actin patches in yeast. *J. Cell Biol.* **132**, 861–870.
Walther, P., Mueller, M., and Schweingruber, M. E. (1984). The ultrastructure of the cell surface and plasma membrane of exponential and stationary phase cells of schizosaccharomyces pombe, grown in different media. *Arch. Microbiol.* **137**, 128–134.
Walther, P., and Ziegler, A. (2002). Freeze substitution of high-pressure frozen samples: The visibility of biological membranes is improved when the substitution medium contains water. *J. Microsc.* **208**, 3–10.

CHAPTER 11

Fission Yeast: A Cellular Model Well Suited for Electron Microscopy Investigations

Hélio Roque *and* Claude Antony
Cell Biology and Biophysics Program, European Molecular Biology Laboratories, Heidelberg 69117, Germany

METHODS IN CELL BIOLOGY, VOL. 96

978-0-12-381007-6
DOI: 10.1016/S0091-679X(10)96011-1

Abstract

The fission yeast *Schizosaccharomyces pombe* has become a prominent model in molecular biology, both in yeast genetics and to investigate the molecular mechanism of the cell cycle. It has also proved to be a suitable model organism for looking at cell architecture and ultrastructure using electron microscopy (EM). Here we discuss what makes *S. pombe* particularly suited to EM and summarize the important discoveries regarding cell organization that have emerged from such studies. We describe the procedures and conventional methods used in EM analysis of fission yeast cells, and lay particular emphasis on cryogenic procedures, which preserve the cell structure in a near-native state, allowing elaborate three-dimensional reconstruction using electron tomography. The chapter also gives several examples of how contemporary EM approaches can be applied to provide a detailed read-out of phenotypes in this versatile cell system. A list of instruments and detailed protocols are provided together with EM-specific reagents required for sample preparation. Finally, potential new avenues of research are discussed, anticipating forthcoming topics in EM as well as new approaches to fission yeast research in the future.

I. Introduction

A. General Facts

Yeast cells are eukaryotic unicellular fungi, of which there are over 1500 species. Most yeasts reproduce asexually by budding, some by fission. The species *Schizosaccharomyces pombe*, a fission yeast, belongs to the class ascomycetes but is variously classified as archaeascomycete (Eukaryota/Fungi/Ascomycota/Archaeascomycetes/Schizosaccharomycetales/Schizosaccharomycetaceae/Schizosaccharomyces). Yeasts are commonly used in baking or fermenting alcoholic beverages and *S. pombe* was originally isolated in millet beer from East Africa. The name *pombe* derives from the Swahili word for beer. The first description of this yeast dates from 1893 (Lindner, 1893).

Both *S. pombe* and *Saccharomyces cerevisiae* (see also Chapter by Buser, this issue, chapter 10) are very popular as model systems since these rapidly growing eukaryotic cells are genetically tractable and thus well suited to molecular manipulation. In addition, many of the basic cell functions and relevant genes are conserved from fungi to humans. For example, many of the genes involved in the regulation of cell division in vertebrates, including humans, also exist with conserved function in yeasts. Since yeasts are simple organisms, essential gene functions can be extracted more easily than in higher eukaryotes, which is a further reason why yeast has been adopted as an ideal model organism. This is exemplified in the work of Paul Nurse in the UK who used *S. pombe* to decipher major conserved molecular mechanisms in the cell cycle for which he was awarded the Nobel Prize together with Lee Hartwell and Tim Hunt in 2001 (Nurse, 2002).

Schizosaccharomyces pombe and *Saccharomyces cerevisiae* are in fact very different cells (Hedges, 2002). These species diverged some 300–600 million years ago.

S. pombe has only three chromosomes whereas *S. cerevisiae* has sixteen. *Schizosaccharomyces pombe* has the smallest sequenced eukaryotic genome (13.8 Mb) (Wood *et al.*, 2002) with 4824 open reading frames (ORFs) compared with *S. cerevisiae* which has about 5600 ORFs. Unlike *S. cerevisiae,* genome duplication is not found in *S. pombe* (Wolfe and Shields, 1997; Wood *et al.*, 2002), hence *S. pombe* has single copy genes, making it advantageous for gene functional studies. *Schizosaccharomyces pombe* remains in the G2 phase for most of the cell cycle and consequently the G2-M transition is under tight control, while *S. cerevisiae* is in G1 phase for most of its cell cycle and as a result the G1-S transition is tightly controlled. *Schizosaccharomyces pombe* has large repetitive centromeres (40–100 kb) similar to mammalian centromeres, while *S. cerevisiae* has smaller centromeres measuring approximately 120 bp. Another feature that distinguishes *S. pombe* from *S. cerevisiae* is its heterochromatin and RNAi machinery genes, which are conserved in vertebrates, but are absent in *S. cerevisiae*, making *S. pombe* an advantageous model system to study cell division.

B. Electron Microscopy

Schizosaccharomyces pombe cells are small rod-shaped cells, about 4–5 μm in diameter and 8–15 μm in length, lined by a cell wall. The minimal number of molecular compounds in yeasts is also reflected in the minimal number of subcellular elements. This presents an advantage for electron microscopy (EM) studies as all subcellular elements can be analyzed and measured. Using electron tomography (ET), for instance, it is possible to visualize large cell volumes, or even whole cells, such that all organelles or cytoskeletal fibers within a single cell can be modeled (Höög and Antony, 2007). Precise quantification is straightforward since most subcellular elements within the cell can be identified. These characteristics make fission yeast an ideal model for investigating cell architecture using EM.

C. EM History of Fission Yeast

Protocols for EM investigations in *S. pombe*, fungi, and filamentous algae were already established in the 1960s. It was Kenji Tanaka who first described fission yeast's morphological features using permanganate fixation and phosphotungstic acid contrasting. His work showed the continuity between the nuclear envelope and the endoplasmic reticulum, the mitochondria structure, and characteristic cristae, and also the Golgi-like tubular membrane structure (Tanaka, 1963). Although fruitful, early investigations from the 1960s using EM on fungi and particularly on *S. pombe* were not extensive. Pickett-Heaps investigated dividing fungi and evolutionary aspects of the geometry of mitosis in various filamentous algae, especially in the Oedogonium gender (Pickett-Heaps and Fowke, 1969). MacLean gave an improved description of *S. pombe* using potassium permanganate and osmium tetroxide (OsO4) as fixatives, which showed the double nuclear envelope, nucleolus, vacuoles and vesicles, and granules (ribosomes) (Maclean, 1964). Other early studies focused on the division of

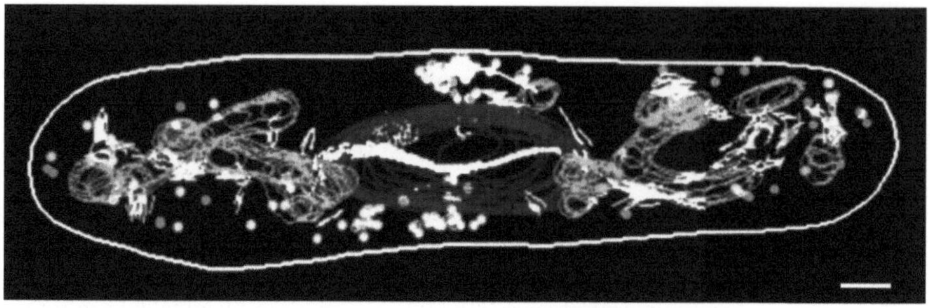

Fig. 1 Three-dimensional reconstruction of a mitotic cell. Three-dimensional reconstruction of a mitotic cell from 28 serial sections of 100 nm each. The sample was plunge-frozen, freeze-substituted in 2% OsO_4 in acetone, and embedded in Epon–Araldite resin. Adapted with permission from *J. Cell. Sci.* **94**(Pt 4), 647–656.

yeast mitochondria using electron microscopy and the replication of mitochondrial DNA (Osumi and Sando, 1969).

In the 1970s, major studies on cell physiology and genetics suggested that fission yeast could be used as a good model system (Leupold, 1970; Mitchison, 1970). Kenji Tanaka also undertook work on *S. pombe* meiosis (Hirata and Tanaka, 1982). Importantly, this latter group was the first to provide three-dimensional reconstruction and modeling of the fission yeast subcellular elements from serial sections (Fig. 1) (Kanbe *et al.*, 1989). The septum formation, a typical feature of the fission yeast, was described by EM as early as 1970 (Oulevey *et al.*, 1970). Following this, an analysis of mitosis in fission yeast was reported for the first time with a comparative approach using light microscopy and EM (McCully and Robinow, 1971). In their work, they showed the nuclear fine structure including the narrow microtubular mitotic spindle with a kinetochore-equivalent region lying in a ribosome-free area. Furthermore, they identified the divided nuclei as they followed the expansion of the nuclear envelope and the marked elongation of the nuclei and their content, which resulted in the typical dumbbell shape in the anaphase and telophase stages. Moreover, they assessed nucleolus partitioning within the daughter cells.

These days, cryofixation by high-pressure freezing (HPF) is the preferred method for studying *S. pombe* since it generates samples with subcellular structures preserved in a near-native state. The frozen samples released in liquid nitrogen are freeze-substituted and embedded in plastic for easy sectioning at room temperature. Sections are then generated for standard 2D imaging or for ET. In the latter case, subcellular volumes or whole cells are reconstructed and modeled in 3D (Höög and Antony, 2007).

II. Rationale

Yeast cell systems offer an efficient and easy means of dissecting the gene function of essential physiological processes using genetic and molecular biology tools.

Schizosaccharomyces pombe is well suited for routine EM and as well as ET investigations since cryogenic techniques and computer-assisted ET can be applied. Indeed, *S. pombe* is easy to vitrify using a HPFreezer as cells can be obtained by filtration in the form of a paste—which contains a minimal amount of water versus a maximum number of cells in its volume—hence minimizing the risk of ice crystal formation. The relative simplicity of the subcellular element organization and their restricted number represent a major advantage for clearly identifying and quantifying elements of interest. Mutations are targeted to a single gene copy, making it clear to analyze specific alterations to structural elements. Moreover, with its short cell cycle, *S. pombe* allows a number of major topics in cell biology to be studied, such as cell division, mitotic apparatus assembly, chromosome segregation, spindle pole body (SPB) duplication or cytokinesis, as well as biochemical aspects of cell cycle events. *Schizosaccharomyces pombe* cells thus offer one of the best systems for combining a molecular and structural analysis in a spatio-temporal frame. Such advantages are reflected in the abundant literature, which increasingly combines molecular mutant dissection with ET.

III. EM Methods

Successful EM analysis of fission yeast requires standardization of the chosen protocol. Here, even small differences in the protocol such as the growing medium or growing conditions can lead to major ultrastructural differences. This is especially important when comparing different mutants to a wild type or comparing different EM protocols. For instance, cells grown in complete medium (YE5S) or in minimal medium (EMM2) vary greatly in the size of a defined type of vacuoles.

A. Conventional Fixation and Plastic Embedding

1. Harvesting

In order to obtain the best possible results it is important to use a harvesting method that keeps cells as intact as possible. In the case of fission yeast we routinely use a vacuum filtering method (McDonald, 1999; McDonald and Muller-Reichert, 2002) for harvesting cells. We normally avoid centrifugation since *S. pombe* is highly sensitive to centrifugation regarding the cytoskeleton and nuclear positioning. Both vacuum filtering and centrifugation are damaging to the cells (see also Chapter by Buser, this issue, chapter 10) and cell prefixing should be considered.

2. Prefixation

Glutaraldehyde (Sabatini *et al.*, 1963) is the fixative of choice for cell prefixation and also for freeze-substitution procedure. In solution, glutaraldehyde will crosslink proteins irreversibly (Dawes, 1979; Hayat, 1981; Robinson *et al.*, 1987). To obtain good results EM-grade glutaraldehyde must be used. Glutaraldehyde is routinely used with a final concentration of 0.5–3%. Another aldehyde fixative is formaldehyde. Formaldehyde is generally used together with glutaraldehyde (Karnovsky, 1965) since formaldehyde penetrates fast but crosslinks weakly, while glutaraldehyde does the reverse. A typical concentration of formaldehyde is 4% (Dawes, 1979; Hayat, 1981; Robinson *et al.*, 1987). Finally, tannic acid possesses good properties as a membrane and microtubule (MT) enhancer (Mizuhira and Futaesaku, 1972). In the preparation of fission yeast samples, tannic acid has been used sparsely as a prefixative at a concentration of 0.2% and is often combined with glutaraldehyde (Kamasaki *et al.*, 2005; Teysset *et al.*, 2003).

1. Prepare enough prefixative for two applications by dissolving glutaraldehyde in 0.1 M sodium phosphate buffer (pH 7.0) to a final concentration of 4% (2× used concentration).
2. Take 5–10 ml of an exponential growing culture ($OD_{595} = 0.4$–0.6) and mix it in a Falcon tube with an equal volume of prefixative. Alternatively, the prefixative can be added directly to the shaking culture (with adapted volumes). Leave at room temperature for 5 min. In the meantime, dissolve the fixative to 1× concentration.
3. Collect cells by centrifugation at 6000 rpm for 5 min in a clinical centrifuge and discard supernatant. Resuspend pellet in 10 ml of 1× fixative.
4. Incubate at room temperature for 30–60 min or at 4°C overnight.

3. Cell-Wall Permebealization/Digestion

Following prefixation, cell-wall permeabilization or its complete digestion should be considered, especially if OsO_4 is to be used as a postfixative due to a poor cell-wall penetration (Wright, 2000). Enzymatic digestion of the cell wall is carried out after a prefixation, but some studies have carried out cell-wall digestion before prefixation, normally with the aim of specially staining MTs or actin (Carazo-Salas *et al.*, 2005; Kamasaki *et al.*, 2005). Zymolyase (Kirin Brewery Co., Ltd., Takasaki, Japan) is the most commonly used enzyme for performing cell-wall digestion at a final concentration of 0.5 mg ml^{-1} 100T. Sorbitol is added to protect the cells from lysis due to osmotic pressure. Cell-wall digestion can be verified by phase-contrast microscopy. The cells lose their halo as cell-wall digestion proceeds. However, one should be aware that cell-wall digestion, even if only partial, may alter the cellular ultrastructure.

1. Wash the cells 3× in 5 ml of buffer by centrifuging at 6000 rpm for 5 min.
2. Resuspend the cells in 2 ml of 0.1 M potassium phosphate, 1 M Sorbitol, and 0.25 mg of Zymolyase 100T. Incubate at 30°C for 30–60 min.

4. Postfixation

Potassium permanganate (Luft, 1956) has been extensively used as a postfixative in *S. pombe* studies. It preserves the lipid membrane bilayer, which appears highly contrasted in electron micrographs. Potassium permanganate leads to the extraction of cytoplasmic components, which is one of the reasons why the membranes appear so clearly (Bradbury and Meek, 1960). Potassium permanganate has been typically used in fission yeast at concentrations of 1–3% in aqueous solution. OsO_4, another common postfixative, is mostly used as a membrane stabilizer (Hayat, 1981). It quickly cross-reacts with lipids and, to some extent, proteins (Griffiths, 1993; Wright, 2000). OsO_4 is normally used in aqueous solution or in cacodylate buffer in concentrations ranging from 0.01% up to 2.5%.

1. Centrifuge the cells at 6000 rpm for 5 min and discard the supernatant. Wash 4–5 times by resuspending in 10 ml of water and centrifuge again. Carefully remove the supernatant to avoid losing the cell pellet. While centrifuging, prepare the postfixation solution: 2% potassium permanganate in aqueous solution (10 ml per sample).
2. After the last wash, remove the supernatant and, together with the residual water, transfer the cell pellet to a round bottom glass tube. Add 2 ml of 2% potassium permanganate and leave at room temperature for 5 min.
3. Pellet the cells at 6000 rpm, 5 min.
4. Replace the solution with 6 ml of 2% potassium permanganate and incubate for 45–60 min at room temperature.
5. Pellet the cells and remove the fixative. Extensively wash by filling the tube with distilled water and removing the water. Gently move the pellet with a laboratory toothpick and let it sediment before removing the water. Repeat until no purple color is visible.

5. En Bloc Staining

"*En bloc*" staining is the process by which a sample is immersed in an aqueous solution of uranyl acetate (0.2–1% in distilled water or ethanol solution) to provide a general background contrast to the sample. Usually, it is used at 4°C to preserve nucleic acid structure. In addition, uranyl acetate provides some degree of fixation without a major effect on protein conformation. This is important for immunocytochemical studies, where *en bloc* staining can replace postfixation (Hayat, 1981).

1. Remove the supernatant and immerse the pellets in 1% uranyl acetate for 1 h at room temperature.
2. Safely discard the uranyl acetate and wash in water 3–5 times.

6. Dehydration

Dehydration is the process by which the sample water is replaced by an organic solvent. It is important to use ice-cold anhydrous ethanol for the 100% steps and incubate the samples on ice to slow down the process of lipid extraction. Finally, gradually increase the ethanol temperature up to room temperature while moving the sample through the graded series to avoid water condensation in the ethanol (Wright, 2000).

1. Resuspend the pellet fragments in 25% ethanol. Leave to settle for 5 min.
2. Exchange by 50% ethanol and leave for 5 min.
3. Repeat step 15 with 75, 95, and 100% ethanol.
4. Incubate 3–5 times in 100% ethanol from a freshly opened bottle. In the final wash mix, add 0.5 ml of Spurr's resin.

7. Infiltration and Embedding

Unless samples are properly infiltrated and embedded in plastic, sections will show holes or even fall apart when they reach the water surface when cut in the diamond cuvette. Indeed, this is one of the most common problems encountered with this method. Yeast is difficult to infiltrate and fission yeast is no exception. Postfixation with permanganate may help to improve infiltration, making it easier than using OsO_4 for postfixation (Wright, 2000). Several resins are commonly used with fission yeast, such as epoxy resins including Epon, Spurr's, Araldite, and Quetol, or alternatively acrylic resins, such as LRWhite and Lowicryl (we recommend using HM20 for fission yeast) (Acetarin *et al.*, 1986; Carlemalm *et al.*, 1985). Epoxy resins can be extensively cross-linked, allowing sections as thin as 50–60 nm to be generated by ultramicrotomy. Sections are therefore highly stable under the electron beam and relatively resistant to electron beam damage (Hayat, 1981). However, the disadvantage of epoxy resins is their hydrophobicity, which, together with heat polymerization, leads to a loss of protein conformation and antigenicity (Griffiths, 1993). By contrast, acrylic resins allow low-temperature resin infiltration. This is considered to be a better method for preserving protein structure. Indeed, the majority of immunolocalization studies in fission yeast use acrylic resins (Kiss and McDonald, 1993).

1. Add a mix of 2:1 ethanol:resin onto the pellet. Transfer the pellets to an eppendorf tube (a plastic pipette is ideal for this transfer). Allow to mix by slow rotation for 2 h.
2. Replace the mix by a 1:1 ethanol:resin mix and allow to mix by rotation overnight.
3. Remove mix and add 100% resin to the eppendorf. Allow to rotate for 1 h.
4. Transfer eppendorfs (with open lid) to a vacuum machine and turn on vacuum for 1 h.
5. Transfer pellets to fresh 100% resin in aluminum weigh boat and turn on the vacuum for 2 h. Repeat once.
6. Transfer a single pellet to the center of a BEEM® capsule.
7. Add 100% resin to the line around the top of the capsule and turn on the vacuum overnight.

8. Polymerization

Finally, after infiltration the resin has to be hardened (i.e., polymerized) to allow sectioning. Epoxy resins are hardened by heat while acrylic resins are hardened by UV light, which allows polymerization at very low temperatures (–45 or –50°C).

1. Incubate capsules filled with Epon resin in the 60°C oven for 48 h.

Alternatively, acrylic resins will polymerize under a UV lamp over 2 days at –45°C followed by another couple of days at room temperature.

B. Cryofixation

Cryofixation or cryoimmobilization is the optimal sample preservation of the cell ultrastructure. By instantaneously "freezing" a cell and all its molecular constituents, it is possible to preserve the subcellular elements structure in a near-native state. Two methods of ultra-rapid freezing are available: plunge-freezing and high-pressure freezing. Both methods were developed in order to freeze samples by vitrifying the intracellular water, avoiding the formation of ice crystals (Moor, 1987). Ice crystals (especially hexagonal ice) are highly destructive and ruin the cell's ultrastructure, and thus constitute a major "enemy" for the electron microscopist.

1. Plunge-Freezing

For plunge-freezing we propose two established protocols, both from the work of Kenji Tanaka (Tanaka and Kanbe, 1986), the only difference being how cell layers are prepared for freezing. Plunge-freezing has the advantage of being relatively cheap and easy to set up in a laboratory. However, one drawback is that plunge-freezing is limited to a sample thickness of just a few micrometers (~5 μm) due to the limited speed of heat transfer; therefore in-depth freezing is inefficient. Despite this limitation, well plunge-frozen cells are indistinguishable from high-pressure frozen ones. It can be useful to add a cryoprotectant to the medium to lower the freezing point of the medium. Such molecules, however, must neither enter the cells nor cause osmotic pressure (Giddings *et al.*, 2001; Gilkey and Staehelin, 1986).

1. Cut 3×3 mm small squares of polycarbonate track-etch (PCTE) membranes (Ayscough *et al.*, 1993). Soak them in growing medium and place in an agar plate.
2. Transfer cells to the membranes and allow to grow for 2–3 h, so cells can adhere to the membrane. Take the membranes with forceps and place them in a plunge-freezing machine.

Alternatively:

1. Culture cells in YE5S to exponential growth.
2. Sandwich a thin layer of cells in between two copper grids.

Common to both protocols:

3. Quickly plunge into liquid ethane or propane cooled by a bath of liquid nitrogen.
4. Transfer the samples to liquid nitrogen. This can be kept until postprocessing. In the case of the membranes, the cells will adhere throughout the whole process of dehydration, infiltration, and embedding.

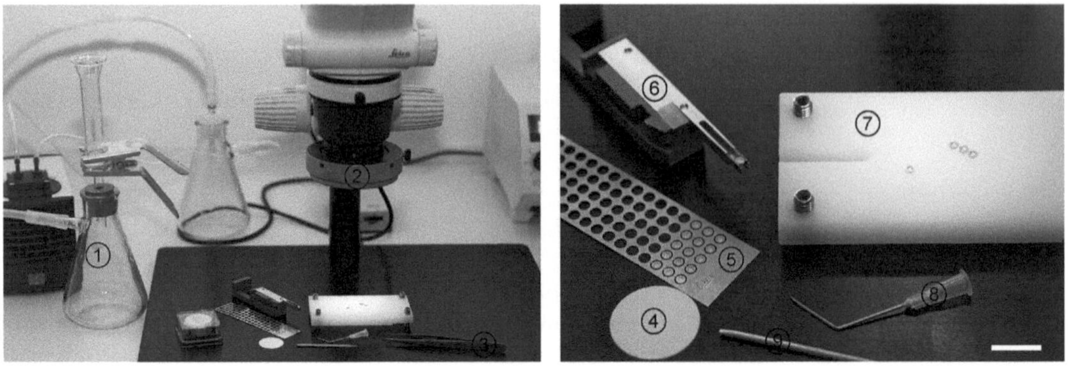

Fig. 2 Set-up for loading samples using the rapid-transfer system (RTS) of the Leica EM PACT2 high-pressure freezer. *Left panel:* 1. Filtration device for preparing yeast paste; 2. Stereomicroscope; 3. Tweezers. *Right panel:* Close-up of the RTS and materials. 4. Millipore® filter (45 μm) for use in filtration; 5. Membrane carriers (Leica); 6. RTS with membrane carrier; 7. RTS loading pad; 8. Needle tip bent 90°; 9. Wooden toothpick with tip cut as a spatula-like end to collect yeast paste. (See Plate no. 10 in the Color Plate Section.)

2. High-Pressure Freezing

The method of choice to vitrify *S. pombe* is high-pressure freezing. There are four types of HPF machines currently available: the Leica EM PACT2 (Leica Microsystems AG, Vienna, Austria); the Leica EM HPM-100 (Leica Microsystems AG, Vienna, Austria,); the Wohlwend Compact HPF-01 (Sennwald, Switzerland, distributed by Technotrade); and the ABRA FLUID HPM-010 (Abra Fluid AG, Widnau, Switzerland). High-pressure freezing is based on the principle that if a high pressure (~2000 bar) is quickly applied to a biological specimen that is then subjected to rapid cooling by liquid nitrogen, the freezing point of water is lowered by about 20 degrees. As a result, the mobility of the water molecules is considerably slowed down and amorphous ice is generated since ice crystals have no time to form (Dahl and Staehelin, 1989; Humbel *et al.*, 2001). One of the main advantages of this method is the possibility of vitrifying reliably large samples up to 200 μm thick (McDonald and Muller-Reichert, 2002; Studer *et al.*, 1993).

1. Prepare the vacuum filtering device. Use a Millipore filter (Type HA, 0.45 μm pore size) (Fig. 2).
2. Take a shaking exponential growing culture ($OD_{595} = 0.4$–0.6) and pour 10–15 ml into the vacuum funnel. It is important to preserve the growing conditions of the culture in between freezing shots.
3. Concentrate the cells by vacuum filtering. When the last visible trace of medium disappears over the filter, rapidly unplug the suction tube that connects the Erlenmayer to the vacuum pump.
4. Unclamp the filter funnel and remove the filter with tweezers. Place the filter on an agar plate to avoid the yeast paste drying out.

5. With a wooden toothpick with a tip cut as a spatula-like end, scrape some of the paste and load it on the carrier (Leica 200 μm deep membrane carrier (McDonald, 2009) or on the bottom carrier (the"*A*" carrier) of the carrier (or "hat") pair (BAL-TEC 200 μm deep hat pair).
6. The paste should completely fill the carrier. Ensure there are no bubbles of air since these can result in poor freezing. Likewise, excess paste also results in poor freezing, so remove all excess material.
7. Use a 45° bent-needle tip to flatten the paste in the carrier to remove any excess paste and clean the edges of the carrier.
8. In the case of Leica EM PACT2, take the rapid-loader with a loaded carrier and freeze the sample. For the BALTEC HPM 010 place the top carrier ("*B*" carrier) over the "*A*" carrier and press down. Close and screw the specimen rod, clamp into position, and freeze. Store the samples in liquid nitrogen.

C. Freeze-Substitution and Embedding

Freeze-substitution (FS) is the equivalent of dehydration, but at very low temperature. A solvent, usually acetone, will progressively substitute the water molecules within cells at –90°C until the sample is fully infused with the solvent. At this temperature the water remains vitrified and thus de-vitrification is prevented (Steinbrecht and Mueller, 1987). Later on, a resin will be added to the solvent and, with increasing concentration, the sample will end up being fully infiltrated with 100% resin. Epoxy or acrylic resins are used, depending on the purpose and advantages (see above).

In our laboratory, we have tried several FS protocols and have come to the conclusion that various FS cocktails provide satisfactory results. The one that has given good and reproducible results for both membranes and cytoskeleton visualization is a mixture of 0.1% GA, 0.01% OsO_4, and 0.25% UA in dry acetone (Muller-Reichert *et al.*, 2003), which we run on the Leica EM AFS1 (Leica Microsystems AG, Vienna, Austria). Finally, we embed our samples in Lowicryl HM20. However, various other FS cocktails have been successfully used in a variety of biological systems (Giddings, 2003; Giddings *et al.*, 2001; McDonald and Muller-Reichert, 2002). Using the specimen holders with flow-through tubes for the AFS2 is perfectly suitable for the AFS1. In addition, both the LEICA and BALTEC carriers can be used in this system.

1. Prepare the freeze-substitution machine and all the materials necessary. Start the program and pause it to cool down the machine.
2. Prepare the FS cocktail (4 ml per sample).
3. Use a small 0.5-ml precooled Eppendorf tube filled of liquid nitrogen to transfer the carrier to the AFS machine. Be sure of using precooled tweezers, preferably with ceramic tips. Slow warm-up from liquid nitrogen to AFS temperature can improve the preservation of the samples (Monaghan *et al.*, 1998).

4. After all specimen carriers are transferred to the AFS machine, release the pause, and start the FS program.
5. Substitution and infiltration take place according to the following protocol:
    a. 56 h at –90°C in the FS cocktail
    b. Raise temperature to –45°C in steps of –5°C per hour (13 h in total)
    c. 72 h at –45°C
        i. When at –45°C remove the FS cocktail and fill the tubes with dry acetone. Leave for 15 min. Repeat this step 3 times.
        ii. Start the infiltration process by removing the acetone and adding HM20 resin in a grade mixture of 3:1 for 1 h, 1:1 for 1 h, 3:1 for 2 h, and 3 times 100% for 2 h, overnight and 2 h.
        iii. Finally replace the resin one last time and start the polymerization.
    d. Place the UV lamp on the AFS and start the polymerization. It should stay at least 48 h at –45°C before raising the temperature.
    e. Raise temperature to +20°C in steps of 10°C per hour.
    f. Continue polymerization for a further 12 h at 20°C.

D. Sectioning and Staining

The type of sectioning depends on what is to be studied. When it is not necessary to follow a cell across several sections, in the case of immunocytochemistry for instance, the sample block should be trimmed in a way that allows the visualization of a large quantity of cells per section. Sections should have a thickness of 50–70 nm. When it is necessary to follow a cell or a structure through several sections, the block face should be trimmed so that the section is wide in the parallel axis to the diamond knife and short in the perpendicular axis to the knife. This increases the number of sections per ribbon and facilitates the search of the same cell across the sections. About 50–70 nm thick sections are required for thin serial sectioning studies or 200–300 nm for tomography purposes. Sections are picked up with a Formvar-coated slot grid and placed in the middle of the slot. Typically, sections are stained in 2% uranyl acetate in 70% methanol for 5 min and rinsed on two subsequent drops of 50% methanol, then on several water droplets. After blotting the excess of water, the grids are then stained in a drop of Reynold's lead citrate for 1 min, followed by 2 min rinse on a drop of water.

E. Immunocytochemistry

A major advantage of EM studies is the possibility of localizing proteins in cells by on-section immunocytochemistry. For proteins to be recognized by antibodies they must retain their antigenicity, which implies an optimal preservation of the cell's fine structure during fixation, dehydration, and plastic embedding (Griffiths, 1993; Jin *et al.*, 2005; Kiss and McDonald, 1993). The best way to accomplish this is cryofixation, followed by freeze-substitution, and hydrophilic resin embedding, as already discussed above. For immunocytochemistry, it is important to avoid OsO_4 in the FS "cocktail" or to keep it at very low percentage (Muller-Reichert *et al.*, 2003).

1. Prepare a blocking solution of 4% nonfat milk in PBST (10 mM of sodium phosphate, pH 7.3, 150 mM sodium chloride, 0.1% Tween).
2. Place the grids with the sections facing the blocking buffer drop. Leave at room temperature for 1–2 h.
3. Dissolve the primary antibody in the blocking buffer. In the absence of specific information, use a trial-and-error approach to discover the best dilution.
4. Incubate sections for 1–2 h at room temperature (possibly overnight at 4°C).
5. Rinse the grids extensively with PBST and incubate with the secondary antibody–gold conjugate dissolved in blocking buffer. Normally the gold size is between 5 and 15 nm (several types of gold-conjugated secondary antibodies are available commercially). Leave at room temperature for 1 h.
6. Rinse the grids extensively with PBST, then with water, and allow drying.
7. Stain the grids to provide the necessary contrast.

The Tokuyasu cryosection method and immunolabeling offers another possibility for immunocytochemistry studies. Detailed information about this technique in general is available in the literature (Griffiths, 1993), as well as particular applications for budding yeast (van Suylekom *et al.*, 2007). It appears that cryosectioning has been used only once for fission yeast (Carmichael *et al.*, 2006).

F. Electron Tomography

Electron tomography is a method of generating 3D images from multiple 2D projection images of a 3D object, obtained over a wide range of viewing directions. To generate a 3D image, a set of 2D projection images are recorded while tilting the object incrementally in the electron microscope. Each 2D image is subsequently back-projected, with the appropriated weighting, to form a 3D-density distribution of the original object (Baumeister *et al.*, 1999).

Electron tomography has been extensively reviewed (Giddings *et al.*, 2001; McIntosh *et al.*, 2005), and our earlier publications explain methods of acquiring, calculating tomograms, and reconstructing large cell volumes in the fission yeast (Fig. 3, and movie S1 at http://www.elsevierdirect.com/companions/9780123810076) (Höög and Antony, 2007; O'Toole *et al.*, 2002). Here we briefly summarize our methods and refer to more recent software advances for both the acquisition and calculation of tomograms.

Before acquiring a tilt-series, we pre-expose the area of acquisition at low magnification (×4200) with 2000 electrons A^{-2}. We normally acquire a tilt-series with a range ±60° with 1° increments in 1 × 3 montage (X–Y) using the acquisition software SerialEM (Mastronarde, 2005). To improve the radial resolution, the sample can be rotated 90° and a tilt-series of the same area acquired. In order to acquire several tilt-series, we use the automated tilt-series acquisition function of SerialEM. We have

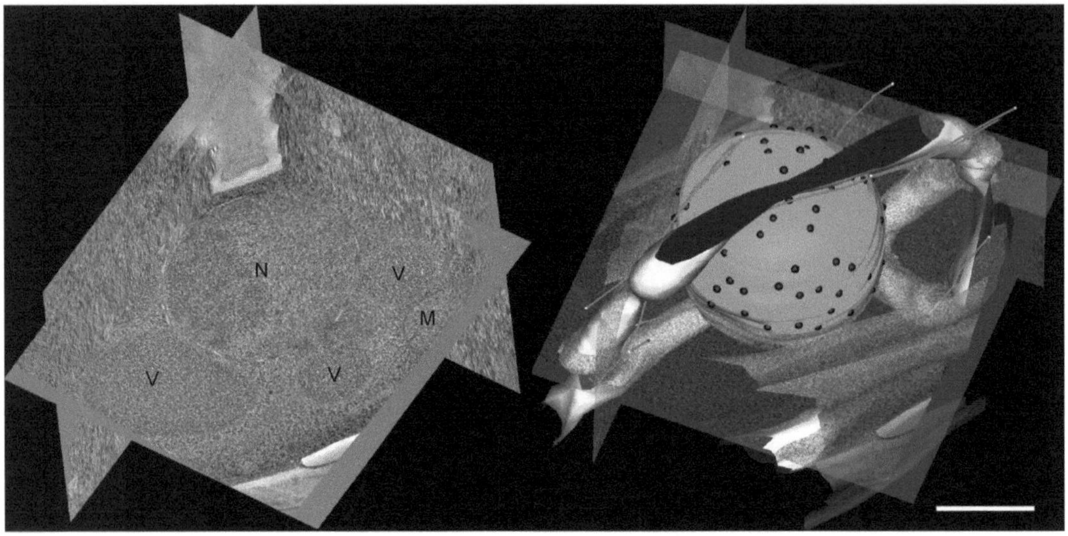

Fig. 3 Reconstruction of a *S. pombe klp2* deleted cell volume. *Left panel:* Tomogram with electronic slices displaying the three axes. N, nucleus; V, vacuoles; M, mitochondria. *Right panel:* superposition of the reconstructed model of this cell volume on the left image. Colors represent the nuclear envelope (turquoise), nucleopores (black), mitochondria (pale brown), plasma membrane (brown), and MTs (green). The sample was processed as indicated in the methods section. The full tomogram is composed of seven joined single frame tomograms (See also movie S1 at http://www.elsevierdirect.com/companions/9780123810076). Scale bar, 500 nm (Hélio Roque unpublished data). (See Plate no. 11 in the Color Plate Section.)

adapted our Dewar microscope support to hold a larger liquid nitrogen Dewar, which allows the microscope to work overnight. Several tilt-series in different sections of the same cell are acquired and the tomograms are later calculated by R-weighted back-projection algorithm using the software package IMOD (Kremer *et al.*, 1996; Mastronarde, 1997). Tomograms of the same area are combined, and those of different sections joined (Höög and Antony, 2007). To improve the joining of sections, we flatten the bent sections using the function Flattenwarp in IMOD (for further information about both SerialEM and IMOD visit http://bio3d.colorado.edu/).

IV. Instrumentation, Materials, and Reagents

A. Conventional Fixation

Instrumentation: Shaking incubator, clinical centrifuge, small tube rotator, vacuum pump, and oven incubator.

Materials: Yeast strains, 15-ml falcon tubes, laboratory toothpicks, plastic pipettes, eppendorfs, aluminum weigh boat, and BEEM capsules.

Reagents: Glutaraldehyde; 0.1 M sodium phosphate buffer; (pH = 7.0); Sorbitol; Zymolase 100T; potassium permanganate; uranyl acetate; Spurr's resin.

B. Cryofixation and Embedding

Instrumentation: Stereomicroscope with light source; plunge-freezing set up; high-pressure freezer (Leica EM PACT2 or ABRA HPM 010; Leica Microsystems AG, Vienna, Austria; or Abra Fluid AG, Widnau, Switzerland); shaking incubator; vacuum pump filtering apparatus; automated freeze-substitution machine (Leica AFS1 or AFS2; Leica Microsystems AG, Vienna, Austria).

Materials: Membrane carriers (200 μm deep); or brass hat pairs (*A* and *B*) (150–200 μm deep); Millipore membrane filters (0.45 μm) (Millipore UK, Ltd.); aluminum weigh boat; tweezers; cocktail sticks; 45° bent needle; YE5S agar plate.

Reagents: Anhydrous acetone (EM grade); glutaraldehyde; uranyl acetate; osmium tetroxide; Lowicryl HM20.

C. Sectioning, Staining, Immunocytochemistry, and Imaging

Instrumentation: Ultramicrotome (Leica UCT, Leica Microsystems AG, Vienna, Austria); electron microscope operated at 100 kV (Biotwin CM120 Philips, FEI, Eindhoven, the Netherlands) equipped with a computer stage.

Materials: Copper–palladium slot grids; Parafilm®.

Reagents: Formvar; methanol; uranyl acetate; Reynold's lead citrate; nonfat milk; phosphate buffer solution with Tween 20.

D. Electron Tomography

Instrumentation: Intermediate-voltage electron microscope operated at 300 kV (we use TECNAI F30 FEG, FEI, Eindhoven, the Netherlands); high-tilt rotating stage (we use Model 2020; Fischione Instruments, Corporate Circle, PA, USA); 4K×4K CCD camera (we use FEI Eagle camera, Eindhoven, the Netherlands); image acquisition software (SerialEM); 3D reconstruction software (IMOD).

Materials: fine tweezers.

Reagents: 15- nm cationic gold (British Biocell, Cardiff, UK).

V. *Schizosaccharomyces pombe*: Some Major Advances using EM

Here we examine some of the major advances that EM investigations have contributed to the field, particularly regarding cell division, microtubular cytoskeleton, cytoskeleton–organelle interactions, and cell-wall formation in *S. pombe*.

A. Mitotic Spindle

Cytological analysis of mitosis in *S. pombe* has disclosed striking features, shared with higher eukaryotes, further strengthening *S. pombe* as a model system in the study of cell division. For example, J.R. McIntosh, working in the Boulder laboratory, achieved the first reconstruction of the whole mitotic spindle of fission yeast

(Ding *et al.*, 1993; McDonald *et al.*, 1996). This laboratory was also the first to introduce HPF to study yeast mitotic apparatus following the pioneering work carried out by Byers and Goetsch (Byers and Goetsch, 1975, 1991). Using cryofixation they optimized the preservation of the spindle ultrastructure and carried out an exemplary three-dimensional modeling of the complete fission yeast mitotic spindle from serial sections and computational reconstruction. To do this, they aligned all serial cross sections, throughout and perpendicularly to the spindle axis, then tracked each individual MT through their lumen from one end to the other, until all elements were assembled as a 3D model. This made it possible to reconstruct up to 12 spindles and to classify and identify all types of MTs with respect to their function (for instance, the kinetochore MTs). Quantitative analysis, such as spacing between the spindle MTs as well as their preferred angles, could also be assessed. Using this method, it was possible to establish a very precise geometry of the spindle, as well as a complete visualization of the spindle in all possible orientations.

B. Determining MT Polarity using ET of Interphase Cells

Applying ET to visualize complex microtubular arrays not only improves the ease and robustness in cell volume reconstruction, but also avoids the laborious tracking of MTs through the serial sections. Moreover, ET provides an unprecedented amount of information on fine morphological features such as MT end structure.

The structure of MT ends was analyzed in a number of studies using *in vitro* systems (for reviews see: Dammermann *et al.*, 2003; Howard and Hyman, 2009). O'Toole *et al.* (1999) performed ET reconstructions of SPBs in early mitotic spindles of *S. cerevisiae* using high voltage and were the first to show MT end morphologies *in vivo*. Morphologically distinct microtubule ends were then visualized in the mitotic centrosome of early *C. elegans* embryos. In this study, open MT minus ends were found associated preferentially with kinetochore-attached MT (O'Toole, 2003). This represented a major step forward, promoting ET as an unprecedented and significant new tool at the beginning of the 21st century for examining detailed features of cell architecture and subcellular elements, such as MTs (McIntosh, 2001). Similarly, an ET approach was applied to study *S. pombe* interphase MT array, combined with a large-scale ET approach (Höög *et al.*, 2007). This latter method was designed to be able to track MTs from one end of the cell to the other and model for the first time cell organelles throughout the full cell volume (Fig. 4). As in previously cited work (O'Toole *et al.*, 1999), the MT polarity was provided by the fine morphology of the MT ends and a color code marked the end types onto the models generated from the tomograms (Höög *et al.*, 2007). Further analysis of organelle features such as MT–mitochondria distribution provided evidence that MTs influence the shape of mitochondria, resulting in a reticulated and stretched pattern of distribution.

C. SPB Duplication

The cycle of SPB duplication, differentiation, and segregation in *S. pombe* also benefits from EM investigation. Taking advantage of the optimal structural

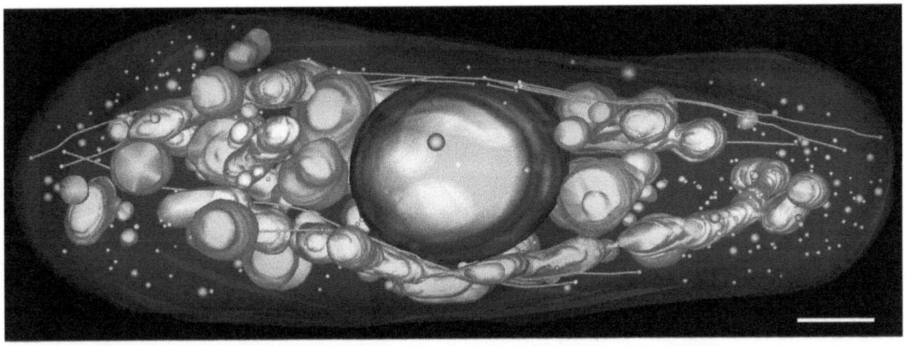

Fig. 4 Reconstruction and modeling of the subcellular elements in fission yeast. Three-dimensional model of a whole fission yeast cell volume reconstructed from 15 montage serial tomograms of 250-nm sections each (1 by 3 montages in XY). The sample was processed as indicated in the methods section. Nucleus, violet; mitochondria, blue; vacuoles, yellow; microtubules, green; various trafficking vesicles (various colors). Scale bar, 1 μm. Adapted with permission from *Dev. Cell* **12,** 349–361. (See Plate no. 12 in the Color Plate Section.)

preservation obtained by HPF/FS, the McIntosh laboratory showed for instance that SPBs in *S. cerevisiae* and *S. pombe* did not behave similarly (Ding *et al.*, 1997). In *S. pombe* the SPB enters and leaves the nuclear envelope (NE) as the cell cycle progresses while in *S. cerevisiae* the SPB always remains anchored within the NE. In fact, the SPB is mostly cytoplasmic in the interphase in *S. pombe*, sitting just on the outside rim of the NE where SPB duplication takes place. The duplicated SPBs then remain connected by an electron-dense bridge until mitosis onset. Upon entry into mitosis, cell-cycle-dependent changes occur. A pocket is formed in the NE, which invaginates and opens up, allowing the SPB to incorporate within the NE. The nucleation of nuclear MTs begins only after NE integration. Interdigitation of nuclear MTs allows SPB separation and spindle assembly. Later, in anaphase, the SPB becomes extruded back to a cytoplasmic localization at the surface of the NE. It is interesting to note that apart from differences in the dynamics of NE insertion, the cycle of SPBs in *S. pombe* is quite similar to the cycle in *S. cerevisiae* (Ding *et al.*, 1997).

D. Immunocytochemistry and the SPB

It is of course fully advantageous to be able to localize main proteins on the fine structural "map" of cell organelles that the 2D thin sections from HPFrozen cells provide. In the case of SPBs several published papers showed clear-cut localization of SPB-bound components. For example, the SPB component *cdc31*, the mutation of which arrests cells in mitosis with a monopolar spindle, was precisely localized to the half-bridge (Fig. 5) (Paoletti *et al.*, 2003). Moreover, EM of HPF/FS-treated *cdc31* mutant cells showed that SPB failed to duplicate in this context, which explains the formation of monopolar spindles. This work demonstrates the powerful and precise information that can be collected using EM combined with mutant analysis on a simple cell system.

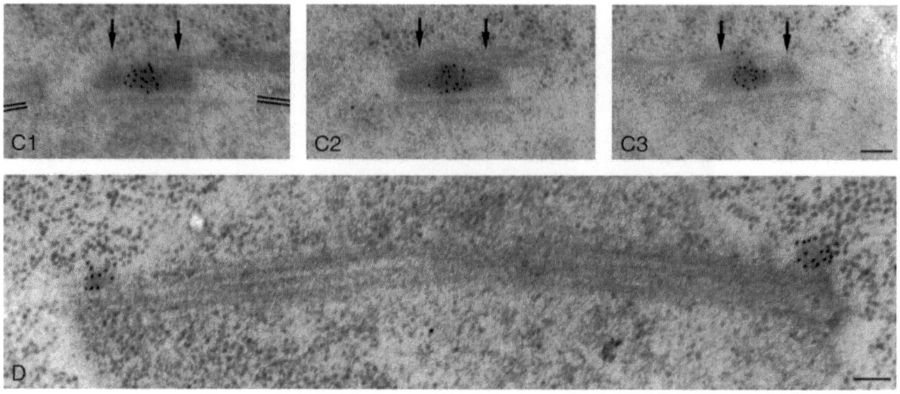

Fig. 5 Immunolocalization of cdc31p, a half-bridge component of the SPB in fission yeast. *Top panels:* Duplicated SPBs (arrows) connected by the half-bridge, where gold particles indicate cdc31p. *Bottom panel:* Section of the entire mitotic spindle; gold-labeling is clearly visible at the "edge" of the SPBs, associated with the half-bridge. Scale bars, 100 nm. Adapted with permission from *Mol. Biol. Cell.* **14,** 2793–2808.

E. Cell-Wall Formation

EM analysis of spheroplasts helped visualize structures such as filasomes and spherical bodies (100–300 nm in diameter), which are involved in cell-wall regeneration (Takagi *et al.*, 2003). Filasomes belong to the F-actin patch structure and contain a single micro-vesicle (35–70 nm in diameter) in the center, surrounded by actin filaments. Filasomes were found not only in the cytoplasm adjacent to the newly formed glucan fibrils (components of the cell wall) but were also found to move from the cytoplasm to the plasma membrane during the process of cell-wall formation. Three-dimensional models were generated from thin sections and these confirmed the distribution of filasomes at the site of glucan network formation in the cell wall. It is interesting to note that cortical actin is also required for the localization of enzymes (i.e., glucan synthases) involved in cell-wall synthesis (Cortés *et al.*, 2002; Humbel *et al.*, 2001). However, the actin network (actin mesh) is not easily detectable by EM in plastic-embedded samples since the background provided by the plastic is too high, burying the actin filament density in the background noise (see Chapter 22 by Resch, this volume).

VI. Discussion and Outlook

Our molecular understanding of cellular processes is evolving fast, aided by cell systems such as fission yeast where molecular biology and genetic techniques can be optimally applied. However, there have been fewer investigations of fission yeast using EM approaches or 3D visualization, including electron tomography studies. One reason for this is that these techniques rely on expensive instruments, such as intermediate- or high-voltage electron microscopes. Moreover, such a pool of equipment requires a team of specialized people: biologists, engineers, and technicians, ready to focus their efforts on cell morphological analysis using tools which belong both to the wet laboratory as well as to the computing office.

Improved instrumentation will further push the boundaries of knowledge regarding cellular structures. The fact that cells can now be easily vitrified with minimal rearrangements of their molecular components suggests that more studies will be done to characterize the subcellular systems at a resolution of about 5–10 nm. ET can now provide details of the 3D fine morphology of subcellular structures, information, which cannot be seen by even the finest 2D snapshots. One exemplary case has been the identification of the MT end structure where the polymer's polarity can be determined (Höög *et al.*, 2007; O'Toole *et al.*, 1999). Furthermore, recent elaborate ET work was able to show slender fibrils connecting the curved protofilaments at the MT tip to the inner kinetochore (McIntosh *et al.*, 2008).

Apart from the morphological analysis of wild-type cells, the mutant phenotypical read-out offers a vast terrain for further ET investigations. In this respect, temperature-sensitive (*ts)* mutants have the advantage of quickly inducing phenotypic alterations, which can be efficiently captured by cryo-immobilization and subsequently analyzed by ET and modeling studies. Secretory and endocytic pathways are topics that would also benefit from combined approaches involving molecular biology and genetics, together with ET analysis. Structural models of compartments and transport intermediates within these pathways are likely to provide further understanding of the spatial organization of the main membrane traffic routes (see Chapter 26 by Verkade *et al.*, this volume).

In the near future, frozen-hydrated sections are likely to provide new information *in vivo* at molecular resolution (Al-Amoudi *et al.*, 2004; Dubochet *et al.*, 2007). For example, in fission yeast the organization of actin cables or bundles has yet to be understood, since it remains very difficult to visualize when using plastic-embedded samples. MT–actin interactions may well be supported by a new pathway as MTs are thought to be able to trigger actin assembly at ectopic sites at the cell cortex under certain conditions (Minc *et al.*, 2009; Terenna *et al.*, 2008). In such a situation, polarity factors (such as bud6p, for3p [forming], and cdc42p) are recruited and actin cables are assembled in a MT-dependent way. Detailed structural information on the interaction between the two networks will also be needed.

Actin-driven membrane dynamics is another important field of interest. For instance, actin is required to support the pulling force necessary to produce endocytic invaginations (Aghamohammadzadeh and Ayscough, 2009). Such compartments have not yet been described by EM/ET methods, and it would be very useful to achieve a better description of these structures, as reported in the case of budding yeast (Idrissi *et al.*, 2008).

Interesting results can be expected as well by investigating molecular complexes *in situ*. It would be helpful for instance to understand how the conformation of MT tips is modified, depending on the binding of the various tip-tracking proteins (e.g., Mal3, Tea2, Tip1), which are known to affect MT dynamics *in vivo* (Busch and Brunner,

2004; Busch *et al.*, 2004). In this case more information is needed from cryoEM tomography (Sandblad *et al.*, 2006). Similarly, other complexes such as the tea1/tea2/mal3 complex, which is transported along MT to the cell tip where polarity markers are delivered, should be further studied by cryoEM methods. We are optimistic that more topics will benefit from further elaborate cryoEM and cryoET investigations on frozen-hydrated sections from fission yeast.

A major issue that concerns future technology development in the field of ET is the design of electron-dense tags that could support protein localization in reconstructed volumes. To our knowledge, there is no available standard method of localizing proteins by ET. However, recently, attempts have been made to generate a clonable tag for EM, based on generating a fusion protein with the metal-binding protein containing chelated metal atoms (Mercogliano and DeRosier, 2007). This technique has been successfully applied to bacteria protein localization using TEM as well as electron tomography (Diestra *et al.*, 2009). Fission yeast would provide an ideal model system in which to try and apply this method in order to localize selected proteins. Furthermore, it should be possible to express a tagged version of a protein of interest, so that all the sites where the protein associates within the 3D models derived from the tomograms can be mapped.

Another recently reported development that is of potential interest to EM specialist is the method of injecting *S. pombe* cells (Riveline and Nurse, 2009). The method circumvents the problem of cell-wall rigidity. The principle is the following: cells are trapped on a topographical patterned substrate forming channels (~5 μm wide) and arranged in a grid pattern. Cells can fit perfectly within such channels. The orthogonal channels can be used to position a needle tip, and by applying enough pressure against the cell, a local shear force can be created, which produces a hole in the rigid wall. A piezo-impact micromanipulator is used to deliver materials flowing out the tip of the needle in the vicinity of the hole, allowing cells to pick up the materials. As an example, fluorescent phalloidin was successfully introduced only within the sheared cells (Riveline and Nurse, 2009). Such selected cells could then be processed for EM fixation and plastic embedding prior to EM or ET visualization and analysis.

Surprisingly, *S. pombe* has recently also acted as a "reactor system" for exogenous protein complex assembly *in vivo*. A striking example showed that *S. pombe* can be modified to express the bacterial tubulin homolog FtsZ (Srinivasan *et al.*, 2008). As a result, this molecule forms a ring-like structure equivalent to the division ring found in normal bacteria. This allows the mechanisms regulating the assembly and organization of the FtsZ ring to be studied in the absence of all bacterial proteins. In this case, the authors showed that an assembly occurred by a process of spooling linear FtsZ filaments. Such assembly would be ideally suited for structural analysis at higher resolution by ET.

Currently available EM technology offers a vast range of possibilities for studying the structural organization of fission yeast, together with gene function dissection. Huge progress has been made over the past 10 years in imaging cellular structures at EM resolution. One of the main targets in cell biology is to try and solve the structure of molecular complexes as well as of supramolecular assemblies *in vivo*. To achieve this goal, and study the main molecular players and their associated functions within the context of the whole cell, cutting-edge EM-based technology—in addition to LM dynamic analysis—will certainly play an essential role in the future.

VII. Summary

In recent years, fission yeast has become a model of choice for addressing various biological questions and has also proved to be an excellent cell system for EM investigation. In this chapter, we provide detailed protocols for several types of EM work that can be performed with fission yeast. Conventional fixation and embedding achieve acceptable results with fission yeast and is a valid approach where expensive material and machines for cryofixation are not readily available. Nonetheless, data interpretation must always take into account the possible negative effects of fixatives on the cell's morphology. The method of choice is HPF followed by FS and low-temperature embedding, which provides the best results while preserving near-to-native cellular structures. This method also allows work to be carried out using both morphology and immunocytochemistry. In addition, electron tomography is most successfully performed with fission yeast cells that have first been cryofixed and embedded in plastic. Finally, fission yeast offers an excellent organism with which to perform large-volume tomography reconstruction and it can be used to analyze the cellular distribution of organelles. The increasing number of high-quality EM publications provides additional evidence of the ongoing success of applying EM techniques to investigate fission yeast wild-type and mutant ultrastructure.

Acknowledgments

We would like to thank the members of our laboratory at the EMBL, Damian Brunner and his group at EMBL for constant support, discussions and generous help with handling fission yeast, and colleagues abroad, particularly Kenji Tanaka (National Institute of Technology and Evaluation, Chiba, Japan), and Anne Paoletti (Institut Curie UMR-144 CNRS, Paris, France) for their useful comments on the manuscript.

References

Acetarin, J. D., Carlemalm, E., and Villiger, W. (1986). Developments of new lowicryl resins for embedding biological specimens at even lower temperatures. *J. Microsc.* **143**, 81–88.

Aghamohammadzadeh, S., and Ayscough, K. (2009). Differential requirements for actin during yeast and mammalian endocytosis. *Nat. Cell Biol.* **11**, 1039–1042.

Al-Amoudi, A., Chang, J., Leforestier, A., Mcdowall, A., Salamin, L., Norlén, L., Richter, K., Blanc, N., Studer, D., and Dubochet, J. (2004). Cryo-electron microscopy of vitreous sections. *EMBO J.* **23**, 3583–3588.

Ayscough, K., Hajibagheri, N. M., Watson, R., and Warren, G. (1993). Stacking of Golgi cisternae in *Schizosaccharomyces pombe* requires intact microtubules. *J. Cell. Sci.* **106**(Pt 4), 1227–1237.

Baumeister, W., Grimm, R., and Walz, J. (1999). Electron tomography of molecules and cells. *Trends Cell Biol.* **9**, 81–85.

Bradbury, S., and Meek, G. A. (1960). A study of potassium permanganate 'fixation' for electron microscopy. *J Microsc. Sci.* **101**, 241–250.

Busch, K. E., and Brunner, D. (2004). The microtubule plus end-tracking proteins mal3p and tip1p cooperate for cell-end targeting of interphase microtubules. *Curr. Biol.* **14**, 548–559.

Busch, K. E., Hayles, J., Nurse, P., and Brunner, D. (2004). Tea2p kinesin is involved in spatial microtubule organization by transporting tip1p on microtubules. *Dev. Cell* **6**, 831–843.

Byers, B., and Goetsch, L. (1975). Behavior of spindles and spindle plaques in the cell cycle and conjugation of *Saccharomyces cerevisiae*. *J. Bacteriol.* **124**, 511–523.

Byers, B., and Goetsch, L. (1991). Preparation of yeast cells for thin-section electron microscopy. *Methods Enzymol.* **194**, 602–608.

Carazo-Salas, R. E., Antony, C., and Nurse, P. (2005). The kinesin klp2 mediates polarization of interphase microtubules in fission yeast. *Science* **309**, 297–300.

Carlemalm, E., Villiger, W., Hobot, J. A., Acetarin, J. D., and Kellenberger, E. (1985). Low temperature embedding with lowicryl resins: Two new formulations and some applications. *J. Microsc.* **140**, 55–63.

Carmichael, J. B., Stoica, C., Parker, H., McCaffery, J. M., Simmonds, A. J., and Hobman, T. C. (2006). RNA interference effector proteins localize to mobile cytoplasmic puncta in *Schizosaccharomyces pombe*. *Traffic* **7**, 1032–1044.

Cortés, J. C., Ishiguro, J., Durán, A., and Ribas, J. C. (2002). Localization of the (1,3)beta-D-glucan synthase catalytic subunit homologue bgs1p/cps1p from fission yeast suggests that it is involved in septation, polarized growth, mating, spore wall formation and spore germination. *J. Cell. Sci.* **115**, 4081–4096.

Dahl, R., and Staehelin, L. A. (1989). High-pressure freezing for the preservation of biological structure: Theory and practice. *J. Electron. Microsc. Tech.* **13**, 165–174.

Dammermann, A., Desai, A., and Oegema, K. (2003). The minus end in sight. *Curr. Biol.* **13**, R614–R624.

Dawes, C. J. (1979). Biological techniques for transmission and scanning electron microscopy. Ladd Research Industries. Burlington.

Diestra, E., Fontana, J., Guichard, P., Marco, S., and Risco, C. (2009). Visualization of proteins in intact cells with a clonable tag for electron microscopy. *J. Struct. Biol.* **165**, 157–168.

Ding, R., McDonald, K. L., and McIntosh, J. R. (1993). Three-dimensional reconstruction and analysis of mitotic spindles from the yeast, *Schizosaccharomyces pombe*. *J. Cell Biol.* **120**, 141–151.

Ding, R., West, R. R., Morphew, D. M., Oakley, B. R., and McIntosh, J. R. (1997). The spindle pole body of *Schizosaccharomyces pombe* enters and leaves the nuclear envelope as the cell cycle proceeds. *Mol. Biol. Cell.* **8**, 1461–1479.

Dubochet, J., Zuber, B., Eltsov, M., Bouchet-Marquis, C., Al-Amoudi, A., and Livolant, F. (2007). How to "read" a vitreous section. *Methods Cell Biol.* **79**, 385–406.

Giddings, T. H. (2003). Freeze-substitution protocols for improved visualization of membranes in high-pressure frozen samples. *J. Microsc.* **212**, 53–61.

Giddings, T. H., Jr., O'Toole, E. T., Morphew, M., Mastronarde, D. N., McIntosh, J. R., and Winey, M. (2001). Using rapid freeze and freeze-substitution for the preparation of yeast cells for electron microscopy and three-dimensional analysis. *Methods Cell Biol.* **67**, 27–42.

Gilkey, J. C., and Staehelin, L. A. (1986). Advances in ultrarapid freezing for the preservation of cellular ultrastructure. *J. Electron. Microsc. Tech.* **3**, 177–210.

Griffiths, G. (1993). "Fine Structure Immunocytochemistry." Springer-Verlag, Heidelberg.

Hayat, M. A. (1981). "Principles and Techniques of Electron Microscopy: Biological Applications". University Park Press, Baltimore.

Hedges, S. B. (2002). The origin and evolution of model organisms. *Nat. Rev. Genet.* **3**, 838–849.

Hirata, A., and Tanaka, K. (1982). Nuclear behaviour during conjugation and meiosis in the fission yeast *Schizosaccharomyces pombe*. *J. Gen. Appl. Microbiol.* **28**, 263–274.

Höög, J. L., and Antony, C. (2007). Whole-cell investigation of microtubule cytoskeleton architecture by electron tomography. *Methods Cell Biol.* **79**, 145–167.

Höög, J. L., Schwartz, C., Noon, A. T., O'Toole, E. T., Mastronarde, D. N., McIntosh, J. R., and Antony, C. (2007). Organization of interphase microtubules in fission yeast analyzed by electron tomography. *Dev. Cell* **12**, 349–361.

Howard, J., and Hyman, A. A. (2009). Growth, fluctuation and switching at microtubule plus ends. *Nat. Rev. Mol. Cell Biol.* **10**, 569–574.

Humbel, B. M., Konomi, M., Takagi, T., Kamasawa, N., Ishijima, S. A., and Osumi, M. (2001). In situ localization of beta-glucans in the cell wall of *Schizosaccharomyces pombe*. *Yeast* **18**, 433–444.

Idrissi, F. Z., Grotsch, H., Fernandez-Golbano, I. M., Presciatto-Baschong, C., Riezman, H., and Geli, M. I. (2008). Distinct acto/myosin-I structures associate with endocytic profiles at the plasma membrane. *J. Cell Biol.* **180**, 1219–1232.

Jin, Y., Mancuso, J. J., Uzawa, S., Cronembold, D., and Cande, W. Z. (2005). The fission yeast homolog of the human transcription factor EAP30 blocks meiotic spindle pole body amplification. *Dev. Cell* **9**, 63–73.

Kamasaki, T., Arai, R., Osumi, M., and Mabuchi, I. (2005). Directionality of F-actin cables changes during the fission yeast cell cycle. *Nat. Cell Biol.* **7**, 916–7.

Kanbe, T., Kobayashi, I., and Tanaka, K. (1989). Dynamics of cytoplasmic organelles in the cell cycle of the fission yeast *Schizosaccharomyces pombe*: Three-dimensional reconstruction from serial sections. *J. Cell. Sci.* **94**(Pt 4), 647–656.

Karnovsky, M. J. (1965). A formaldehyde–glutaraldehyde fixative of high osmolality for use in electron microscopy. *J. Cell Biol.* **27**, 137A–138A.

Kiss, J. Z., and McDonald, K. (1993). Electron microscopy immunocytochemistry following cryofixation and freeze substitution. *Methods Cell Biol.* **37**, 311–341.

Kremer, J. R., Mastronarde, D. N., and McIntosh, J. R. (1996). Computer visualization of three-dimensional image data using IMOD. *J. Struct. Biol.* **116**, 71–76.

Leupold, U. (1970). Genetical methods for *Schizosaccharomyces pombe*. *In* "Methods in Cell Physiology" (D. M. Prescott, , ed.) **Vol. 4**, pp. 169–177. Academic Press, New York.

Lindner, P. (1893). *Wochenschrift für Brauerei* **10**, 1298–1300.

Luft, J. H. (1956). Permanganate; a new fixative for electron microscopy. *J. Biophys. Biochem. Cytol.* **2**, 799–802.

Maclean, N. (1964). Electron microscopy of a fission yeast, *Schizosaccharomyces pombe*. *J. Bacteriol.* **88**, 1459–1466.

Mastronarde, D. N. (1997). Dual-axis tomography: An approach with alignment methods that preserve resolution. *J. Struct. Biol.* **120**, 343–352.

Mastronarde, D. N. (2005). Automated electron microscope tomography using robust prediction of specimen movements. *J. Struct. Biol.* **152**, 36–51.

McCully, E. K., and Robinow, C. F. (1971). Mitosis in the fission yeast *Schizosaccharomyces pombe*: A comparative study with light and electron microscopy. *J. Cell. Sci.* **9**, 475–507.

McDonald, K. (1999). High-pressure freezing for preservation of high resolution fine structure and antigenicity for immunolabeling. *Methods Mol. Biol.* **117**, 77–97.

McDonald, K. L. (2009). A review of high-pressure freezing preparation techniques for correlative light and electron microscopy of the same cells and tissues. *J. Microsc.* **235**, 273–281.

McDonald, K., and Muller-Reichert, T. (2002). Cryomethods for thin section electron microscopy. *Methods Enzymol.* **351**, 96–123.

McDonald, K., O'Toole, E. T., Mastronarde, D. N., Winey, M., and Richard McIntosh, J. (1996). Mapping the three-dimensional organization of microtubules in mitotic spindles of yeast. *Trends Cell Biol.* **6**, 235–239.

McIntosh, J. R. (2001). Electron microscopy of cells: A new beginning for a new century. *J. Cell Biol.* **153**, F25–F32.

McIntosh, J., Grishchuk, E., Morphew, M., Efremov, A., Zhudenkov, K., Volkov, V., Cheeseman, I., Desai, A., Mastronarde, D., and Ataullakhanov, F. (2008). Fibrils connect microtubule tips with kinetochores: A mechanism to couple tubulin dynamics to chromosome motion. *Cell* **135**, 322–333.

McIntosh, R., Nicastro, D., and Mastronarde, D. (2005). New views of cells in 3D: An introduction to electron tomography. *Trends Cell Biol.* **15**, 43–51.

Mercogliano, C. P., and DeRosier, D. J. (2007). Concatenated metallothionein as a clonable gold label for electron microscopy. *J. Struct. Biol.* **160**, 70–82.

Minc, N., Bratman, S. V., Basu, R., and Chang, F. (2009). Establishing new sites of polarization by microtubules. *Curr. Biol.* **19**, 83–94.

Mitchison, J. M. (1970). Physiological and cytological methods for *Schizosaccharomyces pombe*. *In* "Methods in Cell Physiology" (D. M. Prescott, ed.) **Vol. 4**, pp. 131–165. Academic Press, New York.

Mizuhira, V., and Futaesaku, Y. (1972). New fixation method for biological membranes using tannic acids. *Acta Histochem. Cytochem.* **5**, 233–236.

Monaghan, P., Perusinghe, N., and Muller, M. (1998). High-pressure freezing for immunocytochemistry. *J. Microsc.* **192**, 248–258.

Moor, H. (1987). "Theory and Practise of High Pressure Freezing." Springer-Verlag, Berlin, Germany.

Muller-Reichert, T., Sassoon, I., O'Toole, E., Romao, M., Ashford, A. J., Hyman, A. A., and Antony, C. (2003). Analysis of the distribution of the kinetochore protein ndc10p in saccharomyces cerevisiae using 3-D modeling of mitotic spindles. *Chromosoma* **111**, 417–428.

Nurse, P. (2002). Cyclin dependent kinases and cell cycle control (nobel lecture). *ChemBioChem* **3**, 596–603.

Osumi, M., and Sando, N. (1969). Division of yeast mitochondria in synchronous culture. *J. Electron. Microsc. (Tokyo)*. **18**, 47–56.

O'Toole, E. (2003). Morphologically distinct microtubule ends in the mitotic centrosome of caenorhabditis elegans. *J. Cell Biol.* **163**, 451–456.

O'Toole, E. T., Winey, M., and McIntosh, J. R. (1999). High-voltage electron tomography of spindle pole bodies and early mitotic spindles in the yeast *Saccharomyces cerevisiae*. *Mol. Biol. Cell.* **10**, 2017–2031.

O'Toole, E. T., Winey, M., Mcintosh, J., and Mastronarde, D. N. (2002). Electron tomography of yeast cells. *Methods Enzymol.* **351**, 81–95.

Oulevey, N., Deshusses, J., and Turian, G. (1970). Etude de la zone septale de *Schizosaccharomyces pombe* en division a ses etapes succesives. *Protoplasma* **70**, 217–224.

Paoletti, A., Bordes, N., Haddad, R., Schwartz, C. L., Chang, F., and Bornens, M. (2003). Fission yeast cdc31p is a component of the half-bridge and controls SPB duplication. *Mol. Biol. Cell.* **14**, 2793–2808.

Pickett-Heaps, J. D., and Fowke, L. C. (1969). Cell division in oedogonium. *Aust. J. Biol. Sci.* **22**, 857–894.

Riveline, D., and Nurse, P. (2009). 'Injecting' yeast. *Nat. Methods* **6**, 513–4.

Robinson, D., Ehlers, U., Herken, R., Herrmann, B., Mayer, F., and Schurmann, F. -W. (1987). "Methods of Preparation for Electron Microscopy." Springer-Verlag, Berlin.

Sabatini, D. D., Bensch, K., and Barrnett, R. J. (1963). Cytochemistry and electron microscopy. The preservation of cellular ultrastructure and enzymatic activity by aldehyde fixation. *J. Cell Biol.* **17**, 19–58.

Sandblad, L., Busch, K., Tittmann, P., Gross, H., Brunner, D., and Hoenger, A. (2006). The *Schizosaccharomyces pombe* EB1 homolog mal3p binds and stabilizes the microtubule lattice seam. *Cell* **127**, 1415–1424.

Steinbrecht and Mueller (1987). Freeze substitution and freeze drying. *In*: R. A. Steinbrecht, K. Zierold, Eds., Cryotechniques in Biological Electron Microscopy. Springer-Verlag, Berlin, pp. 149–172.

Srinivasan, R., Mishra, M., Wu, L., Yin, Z., and Balasubramanian, M. K. (2008). The bacterial cell division protein FtsZ assembles into cytoplasmic rings in fission yeast. *Genes Dev.* **22**, 1741–1746.

Studer, D., Hennecke, H., and Müller, M. (1993). High-pressure freezing of soybean nodules leads to an improved preservation of ultrastructure. *Planta* **188**, 155–163.

Takagi, T., Ishijima, S. A., Ochi, H., and Osumi, M. (2003). Ultrastructure and behavior of actin cytoskeleton during cell wall formation in the fission yeast *Schizosaccharomyces pombe*. *J. Electron. Microsc. (Tokyo)*. **52**, 161–174.

Tanaka, K. (1963). The fine structure of yeast cell as revealed by ultrathin sectioning and electron microscopy. *Ann. Rept. Inst. Food Microbiol. Chiba Univ.* **15**, 61–70.

Tanaka, K., and Kanbe, T. (1986). Mitosis in the fission yeast *Schizosaccharomyces pombe* as revealed by freeze-substitution electron microscopy. *J. Cell. Sci.* **80**, 253–268.

Terenna, C. R., Makushok, T., Velve-Casquillas, G., Baigl, D., Chen, Y., Bornens, M., Paoletti, A., Piel, M., and Tran, P. T. (2008). Physical mechanisms redirecting cell polarity and cell shape in fission yeast. *Curr. Biol.* **18**, 1748–1753.

Teysset, L., Dang, V. D., Kim, M. K., and Levin, H. L. (2003). A long terminal repeat-containing retrotransposon of *Schizosaccharomyces pombe* expresses a gag-like protein that assembles into virus-like particles which mediate reverse transcription. *J Virol.* **77**, 5451–5463.

van Suylekom, D., van Donselaar, E., Blanchetot, C., Do Ngoc, L. N., Humbel, B. M., and Boonstra, J. (2007). Degradation of the hexose transporter hxt5p in *Saccharomyces cerevisiae*. *Biol. Cell.* **99**, 13–23.

Wolfe, K. H., and Shields, D. C. (1997). Molecular evidence for an ancient duplication of the entire yeast genome. *Nature* **387**, 708–713.

Wood, V., Gwilliam, R., Rajandream, M. A., Lyne, M., Lyne, R., Stewart, A., Sgouros, J., Peat, N., Hayles, J., and Baker, S., et al., , 2002. The genome sequence of *Schizosaccharomyces pombe*. *Nature*. **415**, 871–880.

Wright, R. (2000). Transmission electron microscopy of yeast. *Microsc. Res. Tech.* **51**, 496–510.

CHAPTER 12

Electron Microscopy and High-Pressure Freezing of *Arabidopsis*

Byung-Ho Kang

Microbiology and Cell Science Department, Electron Microscopy and Bioimaging Lab, Interdisciplinary Center for Biotechnology Research, University of Florida, Gainesville, Florida 32611

METHODS IN CELL BIOLOGY, VOL. 96

978-0-12-381007-6
DOI: 10.1016/S0091-679X(10)96012-3

Abstract

In this chapter, we will discuss methods and protocols for high-pressure freezing (HPF) and freeze substitution (FS) to examine *Arabidopsis* tissues by transmission electron microscopy (TEM). By use of HPF in combination with FS, it is possible to obtain *Arabidopsis* samples that are far better preserved for both ultrastructural analysis and immunogold labeling than by conventional chemical fixation. Like other cryofixation methods, ice crystal growth is still a problem in HPF if samples are too thick (> 200 μm) or if their water content is too high. Furthermore, damage done to cells/tissues prior to freezing cannot be "reverted" by HPF. In general, FS of plant tissues is more difficult than that of nonplant tissues because plant cell walls impede removal of water from the enclosed cells as well as from the walls themselves. To overcome these challenges, we describe the details of a HPF, FS, and resin-embedding protocol for *Arabidopsis* tissues here. In addition, the generation of ribbons of serial sections from *Arabidopsis* TEM blocks, three-dimensional (3D) analysis of organelle shapes and distribution within the tissue, and immunogold labeling are also explained. The *Arabidopsis* research community has developed many research tools to investigate gene functions such as knockout mutant lines, antibodies, and transgenic lines expressing epitope-tagged proteins. The TEM techniques explained here have been combined with these tools to elucidate how a particular gene of interest functions in the *Arabidopsis* cell.

I. Introduction

Over the last two decades, plant development and physiology studies have used *Arabidopsis thaliana*, a small weed species of the Cruciferae family, as the primary model organism. *Arabidopsis* has many merits for genetic analyses/manipulations that have helped define biological pathways of plants and identify molecular players in these pathways, using mutations or natural variations. With its full genome sequence and other related tools, identification of the genes responsible for certain mutant phenotypes is relatively easy, compared to other plant species. However, the ability to clone a gene does not help understand how the gene functions in the cell. To this end, it is essential to characterize mutant phenotypes and the protein produced by the gene precisely. Electron microscopy (EM) is a very powerful tool for such a functional characterization of genes and their products.

Characterization of the *knolle* (*kn*) mutant illustrates this point very well. *kn* is an *Arabidopsis* seedling lethal mutant in which plant body organization is disrupted. Light microscopy imaging showed that cells of the mutant plants have incomplete cell walls and are often multinucleated, indicating a cytokinesis defect in the mutant plants (Lukowitz *et al.*, 1996). Microscopic localization of *Kn* gene transcription and of the Kn protein revealed that the gene is transcribed only in dividing cells and that the protein is targeted to the cell plate. Most importantly, it was demonstrated that clusters of vesicles accumulate at the cell plate in the *kn* mutant by transmission electron microscopy (TEM) imaging. The cell plate is a transient organelle that appears only in plant cells undergoing cytokinesis. It is assembled by the fusion of Golgi-derived vesicles and matures into a new cell wall. The Kn protein displays amino acid sequence similarity with syntaxin family proteins that are involved in vesicle fusion events in eukaryotes. The accumulation of vesicles and incomplete formation of the cell wall in the *kn* mutant indicate that the *Kn* protein is likely to be a syntaxin contributing to the fusion of Golgi-derived vesicles at the cell plate (Lauber *et al.*, 1997; Waizenegger *et al.*, 2000).

Cryofixation is the preferred technique over chemical fixation for preserving biological samples for high-resolution TEM. Chemical fixation artifacts have been observed by numerous electron microscopists and many of these artifacts can be avoided by cryofixation (Gilkey and Staehelin, 1986; Mersey and McCully, 1978). However, cryofixation can also lead to artifacts, the most significant of which is ice crystal damage. This damage occurs when ice crystal formation is faster than heat removal from the sample (Echlin, 1991). Several cryofixation techniques, such as plunge freezing, propane jet freezing, and high-pressure freezing (HPF), have been developed to reduce ice crystal formation during freezing. However, the use of both plunge freezing and propane jet freezing is limited to samples thinner than 10 μm. Heat transfer from thicker samples is not fast enough to suppress the growth of ice crystals, to an extent that the ice crystal growth does not distort cellular membranes and cytoskeletal elements. HPF suppresses ice crystal growth by pressurizing samples to as high as 2000 bars as they are frozen by liquid nitrogen. This allows one to freeze samples as thick as 200 μm without noticeable ice crystal damage. Many *Arabidopsis* tissues are either thinner than 200 μm or can be quickly dissected into small pieces with minimal disruption prior to freezing (Bowman, 1994). HPF is currently the best cryofixation method for imaging *Arabidopsis* by EM (Kiss and Staehelin, 1995), as well as for electron microscopic analyses of other model organisms (McDonald, 2007).

HPF has been used in only a "handful" of plant research laboratories since its introduction in the early 1980s, despite its superiority over chemical fixation. The high cost of purchasing and running a high-pressure freezer and the additional equipments required for processing samples at low temperatures is probably the primary reason for its limited use. In addition, freeze substitution (FS) and low-temperature resin embedding of frozen plant samples is more time consuming than conventional room temperature protocols. However, the superiority of HPF preservation was recognized early by several plant cell biologists who published papers in the late 1980s and early 1990s; these papers detailed novel features of the plant endoplasmic reticulum (ER) (Craig and Staehelin, 1988), Golgi stack morphology (Staehelin *et al.*, 1990), ultrastructural differences in plant root tip samples preserved by chemical fixation and HPF (Kiss *et al.*, 1990), and structural polarity of the Chara rhizoid (Kiss and Staehelin, 1993).

Another example of plant ultrastructural studies that have benefitted from HPF is the electron microscopic analysis of cell plate formation. Cell plate formation is a dynamic process involving a cytokinetic organelle, called a phragmoplast, in which vesicle trafficking, microtubule reorganization, and expansion of the cell plate take place (Bednarek and Falbel, 2002). The phragmoplast consists of intricate membranous and cytoskeletal components that can easily be damaged by chemical fixation. By use of HPF, Samuels *et al.* (1995) described how the cell plate arises and expands from a group of spherical vesicles with unprecedented accuracy. Characterization of the complex phragmoplast structure was further benefitted by applying electron tomographic techniques which greatly improves 3D resolution. Combined with HPF/FS, electron tomographic analyses provided novel and detailed morphological information that helped elucidate the mechanism of cell plate formation, expansion, and the 3D organization of the phragmoplast microtubule (Austin *et al.*, 2005; Otegui *et al.*, 2001; Segui-Simarro *et al.*, 2004).

HPF has also enabled TEM imaging of short-lived events or subtle structures that are difficult to capture using chemical fixation techniques in plant cells, such as ER-to-Golgi transport (Kang and Staehelin, 2008), the response of Golgi/TGN (*trans*-Golgi network) complex to brefeldin A treatment (Ritzenthaler *et al.*, 2002), morphological classification of vesicles associated with the plant and algal Golgi (Donohoe *et al.*, 2007), the interaction between statolith and cortical ER in gravity-sensing cells of root tips (Leitz *et al.*, 2009), reorganization of phragmoplast microtubules (Austin *et al.*, 2005) and of protein/membrane trafficking through the TGN (Lam *et al.*, 2007), and multivesicular bodies (Otegui *et al.*, 2006; Tse *et al.*, 2004). In addition, stromal membrane assembly and plastoglobule development in the chloroplast have also been well described by means of HPF/FS and electron tomography (Austin *et al.*, 2006; Shimoni *et al.*, 2005).

II. Rationale

The rationale of this chapter is to explain the technique of HPF and the accompanying TEM methods for serial sectioning and localization of macromolecules by immunogold labeling with *Arabidopsis* specimens (Fig. 1). The discussion will focus on non- or moderately vacuolated cells of *Arabidopsis*, such as cells in the root tip, in the shoot apical meristem, in young leaves, in anthers, and in immortalized liquid-cultured cells. Mature plant cells contain few organelles and they are almost completely filled with vacuoles. By contrast, dividing and growing cells in the above-mentioned tissues are highly active in gene expression, catabolic/anabolic metabolism, and membrane trafficking, making them better suited for protein localization as well as for characterizing mutant phenotypes in their cellular context. Furthermore, cells without large vacuoles are better preserved by HPF because their cytoplasm contains diverse macromolecules that suppress ice crystal growth during HPF. A more general discussion on cryopreservation of plant cell samples including HPF of non-*Arabidopsis* plants can be found in an excellent review by Hess (2007).

III. Methods

A. Plant Material

After surface sterilization, *Arabidopsis* seeds are sprinkled onto solid media, cold-treated at 4°C for a day, and grown under conditions appropriate to each particular

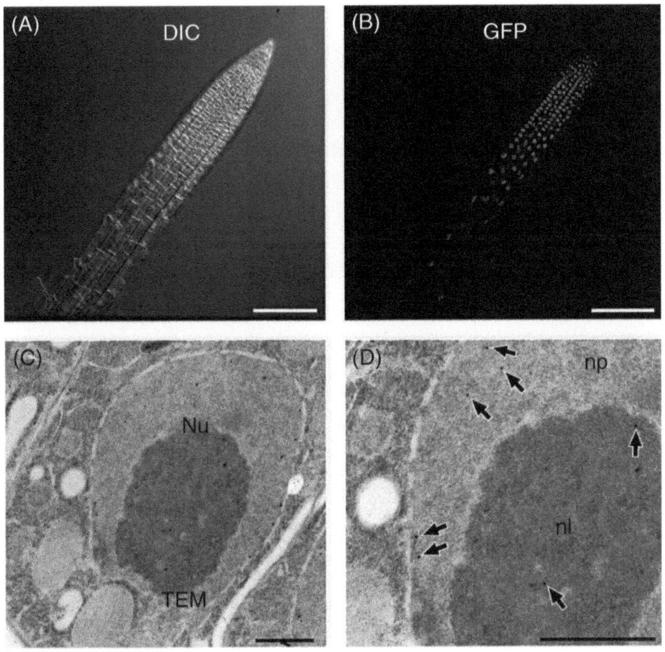

Fig. 1 Localization of green fluorescent protein (GFP) by confocal laser scanning microscopy and by electron microscopy. (A) Differential Interference Contrast (DIC) micrograph of an *Arabidopsis* root tip. (B) Confocal micrograph of the root tip shown in (A) visualizing GFP. This plant expresses a GFP fusion protein targeted to the nucleus. (C) Transmission electron micrograph (TEM) of a meristematic cell in the root tip of the specimen shown in (A) and (B). The root tip was high-pressure frozen, freeze substituted, and embedded in Lowicryl HM20 resin. This section from the root tip sample was labeled with a GFP antibody to localize the GFP fusion protein. Gold particles (15 nm) are localized in the nucleus (Nu). (D) Higher magnification of the image shown in (C). Gold particles (arrows) are seen in the nucleoplasm (np) as well as in the nucleolus (nl). Scale bars in (A) and (B): 100 μm, Scale bars in (C) and (D): 1 μm (See Plate no. 13 in the Color Plate Section.)

experiment. Petri dishes can be held vertically so that root tips grow along but not into the solid media. This will make it easier to recover root tip samples without damage. Root tips or shoot apices are harvested from seedlings grown for 5–7 days after germination. Developing anthers can be isolated from unopened flower buds after removing the petals and sepals.

When harvesting samples for localizing green fluorescent protein (GFP) by immunogold labeling (Fig. 1), it is important to confirm that the parts of the seedlings or plants that are being frozen express GFP. We screen seedlings or plants by comparing them with nontransgenic seedlings or plants using a fluorescence stereomicroscope. This relatively simple task saves tremendous time and effort that could have been wasted working on samples that do not express appropriate amounts of GFP.

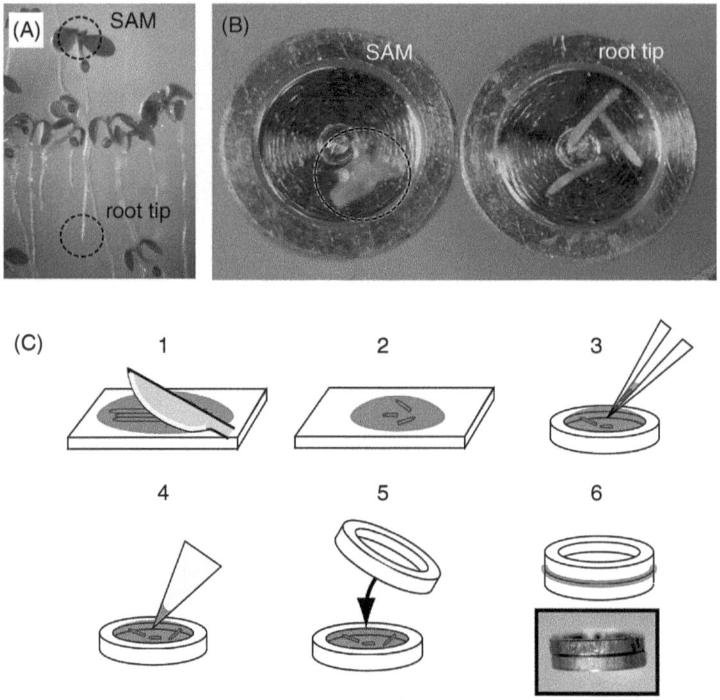

Fig. 2 Dissection of *Arabidopsis* seedlings and loading of shoot apical meristem (SAM) and root tip specimens into HPF planchettes. (A) *Arabidopsis* seedlings grown vertically on MS solid media. (B) An SAM and three root tip samples loaded into planchettes. (C) Steps in loading root tip samples into an HPF planchette. (1) Cut root tips in a 0.15 M sucrose solution with a #10 scalpel blade. For rapid dissection, align multiple root tips and cut them simultaneously. (2–3) Transfer sliced root tip specimens to a planchette filled with the 0.15 M sucrose solution. Root tips can be carried with the sucrose solution retained at the tip of a pair of tweezers, and then released into the planchette as shown in (3). Transfer of root tips in the sucrose solution will avoid mechanical damage that could result from gripping the samples with the tweezers. (4) Removal of excess sucrose solution with a piece of filter paper. (5) Covering of the A-type planchette with a B-type planchette with its flat side down. (6) Planchette sandwich carrying the *Arabidopsis* specimens ready for HPF. An actual planchette sandwich is shown in the inset.

B. High-Pressure Freezing

The procedure for dissecting and freezing *Arabidopsis* root tip samples is illustrated in Fig. 2. Shoot apex and other samples can be dissected out and placed into the planchettes in a similar way. For freezing young anther samples, we make a slit on the pollen sac before freezing to facilitate subsequent FS.

A sucrose solution (0.1–0.15 M) is the most commonly used cryoprotectant/filler for *Arabidopsis* root tip, shoot apex, and pollen sac samples. We use 1-hexadecene for leaf tissues because it permeates into intercellular air spaces readily, owing to its low surface tension. Leaf tissues are split into small pieces in the 1-hexadecene using small-gauge syringe needles and then transferred into the planchettes. Liquid-cultured cells are gently spun down and resuspended in their growth media supplemented with sucrose. After the cells are concentrated again by gentle centrifugation, the cell slurry is transferred to a planchette with a metal spatula.

Arabidopsis root tips are about 0.1 mm thick or slightly thicker. We use the 0.2-mm-deep well of the type A planchette as a carrier for root tips. The loaded planchette is then covered with the flat side of a type B planchette. Using this combination of planchettes, we can reduce the sample volume but still prevent crushing/damage of root specimens. For shoot apex and other specimens thicker than root samples, two type B planchettes are utilized, one for carrying the samples and the other for a lid, with its flat side down[1]. It is important to fill the carrier planchette completely with cryoprotectant/filler. Trapped air within the planchette will interfere with freezing and will collapse during pressurization, thus deforming the samples.

Once the planchette sandwich is ready (Fig. 2C), it is inserted into the HPF machine. If you are using a HPM010, you have to unscrew the planchette holder and separate frozen planchettes with tweezers from the holder while under liquid nitrogen. With the HPM100 machine, each planchette sandwich is placed within a plastic carrier plate. An accessory tool that comes with the HPM100 machine is used for punching the planchettes out of the carrier plate. The planchettes with the frozen samples are then transferred to cryovials containing appropriate FS cocktails or stored in liquid nitrogen for further use. In Fig. 3, morphological indicators of good preservation by HPF/FS as well as signs of ice crystal damage in *Arabidopsis* meristematic cells are shown.

C. Freeze Substitution

During FS, frozen crystalline and noncrystalline water in the samples is replaced with an organic solvent, usually acetone plus fixatives. We describe FS protocols for *Arabidopsis* root tip and shoot apex samples. As shown below, samples for ultrastructural characterization and for immunogold labeling are processed in different FS cocktails and with different protocols. The main difference between the two protocols is that osmium tetroxide (OsO_4) is the cross-linking fixative for ultrastructural samples while low concentrations of glutaraldehyde are used for immunogold labeling samples. This is because proteins are commonly degraded proteolytically by their reaction with OsO_4, and therefore, protein epitopes are destroyed in osmicated samples. However, cell wall polysaccharides can be localized by immunogold labeling using osmicated samples (Lynch and Staehelin, 1992; Moore *et al.*, 1991) probably because polysaccharide epitopes are abundant in the cell wall and/or might be more resistant to degradation by OsO_4.

After FS, ultrastructural samples are warmed up to room temperature for embedding in EPON or Spurr's resin. By contrast, immunogold labeling samples are kept at –50°C for embedding in HM20 resin without temperature changes.

[1] The depth of type B planchettes is 0.3 mm, which is larger than the upper limit of sample thickness for reliable HPF (0.2 mm). Therefore, chances of poor preservation increase. Try to dissect specimens thinner than 0.2 mm if possible.

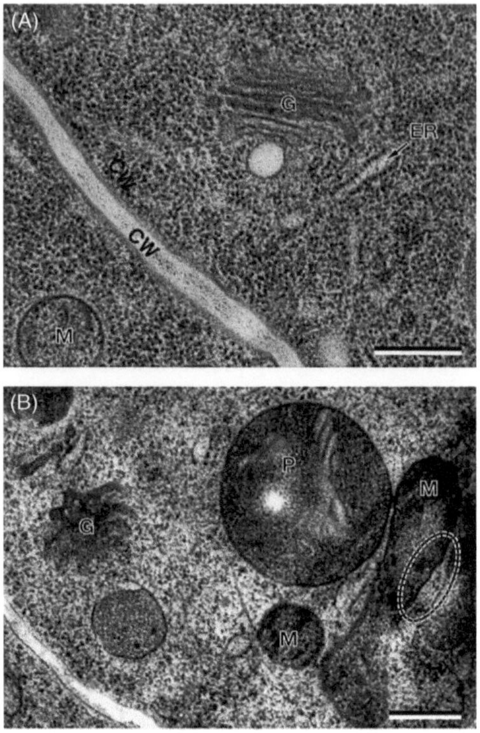

Fig. 3 Electron micrographs (80-nm-thin sections) from high-pressure frozen and freeze-substituted *Arabidopsis* meristematic cells. These cells were freeze substituted in 2% OsO_4–acetone and embedded in Epon resin. (A) A well-preserved cell. All membranes are smooth and tight. The cytoplasm is filled with ribosomes and appears uniformly gray. No blank holes are seen in the well-frozen cells. (B) A cell damaged by ice crystal formation. Staining of the cytoplasm is not homogeneous and ribosomes are shrunken. Membranes are darker than those in panel A. Wavy membranes are often seen in damaged cells (dashed oval). ER: endoplasmic reticulum; G: Golgi; M: mitochondria; P: plastid; CW: cell wall. Scale bars: 500 nm.

1. Freeze Substitution for Ultrastructural Analysis

a. Freeze Substitution Cocktail We use 1–4% OsO_4 dissolved in anhydrous acetone. The OsO_4 concentration can be adjusted to optimize brightness and contrast of structures under study. If a high OsO_4 concentration cocktail is used, nonreacted OsO_4 should be washed out thoroughly with anhydrous acetone before resin embedding. This can be achieved by incubation in acetone for 2–3 h with several changes of the acetone.

b. Freeze Substitution Protocol We use the following protocol as a starting point. Based on fixation and resin infiltration of the resulting samples, the incubation time for each temperature step can be modified:

1. FS at –80°C for 48 h
2. Gradual warming to –20°C over a 24-h period

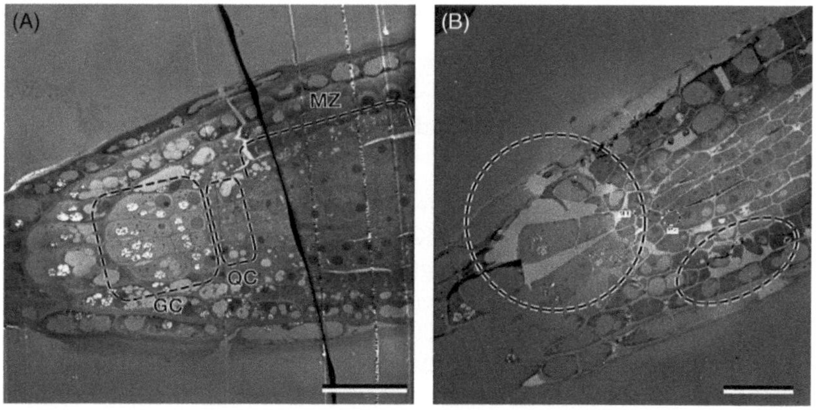

Fig. 4 Low-magnification electron micrographs of medial longitudinal sections of *Arabidopsis* root tips. (A) An *Arabidopsis* root tip showing gravity-sensing columella cells (GC), the quiescent center (QC), and the meristematic zone (MZ). (B) A section from an *Arabidopsis* root tip sample in which FS and resin-embedding quality were unsatisfactory. Columella cells are separating and some cells have been lost (dashed circle). In the dashed oval, cells have been crushed and displaced. Scale bars: 20 μm.

3. Incubation at –20°C for 12 h
4. Gradual warming to 4°C over a 4-h period
5. Washing with anhydrous acetone at room temperature (three changes of acetone)
6. Separation of samples from planchettes if they are still attached to the planchettes.

The plant cell wall is a problem for good FS and impedes substitution of its enclosed cell contents. FS protocols for mammalian cells that involve incubation at –80 to –90° C for only several hours are not appropriate for most plant samples. Such short protocols produce *Arabidopsis* samples that shatter, collapse, or lose cells during sectioning (Fig. 4). We keep frozen *Arabidopsis* samples at –80°C for at least 48 h to remove as much vitreous ice as possible. Samples with especially thick cell walls, such as pollen grains, are freeze substituted at –80°C for as long as 72 h and are warmed up to room temperature slowly over 2 days (Otegui and Staehelin, 2004).

2. Freeze Substitution for Immunogold Labeling

a. Freeze Substitution Cocktail Anhydrous acetone containing 0.25% glutaraldehyde and 0.1% uranyl acetate is used for FS. For efficient removal of water in vacuolated plant cells, dimethoxypropane (DMP) is added to the substitution cocktail up to 8% (Samuels *et al.*, 1995). We have obtained reasonably good FS of tobacco BY-2 suspension cell cultures (Kang and Staehelin, 2008) and *Arabidopsis* leaf samples using the FS cocktail supplemented with DMP (Christopher *et al.*, 2007). It is also possible to add DMP to an osmium-containing FS solution. DMP and OsO_4 react very quickly in acetone at room temperature. Therefore, acetone should be pre-cooled in dry ice before mixing the two chemicals.

b. Freeze Substitution Protocol We use the following protocol:

1. FS at –80°C for 48 h
2. Gradual warming to –50°C over a 30-h period
3. Washing with anhydrous acetone (three changes) and initiating HM20 resin embedding at –50°C.

There are other FS protocols in which frozen samples are warmed up to –35°C and embedded in Lowicryl K4M resin, or warmed up to 0°C and embedded in LR White resin. It is generally agreed that processing at –50°C with Lowicryl HM20 resin provides the best preservation of morphology and of reactive epitopes for immunogold labeling (McDonald, 2007). However, embedding in LR White has been shown to work reliably as well (Dammermann *et al.*, 2004).

D. Resin Embedding and Curing

1. For Ultrastructure Studies

Infiltration with a progressive series of resin to acetone dilutions is carried out at room temperature. We use a hard grade Epon resin mix consisting of 50% Epon resin monomer, 15% DDSA (dodecenyl succinic anhydride), and 35% NMA (nadic methyl anhydride), by weight. Resin embedding begins by incubating specimens with 5% resin in acetone (v/v) overnight. Prolonged incubation with the low concentration resin mix appears to reduce the separation of cytoplasm from the plant cell wall and probably facilitates resin infiltration into the cell wall. After incubation in the 5% dilution, the concentration of the Epon resin increases stepwise, 10, 25, 50, 75, and 100%. At each dilution, specimens are incubated at least several hours, which makes resin embedding a 2–3 day-long procedure. After three changes of 100% resin, samples are transferred to a plastic mold and cured at 60°C for 2 days. Samples are agitated on a low-speed rocking shaker starting at the 75% resin step. We add accelerator, DMP-30 (2,4,6-tri-(dimethylaminemethyl) phenol) or BDMA (benzyldimethylamine), to the 100% resin for the last two resin changes.

2. For Immunogold Labeling Studies

Freeze-substituted samples for immunogold labeling are infiltrated with HM20 acrylic resin mix (Crosslinker D: 5.96 g, Monomer E: 34.04 g, and initiator: 0.20 g). The infiltration is carried out at –50°C with a stepwise increase of resin in acetone from 33 to 66%, and then to 100%. Specimens are infiltrated at least 12 h at each step. After three changes of 100% resin, the samples are removed from the HPF planchettes and transferred either to flat-bottomed BEEM capsules or to a flat embedding mold[2]. It is helpful to use a dissecting microscope to recover small samples like *Arabidopsis* root tips from cryovials and to transfer them to the BEEM capsules or to a flat

[2] HPF planchettes can be removed either before or after resin embedding.

embedding mold[3]. Be sure to remove all the HPF planchettes if they have accidentally been placed in the BEEM capsules or flat embedding mold. The HM20 resin is polymerized under UV irradiation for 24–36 h at –50°C. After the HM20 resin is cured, warm the samples up to room temperature. *Arabidopsis* root tip samples sometimes look yellowish due to uranyl acetate in the FS cocktail, but they should be almost invisible if FS and resin embedding went properly. *Arabidopsis* shoot apex and leaf samples usually retain their green color.

E. Mounting, Trimming, and Ultramicrotomy

After polymerization, Epon-embedded root tip or shoot apical meristem samples are cut out with a jeweler's saw, but HM20 resin blocks are usually too brittle for this procedure. HM20 blocks are fractured into pieces with a surgical prep blade. Root tip samples in HM20 blocks are difficult to locate, so they need to be marked under a dissecting microscope before fracturing. Mounting, trimming, and ultramicrotomy are performed as described in comprehensive books about EM (Bozzola and Russel, 1999; Hagler, 2007). In the following paragraph, some "tips" to facilitate ribbon formation for serial-section analysis are given.

1. Trimming and Coating with Rubber Cement

It is important that the top and bottom edges of the trapezoidal block face are parallel. If not, section ribbons will be curved, and curved ribbons are difficult to handle and cannot be accurately positioned on slot grids easily. Curved section ribbons will also make it more difficult to follow the same field of view in each section in the ribbon and, subsequently, will require more time later for aligning images when they are assembled into a stack for 3D reconstruction. In addition, to facilitate ribbon formation we use rubber cement (Fig. 5). Apply a dab of diluted rubber cement (~10-fold dilution in xylene or toluene) at the base of the trapezoid to bind new sections to the ribbon as they are cut. Wick away excess rubber cement.

2. Formvar Coating of EM Slot Grids

For serial-sectioning analyses, it is essential to retrieve ribbons of sections on slot grids so that structures of interest can be imaged on each section without being masked by the metal bars of mesh grids. Slot grids are also recommended for immunogold labeling because the metal bars may block labeled structures. Before collecting sections on a slot grid, the slot opening must be covered with a supporting film. We usually use 2 × 1 mm copper or gold slot grids, which have been coated with a Formvar (polyvinyl formal) film.[4] Due to limited space, protocols for preparing Formvar-coated grids will not be

[3] A dissecting microscope can be installed on the Leica AFS2 system.

[4] Formvar-coated grids may be purchased from electron microscopy supply vendors. However, in our hands, the pre-coated Formvar film is too thin to survive frequent washing or cycles of incubation and washing during immunogold labeling. In addition, Formvar-coated slot grids (~$2 per grid) are far more expensive than uncoated slot grids (Gilder 2 mm × 1 mm copper grid—~$0.05 per grid).

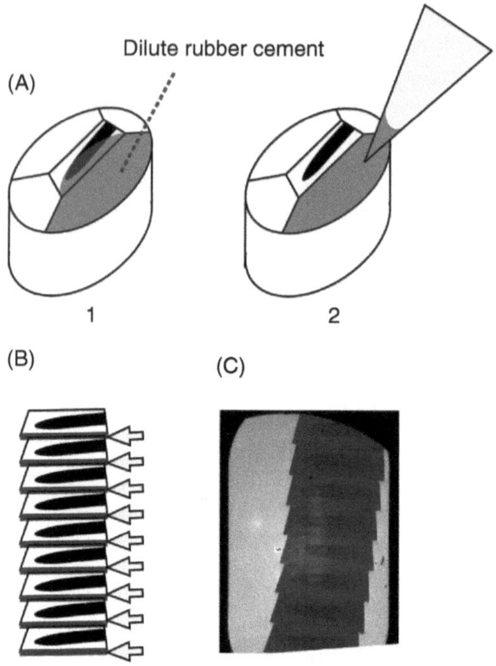

Fig. 5 Preparing a ribbon of serial sections. (A) After trimming and facing the block, a drop of rubber cement diluted in xylene (or toluene) is applied on the base of the trapezoidal face (1). Excess rubber cement dilution is removed with a piece of filter paper (2). If the rubber cement contaminates the block face of the trimmed sample, it should be removed by thick-sectioning with a glass knife. It is difficult to clean rubber cement from a diamond knife edge. (B) Cartoon of a serial-section ribbon held together by intervening rubber cement (arrows). (C) Eight serial sections collected on a Formvar-coated slot grid (2 mm × 1 mm).

covered here. Our protocol is similar to the one described in page 94 of Bozzola and Russel (1999) except that we use a 0.75% solution of Formvar in ethylene dichloride instead of a 0.3–0.5% Formvar solution in chloroform.

F. Immunogold Labeling

On-section immunogold labeling involves multiple incubation steps in various solutions as well as multiple washing steps. The ultrathin sections can be contaminated as they go through all these procedures. Therefore, it is important to use buffers and reagents that are clean and fresh, and forceps and microcentrifuge tubes free of dust and oil. Contamination may also be caused by allowing solutions to "run" onto the back of the grid, where they may dry during subsequent steps.

1. Buffers and Antibody Solutions

All incubation steps are carried out in a humid chamber on Parafilm (to prepare a humid chamber place damp Kimwipes® in the dish; Fig. 6A). Sections should not be

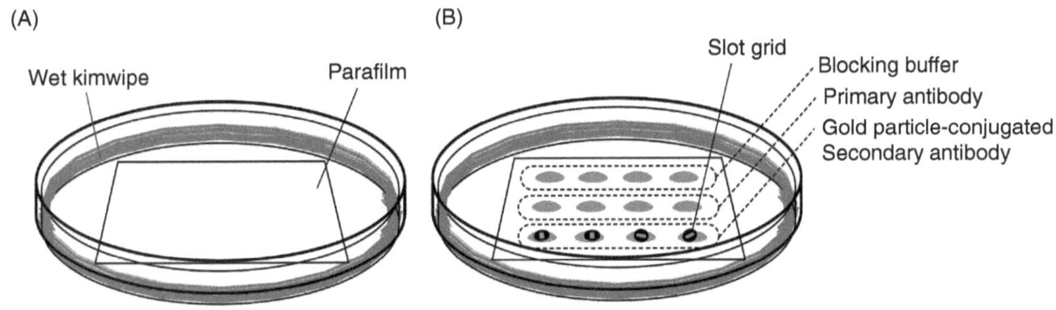

Fig. 6 Setup of a humid chamber for immunogold labeling. (A) The humid chamber is assembled in a Petri dish (10–20 cm in diameter). (B) Grids floating on drops of gold particle-conjugated secondary antibody solutions. Drops of buffers and solutions are placed on the Parafilm immediately prior to use.

allowed to dry out throughout the procedure, until after the final wash with distilled water. All buffers and antibody solutions should be made fresh from stock solutions.

a. PBST (Phosphate-Buffered Saline + Tween 20) Prepare a 1× phosphate-buffered saline (PBS) solution from a 10× PBS stock (80 g NaCl, 2 g KCl, 11.4 g $Na_2HPO_4 \bullet H_2O$, 2g KH_2PO_4, pH 7.3) and add Tween 20 to the final concentration between 0.1 and 0.2%. Higher concentrations of Tween 20 lower background labeling, but concentrations higher than 0.5% are not recommended because high concentrations of detergent inhibit antibody–epitope interaction and EM grids do not float well on drops of PBS with high detergent concentrations.

b. Blocking Buffer Use 2% nonfat milk suspended in the PBST for masking nonspecific binding sites. Sonicate the blocking buffer for 5–10 min after adding the nonfat milk powder. Always use fresh blocking buffer.

c. Primary Antibody Solution Dilute the primary antibody in the blocking buffer or in a 1:1 dilution of blocking buffer in PBST (1% nonfat milk in PBST). It is important to verify that an antibody is detecting a single polypeptide by immunoblot analysis prior to immunogold labeling. For a pilot immunogold labeling run, we test primary antibody concentrations that are 10–20 times higher than those prepared for immunoblot analysis. At an optimal concentration, immunogold particles are associated with specific structures and show only a very low background of randomly scattered gold particles. We have found that many primary antibodies work well for immunogold labeling at concentrations of 1–10 μg/ml. Centrifuge the primary antibody stock (12,000 rpm for 1 min) before preparing dilutions. If the amount of the primary antibody stock is limited, the leftover antibody dilution can be reused. Care must be taken to prevent bacterial contamination during storage.

d. Secondary Antibody Conjugated to Gold Particles We use 5-, 10-, or 15-nm colloidal gold coated with secondary antibodies. After centrifuging at 12,000 rpm for 30 s, secondary antibodies are diluted in the PBST containing 0.5% nonfat milk. Two antigens can be localized simultaneously with a primary antibody solution containing two antibodies prepared from different animal species (e.g., one antibody from a rabbit and the other from a mouse). These two antibodies are differentially detected by gold particles with different sizes onto which secondary antibodies specific to two different host animals are adsorbed (e.g., goat anti-rabbit immunoglobulin G and goat anti-mouse immunoglobulin G).

2. Procedure

We use the following basic protocol for immunolabeling:

1. Set up a humid chamber as shown in Fig. 6A.
2. Prior to the blocking step, float grids on 0.1 N HCl for 5–10 min with the section side in contact with the solution (Fig. 6B). The acid is thought to remove glutaraldehyde from the section surface, enhancing labeling efficiency. However, incubation for more than 10 min seems to increase gold particles in the background.
3. Incubate grids for 30 min at room temperature on the blocking buffer.
4. Blot the grids to remove excess blocking buffer and float grids on the primary antibody solution. Incubate for 1–3 h at room temperature.
5. Rinse grids with PBST by floating grids on three changes of the buffer for 5–10 min each, blotting grids after each rinse.
6. Incubate grids on the secondary antibody solution for 1 h at room temperature.
7. Rinse grids with PBST by floating grids on two changes of the buffer, followed by jet washing with PBST and, finally, thorough jet washing with distilled water.
8. Dry the grids by drawing the water off from the edges with a piece of filter paper and post fix with 0.5% glutaraldehyde. The glutaraldehyde will stabilize bound antibody to help prevent unbinding during post-staining steps.

G. Preparing an Image Stack from Electron Micrographs of Serial Sections

The structural information from a single TEM section is limited because regular thin-sections are only 70–90 nm thick, and that is too thin to convey information about the 3D arrangements of cellular structures. To obtain a comprehensive view of a particular site in the cytoplasm of a cell, electron micrographs from the site along a serial section ribbon are acquired and these serial images can be examined one by one. But it is often more informative to produce a virtual 3D volume in which images from each section are aligned and rendered into an image stack. Structures of interest can be built into 3D models from such a stack of images. From the 3D models, quantitative parameters such as surface area, volume, and density can be calculated (Donohoe *et al.*, 2006).

After determining an area that will be further examined by 3D reconstruction, confirm that the area is not obscured by dust, oil, post-staining crystals, or folding along the serial section ribbon. Take all images at the same magnification and save them as gray scale TIF format files. The TIF files can then be converted into a 3D image stack using the IMOD software package. This procedure involves the following steps:

1. Convert the images into an mrc stack file with the tif2mrc command.
2. With the Midas program, align the images in the stack and save the alignment as an xg file. Use the –g option in the Midas program for global transformation by which each image is aligned to the entire image stack, instead of the default local transformation mode, which is used to align two consecutive images.
3. Once the xg file is ready, make a new image stack from the original mrc stack file using the newstack command with the –xform option. This option will apply the alignment parameters to the new mrc file.

For a detailed explanation of the commands and parameters, consult the website http://bio3d.colorado.edu/imod/#Guides.

Membranous compartments and cytoskeletal components can be more accurately visualized by generating 3D models of them. Figure 7 shows a 3D reconstruction from 23 consecutive 80-nm-thick sections of an *Arabidopsis* root meristematic cell showing a nucleus, plastids, mitochondria, and the cell wall. The 3D model provides a more comprehensive view than any single section electron micrograph does, and provides higher resolution images than confocal microscopy does. The 3D models were constructed from 2D outlines drawn on individual electron micrographs using the imodmesh command.

The 3D models generated from outlines in serial sections are extremely compressed along the *z*-axis (electron beam direction) because the entire thickness (~80 nm) of each section is collapsed onto a single micrograph with a pixel size of 1–5 nm. This problem can be corrected by adjusting the *z*-scale. It is impossible to set the *z*-scale in an image stack of serial sections as accurately as in an electron tomogram. (More details about the *z*-scale and pixel sizes are described at http://bio3d.colorado.edu/imod/doc/3dmodguide.html#SettingZ-scale.) However, we calibrate thickness using round organelles like nuclei[5]. 3D models are stretched along the *z*-axis by changing the *z*-scale value in the Model Header menu item from the Edit menu in the 3dmod. The *z*-scale value varies depending on section thickness and extents of section thinning by electron beam.

Computer-aided 3D reconstructions from serial images like Fig. 7 have been published since the 1990s (Bridge *et al.*, 1998; O'Toole *et al.*, 1997; Winey *et al.*, 1995). Recently, we were able to distinguish differences in mitochondrial morphology between wild-type versus mutant *Caenorhabditis elegans* embryos by using this method (Breckenridge *et al.*, 2009).

[5] This is how we adjust the *z*-scale using a nucleus. First, make a 3D model of a round nucleus. Then the radius of the nucleus model is measured in the *xy* plane. The *z*-scale of the image stack is set to make the nuclear radius be consistent in the *yz* planes. The radius in the two planes should be measured from a single point on the nucleus model at the intersection of the two perpendicular planes.

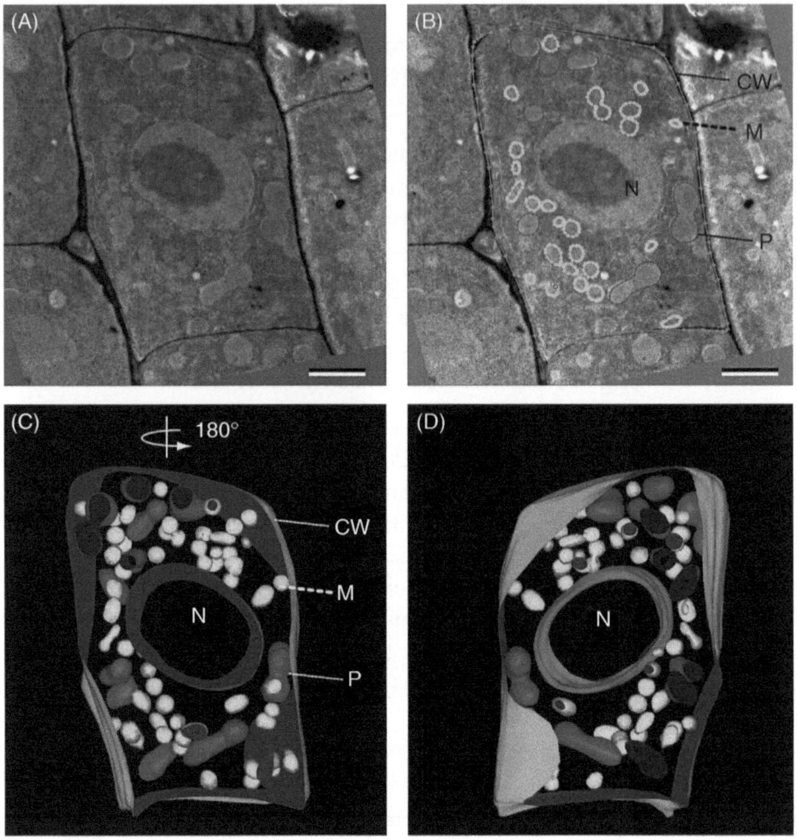

Fig. 7 3D reconstruction of an *Arabidopsis* root meristematic cell from 23 serial sections. (A) Electron micrograph of an *Arabidopsis* root meristematic cell. (B) The same cell as in (A) after segmentation of the cell wall (CW), mitochondria (M), plastids (P), and nucleus (N). (C) 3D model of the organelles and the cell wall from the 23 serial sections. This model view illustrates size and 3D distribution of the organelles. Mitochondria are mostly round while plastids are oblong. Note that plastids are larger, but are fewer, in numbers than the mitochondria in the cell. (D) Rotation of the model around the vertical axis. Scale bars in (A) and (B): 2 μm. (See Plate no. 14 in the Color Plate Section.)

For 3D reconstruction with a *z*-resolution in the nanometer scale, cells have to be examined by electron tomography (McIntosh *et al.*, 2005). Regular TEM sections are about 30–40 times thicker than computationally generated tomographic slices (Staehelin and Kang, 2008). 3D models based on regular TEM micrographs are, therefore, lower in resolution than those obtained from electron tomograms. However, serial-section analysis has an advantage over electron tomography when a larger cell volume needs to be covered. To reconstruct a 1.0 μm^3 area by electron tomography, a tomogram consisting of 400–500 1 μm^2 tomographic slices must be generated. It is possible to study a large volume or even a whole cell by electron

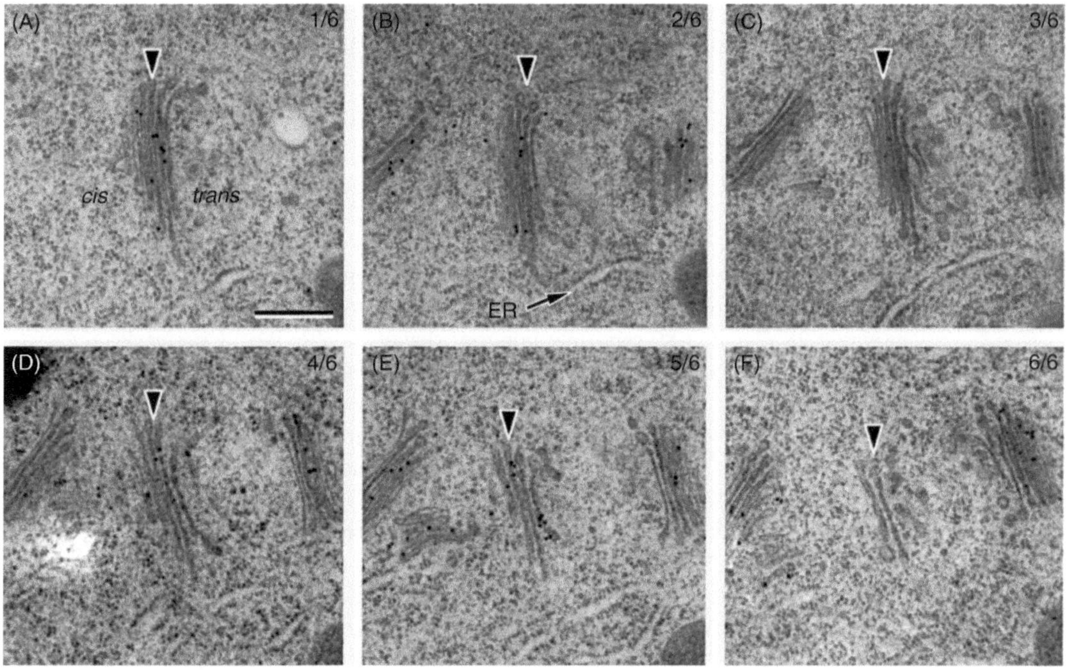

Fig. 8 Six serial sections through Golgi stacks in an *Arabidopsis* meristematic cell. The sections are immunogold labeled with an antibody against a Golgi-localized protein. The Golgi stack in (A) is marked with an arrowhead in each panel. The section in (F) contains the periphery of the Golgi stack. Immunogold particles (15 nm) are associated with *cis*, medial, and *trans* cisternae. Scale bar: 300 nm.

tomography (Noske *et al.*, 2008; Hoog and Antony, 2007), but it requires generation of dozens of tomograms from small areas and then assembly into a gigantic composite tomogram file by stitching and stacking individual tomograms. Conversely, approximately 10 serial sections are enough for enclosing a 1.0 μm^3 volume, and particularly informative sections can be revisited by electron tomography. In addition, 3D reconstruction from serial sections can be carried out with a conventional electron microscope. An intermediate voltage electron microscope equipped with an automatic goniometer-tilting holder is required to obtain high-quality electron tomograms.

Immunogold labeling of serial sections facilitates accurate localization of macromolecules (Hoenger and McIntosh, 2009). In Fig. 8, a Golgi-localized protein has been immunogold labeled in six serial sections. After examining the six sections showing four different Golgi stacks, it is clear that the immunogold particles are specifically localized to the Golgi stack but not to the ER, the vacuoles, or the mitochondria. (Vacuoles and mitochondria are not shown in the Fig. 8.) Within the Golgi stack, the gold particles are associated with all cisternae from *cis* to *trans*. It is possible to determine whether gold particles are preferentially associated with a particular type of cisternae within the Golgi by scanning Golgi stacks across serial sections. Using this method, it was shown that mannosidase I is enriched in the medial Golgi cisternae in *Arabidopsis* meristematic cells (Staehelin and Kang, 2008).

IV. Materials

A. Plant Material

Instrumentation: Arabidopsis seedlings were grown under continuous light at 20°C in an AR-36L2 growth chamber (Percival, Perry, IA). Suspension cultured cells were grown at 28°C and at 80 rpm on a gyratory shaker (VWR, West Chester, PA).

Materials: Arabidopsis plants ecotype Wassilewskija or Columbia were grown as described in Kang *et al.* (Kang et al., 2003). For examining *Arabidopsis* microspore development, anthers from stage 10–13 flowers (Smyth *et al.*, 1990) were collected.

Reagents: Surface-sterilized *Arabidopsis* seeds were germinated and grown vertically on the solid media (2.15 g/L Murashige and Skoog salt, 0.6% phytagar in a Petri dish, pH 5.7) Suspension cultured cells were maintained axenically in liquid Murashige and Skoog salt (4.3 g/L) media supplemented with 0.2 mg/ml 2,4-dichlorophenoxy acetic acid and 1.32 mM KH_2PO_4. Mursashige and Skoog salt, 2,4-dichlorophenoxy acetic acid, and KH_2PO_4 were purchased from Sigma-Aldrich (St. Louis, MO). Phytagar was purchased from Invitrogen (Carlsbad, CA).

B. High-Pressure Freezing

Instrumentation: We have used an HPM010 (Bal-Tec, now sold by RMC, Tucson, AZ) and an HPM100 (Leica, Bannockburn, IL) for freezing various tissues from *Arabidopsis*. We have not observed any obvious differences in freezing or preservation quality between the two machines. However, the HPM100 is a better machine for a core facility-type electron microscopy laboratory that processes a large number of samples because 1) its operation is simpler, making it easier to train new users, 2) the wait time between freezing runs is shorter, 3) it consumes less liquid nitrogen, 4) the parameters of each freezing are automatically saved in the machine, and 5) it is smaller and is portable. Electron micrographs in Figs.1, 3, 7, and 8 were taken from samples frozen with an HPM100.

Materials: There are two types of planchettes, type A and type B, made for the HPM100 high-pressure freezer. Type A planchettes have a 0.1 mm deep well on one side and a 0.2 mm deep well on the other side. Type B planchettes have a 0.3 mm deep well on one side and are flat on the other side. The inner diameter of both planchettes is 2.0 mm. The planchettes were purchased from Technotrade (Manchester, NH) and they can be used with the HPM100 as well as the HPM010.

We recycle planchettes. After FS (or after resin embedding in case of Lowicryl HM20 embedding), empty planchettes are recovered and rinsed with acetone. The planchettes are cleaned by sonication in 10% aqueous solution of Liquinox for an hour. After sonication, the planchettes are rinsed with distilled water and with 95% ethanol.

Reagents: All the reagents were ordered from Sigma-Aldrich (St. Louis, MO) except the Liquinox detergent. Liquinox was bought from Alconox. (White Plains, NY).

C. Freeze Substitution

Instrumentation: FS with OsO_4 for ultrastructural analyses can be carried out in a Styrofoam box filled with dry ice to keep the sample temperature around -80°C. (Dry ice sublimes at -78.5°C at 1 atmospheric pressure.) After FS in dry ice, the samples are slowly warmed up by transferring to a -20°C freezer and then to a 4°C refrigerator. However, It is crucial to use an automatic freeze substitution machine like the AFS2 system (Leica, Bannockburn, IL) when you are processing samples for immunogold labeling by FS and subsequent HM20 resin embedding at -50°C. We have tested the UVC3 Cryochamber system (Ted Pella, Redding, CA) as well but preservation quality by this system was poorer than that obtained with the AFS2 system.

Materials: Dry ice, cryovials.

Reagents: OsO_4 (Ted Pella, Redding, CA), uranyl acetate (Ted Pella, Redding, CA).

D. Resin Embedding and Curing

Instrumentation: Embedding and curing of HM20 resin were carried out in an AFS2 automatic freeze substitution unit (Leica, Bannockburn, IL). Embedding of EMBed resin at room temperature was facilitated by a rocking shaker (VWR, West Chester, PA).

Materials: Flat embedding molds for HM20 resin were constructed by affixing silicone isolators (Grace Bio Lab, Bend, OR) to clean glass slides.

Reagents: We use an EMbed-812 kit (Electron Microscopy Sciences, Hatfield, PA) for osmicated samples. Samples for immunogold labeling experiments are embedded in Lowicryl HM20 resin (Electron Microscopy Sciences, Hatfield, PA). Anhydrous acetone for rinsing and for resin dilution was purchased from Electron Microscopy Sciences (Hatfield, PA) or from Ernest Fullam (Latham, NY).

E. Immunogold Labeling

Instrumentation: None.

Materials: Formvar solutions were purchased from Ted Pella (Redding, CA). 0.75% Formvar solution was prepared by mixing equal volume of 0.5% and 1.0% Formvar solutions.

Reagents: We purchase most immunogold solutions from BioCell (Cardiff, UK). For double-labeling, we use a combination of 10 and 15 nm gold particle-conjugated secondary antibodies. For GFP localization, we purchase GFP-specific antibodies from Santa Cruz Biotechnology (Santa Cruz, CA) or from Rockland (Gilbertsville, PA).

F. Preparing an Image Stack from Electron Micrographs of Serial Sections

Instrumentation: We carried out all of the computer-assisted image analysis using a Macintosh computer (Cuppertino, CA) with the Os X (ver. 1.6.2) operating system.

The IMOD software package for Macintosh computers can be downloaded from http://bio3d.colorado.edu/imod/download.html.

Materials: None.

Reagents: None.

V. Discussion

A. Dissection and Use of Sucrose as a Filler

Even the best high-pressure freezers and electron microscopes cannot "correct" for sample damage that occurs during dissection or during loading into specimen carriers. Dissection and loading must be completed quickly so that *Arabidopsis* seedlings and other tissues are as unperturbed as possible, up to the time of freezing. Before beginning sample collection, make sure that the high-pressure freezer is cooled down and ready for freezing, and keep tools readily accessible. It is also recommended to practice loading the planchettes before attempting to freeze important samples. There are many excellent reviews and book chapters that explain the principles of HPF and provide useful tips/suggestions for proper cryofixation (McDonald, 2007; McDonald *et al.*, 2007; McIntosh *et al.*, 2005; Morphew, 2007).

Plant sugar signaling is influenced by levels of glucose and its catabolic intermediates (Rolland *et al.*, 2006). Since glucose is a hydrolytic product of sucrose, the question of whether the use of a sucrose solution as a cryoprotectant could affect normal metabolism in plant cells has been raised. However, filling the planchettes with a 0.1–0.15 M sucrose solution is not likely to induce artifacts for the following reasons. Wild-type *Arabidopsis* seedlings do not display growth defects when grown in a media containing as high as 5% (~0.15 M) sucrose for a week (Kurepa *et al.*, 1998). Secondly, sucrose cannot readily penetrate into the cytoplasm of cells where sucrose is digested into metabolic intermediates. Glycerol passes through the plasma membrane and is considered to be an intracellular cryoprotectant (McDonald *et al.*, 2007). When compared with glycerol, sucrose has ~100 times slower membrane permeability (Abbott and Romero, 1996). Therefore, sucrose is primarily an extracellular cryoprotectant rather than an intracellular one.

B. Membrane Contrast in Plant Samples Processed by HPF/FS

Membrane contrast in high-pressure frozen/OsO_4 freeze-substituted cells is weaker than in chemically fixed cells. To improve the poor contrast, OsO_4–acetone FS cocktails are supplemented with 0.1% uranyl acetate (McDonald, 2007) or up to 5% distilled water (Walther and Ziegler, 2002). These modified FS cocktails enhanced membrane contrast in yeast cells and in mammalian cells. However, neither of these modified FS cocktails increased membrane contrast in *Arabidopsis* samples in our hands. We observed that adding water to the FS cocktail was detrimental in *Arabidopsis* because it increased ice crystal formation in the cells. Vesicle coats and cytoskeletal components in plant cell samples processed by HPF/FS are better contrasted by staining sections with OsO_4 and uranyl acetate after FS (Murata *et al.*, 2002). Membrane contrast in the HM20 embedded samples is better than in Epon-embedded osmicated samples (Giddings, 2003).

In plant cell samples freeze substituted with a regular 1–2% OsO_4 FS cocktail, the chloroplast stroma is darkly stained. This heavy staining can obscure the thylakoid membranes. To reduce OsO_4 accumulation in the stroma, excessive OsO_4 in the FS media can be rinsed out by washing the samples several times with pre-cooled acetone at –80°C prior to warming.

C. Correlative Light and Electron Microscopy in *Arabidopsis* Meristematic Samples

Correlative light and electron microscopy requires the ability to find the exact same cell or group of cells that were examined by live cell imaging in EM sections (Perkins *et al.*, 2009). It is possible to image a living root tip by light microscopy and recover the exact same root tip after HPF/FS for further TEM imaging (Fig. 1). However, it is very difficult to locate a particular cell in *Arabidopsis* tissues in electron micrographs, especially in the meristematic regions, because they consist of small brick-shaped cells that look almost identical. The cylindrical symmetry of their cell organization makes it harder to orient a sample block in a direction that matches the plane of focus in the light micrographs. Correlative light and electron microscopy with high-pressure frozen samples as carried out with *in vitro* cultured cells (Lanman *et al.*, 2008) and early *C. elegans* embryos (Pelletier *et al.*, 2006) has not yet been successful in *Arabidopsis* tissues.

VI. Concluding Remarks

HPF in combination with FS has greatly contributed to elucidating the mechanisms of cellular processes and gene functions in *Arabidopsis* by preserving short-lived and/or intricate structures that are difficult to capture by conventional chemical fixation. The HPF and FS procedures for *Arabidopsis* samples share many steps and reagents with those for other model organisms. However, the cell walls and large vacuoles of plant cells can be problematic during HPF, FS, and resin embedding. To overcome the problems, we have used prolonged FS and resin-embedding protocols and supplemented the FS media with chemicals that remove water molecules, such as DMP. These modifications have enabled us to obtain well-preserved samples from moderately vacuolated cells. In samples that are preserved by HPF and FS, localization of macromolecules can be determined precisely using immunogold labeling. Furthermore, a comprehensive view of a large cell volume can be acquired using serial-section analysis and computer-aided 3D reconstruction.

Currently, more plant molecular and cellular research is focused on crop plants such as rice, maize, soybean, and others. Crop plants are larger than *Arabidopsis* plants, and the economically important parts of these crop plants often consist of highly differentiated cells with distinctive biochemical compositions of their cytoplasm. These complications pose formidable challenges for HPF/FS because dissecting large plants without inflicting damage and subsequent loading into an HPF machine is more difficult. Special FS cocktails and protocols may be required for different types of differentiated cells. Despite these complications, what we have learned from *Arabidopsis* can certainly serve as a starting point for utilizing these advanced TEM techniques in crop plant research (Kang *et al.*, 2009).

Acknowledgments

I would like to thank Dr. Müller-Reichert (TU Dresden), Dr. Eileen O'Toole (University of Colorado), and Donna Williams (University of Florida) for their careful reading and helpful comments for this chapter. I also thank Dr. Sibum Sung (University of Texas) and Dr. Andreas Nebenführ (University of Tennessee) for the GFP*Arabidopsis* lines shown in the figures. I am grateful to members of the L. Andrew Staehelin laboratory and the Boulder Laboratory for 3D Electron Microscopy of Cells for their support and sharing expertise.

References

Abbott, N. J., and Romero, I. A. (1996). Transporting therapeutics across the blood-brain barrier. *Mol. Med. Today* **2**, 106–113.

Austin, J. R., II, Frost, E., Vidi, P. A., Kessler, F., and Staehelin, L. A. (2006). Plastoglobules are lipoprotein subcompartments of the chloroplast that are permanently coupled to thylakoid membranes and contain biosynthetic enzymes. *Plant Cell* **18**, 1693–1703.

Austin, J. R., II, Segui-Simarro, J. M., and Staehelin, L. A. (2005). Quantitative analysis of changes in spatial distribution and plus-end geometry of microtubules involved in plant-cell cytokinesis. *J. Cell. Sci.* **118**, 3895–3903.

Bednarek, S. Y., and Falbel, T. G. (2002). Membrane trafficking during plant cytokinesis. *Traffic* **3**, 621–629.

Bowman, J. (1994). "*Arabidopsis*, An Atlas of Morphology and Development." Springer-Verlag, Berlin.

Bozzola, J., and Russel, L. (1999). "Electron Microscopy: Principles and Techniques for Biologists." Jones and Bartlett, Sudbury.

Breckenridge, D. G., Kang, B. H., and Xue, D. (2009). Bcl-2 proteins EGL-1 and CED-9 do not regulate mitochondrial fission or fusion in *Caenorhabditis elegans*. *Curr. Biol.* **19**, 768–773.

Bridge, A. J., Morphew, M., Bartlett, R., and Hagan, I. M. (1998). The fission yeast SPB component cut12 links bipolar spindle formation to mitotic control. *Genes Dev.* **12**, 927–942.

Christopher, D. A., Borsics, T., Yuen, C. Y., Ullmer, W., Andeme-Ondzighi, C., Andres, M. A., Kang, B. H., and Staehelin, L. A. (2007). The cyclic nucleotide gated cation channel AtCNGC10 traffics from the ER via Golgi vesicles to the plasma membrane of Arabidopsis root and leaf cells. *BMC Plant Biol.* **7**, 48.

Craig, S., and Staehelin, L. A. (1988). High pressure freezing of intact plant tissues. Evaluation and characterization of novel features of the endoplasmic reticulum and associated membrane systems. *Eur. J. Cell Biol.* **46**, 81–93.

Dammermann, A., Muller-Reichert, T., Pelletier, L., Habermann, B., Desai, A., and Oegema, K. (2004). Centriole assembly requires both centriolar and pericentriolar material proteins. *Dev. Cell* **7**, 815–829.

Donohoe, B. S., Kang, B. H., and Staehelin, L. A. (2007). Identification and characterization of COPIa- and COPIb-type vesicle classes associated with plant and algal Golgi. *Proc. Natl. Acad. Sci. USA* **104**, 163–168.

Donohoe, B. S., Mogelsvang, S., and Staehelin, L. A. (2006). Electron tomography of ER, Golgi and related membrane systems. *Methods* **39**, 154–162.

Echlin, P. (1991). Ice crystal damage and radiation effects in relation to microscopy and analysis at low temperatures. *J Microsc.* **161**, 159–170.

Giddings, T. H. (2003). Freeze-substitution protocols for improved visualization of membranes in high-pressure frozen samples. *J. Microsc.* **212**, 53–61.

Gilkey, J. C., and Staehelin, L. A. (1986). Advances in ultrarapid freezing for the preservation of cellular ultrastructure. *J. Electron Microsc. Tech.* **3**, 177–210.

Hagler, H. K. (2007). Ultramicrotomy for biological electron microscopy. *In* "Electron Microscopy: Methods and Protocol" (J. Cuo, ed.) pp. 67–96. Humana Press, Totowa.

Hess, M. W. (2007). Cryopreservation Methodology for Plant Cell Biology. *In* "Cellular Electron Microscopy" J. McIntosh, ed. (San Diego: Elsevier). pp. 58–101.

Hoenger, A., and McIntosh, J. R. (2009). Probing the macromolecular organization of cells by electron tomography. *Curr. Opin. Cell Biol.* **21**, 89–96.

Hoog, J. L., and Antony, C. (2007). Whole-cell investigation of microtubule cytoskeleton architecture by electron tomography. *Methods Cell Biol.* **79**, 145–167.

Kang, B. H., Busse, J., and Bednarek, S. (2003). Members of the Arabidopsis dynamin-like gene family, ADL1, are essential for plant cytokinesis and polarized cell growth. *Plant Cell* **15**, 899–913.

Kang, B. H., and Staehelin, L. A. (2008). ER-to-Golgi transport by COPII vesicles in *Arabidopsis* involves a ribosome-excluding scaffold that is transferred with the vesicles to the Golgi matrix. *Protoplasma* **234**, 51–64.

Kang, B. H., Xiong, Y., Williams, D., Pozueta-Romero, D., and Chourey, P. (2009). Miniature1-encoded cell wall invertase is essential for assembly and function of wall-in-growth in the maize endosperm transfer cell. *Plant Physiol.* **151**, 1366–1376.

Kiss, J. Z., Giddings, T. H., Jr., Staehelin, L. A., and Sack, F. D. (1990). Comparison of the ultrastructure of conventionally fixed and high pressure frozen/freeze substituted root tips of *Nicotiana* and *Arabidopsis*. *Protoplasma* **157**, 64–74.

Kiss, J. Z., and Staehelin, L. A. (1993). Structural polarity in the Chara rhizoid: a reevaluation. *Am. J. Bot.* **80**, 273–282.

Kiss, J. Z., and Staehelin, L. A. (1995). High pressure freezing. *In* "Rapid Freezing, Freeze Fracture and Deep Etching" (D. M. Shotton, ed.) pp. 89–104. Wiley-Liss, New York.

Kurepa, J., Smalle, J., Van Montagu, M., and Inze, D. (1998). Effects of sucrose supply on growth and paraquat tolerance of the late-flowering gi-3 mutant. *Plant Growth Regul.* **26**, 91–96.

Lam, S. K., Siu, C. L., Hillmer, S., Jang, S., An, G., Robinson, D. G., and Jiang, L. (2007). Rice SCAMP1 defines clathrin-coated, trans-golgi-located tubular-vesicular structures as an early endosome in tobacco BY-2 cells. *Plant Cell* **19**, 296–319.

Lanman, J., Crum, J., Deerinck, T. J., Gaietta, G. M., Schneemann, A., Sosinsky, G. E., Ellisman, M. H., and Johnson, J. E. (2008). Visualizing flock house virus infection in drosophila cells with correlated fluorescence and electron microscopy. *J. Struct. Biol.* **161**, 439–446.

Lauber, M. H., Waizenegger, I., Steinmann, T., Schwarz, H., Mayer, U., Hwang, I., Lukowitz, W., and Jurgens, G. (1997). The *Arabidopsis* KNOLLE protein is a cytokinesis-specific syntaxin. *J. Cell Biol.* **139**, 1485–1493.

Leitz, G., Kang, B. H., Schoenwaelder, M. E., and Staehelin, L. A. (2009). Statolith sedimentation kinetics and force transduction to the cortical endoplasmic reticulum in gravity-sensing *Arabidopsis* columella cells. *Plant Cell* **21**, 843–860.

Lukowitz, W., Mayer, U., and Jurgens, G. (1996). Cytokinesis in the *Arabidopsis* embryo involves the syntaxin-related KNOLLE gene product. *Cell* **84**, 61–71.

Lynch, M. A., and Staehelin, L. A. (1992). Domain-specific and cell type-specific localization of two types of cell wall matrix polysaccharides in the clover root tip. *J. Cell Biol.* **118**, 467–479.

McDonald, K. (2007). Cryopreparation methods for electron microscopy of selected model systems.*Methods Cell Biol.* **79**, 23–56.

McDonald, K., Morphew, M. K., Verkade, P., and Mueller-Reichert, T. (2007). Recent advances in high-pressure freezing: Equipment- and specimen-loading methods. *In* "Electron Microscopy, Methods and Protocols" (J. Cuo, ed.) pp. 143–174. Humana Press, Totowa.

McIntosh, R., Nicastro, D., and Mastronarde, D. (2005). New views of cells in 3d: An introduction to electron tomography. *Trends Cell Biol.* **15**, 43–51.

Mersey, B., and McCully, M. (1978). Monitoring the course of fixation of plant cells. *J. Microsc.* **166**, 43–56.

Moore, P. J., Swords, K. M., Lynch, M. A., and Staehelin, L. A. (1991). Spatial organization of the assembly pathways of glycoproteins and complex polysaccharides in the golgi apparatus of plants. *J. Cell Biol.* **112**, 589–602.

Morphew, M. K. (2007). 3D immunolocalization with plastic sections. *Methods Cell Biol.* **79**, 493–513.

Murata, T., Karahara, I., Kozuka, T., Thomas, H. G., Jr., Staehelin, L. A., and Mineyuki, Y. (2002). Improved method for visualizing coated pits, microfilaments, and microtubules in cryofixed and freeze-substituted plant cells. *J. Electron Microsc. (Tokyo)* **51**, 133–136.

Noske, A. B., Costin, A. J., Morgan, G. P., and Marsh, B. J. (2008). Expedited approaches to whole cell electron tomography and organelle mark-up *in situ* in high-pressure frozen pancreatic islets. *J. Struct. Biol.* **161**, 298–313.

Otegui, M. S., Herder, R., Schulze, J., Jung, R., and Staehelin, L. A. (2006). The proteolytic processing of seed storage proteins in *Arabidopsis* embryo cells starts in the multivesicular bodies. *Plant Cell* **18**, 2567–2581.

Otegui, M. S., Mastronarde, D. N., Kang, B. H., Bednarek, S. Y., and Staehelin, L. A. (2001). Three-dimensional analysis of syncytial-type cell plates during endosperm cellularization visualized by high resolution electron tomography. *Plant Cell* **13**, 2033–2051.

Otegui, M. S., and Staehelin, L. A. (2004). Electron tomographic analysis of post-meiotic cytokinesis during pollen development in *Arabidopsis thaliana*. *Planta* **218**, 501–515.

O'Toole, E., Mastronarde, D., Giddings, T., Jr, Winey, M., Burke, D., and McIntosh, J. (1997). Three-dimensional analysis and ultrastructural design of mitotic spindles from the cdc20 mutant of *Saccharomyces cerevisiae*. *Mol. Biol. Cell* **8**, 1–11.

Pelletier, L., O'Toole, E., Schwager, A., Hyman, A. A., and Muller-Reichert, T. (2006). Centriole assembly in Caenorhabditis elegans. *Nature* **444**, 619–623.

Perkins, G. A., Sun, M. G., and Frey, T. G. (2009). Chapter 2 correlated light and electron microscopy/ electron tomography of mitochondria *in situ*. *Methods Enzymol.* **456**, 29–52.

Ritzenthaler, C., Nebenfuhr, A., Movafeghi, A., Stussi-Garaud, C., Behnia, L., Pimpl, P., Staehelin, L. A., and Robinson, D. G. (2002). Reevaluation of the effects of brefeldin A on plant cells using tobacco bright yellow 2 cells expressing Golgi-targeted green fluorescent protein and COPI antisera. *Plant Cell* **14**, 237–261.

Rolland, F., Baena-Gonzalez, E., and Sheen, J. (2006). Sugar sensing and signaling in plants: Conserved and novel mechanisms. *Annu. Rev. Plant Biol.* **57**, 675–709.

Samuels, A. L., Giddings, T. H., Jr., and Staehelin, L. A. (1995). Cytokinesis in tobacco BY-2 and root tip cells: A new model of cell plate formation in higher plants. *J. Cell Biol.* **130**, 1345–1357.

Segui-Simarro, J. M., Austin, J. R., II, White, E. A., and Staehelin, L. A. (2004). Electron tomographic analysis of somatic cell plate formation in meristematic cells of *Arabidopsis* preserved by high-pressure freezing. *Plant Cell* **16**, 836–856.

Shimoni, E., Rav-Hon, O., Ohad, I., Brumfeld, V., and Reich, Z. (2005). Three-dimensional organization of higher-plant chloroplast thylakoid membranes revealed by electron tomography. *Plant Cell* **17**, 2580–2586.

Smyth, D. R., Bowman, J. L., and Meyerowitz, E. M. (1990). Early flower development in Arabidopsis. *Plant Cell* **2**, 755–767.

Staehelin, L. A., Giddings, T. H., Jr., Kiss, J. Z., and Sack, F. D. (1990). Macromolecular differentiation of Golgi stacks in root tips of *Arabidopsis* and *Nicotiana* seedlings as visualized in high pressure frozen and freeze-substituted samples. *Protoplasma* **157**, 75–91.

Staehelin, L. A., and Kang, B. H. (2008). Nanoscale architecture of endoplasmic reticulum export sites and of Golgi membranes as determined by electron tomography. *Plant Physiol.* **147**, 1454–1468.

Tse, Y. C., Mo, B., Hillmer, S., Zhao, M., Lo, S. W., Robinson, D. G., and Jiang, L. (2004). Identification of multivesicular bodies as prevacuolar compartments in *nicotiana tabacum* BY-2 cells. *Plant Cell* **16**, 672–693.

Waizenegger, I., Lukowitz, W., Assaad, F., Schwarz, H., Jurgens, G., and Mayer, U. (2000). The *Arabidopsis* KNOLLE and KEULE genes interact to promote vesicle fusion during cytokinesis. *Curr. Biol.* **10**, 1371–1374.

Walther, P., and Ziegler, A. (2002). Freeze substitution of high-pressure frozen samples: The visibility of biological membranes is improved when the substitution medium contains water. *J. Microsc.* **208**, 3–10.

Winey, M., Mamay, C. L., O'Toole, E. T., Mastronarde, D. N., Giddings, T. H., Jr., McDonald, K. L., and McIntosh, J. R. (1995). Three-dimensional ultrastructural analysis of the *Saccharomyces cerevisiae* mitotic spindle. *J. Cell Biol.* **129**, 1601–1615.

CHAPTER 13

Preparation Techniques for Transmission Electron Microscopy of *Hydra*

Thomas W. Holstein[*], **Michael W. Hess**[†], *and* **Willi Salvenmoser**[‡]

[*]Institute of Zoology, Heidelberg University, D-69120 Heidelberg, Germany

[†]Division of Histology and Embryology, Innsbruck Medical University, A-6020 Innsbruck, Austria

[‡]Center for Molecular Biosciences, Institute of Zoology, University of Innsbruck, A-6020 Innsbruck, Austria

Abstract

Hydra is a classical model organism in developmental and cell biology with a simple body plan reminiscent of a gastrula with one body axis and a limited number of cell

978-0-12-381007-6
DOI: 10.1016/S0091-679X(10)96013-5

types. This rather simple organism exhibits a regeneration capacity that is unique among all eumetazoans and is largely dependent on the stem cell properties of its epithelial stem cell population. Molecular work in the past few years has revealed an unexpected genetic complexity of these simple animals, making them an interesting model for studying the generation of animal form and regeneration. In addition, *Hydra* has an interstitial stem cell system with a unique population of nematocytes, neuronal cells that are characterized by an explosive exocytotic discharge. Here, we compare classical and modern transmission electron microscopy (TEM) fixation protocols including protocols for TEM immunocytochemistry (post-embedding immunogold labeling). We presume that TEM studies will become an important tool to analyze cell–cell interactions as well as cell matrix interrelationships in *Hydra* in the future.

I. Introduction

The freshwater polyp *Hydra* is a member of the ancient phylum Cnidaria and is famous for its high regeneration capacity. It comprises one of the simplest living metazoans and is an important model for studies of axial patterning (Hobmayer *et al.*, 2000; Meinhardt, 2002; Technau *et al.*, 2000), stem cell biology (Bosch, 2009; Watanabe *et al.*, 2009), and regeneration (Bosch, 2007; Holstein *et al.*, 2003). *Hydra* belongs to the hydrozoans that diverged from anthozoans at least 540 million years ago (Cartwright *et al.*, 2007; Peterson *et al.*, 2008). *Hydra* is different from other hydrozoans (e.g., *Clytia*) in lacking the characteristic larval and medusa stages. Recently, the *Hydra* genome has been published demonstrating an unexpectedly high genetic complexity of these morphologically simple organisms that is similar to lower vertebrates (Chapman *et al.*, 2010).

Hydra polyps propagate asexually, so that they are in a steady state of constant growth and tissue turnover. Both tissue layers, ectoderm and endoderm, are formed by dividing the epithelial stem cells, in which newborn daughter cells are passively displaced upward to form the stinging tentacles, downward to form the foot, or bud off at the sides to make replica animals. An important consequence is that the passively displaced cells have to assess their relative position in the organism continuously. Hence, different from vertebrates, patterning systems necessary to provide this information are continuously active in *Hydra* polyps. These position-specific morphogenetic signals are activated at the site of regeneration (Rentzsch *et al.*, 2007).

The ability to regenerate becomes significantly reduced throughout metazoan evolution. In higher bilaterians, the capacity to regenerate is limited to specific organs or tissues. The very basal metazoans exhibit an almost unlimited regeneration capacity (Bode, 2003; Bosch, 2007; Holstein *et al.*, 2003; Sanchez Alvarado, 2003). In that respect, *Hydra* regeneration shares important similarities with plants despite the fact that multicellularity almost certainly evolved independently in animals and plants. However, *Hydra* polyps can even regenerate from dissociated single cells and thereby serve as a paradigm for *de novo* pattern formation (Gierer *et al.*, 1972). The regeneration stimulus starts the repatterning of the tissue at the site of regeneration. Work over the past years revealed that the Wnt signaling pathway plays a fundamental role in this process (Guder *et al.*, 2006; Hobmayer *et al.*, 2000; Holstein, 2008; Kusserow *et al.*, 2005; Lengfeld *et al.*, 2009; Philipp *et al.*, 2009). The *Wnt* genes are the main constituents of the head organizer and blastoporal organizer formed during embryogenesis (Fritzenwanker *et al.*, 2007; Kusserow *et al.*, 2005; Lengfeld *et al.*, 2009; Momose and Schmid, 2006). The kinetics of *Wnt* gene expression during head regeneration suggests a cascade of consecutive *Wnt* activation with *HyWnt3* at the top of the cascade (Lengfeld *et al.*, 2009; Philipp *et al.*, 2009). Accordingly, HyWnt3 protein rescues the head regeneration deficient mutant strain reg-16. On the molecular level, the *BMP/Chordin* are also upregulated shortly after cutting at the site of regeneration (Rentzsch *et al.*, 2007).

An important feature of the *Hydra*'s biology are the stem cells. Stem cells have been described for animals (Metazoa), fungi, and plants and are probably a basic feature of all multicellular organisms. All stem cell systems share one universal property, in that they continuously reproduce themselves and generate progeny of differentiated cells. Despite the fact that stem cell systems have been identified in all multicellular organisms, so far virtually nothing is known about the molecular mechanisms of these stem cell systems in basal metazoan groups (Sanchez Alvarado, 2007). The best understood stem cell systems so far at the base of metazoan evolution are those of the freshwater polyp *Hydra*.

Three stem cell lines with the capacity for constant self-renewal have been distinguished: the (1) ectodermal epithelial, (2) endodermal epithelial, and (3) interstitial stem cell lineage. Epithelial stem cells are mainly located in the gastric and hypostomal region of the polyp (Steele, 2002). Those epithelial cells that are located at the tentacles and foot are arrested in the G2 phase of the cell cycle (Bosch, 2009; Steele, 2002; Watanabe *et al.*, 2009). The interstitial stem cell lineage is embedded in the interstitial space of the ectodermal epithelial cells. Interstitial stem cells have a shorter cell cycle than epithelial cells (Campbell and David, 1974; David and Campbell, 1972; Holstein and David, 1990) and can be easily removed by drugs affecting the cell cycle (Bode *et al.*, 1976; Campbell, 1976) or by using mutant strains carrying temperature-sensitive interstitial stem cells (Sugiyama and Fujisawa, 1978). The potential of self-renewal and multipotency of interstitial stem cells have been demonstrated in statistical cloning experiments (David and Murphy, 1977) showing that the interstitial cell lineage consists of multipotent interstitial stem cells that differentiate into germ, gland, mucus, and nerve cells including nematocytes. Together with differentiating intermediates and product cells, the interstitial cell system comprises about 75% of all cells in rapidly growing, asexually reproducing animals.

Of particular interest is the differentiation of interstitial stem cells into neuronal cells. Neuronal cells comprise the main cell type produced by pluripotent interstitial stem cells, since about 70% of all differentiation products in the interstitial cell lineage are stinging cells (nematocytes) and neuronal cells that form a simple nerve net (David and Challoner, 1974). The nematocytes are a unique and highly specialized neuronal cell type that is characteristic for all cnidarians (David *et al.*, 2008; Holstein and Tardent, 1984; Hwang *et al.*, 2008; Nuchter *et al.*, 2006). They possess the cnidocyst, which is used for the capture of prey and defense. It consists of a cylindrical capsule, which releases a long tubule upon triggering. Cnidocysts develop inside a giant post-Golgi vesicle by a sequential accumulation of proteins from the Golgi apparatus. During morphogenesis, the capsule and its long tubule form in the cytoplasm within this giant post-Golgi vesicle. At the end of morphogenesis the tubule invaginates and the capsule gets its final form. The mature nematocyst capsule comprises a collagenous polymer with remarkable biophysical properties able to withstand an osmotic pressure of 150 bars. Release of the capsule and discharge is probably initiated by classical exocytosis (for review, see David *et al.*, 2008; Watanabe *et al.*, 2009).

II. Rationale

Hydra has become a very attractive model for analyzing basic mechanisms in cell and developmental biology at the very base of eumetazoan evolution. Here, we present and discuss protocols for thin-section electron microscopy (EM) that can be used in combination with other techniques in molecular cell biology. In addition to standard chemical fixation, including immunocytochemistry, we introduce cryo-based sample processing (high-pressure freezing/freeze substitution, HPF/FS) to *Hydra* research.

III. Methods

A. Chemical Fixation—General Aspects

An adult *Hydra* has around 10^5 cells and measures about 1 cm in length and several millimeters in diameter. This is too large for optimal transmission electron microscopy (TEM) fixation, and suboptimal preservation of organelles, e.g., mitochondria in the endoderm, may occur after fixation of whole animals. For the best results, cross and/or lateral dissection of the animal in *Hydra* culture medium or fixative is therefore recommended; this reduces the size and opens the endodermal space for the fixative (Anne-Kathrin Gorny, personal communication). To avoid muscle contraction during fixation 2% (v/v) urethane should be used as a relaxant for ~1 min prior to fixation. *Hydra* epidermal cells contain large, water-filled vacuoles, which require a careful combination of fixatives and buffer concentration to prevent artifactual shrinkage. Simultaneous application of glutaraldehyde and osmium tetroxide (OsO_4) gives generally better results than their consecutive application (e.g., Campbell, 1987). Acceptable results can be obtained using consecutive application of these reagents, but only for some parts of the tissue (e.g., Lentz, 1966a, b; Westfall *et al.*, 1971). A combination of low concentrations of glutaraldehyde and formaldehyde (prepared from paraformaldehyde) and OsO_4 also yields good ultrastructural preservation (e.g., Wood, 1985; Wood and Novak, 1982). For immunocytochemistry (e.g., post-embedding immunogold labeling), fixation with formaldehyde is recommended. Alternatively, body column pieces from *Hydra* polyps can be fixed in a mixture of glutaraldehyde and buffered formaldehyde. Embedding is performed in acrylic resins, such as Lowicryl K4M or LR White (for reviews on relevant fixatives and embedding procedures, see Griffiths, 1993; Webster *et al.*, 2008).

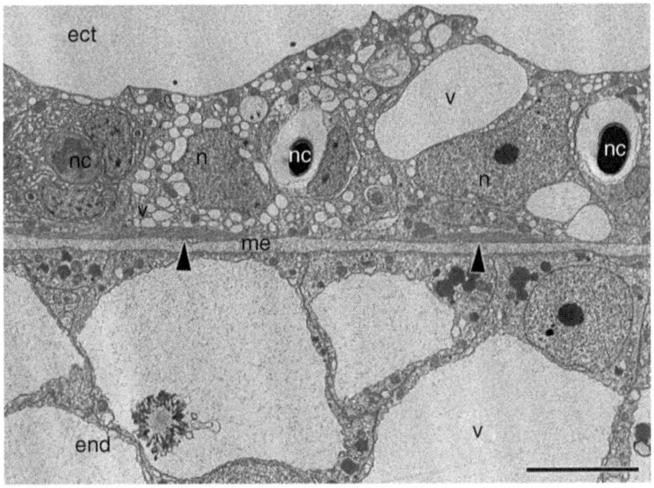

Fig. 1 Overview of the body wall of *Hydra vulgaris* after chemical fixation (Campbell, 1987). Epithelio-muscular cells of the ectoderm and endoderm with numerous water-containing vacuoles and nuclei are present. Note the epithelial processes with myofibrils (myonemes: marked by arrowheads) at the base near the mesoglea. Different stages of nematocytes surround the epithelial cells. Ectoderm (ect), endoderm (end), mesoglea (me), nematocyst (nc), nucleus of epithelio-muscular cell (n), and vacuole (v). Scale bar is 10 µm.

B. Chemical Fixation and Embedding Protocols for Morphology

1. Simultaneous Fixation with Glutaraldehyde and OsO_4 (Campbell, 1987): Figs. 1–3, 4A

Buffer: 0.001 M Tris–HCl, 1 mM $CaCl_2$, 0.1 mM $MgCl_2$, 0.1 mM KCl, 1 mM NaH_2CO_3; pH 7.8 (*Hydra* culture medium)

Fixative: 1% (v/v) glutaraldehyde + 0.2% (w/v) OsO_4 in buffer

Fixation procedure: Relax animals with urethane and fix on ice or at room temperature for at least 1 h. Fixation on ice works slower and therefore it is preferred for smaller animals. Due to the low concentration of OsO_4, the solution remains clear even at room temperature. If the solution becomes brownish during extended fixation times it should be exchanged. After fixation, wash three times in buffer and dehydrate with either a graded series of ethanol or acetone and embed in Epon or Spurr's low viscosity resin according to standard protocols (Glauert and Lewis, 1998).

2. Simultaneous Fixation with Glutaraldehyde and OsO_4 Modified after Shigenaka *et al.* (1971): Figs. 5–6, 7A

Buffer: 0.06 M phosphate buffer pH 7.2

Fixative A: 6% (v/v) glutaraldehyde in <0.05 M phosphate buffer, 2 mM sucrose, 0.02 mM magnesium sulfate; pH 7.2 (i.e., mix 2.4 ml 25% glutaraldehyde, 0.2 ml 0.1 M sucrose, 0.2 ml 1 mM magnesium sulfate with 7.2 ml 0.06 M phosphate buffer; pH 7.2)

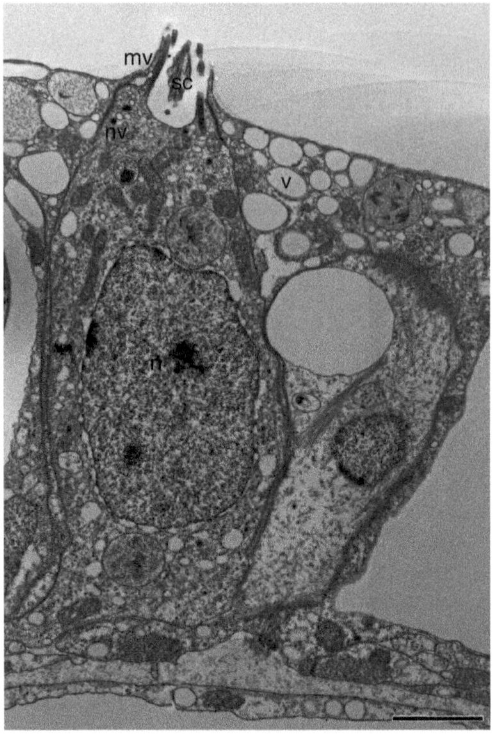

Fig. 2 Nerve cell (N1 cell) of *Hydra* tentacle after chemical fixation (Campbell, 1987) with apical sensory cilium surrounded by microvilli (stereo cilia), as well as neuronal vesicles within the cytoplasm (see Hobmayer *et al.*, 1990). Microvilli (mv), neurosecretory vesicle (nv), nucleus of the nerve cell (n), sensory cilium (sc), and vacuole of the battery cell (v). Scale bar is 2 μm.

Fixative B: 2% (w/v) OsO_4 in distilled water.

Working solution: equal volumes of fixatives A and B.

Fixation procedure: Always prepare a fresh solution. Fix relaxed animals on ice for 1 h. High concentrations of glutaraldehyde and sucrose lead to the reduction of OsO_4. Thus, the fixative should be exchanged once becoming dark. After fixation, do not wash with buffer but start directly with dehydration in a graded series of acetone. Usually the solutions with a low concentration of acetone will begin getting black during the dehydration process, because OsO_4 will be "washed away." Solutions with a high concentration of acetone (~100%) should remain clear. Embed in Epon or Spurr's low viscosity resin.

C. Chemical Fixation and Embedding Protocols for Immunocytochemistry

1. Recommended Fixation Protocol

Buffer: 0.1 M phosphate-buffered saline (PBS, pH 7.2)

Fixative: 4% (w/v) formaldehyde (freshly made from paraformaldehyde) in PBS

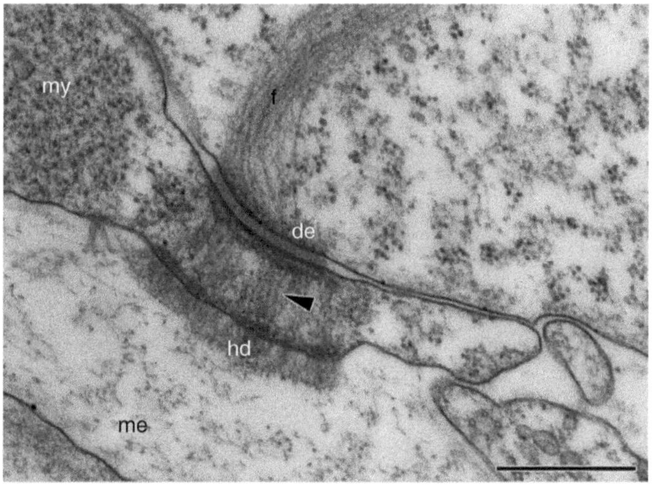

Fig. 3 The nematocyte battery cell mesogloea (NBM) complex of a mounted *Hydra* nematocyte (Wood and Novak, 1982) seen in chemically fixed samples (Campbell, 1987). A desmosome anchors the nematocyte to the epithelial battery cell and a hemidesmosome simultaneously attaches the myofibrils containing processes of the battery cell to the mesoglea. Arrowhead points to microtubules between desmosome and hemidesmosome. Desmosome (de), cytoskeletal filaments of the nematocyte (f), hemidesmosome (hd), mesoglea (me), and myoneme of the battery cell (my). Scale bar is 0.5 μm.

Fixation procedure: Fix animals at +4°C for 1 h to several hours and wash in buffer. Note, however, that the salt component of the PBS buffer may induce muscle contractions during fixation if the animals have been anesthetized with urethane (Roland Aufschnaiter, personal communication). Thus, it is advised to start with a short primary fixation (~5 min) with aldehydes dissolved in *Hydra* culture medium before transferring the animals to PBS-buffered fixatives.

2. Alternative Fixation Protocol: Fig. 9

Buffer: 0.05 M phosphate buffer (pH 7.2)
Fixative: 0.2% (v/v) glutaraldehyde and 2% (w/v) formaldehyde in phosphate buffer
Fixation procedure: Fix animals at +4°C for 1 h to several hours and wash in buffer.

3. Embedding in Acrylic Resins

Dehydrate in dimethylformamid (DMF 50, 70, 90%) and infiltrate with a series of increasing Lowicryl K4M resin concentrations at 4°C (DMF:Lowicryl 2:1 for 15 min, 1:1 for 30 min, 1:2 for 2 h, and 100% Lowicryl K4M for 12 h; for further details, see Glauert and Lewis, 1998). UV polymerization is performed at 0°C for 3 days (Engel *et al.*, 2002).

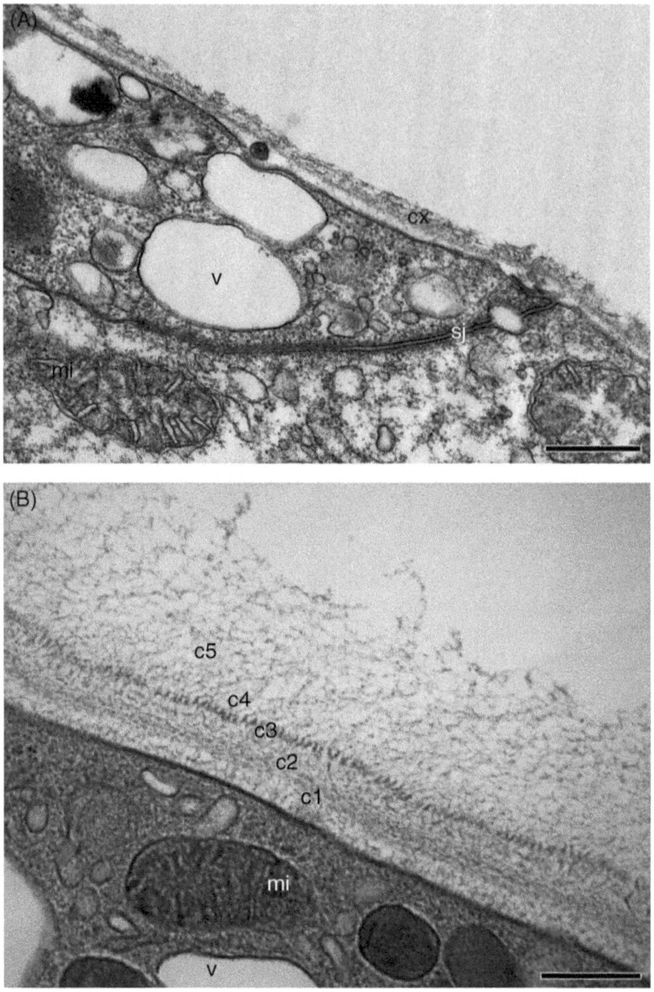

Fig. 4 *Hydra* ectodermal epithelial cells prepared by chemical fixation (A) or HPF/FS (B). (A) Chemical fixation (Campbell, 1987) shows good preservation of mitochondria and vesicles. However, the elaborated glycocalyx ("cuticle layer") of ectodermal epithelial cells is only poorly preserved. (B) HPF/FS provides excellent preservation of the cuticle layer revealing five distinct sheets (c1, c2, c3, c4, c5) not recognizable in chemically fixed samples (A: cx). Mitochondrium (mi), septate junction (sj), and vacuole (v). Scale bar is 2 μm.

D. Cryo-Processing—General Aspects

Rapid-freezing immobilizes subcellular structures and dynamics much quicker and more reliably than conventional chemical fixation at ambient temperatures (Heuser *et al.*, 1979; McIntosh, 2001). HPF (Moor, 1987) is the only freezing method suitable for cryo-immobilization of native, unfixed specimens thicker than ~10 μm. HPF

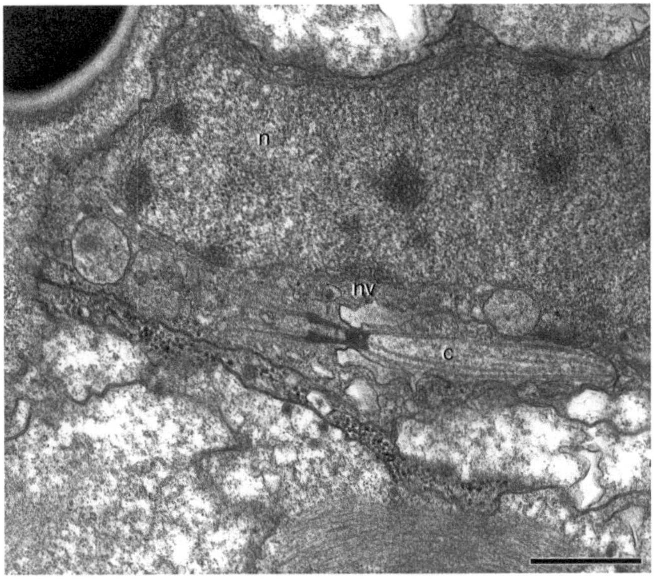

Fig. 5 *Hydra* ganglion cell with "intracellular" cilium after chemical fixation (Shigenaka *et al*, 1971). Cilium (c), nucleus (n), and neuronal vesicle (nv). Scale bar is 1 μm.

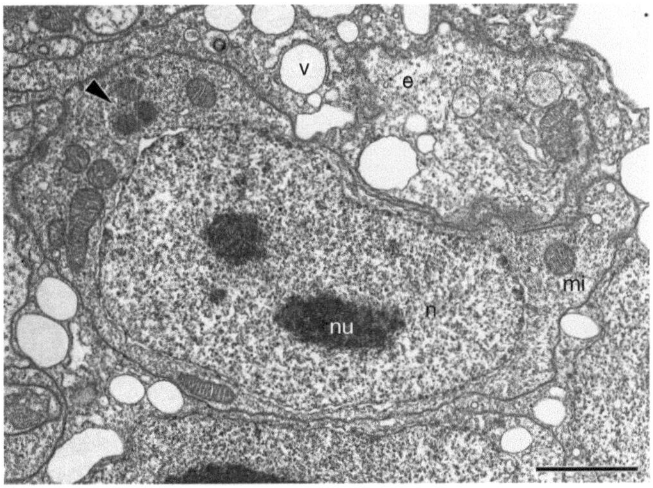

Fig. 6 Putative interstitial stem cell (i-cell) of *Hydra* after chemical fixation (Shigenaka *et al*, 1971) showing nucleus, nucleolus, mitochondria, and a conspicuous nuage/chromatoid body (arrowhead). The i-cell is embedded between epithelial cells. Epithelio-muscular cell (e), mitochondrium (mi), nucleus of i-cell (n), nucleolus (nu), and vacuole of epidermal cell (v). Scale bar is 2 μm.

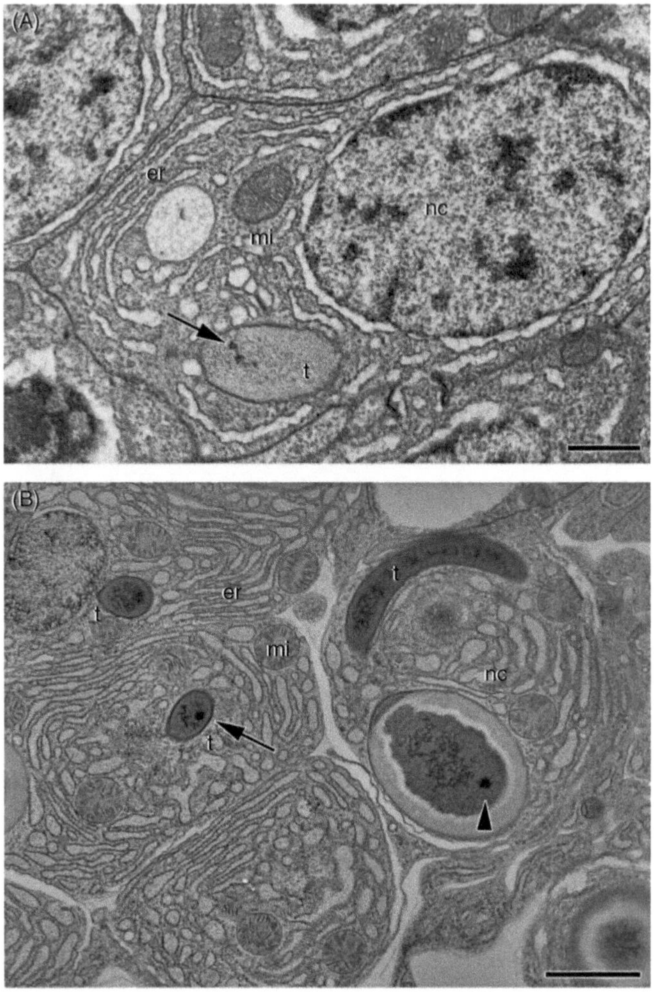

Fig. 7 Differentiating nematocysts of *Hydra* after chemical fixation (Shigenaka *et al*, 1971) (A) and HPF/FS (B). Note the prominent endoplasmic reticulum characteristic for this cell type. (A) High-contrast imaging (HCI) of a differentiating nematocyst (no section poststaining). Note the inclusion bodies in the tubule (arrow). (B) Arrowhead and arrows mark inclusion bodies in the matrix of a forming nematocyst capsule and tubules, respectively. Endoplasmic reticulum (er), mitochondrium (mi), and nematocyst (nc). Scale bars are 1 μm.

lowers the freezing point of water, and ice nucleation is considerably reduced (Müller and Moor, 1984). Thus, bulky animal samples with a thickness of 0.1–0.2 mm, on average, can be frozen with negligible (or even without) ice crystal damage. The two most widely used methods to process rapidly frozen samples for EM are freeze fracturing (Moor and Mühlethaler, 1963; Steere, 1957) and freeze substitution (FS), followed by resin embedding (Van Harreveld and Crowell, 1964). The freeze-fracture replica technique has been successfully applied to *Hydra* since the early days of biological EM (Filshie and Flower, 1977; McDowall and Grimmelikhuijzen, 1980; Wood, 1977). FS too is not a new invention (e.g., Simpson, 1941; Van Harreveld and Crowell, 1964) but with a few exceptions (e.g., Clode and Marshall, 2002) this method has not been used yet for coelenterates/cnidaria, in particular, *Hydra*. FS is a process in which the frozen sample is dehydrated through acetone or other organic solvents and crosslinked at low temperatures, i.e., −90 to −80°C (Humbel and Müller, 1986; Kellenberger, 1991). In parallel, chemical stabilization/crosslinking and staining of the sample are achieved by adding glutaraldehyde and heavy metals to the organic solvent. A standard FS recipe for morphological studies consists of anhydrous acetone plus 1% OsO_4 (Van Harreveld and Crowell, 1964) and 0.1–0.2% uranyl acetate (UA). FS with this popular cocktail is generally combined with embedding in epoxy resins. Alternatively, it was shown that anhydrous acetone, optionally supplemented with UA and/or formaldehyde and/or glutaraldehyde, and followed by embedment in (meth) acrylate resins, is useful for immuno- and lectin-cytochemistry. This approach, however, is not discussed in detail here (for reviews, see Hess, 2007; Nicolas, 1991; Schwarz and Humbel, 2007; and references therein). Typically, FS is performed for ~8 h (Humbel and Müller, 1986) up to several days (Staehelin *et al.*, 1990) at −90 to −80°C with the aid of FS devices or by using dry ice–acetone mixtures (Steinbrecht and Müller, 1987). Likewise, warming-up usually takes ~12–30 h ending with thorough rinsing of the specimens with pure solvents at temperatures between −30 and +20°C (Humbel and Müller, 1986). Finally, specimens are embedded in resin for ultramicrotomy and TEM.

E. Cryo-Processing—Protocols for Morphology

1. Recommended method: Rapid cryo-immobilization by means of HPF followed by FS and epoxy resin embedding: Figs. 4B, 7B, and 8

a. Freeze Substitution Media Anhydrous acetone containing 1% (w/v) OsO_4 (Van Harreveld and Crowell, 1964) plus 0.1–0.2% (w/v) UA (diluted from a 10% (w/v) stock solution of UA in methanol) for morphological studies (for suggestions how to prepare the cocktail, see Hess, 2007; McDonald and Muller-Reichert, 2002).

b. High-Pressure Freezing Procedure *Hydra* polyps immersed in *Hydra* culture medium were quickly dissected with a new, sharp scalpel blade into appropriate pieces to fit into cup-shaped HPF specimen carriers. Standard carriers for instruments of the HPM-010-type have a cavity depth of 0.1, 0.2, or 0.3 mm and an inner diameter of 2 mm (recently, larger carriers with an outer diameter of 6 mm have been made available for Leica HPM-100 machines; Leica Microsystems Wetlzlar, Germany). Tissue pieces are pipetted with *Hydra* culture medium into a 0.2- or 0.3-mm-deep carrier and covered with an additional carrier (flat side facing down). It is crucial to avoid enclosure of air bubbles inside the carrier's cavity since they impede proper freezing. Finally, the obtained sandwiches are cryo-immobilized by HPF. Frozen sandwiches are

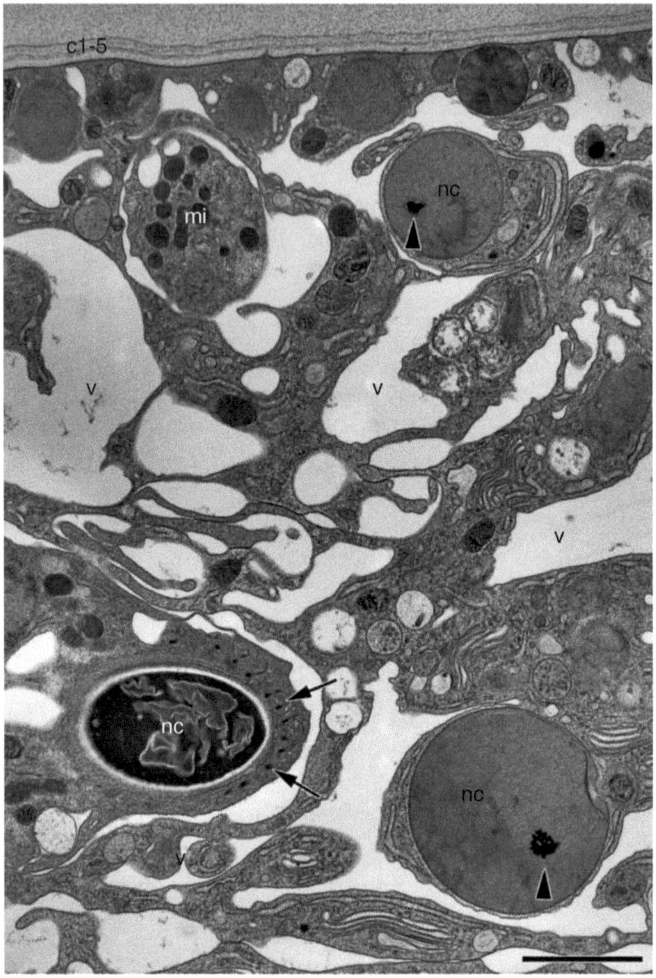

Fig. 8 Overview of epithelial cells and differentiating nematocysts in a HPF/FS *Hydra* sample. Note the pleiomorphic outlines of the complex vacuolar system. Cuticle layer (cl–5); mitochondrium (mi); nematocyst (nc); vacuole (v); arrowheads point to inclusion bodies of nematocysts; arrows mark the cnidocil-associated apparatus. Scale bar is 2 μm.

transferred unopened into appropriate containers for storage in LN_2 for later use or subjected to FS. It is essential to use precooled tweezers/tools and containers in any case and keep the samples under LN_2 whenever possible. Whether it is better to open the frozen sandwich prior to FS or simply wait until the ice dissolves during FS (so that the carriers separate) is not easy to judge. Opening the sandwich frequently leads to artifactual "freeze fracturing," thus damaging the sample. However, it also occurs that sandwiches remain tightly closed until the end of FS and samples become, therefore, poorly infiltrated with the FS media.

c. Freeze Substitution Procedure and Embedding The sandwiches with frozen samples are transferred under LN_2 into cryovials containing frozen FS cocktails. Subsequently, the lids are screwed loosely onto the vials to permit safe evaporation of excess N_2 gas. Finally, the vials are placed into the precooled FS device and after about 1 h the lids are tightened and FS is started. A feasible example for FS reads as follows: FS for at least 8 h (to be extended to ~36 h, if FS is performed over the weekend) at −80 to 90°C; warming up to −55°C at a rate of 5–10°C per hour, subsequent post-fixation and staining at −55°C for ~6 h, followed by warming up to −30°C at a rate of 5–10°C per hour where samples are left for an additional ~3 h. The stepwise warming-up should allow fixatives/stains included in the FS media to stabilize the specimens at low temperatures (for details on the reactivity of selected fixatives at low temperatures, see the following reviews: Hess, 2007; Humbel, 2009). Finally, the specimens are allowed to reach room temperature, rinsed three times with acetone (10 min each), and embedded in Epon and/or Araldite epoxy resin (Mollenhauer, 1964). We avoid embedding FS specimens in Spurr's mixture (Spurr, 1969) as this may solubilize and extract the subcellular constituents (Hess, 1990; note that this constraint does not apply to conventional, chemically fixed samples). A serious problem frequently associated with animal samples is their removal from the HPF carriers after FS or resin infiltration without damage. However, this is conveniently solved by leaving the sample within the aluminum carrier during the whole FS and embedding procedure, including the final resin polymerization. Following polymerization, the carrier's rim is freed from excess resin by thorough trimming (Chapter 14 by Salvenmoser *et al.*, 2010: this volume; Sawaguchi *et al.*, 2003). The sample plus carrier is then immersed in LN_2 so that the embedded sample falls off the carrier (which requires sometimes several freeze–thaw cycles). Finally, the sample may be re-orientated and re-embedded.

F. Microtomy and TEM

Sections are cut with an ultramicrotome preferably using diamond knives. After optional poststaining with UA and/or lead citrate, sections are analyzed in a conventional TEM or an Energy Filter TEM (EFTEM). Note that FS samples containing OsO_4 and UA frequently need no section poststaining when imaged with modern digital cameras. High-contrast images (HCI) can be obtained also from unstained sections of chemically fixed samples by using EFTEM. With this method, electrons with an energy loss of 250 eV are used for imaging (see Fig. 7A) to distinguish electron-dense substances from staining artifacts in sections.

IV. Materials

(Beware that most of the reagents are more or less toxic and/or hazardous to health; for their safe use and disposal, consult the relevant Material Safety Data Sheets).

A. Chemical Fixation for Morphology and Immunocytochemistry

For all studies, we used *Hydra vulgaris* (strain Basel).

Embryo dishes (preferably black, if used for small, quasi-transparent samples) or crystallization dishes (e.g., from Agar, Stansted, U.K.);

0.05–0.2 M cacodylate buffer (contains arsenic compounds!), optionally supplemented with 2 mM sucrose;

0.05–0.1 M phosphate buffer, optionally supplemented with 2 mM sucrose and/or 0.02 mM magnesium sulfate;

0.001 M Tris–HCl, 1 mM $CaCl_2$, 0.1 mM $MgCl_2$, 0.1 mM KCl, 1 mM NaH_2CO_3, pH 7.8 (*Hydra* culture medium);

Urethane was from Sigma-Aldrich, Inc (St Louis, U.S.A.);

Paraformaldehyde (Sigma-Aldrich), glutaraldehyde, OsO_4 crystals, epoxy resins (Epon, Spurr's) were from Sigma-Aldrich (St Louis, MO), Agar-Scientific (Stansted, England), EMS (Hatfield, PA), Polysciences (Warrington, PE), or Ted Pella (Redding, CA).

LR White acrylic resin was from Sigma or London Resin Co. (Woking, Surrey, U.K.), Lowicryl K4M acrylic resin was from Agar, DMF from Merck (Darmstadt, Germany).

B. Cryo-Processing

Embryo dishes or crystallization dishes (e.g., from Agar, Stansted, U.K.).

2 ml cryovials (e.g., Nalgene, from Nalge Nunc International, Rochester, NY; note that these vials are not explicitly LN_2 certified).

OsO_4 crystals, UA, and epoxy resin (Epon) were from Sigma, Agar, EMS (Hatfield, PA), Polysciences (Warrington, PE), or Ted Pella (Redding, CA).

Several types of aluminum carriers for HPF were from Martin Wohlwend Engineering (CH-9466 Sennwald, Switzerland; martin-wohlwend@bluewin.ch) or Leica-Microsystems (Vienna, Austria).

Tweezers, scalpels, slot, or mesh grids (copper, nickel) coated with Formvar and carbon were all from Agar, EMS, Polysciences, or Ted Pella.

Further EM instruments used in this study: dissecting microscope (i.e., Leica-EM-workstation), HPF apparatus HPM-010 (BAL-TEC, Balzers, Liechtenstein; note that this instrument is currently sold by ABRA-Fluid AG, Widnau, Switzerland), automated FS device AFS (Leica), ultramicrotome Ultracut S, diamond knives for ultramicrotomy (Diatome, Biel, Switzerland), transmission EM Libra 120 EFTEM (Zeiss, Oberkochen, Germany), or Philips CM120 TEM (F.E.I, Eindhoven, The Netherlands).

V. Results and Discussion

A. Chemical Fixation for Morphology and Immunocytochemistry

For most applications related to *Hydra* ultrastructure research, chemical fixation yields satisfactory results (Figs. 1–3, 4A, 5, 6, and 7A). Nematoblasts and nematocytes, the most prominent subcellular features in cnidarians, are well preserved (Figs. 1, 7A). By contrast, preservation of the cuticle layer (glycocalyx), for example,

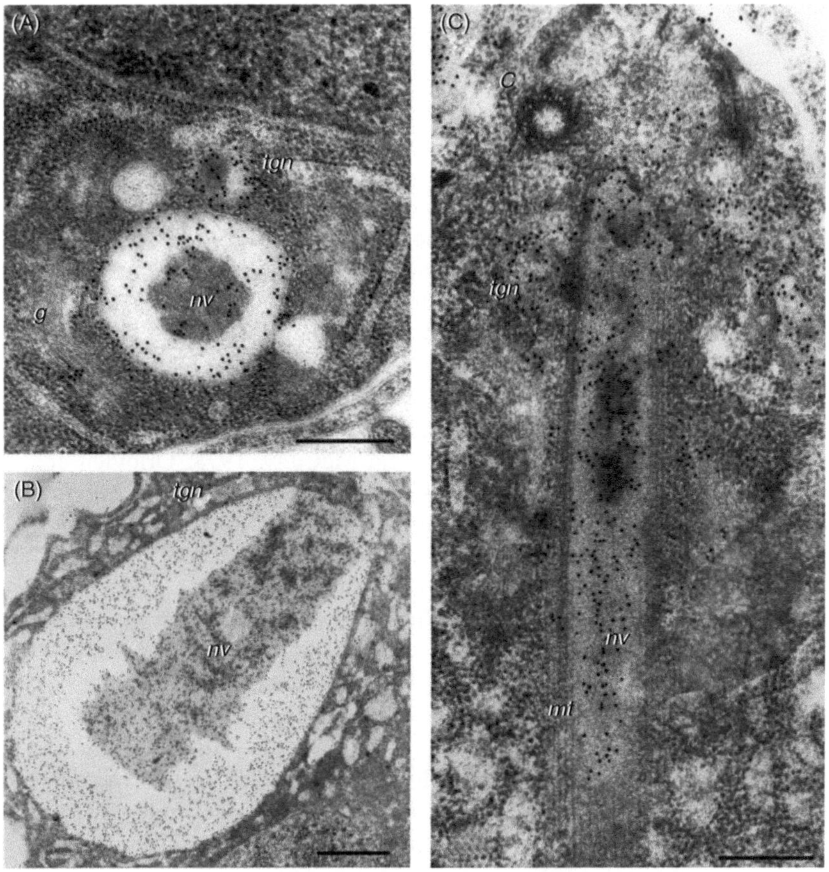

Fig. 9 Post-embedding immunogold labeling of differentiating *Hydra* nematocysts after chemical fixation and Lowicryl embedding showing the distribution of NOWA, a structural protein of nematocysts (Engel *et al.*, 2002). Antibodies used were H22 mouse monoclonal antibody (from Sigma) and goat anti-mouse secondary antibodies coupled to 10-nm colloidal gold (Sigma) for visualization. (A) Early formation of a nematocyst vesicle (nv) as a giant post-Golgi vesicle (reproduced with permission from *J Cell Sci 115(20), 3923*), (B) growth phase of the nematocyst capsule, (C) growth phase showing the formation of the long tubule stabilized by microtubules (mt). Note the specific immunogold labeling at the site of the Golgi apparatus, trans-Golgi network, and capsule. Golgi apparatus (g), trans-Golgi network (tgn). Scale bar is 0.5 μm.

is suboptimal or even poor (Fig. 4A). Moreover, the arrangement of the different vacuoles in epidermal cells (Fig. 1) is better preserved by HPF/FS (see Fig. 8). For TEM, we can recommend the fixation protocol according to Campbell (1987) for standard experiments (Figs. 1–3, and 4A) and the protocol according to Shigenaka *et al.* (1971; see Figs. 5, 6, and 7A) for improved preservation of fibers in the mesoglea. Regarding standard immunocytochemistry, the proposed protocols that include fixation with aldehydes and embedding in Lowicryl or LR-White yielded acceptable results (e.g., Fig. 9). However, depending on the tissue under investigation and the fixatives/acrylic resins employed, the omission of OsO_4 frequently leads to low specimen contrast. Cryo-fixation and FS combined with embedding in acrylic resins (Humbel, 2009) of *Hydra* have not been tested by the authors so far.

B. Cryo-Processing

To our knowledge, we demonstrate here for the first time the feasibility of HPF and FS for the EM analysis of *Hydra* (Figs. 4B, 7B, and 8). The patterns of organelles, endomembranes, and cytoskeleton correlate well in our *Hydra* samples with those reported from other animals, plants, and fungi prepared by HPF/FS (Hess, 2003; Howard and Aist, 1980; Kirschning *et al.*, 1998; Murk *et al.*, 2003; Otegui *et al.*, 2002). The reliable preservation of pleiomorphic vacuoles, a prominent feature of *Hydra* tissues (Fig. 1), is technically challenging for EM, as known from similar endomembranes in plants, fungi, and mammalian nerve growth cones (Dailey and Bridgman, 1993; Mersey and McCully, 1978; Orlovich and Ashford, 1993; Wilson *et al.*, 1990). Rapid freezing and FS, instead of conventional chemical fixation, properly stabilized the irregular shapes of such vacuoles in *Hydra*, thus providing authentic snapshots of their dynamic changes (Fig. 8). Subcellular components specific for *Hydra* and related Cnidaria/Coelenterata, such as nematocysts, are also well preserved by HPF/FS (Figs. 7B, 8). With respect to nematocyst structure and development, it is interesting to note that the data derived from HPF/FS appear quite similar to those from specimens that were chemically fixed with buffered OsO_4 (Slautterback and Fawcett, 1959), but to a lesser extent to those from glutaraldehyde- and OsO_4-fixed materials (Holstein, 1981). Finally, HPF/FS allowed a considerably improved preservation of the cuticle unraveling so far unknown complexity. Up to five distinct layers of variable thicknesses were recognizable after cryo-processing (Fig. 4B) whereas these structures appeared quite "blurred" and partially extracted after aqueous chemical fixation and sample dehydration at room temperature (Fig. 4A). The specific chemical composition of these layers, as well as their relevance/function, remains still open to further research. It is obvious, however, that this elaborated cuticle layer represents an efficient diffusion barrier for bacteria, viruses, and large organic molecules.

We consider the structural preservation achieved with small-sized animals and/or tissue pieces, including buds or daughter polyps, as very good to excellent, and the quality comparable to any other metazoan samples processed by HPF/FS so far. The yield of usable, adequately frozen *Hydra* specimens, however, is still quite moderate. This is likely attributable to the high water content of *Hydra* tissue, as well as to the size of the samples. The physicochemical properties of cells/tissues to be frozen strongly influence the freezing quality, in particular, the presence or absence of natural cryoprotectants within the cells. Furthermore, in HPF, the greatly reduced cooling rate in the center of bulky objects is an inherent problem of specimens thicker than ~200 μm (Sitte *et al.*, 1987; Studer *et al.*, 1995). Thus, optimizing the freezing efficiency is definitely required for *Hydra* samples. Possible improvements here could be a volume reduction of the whole "sandwich" as the mass of the aluminum carriers must be considered as well. For instance, one could use some kind of "mini-aquarium" to contain *Hydra* samples for HPF as introduced by us for flatworms (Chapter 14 by Salvenmoser *et al.*: this volume). Briefly, such a "mini-aquarium" consists of a 0.2- to 0.3-mm-deep carrier and a sapphire disk, usually used for 2D cell cultures, serving as a lid. The final height of this sandwich is adjusted with a suitable spacer ring to meet the dimensions of the sample holder of the HPF apparatus (in analogy to the two-sapphire sandwich described elsewhere: Hawes *et al.*, 2007). Alternatively, one could use custom-made carriers (Craig *et al.*, 1987) with a thinner bottom, similar to the so-called "membrane carriers" introduced for the EMPACT-2 high-pressure freezer (McDonald *et al.*, 2007). Moreover, the use of physiologically appropriate space "fillers" to surround the sample inside the HPF carriers presumably enhances the freezing efficiency. The so-called space "fillers" are liquids with a lower freezing point and/or better heat conductivity than water (Studer *et al.*, 1989). Among the most widely used reagents are concentrated protein or polysaccharide solutions, such as BSA (McDonald *et al.*, 2007) or dextran (Bleck *et al.*, 2010). BSA (20% w/v) is suitable for HPF of nematodes and marine sponges (McDonald *et al.*, 2007) and certain ctenophores (M. Müller-Reichert, personal communication). Dense suspensions of yeast and/or *Escherichia coli* paste (McDonald, 1994; Muller-Reichert *et al.*, 2003) or, as recently reported, cyanobacteria (Daghma *et al.*, 2009) serve the same purpose. Thus, such suspensions appear to be a good candidate to enhance the efficiency of the rapid-freezing process for these challenging samples.

VI. Concluding Remarks

Cryo-processing, such as HPF/FS, has not been established for *Hydra* research so far. We expect that this technique will be extremely useful for this simple organism. Despite the fact that chemical fixation represents a relatively cheap approach, cryo-processing provides superior preservation of *Hydra* tissue. The high spatial and temporal resolution obtained with these methods may be important for unraveling minute ultrastructural details and dynamic features, for example, of the cytoskeleton during changes in planar cell polarity or during the secretion process of nematocysts. In addition, HPF/FS can be linked to low-temperature resin embedding (Humbel and Müller, 1986) as well as to Tokuyasu cryosection immunolabeling (Stierhof and El Kasmi, 2010), opening new perspectives for high-resolution localization of delicate and/or rare antigens. Finally, HPF is the fixation method of choice for correlative microscopy (Schwarz and Humbel, 2007; see also: Robinson and Takizawa, 2009) and EM tomography (Lucic *et al.*, 2008; Marsh *et al.*, 2001).

Acknowledgments

We thank Karin Gutleben for excellent technical assistance; Anne-Kathrin Gorny, Roland Aufschnaiter, and Thomas Müller-Reichert for communicating unpublished observations. Bernhard Egger and Bert Hobmayer are thanked for helpful comments. Fig. 9A is adapted from *J. Cell Sci.* **115**(2), 3923 Engel *et al.*, 2002) with kind permission of the publisher The Company of Biologists Ltd.. T.W. Holstein is supported by the Deutsche Fortschungsgemeinschaft (DFG). M.W. Hess is supported by grants from FWF Austrian Science Funds (P 19486-B12), Oesterreichische Nationalbank-Jubiläumsfonds (P-11050), and Tiroler Wissenschaftsfonds (P-UNI-0404/100).

References

Bleck, C. K., Merz, A., Gutierrez, M. G., Walther, P., Dubochet, J., Zuber, B., and Griffiths, G. (2010). Comparison of different methods for thin section EM analysis of *Mycobacterium smegmatis*. *J. Microsc.* **237**(1), 23–38.

Bode, H. R. (2003). Head regeneration in *Hydra*. *Dev. Dyn.* **226**, 225–236.

Bode, H. R., Flick, K. M., and Smith, G. S. (1976). Regulation of interstitial cell differentiation in *hydra* attenuata. I. Homeostatic control of interstitial cell population size. *J. Cell Sci.* **20**, 29–46.

Bosch, T. C. (2007). Why polyps regenerate and we don't: Towards a cellular and molecular framework for *hydra* regeneration. *Dev. Biol.* **303**, 421–433.

Bosch, T. C. (2009). *Hydra* and the evolution of stem cells. *Bioessays* **31**, 478–486.

Campbell, R. D. (1976). Elimination by *hydra* interstitial and nerve cells by means of colchicine. *J. Cell Sci.* **21**, 1–13.

Campbell, R. D. (1987). Organization of the nematocyst battery in the tentacle of *hydra*: Arrangement of the complex anchoring junctions between nematocytes, epithelial cells, and basement membrane. *Cell Tissue Res.* **249**(3), 647–655.

Campbell, R. D., and David, C. N. (1974). Cell cycle kinetics and development of *Hydra attenuata*. II. Interstitial cells. *J. Cell Sci.* **16**, 349–358.

Cartwright, P., Halgedahl, S. L., Hendricks, J. R., Jarrard, R. D., Marques, A. C., Collins, A. G., and Lieberman, B. S. (2007). Exceptionally preserved jellyfishes from the middle cambrian. *PLoS ONE* **2**(10), e1121.

Chapman, J. A., Kirkness, E. F., Simakov, O., Hampson, S. E., Mitros, T., Weinmaier, T., Rattei, T., Balasubramanian, P. G., Borman, J., Busam, D., Disbennett, K., Pfannkoch, C., et al., (2010). The dynamic genome of *hydra*. *Nature* **464**, 592–596.

Clode, P. L., and Marshall, A. T. (2002). Kalisomes in corals: A novel KCl concentrating organelle? *Tissue Cell* **34**(3), 199–209.

Craig, S., Gilkey, J. C., and Staehelin, L. A. (1987). Improved specimen support cups and auxiliary devices for the balzers high pressure freezing apparatus. *J. Microsc.* **148**(Pt 1), 103-106.

Daghma, D. S., Wiesner, M., Rutte, T., and Melzer, M. (2009). Cyanobacteria as "filler" for loading nitrocellulose capillaries: An innovative method for high pressure freezing of pollen. *In* "MC2009-Graz, Volume 2 Life Sciences" (Pabst A. M., Zellnig G., editors) pp. 161–162. Graz. Verlag der Technischen Universität, Graz.

Dailey, M. E., and Bridgman, P. C. (1993). Vacuole dynamics in growth cones: Correlated EM and video observations. *J. Neurosci.* **13**(8), 3375–3393.

David, C. N., and Campbell, R. D. (1972). Cell cycle kinetics and development of *Hydra attenuata*. I. Epithelial cells. *J. Cell Sci.* **11**, 557–568.

David, C. N., and Challoner, D. (1974). Distribution of interstitial cells and differentiating nematocysts in nests in *Hydra attenuata*. *Am. Zool.* **14**(2), 537–542.

David, C. N., and Murphy, S. (1977). Characterization of interstitial stem cells in *hydra* by cloning. *Dev. Biol.* **58**(2), 372–383.

David, C. N., Ozbek, S., Adamczyk, P., Meier, S., Pauly, B., Chapman, J., Hwang, J. S., Gojobori, T., and Holstein, T. W. (2008). Evolution of complex structures: Minicollagens shape the cnidarian nematocyst. *Trends Genet.* **24**(9), 431–438.

Engel, U., Ozbek, S., Streitwolf-Engel, R., Petri, B., Lottspeich, F., and Holstein, T. W. (2002). Nowa, a novel protein with minicollagen cys-rich domains, is involved in nematocyst formation in *hydra*. *J. Cell Sci.* **115**(Pt 20), 3923–3934.

Filshie, B. K., and Flower, N. E. (1977). Junctional structures in *hydra*. *J. Cell Sci.* **23**, 151–172.

Fritzenwanker, J. H., Genikhovich, G., Kraus, Y., and Technau, U. (2007). Early development and axis specification in the sea anemone *Nematostella vectensis*. *Dev. Biol.* **310**(2), 264–279.

Gierer, A., Berking, S., Bode, H., David, C. N., Flick, K., Hansmann, G., Schaller, H., and Trenkner, E. (1972). Regeneration of *hydra* from reaggregated cells. *Nat. New Biol.* **239**(91), 98–101.

Glauert, A. M., and Lewis, P. R. (1998). "Biological Specimen Preparation for Transmission Electron Microscopy" (A. M. Glauert, , ed.) 326 p. Portland Press, London.

Griffiths, G. (1993). "Fine Structure Immunocytochemistry." Springer, Berlin (371 p).

Guder, C., Philipp, I., Lengfeld, T., Watanabe, H., Hobmayer, B., and Holstein, T. W. (2006). The wnt code: Cnidarians signal the way. *Oncogene* **25**, 7450–7460.

Hawes, P., Netherton, C. L., Mueller, M., Wileman, T., and Monaghan, P. (2007). Rapid freeze-substitution preserves membranes in high-pressure frozen tissue culture cells. *J. Microsc.* **226**(Pt 2), 182–189.

Hess, M. W. (1990). Spurr-resin affects the ultrastructure of freeze-fixed/freeze-substituted pollen grains. *Ultramicroscopy* **34**, 219–221.

Hess, M. W. (2003). Of plants and other pets: Practical aspects of freeze-substitution and resin embedding. *J. Microsc.* **212**(Pt 1), 44-52.

Hess, M. W. (2007). Cryopreparation methodology for plant cell biology. *Methods Cell Biol.* **79**, 57–100.

Heuser, J. E., Reese, T. S., Dennis, M. J., Jan, Y., Jan, L., and Evans, L. (1979). Synaptic vesicle exocytosis captured by quick freezing and correlated with quantal transmitter release. *J. Cell Biol.* **81**(2), 275–300.

Hobmayer, B., Holstein, T. W., and David, C. N. (1990). Tentacle morphogenesis in *hydra*. II. Formation of a complex between a sensory nerve cell and a battery cell. *Development* **109**, 897–904.

Hobmayer, B., Rentzsch, F., Kuhn, K., Happel, C. M., von Laue, C. C., Snyder, P., Rothbacher, U., and Holstein, T. W. (2000). WNT signalling molecules act in axis formation in the diploblastic metazoan *hydra*. *Nature* **407**(6801), 186–189.

Holstein, T. (1981). The morphogenesis of nematocytes in *hydra* and *forskalia*: An ultrastructural study. *J. Ultrastruct. Res.* **75**(3), 276–290.

Holstein, T. W. (2008). Wnt signaling in cnidarians. *Methods Mol. Biol.* **469**, 47–54.

Holstein, T. W., and David, C. N. (1990). Cell cycle length, cell size, and proliferation rate in *hydra* stem cells. *Dev. Biol.* **142**, 392–400.

Holstein, T. W., Hobmayer, E., and Technau, U. (2003). Cnidarians: An evolutionarily conserved model system for regeneration? *Dev. Dyn.* **226**, 257–267.

Holstein, T., and Tardent, P. (1984). An ultrahigh-speed analysis of exocytosis: Nematocyst discharge. *Science* **223**, 830–833.

Howard, R. J., and Aist, J. R. (1980). Cytoplasmic microtubules and fungal morphogenesis: Ultrastructural effects of methyl benzimidazole-2-ylcarbamate determined by freeze-substitution of hyphal tip cells. *J. Cell Biol.* **87**(1), 55–64.

Humbel, B. M. (2009). Freeze-substitution. *In* "Handbook of Cryo-Preparation Methods for Electron Microscopy" (Cavalier A., Spehner D., Humbel B. M., eds.) pp. 320–341. CRC Press, Boca Raton.

Humbel, B., and Müller, M. (1986). Freeze substitution and low temperature embedding. *In* "The Science of Biological Specimen Preparation 1985" (Wolosewick J. J., ed.) pp. 175–183. SEM Inc AMF O'Hare, Chicago.

Hwang, J. S., Takaku, Y., Chapman, J., Ikeo, K., David, C. N., and Gojobori, T. (2008). Cilium evolution: Identification of a novel protein, nematocilin, in the mechanosensory cilium of *hydra* nematocytes. *Mol. Biol. Evol.* **25**(9), 2009–2017.

Kellenberger, E. (1991). The potential of cryofixation and freeze substitution: Observations and theoretical considerations. *J. Microsc.* **161**(Pt 2), 183–203.

Kirschning, E., Rutter, G., and Hohenberg, H. (1998). High-pressure freezing and freeze-substitution of native rat brain: Suitability for preservation and immunoelectron microscopic localization of myelin glycolipids. *J. Neurosci. Res.* **53**(4), 465–474.

Kusserow, A., Pang, K., Sturm, C., Hrouda, M., Lentfer, J., Schmidt, H. A., Technau, U., von Haeseler, A., Hobmayer, B., Martindale, M. Q., and Holstein, T. W. (2005). Unexpected complexity of the wnt gene family in a sea anemone. *Nature* **433**, 156–160.

Lengfeld, T., Watanabe, H., Simakov, O., Lindgens, D., Gee, L., Law, L., Schmidt, H. A., Ozbek, S., Bode, H., and Holstein, T. W. (2009). Multiple wnts are involved in *hydra* organizer formation and regeneration. *Dev. Biol.* **330**, 186–199.

Lentz, T. L. (1966a). "The Cell Biology of *Hydra*." North-Holland Publishing Company, Amsterdam.

Lentz, T. L. (1966b). Intramitochondrial glycogen granules in digestive cells of *hydra*. *J. Cell Biol.* **29**(1), 162–167.

Lucic, V., Leis, A., and Baumeister, W. (2008). Cryo-electron tomography of cells: Connecting structure and function. *Histochem. Cell Biol.* **130**(2), 185–196.

Marsh, B. J., Mastronarde, D. N., Buttle, K. F., Howell, K. E., and McIntosh, J. R. (2001). Organellar relationships in the Golgi region of the pancreatic beta cell line, HIT-T15, visualized by high resolution electron tomography. *Proc. Natl. Acad. Sci. U.S.A.* **98**(5), 2399–2406.

McDonald, K. L. (1994). Electron microscopy and EM immunocytochemistry. *Methods Cell Biol.* **44**, 411–444.

McDonald, K. L., Morphew, M., Verkade, P., and Muller-Reichert, T. (2007). Recent advances in high-pressure freezing: Equipment- and specimen-loading methods. *Methods Mol. Biol.* **369**, 143–173.

McDonald, K., and Muller-Reichert, T. (2002). Cryomethods for thin section electron microscopy. *Methods Enzymol.* **351**, 96–123.

McDowall, A. W., and Grimmelikhuijzen, C. J. (1980). Intercellular junctions in nerve-free *hydra*. *Cell Tissue Res.* **209**(2), 217–224.

McIntosh, J. R. (2001). Electron microscopy of cells: A new beginning for a new century. *J. Cell Biol.* **153**(6), F25–F32.

Meinhardt, H. (2002). The radial-symmetric *hydra* and the evolution of the bilateral body plan: An old body became a young brain. *Bioessays* **24**(2), 185–191.

Mersey, B., and McCully, M. (1978). Monitoring the course of fixation of plant cells. *J. Microsc.* **114**, 49–76.

Mollenhauer, H. H. (1964). Plastic embedding mixtures for use in electron microscopy. *Stain. Technol.* **39**, 111–114.

Momose, T., and Schmid, V. (2006). Animal pole determinants define oral–aboral axis polarity and endodermal cell-fate in hydrozoan jellyfish *Podocoryne carnea*. *Dev. Biol.* **292**, 371–380.

Moor, H. (1987). Theory and practice of high pressure freezing. *In* "Cryotechniques in Biological Electron Microscopy" (Steinbrecht R. A., Zierold K., eds.) pp. 175–191. Springer, Berlin, Heidelberg.

Moor, H., and Mühlethaler, K. (1963). Fine structure in frozen-etched yeast cells. *J. Cell Biol.* **17**, 609–628.

Müller, M., and Moor, H. (1984). Cryofixation of thick specimens by high-pressure freezing. *In* "The Science of Biological Specimen Preparation" (Haggins G. H., ed.) pp. 131–138. SEM Inc. AMF O'Hare, Chicago.

Muller-Reichert, T., Hohenberg, H., O'Toole, E. T., and McDonald, K. (2003). Cryoimmobilization and three-dimensional visualization of *C. Elegans* ultrastructure. *J. Microsc.* **212**(Pt 1), 71-80.

Murk, J. L., Posthuma, G., Koster, A. J., Geuze, H. J., Verkleij, A. J., Kleijmeer, M. J., and Humbel, B. M. (2003). Influence of aldehyde fixation on the morphology of endosomes and lysosomes: Quantitative analysis and electron tomography. *J. Microsc.* **212**(Pt 1), 81–90.

Nicolas, G. (1991). Advantages of fast-freeze fixation followed by freeze-substitution for the preservation of cell integrity. *J. Electron Microsc. Tech.* **18**(4), 395–405.

Nuchter, T., Benoit, M., Engel, U., Ozbek, S., and Holstein, T. W. (2006). Nanosecond-scale kinetics of nematocyst discharge. *Curr. Biol.* **16**, R316–R318.

Orlovich, D. A., and Ashford, A. E. (1993). Polyphosphate granules are an artefact of specimen preparation in the ectomycorrhizal fungus *Pisolithus tinctorius*. *Protoplasma* **173**, 91–102.

Otegui, M. S., Capp, R., and Staehelin, L. A. (2002). Developing seeds of *Arabidopsis* store different minerals in two types of vacuoles and in the endoplasmic reticulum. *Plant Cell* **14**(6), 1311–1327.

Peterson, K. J., Cotton, J. A., Gehling, J. G., and Pisani, D. (2008). The Ediacaran emergence of bilaterians: Congruence between the genetic and the geological fossil records. *Philos. Trans. R Soc. Lond., B, Biol. Sci.* **363**(1496), 1435–1443.

Philipp, I., Aufschnaiter, R., Ozbek, S., Pontasch, S., Jenewein, M., Watanabe, H., Rentzsch, F., Holstein, T. W., and Hobmayer, B. (2009). Wnt/beta-catenin and noncanonical wnt signaling interact in tissue evagination in the simple eumetazoan *hydra*. *Proc. Natl. Acad. Sci. U.S.A.* **106**, 4290–4295.

Rentzsch, F., Guder, C., Vocke, D., Hobmayer, B., and Holstein, T. W. (2007). An ancient chordin-like gene in organizer formation of *hydra*. *Proc. Natl. Acad. Sci. U.S.A.* **104**(9), 3249–3254.

Robinson, J. M., and Takizawa, T. (2009). Correlative fluorescence and electron microscopy in tissues: Immunocytochemistry. *J. Microsc.* **235**(3), 259–272.

Salvenmoser, W., Egger, B., Achatz, J. G., Ladurner, P., and Hess, M. W. (2010). Electron microscopy of flatworms: Standard and cryo-preparation methods. *Methods Cell Biol.* **96**, 289–311.

Sánchez Alvarado, A. (2003). The freshwater planarian *Schmidtea mediterranea*: embryogenesis, stem cells and regeneration. *Curr. Opin. Genet. Dev.* **13**(4), 438–444.

Sanchez Alvarado, A. (2007). Stem cells and the planarian *Schmidtea mediterranea*. *C. R. Biol.* **330**(6–7), 498–503.

Sawaguchi, A., Yao, X., Forte, J. G., and McDonald, K. L. (2003). Direct attachment of cell suspensions to high-pressure freezing specimen planchettes. *J. Microsc.* **212**(Pt 1), 13–20.

Schwarz, H., and Humbel, B. M. (2007). Correlative light and electron microscopy using immunolabeled resin sections. *Methods Mol. Biol.* **369**, 229–256.

Shigenaka, Y., Roth, L. E., and Pihlaja, D. J. (1971). Microtubules in the heliozoan axopodium. 3. Degradation and reformation after dilute urea treatment. *J. Cell Sci.* **8**(1), 127–151.

Simpson, W. L. (1941). An experimental analysis of the Altmann technique of freeze-drying. *Anat. Rec.* **80**, 173–189.

Sitte, H., Edelmann, L., and Neumann, K. (1987). Cryofixation without pretreatment at ambient pressure. *In* "Cryotechniques in Biological Electron Microscopy" (Steinbrecht R. A., Zierold K., eds.) pp. 87–113. Springer, Berlin, Heidelberg.

Slautterback, D. B., and Fawcett, D. W. (1959). The development of the cnidoblasts of *hydra*; an electron microscope study of cell differentiation. *J. Biophys. Biochem. Cytol.* **5**(3), 441–452.

Spurr, A. R. (1969). A low-viscosity epoxy resin embedding medium for electron microscopy. *J. Ultrastruct. Res.* **26**(1), 31–43.

Staehelin, L. A., Giddings, T. H., Kiss, J. Z., and Sack, F. D. (1990). Macromolecular differentiation of Golgi stacks in root tips of *Arabidosis* and *Nicotiana* seedlings as visualized in high pressure frozen and freeze-substituted samples. *Protoplasma* **157**, 75–91.

Steele, R. E. (2002). Developmental signaling in *hydra*: What does it take to build a "simple" animal? *Dev. Biol.* **248**(2), 199–219.

Steere, R. L. (1957). Electron microscopy of structural detail in frozen biological specimens. *J. Biophys. Biochem. Cytol.* **3**(1), 45–60.

Steinbrecht, R. A., and Müller, M. (1987). Freeze-substitution and freeze-drying. *In* "Cryotechniques in Biological Electron Microscopy" (Steinbrecht R. A., K. Zierold, eds.) pp. 149–172. Springer, Berlin, Heidelberg.

Stierhof, Y. D., and El Kasmi, F. (2010). Strategies to improve the antigenicity, ultrastructure preservation and visibility of trafficking compartments in *Arabidopsis* tissue. *Eur. J. Cell Biol.* **89**(2–3), 285–297.

Studer, D., Michel, M., and Muller, M. (1989). High pressure freezing comes of age. *Scanning Microsc. Suppl.* **3**, 253–268, discussion 268–259.

Studer, D., Michel, M., Wohlwend, M., Hunziker, E. B., and Buschmann, M. D. (1995). Vitrification of articular cartilage by high-pressure freezing. *J. Microsc.* **179**(Pt 3), 321–332.

Sugiyama, T., and Fujisawa, T. (1978). Genetic analysis of developmental mechanisms in *Hydra*. V. Cell lineage and development of chimera *hydra*. *J. Cell Sci.* **32**, 215–232.

Technau, U., Cramer von Laue, C., Rentzsch, F., Luft, S., Hobmayer, B., Bode, H. R., and Holstein, T. W. (2000). Parameters of self-organization in *hydra* aggregates. *Proc. Natl. Acad. Sci. U.S.A.* **97**(22), 12127–12131.

Van Harreveld, A., and Crowell, J. (1964). Electron microscopy after rapid freezing on a metal surface and substitution fixation. *Anat. Rec.* **149**, 381–385.

Watanabe, H., Hoang, V. T., Mattner, R., and Holstein, T. W. (2009). Immortality and the base of multicellular life: Lessons from cnidarian stem cells. *Semin. Cell Dev. Biol.* **20**(9), 1114–1125.

Webster, P., Schwarz, H., and Griffiths, G. (2008). Preparation of cells and tissues for immuno EM. *Methods Cell Biol.* **88**, 45–58.

Westfall, J. A., Yamataka, S., and Enos, P. D. (1971). Ultrastructural evidence of polarized synapses in the nerve net of *Hydra*. *J. Cell Biol.* **51**(1), 318–323.

Wilson, T. P., Canny, M. J., McCully, M. E., and Lefkovitch, L. P. (1990). Breakdown of cytoplasmic vacuoles. A model of endoplasmic membrane rearrangement. *Protoplasma* **155**, 144–152.

Wood, R. L. (1977). The cell junctions of *Hydra* as viewed by freeze-fracture replication. *J. Ultrastruct. Res.* **58**(3), 299–315.

Wood, R. L. (1985). The use of *Hydra* for studies of cellular ultrastructure and cell junctions. *Arch. Sci. Geneve* **38**, 371–383.

Wood, R. L., and Novak, P. L. (1982). The anchoring of nematocysts and nematocytes in the tentacles of *Hydra*. *J. Ultrastruct. Res.* **81**(1), 104–116.

CHAPTER 14

Electron Microscopy of Flatworms: Standard and Cryo-Preparation Methods

Willi Salvenmoser*, Bernhard Egger*, Johannes G. Achatz*, Peter Ladurner*, *and* Michael W. Hess†

*Center for Molecular Biosciences, Institute of Zoology, University of Innsbruck, Innsbruck, Austria

†Division of Histology and Embryology, Innsbruck Medical University, Innsbruck, Austria

Dedicated to the memory of our mentor and friend Reinhard Rieger.

978-0-12-381007-6
DOI: 10.1016/S0091-679X(10)96014-7

Abstract

Electron microscopy (EM) has long been indispensable for flatworm research, as most of these worms are microscopic in dimension and provide only a handful of characters recognizable by eye or light microscopy. Therefore, major progress in understanding the histology, systematics, and evolution of this animal group relied on methods capable of visualizing ultrastructure. The rise of molecular and cellular biology renewed interest in such ultrastructural research. In the light of recent developments, we offer a best-practice guide for users of transmission EM and provide a comparison of well-established chemical fixation protocols with cryo-processing methods (high-pressure freezing/freeze-substitution, HPF/FS). The organisms used in this study include the rhabditophorans *Macrostomum lignano, Polycelis nigra* and *Dugesia gonocephala*, as well as the acoel species *Isodiametra pulchra*.

I. Introduction

Flatworms (= Platyhelminthes *sensu lato*) comprise a diverse group of organisms ranging from free-living taxa to human and animal parasitic groups. For a long time, they have been at the center of regeneration research, being called "almost immortal under the edge of the knife" (Dalyell, 1814). Some flatworms are able to regenerate all body parts from tiny tissue fragments—including the head (Egger *et al.*, 2006; Montgomery and Coward, 1974). While not all flatworms are able to regenerate equally well, most species can restore at least the posterior part of the body behind brain or pharynx (Egger *et al.*, 2007).

Regeneration in flatworms is made possible by totipotent stem cells, called neoblasts (Ladurner *et al.*, 2008). These undifferentiated cells have a high nucleus-to-cytoplasm ratio (Bode *et al.*, 2006) and are the only proliferating cells in juvenile and adult flatworms. For studying neoblasts during regeneration, growth, and tissue maintenance, small and transparent flatworms in the "millimeter range", such as *Macrostomum lignano* and *Isodiametra pulchra*, have recently been introduced to research in addition to the traditionally used large (centimeter range) and pigmented triclads, often called "planarians" (De Mulder *et al.*, 2009; Egger *et al.*, 2007).

A. Systematics, Phylogeny, and Morphology

Molecular studies suggest that the Platyhelminthes *sensu strictu*, comprising the sister taxa Catenulida and Rhabditophora, belong to the Lophotrochozoa, whereas the Acoelomorpha represent a separate offshoot at the base of the Bilateria (Hejnol *et al.*, 2009). However, this issue is not entirely settled yet (Egger *et al.*, 2009b). Because all these organisms are processed similarly for EM, the acoels are also included in this chapter. Within the Platyhelminthes the large taxon of the Rhabditophora is characterized by the presence of rhabdite glands and is composed of the Macrostomorpha, Polycladida, and the Neoophora, including among others the Tricladida and the parasitic Neodermata

(Ehlers, 1985). The latter comprises the tapeworms, the flukes (e.g., the blood fluke *Schistosoma mansoni* that causes bilharzia and the liver fluke *Fasciola hepatica* that affects sheep, cattle, and human), and the ectoparasitic monogeneans that cause problems in aquaculture.

Flatworms are acoelomate, bilaterally symmetric animals that primarily lack segmentation, a circulatory system, respiratory organs, and an anus (Rieger *et al.*, 1991). The cephalized nervous system consists of an anteriorly positioned brain and posteriorly extending longitudinal nerve cords, interconnected by transverse commissures. Flatworms possess an elaborate muscle system consisting of longitudinal, circular, and oblique fibers. Small free-living species primarily use their epidermal cilia for locomotion, assisted by muscular contractions in larger forms. The organization of the gut is sometimes reflected in the name of a flatworm clade, as in the stunningly beautiful marine polyclads, which possess a highly branched diverticular gut, the triclads, which exhibit three gut branches, or the rhabdocoels with a rod-shaped gut. Many acoel flatworms have no gut lumen at all but are characterized by a compact central digestive syncytium. The tapeworms (Cestoda) have completely abolished their gut, taking up nutrients through their epidermis. Unpaired (Catenulida) and paired (Rhabditophora) protonephridia with "flame cells" are responsible for osmoregulation, but are absent in the basal group of the Acoelomorpha. Most flatworms are sexually reproducing hermaphrodites (one prominent exception is the diecious *S. mansoni*), while other species are asexual and multiply by fission.

B. Flatworms as Models

Except for stable transgenic cell lines, the whole molecular "toolbox" is available for the flatworm models. Additionally, small flatworms are exceptionally amenable to whole-mount immuno-cytochemistry, i.e., for studying stem cells and other tissues, such as musculature, nervous system, epidermis, gut, and various glands (Egger *et al.*, 2009a; Inoue *et al.*, 2004; Ladurner *et al.*, 2005).

Among acoel flatworms, three species have emerged as model systems, particularly for developmental studies: the small (0.5–1 mm in length) *Isodiametra pulchra* (De Mulder *et al.*, 2009; Ladurner and Rieger, 2000) and the relatively large species of *Convolutriloba longifissura* (up to 6 mm in length) (Åkesson *et al.*, 2001; Gaerber *et al.*, 2007; Hejnol and Martindale, 2008) and *Symsagittifera roscoffensis* (4 mm in length) (Moreno *et al.*, 2009; Semmler *et al.*, 2008), the latter two having symbiotic green algae in the parenchyma.

Unfortunately, in the Catenulida, no model system has been established so far. In the largest group of flatworms, the rhabditophorans, a couple of models have been used extensively, such as *Macrostomum lignano**, *Schmidtea mediterranea**, *Dugesia japonica**, *Girardia tigrina, Polycelis nigra, Echinococcus granulosus**, and *Schistosoma mansoni**. Whole genomes are being sequenced or already available for all species marked with an asterisk.

In recent years, *Schmidtea mediterranea* (Salo *et al.*, 2009) and *Macrostomum lignano* (Ladurner *et al.*, 2005) in particular have emerged as promising model systems for embryonic development (Morris *et al.*, 2004; Sanchez Alvarado, 2003), regeneration (Egger *et al.*, 2006; Forsthoefel and Newmark, 2009; Sanchez Alvarado, 2006), reproductive biology research (Janicke and Scharer, 2009; Newmark *et al.*, 2008), ageing (Mouton *et al.*, 2009), and stem cell research (Ladurner *et al.*, 2008; Palakodeti *et al.*, 2008).

II. Rationale

Here we discuss the most appropriate protocols for thin-section EM for ultrastructure and immuno-cytochemistry studies. All three major systematic groups are covered, including marine and freshwater flatworms. In particular, chemical fixation as well as cryo-based sample processing (high-pressure freezing/freeze-substitution, HPF/FS) is described for the triclads *Polycelis nigra*, *Dugesia gonocephala*, the macrostomorph *Macrostomum lignano*, and the acoel species *Isodiametra pulchra*[1].

III. Methods

A. Chemical Fixation—General Aspects

Simultaneous use of glutaraldehyde and osmium tetroxide (OsO_4), even with low concentrations of OsO_4, gives best results for marine and freshwater species. We recommend cacodylate buffer-based fixatives for marine organisms for avoiding the so-called "fixation pepper" (i.e., artifactual precipitation of electron-dense substances, see Mollenhauer, 1988). The concentration of buffers used is also crucial. For freshwater organisms, buffer concentration should be low (e.g., 0.01–0.05 M) to prevent shrinking artifacts (Gremigni and Falleni, 1991). Some authors, however, recommend higher concentrations for proper buffering (Griffiths, 1993), especially if the primary fixative does not contain OsO_4. For marine organisms, sucrose and/or sodium chloride is added to the cacodylate buffer to increase molarity. Magnesium chloride used for relaxation also provides protection of membranes in marine species (Glauert and Lewis, 1998). However, traces of ions should be added in a millimolar concentration to fixatives for freshwater species. As a complement to aforementioned simultaneous fixation, the consecutive application of glutaraldehyde and OsO_4 also gives acceptable results except for stem cell morphology, and is convenient for "field work". In general, marine organisms are easier to fix than freshwater animals, as long as the fixative's molarity is equivalent to seawater. Another important parameter in addition to the size and the salinity of the natural environment is the presence and amount of water-filled spaces within the body (e.g., chordoid vacuoles in acoels), the latter making

[1] This contribution is meant as a practical instruction for laboratory work. Thus, the literature list is not exhaustive and we apologize to all whose work—though related to the topic—could not be included. For ease of access to relevant literature we cited preferentially original papers and—when necessary—book chapters.

a proper fixation sometimes more difficult to accomplish. Due to their soft body, relaxation of flatworms prior to fixation is mandatory to keep them in a life-like shape and to assure proper orientation for sectioning. Therefore, different relaxation methods are in use to avoid a contraction of the musculature at the beginning of fixation. Magnesium chloride solution isotonic to seawater is used for marine organisms. For freshwater animals, different chemicals have been tested so far (e.g., 2% urethane, 0.1% 1-phenoxy-2-propanol), as well as cooling on ice.

B. Chemical Fixation—Protocols[2]

1. Marine Organisms (e.g., *Macrostomum lignano, Isodiametra pulchra*)

a. Recommended Method Simultaneous fixation with glutaraldehyde and OsO_4 (Eisenman and Alfert, 1982); see Figs. 1A, 2, 4A and 5A.

Buffer A: 0.2 M cacodylate buffer, 0.1 M NaCl; 0.35 M sucrose; pH 7.2
Buffer B: 0.2 M cacodylate buffer, 0.3 M NaCl; pH 7.2
Fixative A: 4% (v/v) glutaraldehyde in 0.2 M cacodylate buffer containing 0.1 M NaCl; 0.35 M sucrose; pH 7.2
Fixative B: 1% (w/v) OsO_4 in 0.2 M cacodylate buffer containing 0.3 M NaCl; pH 7.2
Cocktail: mix 9.5 ml fixative A and 0.5 ml fixative B immediately before use

Fixation procedure:
Fix relaxed animals in cocktail for about 5–10 min on ice. Specimens should become brownish; solution should remain clear. Subsequently, replace cocktail with fixative A. Fix in fixative A at +4°C for 1 h. Wash three times with buffer A and once with buffer B. Fix animals for 1 h with fixative B at +4°C. Wash three times with buffer B, once with buffer B mixed with distilled water (1:1), and once with pure distilled water. Dehydrate specimens in a graded series of either ethanol or acetone (50, 70, 90, and three times 100% at least for 15 min each) and embed in Epon or Spurr's resin.

b. Alternative Method Consecutive fixation with glutaraldehyde and OsO_4.
Buffer A: 0.1 M cacodylate buffer, 9% (w/v) sucrose; pH 7.2–7.4
Buffer B: 0.05 M cacodylate buffer; pH 7.2–7.4
Fixative A: 2.5% (v/v) glutaraldehyde in 0.1 M cacodylate buffer containing 9% (w/v) sucrose; pH 7.2–7.4
Fixative B: 1% (w/v) OsO_4 in 0.05 M cacodylate buffer; pH 7.2–7.4

Fixation procedure:
Fix relaxed animals in fixative A for 1 h at +4°C, wash three times with buffer A (but without sucrose!) 15 min each, postfix in fixative B for 1 h at +4°C, and wash

[2] For ease of reading, we used neutral terms such as "fixative/buffer A,B..." instead of the original nomenclature from literature.

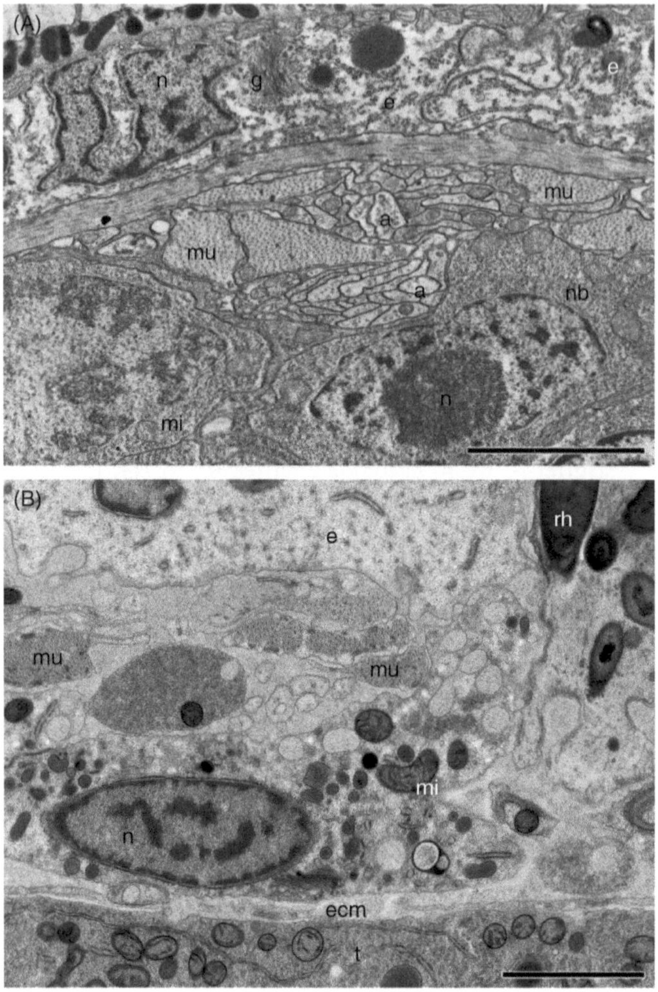

Fig. 1 Comparison of the body wall of *Macrostomum lignano* after chemical fixation and high-pressure freezing (HPF) and freeze-substitution (FS). a, axon; e, epidermal cell; ecm, extracellular matrix; g, Golgi stack; mi, mitochondrion; mu, muscle; n, nucleus; nb, neoblast; rh, rhabdite; t, testis. Scale bar = 2 μm. (A) Chemical fixation (Eisenman and Alfert, 1982). Epidermis, main nerve cord and neoblast, the latter showing the typical morphology with large nucleus and a thin rim of cytoplasm. (B) HPF and FS. Detail of the body wall including proximal parts of the testis.

again three times with buffer B. Dehydrate in an ethanol or acetone series and embed in Epon or Spurr's resin. For "field work", fixation should be carried out up to the last ethanol or acetone dehydration step, followed by sample storage in a refrigerator until resin embedding can be performed.

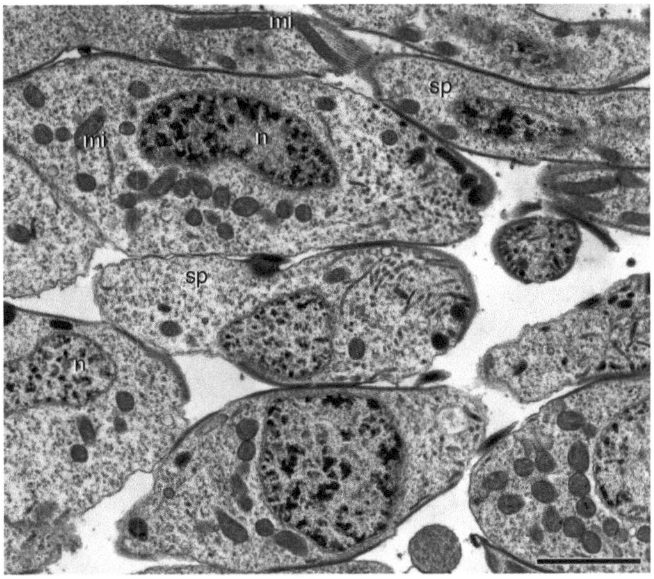

Fig. 2 Early spermatids in the testis of *Macrostomum lignano* after chemical fixation (Eisenman and Alfert, 1982). mi, mitochondrion; n, nucleus; sp, spermatids. Scale bar = 2 μm.

*c. **Fixation for Immuno-Cytochemistry of Marine Organisms*** Buffer: 0.1 M phosphate buffer containing 9% (w/v) sucrose; pH 7.2–7.4

Fixative: 4% (w/v) formaldehyde (freshly prepared from paraformaldehyde) in 0.1 M phosphate buffer containing 9% (w/v) sucrose; pH 7.2–7.4
Fixation procedure:

Fix relaxed animals in formaldehyde fixative for 1 h at +4°C, or on ice, wash three times with phosphate buffer (but without sucrose), dehydrate in alcohol, and embed in an acrylic resin (e.g., LR-White).

2. Freshwater Organisms (triclads, e.g., *Schmidtea mediterranea, Dugesia gonocephala, Polycelis nigra*)

*a. **Recommended Method*** Simultaneous fixation with glutaraldehyde and OsO_4 (Rombout *et al.*, 1978); see Fig. 6.

Buffer: 0.1 M cacodylate buffer; pH 7.2
Fixative A: 2.0% (v/v) glutaraldehyde + 1% (w/v) OsO_4 in 0.1 M cacodylate buffer; pH 7.2
Fixative B: 2% (v/v) glutaraldehyde + 1% (w/v) OsO_4 + 1% (w/v) potassium dichromate in 0.1 M cacodylate buffer; pH 7.2

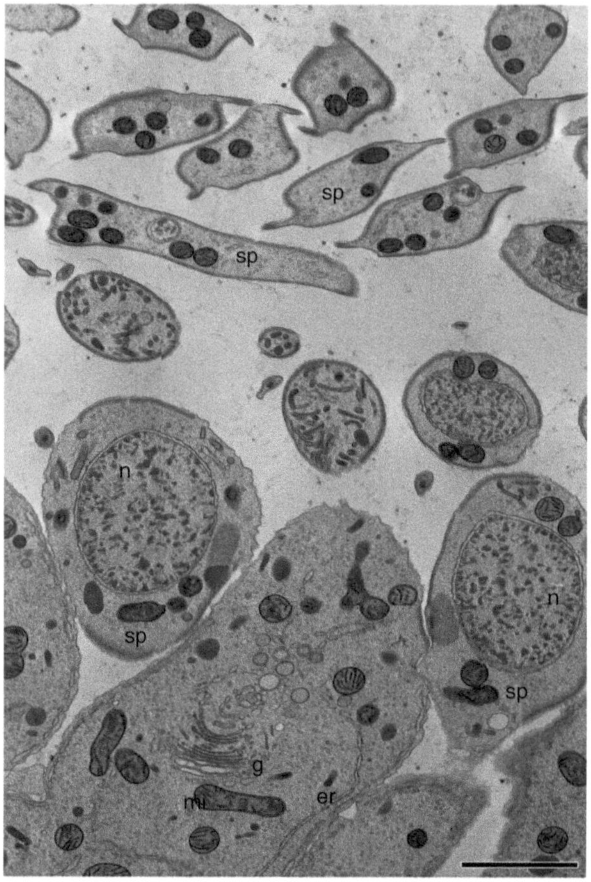

Fig. 3 Testis of *Macrostomum lignano* processed with HPF and FS showing early and late spermatids. er, endoplasmic reticulum; g, Golgi stack; mi, mitochondrion; n nucleus; sp, spermatid. Scale bar = 2 μm.

Fixation procedure:
Fix relaxed animals for 1 h in fixative A on ice. Replace solution without washing with fixative B and fix for 1 h on ice. Wash three times in buffer, dehydrate specimens in an ethanol or acetone series and embed in Epon or Spurr's resin.

*b. **Alternative Method*** Simultaneous fixation with glutaraldehyde and OsO_4 (Campbell, 1987).

Buffer: 0.001 M Tris-HCl, 1 mM $CaCl_2$, 0.1 mM $MgCl_2$, 0.1 mM KCl, 1 mM NaH_2CO_3; pH 7.8 (*Hydra* culture medium)
Fixative: 1% (v/v) glutaraldehyde + 0.2% (w/v) OsO_4 in buffer

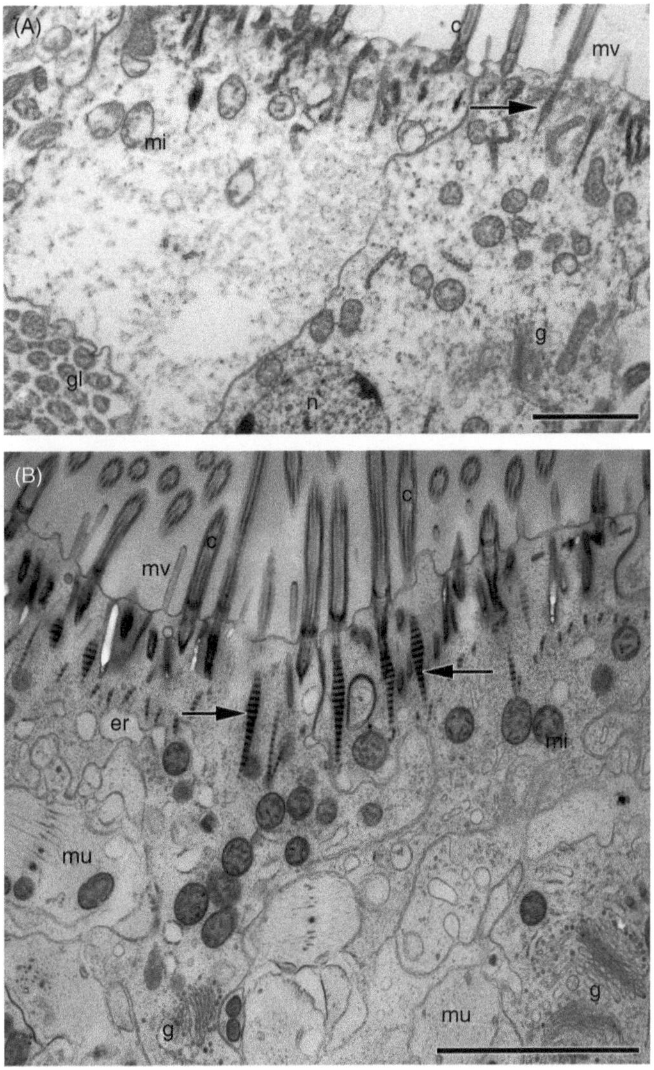

Fig. 4 Body wall of *Isodiametra pulchra* after chemical fixation (A) according to Eisenman and Alfert (1982) and after HPF/FS (B). c, cilium; er, endoplasmic reticulum; g, Golgi stack; gl, gland cell; mi, mitochondrion; mu, muscle; mv, microvilli; n, nucleus. Scale bar = 2 μm. The apical part of the epidermis shows microvilli and basal parts of cilia, the ciliary rootlet system (arrows), and mitochondria. Note in chemically fixed samples (A) the shrinkage of membranes of mitochondria and poor preservation of the Golgi stack as well as artifactual extraction of cytoplasm, as compared to excellent ultrastructure preservation achieved with HPF/FS (B).

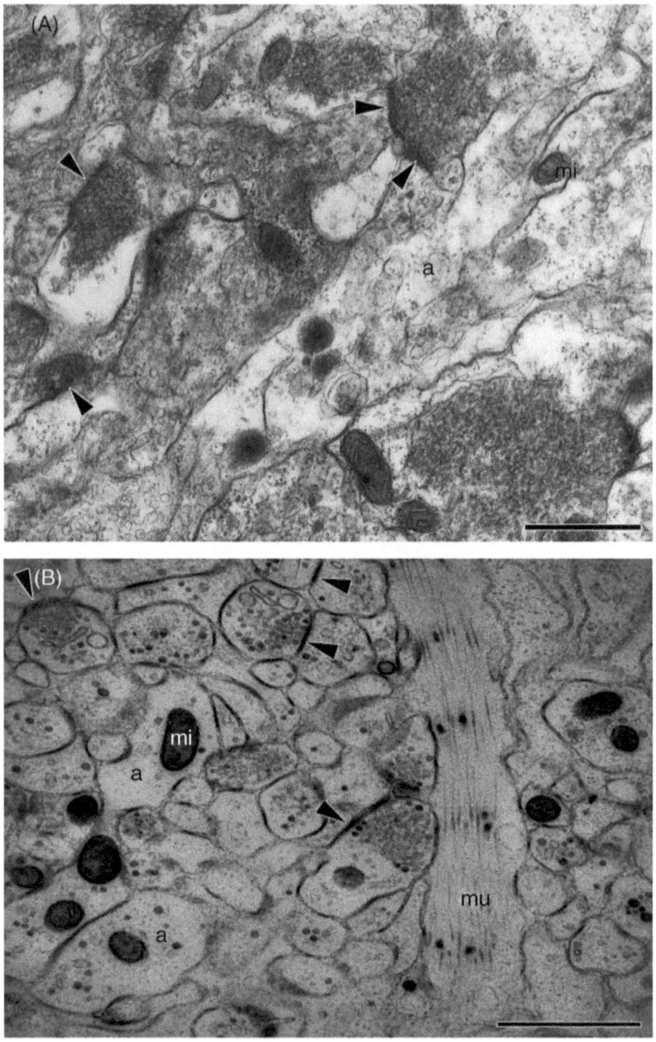

Fig. 5 Detail of the brain of *Isodiametra pulchra* after chemical fixation (A) according to Eisenman and Alfert (1982) and after HPF/FS (B). a, axon; mi, mitochondrion; mu, muscle. Scale bar = 1 μm. Synapses (arrow heads) and clusters of vesicles in their vicinity called synaptic clouds are seen. Note in HPF/FS samples (B) the distinct appearance of the vesicles and the absence of shrinkage of any membranes.

Fixation procedure:

Fix relaxed animals on ice or at room temperature for at least 1 h. Fixation on ice works slower and therefore it is preferred for smaller animals. Due to the low concentration of OsO_4 the solution remains clear even at room temperature. If solution

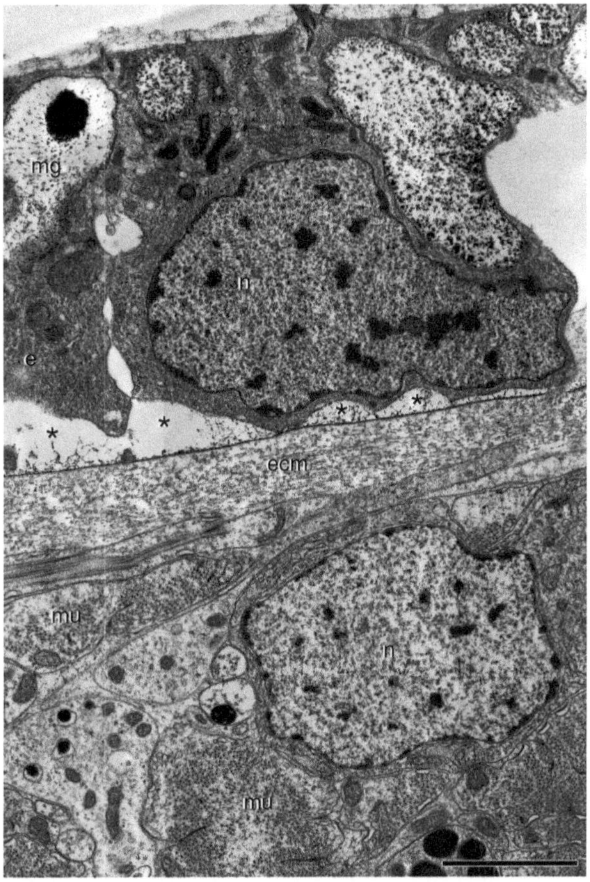

Fig. 6 Juvenile triclad, *Polycelis nigra* after chemical fixation (Rombout *et al.*, 1978). e, epidermal cell; ecm, extracellular matrix; mg, mucous gland; mu, muscle; n, nucleus. Scale bar = 2 μm. Detail of the body wall with a prominent nucleus of an epidermal cell, the extracellular matrix, and a nucleus of a parenchymal cell. Note that soluble components of the extracellular matrix became extracted (asterisks mark these almost empty spaces).

becomes brownish during extended fixation times it should be exchanged. After fixation, wash three times in buffer and dehydrate with either ethanol or acetone and embed in Epon or Spurr's resin.

c. ***Alternative Method*** Simultaneous fixation with glutaraldehyde and OsO_4 modified after Shigenaka *et al.* (1971).

Buffer: 0.06 M phosphate buffer; pH 7.2
Fixative A: 6% (v/v) glutaraldehyde in <0.05 M phosphate buffer, 2 mM sucrose, 0.02 mM magnesium sulfate; pH 7.2 (i.e., mix 2.4 ml 25% glutaraldehyde, 0.2 ml 0.1 M sucrose, 0.2 ml 1 mM magnesium sulfate with 7.2 ml 0.06 M phosphate buffer; pH 7.2)

Fixative B: 2% (w/v) OsO_4 in distilled water
Cocktail: mix equal volumes of fixative A and B immediately before use

Fixation procedure:

Always prepare fresh solutions. Fix relaxed animals on ice for 1 h. High concentrations of glutaraldehyde and sucrose lead to the reduction of OsO_4. Thus, the fixative should be exchanged once becoming dark. After fixation do not wash with buffer but start directly with dehydration in a standard graded series of acetone. Usually the solutions with low concentration of acetone will start to get black during the dehydration process, but solutions with a high concentration of acetone (~100%) should remain clear. Embed in Epon or Spurr's resin.

d. Fixation for Immuno-Cytochemistry of Freshwater Organisms Buffer: 0.05 M phosphate buffer; pH 7.2–7.4

Fixative: 4% (w/v) formaldehyde (freshly prepared from paraformaldehyde) in phosphate buffer

Fixation procedure:

Fix relaxed animals in formaldehyde fixative for 1 h at +4°C, or on ice, wash three times with phosphate buffer (but without sucrose), dehydrate in ethanol, and embed in an acrylic resin.

C. Cryo-Processing—General Aspects

Rapid freezing undoubtedly immobilizes subcellular structures and dynamics much quicker and more reliably than conventional chemical fixation at ambient temperatures (Heuser *et al.*, 1979; McIntosh, 2001; Muller *et al.*, 1980; Sitte *et al.*, 1987). The only freezing method that is suitable for cryo-immobilization of bulky specimens thicker than ~10 μm is HPF (Moor and Riehle, 1968). HPF lowers the freezing point of water, and both crystal growth and nucleation rates are markedly reduced (Müller and Moor, 1984). The maximum thickness of metazoan samples that can be high-pressure frozen without or with negligible amounts of disturbing ice crystal damage ranges from 0.1 to 0.2 mm. This is primarily dependent on the physicochemical and physiological properties of the samples, e.g., their contents of water and natural "cryo-protectants" within the cytoplasm. One of the most feasible ways to process cryo-immobilized samples for cellular EM comprises FS followed by resin embedding at either room or low temperatures; for recent reviews on various other cryo-methods, see Hess (2007) and Mobius (2009). FS of the sample is accomplished by dehydration at very low temperatures through organic solvents, such as ethanol or acetone. For chemical stabilization and staining of the biological sample, fixatives are usually added to the organic solvent. Out of a wide variety of published FS recipes, we consider FS with anhydrous acetone containing 1% OsO_4 (Van Harreveld and Crowell, 1964) plus 0.1–0.2% uranyl acetate (UA) as optimal for morphological studies. This popular cocktail may be further supplemented with 1–4% water (Buser and Walther, 2008; Walther and Ziegler, 2002) for improving the contrast of certain membranes and the cytoskeleton. FS with pure, anhydrous acetone, optionally supplemented with UA, followed by embedment in (meth)acrylate resins can be used for immuno- and lectin-cytochemistry (this aspect is not discussed here; for reviews on this topic, see Hess, 2007; Humbel and Müller, 1986; Schwarz and Humbel, 2007). Typically, FS is performed for ~8 h (Humbel and Müller, 1986) up to several days (Zhang *et al.*, 1993) at −90 to −80°C with the aid of commercial or homemade FS devices or by using dry ice–acetone mixtures (Steinbrecht and Müller, 1987). Likewise, warming-up usually takes ~12–30 h ending with thorough rinsing of the specimens with pure solvent at temperatures between −30 to +20°C (Humbel and Müller, 1986). Finally, specimens are embedded in resin for ultramicrotomy and transmission EM. According to our experiences and the published literature from various cell types and organisms, short and extended protocols work equally well (plants, fungi, mammalian cell cultures, e.g., Hess, 2003 and our unpublished data). A highly similar quality in terms of ultrastructure preservation is obtained, provided the specimens have been cryo-immobilized properly.

D. Cryo-Processing—Protocols

1. Marine and Freshwater Organisms

a. Recommended Method Rapid cryo-immobilization by means of HPF followed by FS and epoxy resin embedding; see Figs. 1B, 3, 4B, 5B, 7 and 8.

Freeze-substitution media Recipe I: Anhydrous acetone containing 1% (w/v) OsO_4 (Van Harreveld and Crowell, 1964) plus 0.1–0.2% (w/v) UA (diluted from a 10% (w/v) stock solution of UA in methanol) is optimal for morphological studies.

Recipe II: FS medium of recipe I, further supplemented with 4% (v/v) water (Walther and Ziegler, 2002).

The fixatives (prechilled OsO_4 crystals, UA stock) are carefully added to the acetone in a beaker at room temperature followed by a few seconds of gentle mixing. Subsequently, cryovials are quickly filled with 1.2–1.5 ml FS media, closed and immediately frozen in upright position with LN_2 to prevent premature degradation. We prefer 2 ml cryovials with screw-caps, plus the appropriate cryoracks with locking bottoms that match the cryovials since they permit one-hand operation (for detailed instructions how to make the cocktail see Hess, 2007; McDonald and Muller-Reichert, 2002).

High-pressure freezing procedure Worms are pipetted with their natural medium (seawater, freshwater) into an embryo dish and subsequently kept on ice for 15–30 min to slow down the animals' activity. Next, the cup-shaped aluminum HPF specimen carriers (cavity depth: 0.1–0.2 mm) are slightly overfilled with a droplet of freshwater or seawater containing worms[3]. A sapphire coverslip (diameter 3 mm, thickness 50 μm) is

[3] Our experience and, thus, recommendations given here relate to "BAL-TEC HPM-010-type" HPF machines but not to the "EM-PACT-type" instruments, which have slightly different and smaller specimen carriers.

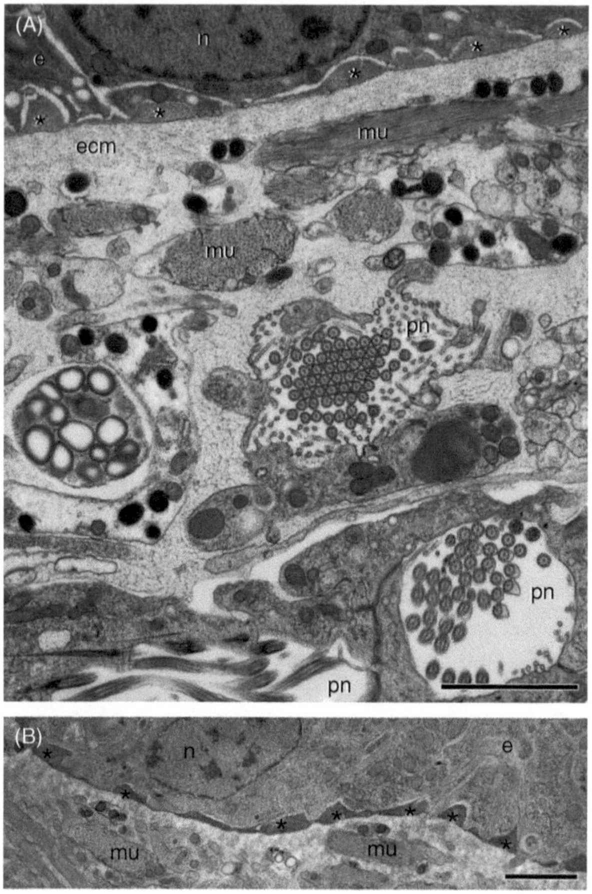

Fig. 7 Body wall of triclads processed with HPF and FS. e, epidermal cell; ecm, extracellular matrix; mu, muscle; n, nucleus; pn, protonephridium. Scale bar = 2 μm. Asterisks mark components of the extracellular matrix, which are usually extracted by aqueous chemical fixation and/or dehydration at ambient temperature (see Fig. 6) but remained well preserved by HPF/FS; note their characteristic pattern at the base of the epidermal cells. (A) *Polycelis nigra*, juvenile. (B) *Dugesia gonocephala.*

immediately placed on top of the droplet and excess water is soaked with filter paper so that the specimen carrier is sealed. This "mini-aquarium" is then transferred with tweezers into the sample holder of the HPF apparatus. A wetted, 0.3-mm-deep aluminum carrier is placed with its flat side down on top of the sapphire[4]. Finally, the sample holder

[4] Clearly, the additional sapphire disk does increase the thickness of the sandwich, and thus slightly delays rapid freezing. However, the transparent sapphire lid allows for full visual control of the whole loading process so that the light-shunning animals cannot escape unobserved from the carrier. Moreover, accidentally trapped air bubbles, that might impede proper freezing or damage the sample, are easy to detect and loading can be repeated.

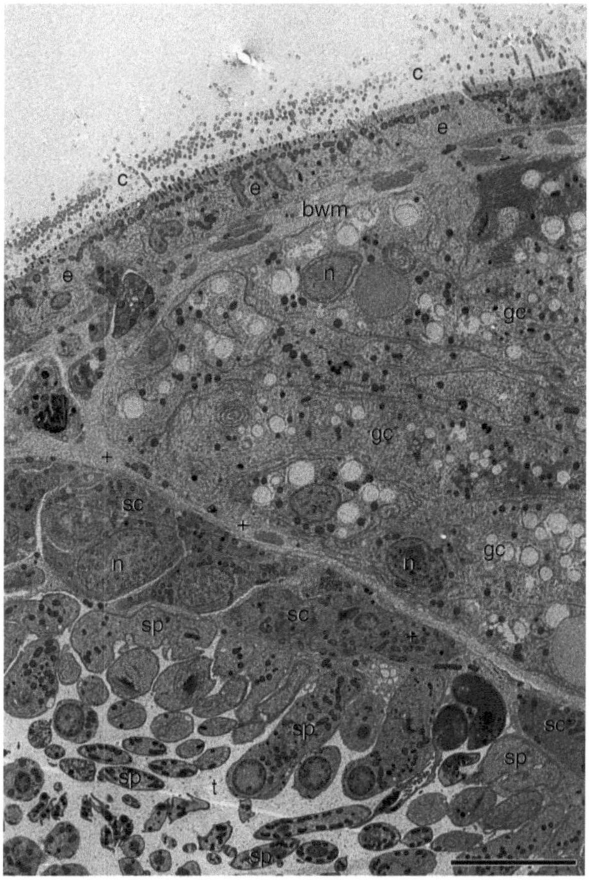

Fig. 8 *Macrostomum lignano* (HPF/FS). Gross anatomy of an adult animal showing epidermis, musculature, gut, and testis in a low magnification electron photomicrograph. The testis is ventral and the gut dorsal to the extracellular matrix that is marked by crosses (+). Scale bar = 10 μm. bwm, body wall musculature; c, cilia; e, epidermis; gc, gut cell; n, nucleus; sc, spermatocyte; sp, spermatid; t, testis lumen.

is closed and the "sandwich" is immediately cryo-immobilized by HPF. Frozen sandwiches are subjected to FS or stored in LN_2 for later use. (Larger animals, e.g., freshwater triclads must be cut with a sharp scalpel blade into appropriate pieces to fit into the HPF carriers.)

Freeze-substitution procedure and embedding The cryo-immobilized samples are transferred under LN_2 with precooled tweezers into the cryovials containing the frozen FS cocktails. Subsequently, the lids are loosely screwed onto the vials to permit safe evaporation of excess N_2 gas. Finally, the vials are placed into the precooled FS container and after about 1 h the lids are tightened and FS is started.

A feasible example for FS (to be performed over the weekend) reads as follows: FS for 36 h at −90°C (or at −80°C, if a dry ice–acetone mixture is used instead of an FS device (Steinbrecht and Müller, 1987)), warming up to −55°C at a rate of 5°C per hour, subsequently postfixation and staining at −55°C for 6 h, followed by warming up to −30°C at a rate of 5°C per hour where samples are left for additional ~8–12 h. The step-wise warming-up should allow fixatives/stains included in the FS media to stabilize the specimens at low temperatures (for details on the reactivity of selected fixatives at low temperatures see the following reviews: Hess, 2007; Humbel, 2009). Finally, the specimens are allowed to reach room temperature, rinsed three times with pure solvent (10 min each), and embedded in any epoxy resin. We avoid embedding in Spurr's mixture (Spurr, 1969) as this may solubilize and extract subcellular constitutents of FS specimens (Hess, 1990). Ultrathin sections of FS samples were viewed without poststaining, whereas chemically fixed samples were poststained with UA and lead.

IV. Materials

A. Animals

The rhabditophoran flatworm *Macrostomum lignano*, originally sampled in the Mediterranean (Ladurner *et al.*, 2005), and the acoel flatworm *Isodiametra pulchra* from the Atlantic coast of Maine (Smith and Bush, 1991) are kept in permanent laboratory cultures at the University of Innsbruck since 1995 and 2003, respectively, and are being fed with the diatom *Nitzschia curvilineata*.

The freshwater tricladid rhabditophorans *Polycelis nigra* and *Dugesia gonocephala* were sampled in creeks and ponds in Lower Austria in May 2009 and in a creek near the University of Innsbruck in spring 2009, respectively, and have since been kept in laboratory cultures on veal liver.

B. Chemical Fixation

1. Chemicals (be aware that most of the reagents are more or less toxic and/or hazardous to health; for their safe use and disposal consult the relevant Material Safety Data Sheets)

0.05–0.2 M cacodylate buffer (contains arsenic compounds!), optionally supplemented with varying concentrations of sucrose.

0.05–0.1 M phosphate buffer, optionally supplemented with varying concentrations of sucrose and/or magnesium sulfate.

Hydra culture medium: 0.001 M Tris-HCl, 1 mM $CaCl_2$, 0.1 mM $MgCl_2$, 0.1 mM KCl, 1 mM NaH_2CO_3; pH 7.8 (all Sigma-Aldrich or Merck).

Urethane and 1-phenoxy-2-propanol (Sigma-Aldrich).

Paraformaldehyde, potassium dichromate, as well as glutaraldehyde, OsO_4 crystals, epoxy resins (Epon, Spurr's) from Sigma-Aldrich, Agar (Stansted, U.K.), EMS (Hatfield, PA, U.S.A.), Polysciences (Warrington, PE, U.S.A.), or Ted Pella (Redding, CA,

U.S.A.), LR-White acrylic resin from Sigma-Aldrich or London Resin Co (Woking, Surrey, U.K.).

2. Tools

For specimen handling, embryo dishes (preferably black, if used for small, quasi-transparent species) or crystallization dishes (e.g., from Agar) and fine glass or small plastic pipettes were used.

C. Cryo-Processing

1. Chemicals

OsO_4 crystals, uranyl acetate, epoxy resin (Epon) from Sigma-Aldrich, Agar, EMS, Polysciences, or Ted Pella.

2. Tools

Embryo dishes or crystallization dishes and fine glass pipettes were used for animal handling. Sapphire coverslips (diameter 2 mm; usually used for HPF of adherent 2D cell cultures, see Hess *et al.*, 2010, Chapter 27, this volume) were from Engineering Office Martin Wohlwend (CH-9466 Sennwald, Switzerland; martin-wohlwend@bluewin.ch) and are also available from Leica Microsystems (Vienna, Austria).

Several types of aluminum carriers for HPF were from Martin Wohlwend Engineering or Leica.

For FS, we recommend 2 ml cryovials (e.g., Nalgene, from Nalge Nunc International, Rochester, NY; note that these vials are not explicitly LN_2-certified!).

Tweezers, scalpels, slot, or mesh grids (copper) coated with formvar and carbon were all from Agar, EMS, Polysciences, or Ted Pella.

3. Instruments

Further EM instruments used in this study: dissecting microscope (i.e., Leica EM workstation), HPF apparatus HPM-010 (from BAL-TEC, Balzers, Liechtenstein; note that this instrument is currently sold by ABRA-Fluid AG, Widnau, Switzerland), automated FS device AFS (Leica), ultra-microtome Ultracut S, diamond knives for ultra-microtomy (from DIATOME, Biel, Switzerland), transmission EM Libra 120 EFTEM (from Zeiss, Oberkochen, Germany), or Philips CM120 TEM (from F.E.I., Eindhoven, The Netherlands).

V. Results and Discussion

A. Chemical Fixation

In our hands, simultaneous fixation with glutaraldehyde and OsO_4 gives the best results, especially for stem cell research. The generally accepted criteria for a good fixation for EM (Hayat, 2000) are: clearly contrasted membranes, including the nuclear envelope and mitochondrial membranes, and a so-called "full cytoplasm" (e.g., Figs. 1A and 2). Inadequate fixation causes membrane undulations or interruptions, shrinkage or swelling of mitochondria or endomembrane compartments, such as the Golgi apparatus (e.g., Fig. 4A vs. Fig. 4B), as well as the (partial) loss of the matrix of organelles such as mitochondria or peroxisomes (not shown). Extraction of easily soluble compounds of the cytoplasm often leads to the appearance of an "empty cytoplasm" (Fig. 4A).

A proper chemical fixation allows the study of the gross anatomy of both marine and freshwater flatworms at the ultrastructural level, including more voluminous species (Figs. 1A, 2, and 6). Research on stem cells and regeneration as well as studies on changes of subcellular structures after gene silencing by RNAi (RNA interference) are routinely possible (De Mulder *et al.*, 2009). However, the preservation of certain structures still remains unsatisfactory especially in comparison to cryo-processed samples. Cellular structures with high concentrations of chemically bound water, such as extracellular matrix, show alterations and shrinkage (Fig. 6 vs. Fig. 7). The amorphous "ground substance" only is affected, but not the fibers. Depicting the structures underlying certain physiological processes, such as liquid transport by vesicles, is not always possible. In many cases, these processes are too fast to be properly immobilized by chemical fixation. The preservation of the nervous system is also frequently suboptimal. The ultrastructure of the different vesicle types, like dense core and lucent vesicles but also synaptic sites, are adequately preserved thus allowing distinction and functional analysis of the different axons types. However, the shape of axons and the composition of the microtubular and neuronal fiber network differ in comparison to cryo-processed specimens. As an example, axons in chemically fixed samples do not appear round (Figs. 1A and 5A vs. Fig. 5B) and usually microtubules and neuronal fibers are not equally distributed.

Out of the various immuno-cytochemical methods available, immuno-gold labeling of resin sections has already been successfully applied to flatworm research (Bode *et al.*, 2006; Pascolini *et al.*, 1992). Depending on the antigen and the quality of the antibody, even simultaneous or consecutive fixation with glutaraldehyde and OsO_4 is occasionally possible (e.g., the immuno-labeling of BrdU (Bode *et al.*, 2006)). Usually, formaldehyde fixation and embedding in an acrylic resin gives satisfactory results. For optimal tissue preservation and antigen localization, cryo-processing should be considered.

B. Cryo-Processing

"Representatives" from all three major systematic groups of nonparasitic flatworms were processed by means of HPF and FS. In part, excellent or at least adequate results were obtained (Figs. 1B, 3, 4B, 5B and 8, and Fig. 7, respectively). The ultrastructural patterns observed were fully consistent with those reported in the literature from cryo-processed nematodes, e.g., *C. elegans* (McDonald, 2007; Muller-Reichert *et al.*, 2003; O'Toole *et al.*, 2003), as well as other metazoa including mammalian systems. Briefly, the outlines of any cell type under investigation appeared smooth and not undulated, and so were the membranes of the generally turgescent organelles. In case of suboptimal freezing, however, membranes showed somewhat slightly bent or "zigzag" patterns, characteristic for minor ice crystal damage. Whether an artifact is acceptable or not depends on its intensity and on the scientific question under investigation. Matrix constituents (i.e., extracellular matrix, mitochondrial matrix, contents of Golgi cistern/vesicles, secretory vesicles of various glands) were generally much better preserved than after conventional chemical fixation, thus, stained more intensively (e.g., Figs. 4B and 7 vs. Figs. 4A and 6). The cytoskeleton, too, proved well preserved and clearly contrasted in our preparations. No major differences between anhydrous and "watery" FS media were apparent, except that the triclad *P. nigra* showed very low specimen contrast after anhydrous FS (not shown). Adequate staining of membranes and cytoskeletons was achieved here by adding water to the FS media (e.g., Fig. 7A). Taken together, we consider the preservation quality and the yield of usable specimens obtained with small-sized animals, such as *M. lignano* and *I. pulchra* as very good (Figs. 1B, 3, 4B, 5B and 8).

Concerning, however, the preparation of more voluminous flatworms, the following steps need further optimization. Most importantly, the freezing quality achieved with thicker animals, such as triclads, is not perfect yet (e.g., Fig. 7). The greatly reduced cooling rate in the center of bulky objects is an inherent problem of specimens thicker than ~200 μm (Studer *et al.*, 2008). Nevertheless, improvements should be possible by reducing the volume of the whole sandwich since the mass of aluminum specimen carriers counts as well. For the "mini-aquarium" described here, for example, one could use a suitable spacer ring instead of the solid lid on top of the sapphire disk, in analogy to the two-sapphire sandwich described elsewhere (Hawes *et al.*, 2007). An alternative could be the use of custom-made specimen carriers (Craig *et al.*, 1987). Standard carriers for the BAL-TEC HPM010 (and similar instruments) have a 200-μm-thick bottom. Carriers with a thinner bottom, so-called "membrane carriers", were developed for the EM PACT2 (McDonald, 2007) and quite thin steel foils may be suitable as well (Sawaguchi *et al.*, 2008). Furthermore, it remains to be tested whether it is possible to enclose the animals within the sandwich in physiologically appropriate media other than water. Usually, liquids with a lower freezing point and/or better heat conductivity than water or buffer serve as so-called space "fillers" to improve the rapid freezing process (Studer *et al.*, 1989). Hexadecene (a paraffin oil) is widely used, especially for plants and fungi (Hess, 2007; Studer *et al.*, 1989) but is definitely not suitable for flatworms. Hexadecene quickly disintegrates the flatworms' epidermis and parenchyma (our unpublished data) similar to its negative effects on nematodes (Hohenberg *et al.*, 1994). Concentrated bovine serum albumine solutions (~20% w/v), for example (McDonald *et al.*, 2007), has proven as to be apt for various cell types/organisms, including marine sponges (McDonald *et al.*, 2007) and ctenophores (T. Müller-Reichert, pers. comm.). However, negative effects of this reagent were also observed, in particular with certain triclads (T. Müller-Reichert, pers. comm.). Thus, a suitable filler remains to be found by systematic testing the various reagents employed for other organisms. Finally, it has to be mentioned that large-sized flatworms frequently stick firmly within the cavity of the specimen carrier. This leads to accidental freeze-fracturing (especially at the level of the epidermis) when the sandwich is removed from the sample holder or subsequently opened. Likewise, it is almost impossible to peel stuck worms off the carrier's cavity after FS or resin infiltration without damaging the sample. Polymerizing the sample still contained within the aluminum carrier solves this problem. After polymerization, one must only free the carrier's rim from excess resin by thorough, careful trimming with a razor blade, followed by immersion in liquid nitrogen. Sometimes it takes several freeze–thaw cycles until the resin block with the sample falls off the carrier. Finally, the sample is re-embedded (see also Sawaguchi *et al.*, 2003 in this context). Alternatively, coating the carriers and the sapphire with lecithin (Craig *et al.*, 1987) is worth a try so that the sample is released already during FS.

VI. Concluding Remarks

The major advantages of chemical fixation are its low costs and ease of handling. It is the method of choice for routine work, "fieldwork", and for the fixation of large specimens. The most serious disadvantage of chemical fixation is the well-known problem of fixation artifacts (Bowers and Maser, 1988; Strangeways and Canti, 1927). The degree of damage depends on the type and physiological condition of the cells/tissue/organism under investigation.

Cryo-preparation techniques, namely, HPF and FS, were only recently introduced to flatworm ultrastructure research (Egger *et al.*, 2009a). Thus, we are at the very beginning of exploiting their potential for this group of organisms. A well-established, straightforward approach for (immuno-)cytochemistry that should be applied to flatworms is FS followed by embedding in acrylic resin, as described for various organisms (Humbel and Müller, 1986; Schwarz and Humbel, 2007). Another promising method serving the same purpose is HPF/FS followed by sample rehydration and Tokuyasu cryosection immuno-labeling (van Donselaar *et al.*, 2007); this somewhat more laborious method has meanwhile been optimized and also employed for flies and nematodes (Ripper *et al.*, 2008; Stierhof *et al.*, 2009). Finally, it is conceivable that SDS-digested freeze-fracture replica labeling (Fujimoto, 1995; Kaufmann *et al.*, 2009) is suitable for flatworm research as well, as made likely by classical freeze-fracture studies (Rieger *et al.*, 1991).

Acknowledgments

We thank Karin Gutleben for the excellent technical assistance and Thomas Müller-Reichert for communicating unpublished observations. P. Ladurner is supported by grant P-18099 from FWF Austrian Science Funds; B. Egger is supported by a Sparkling Science grant (funded by the Austrian Ministry of Science and Research); and M.W. Hess is supported by grants from FWF Austrian Science Funds (P-19486-B12), Österreichische Nationalbank-Jubiläumsfonds (P-11050), and Tiroler Wissenschaftsfonds (P-UNI-0404/100).

References

Åkesson, B., Gschwentner, R., Hendelberg, J., Ladurner, P., Müller, H., and Rieger, R. (2001). Fission in *Convolutriloba longifissura*: Asexual reproduction in acoelous turbellarians revisited. *Acta Zool. (Stockh)* **82**, 231–239.

Bode, A., Salvenmoser, W., Nimeth, K., Mahlknecht, M., Adamski, Z., Rieger, R. M., Peter, R., and Ladurner, P. (2006). Immunogold-labeled S-phase neoblasts, total neoblast number, their distribution, and evidence for arrested neoblasts in *Macrostomum lignano* (Platyhelminthes, Rhabditophora. *Cell Tissue Res.* **325**(3), 577–587.

Bowers, B., and Maser, M. (1988). Artifacts in fixation for electron microscopy. *In* "Artifacts in Biological Electron Microscopy." (R. F. E. Crang, and K. L. Klomparens, eds.) pp. 13–42. Plenum Press, NY, London.

Buser, C., and Walther, P. (2008). Freeze-substitution: The addition of water to polar solvents enhances the retention of structure and acts at temperatures around −60 degrees C. *J. Microsc.* **230**(Pt 2), 268–277.

Campbell, R. D. (1987). Organization of the nematocyst battery in the tentacle of *Hydra*: Arrangement of the complex anchoring junctions between nematocytes, epithelial cells, and basement membrane. *Cell Tissue Res.* **249**, 647–655.

Craig, S., Gilkey, J. C., and Staehelin, L. A. (1987). Improved specimen support cups and auxiliary devices for the balzers high pressure freezing apparatus. *J. Microsc.* **148**(Pt 1), 103–106.

Dalyell, J. G. (1814). "Observations on Some Interesting Phenomena in Animal Physiology Exhibited by Several Species of Planariae." Archibald Constable & Co., Edinburgh.

De Mulder, K., Kuales, G., Pfister, D., Willems, M., Egger, B., Salvenmoser, W., Thaler, M., Gorny, A., Hrouda, M., Borgonie, G., and Ladurner, P. (2009). Characterization of the stem cell system of the acoel *Isodiametra pulchra*. *BMC Dev. Biol.* **9**, 69.

Egger, B., Gschwentner, R., Hess, M. W., Nimeth, K. T., Adamski, Z., Willems, M., Rieger, R., and Salvenmoser, W. (2009a). The caudal regeneration blastema is an accumulation of rapidly proliferating stem cells in the flatworm *Macrostomum lignano*. *BMC Dev. Biol.* **9**, 41.

Egger, B., Gschwentner, R., and Rieger, R. (2007). Free-living flatworms under the knife: Past and present. *Dev. Genes Evol.* **217**(2), 89–104.

Egger, B., Ladurner, P., Nimeth, K., Gschwentner, R., and Rieger, R. (2006). The regeneration capacity of the flatworm *Macrostomum lignano*—on repeated regeneration, rejuvenation, and the minimal size needed for regeneration. *Dev. Genes Evol.* **216**(10), 565–577.

Egger, B., Steinke, D., Tarui, H., De Mulder, K., Arendt, D., Borgonie, G., Funayama, N., Gschwentner, R., Hartenstein, V., Hobmayer, B., Hooge, M., Hrouda, M., et al., (2009b). To be or not to be a flatworm: The acoel controversy. *PLoS One* **4**(5), e5502.

Ehlers, U. (1985). "Das Phylogenetische System der Plathelminthes." Gustav Fischer, Stuttgart.

Eisenman, E. A., and Alfert, M. (1982). A new fixation procedure for preserving the ultrastructure of marine invertebrate tissues. *J. Microsc.* **125**, 117–120.

Forsthoefel, D. J., and Newmark, P. A. (2009). Emerging patterns in planarian regeneration. *Curr. Opin. Genet. Dev.* **19**(4), 412–420.

Fujimoto, K. (1995). Freeze-fracture replica electron microscopy combined with SDS digestion for cytochemical labeling of integral membrane proteins. Application to the immunogold labeling of intercellular junctional complexes. *J. Cell. Sci.* **108**(Pt 11), 3443–3449.

Gaerber, C. W., Salvenmoser, W., Rieger, R. M., and Gschwentner, R. (2007). The nervous system of *Convolutriloba* (acoela) and its patterning during regeneration after asexual reproduction. *Zoomorphology* **126**, 73–87.

Glauert, A. M., and Lewis, P. R. (1998). "Biological Specimen Preparation for Transmission Electron Microscopy." Portland Press, London.

Gremigni, V., and Falleni, A. (1991). Ultrastructural features of cocoon-shell globules in the vitelline cells of neoophoran platyhelminths. *Hydrobiologia* **227**, 105–111.

Griffiths, G. (1993). "Fine Structure Immunocytochemistry." Springer, Berlin.

Hawes, P., Netherton, C. L., Mueller, M., Wileman, T., and Monaghan, P. (2007). Rapid freeze-substitution preserves membranes in high-pressure frozen tissue culture cells. *J. Microsc.* **226**(Pt 2), 182–189.

Hayat, M. A. (2000). "Principles and Techniques of Electron Microscopy. Biological Applications." Cambridge University Press, Cambridge.

Hejnol, A., and Martindale, M. Q. (2008). Acoel development supports a simple planula-like urbilaterian. *Philos. Trans. R. Soc. Lond., B, Biol. Sci.* **363**(1496), 1493–1501.

Hejnol, A., Obst, M., Stamatakis, A., Ott, M., Rouse, G. W., Edgecombe, G. D., Martinez, P., Baguna, J., Bailly, X., Jondelius, U., Wiens, M., Muller, W. E., et al., (2009). Assessing the root of bilaterian animals with scalable phylogenomic methods. *Proc. Biol. Sci.* **276**(1677), 4261–4270.

Hess, M. W. (1990). Spurr-resin affects the ultrastructure of freeze-fixed/freeze-substituted pollen grains. *Ultramicroscopy* **34**, 219–221.

Hess, M. W. (2003). Of plants and other pets: Practical aspects of freeze-substitution and resin embedding. *J. Microsc.* **212**(Pt 1), 44–52.

Hess, M. W. (2007). Cryopreparation methodology for plant cell biology. *Methods Cell Biol.* **79**, 57–100.

Hess, M. W., Pfaller, K., Ebner, H. L., Beer, B., Hekl, D., and Seppi, T. (2010). 3D versus 2d cell culture: Implications for electron microscopy. *Methods Cell Biol*, **96**, 617–637.

Heuser, J. E., Reese, T. S., Dennis, M. J., Jan, Y., Jan, L., and Evans, L. (1979). Synaptic vesicle exocytosis captured by quick freezing and correlated with quantal transmitter release. *J. Cell Biol.* **81**(2), 275–300.

Hohenberg, H., Mannweiler, K., and Muller, M. (1994). High-pressure freezing of cell suspensions in cellulose capillary tubes. *J. Microsc.* **175**(Pt 1), 34–43.

Humbel, B. M. (2009). Freeze-substitution. *In* "Handbook of Cryo-Preparation Methods for Electron Microscopy." (A. Cavalier, D. Spehner, and B. M. Humbel, eds.) pp. 320–341. CRC Press, Boca Raton, FL.

Humbel, B., and Müller, M. (1986). Freeze substitution and low temperature embedding. *In* "The Science of Biological Specimen Preparation 1985." (J. J. Wolosewick, ed.) pp. 175–183. SEM Inc AMF O'Hare, Chicago.

Inoue, T., Kumamoto, H., Okamoto, K., Umesono, Y., Sakai, M., Sanchez Alvarado, A., and Agata, K. (2004). Morphological and functional recovery of the planarian photosensing system during head regeneration. *Zoolog. Sci.* **21**(3), 275–283.

Janicke, T., and Scharer, L. (2009). Determinants of mating and sperm-transfer success in a simultaneous hermaphrodite. *J. Evol. Biol.* **22**(2), 405–415.

Kaufmann, W. A., Ferraguti, F., Fukazawa, Y., Kasugai, Y., Shigemoto, R., Laake, P., Sexton, J. A., Ruth, P., Wietzorrek, G., Knaus, H. G., Storm, J. F., and Ottersen, O. P. (2009). Large-conductance calcium-activated potassium channels in Purkinje cell plasma membranes are clustered at sites of hypolemmal microdomains. *J. Comp. Neurol.* **515**(2), 215–230.

Ladurner, P., Egger, B., De Mulder, K., Pfister, D., Kuales, G., Salvenmoser, W., and Schärer, L. (2008). The stem cell system of the basal flatworm *Macrostomum lignano*. *In* "Stem Cells: From Hydra to Man." (T. C. Bosch, ed.) pp. 75–94. Springer Science + Business Media B.V, Heidelberg.

Ladurner, P., and Rieger, R. (2000). Embryonic muscle development of *Convoluta pulchra* (turbellaria-acoelomorpha, platyhelminthes). *Dev. Biol.* **222**(2), 359–375.

Ladurner, P., Scharer, L., Salvenmoser, W., and Rieger, R. M. (2005). *Macrostomum lignano*, n. sp. (Rhabditophora, Macrostomorpha): A new model organism among the lower Bilateria and the use of digital video microscopy in taxonomy of meiobenthic Platyhelminthes. *J. Zool. Syst. Evol. Res.* **43**(2), 114–126.

McDonald, K. (2007). Cryopreparation methods for electron microscopy of selected model systems. *Methods Cell Biol.* **79**, 23–56.

McDonald, K. L., Morphew, M., Verkade, P., and Muller-Reichert, T. (2007). Recent advances in high-pressure freezing: Equipment- and specimen-loading methods. *Methods Mol. Biol.* **369**, 143–173.

McDonald, K., and Muller-Reichert, T. (2002). Cryomethods for thin section electron microscopy. *Methods Enzymol.* **351**, 96–123.

McIntosh, J. R. (2001). Electron microscopy of cells: A new beginning for a new century. *J. Cell Biol.* **153**(6), F25–F32.

Mobius, W. (2009). Cryopreparation of biological specimens for immunoelectron microscopy. *Ann. Anat.* **191**(3), 231–247.

Mollenhauer, H. H. (1988). Artifacts caused by dehydration and epoxy embedding in transmission electron microscopy. *In* "Artifacts in Biological Electron Microscopy." (R. F.E. Crang, and K. L. Klomparens, eds.) pp. 43–64. Plenum Press, NY, London.

Montgomery, J. R., and Coward, S. J. (1974). On the minimal size of a planarian capable of regeneration. *Trans. Am. Microsc. Soc.* **93**(3), 386–391.

Moor, H., and Riehle, U. (1968). Snap-freezing under high pressure: A new fixation technique for freeze etching. *In* "Electron Microscopy 1968. Pre-Congress Abstracts of Papers Presented at the Fourth European Regional Conference. Rome, September 1–7. Vol. II." (D. A. Bocciarelli, ed.) pp. 33–34. Tipografia Poliglotta Vaticana, Rome.

Moreno, E., Nadal, M., Baguna, J., and Martinez, P. (2009). Tracking the origins of the bilaterian Hox patterning system: Insights from the acoel flatworm *Symsagittifera roscoffensis*. *Evol. Dev.* **11**(5), 574–581.

Morris, J., Nallur, R., Ladurner, P., Egger, B., Rieger, R., and Hartenstein, V. (2004). The embryonic development of the flatworm *Macrostomum* sp. *Dev. Genes Evol.* **214**(5), 220–239.

Mouton, S., Willems, M., Braeckman, B. P., Egger, B., Ladurner, P., Scharer, L., and Borgonie, G. (2009). The free-living flatworm *Macrostomum lignano*: A new model organism for ageing research. *Exp. Gerontol.* **44**(4), 243–249.

Muller, M., Meister, N., and Moor, H. (1980). Freezing in a propane jet and its application in freeze-fracturing. *Mikroskopie* **36**(5-6), 129–140.

Muller-Reichert, T., Hohenberg, H., O'Toole, E. T., and McDonald, K. (2003). Cryoimmobilization and three-dimensional visualization of *C. elegans* ultrastructure. *J. Microsc.* **212**(Pt 1), 71–80.

Müller, M., and Moor, H. (1984). Cryofixation of thick specimens by high-pressure freezing. *In* "The Science of Biological Specimen Preparation." (G. H. Haggins, ed.) pp. 131–138. SEM Inc. AMF O'Hare, Chicago.

Newmark, P. A., Wang, Y., and Chong, T. (2008). Germ cell specification and regeneration in planarians. *Cold Spring Harb. Symp. Quant. Biol.* **73**, 573–581.

O'Toole, E. T., McDonald, K. L., Mantler, J., McIntosh, J. R., Hyman, A. A., and Muller-Reichert, T. (2003). Morphologically distinct microtubule ends in the mitotic centrosome of *Caenorhabditis elegans*. *J. Cell Biol.* **163**(3), 451–456.

Palakodeti, D., Smielewska, M., Lu, Y. C., Yeo, G. W., and Graveley, B. R. (2008). The PIWI proteins SMEDWI-2 and SMEDWI-3 are required for stem cell function and piRNA expression in planarians. *RNA* **14**(6), 1174–1186.

Pascolini, R., Panara, F., Di Rosa, I., Fagotti, A., and Lorvik, S. (1992). Characterization and fine-structural localization of actin- and fibronectin-like proteins in planaria (*Dugesia lugubris* s.l.). *Cell Tissue Res.* **267** (3), 499–506.

Rieger, R. M., Tyler, S., Smith, J. P.S. I., and Rieger, G. E. (1991). Platyhelminthes: Turbellaria. *In* "Microscopic Anatomy of Invertebrates. Vol. 3: Platyhelminthes and Nemertinea." (F. W. Harrison, and B. J. Bogitsh, eds.) pp. 7–140. Wiley-Liss, NY.

Ripper, D., Schwarz, H., and Stierhof, Y. D. (2008). Cryo-section immunolabelling of difficult to preserve specimens: Advantages of cryofixation, freeze-substitution and rehydration. *Biol. Cell* **100** (2), 109–123.

Rombout, J. H., Lamers, C. H., and Hanstede, J. G. (1978). Enteroendocrine APUD cells in the digestive tract of larval *Barbus conchonius* (Teleostei, Cyprinidae). *J. Embryol. Exp. Morphol.* **47**, 121–135.

Salo, E., Abril, J. F., Adell, T., Cebria, F., Eckelt, K., Fernandez-Taboada, E., Handberg-Thorsager, M., Iglesias, M., Molina, M. D., and Rodriguez-Esteban, G. (2009). Planarian regeneration: Achievements and future directions after 20 years of research. *Int. J. Dev. Biol.* **53**, 1317–1327.

Sanchez Alvarado, A. (2003). The freshwater planarian *Schmidtea mediterranea*: Embryogenesis, stem cells and regeneration. *Curr. Opin. Genet. Dev.* **13**(4), 438–444.

Sanchez Alvarado, A. (2006). Planarian regeneration: Its end is its beginning. *Cell* **124**(2), 241–245.

Sawaguchi, A., Aoyama, F., Ide, S., Goto, Y., and Suganuma, T. (2008). A new device for high-pressure freezing of cultured cell monolayer using 10-microm-thin stainless discs as both culture plate and specimen carrier. *J. Electron Microsc. (Tokyo)* **57**(6), 203–206.

Sawaguchi, A., Yao, X., Forte, J. G., and McDonald, K. L. (2003). Direct attachment of cell suspensions to high-pressure freezing specimen planchettes. *J. Microsc.* **212**(Pt 1), 13–20.

Schwarz, H., and Humbel, B. M. (2007). Correlative light and electron microscopy using immunolabeled resin sections. *Methods Mol. Biol.* **369**, 229–256.

Semmler, H., Bailly, X., and Wanninger, A. (2008). Myogenesis in the basal bilaterian *Symsagittifera roscoffensis* (Acoela). *Front. Zool.* **5**, 14.

Shigenaka, Y., Roth, L. E., and Pihlaja, D. J. (1971). Microtubules in the heliozoan axopodium. 3. Degradation and reformation after dilute urea treatment. *J. Cell Sci.* **8**(1), 127–151.

Sitte, H., Edelmann, L., and Neumann, K. (1987). Cryofixation without pretreatment at ambient pressure. *In* "Cryotechniques in Biological Electron Microscopy." (R. A. Steinbrecht, and K. Zierold, eds.) pp. 87–113. Springer, Berlin, Heidelberg.

Smith, J., and Bush, L. (1991). *Convoluta pulchra* n. sp. (Turbellaria: Acoela) from the East Coast of North America. *Trans. Am. Microsc. Soc.* **110**, 12–26.

Spurr, A. R. (1969). A low-viscosity epoxy resin embedding medium for electron microscopy. *J. Ultrastruct. Res.* **26**(1), 31–43.

Steinbrecht, R. A., and Müller, M. (1987). Freeze-substitution and freeze-drying. *In* "Cryotechniques in Biological Electron Microscopy." (R. A. Steinbrecht, and K. Zierold, eds.) pp. 149–172. Springer, Berlin, Heidelberg.

Stierhof, Y. D., van Donselaar, E., Schwarz, H., and Humbel, B. M. (2009). Cryo-fixation, freeze-substitution rehydration and tokuyasu cryo-sectioning. *In* "Handbook of Cryo-Preparation Methods for Electron Microscopy." (A. Cavalier, D. Spehner, and B. M. Humbel, eds.) pp. 345–365. CRC Press, Boca Raton, FL.

Strangeways, T. S.P., and Canti, R. G. (1927). The living cell *in vitro* as shown by dark-ground illumination and the changes induced in such cells by fixing reagents. *Q. J. Microsc. Sci.* **71**, 1–14.

Studer, D., Humbel, B. M., and Chiquet, M. (2008). Electron microscopy of high pressure frozen samples: Bridging the gap between cellular ultrastructure and atomic resolution. *Histochem. Cell Biol.* **130**(5), 877–889.

Studer, D., Michel, M., and Muller, M. (1989). High pressure freezing comes of age. *Scanning Microsc. Suppl.* **3**, 253–268; discussion 268–269.

van Donselaar, E., Posthuma, G., Zeuschner, D., Humbel, B. M., and Slot, J. W. (2007). Immunogold labeling of cryosections from high-pressure frozen cells. *Traffic* **8**(5), 471–485.

Van Harreveld, A., and Crowell, J. (1964). Electron microscopy after rapid freezing on a metal surface and substitution fixation. *Anat. Rec.* **149**, 381–385.

Walther, P., and Ziegler, A. (2002). Freeze substitution of high-pressure frozen samples: The visibility of biological membranes is improved when the substitution medium contains water. *J. Microsc.* **208** (Pt 1), 3–10.

Zhang, G. F., Driouich, A., and Staehelin, L. A. (1993). Effect of monensin on plant Golgi: Re-examination of the monensin-induced changes in cisternal architecture and functional activities of the Golgi apparatus of sycamore suspension-cultured cells. *J. Cell Sci.* **104**(Pt 3), 819–831.

CHAPTER 15

Three-Dimensional Reconstruction Methods for *Caenorhabditis elegans* Ultrastructure

Thomas Müller-Reichert*, **Joel Mancuso**†, **Ben Lich**‡, *and* **Kent McDonald**§

*Medical Theoretical Center, TU Dresden, 01307 Dresden, Germany

†Gatan, Inc., Pleasanton, California 94588

‡FEI Company, 5651 Eindhoven, The Netherlands

§Electron Microscope Laboratory, University of California, Berkeley, California 94720

METHODS IN CELL BIOLOGY, VOL. 96

978-0-12-381007-6
DOI: 10.1016/S0091-679X(10)96015-9

Abstract

The roundworm *Caenorhabditis elegans* is one of the major model organisms in modern cell and developmental biology. Here, we present methods for the three-dimensional (3D) reconstruction of the worm ultrastructure. We describe the use of (1) serial-section analysis, (2) electron tomography, and (3) serial block face imaging by scanning electron microscopy (SEM). Sample preparation for high-pressure freezing/freeze substitution (HPF/FS) has been extensively covered in a previous volume of this "Methods in Cell Biology" series and will only be described briefly. We will discuss these 3D methods in light of recent research activities related to worm and early embryo biology.

I. Introduction

A. The Worm and Electron Microscopy—Some Early History

The history of *Caenorhabditis elegans* as a model system is also a history of using electron microscopy (EM) to understand three-dimensional (3D) cell relationships. From the beginning, EM was part of the plan for the worm as a model system (for historical perspective, see Brenner, 2009). The use of computers to aid in the reconstruction and analysis was planned as well, but the idea was mostly too far ahead of the hardware and software available at the time. Finally, the students, postdocs, and other associates working with Sydney Brenner used a combination of computer and manual methods to produce detailed reconstructions that were heroic in their scope and effort (Ward *et al.*, 1975; Ware *et al.*, 1975; White *et al.*, 1976). These culminated in the paper by White *et al.*, sometimes referred to as "The Mind of the Worm" (White *et al.*, 1986). This publication included 600 micrographs and 175 drawings of the *C. elegans* nervous system. Other publications followed, including David Hall's work on the EM reconstruction of the posterior nervous system that began as a thesis work at Cal Tech about this same period, and then published later (Hall and Russell, 1991).

However, with a few exceptions, EM was not to continue as a regular part of worm research. The pull of molecular genetics coupled with problems of development, physiology, neurobiology, and behavior took most worm researchers into areas where EM was not deemed essential. On the other hand, the worm is an excellent subject for light microscopy, and as light microscopes and reporter molecules evolved toward their current state, these tools were used to document nearly every aspect of worm structure and development that could be imagined. But while some thought (and some still think) that light microscopy would make EM obsolete, it has had precisely the opposite effect. Among all the many light microscope images of the worm that have accrued, there is a significant subset that comes up against the reality of resolution limits. True, there are some new developments in LM technology that are pushing beyond the old limits of 200 nm or so, but they will never get close to the resolution of the electron microscope. Fortunately, it is not an "either/or" type of situation, but one we could call "both/and." Today, we need to use both light and electron microscopy to fully understand the molecular basis for many biological events. More about this later, but we first need to understand why the specimen preparation for worm EM is so much better today than it was previously.

B. Worms Are Hard to Fix for EM but There Is a Solution

The role of EM may have continued to have been more integrated into the whole of worm research if it were not for the fact that the worm is extraordinarily difficult to fix well for ultrastructural studies. It is noteworthy that all the early studies were done with osmium as the primary fixative, even though glutaraldehyde was the general fixative of choice by that time. This may have actually been an advantage for cell lineage studies because the combination of extraction and high membrane contrast must have made it easier to follow cell connections. Later researchers found that by cutting open worms in the presence of glutaraldehyde, or glutaraldehyde plus osmium, they could get reasonable preservation of ultrastructure with less cytoplasm extraction (Hall and Russell, 1991). A summary of these fixation methods plus details of serial sectioning and 3D reconstruction methods was published by David Hall, and this work remains a valuable resource for anyone contemplating doing EM work on *C. elegans* (Hall, 1995). However, we now know that for the best possible preservation for EM studies, ultra-rapid freezing under pressures of about 2000 bar is the best choice. This technique is known as "high-pressure freezing" (HPF).

HPF was developed by Hans Moor at the Eidgenössische Technische Hochschule (ETH) in Zürich in the late 1960s (Moor and Riehle, 1968) and turned into a commercially available product in the mid/late 1980s (Moor, 1987). Fast-freezing methods for cytological preservation were well known in the 1960s and had their best applications in freeze-fracture studies that elucidated the basic structure of cell membranes. To prepare frozen samples for thin-section EM, it is usual to go through a process called "freeze substitution" (FS), whereby samples are dehydrated and fixed at low temperatures before warming up to higher temperatures for resin embedding and polymerization prior to sectioning. Martin Müller at the ETH pioneered in using HPF in combination with FS to visualize cell

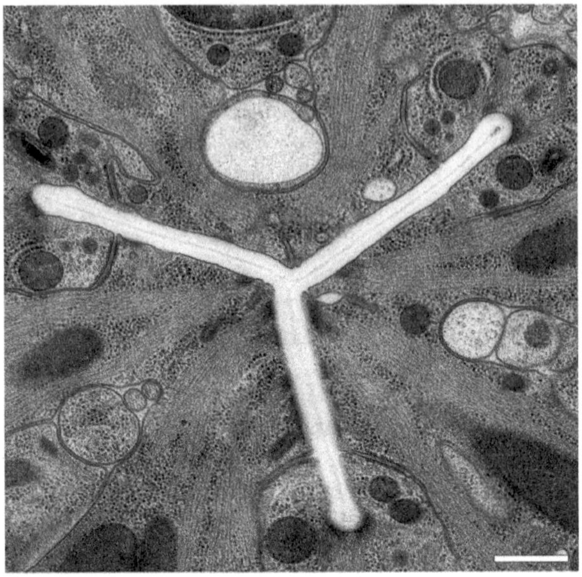

Fig. 1 Thin-section EM of a pharyngeal region of a wild-type dauer larva of *C. elegans*. The figure illustrates the quality of sample preparation after high-pressure freezing and freeze substitution in acetone containing 1% osmium tetroxide and 0.1% uranyl acetate. (Project with T. Kurzchalia, MPI-CBG, Dresden). Scale bar = 500 nm.

ultrastructure in thin resin sections (Müller and Moor, 1984; Humbel and Müller, 1986). Since the commercial availability of the BAL-TEC HPM 010 high-pressure freezer in 1985, there has been a slow but steady increase in the studies using HPF-FS to prepare samples for EM analysis (for a recent summary on HPF/FS, see: Humbel, 2009). The first studies to use HPF-FS for the study of worm ultrastructure and EM immunolabeling also came from the Müller laboratory (Favre *et al.*, 1995, 1998; Hohenberg *et al.*, 1994). For a review of how HPF has been applied to worm EM more recently, see the articles by Müller-Reichert (Muller-Reichert *et al.*, 2003, 2008). The quality of ultrastructural preservation one can get with *C. elegans* samples is illustrated in Fig. 1.

At this point we must mention an alternative to HPF that can be used to preserve worm ultrastructure well. It is a technique developed recently by Leunissen and Yi (Leunissen and Yi, 2009) that they call "self-pressurized rapid freezing," or SPRF. In this method, worms are sealed in a small copper tube (16 mm long by 350 μm inner diameter) and plunged into liquid nitrogen, or cooled liquid propane. By a mechanism not well understood, a reasonable percentage of worms can be frozen without significant ice damage. SPRF might be a very inexpensive option for those worm laboratories wishing to carry out EM, but which do not have access to a high-pressure freezer. It should be mentioned that high-pressure freezers cost between US$ 140,000 and 240,000. On the other hand, the percentage of well-frozen worms possible by HPF is quite high and one can follow the same worm in LM and EM (Muller-Reichert *et al.*, 2003).

C. Combining High-Pressure Freezing, Light Microscopy, and Electron Microscopy

Now that we know the best way to preserve worm ultrastructure, let us return to the subject of correlative light and electron microscopy (CLEM) as a powerful approach to answering basic questions about the molecular biology of the worm. There are different ways to define CLEM (McDonald, 2009), but for our purposes we mean that the same cells in the worm followed by LM are then observed in the EM. The next question is, how do we use HPF as the fixation tool in between the LM and EM observations? One answer is a relatively new type of HPF machine called the Leica EM PACT2+RTS. The RTS part of the name refers to the Rapid Transfer System, developed by Paul Verkade (University of Bristol, UK) to transfer a living sample from a light microscope to being high-pressure frozen within less than 5 s (Verkade, 2008). Applications to worm EM have been published previously (Muller-Reichert *et al.*, 2007, 2008; Pelletier *et al.*, 2006). CLEM of worms not requiring rapid transfer times can be done with any HPF machine as exemplified by the study of Sims and Hardin (2007). While CLEM has obviously the benefit of studying a tissue/or early embryo whose history is known and filmed, there is the added advantage that it greatly increases the sample size for EM studies. This is especially important if the subject of the study is of rare occurrence either spatially and/or temporally. With CLEM you can go to the electron microscope with confidence that you will be able to see the structure that you want to see.

D. A New Generation of 3D Reconstruction Tools for EM

Brenner originally included EM as part of his worm studies, but also mapping structures in three dimensions (see above). We have already discussed the fact that he was ahead of his time in using computers for making 3D reconstructions, but happily, computer hardware and software are now up to the task. Furthermore, we now have digital cameras of sufficient resolution that for most projects film is unnecessary. Without film we lose some resolution, but gain a lot of time between data collection and model building. What kinds of 3D imaging tools are available to the current generation of worm researchers interested in EM? We present four in this chapter: conventional serial-section reconstruction (Hall, 1995), electron tomography (ET; Frank, 1992), serial block face (SBF) imaging "a la Denk" (Denk and Horstmann, 2004), and focused ion beam (FIB) milling and imaging (Knott *et al.*, 2008). The latter two methods use high-resolution scanning EM for imaging. Therefore, the resolution is less than transmission EM but for certain projects it is perfectly adequate and much faster. In summary, the two factors you need to consider when choosing a 3D reconstruction methods are (1) the resolution you need, and (2) the volume of structure you need to analyze.

E. 3D from Serial Thin Sections

As can be seen in Fig. 2, the *X–Y* resolution one can achieve with the different methods goes from the highest with ET and serial thin sectioning TEM, followed by FIB scanning electron microscopy (FIB-SEM) and SBF-SEM in the order of descending resolution. This figure includes a ranking for *Z*-resolution as well, but these

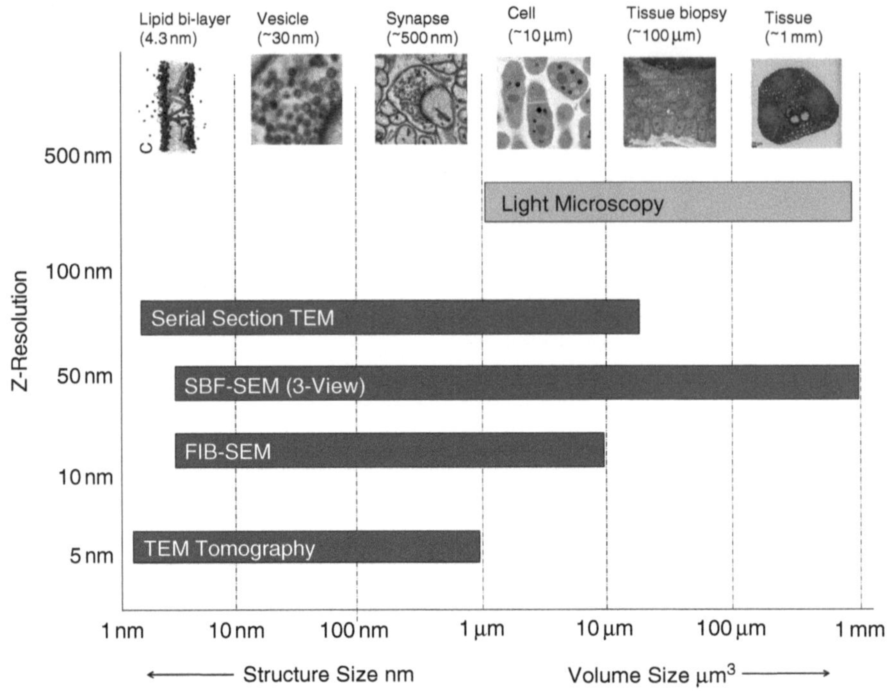

Fig. 2 Comparison of imaging methods for 3D analysis of biological samples. The methods basically differ in the volume that can be reconstructed and in the *Z*-resolution that can be obtained. (See Plate no. 15 in the Color Plate Section.)

numbers should be understood as approximations only. They will vary from sample to sample depending on factors, such as the section thickness chosen by the investigator, the density of the cytoplasm, or the quality of the microscope being used, among others. Within the SEM modes, the FIB-SEM can have the best *Z*-resolution because it can take such thin slices and the nature of the microscopes used, but the total volume reconstructed will be relatively small. The strength of SBF-SEM lies with the ability to reconstruct very large volumes.

Within the TEM modes, it is always best to use ET if possible, but there are situations where old-style serial-section reconstruction methods are preferable. The instance where this is most obvious is when one wants to reconstruct a structure that has a high axial ratio, i.e., the proportion of length to width. The worm is a perfect example of an organism with a high length-to-width ratio, and there are numerous examples of cellular components that can be tracked over distances of many hundreds of microns. The nervous system is historically the best example (see Brenner, 2009 and other citations given in the Introduction) but, as we will see, other cell types fit this criterion as well. While whole cells can be analyzed in 3D by serial ET (Hoog *et al.*, 2007; Marsh *et al.*, 2001), this approach is very labor intensive and typically can take months per single dataset. ET is in fact better suited for subcellular volumes that are at the level of organelles, such as the centrosome (O'Toole *et al.*, 2003; Pelletier *et al.*, 2006).

No discussion of serial-section analysis (SSA) of worms would be complete without mentioning the paper by Hall on this subject (Hall, 1995). Here the reader will find valuable tips on trimming, sectioning, picking up sections, poststaining, and how to recognize where you are when you are looking at the serial sections in the electron microscope. We will offer some alternative techniques for these subjects, but this paper must be read by anyone choosing to use serial thin sections for 3D reconstruction. Since 1995, a great deal has improved on specimen fixation as well as computer reconstruction methods, so these aspects of the Hall paper are a bit dated. The objective of this chapter is to bring these subjects up to date with the sections on HPF/FS and 3D reconstruction methods based on contemporary computer hardware and software. Fortunately, the software that has been developed for aligning and modeling tilt series in ET can be adapted for the same functions with serial thin sections.

It is worth mentioning at this point that serial sections are useful for things other than making 3D models. They can be used to quantify the numbers of organelles or other structures within a volume of the worm, or just to identify their presence or absence within. In addition, serial sections are useful for immuno-EM studies to check the specificity of the gold labeling.

F. Electron Tomography for 3D Modeling and Analysis

Electron tomography (ET) is a powerful method for 3D analysis, especially if one is interested in subcellular organelles at high resolution (Frank *et al.*, 2002). Despite the ambitions of some researchers (Marsh *et al.*, 2001; Noske *et al.*, 2008; others), ET works best on volumes of a few square microns. There are some excellent recent papers on ET of the worm that illustrate this idea (O'Toole *et al.*, 2003; Pelletier *et al.*, 2006; Srayko *et al.*, 2006; reviewed by Muller-Reichert *et al.*, 2010). To perform ET at this level you will need an intermediate-voltage electron microscope, a good digital camera, and the appropriate computer hardware and software. Fortunately, most transmission EMs sold these days are outfitted for ET. Automated image acquisition programs are available from commercial vendors (FEI Company, Eindhoven, The Netherlands; Tietz Video and Image Processing Systems (TVIPS), Gauting, Germany). Several software packages are freely available (TOM, UCSF Tomography, and SerialEM, Boulder Laboratory for 3D Electron Microscopy of Cells). Depending on the organelle that you wish to reconstruct, there may or may not be software to help in the modeling phase of ET. This "segmentation" step remains a labor-intensive task for many projects.

G. Serial Block Face Reconstruction using Scanning EM (SBF-SEM)

To reconstruct local neural circuits, the SBF method was developed by Winfried Denk and Heinz Horstmann at the Max Planck Institute for Medical Research (Heidelberg, Germany) (Denk and Horstmann, 2004). A custom microtome is placed inside the specimen chamber of a scanning EM and a back-scatter detector is used to collect images of the block face (Fig. 3A). Then the block is sectioned in 30–200 nm

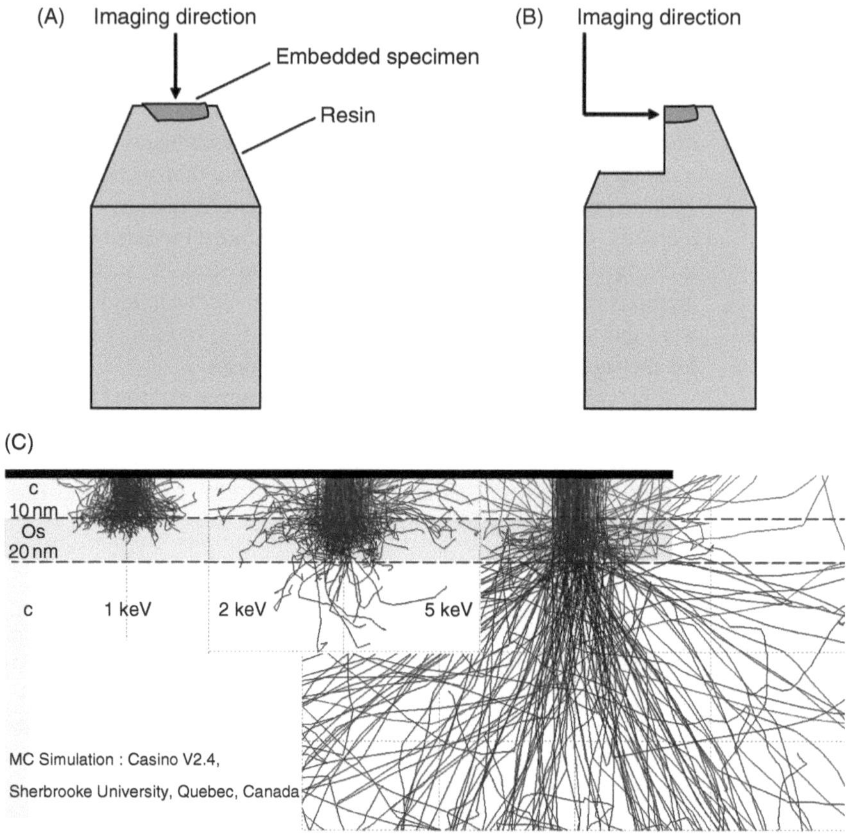

Fig. 3 Schematic diagram illustrating specimen orientation for block face imaging. (A) Serial block face SEM (SBF-SEM) using "3-View." The directions of sectioning and imaging are identical. (B) Focused ion beam SEM (FIB-SEM). The directions for FIB milling and imaging are perpendicular to each other. (C) Interaction of a 5-, 2-, and 1-keV primary electron beam with a carbon substrate as simulated using Casino v2.4.

increments, the section discarded, and the block face imaged again. There is currently a commercially available package for sectioning and data acquisition from Gatan, Inc., called "3-View," and we used this package for the work on SBF-SEM reported here.

Fixation strategies for SEM are the same as for TEM sample preparation. In principle, both HPF and traditional chemical fixation techniques are suitable for SBF-SEM. As mentioned above, HPF, however, is clearly the method of choice for the fixation of worm samples. Unlike conventional serial sectioning, SBF-SEM images the block face instead of the sections. Consequently, the SFB-SEM technique does not allow one to take advantage of postsectional staining, such as uranyl acetate or lead citrate. "*En bloc*" staining strategies have to be developed to maximize the back-scatter signal, thus improving resolution.

"3-View" can accommodate block faces up to 1 × 1 mm and can image large areas of the block at high resolution with a montage feature. For example, creating a 3 × 2 montage of 8k × 8k images will give a field of view of 120 × 80 μm with a pixel size of 5 nm allowing the user to resolve mitochondrial cristae, microtubules, and synaptic vesicles.

Unlike other 3D imaging strategies, SBF-SEM raw images are perfectly aligned. Because the block face is imaged, and not the section itself, compression distortions, common in serial sectioning, are absent. In addition, because the block face itself is the focal plane of the microscope, no corrective focusing algorithms are necessary. Aligned sections can be viewed and scrolled through live during acquisition. "3-View" can acquire 1000–2000 1k × 1k serial images overnight using a standard beam current. Therefore, reconstructing large volumes from thousands of images becomes a streamlined and automated process when compared to manual techniques, such as serial-section TEM.

H. The Dual-Beam Focused Ion Beam Method (FIB-SEM) of 3D Modeling

Similar to the previous section, this is another method that uses SEM as the imaging tool, but rather than sectioning away material with an in-column microtome, an FIB is used to mill away material in between imaging steps. Compared to SBF-SEM imaging, the amount of material that can be removed per cycle is much less, and therefore the resolution in *Z* is improved accordingly. But like ET compared with serial-section reconstruction, the area that can be analyzed is relatively small. The achievable reconstruction volume for this technology is about one order of magnitude smaller than for serial sectioning, but about three orders of magnitude higher than for ET. The maximum surface using image tiling is approximately 100 × 50 μm with a pixel size as small as 2 nm. In the *Z*-direction, the number of slices is only time limited. The combination of ultrathin serial sectioning with high-resolution back-scattered electron imaging enables acquiring datasets with isotropic voxels as small as 5 nm, allowing reconstructing a whole cell or multiple cells with adequate resolution for subcellular organelles.

The FIB in combination with a high-resolution scanning electron microscope (SEM) has proven to be very useful for obtaining high-resolution 3D data in an automated process (Knott *et al.*, 2008). The principle of FIB-SEM is to remove serial slices of material and to image the block face as illustrated in Fig. 3B. The FIB is used for sectioning and allows cutting of ultra-thin sections that can be an order of magnitude thinner than sections cut by traditional ultramicrotomy. The FIB uses Ga^{2+} ions that are accelerated toward the sample. The gallium ion beam is scanned like in an SEM, and due to the mass and velocity of these ions, the material is removed by sputtering. By precisely scanning the FIB, the area of material removal can be controlled almost to the nanometer.

The SEM is used for imaging through back-scattered electrons. Importantly, in a biological specimen (prepared for TEM imaging) the contrast is created/increased by using heavy metals, such as osmium tetroxide, during the sample preparation process. The interaction of a 5-, 2-, and 1-keV primary electron beam with a carbon substrate is simulated using Casino v2.4. These simulations show that a primary beam of lower energy yields a smaller interaction volume. Hence, the resolution that can be obtained from back-scatter imaging will improve with lower electron beam-accelerating voltages. Higher kV electron beam generates more signals from deeper layers in the specimen and hence "blurs" the structures close to the cross-sectional surface (Fig. 3C).

II. Rationale

Rationale of this chapter is to present and discuss methods for the 3D reconstruction of *C. elegans* ultrastructure. Specimen loading for HPF in combination with FS has been described in detail previously. Here, we give protocols for thin-layer embedding, CLEM, and the 3D reconstruction of cellular structures using SSA, ET, or SBF imaging by SEM.

III. Methods

A. High-Pressure Freezing and Freeze Substitution

As mentioned, these aspects of specimen preparation for HPF have been covered in detail in a number of recent publications (McDonald *et al.*, 2007; Muller-Reichert *et al.*, 2003, 2007, 2008). We will not repeat here what has been exhaustively covered previously. In brief, we routinely use 1% osmium tetroxide plus 0.1% uranyl acetate in anhydrous acetone for morphological studies (McDonald and Muller-Reichert, 2002). It is also possible to add up to 5% of water to the FS "cocktail" to increase membrane contrast (Buser and Walther, 2008).

In the following sections, we will, however, include those methods that are particularly useful for CLEM work and/or precisely oriented sectioning.

B. Thin-Layer Embedding

Labeled microscope slides are wiped clean with a soft cloth, submersed in a Teflon solution (MS-143V TFE Release Agent, Miller-Stephenson Chemical Co., Inc., Danbury, CT, USA), and allowed to dry. The slides are then polished to get a clean transparent surface, and two layers of Parafilm are placed on the margins of the glass slide to serve as spacers (Fig. 4A). About 200–300 μl of Epon/Araldite are placed on the microscope slide and allowed to disperse on the surface of the slide. Subsequently, specimen holders are placed in the resin droplet and held tightly with appropriate tweezers. Specimens are then removed from the sample holder (i.e., a membrane carrier for the EMPACT2 high-pressure freezer) with a sharpened tungsten needle or a comparable tool. Importantly, specimens tend to be brittle after FS, and care must be taken to avoid any mechanical damage to the embryo during these procedures. After removal of the empty membrane carrier, specimens in their final orientation are then softly pressed down to the surface of the glass slide, and

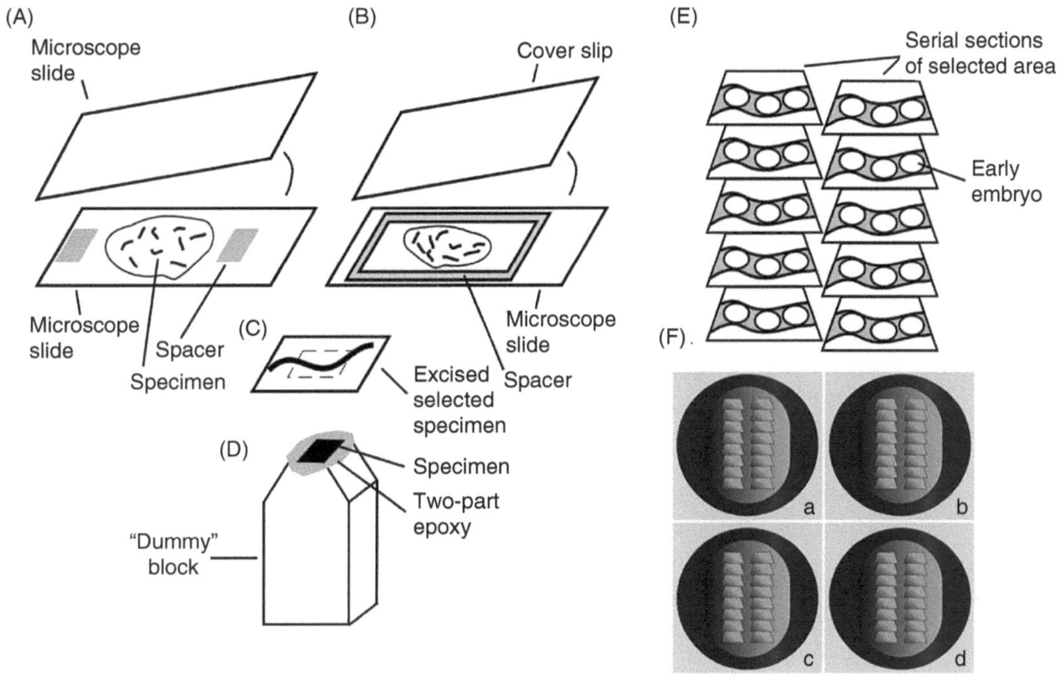

Fig. 4 Selection of thin-layer embedded *C. elegans* samples for electron microscopy. (A) Epon/Araldite-infiltrated samples are sandwiched between two Teflon®-coated microscope slides. Spacers are used to hold the slides apart. (B) For immuno-labeling, samples infiltrated with LR White resin are placed in slide molds made of a microscope slide and a Thermanox® spacer. Slide molds are closed using an Aclar® cover slip. (C) After polymerization of the resin, specimens are selected and excised. (D) Selected worms or early embryos are remounted on "dummy" blocks. (E) Serial sectioning of worms in longitudinal orientation. (F) Pairs of ribbons of serial sections can be placed on single EM grids. A whole series of serial sections (grids a–d) can be collected (Muller-Reichert *et al.*, 2003). (See Plate no. 16 in the Color Plate Section.)

a second, coated and cleaned glass slide is put on top of the resin samples. This "sandwich" is placed on an appropriate support in the oven (about 48 h at 60°C) to polymerize the resin. One of the microscope slides is removed after resin polymerization by pressing a razor blade between the resin and one of the glass slides. At this stage, the microscope slide is suitable for screening in a light microscope and the specimens are ready for remounting on "dummy" blocks for ultramicrotomy (Muller-Reichert *et al.*, 2003). A similar chamber can be made for samples infiltrated with LR White. Such samples are used for immuno-labeling experiments (Fig. 4B).

C. Screening and Remounting for Oriented Thin and Semi-Thin Sectioning

Polymerized samples are observed with a light microscope in bright field or with phase optics to relocate the specimens, whole worms, or capillary tubes with early embryos, for example, on the microscope slide (Muller-Reichert *et al.*, 2003, 2007). An objective marker with a diamond tip is used to mark a circle around the specimen. The purpose of this circling is threefold: (1) Whole worms or embryos in capillary tubes are clearly marked for handling in subsequent steps (i.e., during remounting). (2) The circle is used to determine the position of the specimen within the resin. By focusing up and down at high magnification, it can easily be determined on which surface of the thin resin layer the worm is located. Because most objects sink to the bottom of the resin layer prior to polymerization, they will always be at or near the surface and this must be known before remounting. By remounting the cells so they are at the upper surface, you can be cutting biological material in your first few sections. If you are cutting embryos inside the tubing, you can use the calibration of the *Z* movement on your microscope to estimate how far in the tubing you have to section before coming to the embryo. (3) The scribed circle can be used as an orientation aid. Marks can be scratched through the circle with a No. 11 scalpel blade to indicate particular orientations or subregions of the worm to be sectioned. Marks indicating where particular regions are located along the worm's long axis can help with rough trimming for serial cross sections (Muller-Reichert *et al.*, 2007).

Using a scalpel, small squares of resin about 1×1 mm containing the specimen are cut out under a stereomicroscope. Miller-Stephenson (M-S) 907 Epoxy glue is used to remount the samples on "dummy" blocks. For serial longitudinal sections we mount the samples facing up (Fig. 4C–D); for serial cross sections we prefer to use dummy blocks from latex flat molds and glue the worm or early embryo on the side, rather than on top of the block. We prefer the M-S 907 glue because it can be sectioned and is stable in the electron beam. Polymerization of the glue at 60°C is completed within about 30 min. Remounted samples are then trimmed for thin sectioning.

D. Conventional Serial Thin Sectioning for 3D Reconstruction

1. Trimming Blocks for Conventional Serial Sectioning

a. Starting with the remounted sample(s) from HPF/FS-treated and thin-layer embedded material, the choice will be to trim for longitudinal or cross sections of the worm. Since Hall covered the cross-sectional approach so well, we will concentrate only on longitudinal sections of the worm (Hall, 1995). The instructions below are for sections of whole worms. Of course, if there are subsections along the length that are of particular interest, e.g., the spermatheca, or embryos at particular stages, then one can trim away the parts of the head and tail that are not of interest.
b. Rough trim with a single-edge razor blade, or even a jeweler's saw until there is an area of several hundred micrometers around each side of the worm.
c. For the final trimming switch to a very thin, double-edged razor blade such as the double-edged, stainless steel blades from Electron Microscopy Sciences, Cat. No. 72002 (Hatfield, PA, USA). These can be broken in half to form two single-edged blades by carefully folding the blade lengthwise while still in its wrapper until it snaps in two. Carefully remove the half blade from the wrapper and wipe the edge with ethanol-soaked laboratory tissue to remove grease from the manufacturer.

d. Under a stereomicroscope, either on the microtome or otherwise, trim away very thin (20 μm or so thick) slices parallel to the long axis of the worm until you are as close as you can get while still comfortable that you won't cut the worm itself. We would typically get within 50 μm or less. Pay careful attention to the reflectivity of the surface of this side of the block. It should be shiny and completely transparent so that you can see into the block. If it is opaque or rough in any way, then the blade is not sharp enough and you should move to an unused portion of the blade, especially for the final cut for this particular side.
e. Rotate the block 180° and repeat step 4 for the opposite side of the block. You should end up with parallel trimmed surfaces on either side of the worm. The final width will be determined by how much the body of the worm bends, if any.
f. Rotate 90° in either direction to trim the sides of what will be your final block face. We prefer to taper the sides so the final shape is an elongated trapezoid. Having these long top and bottom edges will help the subsequent serial sections stick to each other and form a ribbon.

2. Sectioning

g. Serial sections can be cut on almost any microtome, but we find that the Reichert-Jung (now Leica Microsystems, Vienna, Austria) Ultracut E to be particularly useful. This is because the sectioning arm and frame of the microtome are not connected, so this means you can rest your arm on the frame while the arm is cutting and it will not affect the sections. These microtomes come with a plastic shield that rests on the front platform, and we find this the best place to place our hand during sectioning. Other microtomes, including the newest models, tend to have mechanical connections between the frame and the arm so that any movement of the frame will affect the sections. Some have separate armrests that are connected only to the table, and if you do not have one of these models, consider making such an armrest.
h. Prepare Formvar-coated slot grids by whatever method you prefer. We like to make the films slightly thicker than we would use for mesh grids. The reflectance color on water should be solid silver, perhaps with a faint tinge of gold. Do not carbon-coat the filmed grids. This will only make them prone to becoming hydrophobic and very brittle. When carbon-coated slot grids are punctured, they will propagate a long split in the film and usually most sections are lost. With non-carbon-coated films, a puncture (say with a forceps tip) will just leave a hole and that hole can be "patched" by applying another, thinner layer of film over the hole and the sections can be saved. If grids are not to be used right away (within a day), store them in a refrigerator. This will keep them hydrophilic for several weeks in our experience. The exact duration varies with the humidity of the refrigerator, with more humidity working for longer times. We highly recommend using the Maxtaform copper–rhodium slot grids because each side is a different color. Put the film on the copper side.

i. Start sectioning the block at the preferred thickness. If the first few sections do not stick together, or if they are not coming off in a straight ribbon, then they must be adjusted accordingly. Retrimming the parallel sides should give straight ribbons. If the sections do not stick together, then there may be several reasons and remedies: (1) The parallel sides are not smooth enough due to a dull razor blade. Retrim with a new blade. (2) The water in the boat is not "conditioned." We have found that you sometimes have to break the surface tension in the boat by inserting a toothpick or another object into the boat. Watch the sections that are floating around when you do this, and if you see them rush forward toward the knife edge when you insert the toothpick, then you probably have corrected the problem. Often this has to be done twice or more before you see the sections move. Try not to introduce too much dirt with the stick. Cut a few more sections to see if they now come off as a ribbon. (3) If your microtome is equipped with an ionizer unit, try turning that on during sectioning. You can also place a polonium strip ionizer (GE Osmonics Labstore, Part. No. 1215174, 1-800-444-8212 in the U.S., and 1-952-988-6665 for international distributor information) near the knife and that may help.
j. By using an ocular with reticle, or by cutting a test ribbon, determine how many sections will fit within the 2-mm length of the oval slot. Also see if you can fit only one ribbon or more within the 1-mm width of the grid. All of this will vary, of course, depending on whether you are cutting longitudinal or cross sections and the size of the worm you are cutting.
k. When you reach the region of the worm where you are ready to start collecting sections, cut ribbons in lengths appropriate for the slot grid. As each ribbon comes to the proper length, use an eyelash or hair tool to dislodge it from the knife edge by gently touching the section just behind the one on the knife edge. Move the ribbon over to the side and continue cutting the next ribbon. If you rest your hand on the armrest you can do this without stopping and starting the microtome. As soon as you have enough sections/ribbons for one grid, move them over to the side of the boat. Continue like this until you have enough for four grids worth before stopping the microtome. The block should be stopped when it is below the knife edge but before it starts the return cycle. Using the fine control for retracting the arm, back it up 1–2 μm. Arrange the ribbons in pairs (Fig. 4E).
l. Next to the microtome you should have five pairs of forceps holding the slot grids. These should be numbered or marked so that you know the order they are in. Use forceps #1 to pick up the first set of sections cut. Insert the slot grid into the water at about a 45° angle and with the other hand use a hair tool to position the sections over the hole. Try to have the meniscus of the water at one edge of the slot and attach one edge of the ribbon raft to that part of the film. Then with a rolling and lifting motion, bring the grid out of the water so that the sections are over the slot. Carefully blot off excess water from underneath the grid with a laboratory tissue wrapped around a finger. Set this grid aside to dry and repeat with the remaining grids. Put them in a grid box in a known order after they dry (Fig. 4F).
m. Prepare five more slot grids in the forceps and have them next to the microtome. Resume sectioning and watch carefully as the block advances toward the knife. If

you see too much gap, advance the block manually, but as it gets close to cutting sections again just be patient and let it go until sections appear. With luck, you will continue to cut ribbons without a break from the previous series. If you do lose a section or two, make a note of it. Cut five more grids worth before repeating the last two steps.

3. Poststaining Grids

n. Stain sections using 1% uranyl acetate in 70% methanol for 10 min and Reynolds' lead citrate for 5 min. These times may vary depending on resin type and section thickness. We like to stain more than one grid at a time using a multiple grid staining device, such as the Pelco™22510 Grid Staining Matrix System (Ted Pella, Inc., Redding, CA, USA). We find most other devices for staining multiple grids to be inadequate because they tend to break the slots during loading and unloading.

4. Transmission Electron Microscopy

o. The first thing you must know when looking at serial sections on the microscope is the orientation, i.e., where are the first and the last sections. If you only have one ribbon and you have trimmed a trapezoid shape then there is no problem. With multiple ribbons you will have to determine which was cut first and last. You cannot rely on the arrangement that was in the boat, e.g., first ribbon on the right and second on the left with trapezoid pointing up, because different microscopes may or may not reverse that order as you see it on the screen or monitor. You will just have to check the continuity between the top section of one ribbon and the bottom section of its neighbor. When you get the correct order you may want to make a diagram that you can refer to in subsequent microscope sessions.
p. Grids without a carbon coat can be subject to charging and drift. However, there is something you can do that may stabilize the film. Go to the lowest magnification where you can still see the grid but not the objective aperture. Spread the beam to fill the field and use stage controls to irradiate all around the edge of the slot. Then, make several passes from one side of the slot to the other. If your magnification allows you to see the whole slot, then just leave the beam spread for several minutes as it irradiates the grid. This seems to stabilize the film and reduce problems with drift. If you have persistent drift problems, then you may have to carbon-coat the film, despite the other problems this might introduce.
q. It is a good idea to make a grid map, i.e., a diagram or a low-magnification image that shows the number of sections in each ribbon and the arrangement of ribbons. It is especially important if your ribbons are not perfectly aligned or if they have spaces between sections along their length.
r. Find areas of interest and start collecting serial images.

5. Three-Dimensional Reconstruction

s. We use the IMOD software package (Kremer *et al.*, 1996) to segment features of interest. The package allows the superposition of 2D traces to create a 3D representation of the subcellular structures. In addition, IMOD allows to correct some image distortions that arise during ultramicrotomy and imaging (Mastronarde *et al.*, 1993; McDonald *et al.*, 1992).

An example of 3D modeling from serial sections is given in Fig. 5 (Evans *et al.*, 2006). This figure shows a 3D reconstruction of amphid channel cilia in the worm "head".

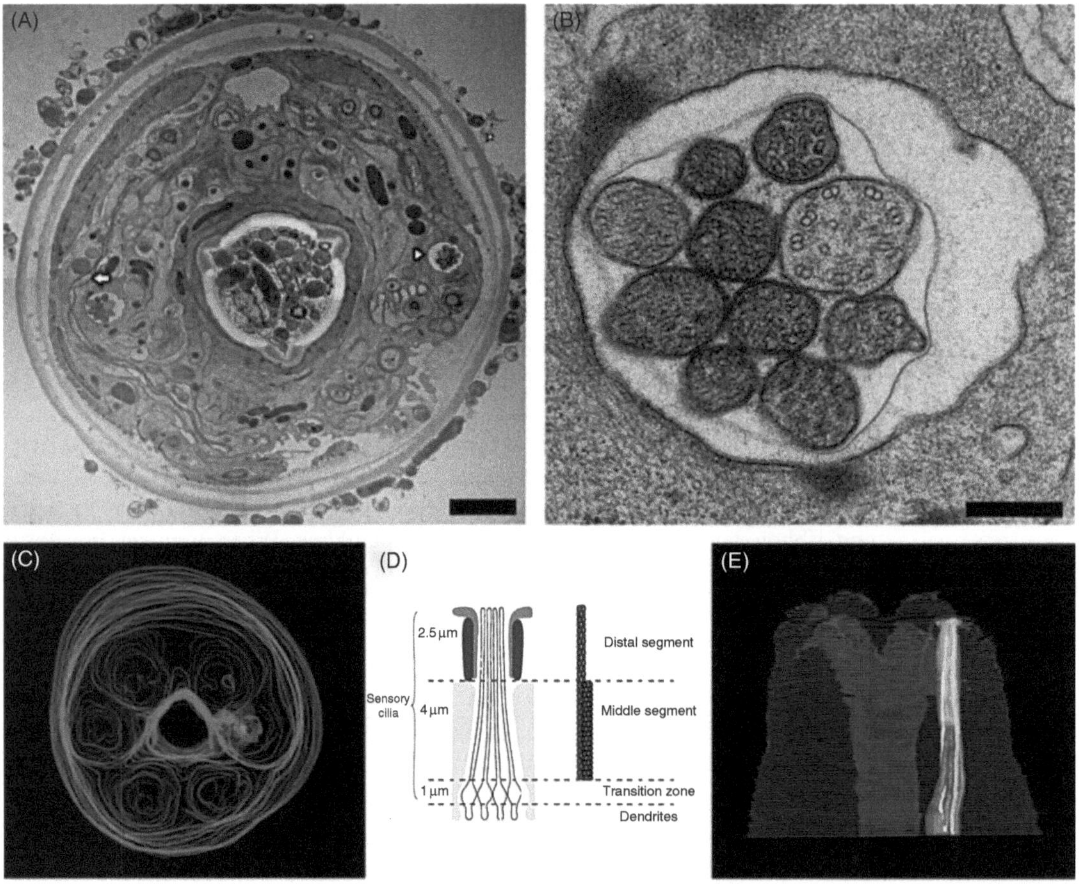

Fig. 5 Serial-section analysis of wild-type *C. elegans*. (A) Cross section through the head of a high-pressure-frozen/freeze substituted specimen embedded in Epon/Araldite. (B) Ultrastructure of the right amphid. (C and E) Top and side views of a 3D reconstruction from aligned serial sections showing overview of head and location of amphid channel cilia. Three-dimensional reconstruction was carried out using the IMOD software package. (D) Schematic illustration of channel cilia (from Evans *et al.*, 2006, with permission). (See Plate no. 17 in the Color Plate Section.)

E. Electron Tomography

All of the operations described above in sections III-D1–3, (Trimming blocks – post staining grids) can be used for preparing samples for ET with some slight variations. For HPF of early embryos, we use a correlative microscopy approach using a Leica EM PACT2+RTS high-pressure freezer. The method has been explained in detail previously (Muller-Reichert *et al.*, 2007). For sectioning, we also use serial sections processed as described above, but we cut sections about 300 nm thick and usually put two ribbons in the center of a Formvar-coated copper 1 × 2 mm slot grid. The Formvar-film should have a "goldish" color. Using such prepared samples, we apply the following procedure:

1. Acquisition of Tomographic Data

a. Image a tilt series of semi-thick sections in an intermediate voltage (200–300 kV) EM equipped with a eucentric tilting stage. Collect serial, tilted views of the section every degree over a ±60° or 65° range.
b. After the first tilt series has been acquired, rotate the grid 90° to image a second tilt series over a ±60° or 65° range.
c. Use both tilt series to calculate a double-tilt tomogram as described (Mastronarde, 1997).

2. Segmentation and 3D Modeling of Microtubules

In the following paragraph, we briefly describe how to model microtubules in tomograms (O'Toole and Muller-Reichert, 2009). For the segmentation of other features, the reader is referred to published procedures (Ladinsky *et al.*, 1999; Marsh *et al.*, 2001).

d. Open the reconstructed volume using "3dmod" and go into model mode.
e. Create a model object and edit the object type as an open contour.
f. Open the slicer window and choose a microtubule by clicking the left mouse button. Rotate the *x, y*, and *z* sliders at the top of the slicer window to orient the microtubule along its long axis. Deposit model points along the microtubule using the middle mouse button. Each new microtubule modeled is a new contour in the object. In addition, create new objects for "open" and "closed" microtubule plus- and minus-ends. Deposit model points at each microtubule end. A projection of the 3D model can then be opened.
g. Different classes of microtubules (i.e., kinetochore microtubules, astral microtubules) are organized as separate objects and can be distinguished using different colors.
h. Once the microtubules in the volume have been modeled, a program can be run, mtrotlong, which will extract a series of subvolumes that contain the microtubules in a longitudinal orientation. The operator can then step through successive tomographic slices of the subvolumes to analyze the microtubule end morphology in detail.
i. Model objects can be displayed together or subsets of objects can be turned off or on to highlight the 3D relationships of particular features within the cell.

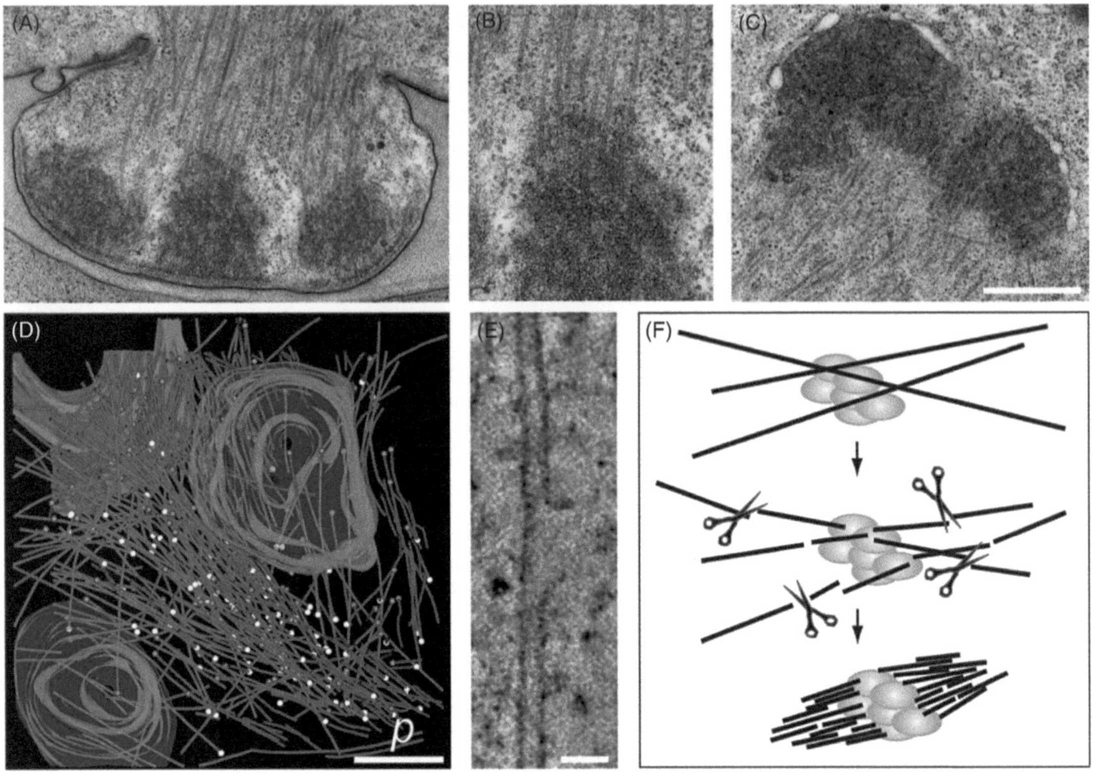

Fig. 6 Tomographic reconstruction of meiotic spindle assembly in *C. elegans*. (A–C) Thin-section EM showing a polar body, a kinetochore region with microtubules, and the formation of the female pronucleus. Scale bar = 500 nm. (D) Three-dimensional reconstruction of one half of a meiotic wild-type spindle (microtubules in red, pole-proximal ends as white spheres, pole-distal ends as blue spheres, chromatin in green, *p* is spindle pole) (Srayko *et al.*, 2006). Scale bar = 500 nm. (E) Tomographic slice showing lateral disruption of the microtubule lattice (arrow). Scale bar = 50 nm. (F) Model explaining the role of *C. elegans* katanin in female meiotic spindle assembly (modified from Muller-Reichert *et al.*, 2010). (See Plate no. 18 in the Color Plate Section.)

An example of the tomographic reconstruction of spindle microtubules in the early *C. elegans* embryo is given in Fig. 6. For details, the reader is referred to Srayko *et al.* (2006).

F. Serial Block Face SEM (SBF-SEM)

1. Specimen Preparation

Many of the procedures described above up until the point where the sample block is mounted for sectioning III-D1 (Trimming Blocks for Conventional Serial Sectioning) apply to sample preparation for SBF-SEM. What is different is that the contrasting in the material must be done prior to sectioning, and this may require some new "*en bloc*" methods. We applied contrasting during FS in 1% osmium tetroxide and 0.1% uranyl acetate in anhydrous acetone (Muller-Reichert *et al*., 2003), but further protocols need to be developed (see discussion).

2. Specimen Trimming

Depending on the sample, the researcher may want to pre-screen the resin block by cutting thick sections for light microscopy to determine where in the biological material they want to start imaging in "3-View." Trimming should be such that the area of interest is surrounded by a large buffer zone, especially if the structure may move laterally as one proceeds in the *Z*-direction. For worms, it is simply a matter of leaving enough area of blank resin around the body of the worm that the worm itself will always be within the boundaries of the section.

3. Setting up the Instrument for Data Acquisition using "3-View"

Users will find setting up "3-View" to be familiar and almost identical to setting up a conventional ultramicrotome. Approaching the diamond knife to the 3-View sample resembles the same strategy as in a conventional ultramicrotome, using light reflections and shadows to align the diamond knife to a sample. Once the sample is within a few microns, the microtome is initialized to begin taking 200-nm cuts until the sample begins to cut. Once the microtome has begun to cut the surface, the chamber is pumped down and the user begins to search for an area of interest. 3-View software can be set up to automatically acquire an aligned 3D dataset. The user can choose the slice thickness, number of images, and the image size. Since the acquired images are aligned into stacks of images as they are acquired, the user can browse the 3D data life during the experiment.

SBF-SEM of worm samples is illustrated in Fig. 7 (see also Suppl. Movie 1 at http://www.elsevierdirect.com/companions/9780123810076). A dataset of 400 images was acquired with "3-View" in about 6 h. Figure 8 shows a manual segmentation of individual neurons from the dorsal nerve cord of *C. elegans*. For this particular dataset, segmentation was carried out using the Imaris software package (Bitplane, Zurich, Switzerland).

G. Focused Ion Beam SEM (FIB-SEM)

1. Sample Preparation for FIB-SEM

For the automated 3D data acquisition, the sample preparation can be kept almost identical to the sample preparation process for serial-section TEM analysis—up to the point where the sections are cut with the ultramicrotome (III-D1—Trimming Blocks for Conventional Serial Sectioning). The samples are processed for HPF/FS, thin-layer embedded, and mounted on "dummy" blocks. However, there is one modification to our standard embedding protocol: instead of embedding in Epon/Araldite, we follow the protocol by Knott for sample embedding in Durcupan (Knott *et al*., 2008). The FIB will cut the specimen perpendicular to the surface of the block (see above, Fig. 3B) and hence will create a cross-sectional surface for 3D imaging perpendicular to the plane that one would see with the TEM after taking ultra-thin microtome sections. Therefore, it is

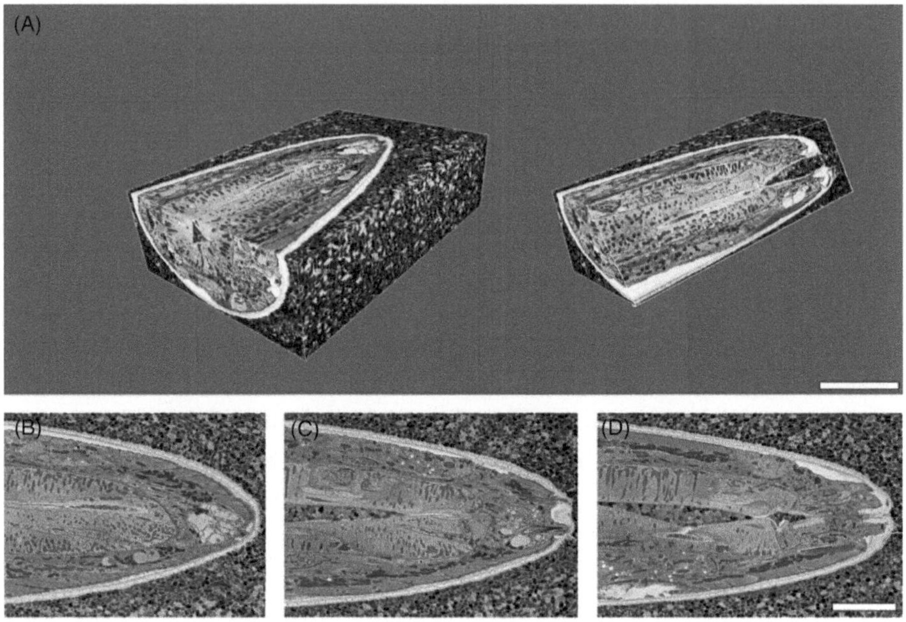

Fig. 7 Three-dimensional serial block face dataset of 400 images acquired with "3-View" of *C. elegans* in about 6 h. The slice thickness is 50 nm and the pixel size 50 nm as well, producing 50 nm isotropic voxels. 3D visualization and rendering was done using Imaris Bitplane software package. Scale bar = 250 μm. (B–D) Individual serial images of from the 3D data stack. Scale bars = 10 μm.

recommended to realize before mounting the specimen to the resin support what the ideal cut direction is and mount the specimen accordingly. The one step that differs for the sample preparation process for the Dual Beam is the removal of the mounting resin and some of the specimen, so that the volume of interest is as close as possible to the specimen surface (Fig. 9A–B). This is typically done using an ultramicrotome.

2. Setting up the Slice-and-View Process

The Dual Beam or FIB-SEM can be set up to automatically collect the 3D data by using the so-called "slice and view" process. This automated routine allows to set up (1) the slice thickness, (2) the amount of slices required, and (3) the field of view with the desired resolution. This automatic process allows overnight acquisition of data. A typical setup applied for studying the 3D architecture of subcellular organelles is using approximately 3 nm pixels in *X–Y* and a slice thickness of about 10 nm. This yields voxels of the size of $3 \times 3 \times 10$ nm. The "slice and view" process automatically prompts the user to go through all the required steps to get a successful dataset. It will create the FIB-removed trenches adjacent to the volume of interest for removed material to redeposit. Also it will prompt the system to deposit a layer of platinum on the top of the specimen for creating better cross sections with the FIB.

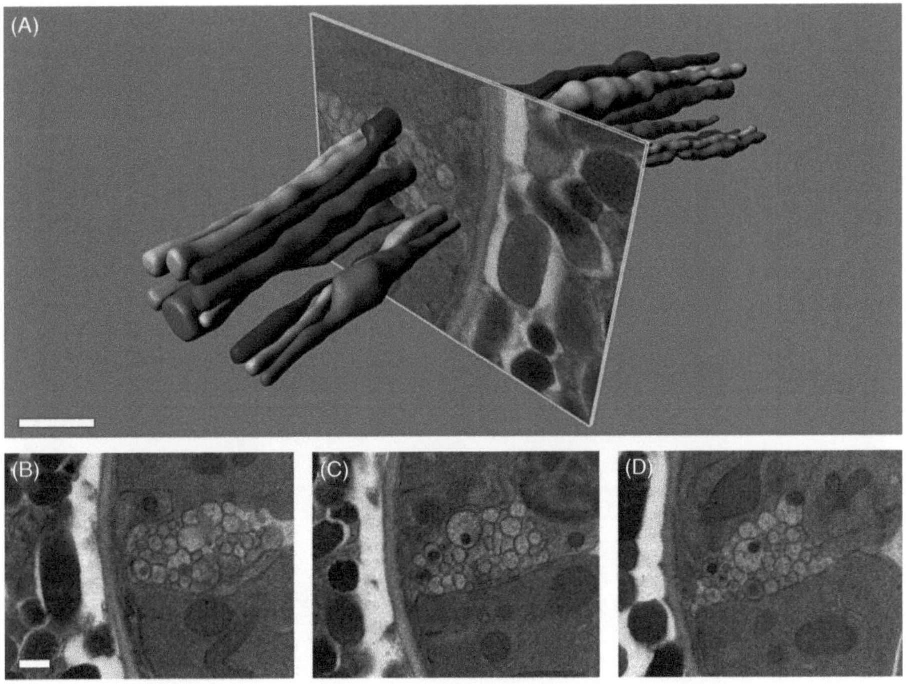

Fig. 8 Three-dimensional reconstruction by SBF-SEM using a Quanta 600 (FEI). (A) Small subset of 200 serial images visualized from a large 3D volume consisting of 1000 serial images. Individual neurons from the dorsal nerve cord of *C. elegans* have been manually segmentated using the Imaris Bitplane software package. Scale bar = 1 μm. (B–D) Individual serial images from the 3D data stack. Scale bar = 0.5 μm. (See Plate no. 19 in the Color Plate Section.)

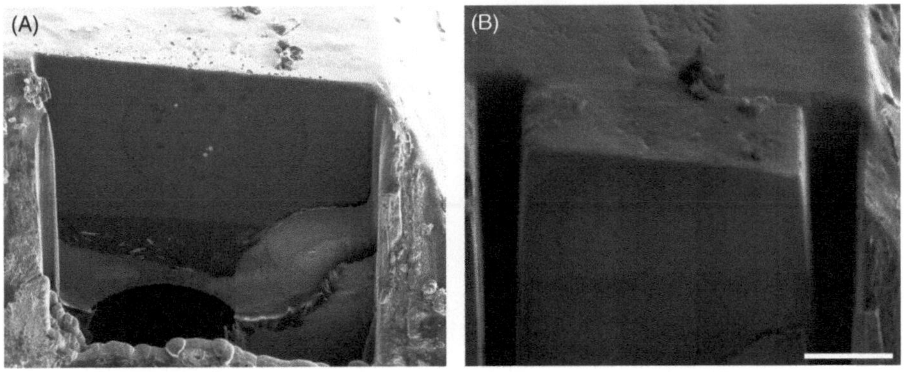

Fig. 9 Sample mounting for FIB-SEM. (A) The sample needs to be close to the resin surface. (B) Field of view at higher magnification. Scale bar = 10 μm.

An example of FIB-SEM is shown in Fig. 10. A larval worm (L3) was "sliced and viewed" in cross section. In this case, a layer of resin of 50 nm was removed per slice/view cycle, the voxel size was 13 × 13 × 50 nm, and the sample was imaged at 2 kV.

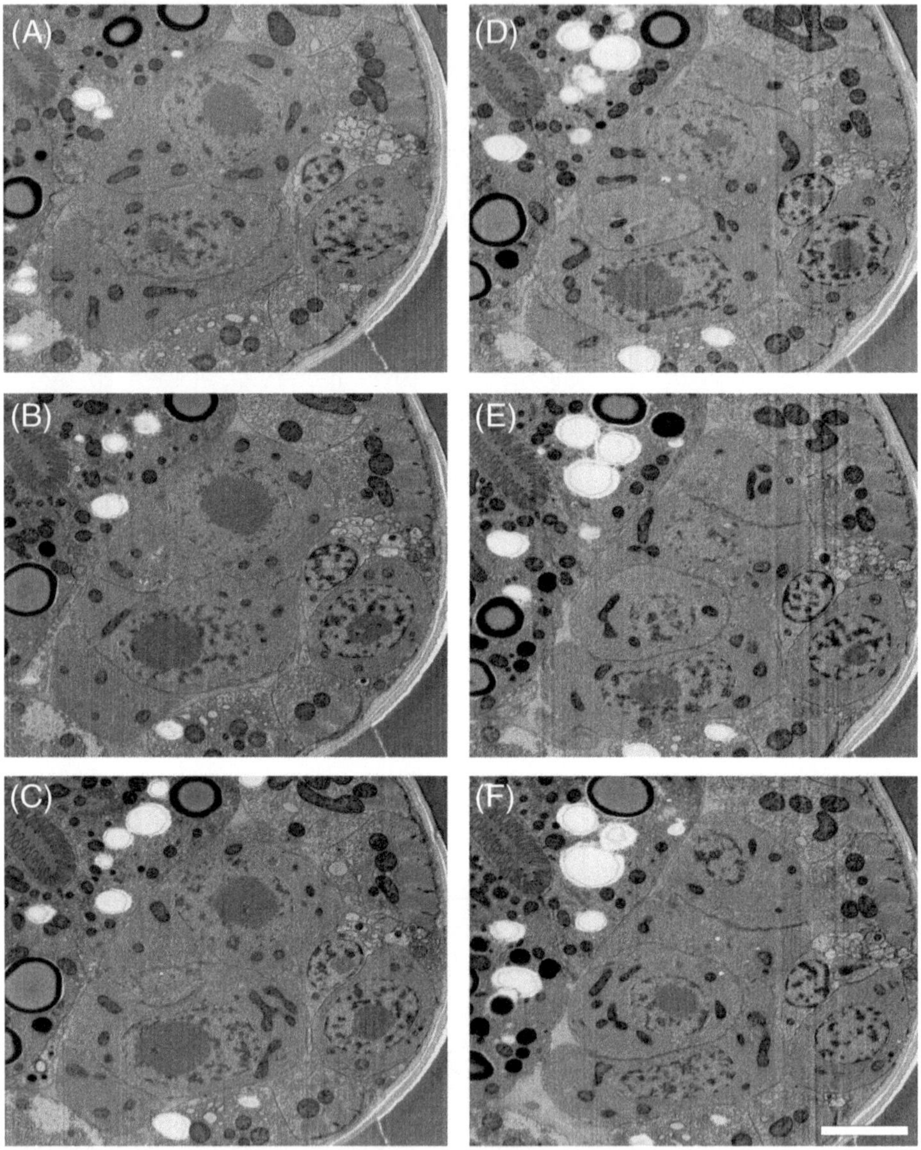

Fig. 10 Serial FIB-SEM slices through an L3 larva of *C. elegans*. This figure shows six images out of a series of 200 images. The sample was images with a Helios Nanolab 600 SEM operated at 2 kV. Slices of 50 nm were removed per cycle. Scale bar = 2.5 μm.

3. Three-Dimensional Reconstruction

As described above, the IMOD software package can be used to segment and model the features of interest (Kremer *et al.*, 1996).

IV. Materials and Instrumentation

A. High-Pressure Freezing

Instrumentation: The currently available commercial high-pressure freezers include the ABRA HPM 010 (RMC-Boeckler, Tuscon, AZ, USA); the Wohlwend Compact HPF 01 (Wohlwend Engineering, Sennwald, Switzerland); the Leica EM PACT2 with or without the rapid transfer system (RTS) for specimen loading; and the Leica HPM 100 (Leica Microsystems).

Materials: For the EM PACT2 we use 100-μm-deep "membrane" carriers available from Leica Microsystems (McDonald *et al.*, 2007). For the other instruments the least expensive source of specimen carriers is Wohlwend Engineering. One can get specimen carriers ranging in depths from 25 to 300 μm and we use the depth that most closely approximates the thickness of the worm generation that we are freezing. One can also use 3-mm EM grids as variable depth spacers as explained in detail elsewhere (McDonald *et al.*, 2007, 2010). Tools for handling worms and specimen carriers include worm picks and an alcohol lamp, fine-tipped (sizes 0–00) paint brushes, micropipettors and tips, bibulous paper for wicking off excess liquid, and fine forceps for handling the specimen carriers during loading. An inexpensive, lower wattage dryer is best for warming and drying of specimen loaders.

Reagents: M-9 buffer (22 mM potassium phosphate monobasic [KH_2PO_4], 19 mM NH_4Cl, 48 mM sodium phosphate dibasic [Na_2HPO_4], 9 mM NaCl), bovine serum albumin (BSA) (Sigma-Aldrich, Steinheim, Germany). Air space in the carriers will cause poor freezing. To fill in the spaces between worms, we tend to use either thick *E. coli* paste from worm food plates (Muller-Reichert *et al.*, 2003) or 20% BSA made up in M-9 buffer (Muller-Reichert *et al.*, 2007).

B. Freeze Substitution

Instrumentation: Automatic FS device. Currently, one can get the AFS2 from Leica Microsystems or the EMS-002 Rapid Immersion Freezer with Freeze Substitution from Electron Microscopy Sciences. Alternatively, FS can be carried out on dry ice (for details, see McDonald, 1994).

Materials: Cryotubes.

Reagents: Anhydrous acetone (EM grade), osmium tetroxide, and uranyl acetate. The FS "cocktail" is made up ahead of time in cryotubes and frozen in LN_2, and the samples are added to the frozen fixative for either storage or just before starting FS. Details on making up fixatives can be found elsewhere (McDonald, 1999).

C. Thin-Layer Embedding

Instrumentation: Oven set at 60°C for polymerization.

Materials: Plastic beakers, transfer pipettes, a digital scale for weighing out components, magnetic stirrer, stir bars for mixing, waste containers, laboratory tissues, glass slides, polytetrafluoroethylene (PTFE) release agent (MS-143V, Miller-Stephenson Chemical Co., Inc.), lint-free cloth, air can, parafilm, and scissors.

Reagents: Epon 812 substitute, or Epon/Araldite mix, anhydrous acetone for making a graded infiltration series.

D. Screening, Remounting for Oriented Thin and Semi-Thin Sectioning

Instrumentation: Light microscopy equipped with phase optics to screen for samples, scribing tool that mounts on a light microscope objective turret (i.e., marker with a diamond tip, Leica), stereo microscope for remounting selected specimens, and ultramicrotome (preferably a recent model that will cut sections of consistent thickness). We use either an Ultracut E or an UCT ultramicrotome (Leica Microsystems).

Materials: Scalpel with No. 11 blade, remount glue (Epoxy 907 Adhesive System, Miller-Stephenson Chemical Co., Inc.), "dummy blocks", razor blades for trimming blocks, self-closing forceps (five pairs; Electron Microscopy Sciences, Cat. No. 72864-D), slot grids (Maxtaform copper–rhodium, 1 × 2 mm; Electron Microscopy Sciences, Cat. No. 2010-CR), and grid storage boxes.

Reagents: Formvar solution (0.5%; Electron Microscopy Sciences, Cat. No. 15820) and film casting accessories, uranyl acetate, lead citrate, and methanol.

E. Electron Tomography

Instrumentation: Intermediate-voltage electron microscope operated at 300 kV (we use a TECNAI F30 FEG, FEI Company, Eindhoven, The Netherlands), high-tilt rotating stage (Gatan model 650, Pleasanton, CA), 2K × 2K CCD camera (e.g., Gatan), image acquisition software package (SerialEM), and 3D reconstruction software (IMOD). Details can be found at http://bio3d.colorado.edu. In principle, ET can be done on most electron microscopes that are computer-controlled and fitted with digital cameras. As the acceleration voltage of the microscope goes up, the sample thickness can be thicker, and most ET is done these days with 200–400 kV instruments. The general "rule" is that the thickness of the sample (usually a resin section) in nanometers is roughly comparable to the kV of the microscope. Therefore, a 120-kV microscope could look at sections up to about 120 nm, a 300-kV scope could view 300 nm sections, and so on. The exact thickness of the sample will vary with the density of the sample, so low-density structures might be viewed in thicker sections.

Materials: Parafilm®, fine-tipped tweezers.

Reagents: 10- or 15-nm-colloidal gold (Ted Pella, Inc.).

F. Serial Block Face SEM (SBF-SEM)

Instrumentation: Scanning electron microscope. We used a Quanta 600 (FEI). "3-View" (Gatan) and Imaris software package (Bitplane, Saint Paul) was used.

G. Focused Ion Beam SEM (FIB-SEM)

Instrumentation: Scanning electron microscope. We used the Helios Nanolab 600 (FEI) for imaging.

V. Discussion

The choice of a 3D method is strongly dependent on the scientific question and, therefore, it is not really possible to give a general advice here on when to use which method for which biological project. We will instead give some method-related comments, which might help the reader to find the right method for her/his project.

1. Serial-Section Analysis

SSA is a rather "old" method. As mentioned earlier, it has been applied in the early days of *C. elegans* research to reconstruct the nervous system (Brenner, 2009), and the only complete wiring diagram of a neural circuit in existence is that of *C. elegans* (reviewed by Mishchenko, 2009). SSA is still very useful for reconstructing *C. elegans* ultrastructure (Evans *et al.*, 2006), as well as that of other nematode species (Ragsdale *et al.*, 2009). SSA is mainly applied when rather large structures have to be followed within the worm and modeled in 3D. In particular, SSA is useful when the length-to-width ratio for a structure is high. While it is possible to carry out serial ET through several sections (Hoog *et al.*, 2007; Marsh *et al.*, 2001; Noske *et al.*, 2008), SSA is a more reasonable option when the number of sections required is tens to hundreds (see also Chapter 12 by Kang, 2010; this MCB volume). For the handling of large datasets, an interesting software package, called Collaborative Annotation Toolkit for Massive Amounts of Image Data (CATMAID), has been developed (Saalfeld *et al.*, 2009).

SSA works particularly well when worms are mounted in a way that can be cut and imaged in cross section. We usually collect the ribbons for serial sectioning in pairs on filmed 1×2 mm slot grids. In general, the sectioning through whole worms is technically demanding and time consuming. In addition, there is always the danger of losing sections during ultramicrotomy and ribbon uptake, which results in "gaps" in the 3D reconstructions. To avoid loss of sections, automated high-throughput serial sectioning using the ATLUM approach (Automatic Tape-Collecting Lathe Ultramicrotome) has been developed. The ATLUM operation allows a spiral cut through a sample block, yielding a continuous ribbon of the biological material in the knife's water boat. The ribbons are collected by a submerged conveyor belt, allowing the production of ultra-thin section libraries for imaging by SEM. For more information, the reader is referred to http://www.mcb.harvard.edu/lichtman/ATLUM/ATLUM_Web.htm. To our knowledge, this method has not been applied yet for either *C. elegans* worms or early embryos.

A different aspect that cannot be solved by continuous sectioning, however, is the fact that serial sections never fit perfectly when stacked *in silico*. The sectioning process itself induces some degree of compression, and the interaction of the electron beam with the plastic section causes both "global" and "local" distortions. Software packages, such as IMOD, allow to correct some of these distortions computationally (Mastronarde *et al.*, 1993; McDonald *et al.*, 1992). These distortions can be a serious problem when large areas have to be reconstructed in 3D.

2. Electron Tomography

The fact that the nematode worm develops in an amazingly stereotypical manner (Sulston and Horvitz, 1977) and the availability of a genome-wide RNAi screen, which allowed a classification of cellular defects from a collection of over 100,000 videos (Sonnichsen *et al.*, 2005), made the early *C. elegans* embryo a popular model system for a number of cell biological questions. In recent years, we have used ET of plastic-embedded material for systematic structure and function analyses (reviewed by Muller-Reichert *et al.*, 2010). In particular, we have analyzed the meiotic and mitotic spindle organization (O'Toole *et al.*, 2003; Ozlu *et al.*, 2005; Schlaitz *et al.*, 2007; Srayko *et al.*, 2006), the interaction of microtubules with kinetochores and centrosomes (O'Toole *et al.*, 2003), and structural intermediates of centriole duplication (Pelletier *et al.*, 2006). As mentioned, the volume that can be reconstructed by ET is rather small. In our studies, we have applied "montaging" of several tomograms in combination with serial tomography to enlarge the reconstructed volume, but we did not intend to reconstruct a whole mitotic spindle by ET (O'Toole *et al.*, 2003). It is worth mentioning here that STEM tomography can be applied to use plastic sections with a thickness of up to 600 nm (Chapter 25 by Walther *et al.*, 2010, this volume). To our knowledge, this method has not been applied so far for *C. elegans* samples.

It is important to note here that ET can also be applied to visualize frozen-hydrated samples in the TEM at liquid nitrogen temperature (Al-Amoudi *et al.*, 2007; Beck *et al.*, 2007; Medalia *et al.*, 2002). Such samples for cryo-ET do not go through dehydration, fixation, or staining procedures as described for plastic embedding and have the potential to deliver images with higher resolution. Imaging of frozen-hydrated samples, however, has to be achieved at low-dose conditions (Al-Amoudi *et al.*, 2004). Searching for specific "spots" within low-contrast cryo-sections of cells, however, is/remains a general problem of this method. Recently, it was hoped that tagging selected proteins by green fluorescent protein (GFP) might facilitate the identification of organelles in cryo-sections as visualized by deep-temperature light microscopy prior to ET (Gruska *et al.*, 2008; Schwartz *et al.*, 2007). It has to be kept in mind, however, that GFP signals in cells can be rather "weak" and that cryo-sections are usually much thinner compared to plastic sections. In addition, serial cryo-sectioning is not a trivial task. When searching for a specific structure, such as the centrosome in the "huge" early embryo, is necessary, ET of plastic sections has certainly an advantage over cryo-ET.

We routinely use the IMOD software package, which contains all of the programs needed for calculating tomograms and for the display and modeling of subcellular features within the reconstructed volume (Kremer *et al.*, 1996). The IMOD software package runs on multiple platforms, including Linux, Mac OSX, and Windows. The programs used for tomographic reconstruction are managed by a graphical user interface, eTomo. The eTomo interface facilitates the ease with which users go through the various steps of the process, much like a flow chart. Image display and modeling are carried out with the "3dmod" viewing program from the IMOD software package. This program can be run by command line; it contains windows for image display and the slicer tool for rotating slices of image data and for modeling features of interest in the reconstruction. Importantly, the IMOD software can also be used to model and reconstruct data obtained from either serial sectioning or block face imaging using SEM.

EM, using either of the methods presented above, delivers a high-resolution snapshot of biological samples, and this snapshot is ideally obtained by HPF of either whole worms (Muller-Reichert *et al.*, 2003) or isolated, staged early embryos (Muller-Reichert *et al.*, 2008). Biological processes, however, are dynamic and it is very powerful to combine the strength of the light microscope with the potential of the electron microscope. The general advantages of such a CLEM approach in combination with HPF have been highlighted recently (McDonald, 2009; Verkade, 2008). One of the most important, and often overlooked, aspects of CLEM is the ability to increase the sample size for EM analysis. Using CLEM methods means that you always know that your EM sample will contain exactly what you want to image. But if the feature of interest is small in size it may require pre-screening of serial semi-thick sections prior to carrying out ET (Muller-Reichert *et al.*, 2007).

3. Block Face Imaging

As mentioned above, 3D reconstruction of whole worms can be achieved by SSA with the disadvantages that cutting of thin sections (i.e., 50–80 nm) is time consuming and demanding, and that images of serial sections do not allow a perfect superposition due to distortions. ET can be considered to take a somewhat "opposite position" to this. Very thin tomographic slices can be produced computationally (i.e., in the range of 2.5 nm). As mentioned, the volume of reconstruction is rather low, but can be increased by combining several datasets and applying "serial" ET (Hoog *et al.*, 2007; Marsh *et al.*, 2001; Noske *et al.*, 2008). One would therefore hope to use a microscope where serial sectioning of rather thin slices is automated, and where the disadvantageous distortions due to the sectioning process can be avoided. Block face imaging by SEM is considered to occupy this "intermediate position." Using either SBF-SEM or FIB-SEM, the major advantage of this method is that imaging of the block face and not the section itself eliminates the creation of global and local distortions. The *Z*-resolution of this technique basically depends on the thickness of the material that can be removed per cutting cycle, and it appears that the removal of the resin by a focused ion beam has an advantage over the use of an in-column microtome.

A problem that might occur with block face imaging is a rather low contrast of the biological specimen. In general, contrasting of samples as carried out for serial sections cannot be applied because the material is removed during each slicing step. We found that the contrast of samples after FS in acetone containing 1% osmium tetroxide and 0.1% uranyl acetate was acceptable for the imaging of cross sections of L3 larval worms. However, the contrast was not sufficient to visualize spindle microtubules in early embryos. Addition of 0.1% tannic acid to the FS "cocktail" obviously increases the contrast in mouse brain tissue prepared for block face imaging (Moebius *et al.*, 2010, this volume). Further staining methods need to be developed in the future and will be critical for the success of block face imaging in general. We have also noticed that our routine Epon/Araldite recipe is not ideal for block face imaging by FIB-SEM. Instead, we followed the protocol published by Knott *et al.* to embed samples in Durcupan (Knott *et al.*, 2008). Such prepared samples turned out to be harder and advantageous for the milling of material.

Segmentation of data obtained from SSA, ET, or block face imaging is currently the "bottleneck" in 3D reconstruction. When modeling spindle microtubules in tomograms, for example, the path of individual microtubules has to be modeled manually using the "slicer window" in IMOD (O'Toole and Muller-Reichert, 2009). It will be crucial for future studies to develop software packages that will allow the segmentation of organelles and even whole cells automatically (Sandberg, 2007). Efforts toward automation of segmentation are most likely the most important steps for future progress in the 3D EM field.

Acknowledgments

The authors would like to thank Jana Mäntler (MPI-CBG, Dresden) for excellent technical assistance. The lab of TMR is supported by the Deutsche Forschungsgemeinschaft (DFG MU 1423/2-1 and 1423/3-1).

References

Al-Amoudi, A., Diez, D. C., Betts, M. J., and Frangakis, A. S. (2007). The molecular architecture of cadherins in native epidermal desmosomes. *Nature* **450**, 832–837.

Al-Amoudi, A., Norlen, L. P., and Dubochet, J. (2004). Cryo-electron microscopy of vitreous sections of native biological cells and tissues. *J. Struct. Biol.* **148**, 131–135.

Beck, M., Lucic, V., Forster, F., Baumeister, W., and Medalia, O. (2007). Snapshots of nuclear pore complexes in action captured by cryo-electron tomography. *Nature* **449**, 611–615.

Brenner, S. (2009). In the beginning was the worm. *Genetics* **182**, 413–415.

Buser, C., and Walther, P. (2008). Freeze-substitution: The addition of water to polar solvents enhances the retention of structure and acts at temperatures around −60 degrees C. *J. Microsc.* **230**, 268–277.

Denk, W., and Horstmann, H. (2004). Serial block-face scanning electron microscopy to reconstruct three-dimensional tissue nanostructure. *PLoS Biol.* **2**, e329.

Evans, J.E., Snow, J. J., Gunnarson, A. L., Ou, G., Stahlberg, H., McDonald, K. L., and Scholey, J. M. (2006). Functional modulation of IFT kinesins extends the sensory repertoire of ciliated neurons in *Caenorhabditis elegans*. *J. Cell Biol.* **172**, 663–669.

Favre, R., Cermola, M., Nunes, C. P., Hermann, R., Muller, M., and Bazzicalupo, P. (1998). Immuno-cross-reactivity of CUT-1 and cuticlin epitopes between *Ascaris lumbricoides, Caenorhabditis elegans*, and *Heterorhabditis*. *J. Struct. Biol.* **123**, 1–7.

Favre, R., Hermann, R., Cermola, M., Hohenberg, H., Muller, M., and Bazzicalupo, P. (1995). Immuno-gold-labelling of CUT-1, CUT-2 and cuticlin epitopes in *Caenorhabditis elegans* and *Heterorhabditis sp.* Processed by high pressure freezing and freeze-substitution. *J. Submicrosc. Cytol. Pathol.* **27**, 341–347.

Frank, J. (1992). "Electron Tomography" Plenum Press, New York.

Frank, J., Wagenknecht, T., McEwen, B. F., Marko, M., Hsieh, C. E., and Mannella, C. A. (2002). Three-dimensional imaging of biological complexity. *J. Struct. Biol.* **138**, 85–91.

Gruska, M., Medalia, O., Baumeister, W., and Leis, A. (2008). Electron tomography of vitreous sections from cultured mammalian cells. *J. Struct. Biol.* **161**, 384–392.

Hall, D. H. (1995). Electron microscopy and three-dimensional image reconstruction. *Methods Cell Biol.* **48**, 395–436.

Hall, D. H., and Russell, R. L. (1991). The posterior nervous system of the nematode *Caenorhabditis elegans*: serial reconstruction of identified neurons and complete pattern of synaptic interactions. *J. Neurosci.* **11**, 1–22.

Hohenberg, H., Mannweiler, K., and Muller, M. (1994). High-pressure freezing of cell suspensions in cellulose capillary tubes. *J. Microsc.* **175**, 34–43.

Hoog, J. L., Schwartz, C., Noon, A. T., O'Toole, E. T., Mastronarde, D. N., McIntosh, J. R., and Antony, C. (2007). Organization of interphase microtubules in fission yeast analyzed by electron tomography. *Dev. Cell* **12**, 349–361.

Humbel, B. (2009). Freeze-substitution. *In* "Handbook of Cryo-preparation Methods for Electron Microscopy" (A. Cavalier, D. Spehner, and B. M. Humbel, eds.), pp. 320–341. CRC Press, Boca Raton.

Humbel, B., and Müller, M. (1986). Freeze substitution and low temperature embedding. *In* "The Science of Biological Specimen Preparation 1985" (J. J. Wolosewick, ed.), pp. 175–183. SEM, Inc., AMF O'Hare, Chicago, IL.

Kang, B.-H. (2010). Electron Microscopy and high-pressure freezing of *Arabidopsis*. *Methods Cell Biol.* **96**, 245–268.

Knott, G., Marchman, H., Wall, D., and Lich, B. (2008). Serial section scanning electron microscopy of adult brain tissue using focused ion beam milling. *J. Neurosci.* **28**, 2959–2964.

Kremer, J. R., Mastronarde, D. N., and McIntosh, J. R. (1996). Computer visualization of three-dimensional image data using IMOD. *J. Struct. Biol.* **116**, 71–76.

Ladinsky, M. S., Mastronarde, D. N., McIntosh, J. R., Howell, K. E., and Staehelin, L. A. (1999). Golgi structure in three dimensions: Functional insights from the normal rat kidney cell. *J. Cell Biol.* **144**, 1135–1149.

Leunissen, J. L., and Yi, H. (2009). Self-pressurized rapid freezing (SPRF): A novel cryofixation method for specimen preparation in electron microscopy. *J. Microsc.* **235**, 25–35.

Marsh, B. J., Mastronarde, D. N., Buttle, K. F., Howell, K. E., and McIntosh, J. R. (2001). Organellar relationships in the Golgi region of the pancreatic beta cell line, HIT-T15, visualized by high resolution electron tomography. *Proc. Natl. Acad. Sci. U. S.A.* **98**, 2399–2406.

Mastronarde, D. N. (1997). Dual-axis tomography: An approach with alignment methods that preserve resolution. *J. Struct. Biol.* **120**, 343–352.

Mastronarde, D. N., McDonald, K. L., Ding, R., and McIntosh, J. R. (1993). Interpolar spindle microtubules in PTK cells. *J. Cell Biol.* **123**, 1475–1489.

McDonald, K. L. (1994). Electron microscopy and EM immunocytochemistry. *Methods Cell Biol.* **44**, 411–444.

McDonald, K. (1999). High-pressure freezing for preservation of high resolution fine structure and antigenicity for immunolabeling. *Methods Mol. Biol.* **117**, 77–97.

McDonald, K. L. (2009). A review of high-pressure freezing preparation techniques for correlative light and electron microscopy of the same cells and tissues. *J. Microsc.* **235**, 273–281.

McDonald, K., Schwarz, H., Muller-Reichert, T., Webb, R., Buser, C., and Morphew, M. (2010). 'Tips and tricks' for high-pressure freezing of model systems. *Methods Cell Biol.* **96**, 639–660.

McDonald, K. L., Morphew, M., Verkade, P., and Muller-Reichert, T. (2007). Recent advances in high-pressure freezing: Equipment- and specimen-loading methods. *Methods Mol. Biol.* **369**, 143–173.

McDonald, K., and Muller-Reichert, T. (2002). Cryomethods for thin section electron microscopy. *Methods Enzymol.* **351**, 96–123.

McDonald, K. L., O'Toole, E. T., Mastronarde, D. N., and McIntosh, J. R. (1992). Kinetochore microtubules in PTK cells. *J. Cell Biol.* **118**, 369–383.

Medalia, O., Weber, I., Frangakis, A. S., Nicastro, D., Gerisch, G., and Baumeister, W. (2002). Macromolecular architecture in eukaryotic cells visualized by cryoelectron tomography. *Science* **298**, 1209–1213.

Mishchenko, Y. (2009). Automation of 3D reconstruction of neural tissue from large volume of conventional serial section transmission electron micrographs. *J. Neurosci. Methods* **176**, 276–289.

Möbius, W., Cooper, B., Kaufmann, W. A., Imig, C., Ruhwedel, T., Snaidero, N., Saab, A. S.., and Varoqueaux, F. (2010). Electron microscopy of the mouse central nervous system. *Methods Cell Biol.* **96**, 453–588.

Moor, H., and Riehle, U. (1968). Snap-freezing under high pressure: A new fixation technique for freeze etching. *In* "Electron Microscopy 1968. Pre-Congress Abstracts of Papers Presented at the Fourth European Regional Conference, September 1-7" (D. A. Bocciarelli, ed.), pp. 33–34. Tipografia Poliglotta Vaticana, Rome.

Moor, H. (1987). Theory and practice of high pressure freezing. *In* "Cryotechniques in Biological Electron Microscopy" (K. Zierold, ed.), pp. 175–191. Springer, Berlin, Heidelberg.

Müller, M., and Moor, H. (1984). Cryofixation of thick specimens by high-pressure freezing. *In* "The science of biological specimen preparation" (G.H. Haggins, ed.), pp. 131–138. SEM, Inc., AMF O'Hare, chicago, IL.

Muller-Reichert, T., Greenan, G., O'Toole, E., and Srayko, M. (2010). The *elegans* of spindle assembly. *Cell Mol. Life Sci.* **67**, 2195–2213.

Muller-Reichert, T., Hohenberg, H., O'Toole, E. T., and McDonald, K. (2003). Cryoimmobilization and three-dimensional visualization of *C. elegans* ultrastructure. *J. Microsc.* **212**, 71–80.

Muller-Reichert, T., Mantler, J., Srayko, M., and O'Toole, E. (2008). Electron microscopy of the early *Caenorhabditis elegans* embryo. *J. Microsc.* **230**, 297–307.

Muller-Reichert, T., Srayko, M., Hyman, A., O'Toole, E. T., and McDonald, K. (2007). Correlative light and electron microscopy of early *Caenorhabditis elegans* embryos in mitosis. *Methods Cell Biol.* **79**, 101–119.

Noske, A. B., Costin, A. J., Morgan, G. P., and Marsh, B. J. (2008). Expedited approaches to whole cell electron tomography and organelle mark-up *in situ* in high-pressure frozen pancreatic islets. *J. Struct. Biol.* **161**, 298–313.

O'Toole, E. T., McDonald, K. L., Mantler, J., McIntosh, J. R., Hyman, A. A., and Muller-Reichert, T. (2003). Morphologically distinct microtubule ends in the mitotic centrosome of *Caenorhabditis elegans*. *J. Cell Biol.* **163**, 451–456.

O'Toole, E., and Muller-Reichert, T. (2009). Electron tomography of microtubule end-morphologies in *C. elegans* embryos. *Methods Mol. Biol.* **545**, 135–144.

Ozlu, N., Srayko, M., Kinoshita, K., Habermann, B., O'Toole, E. T., Muller-Reichert, T., Schmalz, N., Desai, A., and Hyman, A. A. (2005). An essential function of the *C. elegans* ortholog of TPX2 is to localize activated aurora A kinase to mitotic spindles. *Dev. Cell* **9**, 237-248.

Pelletier, L., O'Toole, E., Schwager, A., Hyman, A. A., and Muller-Reichert, T. (2006). Centriole assembly in *Caenorhabditis elegans*. *Nature* **444**, 619–623.

Ragsdale, E. J., Ngo, P. T., Crum, J., Ellisman, M. H., and Baldwin, J. G. (2009). Comparative, three-dimensional anterior sensory reconstruction of *Aphelenchus avenae* (nematoda: Tylenchomorpha). *J. Comp. Neurol.* **517**, 616–632.

Saalfeld, S., Cardona, A., Hartenstein, V., and Tomancák, P. (2009). CATMAID: collaborative annotation toolkit for massive amounts of image data. *Bioinformatics*. **25**, 1984–1986.

Sandberg, K. (2007). Methods for image segmentation in cellular tomography. *Methods Cell Biol.* **79**, 770–799.

Schlaitz, A. L., Srayko, M., Dammermann, A., Quintin, S., Wielsch, N., MacLeod, I., de Robillard, Q., Zinke, A., Yates, J. R. III, Muller-Reichert, T., *et al.,* (2007). The *C. elegans* RSA complex localizes protein phosphatase 2a to centrosomes and regulates mitotic spindle assembly. *Cell* **128**, 115–127.

Schwartz, C. L., Sarbash, V. I., Ataullakhanov, F. I., McIntosh, J. R., and Nicastro, D. (2007). Cryo-fluorescence microscopy facilitates correlations between light and cryo-electron microscopy and reduces the rate of photobleaching. *J. Microsc.* **227**, 98–109.

Sims, P. A., and Hardin, J. D. (2007). Fluorescence-integrated transmission electron microscopy images: Integrating fluorescence microscopy with transmission electron microscopy. *Methods Mol. Biol.* **369**, 291–308.

Sonnichsen, B., Koski, L. B., Walsh, A., Marschall, P., Neumann, B., Brehm, M., Alleaume, A. M., Artelt, J., Bettencourt, P., Cassin, E., *et al.,* (2005). Full-genome RNAi profiling of early embryogenesis in *Caenorhabditis elegans. Nature* **434**, 462–469.

Srayko, M., O'Toole, E. T., Hyman, A. A., and Muller-Reichert, T. (2006). Katanin disrupts the microtubule lattice and increases polymer number in *C. elegans* meiosis. *Curr. Biol.* **16**, 1944–1949.

Sulston, J. E., and Horvitz, H. R. (1977). Post-embryonic cell lineages of the nematode, *Caenorhabditis elegans. Dev. Biol.* **56**, 110–156.

Verkade, P. (2008). Moving EM: The rapid transfer system as a new tool for correlative light and electron microscopy and high throughput for high-pressure freezing. *J. Microsc.* **230**, 317–328.

Walther, P., Wang, L., Liebem, S., and Frascaroli, G. (2010). Viral infection of cells in culture: Approaches for electron microscopy. *Methods Cell Biol.* **96**, 573–588.

Ward, S., Thomson, N., White, J. G., and Brenner, S. (1975). Electron microscopical reconstruction of the anterior sensory anatomy of the nematode *Caenorhabditis elegans. J. Comp. Neurol.* **160**, 313–337.

Ware, R. W., Clark, D., Crossland, K., and Russel, R. L. (1975). The nerve ring of the nematode *Caenorhabditis elegans*: Sensory input and motor output. *J. Comp. Neurol.* **162**, 71–110.

White, J. G., Southgate, E., Thomson, J. N., and Brenner, S. (1976). The structure of the ventral nerve cord of *Caenorhabditis elegans. Philos. Trans. R. Soc. Lond., B, Biol. Sci.* **275**, 327–348.

White, J. G., Southgate, E., Thomson, J. N., and Brenner, S. (1986). The structure of the nervous system of *Caenorhabditis elegans. Philos. Trans. R. Soc. Lond. [Biol.]* **314**, 1–340.

CHAPTER 16

Insects as Model Systems in Cell Biology

Thomas A. Keil* ***and*** **R. Alexander Steinbrecht†**

*Department Molecular Structural Biology, Max-Planck-Institute for Biochemistry, D-82152 Martinsried, Germany

†Max Planck Institute for Ornithology, D-82319 Seewiesen, Germany

METHODS IN CELL BIOLOGY, VOL. 96

978-0-12-381007-6
DOI: 10.1016/S0091-679X(10)96016-0

Abstract

For almost 100 years, insects have been favorable "model systems" in biology. Just to mention a few examples: fruit flies in genetics and developmental biology; bugs and caterpillars in hormone research; houseflies, blowflies, and locusts in neurobiology; silk moths in pheromone research; honeybees and crickets in neuroethology. For more than 50 years the electron microscope (EM) has been a valuable tool in analyzing the structure of cells and organs of these creatures. However, progress in specimen preparation was relatively slow compared with mammalian material and, in 1970, it was taken for granted that insects were much more difficult to fix than mammals. Since then, methods have dramatically improved, and satisfactory results can now be obtained routinely with chemical as well as cryofixation. In this chapter we briefly demonstrate what can be achieved with insect material, and help the researcher to find the most appropriate method for her/his systems and scientific questions.

I. Introduction

Insects have attracted the attention of early electron microscopists because of the high stability of their chitinous exoskeleton: The wing of the housefly was among the first biological objects to enter the high vacuum of an electron microscope (EM) (Driest and Müller, 1935). The fine structure of the insect cuticle was addressed in a series of papers by A. Glenn Richards together with T. F. Anderson in 1942, in which the secret of the lepidopteran wing scales' iridescence was also revealed. The latter topic was independently addressed in Germany at the same time (Anderson and Richards, 1942; Richards and Anderson, 1942; Gentil, 1942; Kinder and Süffert, 1943; Kühn, 1946).

During the 1950s and 1960s, insects gained increasing importance as experimental animals. This is especially true for research fields such as neuroscience, where their anatomical construction turned out to be advantageous for electrophysiological recording. Pioneering work was started during this time on the senses of taste (Dethier, 1955; Hodgson and Roeder, 1956), olfaction (Schneider, 1957, 1962), vision (Autrum and Burkhardt, 1960; Burkhardt and Autrum, 1960), and mechanoreception (Wolbarsht, 1960; Thurm, 1964, 1965). Consequently, the interest in the microanatomy of insect sensory organs grew. EM investigation of their structure started with the work of Eleanor H. Slifer and coworkers (1957, 1959) being mainly concerned with chemoreception. Yet it must be admitted that the quality of specimen preparation left much to be desired. A thorough analysis of the insect ear by E. G. Gray followed in 1960. H. Fernandez-Moran (1958) and O. Trujillo-Cenoz (1965) addressed the fine structure of the insect eye. A good example of the usefulness of the EM is the analysis of the mechanotransduction mechanism, started by U. Thurm in the 1960s and finally leading to a molecular model over 30 years later (Howard and Bechstedt, 2004).

However, we must avoid the impression that only neurobiologists recognized the usefulness of insects as experimental animals. The problem of osmoregulation in mosquito larvae was addressed by the Grand Old Man of insect physiology, Vincent B. Wigglesworth, in 1933 (Wigglesworth, 1933). The cells active in this process were localized on the "anal gills." Thirty years later, an EM analysis of these cells led Copeland (1964) to formulate his theory of the "mitochondrial pump." Gupta and Berridge (1966) came to the conclusion that "a coat of repeating subunits" on the cytoplasmic face of the membrane was active in ion transport. The same coat was found in the *Cecropia* midgut (Anderson and Harvey, 1966) and in ion-transporting cells of insect sensilla (Thurm and Küppers, 1980). These subunits, called "portasomes" by Harvey (1980), were later identified as V-ATPases and intensively studied (reviewed by Harvey, 2009). On the other hand, it could be shown that their transport activity is hormonally controlled by 5-hydroxytryptamine and cyclic AMP (Berridge, 1970). All these findings were made possible by careful EM analysis in the first place.

Mosquitoes were the subject of another EM work leading to important insights into cellular processes of general biological importance: Roth and Porter (1964) investigated the uptake of yolk proteins by the oocyte and discovered the structures that are now well-known as "coated vesicles."

As already mentioned, electron microscopic work on the insect cuticle had started with A. G. Richards in 1942. Careful EM investigations, in combination with biochemical and molecular biological methods, made it clear that cuticle is a natural fibrous composite material, therefore being an object of interest for materials scientists (reviewed by Neville, 1993). "Insects for studying fundamental problems in biology" is the title of a famous lecture of Michael Locke (2001), who throughout his career enthusiastically promoted the use of insects for a great variety of cell biological questions.

The interest in insects grew beyond their advantage as experimental animals. Insects are of great medical (mosquitoes, flies, bugs) and economic (honeybee, silk moth) importance, many species are serious pests in agriculture (beetles, moths), and their ecological importance becomes more and more clear (honeybee decline). And of course developmental genetics could not be imagined without the fruit fly, *Drosophila melanogaster*. Especially in this field, the high resolution of the EM is needed for a proper interpretation of the results (e.g., Avidor-Reiss *et al.*, 2004).

Another point to be considered is that insects are extremely useful for teaching purposes. Most aspects of cell and tissue biology can be covered with examples from the "insect world." Insect tissues can be easily excised with a little skill, but always without bloodshed. These few examples were chosen to illustrate that electron microscopical investigation of insect material had, and still has, an impact on biomedical as well as biomaterials research far beyond the entomologist's horizon.

The major problems for EM preparation were, and still are, the impermeability of the hydrophobous insect cuticle for fixatives as well as the difficulty of producing ultrathin sections of the tough cuticle, which can only be overcome by using diamond knives. A major breakthrough came with the work of Michael Locke (1961) and also after the introduction of glutaraldehyde (GA) fixation (Sabatini *et al.*, 1963). A certain standard was reached with the publication of "Insect Cells" by David S. Smith in 1968 (Smith, 1968). Later, the state of the art in insect ultrastructure research was summarized in the two-volume "Insect Ultrastructure" by King and Akai (1982, 1984) and finally in three volumes of "Microscopic Anatomy of Invertebrates," edited by Harrison and Locke (1998).

Soft tissue and cells, such as muscle, gut, Malphigian tubules, salivary glands, fat body, and hemocytes, no longer pose a problem, whereas the insect brain with its efficient blood–brain barrier is still a challenge. Good preservation of tissues and organs covered or encapsulated by cuticle, e.g., insect sensilla, is difficult with conventional fixation techniques, but can be achieved by the application of rapid freezing and cryosubstitution (Steinbrecht, 1976, 1980).

II. Rationale

For many years, we have worked with a number of insect species, namely, the blowfly *Calliphora vicina*, the fruit fly *D. melanogaster*, the mosquito *Aedes aegypti*, the silk moths *Antheraea polyphemus* (adult and pupal stage) and *Bombyx mori*, the hawk moth *Manduca sexta*, the house cricket *Acheta domesticus*, the louse *Pediculus humanus*, and the bedbug *Cimex lectularius*. Here we present reliable standard protocols for the preparation of insect material, yielding satisfactory results without highly sophisticated and expensive equipment. We also describe some methods that can be applied to address special cell biological problems, such as enhancing the contrast of cell surfaces and extracellular matrix, detection of extracellular pathways, and immunolocalization of proteins. In the following section, the chemical fixation methods will be dealt with by Thomas A. Keil, whereas R. Alexander Steinbrecht will deal with cryopreparation and immunocytochemistry.

III. Methods

A. Standard Protocols for Chemical Fixation

The impermeability of the hydrophobous insect cuticle remains a problem for fixation with aqueous solutions. This is especially true for small, tubular structures, such as olfactory sensilla. For the latter, rapid freezing in combination with freeze substitution can be an alternative (discussed below). For chemical fixation, care must always be taken that access from the haemolymph side is possible. Larger tubular structures, such as antennae, cerci, or legs, must be cut into short pieces and, if possible, opened at both ends. Long and thin tubular structures, such as olfactory sensilla, can be pinched open at their tips before immersion in fixative (Keil, 1982).

A reliable method for EM investigation of insect tissues is fixation in GA, followed by post-fixation in osmium tetroxide, all at low temperatures (4°C). This can be followed by *en bloc* staining in "hot" uranyl acetate, as introduced by Locke *et al.* (1971). The advantage of the last step is that post-staining of sections usually can be omitted, and thus, additional contamination is avoided. Moreover, time is saved, as the

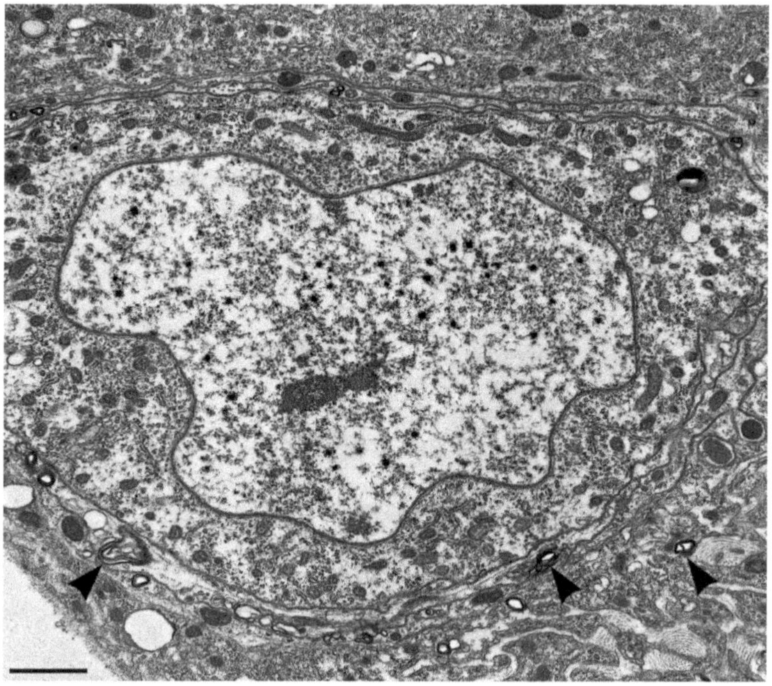

Fig. 1 Mechanosensory neuron (of a macrochaeta) in the thoracal epidermis of the blowfly *Calliphora vicina*. Note that the cell membranes do not appear smooth and that myelin bodies (arrowheads) are induced by chemical fixation. Standard fixation without tannic acid. Scale bar is 1 μm.

dried sections can be investigated immediately after ultramicrotomy. A typical example is shown in Fig. 1.

Procedure:

1. 2.5% GA (opened bottles can be stored for a long time under an inert gas, such as nitrogen or argon) in 0.1 M Na-cacodylate buffer, containing 5% sucrose, pH 7.2, 2–4 hr, 4°C.
2. Post-fixation in 1–2% OsO_4 in the same buffer, 1–2 h, 4°C.
3. En bloc staining in 1% aqueous uranyl acetate at 60°C overnight (Locke *et al.* (1971).
4. Dehydration in ethanol, embedding in Spurr's medium (optional), polymerization overnight at 70°C. It is highly recommended to use longer infiltration times as in Spurr's original recipe for cuticle material.
5. Normally, the contrast of the specimen will be sufficient after application of en bloc staining. However, it can be increased by additional staining of thin sections with 1% aqueous uranyl acetate and lead citrate (Reynolds, 1963) at room temperature.

B. Additional "Tricks"

Chemical fixation has the advantage that a number of additional protocols can be applied. Tannic acid (TA) and ruthenium red (RR) have proven useful for improving the appearance of membranous structures, certain proteins, extracellular matrix, and lymph (e.g., Luft, 1971a, b; Wagner, 1976). The delineation of intercellular clefts and paracellular pathways is possible by the application of lanthanum, but also RR. In the following sections, the use of these substances will be described in detail.

1. Tannic Acid

TA is used for enhancing the contrast of cell membranes, extracellular matrix compounds, and also microtubules. This improved contrast is reached by the strong binding of osmium. Examples are shown in Figs. 2–5 and S1–S4.

The solution has to be prepared fresh and filtered shortly before use. It can be applied either initially in combination with GA or, if you want to be on the safe side, after a pre-fixation with GA alone followed by a combination of GA and TA, always at 4°C. Find out yourself which concentration of TA is most convenient for your specimen. The incubation time is more critical: TA tends to precipitate when mixed with GA.

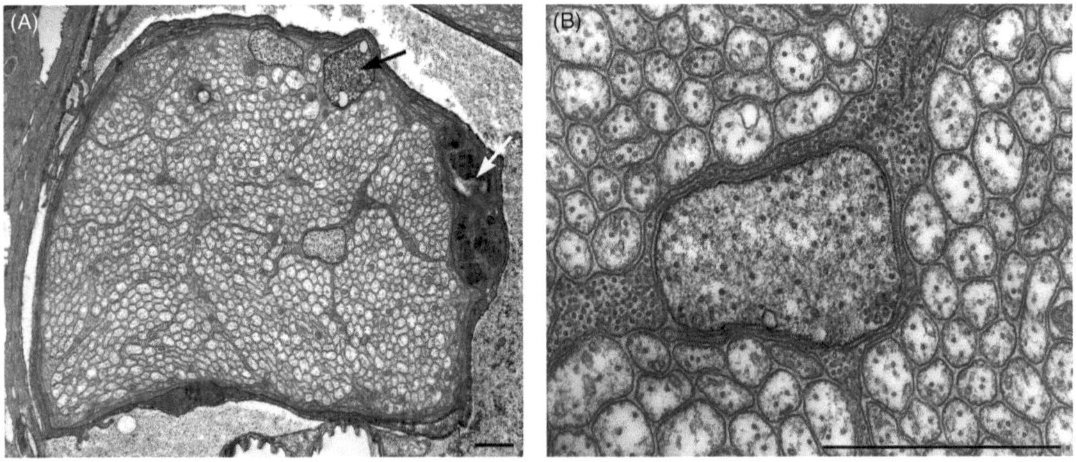

Fig. 2 Cross section of the antennal nerve of the silk moth *Antheraea polyphemus*, standard fixation with tannic acid (TA). Note uniform diameter (approximately 0.2 μm) of olfactory axons. (A) The neural lamella and the underlying glial envelope (white arrow) show dark staining due to the TA treatment, and the thick axon in the upper region has been heavily penetrated by TA (black arrow). (B) Axons and glial envelope at higher magnification. Note that the cytoplasm, microtubules, and membranes of the larger axon and the enveloping glial process are more heavily stained than those of the small olfactory axons. Scale bars are 1 μm.

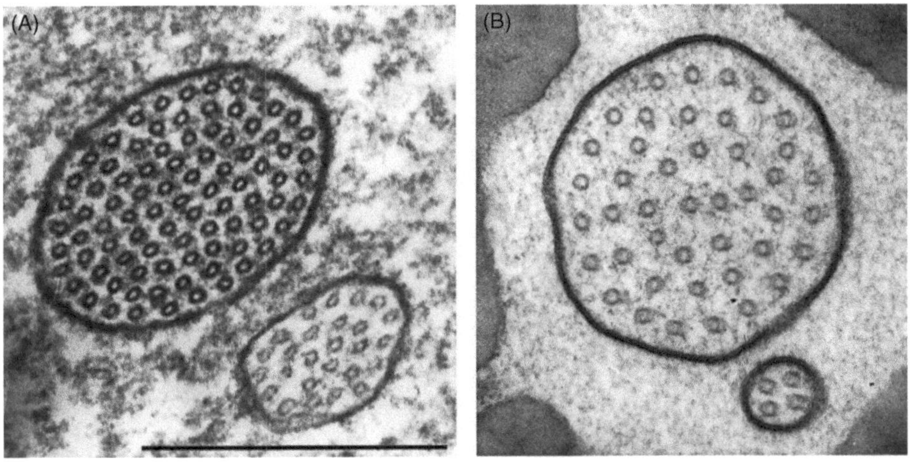

Fig. 3 Cross sections of pheromone-sensitive sensilla trichodea of the silk moths *Antheraea pernyi* (A) and *Bombyx mori* (B) showing two sensory dendrites of different diameter with their microtubular skeleton. (A) The specimen has been treated with tannic acid (TA), but not stained *en bloc* with uranyl acetate (UA). Instead, the section was stained with UA and Pb. If compared with the specimen stained *en bloc*, no differences can be seen. TA has penetrated only the thicker dendrite. The microtubules show heavy binding and a quite regular arrangement. Note the rather faint and more irregular shape of the microtubules in the thinner dendrite, which has not been penetrated by TA. However, the membranes of both dendrites show binding. Also the sensillum lymph shows a dark, flocculent precipitation. (B) Specimen cryosubstituted in acetone containing OsO_4. Note smooth dendritic membrane, evenly spaced microtubules, and homogeneous sensillum lymph around dendrites. Scale bar is 0.5 μm.

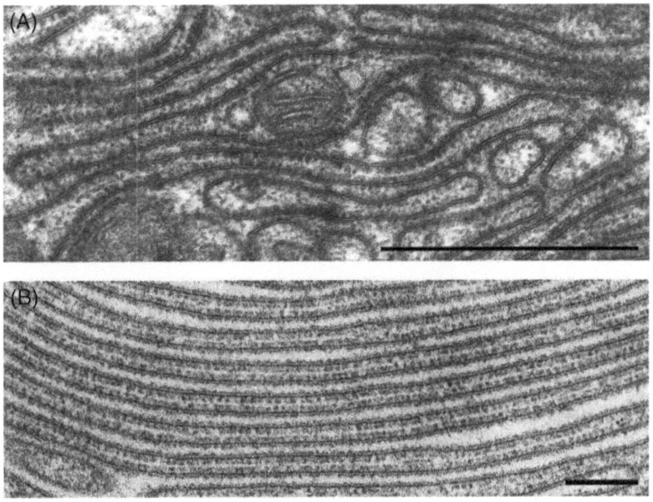

Fig. 4 Portasomes on the apical folds of sensillar auxiliary cells. These regular-spaced particles on the cytoplasmatic face of the lamellae represent a V-ATPase involved in ion transport into the sensillum lymph surrounding the receptor cell dendrites (see also Introduction). (A) Cercal mechanosensitive sensillum of *Achaeta domesticus* after glutaraldehyde fixation and TA treatment. (B) Olfactory sensillum on the antenna of *Bombyx mori* after cryosubstitution. Scale bars are 0.5 μm in A and 100 nm in B.

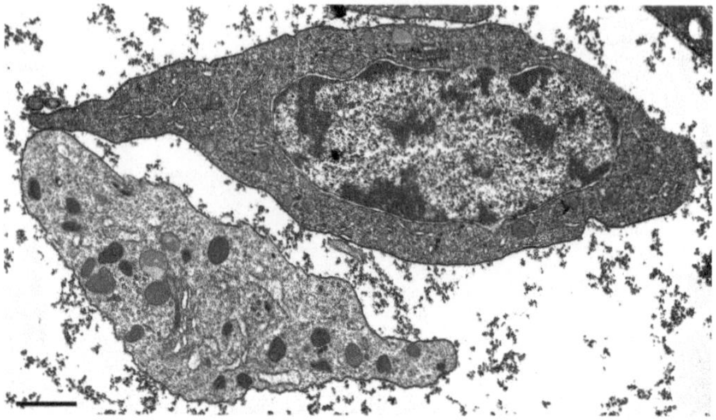

Fig. 5 Haemocytes in the cercus of *Acheta domesticus*, fixed *in situ*. Tannic acid has penetrated the cells to different degrees; the cytoplasm of the upper cell appears much more electron dense. There is precipitation on the cell membrane and in the hemolymph space. Scale bar is 1 μm.

Procedure:

1. Initial fixation in 2.5% GA in 0.1 M Na-cacodylate buffer, containing 5% sucrose, pH 7.2, 2–4 h, at 4°C.
2. A fresh and filtered aqueous solution of 2–8% TA is mixed 1:1 with 5% GA in 0.2 M Na-cacodylate buffer containing 10% sucrose, pH 7.2, immediately before use, and then the specimen is transferred from the initial fixative to this solution for approximately 1 h at 4°C.
3. Post-fixation in 1–2% OsO_4 in the same buffer, 1–2 h, at 4°C.
4. En bloc staining in 1% aqueous uranyl acetate at 60°C overnight.
5. Dehydration in ethanol, embedding in Spurr's medium, polymerization at 70°C.

2. Ruthenium Red

RR was introduced by Luft (1971a, b) for application in electron microscopy. It shows a strong binding to negatively charged substances. Therefore, extracellular matrix as well as the surface coat of the cell membranes is stained even more intensely than after the application of TA. Examples are shown in Fig. 6. Following Luft's original schedule, RR is applied directly with the initial fixation.

Procedure with the following mixture used by Keil (1979):

1. Initial fixation for approximately 20 h at 4°C in the following solution:

   3.6% GA
   0.5 M Na-cacodylate, pH 7.3
   0.6% RR
   at equal volumes.

2. Post-fixation:

   5% OsO_4 in H_2O
   0.2 M cacodylate buffer, pH 7.2
   0.6% RR in H_2O
   at equal volumes.

It had been shown that a simplification of Luft's original schedule, omitting the RR in the initial fixative, also yielded good ultrastructural preservation of cell structures, such as olfactory dendrites (Keil, 1984). The following procedure is therefore recommended:

1. Initial fixation in 2.5% GA in 0.1 M Na-cacodylate buffer, containing 5% sucrose, pH 7.2, 1 h, 4°C.
2. Incubation and post-fixation for approximately 3 h at room temperature in the following solution:

   4% OsO_4 in H_2O
   0.2 M cacodylate buffer, pH 7.2
   0.6% RR in H_2O
   at equal volumes.

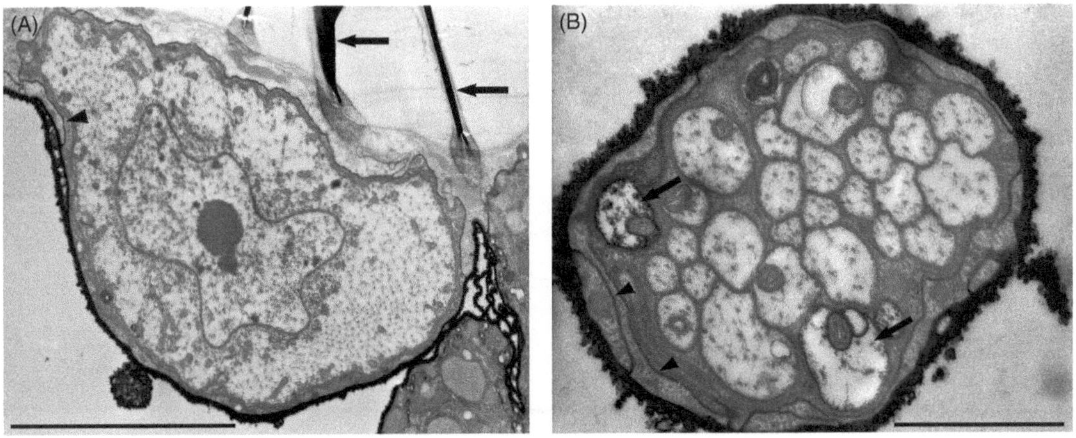

Fig. 6 Thorax of *Calliphora vicina* treated with ruthenium red (RR). (A) A similar mechanosensory neuron as in Fig. 1. RR binds strongly to the extracellular matrix (ECM) (=basal lamina) and to some degree the extracellular clefts (arrowhead), but has not entered the cell. Note the folds in the cuticle (arrows). Scale bar is 5 μm. (B) Thorax of *Calliphora vicina* treated with RR. Cross section of a small thoracal nerve, containg axons from thoracal bristles. Again, the ECM (=neural lamella) shows heavy staining by RR as well as some extracellular clefts (arrowheads). Note that RR has entered at least two of the axons, which therefore show dense staining of the mitochondrial membranes (arrows). Scale bar is 1 μm.

3. Lanthanum

Lanthanum is an excellent tracer substance, which diffuses readily into intercellular clefts. It has a highly charged cation, which binds to polyanions and can be precipitated as electron-dense phosphate; therefore, phosphate buffer has to be used in the fixation medium. Examples are shown in Figs. 7 and S5.

Procedure:

1a. Incubation (perfusion or superfusion) *in vivo*:

5 mM $LaCl_3$ or $La(NO_3)_3$ in the appropriate Ringer solution (e.g., hemolymph Ringer) or other buffer solution (e.g., 200 mM KCl, 50 mM NaCl, 5 mM Tris, pH 7.2) at room temperature for 10–60 min.

1b. Incubation after pre-fixation:

Initial fixation in 2.5% GA in 0.1 M Na-cacodylate buffer with 5% sucrose, pH 7.2, approximately 1 h, 4°C. Incubation with La^{+++} as in 1a. The tracer should now have access from the apical as well as the basal side of epithelia.

2. Fixation in 2.5% GA in 100 mM Sörensen's phosphate buffer, containing 5% sucrose, pH 7.2, 4°C for approximately 1 h. The phosphate buffer is necessary to precipitate the La^{+++}.
3. Post-fixation in 2% OsO_4 in the same buffer.

C. Cryofixation and Cryosubstitution (Freeze Substitution)

Cryofixation at atmospheric pressure is performed either by slamming the specimen onto a cold metal surface or by plunging the specimens into liquid propane cooled by liquid nitrogen (LN_2); liquid propane may also be jetted onto the specimen, while

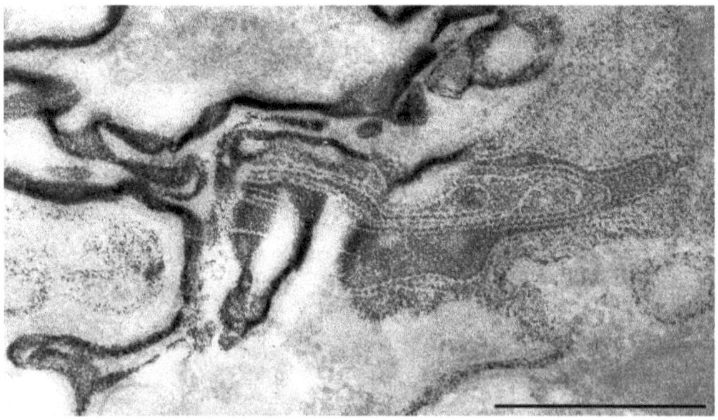

Fig. 7 Thoracal epidermis of *Calliphora vicina*, incubated with $LaCl_3$ from the hemolymph side and precipitated with phosphate buffer. The tracer has entered the convoluted intercellular clefts and clearly outlines the septate junctions in negative contrast, showing the irregular course and the zigzag shape of the septa. Scale bar is 0.5 μm

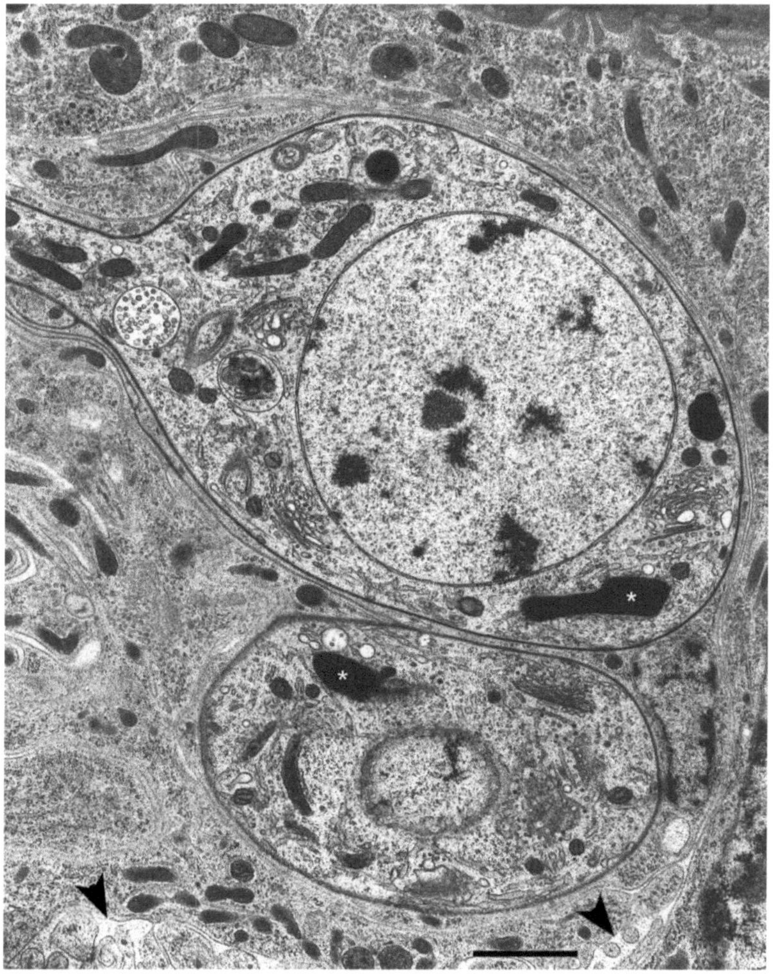

Fig. 8 Two pheromone receptor cells of the silk moth *Bombyx mori* surrounded by auxiliary cells; specimen plunge-frozen and cryosubstituted in acetone containing OsO_4. Note that the contours of the cell borders are smooth and free from wrinkles; mitochondria display a very dense matrix; extracellular space around the basal infoldings is abundant (arrowheads). Asterisks mark paracrystalline inclusions. Scale bar is 1 μm.

suspensions of minute specimens may be sprayed into the cold propane (for references, see Steinbrecht and Zierold, 1987 or Steinbrecht, 1993b).

A totally different technique is cryofixation at high pressure (around 2000 bar), called high-pressure freezing (HPF), which allows for cryofixing specimens of a thickness of up to 200 μm (Moor and Riehle, 1968; Shimoni and Müller, 1998). The application of HPF has been treated in detail elsewhere (see McDonald, 2007, and references therein).

For small insect appendages, such as antennae, cerci, wings, etc., plunging into propane proved the most practical approach. Examples are presented in Figs. 3, 4, 8, 9,

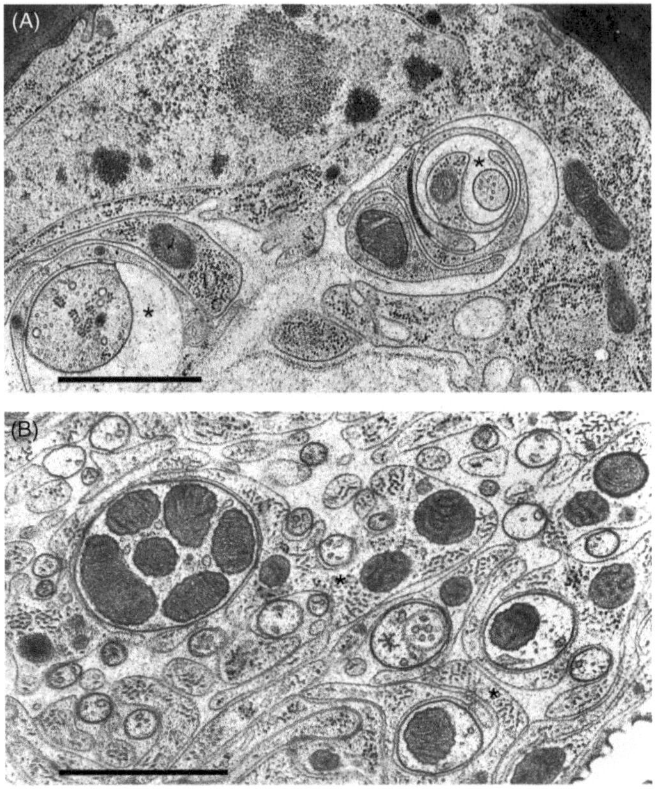

Fig. 9 Freezing damage in cryosubstituted specimens strongly depends on the size of the specimen and on the distance of the structures studied to the specimen surface. (A) *Drosophila melanogaster*, cross section through vein of wing. This tissue that is close to the surface is practically free from freezing damage; the outer mitochondrial membrane is smooth and the cristae are open. All cytomembranes are smooth; no ice crystal ghosts are detectable. Asterisks mark abundant extracellular space around two axons. (B) Antennal nerve of *Bombyx mori*. This sample is from approximately 40 μm below the antennal surface and shows freezing damage: mitochondria are indented by ice crystal ghosts (compare also with Fig. S9) and glial microtubuli are collapsed (asterisks). Nevertheless, tissue preservation may still be sufficient for morphological and morphometric studies, because the overall size and shape of the axon and glial profiles are unchanged. Note the abundant extracellular space as compared to conventionally fixed specimens (cf. Fig. 3). Scale bars are 1 μm.

S6–S9. Technical propane (bottle gas) can be supercooled almost to the temperature of LN_2 (−196°C) without solidifying. The Steinbrecht lab used a simple self-built apparatus, which can be easily assembled in any workshop (Fig. 10A). Propane is contained in a cylindrical aluminum vessel that is 13 mm wide and 80 mm deep, which is immersed into LN_2 for 30–50 mm. The narrow metal cylinder sustains a temperature of ≤−180°C, even close to the surface, without stirring; it is covered with a Teflon lid, except at the moment of plunging. For safety reasons, it is important to note here that all work with propane has to be done in a fume

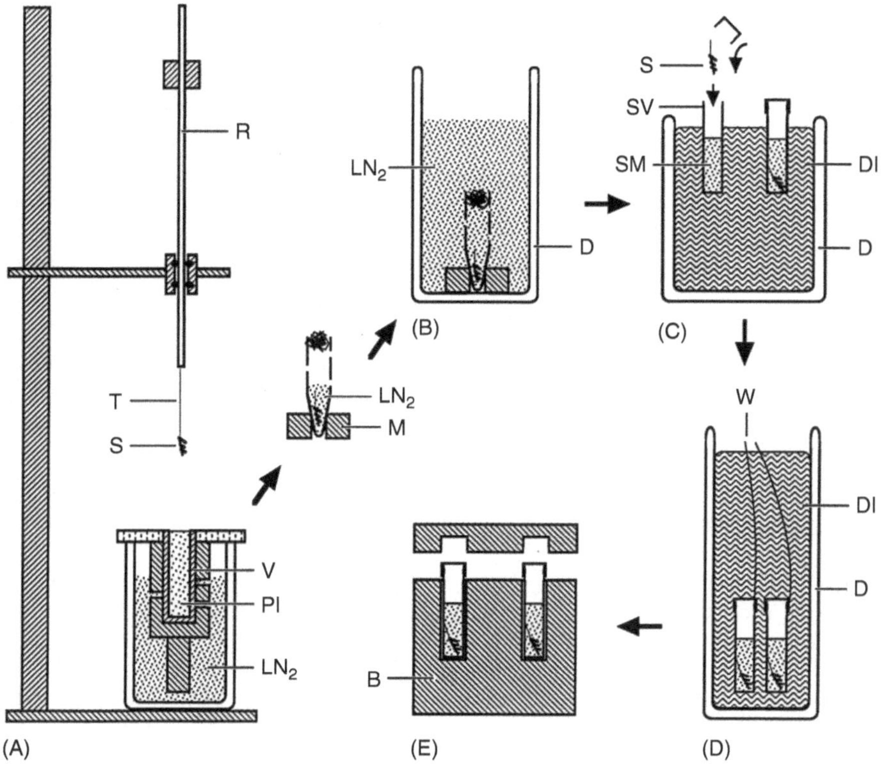

Fig. 10 Simple cryofixation apparatus and freeze substitution (FS) procedure, as used earlier in Steinbrecht's lab. (A) Schematic drawing of the plunging device. Parts made of aluminum are crosshatched, and those made of Teflon are stippled. The specimen (S) is mounted on a tungsten wire (T) fixed to a steel rod (R) guided by a ball bearing and is immersed by gravity into liquid propane (PI), contained in a narrow aluminum vessel (V), which in turn is cooled by liquid nitrogen (LN_2). The vessel is surrounded by an aluminum sleeve with holes so that LN_2 can reach up to half its length. By falling 25 cm the specimen accelerates to approximately 2 m/s. (B) Capsule for storage of cryofixed specimens under LN_2. M, metal nut. (C) Substitution vials (SV) with substitution medium (SM) cooled by dry ice (DI) are loaded with cryofixed specimens. (D) FS is carried out in a deep Dewar (D) filled with dry ice–acetone slush. W, wire for recovery of vial. (E) Slow rewarming after FS is achieved by placing the substitution vials into a large precooled aluminum block (B; approximately 3.5 kg) with tightly fitting holes. (From Steinbrecht, 1993a, *Microsc. Res. Tech.* **24**, 488–504 with permission of Longman Group Ltd.)

chamber, as propane is extremely flammable. The gravity-driven device, as described here, allowed immersion speeds of up to 5 m/s. However, velocities above 2 m/s produced no improvement. It is mandatory though that the specimen keeps moving rapidly in the cryogen during the critical first milliseconds (forced convection). This is achieved by mounting the specimen onto the tip of a long wire, which is fully immersed during plunging. We used the same tungsten wire (0.1 mm thick, 70 mm long) that is also used for electrophysiological recording. Tungsten wire can be electrolytically sharpened if necessary for impaling smaller specimens. Very small specimens, such as the antennae or wings of *Drosophila*, were left on the head and thorax, respectively, and the wire pierced the latter.

After cryofixation the specimens are transferred either directly to the substitution medium or to LN_2 for storage. The wire remains with the specimen to form a convenient handle after shortening to about 5 mm. Transfer by hand, if done quickly (<1 s), is usually without deleterious effects. With particularly sensitive specimens the method of Zalokar (1966) can be applied, in which the substitution medium and the cryogen for freezing are contained in the same vessel. A convenient storage vial for keeping specimens under LN_2 can be made from Eppendorf centrifuge tubes (1.5 ml) by screwing their conical end into metal nuts. Some holes in the upper part and a stopper made of cotton wool allow for the exchange of liquid and gaseous nitrogen (Fig. 10B). The weight of the metal nut keeps these "storage capsules" standing upright under LN_2 and its heat capacity prevents the LN_2 inside from boiling too much during transfer. Cryosubstitution is done in plastic vials that have tight closing lids. These vials can be immersed in dry ice (−79°C) mixed with technical-grade acetone for a good thermal contact (Fig. 10C and D). The following substitution media are recommended and have given the best results in the Steinbrecht lab:

1. A freshly prepared and quickly cooled solution of 2% osmium tetroxide in analytical-grade acetone (water content <0.1%) for morphology.
2. Analytical-grade acetone with 0.1–3% GA (depending on the antigen) and molecular sieve (Linde 0.3 nm) for immunocytochemistry.

Molecular sieve is added in order to remove the water introduced with GA but only when preparing the substitution medium (about 20% v/v); for the actual substitution the solution is decanted. On the other hand, Walther and Ziegler (2002) deliberately add water (1–3%) to a substitution medium consisting of acetone and 1–2% osmium tetroxide in order to increase the visibility of cytomembranes after HPF.

After some 48 h (minimum duration depends on the specimen and usually is grossly exceeded to be on the safe side) at −79°C specimens are warmed to room temperature; the latter process may be slowed by inserting the vials into a precooled aluminum block with a large heat capacity (warming halftime, 3 h; Fig. 10E). At about 0°C the medium is changed to pure acetone, and embedding in Epon is performed at room temperature, with subsequent heat polymerization at 60°C.

Instead of in dry ice–acetone slush, cryosubstitution can be also carried out in a deep freezer at −80°C or in a commercial cryosubstitution machine. These machines allow free choice of various temperature/time protocols and are particularly convenient for low-temperature embedding. A self-made substitution apparatus is described by Marshall and Kent (1991) and has proven useful in the Steinbrecht laboratory for many years. Two protocol examples used for insect antennae are shown in Fig. 11.

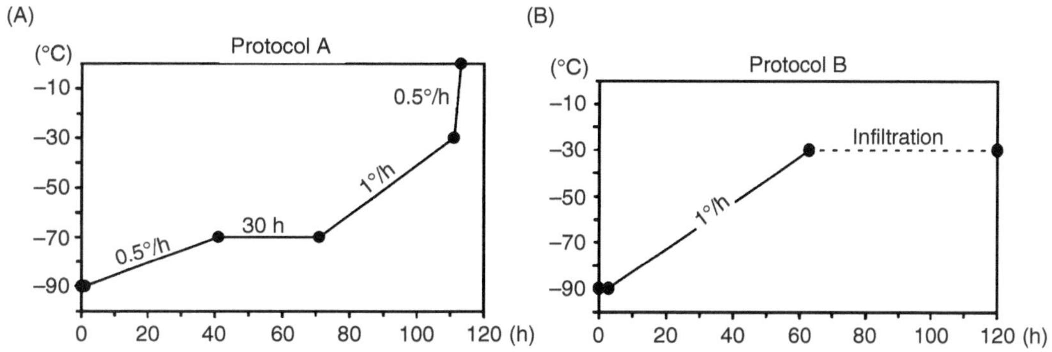

Fig. 11 Two substitution protocols as used in the apparatus according to Marshall and Kent (1991). Protocol A ends at room temperature and is followed by infiltration with Epon or LRWhite. During the plateau at –70° acetone containing osmium tetroxide was exchanged against pure acetone. Protocol B was used for low-temperature embedding at –30° C in Lowicryl followed by UV polymerization at –30°. (Figure courtesy Michael Laue)

D. Immunocytochemistry

In this section we do not intend to describe labeling techniques but rather the preparation methods that are suitable for optimal preservation of the proteins to be labeled.

In general, pre-embedding labeling and post-embedding labeling have to be distinguished. In the first group the antibody reacts with the tissue *en bloc* before fixation and preparation and, therefore, it is of paramount importance that the antibody can permeate the tissue and reach its site of reaction. Cuticle as well as tracheae might efficiently block the entrance of antibodies, so that it depends on specimen and preparation whether such pre-embedding methods are possible.

Post-embedding labeling is done on the sections essentially as with a very specific staining method. It is important that the antigens to be labeled are neither extracted nor translocated by the preceding preparation steps. If feasible, the optimal method is cryofixation (by plunge-freezing or HPF) followed by cryosubstitution and embedding at low temperature. For immunocytochemistry, cryosubstitution is carried out in pure acetone or acetone containing GA (0.1–3%) and molecular sieve. This is followed by low-temperature embedding in Lowicryl K4M or HM20 with subsequent UV polymerization at −30°C (Humbel and Müller, 1986). Alternatively, the specimens are embedded in LRWhite at room temperature with heat polymerization at 60°C.

"Mild" chemical fixation with 2–4% paraformaldehyde and/or 0.5–2% GA followed by embedding in LRWhite or Lowicryl can be used, if cryofixation is not possible (see e.g., Jin *et al.*, 2006). As antigens differ drastically as to their susceptibility to fixation damage, the optimal compromise between a reliable labeling signal and sufficient tissue preservation has to be found experimentally in every instance.

Alternatively, specimens might be gently fixed, impregnated by sugar in order to avoid freezing damage, and then cryofixed and cryosectioned for section labeling as recommended by Tokuyasu (1973, 1986).

E. Sectioning

Different embedding media shrink with different degrees during polymerization. This is not a problem for soft tissues, in which the degree of shrinkage is quite even. However, the tough cuticle shrinks much less, or not at all. Also, embedding media do not adhere properly to the hydrophobic cuticular surface, which means that both might separate during sectioning. Whereas the plastic as well as the embedded soft tissue is permanently compressed by the sectioning process, the cuticle remains elastic, so that small structures tend to pop out of the section. Moreover, at the epidermis–cuticle interface, the latter can show folds perpendicular to the direction of sectioning (Fig. 6A).

Glass knives can be used for soft insect tissues, but the hardness of the cuticle usually leads to heavy sectioning artefacts. The use of diamond knives as well as hard embedding media are thus prerequisites for obtaining good sections of cuticular material. This is especially true for the walls and sockets of insect sensilla with their rather complicated structure (cf. Keil, 1997).

The embedding media that we have used are Epon and ERL (Spurr, 1969) for standard morphology and LRWhite and Lowicryl K4M for immunocytochemistry. Spurr's medium has the advantage of low viscosity and miscibility with ethanol, so the intermediate step with propylene oxide can be omitted. However, it shrinks to a higher degree than Epon.

IV. Materials

Most materials listed in this section can be obtained from different sources. Some examples are given here.

A. Chemical Fixation, Standard Protocol

GA: obtained in 25, 50, and 70% solutions as well as frozen in vials. However, for standard applications it is perfect to use a 25% stock solution, which can be kept under an inert gas, such as nitrogen or argon, in the refrigerator for years (Plano, Wetzlar, Germany; Science Services, Munich, Germany; Serva, Heidelberg, Germany; Sigma, Taufkirchen, Germany).

OsO_4: obtained as crystals as well as a 2 or 4% aqueous solution. Crystals need one day to dissolve in water, so it is easier to start directly with the liquid form (Plano, Wetzlar, Germany; Science Services, Munich, Germany; Sigma, Taufkirchen, Germany).

Sodium cacodylate (Sodium dimethyl arsenate, Dimethylarsinsäure Natriumsalz-Trihydrat; Plano, Wetzlar, Germany; Science Services, Munich, Germany; Merck, Darmstadt, Germany)

Uranyl acetate (Plano, Wetzlar, Germany; Science Services, Munich, Germany)
Sucrose (Science Services, Munich, Germany; Sigma, Taufkirchen, Germany; Merck, Darmstadt, Germany)
Lead citrate (Plano, Wetzlar, Germany; Science Services, Munich, Germany)

B. Additional "Tricks"

Tannic acid (Science Services, Munich, Germany; Merck, Darmstadt, Germany)
Ruthenium red (Plano, Wetzlar, Germany; Science Services, Munich, Germany; Merck, Darmstadt, Germany)
$LaCl_3$, $La(NO_3)_3$ (Merck, Darmstadt, Germany)
Na-phosphate, mono- and dibasic (Science Services, Munich, Germany; Merck, Darmstadt, Germany)

C. Cryofixation and Cryosubstitution

Plunging device: custom-built in workshop
Tungsten wire: 0.1 mm, Clark Electromedical Instruments, Reading, UK
Liquid nitrogen: now available in any larger city
Propane: industrial propane (bottle gas) from local suppliers
Plastic vials (Sarstedt, Numbrecht-Rommelsdorf, Germany)
Acetone: analytical grade (Merck, Darmstadt, Germany)
Molecular sieve: Linde 0.3 nm (Serva, Heidelberg, Germany)

D. Immunocytochemistry

Embedding resin (LRWhite, Lowicryl K4M) (Plano, Wetzlar, Germany; Science Services, Munich, Germany; Polysciences, Eppelheim, Germany)

E. Sectioning

Embedding resin (Spurr's medium, Epon) (Plano, Wetzlar, Germany; Science Services, Munich, Germany)
Diamond knife (Diatome, Biel, Switzerland; Drukker, Cuijk, The Netherlands)

V. Results and Discussion

In this section we present results obtained with the methods described above, illustrated with a few exemplary figures. We also discuss possible artefacts, in particular of cryofixation and cryosubstitution.

Between 1964 and 2001, our focus of research lay on the peripheral sensory organs and nervous system of insects, which explains the choice of illustrations. R. A. S. specialized on the antennal sensory organs of various insects with emphasis on the silk moth *Bombyx mori*. As the standard chemical fixation protocols yielded no satisfactory results, he switched over to cryofixation, which turned out to be the method of choice for these specimens. T. K. specialized first on insect mechanoreceptors (reviewed by Keil, 1997), in which chemical fixation yielded satisfactory results, provided the fixative has good access to the structures of interest (Fig. 1). It can be said that chemical fixation works reliably and rather fast in most insect tissues. Cryofixation, in combination with cryosubstitution (freeze substitution), though more time-consuming, can yield superior results provided the size and geometry of the specimen are suitable (Fig. 8), and is particularly useful with immunocytochemical investigations.

A. Chemical Fixation Protocols

Chemical fixation is the method of choice for tissues and cells farther from the surface, such as nerves (Fig. 2), blood cells (Fig. 5), or muscles. It works well in developmental stages, such as the pupa (Fig. S10; Keil, 1992). Of course, a number of artefacts must be considered if applying chemical fixation followed by dehydration and plastic embedding. The problem of "aggregation artefacts," i.e., the tendency of small particles and fine fibrils to form aggregates and to adhere to larger and more stable cellular structures during fixation and dehydration, has been discussed in detail by Kellenberger (1987) and Dubochet and Sartori Blanc (2001). Another problem, clearly shown in Fig. 1, is the tendency of membranes to become wrinkled, obviously due to shrinkage, and to form myelin-like structures.

When choosing the buffer system for chemical fixation, certain facts must be considered. Na-cacodylate buffer works satisfactorily with both GA and osmium tetroxide and can be stored for months without fungal growth. However, its disadvantage is its content of arsenic. Importantly, working with this substance can lead to basalioma, a curable form of skin cancer, after many years. This buffer has to be handled with special care. Alternatives are given in detail by Glauert (1974). These are phosphate, veronal-acetate, and collidine buffers, all of which have their disadvantages. Phosphate buffers are considered to be the "most physiological" buffers, but support the growth of microorganisms, and can form fine precipitations with calcium. Veronal-acetate buffers turned out to be "useless" at physiological pH, and collidine buffers are very toxic and not recommended by Glauert (1974). So the best alternative to cacodylate seems to be Sörensen's phosphate buffer (see Glauert, 1974).

The routine treatment *en bloc* in hot uranyl acetate has proven to be of particular value, as it gives good contrast to cell membranes and cell surfaces and eliminates the need to stain sections. Typical examples of this preparation are shown in Figs. 1, 2, 4, 5, S1–S4, S10. Uranyl acetate has a strong affinity to nucleic acids (Huxley and Zubay, 1961). These become very prominent within the nucleus, and the ribosomes show good staining as well. It must also be pointed out that the mitochondrial matrix appears very electron-dense after this procedure (Fig. S3). A slight disadvantage of the method is an occasional occurring of fine grainy deposits in the cytoplasm, which become visible only at high magnifications. Larger needle-like precipitations sometimes appear in extracellular regions, but are not disturbing.

TA was first used to show that microtubules are composed of 13 protofilaments (Burton *et al.*, 1975; Tilney *et al.*, 1973). Later, its application was extended to membrane surfaces, vesicles, and the extracellular matrix. In combination with GA, it binds to the outer surface of the cell membranes that, therefore, show an increased thickness (Figs. 3a and 5). Importantly, an increase in contrast is reached only in combination with osmium, which has a strong affinity to TA. Depending on the method of application, cellular membrane systems might be marked in differing intensities (discussed by Wagner, 1976). Normally, TA does not penetrate to a larger extent into the cells. Application of TA following OsO_4 was recommended by Simionescu and Simionescu (1976a, b) in connection with Pb poststaining. This treatment allows penetration of TA into the cells and yields enhanced contrast of microtubules and membranes. Damage of the outer cell membrane leads to heavy penetration, and cellular structures appear blackened throughout. However, penetration can happen very unevenly within a tissue: one cell might show heavy staining, and the neighboring cell none at all (Figs. 2A and B; 3A). The extracellular matrix (basal lamina, neural lamella) always shows strong staining after TA treatment (Figs. 2A and S2).

RR has been in use since 1893 as a light microscopical stain for gums, mucilages, and pectic substances in plant tissues (Mangin, 1893; reviewed by Luft, 1971a). Luft published two extensive papers on the application of RR to animal cells and tissues in the EM and showed its usefulness as an electron-dense stain for polyanionic macromolecules, such as acidic mucosubstances (Luft, 1971a, b). It is bound to anionic groups by electrostatic forces and yields high contrast in combination with OsO_4 (Blanquet, 1976a, b). It has been applied to insect tissues by Dybowska and Dutkowski (1977) and later by Keil (1979, 1984). Simplification of Luft's original schedule showed no negative effect on ultrastructural preservation. Intense staining is found on the extracellular matrix, as well as all cell surfaces directly accessible from the hemolymph side, such as intercellular clefts and the basal labyrinth of epidermal cells (Fig. 6A and B). RR can penetrate into the cells, where it binds to membranes, microtubules, and filamentous structures (Fig. 6B). As other electron-dense tracers, RR does not penetrate the cuticle. If incubation from the apical face of the epidermis is intended, the cuticle has to be opened mechanically.

Lanthanum has been used as a tracer for paracellular diffusion pathways and for marking the cell coat (Doggenweiler and Frenk, 1965; Keil and Steinbrecht, 1987; Lane and Treherne, 1972, Shaklai and Tavassoli, 1982; Skaer *et al.*, 1987). If applied *in vivo* prior to fixation, either from the apical or basal side of an epithelium, the tightness of the latter can be tested. This method has been used to identify the diffusion barriers: it diffuses into intercellular spaces and enters the junctional complexes, but stops finally at the barrier (Lane and Treherne, 1972; Lane *et al.*, 1977; Keil and Steinbrecht, 1987; Skaer *et al.*, 1987). For this purpose, it is necessary to dissolve the lanthanum salt in a phosphate-free Ringer solution (see Methods section). If viewed in tangential sections, it very clearly outlines the junctional complexes, such as septate and gap junctions (Figs. 7, S5). It acts as a strong Ca competitor (Weiss, 1974) and therefore binds to Ca-binding sites. If applied before fixation, it is also taken up by endocytotic vesicles (Keil and Steinbrecht, 1987).

B. Cryofixation and Cryosubstitution

1. Artefacts

Freezing damage is the primary artefact of any specimen preparation using cryofixation. It results from cooling rates that are too low and is revealed by segregation compartments (ice-crystal ghosts). The thermodynamic equilibrium of aqueous solutions is fundamentally different above and below the freezing point: upon freezing the solute is not incorporated into the ice lattice; pure ice crystals are growing, while the solute is concentrated in the interfaces between them. "If the ice lattice would tend to form 'solid solutions', one would not have to worry about segregation artefacts" (Bachmann and Mayer, 1987).

The only known means of avoiding segregation is to extract the heat from the specimen so quickly that it does not have time to adjust to the new equilibrium before arriving at the "frozen-in" state, where the thermal energy of the specimen is too low to permit any changes. As the low heat conduction within biological specimens cannot be altered, specimen size is the main limiting factor for cryofixation (Fig. S6) and freezing damage is usually absent only within the outer 10 μm of a specimen. On the other hand, it depends on the biological questions of the investigator and the resolution desired, how much freezing damage is tolerable; for example, extracellular space in nerves is faithfully represented even after moderate freezing damage (Fig. 9B). Freezing damage resulting from the low cooling rates within larger biological specimens can, however, be avoided or reduced if high pressure (2000 bar) is applied during cooling (Moor and Riehle, 1968).

Of course, the process of segregation will start again whenever the specimen is warmed and the "safe" temperatures are left (for a comprehensive treatment, see Bachmann and Mayer, 1987). Therefore, conceptually we can distinguish between *primary freezing damage*, which occurs during the primary cooling of the specimen, and *secondary freezing damage*, which will happen when a frozen specimen is warmed above the recrystallization temperature before the water has been removed during cryosubstitution.

The sensory hairs of antennal sensilla of *Bombyx mori*, and many other insects, have diameters of 1–5 μm and, therefore, represent ideal test specimens for cryotechniques in electron microscopy. After routine plunging into propane, ice-crystal ghosts are usually absent or smaller than 5 nm in the sensillum lymph and the sensory dendrites (Steinbrecht, 1976, 1980). Earlier, one of the main arguments against cryosubstitution was that—due to the melting points of the solvents used—it had to be carried out at temperatures far above the recrystallization temperature of amorphous ice (−130 to −120°C; Bachmann and Mayer, 1987) and, therefore, secondary freezing damage should be inevitable. Meanwhile, it has been shown, by comparing freeze-substituted specimens with freeze-etched specimens after identical cryofixation, that their morphology is practically identical (Figs. S8, S9) (Steinbrecht, 1976, 1980; for non-insect material, see Steinbrecht and Müller, 1987).

The reason for the substantially elevated recrystallization temperatures of biological specimens in comparison to pure water are the solutes and, in particular, the macromolecules therein. The sensillum lymph in the sensilla trichodea of *Bombyx mori*, for example, contains a pheromone-binding protein at a concentration of 10 mMol. Taking advantage of the low probability of primary freezing damage in these sensory hairs, Steinbrecht (1985) experimentally assessed the recrystallization temperature of these specimens by deliberately warming cryofixed antennae in propane prior to cryosubstitution. Surprisingly, secondary freezing damage was negligible even after 10 min at −43°C and became significant only with a very long incubation (45 min at −43°C) (Fig. S8B) or further elevated temperatures (10 min at −25°C).

A promising way to cope with freezing damage in insect specimens should be the utilization of frost-hardy species. Many insects have developed strategies to survive periods of subzero temperatures and some species incorporate larger amounts of cryoprotectants in their blood and tissue (for reviews, see Sømme, 1982; Storey and Storey, 1988). We have not pursued this line systematically, but in two "winter species" studied, the moth, *Poecilocampa populi*, and the snow flea, *Boreus hiemalis*, good tissue preservation was observed deeper below the surface than in the moths, *Bombyx* and *Antheraea*, or in the fruit fly, *Drosophila* (for figures, see Steinbrecht, 1993a).

In general, freezing damage is inversely proportional to the density of the respective tissue compartment. For instance, larger ice-crystal ghosts are usually found in extracellular rather than in intracellular compartments, in the karyoplasm rather than in the cytoplasm of a given cell, and in the more electron-lucid cytoplasm of receptor rather than in the denser glial cells, etc. (Fig. S6). This can be readily explained, since freezing damage is correlated with the water content. One of the most sensitive indicators of freezing damage is the nuclear chromatin, which easily becomes flocculent and loses its compactness and clear delineation. Extremely susceptible to freezing damage are also the microtubules, which may "collapse" to strands, and the mitochondria, which often are surrounded by a halo of segregation compartments that are much larger than those in the surrounding ground cytoplasm (Figs. 9B and S9A). Such damaged mitochondria display collapsed cristae and an indented outline, and the outer mitochondrial membrane is no longer discernible (Steinbrecht, 1993a).

It should be mentioned in this context that improving our plunging procedure did not improve the quality of our best results, which, sometimes, were achieved by simply dropping the cut antennae into super-cooled propane. However, the yield of well-frozen specimens was substantially increased by guided immersion at high speed (~2 m/s) and forced convection during the first 30 ms of cooling. For artefacts of HPF, see McDonald (2007).

Apart from freezing damage, cryosubstitution—as well as the following steps of warming up, resin embedding, and polymerization—may also produce further artefacts. Extraction of specimen components increases with time and temperature. Prolonged exposure to acetone at room temperature may even lead to the total loss of membranes. The duration of the substitution, therefore, has to be adjusted to the specimen size and the solvent used. Methanol is even more aggressive and is hardly used anymore. The addition of chemical fixing agents, such as osmium tetroxide, uranyl acetate, or GA, to the substitution medium is a means of preventing and/or reducing the extraction of components during the later preparation steps (Humbel, 2008). The fixatives are introduced during freeze substitution at a temperature at which they do not react with the tissue and later, upon rewarming, may react simultaneously at all tissue sites. Osmium tetroxide allegedly starts to react at −70°C and GA at −50°C (for references, see Steinbrecht and Müller, 1987); therefore, it is recommended to raise the temperature slowly after substitution or to use a protocol with temperature steps (see Fig. 6 for examples). Adding water to the substitution medium may result in shifting the onset of substitution to temperatures where osmium tetroxide starts to react (Walther and Ziegler, 2002).

2. Achievements

Although being a highly complex hybrid technique, cryofixation in combination with cryosubstitution has been remarkably successful. This becomes even more evident when compared with the "pure" follow-up techniques, such as freeze-etching or cryosectioning. With insect work, the dual advantage of cryosubstitution becomes especially apparent: it offers the benefits of cryofixation, but at the same time allows for conventional ultramicrotomy at room temperature, an immense help with complex tissues. The superior tissue preservation of cryofixation/cryosubstitution compared to conventional fixation techniques has been demonstrated for a variety of insect tissues. Most papers deal with arthropod antennal sense organs (e.g., Ameismeier, 1987; Crouau, 1990; Haug, 1985; Haug and Altner, 1984; Shanbhag *et al.*, 1999a, b; Steinbrecht, 1980, 1989; Steinbrecht and Gnatzy, 1984; Wolfrum, 1990, 1991), but halteres (Toh, 1985; Völker *et al.*, 1984), tarsi, and ovipositors (Anderson and Hallberg, 1990; Shanbhag *et al.*, 2001) have also been studied. A further example is given with the wing veins of *Drosophila* (Steinbrecht, 1993a). Even compound eyes have been successfully cryofixed (Meinecke and Langer, 1984). In the latter, however, the yield of well-frozen specimens was reportedly very low.

Where cryofixation at atmospheric pressure was insufficient, HPF was applied in several cases. For insect material, this technique was used first by Wolfrum (1990, 1991: adult antennae) and Zimmermann (1991: pupal antennae). It is applied routinely now for the cryofixation of *Drosophila* embryos, also in combination with immunolabeling (e.g., Bardsley *et al.*, 1993; McDonald, 1994, 2007, 2009; McDonald and Morphew, 1993; McDonald *et al.*, 2000; Sharp *et al.*, 1999).

Among the benefits of cryofixation for insect studies the following are particularly evident: The contours of the cell borders are clear and free from wrinkles so that even complex tissues can be reconstructed at low magnification (Figs. 8, 9). There is little extraction of proteins, and the various compartments show a varying density of the ground cytoplasm. In particular, the mitochondrial matrix is very dense (Figs. 8, 9). Occasionally, the F_0/F_1 particles of mitochondrial ATPase can be seen (Steinbrecht, 1986), which are usually visible only in negative stain preparations. Freeze substitution also proved to be very useful in studies of membrane substructures, such as ATPase particles (portasomes) in the apical auxiliary cell membrane (Fig. 4B) or the bossed orthogonal surface substructures (BOSSes) of thermoreceptor membranes (Fig. S7)

(Steinbrecht, 1988). From its first successful application in electron microscopy, freeze substitution has been a valuable tool for the faithful representation of extracellular fluid spaces and the adjacent tissues, which suffer most from slow and inadequate fixation (Van Harreveld *et al.*, 1965). In insect sensilla, the dendrites, microvilli, and microlamellae extending into the extracellular sensillum lymph are preserved without swelling or shrinking (Figs. 3B, 4B, and S7; see also Steinbrecht, 1980). The same holds for intercellular fluid spaces between cells and for the extracellular space around the axons (Fig. 9). In several studies undertaken to link insect sensory morphology and physiology (Gnatzy *et al.*, 1984; Haug, 1985; Steinbrecht, 1973; Steinbrecht *et al.*, 1989), the usefulness of rapid freezing in combination with freeze substitution for obtaining reliable morphometric data has been convincingly shown. Even adaptation processes in insect hygroreceptors were investigated by freeze substitution, because cryofixation enabled the study of dry-adapted hygroreceptors without the need to immerse them in an aqueous fixing solution (Steinbrecht and Müller, 1991). In principle, cryofixation is capable of resolving even fast dynamic processes (for review, see Knoll *et al.*, 1987), e.g., the exocytosis of synaptic vesicles (Heuser *et al.*, 1979), but in insects such studies have not yet been attempted.

C. Immunocytochemistry

The extraction of specimen components becomes a question of paramount importance when freeze substitution is being considered as a method of preparation for immunocytochemistry. As for morphology and morphometry, good tissue preservation is essential. However, if immunolabeling is to be performed on tissue sections (post-section staining technique), the antigen must still retain its reactivity despite fixation, dehydration, embedding, etc. Moreover, it must be made sure that the antigen has not been translocated or even extracted during these various preparation steps.

Preservation of antigenicity is most probably better after cryofixation than after conventional fixation, although each antigen differs in its susceptibility to damage. Also dehydration and embedding at room temperature followed by heat polymerization may have deleterious effects. Therefore, the combination of freeze substitution with low-temperature embedding (Carlemalm *et al.*, 1986; Humbel, 2008; Humbel *et al.*, 1983; Schwarz and Humbel, 1989) appears to be the most promising way to retain as much of the reactivity of the antigen as possible. Nevertheless, the danger of extraction of soluble antigens increases with the omission of chemical crosslinkers (Schwarz and Humbel, 1988; Steinbrecht *et al.*, 1992). Thus, for faithful localization of soluble antigens a compromise has to be made by introducing chemical fixing agents during freeze substitution (see Methods section).

The number of immunocytochemical studies in insects using freeze substitution for specimen preparation is continually increasing. In the beginning, studies on the expression of soluble and membrane-bound proteins in insect sensilla were predominant. Steinbrecht *et al.* (1992, 1995) labeled the pheromone-binding protein and general odorant-binding protein in sensilla trichodea and basiconica of *A. polyphemus* and *Bombyx mori* (Fig. 12A). These are soluble proteins that occur in high

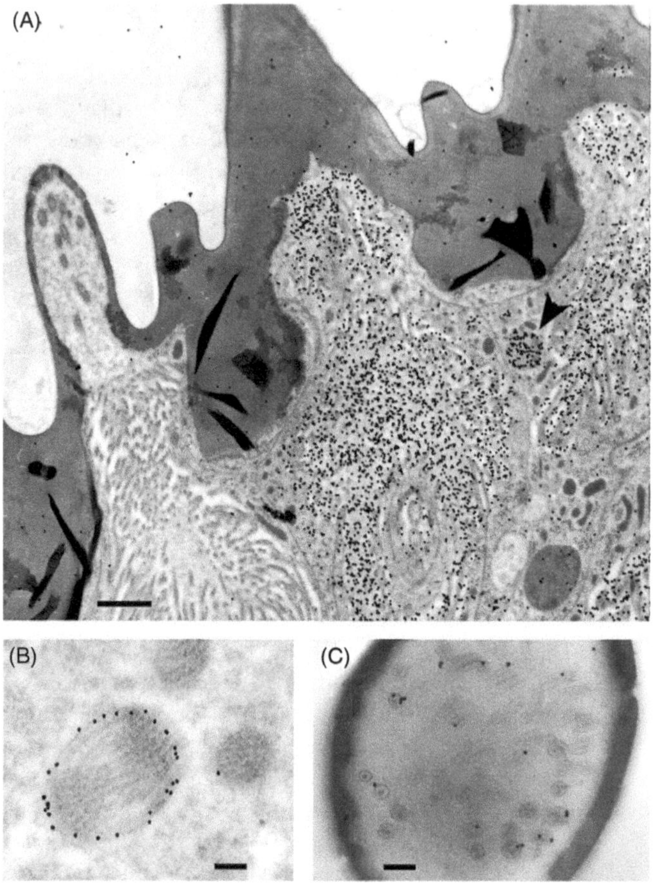

Fig. 12 Immunocytochemical labeling of soluble (A) and membrane-bound (B, C) proteins involved in insect olfactory signal transduction. (A) *Bombyx mori*, oblique section through base of two pheromone-sensitive sensilla trichodea (middle and right) and a plant odor-sensitive sensillum basiconicum at the left. Pheromone-binding protein (PBP) is extensively expressed in the sensillum lymph of two sensilla trichodea, while the sensillum lymph space of the neighboring basiconic sensillum is unlabeled. Arrowhead marks secretory granule containing PBP. Specimen freeze-substituted in acetone containing 2% glutaraldehyde and embedded in LRWhite at room temperature (from Steinbrecht *et al.*, 1992 *Cell Tissue Res.* **270**, 287–303 with permission of Springer Verlag). (B) Sensory neuron membrane protein 1 (SNMP-1) is localized in the membrane of the sensory dendrites of olfactory sensilla of *Antheraea polyphemus* (see also Rogers *et al.*, 2001). Specimen freeze-substituted in pure acetone and embedded in LRWhite at room temperature. (C) Expression of the olfactory receptor protein Or22a/b in the membrane of dendritic branches of a large sensillum basiconicum of *Drosophila melanogaster* (see also Dobritsa *et al.*, 2003). Specimen freeze-substituted in pure acetone and embedded in Lowicryl K4M at –30°C. Note that the density of the label does not permit conclusions on the density of the expressed protein. Scale bars are 1 μm in (A), 100 nm in (B) and (C).

concentrations in the sensillum lymph and are thought to function as carriers for lipophilic odorants through this aqueous medium. Evidence from electrophysiological experiments points to an additional contribution of the pheromone-binding proteins of the mentioned species to stimulus specificity (Pophof, 2002, 2004). Meanwhile, the expression mosaic of several such odorant-binding proteins was mapped in various moth species (Zhang *et al.*, 2001) but also in *Drosophila* (Shanbhag *et al.*, 2005) and the desert locust (Jin *et al.*, 2006). With insect odorant-binding proteins, rapid freezing/freeze substitution in pure acetone followed by low-temperature embedding was insufficient to "fix" these antigens in the sensillum lymph. However, adding 3% GA (plus molecular sieve) to the substitution acetone solved the problem without losing much of the labeling signal. Immunocytochemistry on chemically fixed specimens produced similar labeling results; however, there was inferior tissue preservation, and therefore, localization of the label was less accurate.

In a few cases, cryosubstitution also allowed successful localization of membrane-bound proteins. Klein and Zimmermann (1991) immunolabeled the membrane-bound vacuolar-type ATPase in the apical membrane of sensillar auxiliary cells after freeze substitution in acetone containing 0.1% GA and low-temperature embedding in Lowicryl K4M. The immunolocalization of Na, K-ATPase on the receptor cell membrane in these specimens was reported by Zimmermann (1992). Wolfrum (1991) demonstrated the colocalization of tropomyosin and actin in the scolopale of insect mechanoreceptors using monoclonal antibodies on Lowicryl sections after freeze substitution and low-temperature embedding. Rogers *et al.* (2001) localized the membrane protein SNMP2 (sensory neuron membrane protein 2) to the dendritic membrane of pheromone-sensitive olfactory neurons in sensilla trichodea of *A. polyphemus* (Fig. 12B). In *Drosophila*, SNMP2 meanwhile has been shown to be indispensable for stimulus transduction of the pheromone cis-vaccenyl acetate (Benton *et al.*, 2007). Even the olfactory receptor protein OR22b could be localized on the sensory dendrites of a subclass of sensilla basiconica of *D. melanogaster* by immuno-electron microscopy of freeze-substituted specimens (Fig. 12C) (Dobritsa *et al.*, 2003).

Recently, results of applying the Tokuyasu method with insect specimens (*Drosophila* embryos) have also been reported (e.g., Delanoue *et al.*, 2008; Wucherpfennig *et al.*, 2003).

VI. Concluding Remarks

In this chapter we have presented standard protocols using chemical fixation that may give satisfactory results for transmission electron microscopy of almost any type of insect tissue. In addition, we have demonstrated cryopreparation methods, which may be particularly useful for selected specimens and tissues. When now returning to the general topic—insects as model specimens for cell biology—we have to consider that transmission electron microscopy is only one of the numerous approaches in modern cell biology. The "renaissance" of the light microscope with its developments in fluorescence, confocal laser, bi-photon microscopy, total internal reflection fluorescence (TIRF), structured illumination (SIM), and others offers the potential to combine dynamics with the still unsurpassed resolution of the EM. Such correlative LM and EM approaches will help in answering questions of today's cell biology (see Chapter 24 by Guizetti *et al.* and Chapter 26 by Verkade *et al.*, this volume). In this context the choice of insects as model systems is even more advisable due to the easy culture, short generation span of many insect species, and—last but not least—the greater ease to conform to principles of animal care and welfare. With the ever-growing number of insect species with a fully sequenced genome, many molecular methods are no longer restricted to *Drosophila*. As the basal processes of cell biology are surprisingly similar in all metazoans, cell biologists should feel free to choose the specimen most suitable to test their hypotheses. For microscopists, and in particular those working with the EM, the small size of most insects is an asset, not a disadvantage. After all, these small creatures have colonized planet Earth at least as successfully as mammals, including humans.

VII. Summary

Besides their importance as model organisms in biology, insects have proven to be useful and easy-to-handle objects for electron microscopy. Chemical fixation with glutaraldehyde, osmium tetroxide, and uranyl acetate yields good results in most cases. The use of tannic acid, ruthenium red, and lanthanum is advantageous for cell surfaces, the extracellular matrix, extracellular spaces, and transcellular pathways. Cryofixation/cryosubstitution is particularly useful for peripheral tissues, such as epidermis and sensilla, where it can be superior to chemical methods, and is often required for immunocytochemical investigations. However, the investigator must be aware of the artefacts specific for chemical and cryo-methods, respectively.

References

Ameismeier, F. (1987). Ultrastructure of the chemosensitive basiconic single-walled wall-pore sensilla on the antennae in adults and embryonic stages of *Locusta migratoria* L. (Insecta, Orthoptera). *Cell Tissue Res.* **247**, 605–612.

Anderson, P., and Hallberg, E. (1990). Structure and distribution of tactile and bimodal taste tactile sensilla on the ovipositor, tarsi, and antennae of the flour moth *Ephestia kuehniella* (Zeller) (Lepidoptera Pyralidae). *Int. J. Insect Morphol. Embryol.* **19**,13–23.

Anderson, E., and Harvey, W. R. (1966). Active transport by the *Cecropia* midgut. II. Fine structure of the midgut epithelium. *J. Cell Biol.* **31**, 107–134.

Anderson, T. F., and Richards, A. G. (1942). An electron microscope study of some structural colors in insects. *J. Appl. Phys.* **13**, 748–758.

Autrum, H., and Burkhardt, D, (1960). Die spektrale Empfindlichkeit einzelner Sehzellen. *Naturwissenschaften* **47**, 527.

Avidor-Reiss, T., Maer, A. M., Koundakjian, E., Polyanovsky, A., Keil, T., Subramaniam, S., and Zuker, C. (2004). Decoding cilia function: Defining specialized genes required for compartmentalized cilia biogenesis. *Cell* **117**, 527–539.

Bachmann, L., and Mayer, E. (1987). Physics of water and ice: Implications for cryofixation. *In* "Cryotechniques in Biological Electron Microscopy" (R. A. Steinbrecht and K. Zierold, eds.) pp. 3–34. Springer Verlag, Berlin.

Bardsley, A., McDonald, K., and Boswell, R. E. (1993). Distribution of tudor protein in the *Drosophila* embryo suggests separation of functions based on site of localization. *Development* **119**, 207–219.

Benton, R., Vannice, K. S., and Vosshall, L. B. (2007). An essential role for a CD36-related receptor in pheromone detection in *Drosophila*. *Nature* **450**, 289–295.

Berridge, M. J. (1970). The role of 5-hydroxytryptamine and cyclic AMP in the control of fluid secretion by isolated salivary glands. *J. Exp. Biol.* **53**, 171–186.

Blanquet, P. R. (1976a). Ultrahistochemical study on the ruthenium red surface staining. I. Processes which give rise to electron-dense marker. *Histochemistry* **47**, 63–78.

Blanquet, P. R. (1976b). Ultrahistochemical study on the ruthenium red surface staining. II. Nature and affinity of the electron dense marker. *Histochemistry* **47**, 175–189.

Burkhardt, D., and Autrum, H, (1960). Die Belichtungspotentiale einzelner Sehzellen von *Calliphora erythrocephala*. *Z. Naturforsch.* **15b**, 612–616.

Burton, P. R., Hinkley, R. E., and Pierson, G. B. (1975). Tannic acid-stained microtubules with 12, 13, and 15 protofilaments. *J. Cell Biol.* **65**, 227–233.

Carlemalm, E., Villiger, W., Acetarin, J.-D., and Kellenberger, E. (1986). Low temperature embedding. *In* "The Science of Biological Specimen Preparation 1985" (M. Müller, R. P. Becker, A. Boyde, and J. T. Wolosewick, eds.) Scanning Electron Microscopy, pp. 147–154. AMF O'Hare, Chicago, IL

Copeland, E. (1964). A mitochondrial pump in the cells of the anal papillae of mosquito larvae. *J. Cell Biol.* **23**, 253–263.

Crouau, Y. (1990). Sensory organs in a cavernicolous insect after cryofixation without cryoprotectants: A comparative ultrastructural study. *Tissue Cell* **22**, 245–258.

Delanoue, R., Herpers, B., Soetaert, J., Davis, I., and Rabouille, C. (2008). *Drosophila* Squid/hnRNP helps dynein switch from a *Gurken* mRNA transport motor to an ultrastructural static anchor in sponge bodies. *Dev. Cell* **13**, 523–538.

Dethier, V. G. (1955). The physiology and histology of the contact chemoreceptors of the blowfly. *Q. Rev. Biol.* **30**, 348–371.

Dobritsa, A. A., van der Goes van Naters, W., Warr, C. G., Steinbrecht, R. A., and Carlson, J. R. (2003). Integrating the molecular and cellular basis of odor coding in the *Drosophila* antenna. *Neuron* **37**, 827–841.

Doggenweiler, C. F., and Frenk, S. (1965). Staining properties of lanthanum on cell membranes. *Proc. Natl. Acad. Sci. U.S.A.* **53**, 425–430.

Driest, E., and Müller, H. O. (1935). Elektronenmikroskopische Aufnahmen (Elektronenmikrogramme) von Chitinobjekten. *Z. wiss. Mikrosk.* **52**, 53–57.

Dubochet, J., and Sartori Blanc, N. (2001). The cell in absence of aggregation artefacts. *Micron* **32**, 91–99.

Dybowska, H. E., and Dutkowski, A. B. (1977). Ruthenium red staining of the neural lamella of the brain of *Galleria mellonella*. *Cell Tissue Res.* **176**, 275–284.

Fernandez-Moran, H. (1958). Fine structure of the light receptors in the compound eyes of insects. *Exp. Cell Res. Suppl.* **5**, 586–644.

Gentil, K. (1942). Elektronenmikroskopische Untersuchung des Feinbaues schillernder Leisten von Morpho-Schuppen. *Z. Morph. Ökol. Tiere* **38**, 344–355.

Glauert, A. M. (1974). "Practical Methods in Electron Microscopy." Vol. 3. North Holland Publishing Company, Amsterdam.

Gnatzy, W., Mohren, W., and Steinbrecht, R. A. (1984). Pheromone receptors in *Bombyx mori* and *Antheraea pernyi*. II. Morphometric analysis. *Cell Tissue Res.* **235**, 35–42.

Gray, E. G. (1960). The fine structure of the insect ear. *Philos. Trans. R. Soc. Lond. B.* **243**, 75–94.

Gupta, B. L. and Berridge, M. J. (1966). A coat of repeating subunits on the cytoplasmic surface of the plasma membrane in the rectal papillae of the blowfly, *Calliphora erythrocephala* (Meig.), studied in situ by electron microscopy. *J. Cell Biol.* **29**, 376–382.

Harrison, F. W., and Locke, M. (1998). "Microscopic Anatomy of Invertebrates Vol. 11 A–C: Insecta." Wiley-Liss, New York.

Harvey, W. R. (1980). Water and ions in the gut. *In* "Insect Biology in the Future" (M. Locke and D. S. Smith, eds.) pp. 105–124. Academic Press, New York.

Harvey, W. R. (2009). Voltage coupling of primary H^+ V-ATPases to secondary Na^+- or K^+-dependent transporters. *J. Exp. Biol.* **212**, 1620–1629.

Haug, T. (1985). Ultrastructure of the dendritic outer segments of sensory cells in poreless ("no-pore") sensilla of insects. A cryofixation study. *Cell Tissue Res.* **242**, 313–322.

Haug, T., and Altner, H. (1984). A cryofixation study of presumptive hygroreceptors on the antennule of a terrestrial isopod. *Tissue Cell* **16**, 377–391.

Heuser, J. E., Reese, T. S., Dennis, M. J., Jan, Y., Jan, L., and Evans, L. (1979). Synaptic vesicle exocytosis captured by quick freezing and correlated with quantal transmitter release. *J. Cell Biol.* **81**, 275–300.

Hodgson, E. S., and Roeder, K. D. (1956). Electrophysiological studies of arthropod chemoreception. I. General properties of the labellar chemoreceptors of Diptera. *J. Cell. Comp. Physiol.* **48**, 51–75.

Howard, J. and Bechstedt, S. (2004). Hypothesis: A helix of ankyrin repeats of the NOMPC-TRP ion channel is the gating spring of mechanoreceptors. *Curr. Biol.* **14**, R224–R226.

Humbel, B. M. (2008). Freeze-substitution. *In* "Handbook of Cryo-Preparation Methods for Electron Microscopy" (A. Cavalier, D. Spehner, and B. M. Humbel, eds.) pp. 314–336. CRC Press Inc, Boca Raton, FL

Humbel, B. M., Marti, T. H., and Müller, M. (1983). Improved structural preservation by combining freeze substitution and low temperature embedding. *Beitr. Elektronenmikrosk. Direktabb. Oberfl.* **16**, 585–594.

Humbel, B. M., and Müller, M. (1986). Freeze substitution and low temperature embedding. *In* "The Science of Biological Specimen Preparation 1985" (M. Müller, R. P. Becker, A. Boyde, and J. J. Wolosewick, eds.) Scanning Electron Microscopy, pp. 175–183. AMF O'Hare, Chicago, IL.

Huxley, H. E. and Zubay, G. (1961). Preferential staining of nucleic acid-containing structures for electron microscopy. *J. Cell Biol.* **11**, 273–296.

Jin, X., Zhang, S.-G., and Zhang, L. (2006). Expression of odorant-binding and chemosensory proteins and spatial map of chemosensilla on labial palps of *Locusta migratoria* (Orthoptera: Agrididae). *Arthropod. Struct. Dev.* **35**, 47–56.

Keil, T. A. (1979). Rutheniumrot-Färbung sensorischer Einheiten der Insekten-Epidermis. *Eur. J. Cell Biol.* **19**, 78–82.

Keil, T. A. (1982). Contacts of pore tubules and sensory dendrites in antennal chemosensilla of a silkmoth: Demonstration of a possible pathway for olfactory molecules. *Tissue Cell* **14**, 451–462.

Keil, T. A. (1984). Surface coats of pore tubules and olfactory sensory dendrites of a silkmoth revealed by cationic markers. *Tissue Cell* **16**, 705–717.

Keil, T. A. (1992). Fine structure of a developing insect olfactory organ: Morphogenesis of the silkmoth antenna. *Microsc. Res. Tech.* **22**, 351–371.

Keil, T. A. (1997). Functional morphology of insect mechanoreceptors. *Microsc. Res. Tech.* **39**, 506–531.

Keil, T. A., and Steinbrecht, R. A. (1987). Diffusion barriers in silkmoth sensory epithelia: Application of lanthanum tracer to olfactory sensilla of *Antheraea polyphemus* and *Bombyx mori*. *Tissue Cell* **19**, 119–134.

Kellenberger, E. (1987). The response of biological macromolecules and supramolecular structures to the physics of specimen cryopreparation. *In* "Cryotechniques in Biological Electron Microscopy" (R. A. Steinbrecht and R. Zierold, eds.) pp. 35–63. Springer, Berlin.

Kinder, E., and Süffert, F. (1943). Über den Feinbau schillernder Schmetterlingsschuppen vom Morpho-Typ. *Biol. Zbl.* **63**, 268–288.

King, R. C. and Akai, H. (1982/1984). "Insect Ultrastructure." Vol. 1–2. Plenum Press, New York.

Klein, U., and Zimmermann, B. (1991). The vacuolar-type ATPase from insect plasma membrane: Immunocytochemical localization in insect sensilla. *Cell Tissue Res.* **266**, 265–273.

Knoll, G., Verkleij, A. J., and Plattner, H. (1987). Cryofixation of dynamic processes in cells and organelles. *In* "Cryotechniques in Biological Electron Microscopy" (R. A. Steinbrecht and K. Zierold, eds.) pp. 258–271. Springer Verlag, Berlin.

Kühn, A. (1946). Konstruktionsprinzipien von Schmetterlingsschuppen nach elektronenmikroskopischen Aufnahmen. *Z. Naturforsch.* **1**, 348–351.

Lane, N. J., Swales, L. S., and Abbott, N. J. (1977). Lanthanum penetration in crayfish nervous system: Observations on intact and "desheathed" preparations. *J. Cell Sci.* **23**, 315–324.

Lane, N. J., and Treherne, J. E. (1972). Studies on perineurial junctional complexes and the sites of uptake of microperoxidase and lanthanum in the cockroach central nervous system. *Tissue Cell* **4**, 427–436.

Locke, M. (1961). Pore canals and related structures in insect cuticle. *J. Biophys. Biochem. Cytol.* **10**, 589–618.

Locke, M. (2001). The Wigglesworth Lecture: Insects for studying fundamental problems in biology. *J. Insect Physiol.* **47**, 495–507.

Locke, M., Krishnan, N., and McMahon, J. T. (1971). A routine method for obtaining high contrast without staining sections. *J. Cell Biol.* **50**, 540–544.

Luft, J. H. (1971a). Ruthenium red and violet. I. Chemistry, purification, methods of use for electron microscopy and mechanism of action. *Anat. Rec.* **171**, 347–368.

Luft, J. H. (1971b). Ruthenium red and violet. II. Fine structural localization in animal tissue. *Anat. Rec.* **171**, 369–415.

Mangin, L. (1893). Sur ĺemploi du rouge de ruthenium en anatomie végétale. *C. R. Acad. Sci.* **116**, 653–656.

Marshall, A. T., and Kent, M. (1991). A device for freeze-substituting a large number of samples under controlled conditions. *J. Microsc.* (*Oxford*) **162**, 335–340.

McDonald, K. (1994). Electron microscopy and EM immunocytochemistry. *In* "*Drosophila melanogaster*: Practical Uses in Cell and Molecular Biology" (L. S. B. Goldstein and E. Fyrberg, eds.) Methods in Cell Biology. Vol. 44, pp. 411–444. Academic Press, San Diego, CA.

McDonald, K. (2007). Cryopreparation methods for electron microscopy of selected model systems. *In* "Cellular Electron Microscopy" (J. R. McIntosh, ed.) Vol. 79, pp. 23–56. Elsevier, Amsterdam.

McDonald, K. L. (2009). A review of high-pressure freezing preparation techniques for correlative light and electron microscopy of the same cells and tissues. *J. Microsc.* **235**, 273–281.

McDonald, K., and Morphew, M. (1993). Improved preservation of ultrastructure in difficult-to-fix organisms by high pressure freezing and freeze substitution: I. *Drosophila melanogaster* and *Strongylocentrotus purpuratus* embryos: *Microsc. Res. Tech.* **24**, 465–473.

McDonald, K. L., Sharp, D. J., and Rickoll, W. (2000). Preparation of thin sections of *Drosophila* for examination by transmission electron microscopy. *In* "*Drosophila* Protocols" (W. Sullivan, M. Ashburner, and R. S. Hawley, eds.) pp.245–271. CSHL Press, Cold Spring Harbor, NY.

Meinecke, C.-C., and Langer, H. (1984). Localization of visual pigments within rhabdoms of the compound eye of *Spodoptera exempta* (Insecta, Noctuidae). *Cell Tissue Res.* **238**, 359–368.

Moor, H., and Riehle, U. (1968). Snap-freezing under high pressure: A new fixation technique for freeze-etching. *Proceedings 4th European Reg. Conf. Electron Microsc.* **2**, 33–34.

Müller, M., Marti, Th., and Kriz, S. (1980). Improved structural preservation by freeze substitution. *In* "Electron Microscopy 1980" (P. Brederoo and W. de Priester, eds.) Vol. II. pp. 720–721. Proceedings 7th European Congress Electron Microscopy Leiden.

Neville, A. C. (1993). "Biology of Fibrous Composites." Cambridge University Press, Cambridge, UK.

Pophof, B. (2002). Moth pheromone binding proteins contribute to the excitation of olfactory receptor cells. *Naturwissenschaften* **89**, 515–518.

Pophof, B. (2004). Pheromone-binding proteins contribute to the activation of olfactory receptor neurons in the silkmoths *Antheraea polyphemus* and *Bombyx mori*. *Chem. Senses* **29**, 117–126.

Reynolds, E. S. (1963). The use of lead citrate at high pH as an electron-opaque stain in electron microscopy. *J. Cell Biol.* **17**, 208–212.

Richards, A. G., and Anderson, T. F. (1942). Electron microscope studies of insect cuticle, with a discussion of the application of electron optics to this problem. *J. Morph.* **71**, 135–183.

Rogers, M. E., Steinbrecht, R. A., and Vogt, R. G. (2001). Expression of SNMP-1 in olfactory neurons of male and female antennae of the silkmoth *Antheraea polyphemus*. *Cell Tissue Res.* **303**, 433–446.

Roth, T. F., and Porter, K. R. (1964). Yolk protein uptake in the oocyte of the mosquito *Aedes aegypti*. L. *J. Cell Biol.* **20**, 313–332.

Sabatini, D. D., Bensch, K., and Barrnett, R. J. (1963). Cytochemistry and electron microscopy. The preservation of cellular ultrastructure and enzymatic activity by aldehyde fixation. *J. Cell Biol.* **17**, 19–58.

Schneider, D. (1957). Elektrophysiologische Untersuchungen von Chemo- und Mechanorezeptoren der Antenne des Seidenspinners *Bombyx mori* L. *Z. vergl. Physiol.* **40**, 8–41.

Schneider, D. (1962). Electrophysiological investigations on the olfactory specificity of sexual attracting substances in different species of moths. *J. Insect Physiol.* **8**, 15–30.

Schwarz, H., and Humbel, B. M. (1988). Freeze-substitution and immunolabelling. EUREM 88, York, Inst. Phys. Conf. Ser. No. 93.

Schwarz, H., and Humbel, B. M. (1989). Influence of fixatives and embedding media on immunolabelling of freeze-substituted cells. *In* "The Science of Biological Specimen Preparation 1988." Scanning Microscopy, 1989, pp. 57–64.

Shaklai, M., and Tavassoli, M. (1982). Lanthanum as an electron microscopic stain. *J. Histochem. Cytochem.* **30**, 1325–1330.

Shanbhag, S. R., Müller, B., and Steinbrecht, R. A. (1999a). Atlas of olfactory organs of *Drosophila melanogaster.* 1. Types, external organization, innervation and distribution of olfactory sensilla. *Int. J. Insect Morphol. Embrol.* **28**, 377–397.

Shanbhag, S. R., Müller, B., and Steinbrecht, R. A. (1999b). Atlas of olfactory organs of *Drosophila melanogaster.* 2. Internal organization and cellular architecture of olfactory sensilla. *Arthropod Struct. Dev.* **29**, 211–229.

Shanbhag, S. R., Park, S.-K., Pikielny, C. W., and Steinbrecht, R. A. (2001). Gustatory organs of *Drosophila melanogaster*: Fine structure and expression of the putative odorant-binding protein PBPRP2.*Cell Tissue Res.* **304**, 423–437.

Shanbhag, S. R., Smith, D. P., and Steinbrecht, R. A. (2005). Three odorant-binding proteins are co-expressed in sensilla trichodea of *Drosophila melanogaster. Arthropod Struct. Dev.* **34**, 153–165.

Sharp, D. J., McDonald, K. L., Brown, H. M., Matthies, H. J., Walczak, C., Vale, R. D., Mitchison, T. J., and Scholey, J. M. (1999). The bipolar kinesin, KLP61F, cross-links microtubules within interpolar microtubule bundles of *Drosophila* embryonic mitotic spindles. *J. Cell Biol.* **144**, 125–138.

Shimoni, E., and Müller, M. (1998) On optimizing high-pressure freezing:from heat transfer theory to a new microbiopsy device. *J. Microsc.* **192**, 236–247.

Simionescu, N., and Simionescu, M. (1976a). Galloylglucoses of low molecular weight as mordant in electron microscopy. I. Procedure, and evidence for mordanting effect. *J. Cell Biol.* **70**, 608–621.

Simionescu, N., and Simionescu, M. (1976b). Galloylglucoses of low molecular weight as mordant in electron microscopy. II. The moiety and functional groups possibly involved in the mordanting effect. *J. Cell Biol.* **70**, 622–633.

Skaer, H.Le.B.S. H. P. Maddrell, and J. B., Harrison, (1987). The permeability properties of septate junctions in Malpighian tubules of *Rhodnius. J. Cell Sci.* **88**, 251–265.

Slifer, E. H., Prestage, J. J., and Beams, H. W. (1957). The fine structure of the long basiconic sensory pegs of the grasshopper (Orthoptera, Acrididae) with special reference to those on the antenna. *J. Morph.* **101**, 359–396.

Slifer, E. H., Prestage, J. J., and Beams, H. W. (1959). The chemoreceptors and other sense organs on the antennal flagellum of the grasshopper (Orthoptera, Acrididae). *J. Morph.* **105**, 145–192.

Smith, D. S. (1968). "Insect Cells. Their Structure and Function." Oliver and Boyd, Edinburgh.

Sømme, L. (1982). Supercooling and winter survival in terrestrial arthropods. *Comp. Biochem. Physiol.* **73A**, 519–543.

Spurr, A. R. (1969). A low-viscosity epoxy resin embedding medium for electron microscopy. *J. Ultrastruct. Res.* **26**, 31–43.

Steinbrecht, R. A. (1973). Der Feinbau olfaktorischer Sensillen des Seidenspinners (Insecta, Lepidoptera). Rezeptorfortsätze und reizleitender Apparat. *Z. Zellforsch.* **139**, 533–565.

Steinbrecht, R. A. (1976). Freeze substitution and freeze fracturing of insect sensilla without cryoprotectants. *In* "Electron Microscopy 1976" (Y. Ben-Shaul, ed.) Vol. 2. pp. 111–113. Isr. Society Electron Microscopy. Intertal, Givatayim, Jerusalem.

Steinbrecht, R. A. (1980). Cryofixation without cryoprotectants. Freeze substitution and freeze etching of an insect olfactory receptor. *Tissue Cell* **12**, 73–100.

Steinbrecht, R. A. (1985). Recrystallization and ice-crystal growth in a biological specimen, as shown by a simple freeze substitution method. *J. Microsc.* (Oxford) **140**, 41–46.

Steinbrecht, R. A. (1986). ATPase particles (portasomes) on mitochondrial cristae and the plasma membrane of an insect as demonstrated by freeze substitution. *Naturwissenschaften* **73**, 275–276.

Steinbrecht, R. A. (1988). BOSS-a new membrane substructure in insect thermoreceptors (goniometer analysis of cryosubstituted specimens). EUREM 88, York, Inst. Phys. Conf. Ser., No. 93, Vol. 3, pp. 475–476.

Steinbrecht, R. A. (1989). The fine structure of thermo-hygrosensitive sensilla in the silkmoth *Bombyx* mori: Receptor membrane substructure and sensory cell contacts. *Cell Tissue Res.* **255**, 49–57.

Steinbrecht, R. A. (1993a). Freeze-substitution for morphological and immunocytochemical studies in insects. *Microsc. Res. Tech.* **24**, 488–504.

Steinbrecht, R. A. (1993b). Frost statt Gift—Kryotechniken in der biologischen Elektronenmikroskopie. *Verh. Dtsch. Zool. Ges.* **86**(2), 213–228.

Steinbrecht, R. A., and Gnatzy, W. (1984). Pheromone receptors in *Bombyx mori* and *Antheraea pernyi*. I. Reconstruction of the cellular organization of the sensilla trichodea. *Cell Tissue Res.* **235**, 25–34.

Steinbrecht, R. A., Laue, M., and Ziegelberger, G. (1995). Immunolocalization of pheromone-binding protein and general odorant-binding protein in olfactory sensilla of the silk moths *Antheraea* and *Bombyx*. *Cell Tissue Res.* **282**, 203–217.

Steinbrecht, R. A., Lee, J.-K., Altner, H., and Zimmermann, B. (1989). Volume and surface of receptor and auxiliary cells in hygro-thermoreceptive sensilla of moths *(Bombyx mori, Antheraea pernyi*, and *A. polyphemus)*. *Cell Tissue Res.* **255**, 59–67.

Steinbrecht, R. A., and Müller, M. (1987). Freeze-substitution and freeze-drying. *In* "Cryotechniques in Biological Electron Microscopy" (R. A. Steinbrecht and K. Zierold, eds.) pp. 149–172. Springer Verlag, Berlin.

Steinbrecht, R. A., and Müller, B. (1991). The thermo-hygrosensitive sensilla of the silkmoth, *Bombyx mori*: Morphological changes after dry- and moist-adaptation. *Cell Tissue Res.* **266**, 441–456.

Steinbrecht, R. A., Ozaki, M., and Ziegelberger, G. (1992). Immunocytochemical localization of pheromone-binding protein in moth antennae. *Cell Tissue Res.* **270**, 287–303.

Steinbrecht, R. A., and Zierold, K. (1987). "Cryotechniques in Biological Electron Microscopy." Springer Verlag, Berlin.

Storey, K. B., and Storey, J. M. (1988). Freeze tolerance in animals. *Physiol. Rev.* **68**, 27–84.

Thurm, U. (1964). Mechanoreceptors in the cuticle of the honey bee. *Science* **145**, 1063–1065.

Thurm, U. (1965). An insect mechanoreceptor. I. Fine structure and adequate stimulus. *Cold Spring Harb. Symp. on Quant. Biol.* **30**, 75–82.

Thurm, U., and Küppers, J. (1980). Epithelial physiology of insect sensilla. *In* "Insect Biology in the Future" (M. Locke and D. S. Smith, eds.) pp. 735–763. Academic Press, New York.

Tilney, L. G., Bryan, J., Bush, D. J., Fujiwara, K., Mooseker, M. S., Murphy, D. B., and Snyder, D. H. (1973). Microtubules: Evidence for 13 protofilaments. *J. Cell Biol.* **59**, 267–275.

Toh, Y. (1985). Structure of campaniform sensilla on the haltere of *Drosophila* prepared by cryofixation. *J. Ultrastruct. Res.* **93**, 92–100.

Tokuyasu, K. T. (1973). A technique for ultramicrotomy of cell suspensions and tissues. *J. Cell Biol.* **57**, 551–565.

Tokuyasu, K. T. (1986). Application of cryoultramicrotomy to immunocytochemistry. *J. Microsc.* **143**, 139–149.

Trujillo-Cenoz, O. (1965). Some aspects of the structural organization of the intermediate retina of Dipterans. *J. Ultrastruct. Res.* **13**, 1–33.

Van Harreveld, A, Crowell, J., and Malhotra, S. K. (1965). **A** study of extracellular space in central nervous tissue by freeze-substitution. *J. Cell Biol.* **25**, 117–137.

Völker, W., Meschede, H., Weil, J., and Thurm, U. (1984). Immersionfreezing and freeze-substitution of irregularly shaped specimens with the aid of a cryo-injector. *Mikroskopie* (Wien) **41**, 351–359.

Wagner, R. C. (1976). The effect of Tannic acid on electron images of capillary endothelial cell membranes. *J. Ultrastruct. Res.* **57**, 132–139.

Walther, P., and Ziegler, A. (2002). Freeze substitution of high-pressure frozen samples: The visibility of biological membranes is improved when the substitution medium contains water. *J. Microsc.* **208**, 3–10.

Weiss, G. B. (1974). Cellular pharmacology of lanthanum. *Annu. Rev. Pharmacol.* 14, 343–354.

Wigglesworth, V. B. (1933). The function of the anal gills of the mosquito larva. *J. Exp. Biol.* **10**, 16–26.

Wolbarsht, M. L. (1960). Electrical characteristics of insect mechanoreceptors. *J. Gen. Physiol.* **44**, 105–122.

Wolfrum, U. (1990). Actin filaments: The main components of the scolopale in insect sensilla. *Cell Tissue Res.* **261**, 85–96.

Wolfrum, U. (1991). Tropomyosin is co-localized with the actin filaments of the scolopale in insect sensilla. *Cell Tissue Res.* **265**, 11–17.

Wucherpfennig, T., Wilsch-Bräuninger, M., and Gonzaléz-Gaitán, M. (2003). Role of *Drosophila* Rab5 during endosomal trafficking at the synapse and evoked neurotransmitter release. *J. Cell Biol.* **161**, 609–624.

Zalokar, M. (1966). **A** simple freeze-substitution method for electron microscopy. *J. Ultrastruct. Res.* **15**, 469–479.

Zhang, S.-G., Maida, R., and Steinbrecht, R. A. (2001). Immunolocalization of odorant-binding proteins in noctuid moths (Insecta, Lepidoptera). *Chem. Senses* **26**, 885–896.

Zimmermann, B. (1991). Differentiation of the thermo-/hygrosensitive (no-pore) sensilla on the antenna of *Antheraea pernyi* (Lepidoptera, Saturniidae): A study of cryofixed material. *Cell Tissue Res.* **266**, 427–440.

Zimmermann, B. (1992). Antennal thermosensitive and hygrosensitive sensilla in *Antheraea perny* (Lepidoptera, Saturniidae): Ultrastructure and immunohistochemical localization of Na^{+}, K^{+}-ATPase. *Cell Tissue Res.* **270**, 365–376.

CHAPTER 17

Electron Microscopy of the Amphibian Model Systems *Xenopus laevis* and *Ambystoma mexicanum*

Thomas Kurth[*], Jürgen Berger[†], Michaela Wilsch-Bräuninger[‡], Susanne Kretschmar[*], Robert Cerny[§], Heinz Schwarz[†], Jan Löfberg[¶], Thomas Piendl[#], *and* Hans H. Epperlein[||]

[*]Center for Regenerative Therapies, TU Dresden, D-01307 Dresden, Germany

[†]Max-Planck Institute of Developmental Biology, D-72076 Tübingen, Germany

[‡]Max-Planck Institute of Molecular Cell Biology and Genetics, D-01307 Dresden, Germany

[§]Department of Zoology, Charles University Prague, CZ-128 44 Prague, Czech Republic

[¶]Section of Animal Development and Genetics, Uppsala University, S-752 36 Uppsala, Sweden

[#]NET-Network for Educational Technology, ETH Zürich, CH-8092 Zürich, Switzerland

[||]Department of Anatomy, TU Dresden, D-01307 Dresden, Germany

978-0-12-381007-6
DOI: 10.1016/S0091-679X(10)96017-2

Abstract

In this chapter we provide a set of different protocols for the ultrastructural analysis of amphibian (*Xenopus*, axolotl) tissues, mostly of embryonic origin. For *Xenopus* these methods include: (1) embedding gastrulae and tailbud embryos into Spurr's resin for TEM, (2) post-embedding labeling of methacrylate (K4M) and cryosections through adult and embryonic epithelia for correlative LM and TEM, and (3) pre-embedding labeling of embryonic tissues with silver-enhanced nanogold. For the axolotl (*Ambystoma mexicanum*) we present the following methods: (1) SEM of migrating neural crest (NC) cells; (2) SEM and TEM of extracellular matrix (ECM) material; (3) Cryo-SEM of extracellular matrix (ECM) material after cryoimmobilization; and (4) TEM analysis of hyaluronan using high-pressure freezing and HABP labeling. These methods provide exemplary approaches for a variety of questions in the field of amphibian development and regeneration, and focus on cell biological issues that can only be answered with fine structural imaging methods, such as electron microscopy.

I. Introduction

Amphibians are classic model organisms for vertebrates in cell and developmental biology. Urodeles (newts and salamanders, such as *Triturus* and *Ambystoma*) and anurans (frogs and toads, such as *Rana, Bufo, Xenopus*) are the model species studied most intensively since the late 19th century. In the early days of embryology, many studies on developmental processes were performed especially in amphibians. The main advantages of amphibian models are the generation of vast amounts of embryos, the external development, as well as the robustness of the embryos tolerating experimental manipulations, such as tissue grafting (explantation, implantation, or transplantation), even across species. Many important concepts, such as cell fate determination or embryonic induction, were formulated in these times (Hamburger, 1988), mainly based on studies with urodele embryos. Later, the South African clawed frog, *Xenopus laevis*, was introduced as the most important amphibian model, due to practical considerations such as the reliable induction of ovulation by injection of hormones or the easy chemical removal of the jelly coat around the eggs (Sive *et al.*, 2000). Over the past 100 years, some of the hallmark experiments in developmental biology were performed with amphibians as model organisms, for example, the "organizer experiment." In this experiment, an area in the future dorsal region of the embryo was defined that, upon transplantation into a host embryo, was able to instruct the formation of a complete secondary body axis from graft and host tissues (Spemann and Mangold, 1924). Further important experiments were the first cloning of a vertebrate by microinjection of nuclei from differentiated cells into enucleated eggs (Gurdon *et al.*, 1975), the identification of the first organizer gene, the homeobox gene *goosecoid* (after more than 60 years of frustrating efforts; Cho *et al.*, 1991), followed by the subsequent molecular characterization of the organizer (reviewed in De Robertis, 2009), and recently, the introduction of antisense-Morpholinos for specific gene knock-down in embryos (Heasman *et al.*, 2000).

The major disadvantage of *Xenopus laevis* as a model for genetic studies is its pseudotetraploid genome and the long generation time (1–2 years). Therefore, the interest had shifted to other model organisms, such as *Drosophila melanogaster, Caenorhabditis elegans*, or the zebrafish *Danio rerio*. But *Xenopus* laevis and the axolotl, *Ambystoma mexicanum*, persisted as important models in developmental and regeneration biology. In fact, *Xenopus laevis* still is the best-understood model for mesoderm induction and patterning, or the regulation of cell movements in the early gastrulating or neurulating embryo (Heasman, 2006; Keller *et al.*, 2003; Kimelman, 2006). Recently, a close relative, *Xenopus tropicalis,* emerged as a new amphibian model organism with a diploid genome and a shorter generation time of 3–6 months (Amaya *et al.*, 1998).

In addition, the *Xenopus* egg is a powerful system for biochemical studies using egg extracts (Crane and Ruderman, 2006; Cross and Powers, 2009; Mandato *et al.*, 2001; Maresca and Heald, 2006) and for patch-clamp studies of ion-channels that are ectopically expressed in the oocyte membrane (Tammaro *et al.*, 2008). The axolotl also continues to be an important model in a variety of research areas, from physiology to development and regeneration (Voss *et al.*, 2009).

In the past few years, the regenerative potential of both *Xenopus* and the axolotl has gained increased interest (Slack *et al.*, 2008; Voss *et al.*, 2009). In this respect, the neoteny of the axolotl is especially remarkable as it allows the animal to regenerate complex body structures, such as limbs, tail, spinal cord, heart, or jaws (e.g., Kragl *et al.*, 2009; Mchedlishvili *et al.*, 2007; Sobkow *et al.*, 2006). Recently, it has become possible to study the genomic background for this regenerative potential (Habermann *et al.*, 2004; Putta *et al.*, 2004), which opens the door for deeper insights into the underlying processes, especially with respect to possible medical implications, and a potential to stimulate tissue regeneration in humans.

Many of the questions that arise during developmental or regeneration studies finally focus on basic cell biological questions concerning subcellular protein localization, macromolecular dynamics, vesicle trafficking, signaling regulation, etc. Often these cell biological questions can only be answered by using fine structural imaging methods, including electron microscopy.

II. Rationale

Our goal is to provide a set of different methods for the ultrastructural analysis of amphibian tissues, mostly of embryonic origin. The protocols include ultrastructural and immunolabeling approaches for TEM and SEM. In general, embryos are considered as “problematic samples” due to their yolk content, the concentration of very different tissues within a small specimen and, in many cases, the impossibility of dissecting them before fixation. Here, most of the presented methods are modified standard preparations that help to yield reasonable ultrastructure and/or protein localization. It is well accepted that cryofixation, especially high-pressure freezing (HPF), is currently the method of choice for optimal ultrastructural preservation (Moor, 1987; Studer *et al.*, 2001). Unfortunately, both *Xenopus* and axolotl embryos are too big to be high-pressure frozen properly *in toto*, the only possible exception being small explants (see Section III.B). Therefore, most of the protocols presented here still rely on chemical fixation and gelatin or resin embedding.

III. Methods

A. Xenopus

Although *Xenopus* is widely used as a laboratory animal, information on the histology of embryonic stages is rather limited. There is a detailed light microscopical description of tissue sections through early embryos (Hausen and Riebesell, 1991), but later stages, from tailbud stage 27 onward, and the histogenesis of tissues and organs, are poorly documented. The current knowledge about these later events is deduced mainly from low-resolution gene expression patterns (Sive *et al.*, 2000). Moreover, EM studies on *Xenopus* embryos are relatively rare and the *Xenopus* Laboratory Manual (Sive *et al.*, 2000) does not even contain a paragraph on specimen preparation for EM. However, studies involving EM analysis were performed in many fields of cell and developmental biology, where *X. laevis* and *X. tropicalis* are important vertebrate models. The topics include egg development and early cleavage (e.g., Cardellini *et al.*, 2007; Danilchik and Brown, 2008; Danilchik *et al.*, 2003; Fesenko *et al.*, 2000; Kalt, 1971; Merzdorf *et al.*, 1998; Müller and Hausen, 1995), gastrulation (e.g., Hardin and Keller, 1988; Keller, 1980, 1981; Kurth and Hausen, 2000; Kurth *et al.*, 1999; Nakatsuji *et al.*, 1985; Reintsch and Hausen, 2001; Shook *et al.*, 2004; Winklbauer and Nagel, 1991; Winklbauer *et al.*, 1992), neurulation and neural crest (NC) cell migration (e.g., De Calisto *et al.*, 2005; Sadaghiani and Thiebaud, 1987; Schroeder, 1970; Youn and Malacinski, 1981), and organogenesis (Giacomello *et al.*, 2002; Moriya *et al.*, 2000; Osafune *et al.*, 2002; Youn and Malacinski, 1981; Youn *et al.*, 1980). Most of these studies include SEM observations, but rarely TEM or immuno-EM.

In addition to studies on embryonic development, several cell biological EM studies on the cytoskeleton or nuclear structures were performed with *Xenopus* as a model organism (e.g., Allen *et al.*, 2007a, b; Gajewski and Krohne, 1999; Lourim and Krohne, 1993; Schwab *et al.*, 1998; Stick and Schwarz, 1983; Coughlin *et al.*, 2007; Goldberg *et al.*, 2008; Walther, 2008). The 3D structure of the nuclear pore complex (NPC), for example, was elucidated using *Xenopus* NPCs (Akey and Radermacher, 1993; Maco *et al.*, 2006; Stoffler *et al.*, 2003).

The focus of current *Xenopus* research lies on the genetic control of early development (axis determination, mesoderm induction and patterning, mesoderm and NC cell movements) or organogenesis (somitogenesis, NC derivatives, formation of organ anlagen), which is studied in embryos ranging from cleaving to tailbud stages (De Robertis, 2009; Heasman, 2006; Kimelman, 2006).

In this paragraph, we describe a standard embedding protocol for the ultrastructural analysis of embryos as well as pre- and post-embedding immunolabeling protocols for adult and embryonic tissues.

1. Embedding of Gastrula and Tailbud Embryos into Spurr's Resin for TEM

Cells of early *Xenopus* embryos are "filled" with yolk platelets that contain lipo-glyco-phosphoprotein crystals (Hausen and Riebesell, 1991). This material is difficult to fix, infiltrate, and cut properly (Kalt and Tandler, 1971; Kurth, 2003). To balance extraction, infiltration efficiency, and cutting properties with ultrastructural preservation, the following protocol is used: fixation with a glutaraldehyde–paraformaldehyde-(GA/PFA)-mixture modified after Karnovsky (1965) and embedding in Spurr's low-viscosity resin (Spurr, 1969).

Embryos are kept in modified Barth's solution (MBSH, Fey and Hausen, 1990) and staged according to Nieuwkoop and Faber (1967). Embryos are fixed at least overnight at 4°C in modified Karnovsky's fixative (2.5% GA, 2% PFA in 50 mM HEPES, pH 7.4). After several washes (6 × 10 min in 100 mM HEPES, 2 × 10 min in PBS) samples are embedded in 3% low-melting-point agarose. The agarose block is trimmed asymmetrically to a trapezoid pyramid, which allows orientation of the sample at a later stage during embedding into Spurr's resin. Agarose blocks enclosing the specimens are then postfixed with 1% OsO_4/PBS on ice for at least 4 h to overnight, followed by washes in PBS (5 × 5 min) and water (4 × 5 min), *en bloc* contrasting in 1% aqueous uranyl acetate (2 h on ice) and washes in water (4 × 5 min). Dehydration occurs in a graded series of ethanol in water (30, 50, 70, 90, 96%, 30 min each) with two final steps in water-free ethanol (100% ethanol on a molecular sieve, 30 min each). Subsequently, the samples are infiltrated in mixtures of Spurr's and water-free ethanol (1:2; 1:1; 2:1), 1.5 h each, and finally in pure Spurr's resin for 2 h and overnight. The blocks are embedded in flat embedding molds (EMS), oriented, and cured in an oven at 65°C overnight. Ultrathin sections are contrasted with lead citrate and uranyl acetate after Venable and Coggeshall (1965).

The ultrastructure of *Xenopus* embryos prepared in that way is shown in Fig. 1. The amphibian blastula is surrounded by an outer layer of polarized epithelial cells (Müller and Hausen, 1995). This so-called primary epithelium is the first cell population displaying signs of differentiation. At the beginning of gastrulation, a subpopulation of these outer epithelial cells, at the border of vegetal mass and marginal zone, undergoes apical constriction, thereby acquiring a typical flask-like shape (Figs. 1A and B; Hardin and Keller, 1988; Keller, 1981). This process involves both the actin cytoskeleton and microtubules (Lee and Harland, 2007). At the ultrastructural level, these so-called bottle cells are characterized by apical villi-like surface extensions (Figs. 1B and C), an electron-dense subapical layer, vesicles underneath the apical cell surface, and accumulated pigment granules (Figs. 1B and C). Later on, bottle cells respread and form part of the anterior pharynx epithelium

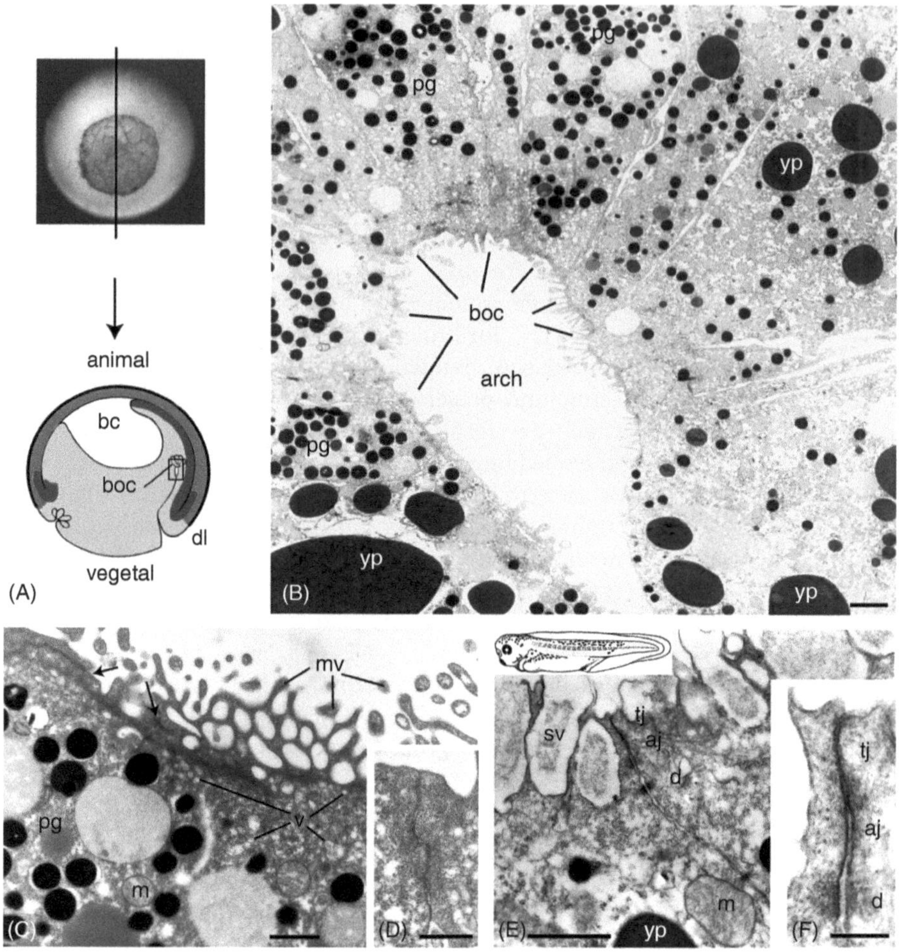

Fig. 1 Ultrathin Epon sections through *Xenopus* gastrula (stage 11, A–D) and tailbud (stage 37, E, F) embryos. (A) Vegetal view of the blastopore (top) and schematic of a median section (bottom, sectioning plane indicated by the black line in the embryo on top). Constricted epithelial cells (bottle cells, boc) are indicated; bc, blastocoel; dl, dorsal lip. The square indicates the region displayed in (B). (B) Apical, constricted parts of several bottle cells at the tip of the archenteron (arch) with accumulated pigment granules (pg). (C) Cellular details of bottle cells: microvilli (mv), subapical vesicles (v), pigment granules, and junctions between the cells (arrows). (D) Typical bottle cell contact at higher magnification. (E) Epidermal cells of a stage 37 tailbud embryo with a junctional complex: tight junction (tj), adherens junction (aj), desmosome (d). m, mitochondrium; sv, secretory vesicle; yp, yolk platelet. (F) Junctional complex at higher magnification. Scale bars: B, 2 μm, C,E, 1 μm, D, 0.5 μm, F, 0.2 μm. (See Plate no. 20 in the Color Plate Section.)

(Keller, 1981). Apical junctions are visible between bottle cells (arrows in Fig. 1C). However, a clear distinction of tight junctions, adherens junctions, and desmosomes, as shown in the outer epithelium of a stage 37 tailbud embryo (Figs. 1E and F), is often not possible.

2. Post-embedding Labeling of Methacrylate Sections (K4M) and Cryosections Through Adult and Embryonic Epithelia for Correlative Light and Electron Microscopy (CLEM)

On-section or post-embedding labeling of resin or cryosections is a powerful method for light and electron microscopy, yielding both ultrastructural and immunolabeling information (Schwarz and Humbel, 2007; Slot and Geuze, 2007). The major disadvantage of this method is its relatively low labeling efficiency, which is often critical in the early amphibian embryo (Kurth, 2003).

a. Fixation Tissue blocks not larger than 1 mm^3 are dissected in 2% PFA in 0.1 M phosphate buffer (PB), transferred immediately into 4% PFA/PB or 4% PFA/ 0.1% GA/PB, and fixed at least overnight at 4°C.

b. Progressive Low Temperature (PLT)-Embedding into K4M and Sectioning Specimens are transferred to 4-ml glass vials with screw caps or 0.5-ml Eppendorf microtubes, washed for 5 × 5 min in PB, and dehydrated in a graded series of ethanol/water: 30%, 50% (20 min each at room temperature), 70% (30 min at 4° C), 80% and 90% (30 min each at −20°C), 96% and 2 × 100% (at −35°C, 1 h each). Infiltration occurs at −35°C in Lowicryl K4M/ethanol mixtures (1:2, 1:1, 2:1) for 1 h each, followed by two steps in pure K4M for 2 h and overnight. The samples are then polymerized by UV light at −35°C for 24 h (e.g., in a Leica AFS2). For immunofluorescence, 200-nm-thick sections are collected from the water trough with a perfect loop, mounted on polylysine-coated microscope slides, and encircled with a water-repellent silicon pen. Ultrathin sections (50–70 nm) are collected on 50–100 mesh grids for immunogold labeling.

c. Embedding into Gelatin/Sucrose and Cryosectioning after Tokuyasu Fixed samples are washed for 5 × 5 min in PB, infiltrated stepwise with 2% gelatin (10 min at room temperature), 5, 10% (10 min each at 37°C), and cooled down on ice. Blocks (1.5 × 1.5 mm future cutting surface or smaller) are cut on ice and transferred to 2.3 M sucrose in water for at least overnight at 4°C. Sucrose-infiltrated blocks are mounted on specimen carriers (pins) and plunge-frozen in liquid nitrogen. Specimens can be stored either in sucrose solution or in LN_2 for at least several months. Semi-thin sections (300 nm) are picked up in methyl cellulose sucrose (MCS, 1 part 2% MC + 1 part 2.3 M sucrose) and mounted on polylysine-coated microscope slides. Ultrathin sections (70 nm) are picked up in MCS and transferred with a perfect loop to 50 mesh copper grids. Sections can be stored at 4°C in a humid chamber until staining (Griffith and Posthuma, 2002).

d. Correlative Fluorescence- and Immunogold-Labeling of K4M-Sections Resin sections are "rehydrated" for 2 × 5 min in PBS, blocked with 0.2% gelatin + 0.5% BSA in PBS (PBG, 2 × 10 min), and incubated with primary antibody (approximately 1–10 μg/ml) for 30 min to 2 h. All antibody/protein A gold solutions are centrifuged before application (5000 U/min for 5 min). After washes in PBG (6 × 3 min) sections are incubated with secondary antibody (e.g., fluorophore-labeled or gold-conjugated antibodies, protein A gold) for 30 min to 2 h. For fluorescence, sections are washed for 6 × 3 min in PBS, counterstained with DAPI (0.4 μg/ml) for 5 min, washed for 2 × 3 min in PBS, and mounted with Mowiol/DABCO or ProLong Gold. For EM, gold-labeled sections are washed for 6 × 3 min in PBG, 2 × 3 min in PBS, and 5 × 2 min in distilled water, before being contrasted with 1–2% aqueous uranyl acetate for 5 min. After final short washes in distilled water the sections dried and ready for investigation. If desired, lead citrate staining can be performed according to Venable and Coggeshall (1965).

e. Correlative Fluorescence- and Immunogold-Labeling of Cryosections For fluorescence microscopy the sections are washed for 2 × 5 min with PBS to get rid of the sucrose and 4 × 5 min (at 37°C) and 5 × 3 min (at room temperature) with PBS to melt and remove the gelatin. Sections are permeabilized with 0.1% Tween 20 in PBS for 5 min, washed with PBS (5 × 2 min), blocked with 0.1% glycin/PBS (2 × 3 min) and 1% BSA/PBS (2 × 5 min), and incubated with primary antibody (approximately 1–10 μg/ml in BSA/PBS, 30 min – 1 h). Sections are washed with BSA/PBS (5 × 2 min), incubated with fluorophore-labeled antibodies (Alexa-conjugated: 1:200, Cy3-conjugated: 1:400; from here on sections have to be protected from light) in BSA/PBS for 30 min to 1 h, washed in PBS (5 × 3 min), counterstained with DAPI (see above), washed (2 × 3 min PBS), and mounted with ProLong Gold.

For EM the grids are put upside down on 2% phosphate-buffered gelatin in a 3-cm petri dish. The gelatin is melted under a lamp (approximately 40°C), transferred to a 37°C-incubator for 20 min, washed with 0.1% glycin/PBS (5 × 1 min), blocked with 1% BSA/PBS (2 × 5 min), and incubated with primary antibody for 30 min to 1 h. After washing in PBS (4 × 2 min), the sections are incubated with gold-labeled secondary antibodies or Protein A gold (30 min-1 h), in PBS (3 × 5s, 4 × 2 min), stabilized with 1% glutaraldehyde (5 min), and washed extensively with distilled water (10 × 1 min) to get rid of the phosphates. With mouse IgGs as primary antibodies, and Prot A gold as marker, a bridging rabbit anti-mouse antibody is used because Prot A does not bind mouse IgGs. Then, sections are stained with neutral uranyl oxalate (2% UA in 0.15 M oxalic acid pH 7.0) for 5 min, washed for a short time in water, and incubated in methyl cellulose uranyl acetate (9 parts 2% MC + 1 part 4% UA pH 4) on ice for 5 min. The uranyl oxalate step is optional. It stabilizes membrane lipids and enhances the contrast in some cases (Griffiths *et al.*, 1984; Tokuyasu, 1976). Finally, grids are looped out and the MC/UA film is reduced to an even and thin film and air dried.

The examples shown in Figs. 2A–C (K4M sections) and Figs. 2 D and E (cryosections) show adult *Xenopus* enterocytes stained with anti-β-catenin (P14L, Schneider *et al.*, 1996). The protein is localized in the basolateral membrane and concentrated in adherens junctions. Since on-section-labeling is highly dependent on the copy number of a given protein, it is in many cases not sensitive enough for labeling embryonic

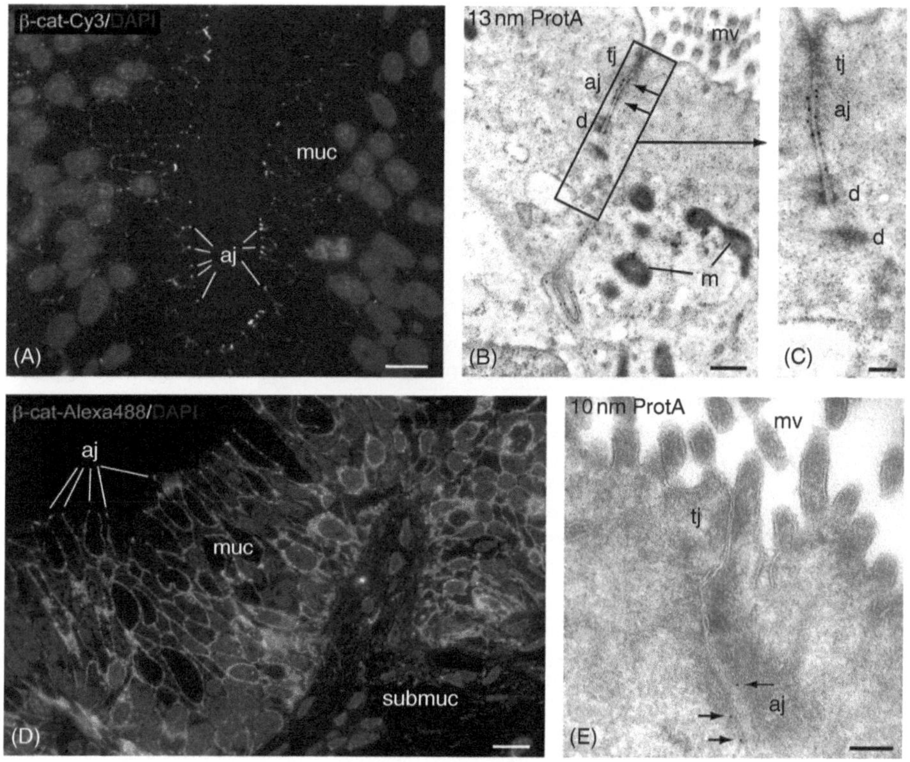

Fig. 2 Correlative on-section labeling of thin K4M sections (A–C) and cryosections (D,E) through mucosal (muc) and submucosal (submuc) layers of the adult *Xenopus* intestine. Sections are stained with anti-β-catenin and fluorescent secondary antibodies or Protein A Gold as indicated. For fluorescence, sections were counterstained with DAPI. The signal is visible in the basolateral membrane domains especially in the adherens junctions (aj); d, desmosome; m, mitochondrium; mv, microvilli; tj, tight junction. Scale bars: A and D, 10 μm; B, 0.5 μm, C and E, 200 nm. (See Plate no. 21 in the Color Plate Section.)

tissues (not shown). The staining is usually stronger in cryosections, which, therefore, might be a promising alternative for embryologists. In Fig. 3, junctions are labeled in cryosections through the epidermis of tails in tailbud embryos. As a proof of principle, the tight junction marker ZO-1 (mouse monoclonal ZO-1-1A12, Invitrogen) and β-catenin are shown. Many developmental biologists focus their interest on early embryos, such as blastula, gastrula, or neurula stages, where many important developmental decisions are made. As mentioned earlier, these embryos, however, are extremely difficult to cut properly because of their high yolk content.

3. Pre-Embedding Labeling of Embryonic Tissues using Silver-Enhanced Nanogold

For the early embryo, an alternative labeling method is needed when post-embedding labeling of resin sections is not sensitive enough and cryosections are difficult to obtain.

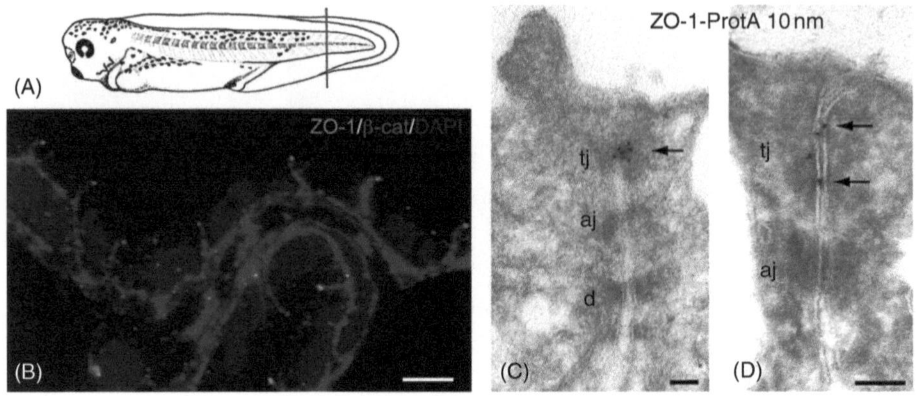

Fig. 3 Correlative on section-labeling of cryo-sections through a stage 37 *Xenopus* embryo. (A) Embryo with the sectioning plane indicated. Immunofluorescence (B) and immunogold-labeling (C,D) of thin cryosections through the epidermis stained with anti-β-catenin (+ goat-anti-rabbit-Alexa555) and anti ZO-1 (+ goat-anti-mouse Alexa488, B) or with anti-ZO-1 followed by Protein A 10 nm gold (C,D). Anti-ZO-1 marks the tight junctions (tj) between the epithelial cells; aj, adherens junction; d, desmosome. Scale bars: B, 5 μm, C and D, 100 nm. (See Plate no. 22 in the Color Plate Section.)

In such cases, we use a pre-embedding approach with silver-enhanced nanogold markers (Fig. 4; Kurth, 2003).

a. Fixation and Preparation of Vibratome Sections Embryos are fixed as described in paragraph b, washed in 100 mM HEPES (6 × 5 min) and PBS (2 × 5 min), and embedded in 2% low-melting-point agarose in PBS. Sections (50 μm) are cut with a vibrating razor blade on a vibratome (e.g., Vibratome series 1000, Pelco, speed 2.5, amplitudes 6–7).

b. Antibody Labeling The sections are blocked with 20% normal goat serum (NGS) in PBS containing 0.1% saponin for permeabilization (2 h at room temperature) and incubated with primary antibodies diluted in 20% NGS/PBS/0.05% saponin for at least overnight at 4°C. After washes in PBS (short, 5, 10, 15, 30 min, 2 × 1 h, 1 × 2 h), nanogold-conjugated antibodies (Fab-fragments, e.g., Aurion/EMS) are applied over-night at 4°C, the sections are washed in PBS (short, 5, 10, 15, 30 min, 2 × 1 h, 1 × 2 h), postfixed in 2% glutaraldehyde/PBS (1 h at room temperature), and again washed in PBS (5 × 5 min).

c. Silver Enhancement After washes in water (4 × 5 min), to get rid of the phosphates, silver enhancement is performed after Danscher (1981; See also Lah *et al.*, 1990; Stierhof, 2009) in the dark for 30 min to 1 h. Alternatively, the R-Gent silver enhancement Kit from EMS/Aurion can be used. After enhancement the specimens are washed in water (4 × 5 min), postfixed in 1% aqueous uranyl acetate for 1h on ice, and washed again for 4 × 5 min in water.

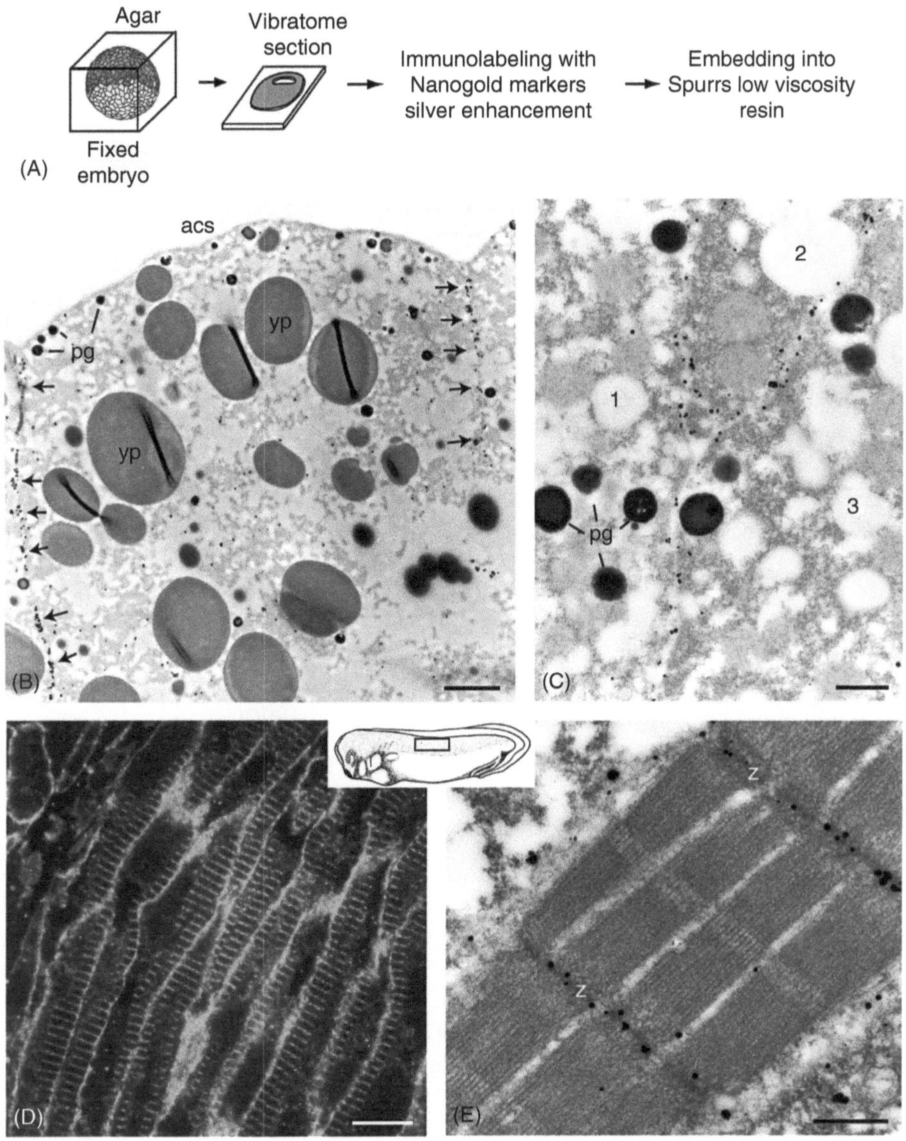

Fig. 4 Pre-embedding labeling of β-catenin in *Xenopus* embryos using nanogold and silver enhancement. (A) Schematic drawing of the procedure. (B) Animal cap cell of a blastula (stage 7) with β-catenin in the basolateral membrane (arrows); acs, apical cell surface; pg, pigment granules; yp, yolk platelet. (C) In the deeper layers the protein is expressed in all membrane contacts between cells (1–3). (D,E) Indirect correlative labeling of sagittal vibratome sections through the myotome of a tailbud embryo (stage 30). (D) Fluorescent staining of β-catenin in a striped pattern (anti-β-catenin, Goat-anti-rabbit-Cy3). (E) Labeling of the Z-disks at the EM level (Z), vibratome section embedded in Spurr's resin. Scale bars: B, 2 μm; C and E 0.5 μm; D, 20 μm.

d. Dehydration and Embedding The samples are dehydrated in a graded series of ethanol/water (30, 50, 70, 80, 90, 96%, 2× 100% ethanol, 20 min each), and samples are infiltrated in Spurr's low-viscosity resin via a graded series of Spurr–ethanol mixtures (1:2, 1:1, 2:1; 1–2 h each) followed by incubation in pure Spurr's resin for 2 h and overnight. The flat-embedding is done between two foils in a slide-embedding mold. The upper foils can be fixed with two to three small coins. One must be careful at this stage as early embryos are extremely brittle and can easily be destroyed, therefore excess pressure must not be applied. Polymerization occurs at 65°C. Later, the specimens are cut out and mounted on an Epon dummy block for sectioning. Sections can be contrasted with uranyl acetate and lead citrate, as described above.

e. Immunofluorescence In parallel, adjacent vibratome sections can be processed for immunofluorescence. Here the sections are stained with fluorophore-labeled secondary antibodies, postfixed in 4% PFA/PBS, dehydrated in a graded series of ethanol, and flat-embedded in either Technovit 7100 or in Spurr's resin. Technovit can be sectioned 1 μm thin, Spurr even thinner, resulting in high-resolution fluorescence (Kurth, 2003; Kurth *et al.*, 1996). Alternatively, silver-enhanced nanogold can be visualized by polarized light on semi-thin sections (Kurth, 2003).

The localization of β-catenin in the basolateral membranes of animal cap cells (blastula embryo, stage 7) is demonstrated (Figs. 4A–C). The protein can also be found in the Z-discs of growing myofibers in the myotome of stage 30 embryos (Figs. 4 D and E). Z-disc localization of β-catenin has been described also for adult heart and skeletal muscle (Kurth *et al.*, 1996).

B. Axolotl

Important issues studied with EM in axolotl include spermatogenesis (*TEM*; Miltner and Armstrong, 1983), somite segmentation (*SEM*; Armstrong and Graveson, 1988), pronephric duct migration (*SEM*; Gillespie *et al.*, 1985; Zackson and Steinberg, 1987), pigment cell migration and development, neural crest (NC) migration, and extracellular matrix (ECM) organization (*TEM, SEM*; Cerny *et al.*, 2004; Epperlein and Löfberg, 1990; Epperlein *et al.*, 1997, 2000a, 2000b, 2007; Falck *et al.*, 2002; Löfberg *et al.*, 1985, 1989a). For the staging of axolotl embryos and larvae up to stage 45, the normal table of Bordzilovskaya *et al.* (1989) is used. The development of limbs, the buds of which start to protrude at stage 40, can be staged from stage 45 onward in a special table (Nye *et al.*, 2003). In this section, EM protocols for SEM and TEM preparations and TEM immunohistochemistry will be presented. They focus on NC cell migration and ECM organization during development, key topics of research in axolotl and other vertebrates.

Axolotl embryos are much larger, but otherwise similar to *Xenopus* embryos. Thus, when transferring the methods applied in *Xenopus* to axolotl, the large size of these embryos has to be considered. In addition, different topics were studied in the axolotl field and a somewhat different set of EM-preparation protocols has been developed.

1. SEM Analysis of Migrating Neural Crest Cells

In axolotl embryos, from which the epidermis is removed, premigratory and migrating neural crest (NC) cells can vaguely be visualized with LM, but very well with SEM. The NC is formed during neurulation. Prospective NC cells are contained in the apical parts of the neural folds bordering either side of the neural plate. After the neural folds have converged dorsally and fused into the neural tube, NC cells leave the association of neural tube cells and lie on top of the neural tube. On the cranial neural tube the NC forms a flat epithelium whereas on the trunk tube it is organized as a narrow, multi-layered cell string (Fig. 5). These embryos were fixed in modified Karnovsky's fixative (2.5% GA and 2% PFA in 0.1 M phosphate buffer, pH 7.5, 4°C), followed by a washing step in buffer and postfixation in 1% OsO_4 in 0.1 M potassium–phosphate buffer, pH 7.5, 4°C (for details see Cerny *et al.*, 2004; Epperlein and Löfberg, 1990). From their site of origin on top of the neural tube, NC cells migrate out into different body sites on stereotyped routes through an ECM-rich environment and give rise to a wide variety of derivatives (Hall and Hörstadius, 1988; Le Douarin and Kalcheim, 1999). In the head, NC cells form several discrete and coherent cell streams. These originate from specific hindbrain levels of the NC and are directed laterally (Fig. 5A). In the trunk, NC cells do not form streams but migrate out laterally as single cells (Fig. 5C) between somites and epidermis, and medially between somites and the neural tube. Lateral NC cells give rise to pigment cells and medial ones form neurons and glia of the peripheral nervous system.

a. Tissue Preparation and Fixation of Embryos for SEM Analysis of NC Cells Embryos are kept in Steinberg solution (Steinberg, 1957) either at room temperature or at 7–8°C to slow down embryonic development. To partially remove the epidermis and expose areas of interest, embryos are transferred into an agar dish filled with cold Steinberg solution (7–8°C). The epidermis is removed with tungsten needles and the embryos are then transferred into modified Karnovsky's fixative in 0.1 M potassium–phosphate buffer (pH 7.5, 4°C) and kept at 4°C overnight. Following washes in PBS (30 min, three changes), fixed embryos are transferred to 1% OsO_4 in 0.1 M potassium–phosphate buffer (pH 7.5, 4°C), and kept overnight at 4° C. Then the embryos are washed in PBS and dehydrated through a graded series of ethanol in water: 30, 50% (30 min each at room temperature), 70% overnight at 4°C, 85%, 96%, 3× 100% ethanol (30–60 min each at room temperature). The slow dehydration process helps to prevent the formation of cracks on the surface of the specimen or in the yolk platelets. Subsequently, critical-point drying with liquid CO_2 is performed using a Polaron critical-point dryer. The exchange of ethanol and liquid CO_2 should happen slowly for at least 2–3 h. Finally, the specimens are sputter-coated with gold–palladium and observed with SEM (Fig. 5A, cranial NC; Fig. 5B, trunk NC).

2. SEM and TEM Analysis of Extracellular Matrix (ECM) Material

The ECM consists of proteoglycans, glycoproteins, and glycosaminoglycans, such as hyaluronan. It includes all secreted molecules outside the cells, insoluble fibrils (like

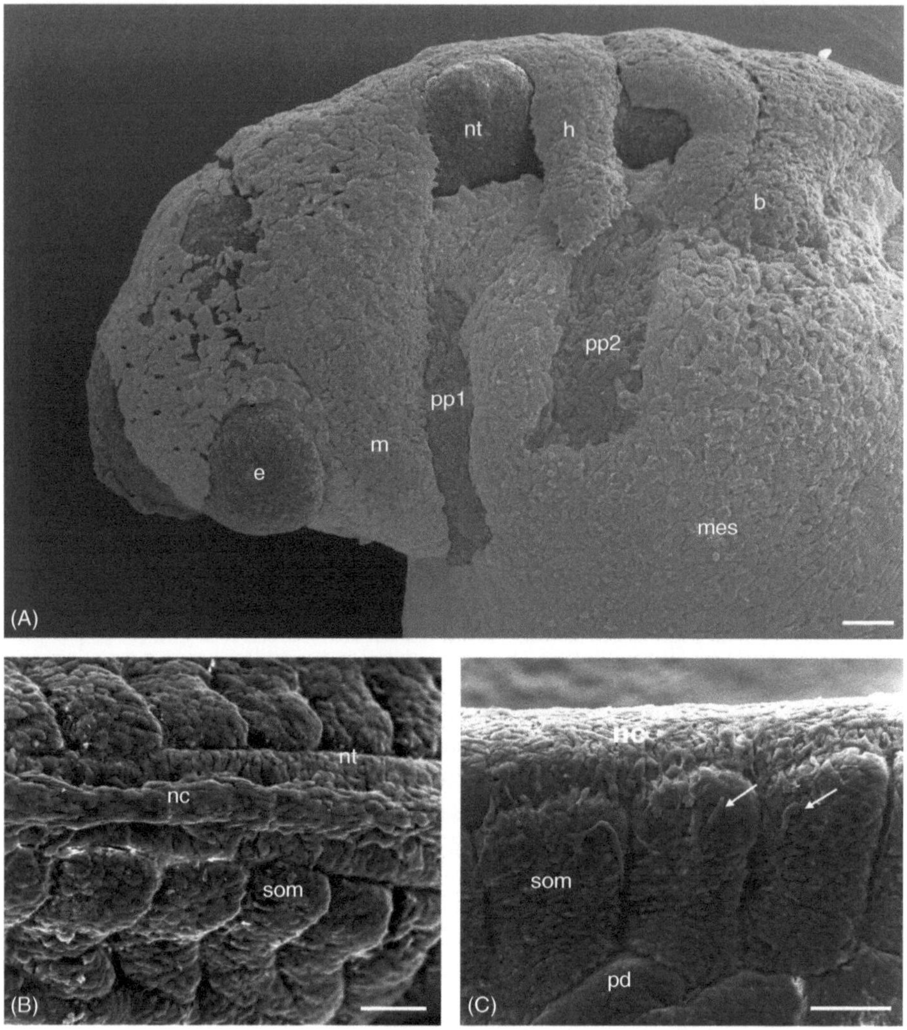

Fig. 5 SEM of axolotl neural crest (NC). (A) Cranial NC streams of an epidermis-free axolotl embryo (stage 25) migrating out laterally on head mesoderm (mes). Prospective pharyngeal pouches (pp 1 and pp 2); mandibular NC stream (m), forming the upper and lower jaw; hyoid NC stream (h), forming the hyoid; common branchial NC stream (b) segregating into NC streams 3–6 later. e, eye; nt, neural tube. (B,C) Midrunk NC. (B) premigratory NC (stage 30/31); NC string (nc) on top of the trunk neural tube (nt). (C) migratory NC (stage 33); single NC cells (arrows) migrate out laterally over the dorsolateral neural tube toward the upper margin of the somites (som). pd, pronephric duct. Scale bars: A, 100 μm, B and C, 50 μm. (See Plate no. 23 in the Color Plate Section.)

collagen or fibronectin), or highly hydrated soluble polymers (like hyaluronan and proteoglycans). The ECM regulates the development and function of almost all cells in metazoan organisms (Reichardt, 1993). In particular, subepidermal ECM from dark

(D/-) axolotl embryos was shown to stimulate NC cell migration in the white (d/d) mutant (Löfberg *et al.*, 1985, 1989a,b). Our knowledge of the ultrastructure of the ECM is very limited and even artifacts have been published. During standard chemical fixation (modified Karnovsky: 2.5% GA and 2% PFA in 0.1 M phosphate buffer) and the following dehydration steps, water is extracted from the ECM, which results in shrinkage and structural damage. For example, wide spaces become visible between fibrocytes in the dorsal fin of axolotl larvae observed in the SEM, and the cells seem to lie in an empty environment (Epperlein *et al.*, 1997). Precipitates of the ECM are retained as fibrillar networks between the cells and on their surfaces. By contrast, living fibrocytes, studied in the confocal microscope, are embedded in their natural interstitial ECM (Epperlein *et al.*, 1997). Here the cells appear voluminous, send out branching cell processes, and lie in an amorphous extracellular ground substance with interspersed fibrils. In order to find components of the ECM that specifically stimulate lateral NC (pigment) cell migration in the wild-type (dark, D/d) axolotl embryo or inhibit this migration in the white mutant (d/d), detailed investigations on ECM components between dark and white embryos were carried out. Differences could be observed in the distribution of collagen type III (Perris *et al.*, 1990) and keratan sulfate (Olsson *et al.*, 1996a). Functional tests with collagen type III, however, could not show any influences on pigment cell distribution (Olsson *et al.*, 1996b), and the effect of keratan sulfate is still unknown. Furthermore, the analysis of the distribution of hyaluronan (HA) did not reveal differences in dark or white embryos (Epperlein *et al.*, 2000a). The ultrastructure of HA is particularly difficult to demonstrate. Because of its hydration HA resists standard chemical fixation and can only be shown with difficulty, even after cryofixation (see below). More recently, the TGF-β growth factor BMP-4 was shown to exert an inhibitory effect on pigment cell (melanophore) distribution in dark embryos whereas Noggin, its antagonist, could stimulate melanophore migration in the white mutant (Hess *et al.*, 2008).

b. Tissue Preparation and Chemical Fixation for SEM and TEM Analysis of ECM To demonstrate ECM material in the SEM or TEM, standard chemical fixation is insufficient. Reagents like ruthenium red (RR) (Luft, 1971a, b) have to be added to the fixative in order to precipitate ECM components and enhance their contrast. The preparation and fixation (including RR) of embryos is as follows: Embryos are fixed in modified Karnovsky's fixative containing 1.5% GA +1.5% PFA in 0.1 M phosphate buffer (pH 7.4) for at least several hours at 4°C, rinsed in PBS, and postfixed in 1% OsO4 in 0.1 M cacodylate buffer for 4 h. During part of the fixation time (Karnovsky: 30 min, OsO_4: 4 h) both fixatives contain 500 ppm RR. Alternatively, 0.05% tannic acid can be added to Karnovsky's fixative only. After several washes in cacodylate buffer, the specimens are dehydrated in a graded series of ethanol, critical-point dried, and sputtered with gold palladium (SEM; Löfberg *et al.*, 1985; see Section 2a), or infiltrated with Epon for ultrathin sectioning (TEM; Löfberg *et al.*, 1985). Sections are stained with lead citrate and uranyl actetate as described above, Section IIIA).

Subepidermal ECM of an axolotl embryo (stage 34) fixed with Karnovsky's fixative + RR is shown in Fig. 6A (SEM) and Fig. 6B (TEM). ECM precipitated on a

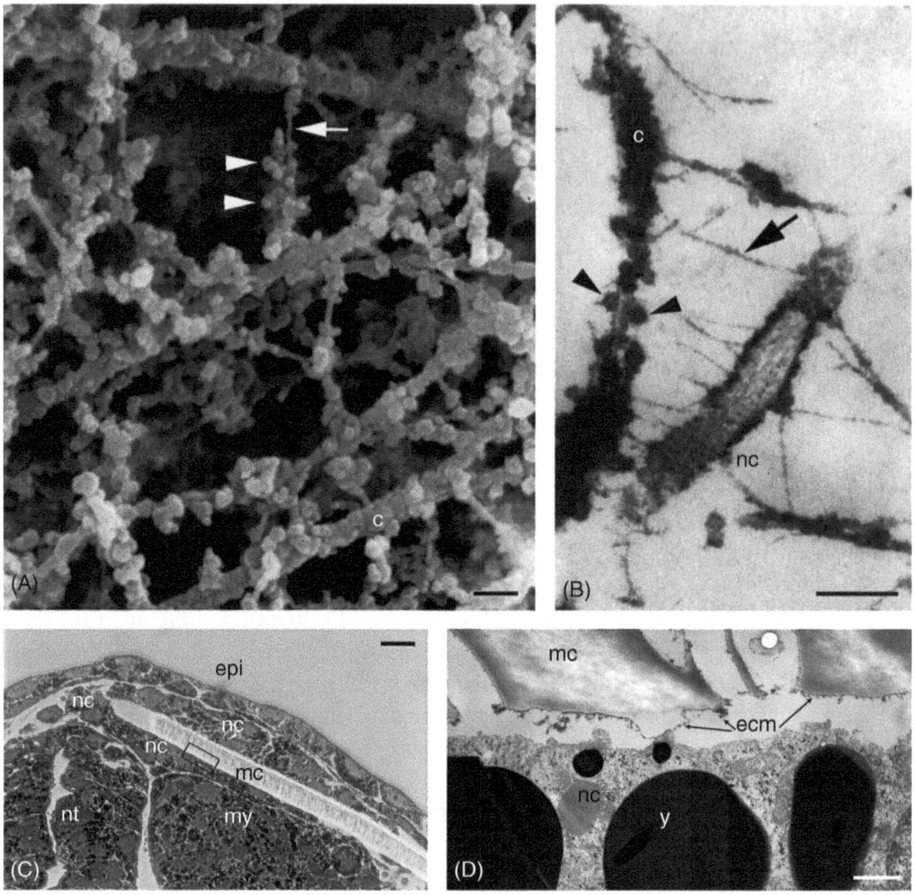

Fig. 6 Subepidermal ECM in the electron microscope. SEM (A) and TEM (B) micrographs of subepidermal ECM from the dorsolateral trunk of axolotl embryos (stage 34) after fixation with Karnovsky's fixative including RR. The thicker fibrils consist of collagen (c), the thinner filaments (arrows) consisting very likely of hyaluronan. Both are decorated with granular precipitates, possibly proteoglycans (arrowheads). RR enhances the preservation of thin filaments and ECM precipitates. In B part of a NC cell (nc) is visible and contacted by matrix filaments and fibrils (C,D) Stimulation of trunk neural crest cell migration in wild-type (dark; D/d) axolotl embryos with dark subepidermal ECM. (C) LM micrograph showing transverse section through the midtrunk region of a wild-type embryo (stage 30) containing a grafted microcarrier (mc, Nuclepore filter) conditioned with ECM of a dark donor embryo. (D) TEM micrograph showing an enlarged contact area (inset in C) between the microcarrier (mc) and a host NC cell whose lateral migration was stimulated by adsorbed dark ECM. ECM material (ecm, arrows) is precipitated following fixation with Karnovsky's fixative containing RR; epi, epidermis; nt, neural tube; my, myotome; ecm, extracellular matrix; y, yolk platelet. Bars: A and B, 0.2 μm, C, 25 μm, D, 0.1 μm. A and B reprinted with permission from Löfberg *et al.* (1989b).

microcarrier using Karnovsky+RR is shown in Figs. 6C and D. Microcarriers (0.1 × 0.5 mm) cut from Nuclepore filter material (Nuclepore Inc.) were implanted into the dorsolateral trunk of dark (D/-) axolotl embryos (stage 25) with their matt side facing the subepidermal ECM. For adsorbing "dark" ECM, microcarriers were left in the embryo for about 20 h (until stage 30), then they were implanted into dark (D/d) or white (d/d) stage 25 hosts and stayed there until stage 30, where they could stimulate lateral migration of host NC cells (Löfberg *et al.*, 1985).

3. Cryo-SEM Analysis of Extracellular Matrix (ECM) Material After Cryofixation

An alternative to chemical fixation is ultrarapid freezing. Several methods of cryofixation, such as plunge-freezing, metal mirror freezing, and high-pressure freezing (HPF), have been introduced over the past decades (Moor, 1987). Tissues that are difficult to fix by conventional means particularly benefit from rapid freezing (i.e., encapsulated tissues, plants, and tissues with a high content of water, such as cartilage; see several chapters of this volume). Therefore, cryofixation is a promising option for the analysis of ECM material.

a. Tissue Preparation and Cryofixation for Cryo-SEM Analysis of ECM Epidermal grafts from the dorsolateral trunk of dark (D/d) anesthetized (MS 222, Sandoz) larvae (stage 40) are transferred into an agar dish containing Steinberg solution (Steinberg, 1957) and quickly arrested between the two halves of a lockable double grid (precleaned by sonication in ethanol/acetone) to prevent curling. The matrix side of the graft has to face the side with the grid slot. The grid containing the epidermal graft is directly plunge-frozen in liquid ethane (plunge velocity 2 m/s), mounted on a cryoholder (Gatan Inc., Pittsburgh, PA) and freeze-dried in a Balzers' MED 010 device (pressure 6.7×10^{-7}mbar; 3 h at −110°C, 1 h at −100°C). Sputter-coating with 3-nm tungsten at the same temperature follows and then the samples are cooled down to −168°C, transferred to the cryo-SEM, and imaged at −168°C (Epperlein *et al.*, 1997).

Subepidermal ECM in epidermal grafts from axolotl larvae (stage 40), fixed in modified Karnovsky solution with the addition of CPC (See section IV), followed by plunge-freezing and freeze-drying, shows only unsatisfactory results in the cryo-SEM. The ECM fibrils, consisting mainly of collagen, form a random network with large empty spaces. The fibrils are decorated with only a few, small granular precipitates, which are probably proteoglycans (Fig. 7A). Better structural preservation is obtained by cryofixation (plunge-freezing and freeze-drying). When observed in the cryo-SEM, the ECM fibrils form a dense and relatively even matrix that contains many small clefts and some larger holes. The fibrils are densely covered with precipitates (Fig. 7B).

4. TEM Analysis of the ECM Component Hyaluronan by Cryofixation and HABP Labeling

Hyaluronan (HA) is an extracellular and cell-surface-associated polysaccharide (Toole, 2004). It belongs to the family of glycosaminoglycans (GAGs) that consist

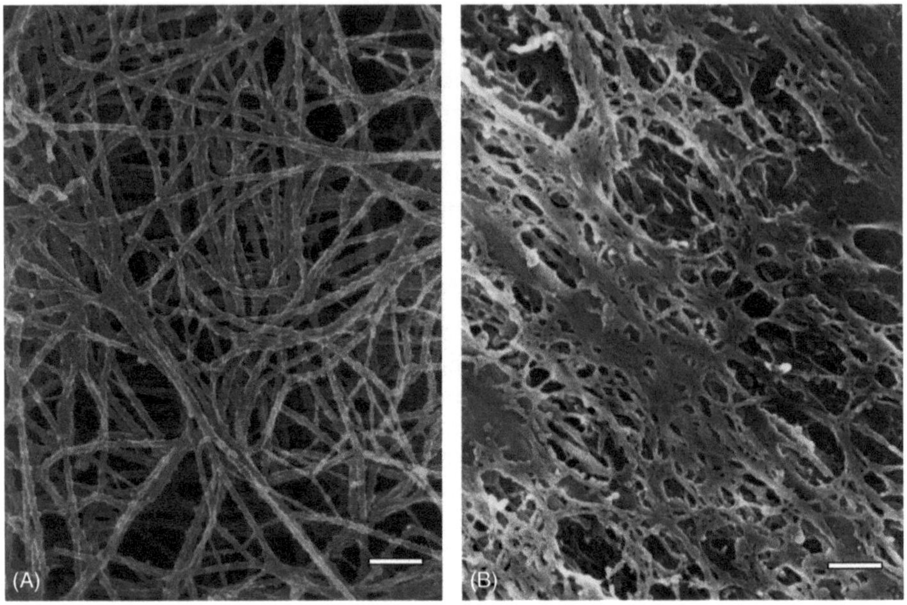

Fig. 7 SEM micrographs of subepidermal ECM observed in the cryo-SEM at -168°C following (A) standard chemical fixation and (B) cryofixation. (A) standard chemical fixation, followed by plunge-freezing and freeze-drying. Collagen fibrils form a random network with large empty spaces and are decorated with only few granular precipitates. (B) Direct plunge-freezing and freeze-drying. Collagen fibrils are densely covered with precipitates and form a dense, even matrix with numerous small clefts. Scale bars: 0.2 μm. Reprinted with permission from Epperlein et al. (1997).

of repeating disaccharides of glucuronic acid and *N*-acetylglucosamine. The number of disaccharide repeats can be 10,000 or more, and thus reach 10 μm if stretched (Hascall and Laurent, 1997). HA forms highly hydrated randomly kinked coils in solution and starts to entangle at concentrations of less than 1 mg/ml (Laurent *et al.*, 1996). Within the matrices of the ECM, HA mainly interacts with proteoglycans via linker proteins. At cell surfaces it binds with hyaluronan-binding proteins (HABPs), such as CD44 and RHAMM (Reichardt, 1993; Toole, 2004). It is present predominantly in the epidermis and dermis of the skin (Hascall and Laurent, 1997), and also, for example, in the vitreous of the eye, in the rooster comb, or in Wharton's jelly of the umbilical cord (Scott *et al.*, 1991). HA is involved in the control of cell proliferation, NC and tumor cell migration, wound repair, and various physiological processes (Chen and Abatangelo, 1999; Hayen *et al.*, 1999; Whight *et al.*, 1991). Furthermore, it plays a role in tumorigenesis, signal transduction, multidrug resistance, and angiogenesis (Toole, 2004).

The molecular model of HA in solution is not clear because it partially relies on TEM observations of HA solutions (Scott *et al.*, 1991) that were not optimally freeze-dried. According to these observations, HA is a network-forming polymer with the ability to form 3D honeycomb-like structures. A more recent characterization describes HA molecules in solution as expanded random coils that may form networks with pores that continuously change in size (Hascall and Laurent, 1997). Because the appearance of HA in natural biological samples may be different from such theoretical models based on HA in solution, the fibrillar networks (Whight *et al.*, 1991), honey-comb-like structures (Scott *et al.*, 1991), or porous networks (Hascall and Laurent, 1997) observed may be due to segregation patterns obtained by fixation artifacts during standard chemical fixation and/or ice crystal formation during cryofixation. In this paragraph we use various cryofixation protocols to demonstrate the structure of HA in axolotl embryos (stage 35) with TEM immunohistochemistry of HABP (Fig. 8). When the cryofixed embryonic axolotl ECM is labeled with HABP, the structural preservation of HA appears less regular (Figs. 8D,D', with UA; Figs. 8E,E', without UA) than after standard chemical fixation and freeze-substitution (Fig. 8C,C') where HA seems to predominate in honeycomb-like patterns, especially when the images are compared at higher magnification. Thus, at present, our TEM observations reveal a less regular honeycomb-like pattern (Figs. 8E and E') as the optimal result for the structure of natural HA in the subepidermal ECM of axolotl embryos (stage 35) following HPF/FS, fixation with 0.05% GA, no UA, and HABP immunohistochemistry (Epperlein *et al.*, 2000a). In addition, the labeling efficiency is much better after HPF (Figs. 8E,E') than after chemical fixation (Figs. 8C, C').

a. Standard Chemical Fixation (PFA/GA) and Freeze-Substitution (FS) Excised trunks of axolotl embryos (stages 33–35) were chemically fixed in 4% PFA in 0.1 M phosphate buffer at pH 7.4 overnight at 4°C. The tissue blocks were then infiltrated with 2.3 M sucrose overnight, frozen in liquid nitrogen, and freeze-substituted (FS) in a Leica AFS device, in pure methanol for 70 h at −80°C, followed by infiltration with Lowicryl HM20 and UV polymerization at −35°C (See Fig. 8C)

b. High-Pressure Freezing (HPF) Small pieces of dorsolateral trunk epidermis (length 2 mm, width 1 mm), including subepidermal ECM of embryos at stages 33–35, are excised with tungsten needles, transferred to cryoprotectants (a mixture of 10% polyvinyl pyrrolidone and 7% methanol in 0.1 M potassium–phosphate buffer; Epperlein *et al.*, 1997), and placed between two cylindrical aluminum platelets that form a double cup with a cavity of 200 μm depth. The tissue samples are then frozen with a high-pressure freezer (BAL-TEC HPM 010, Balzers, Liechtenstein). The platelets are carefully opened under liquid nitrogen and transferred to 2-ml microtubes with screw caps containing the precooled (–90°C) substitution medium consisting of 0.05% glutaraldehyde in acetone either with (Figs. 8D and D') or without (Figs. 8E and E') 0.5% uranyl acetate. The samples are freeze-substituted for 24 h at −90°C, 4 h at −60°C, and 3 h at −30°C in a freeze-substitution device (Bal-Tec FSU 010), washed with pure ethanol at −30°C, and infiltrated with ethanol/HM20. Embedding in Lowicryl HM20 and UV polymerization is performed at −35°C.

c. TEM Immunocytochemistry with HABP Ultrathin sections (70 nm) through conventionally fixed, or HPF-frozen samples (embryos at stages 33–35) embedded in

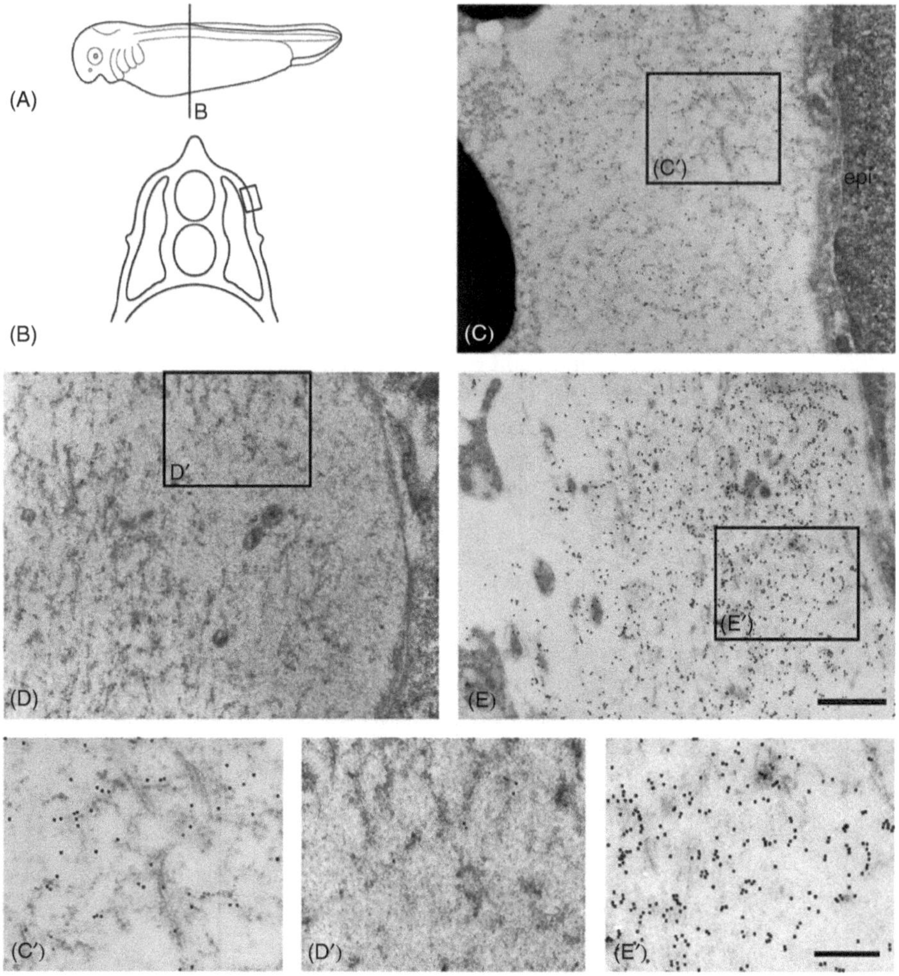

Fig. 8 Post-embedding TEM labeling of hyaluronan with HABP on ultrathin Lowicryl (HM20) sections through trunk epidermis/subepidermal ECM of dark (D/d) axolotl embryos. (A) Schematic drawing of an axolotl embryo at stage 35. The trunk level of subepidermal ECM is boxed. (B) Schematic transverse section through mid trunk area marked in A with inset indicating ECM areas displayed in C–D. (C) Fixation with 4% PFA, freeze-substitution in methanol, no *en bloc* staining with UA. The ECM is evenly labeled, its material present predominantly as "honeycomb"-like nets. (D) Fixation by HPF, freeze-substitution in acetone containing 0.05% GA and 0.5% UA. The ECM is visible as flocculent, irregular net-like and fibrillar material. The labeling intensity is very low. (E) Fixation by HPF, freeze-substitution in acetone containing 0.05% GA, no UA. The ECM appears as flocculent material, forming less regular honeycomb-like nets and fibrils than in C. Strong labeling with HABP. The contrast of cell and matrix structures is lower than in D due to the lack of UA. (C'–E') show framed regions in C–E at higher magnification. Scale bars: 1 μm (C–E), 0.5 μm (C'–E').

Lowicryl HM20, are blocked with PBS containing 0.2% gelatin. The samples are incubated with biotinylated HABP (1:1000) overnight at 4°C in PBS/gelatin. After washing with PBS/gelatin, the samples are incubated with rabbit–anti-biotin antibodies (1:50 in PBS/gelatin) for 1 h at room temperature. Washing in PBS/gelatin follows. The samples are incubated with a 10-nm gold-coupled goat anti-rabbit-IgG for 2 h at room temperature. After washing with PBS/gelatin, PBS and water samples are poststained with aqueous uranyl acetate and lead citrate according to standard protocols prior to investigation.

IV. Materials

- **Embryo culture media:** For *Xenopus*: MBSH (88 mM NaCl, 1 mM KCl, 2.4 mM $NaHCO_3$, 0.82 mM $MgSO_4$, 0.41 mM $CaCl_2$, 0.33 mM $Ca(NO_3)_2$, 10 mM HEPES pH 7.4; 10 mg/ml streptomycinsulfate and penicillin. For *Ambystoma*: Steinberg solution, 1000 ml (Steinberg, 1957): *Solution A*: 3.4 g NaCl, 0.05 g KCl, 0.205 g $MgSO_4 \times 7$ H2O, 4 ml 1 N HCl (about 9 ml 1N HCl in 100 ml aqua bidest), 0.56 g TRIS (Sigma) and 900 ml aqua bidest; plus *Solution B*: 0.08 g $Ca(NO_3)_2 \times 4H_2O$ and 100 ml aqua bidest.
- **Fixatives:** For ultrastructure/SEM: Modified Karnovsky's fixative (Karnovsky, 1965): 2.5% glutaraldehyde (GA), and 2% paraformaldehyde (PFA) in either 50 mM HEPES or in 0.1 M phosphate buffer, or modified Karnovsky's fixative containing 1.5% GA +1.5% PFA when RR, CPC, or tannic acid was added for precipitating ECM material. Secondary fixative: 1% OsO_4 in 0.1 M cacodylate buffer or PBS. For immunocytochemistry: 4% paraformaldehyde ±0.1% glutaraldehyde in 0.1 M phosphate buffer.
- **Embedding media**: For light microscopy: Technovit 7100 (Heraeus-Kulzer). For TEM: Spurr's resin kit (EMS # 14300), Epon 812 or EMbed-812 kit (EMS, # 14120), Lowicryl K4M kit (EMS # 14330) or HM20 kit (EMS # 14340).
- **Chemicals:** BSA (Merck, # 1.12018), gelatin (local food brand), glutaraldehyde (25% in water, EMS # 16220), HEPES (EMS # 16782), lead citrate (EMS # 17810), low-melting-point agarose (e.g., Sigma A9414), methyl cellulose (25 centipoises, Sigma M-6385), OsO_4 (4% in water, EMS # 19170), paraformaldehyde (granular, EMS # 19208), sucrose (EMS # 21600), uranyl acetate (Polysciences # 21447).
- **Secondary antibodies and detection systems:** We use Alexa488- and Alexa555-conjugated secondary antibodies for immunofluorescence, DAPI (Invitrogen) as a counterstain and ProLong Gold (Invitrogen P36930) as mounting medium. For EM, we use protein A conjugated to 10-nm or 15-nm gold particles (Aurion). Alternatively, secondary antibodies conjugated to gold particles from several companies (Aurion, Jackson ImmunoResearch, British Biocell) can be used. Ultrasmall gold-conjugated antibodies are purchased from EMS/Aurion as well as the R-Gent-SE-EM silver enhancement kit (EMS/Aurion # 25521)
- **Tools and materials:** Flat embedding mold (EMS # 70901), lockable double grids (type 50/50, Embra, The Netherlands), microtubes with screw cap (Sarstedt #72.694),

perfect loop (Plano), pins for cryo-sectioning (Leica specimen carrier # 16701950), polylysin-coated microscope slides (Menzel/VWR # 631-1349), slide embedding mold (EMS # 70172), water-repellent silicon pen (Pap-pen, Plano # 22304).

V. Discussion and Outlook

For more than a century, amphibian embryos have been classic models in cell and developmental biology. They were at the forefront of biological sciences in the early 20th century, culminating in the most influential organizer experiment (Spemann and Mangold, 1924). However, for more than 50 years the molecular nature of many discovered processes remained mysterious. Many researchers switched to models, such as *Drosophila* or the zebrafish that seemed to be more suitable for genetic analysis. Only after the discovery and characterization of the first organizer genes, and different embryonic inducers, embryology of amphibians has regained interest and is now successfully combined with molecular biology. Many long-standing developmental puzzles were solved over the past 20 years (De Robertis, 2009; Heasman, 2006; Kimelman, 2006). A plethora of developmental regulators, such as adhesion molecules and signal molecules, receptors, or transcription factors, has been discovered in many different screens where embryonic assays were combined with molecular biology. Various roles of such factors during development are currently being studied. A deeper understanding of the cell biological functions of these factors can, however, only be achieved if their subcellular distribution and trafficking are known, as well as their interaction with other proteins and their involvement in different subcellular processes. In some cases, cell biological processes can only be analyzed by ultrastructural analysis in the electron microscope. In particular, the correlation of fluorescent signals detected at the light microscopical level with subcellular structures, observed in the TEM, is a demanding task. Therefore, preparation methods for correlative LM/EM of amphibian embryos (mainly *Xenopus* and *Ambystoma*) are necessary.

Most of the protocols described here include chemical fixation and, for *Xenopus* embryos, embedding into Spurr's resin. This approach is generally considered "second best" if compared to high-pressure freezing followed by freeze-substitution and Epon embedding (McDonald, 2007; Moor, 1987). Furthermore, early *Xenopus* embryos are immunolabeled prior to embedding (i.e., pre-embedding labeling), which yields only poor ultrastructural preservation compared to post-embedding labeling (Kurth, 2003; Schwarz and Humbel, 2007). For amphibian embryos, these protocols generally still provide the best compromise for obtaining reasonable ultrastructural data. In the future, the systematic adaptation of additional approaches, such as cryopreparation techniques (HPF, Tokuyasu), or microwave-assisted processing, might help to further analyze many cell biological processes in amphibian embryos.

A. Chemical Fixation vs. HPF

High-pressure freezing followed by freeze-substitution is considered the "gold standard" for obtaining optimal ultrastructural preservation of cells and tissues (Moor, 1987; Studer *et al.*, 2001, McDonald, 2007) and was recently successfully applied to adult *Xenopus* tissues (Van Herp *et al.*, 2005; Wang *et al.*, 2005). However, for optimal freezing results, a specimen should not be thicker than 0.2 mm (Studer *et al.*, 2008). *Xenopus* eggs and early embryos are spheres of about 1–1.3 mm in diameter, axolotl eggs about 2 mm. Therefore, only parts of embryos, such as tissue explants or isolated cells, can be frozen successfully. The ECM preparations presented in Figs. 7 and 8 demonstrate that, in such cases, ultrarapid freezing is an attractive option.

Explants and isolated cells are important models for many early developmental processes (e.g., the animal cap assay or Keller explants for the analysis of convergent extension, cultivation of isolated embryonic cells, or organ anlagen, Sive *et al.*, 2000). Those explants and cells in culture might be quite suitable for HPF. Alternatively, chemical fixation might be improved with microwave-assisted fixation protocols, using either the microwave device from Ted Pella (Milestone rapid electron microscopy (REM) histoprocessor) or the Leica microwave-based tissue processor (Leica EM AMW, for examples, see Schroeder *et al.*, 2006). To our knowledge, this has not been tried for amphibian embryos so far. The application and/or combination of HPF and microwave fixation/processing might be a promising option for future experiments (see also Chapter 18 by Webb/Parton, this volume).

B. Spurr's Resin As an Embedding Medium

The use of Spurr's resin as an embedding medium leads to stronger extraction compared with Epon. For early *Xenopus* and axolotl embryos, however, a low-viscosity resin proved to be essential because of the massive amount of intracellular yolk granules and the problems caused by them during infiltration and sectioning (Kurth, 2003; Müller and Hausen, 1995). Again, the microwave option might help to improve resin infiltration and offer the possibility to switch from Spurr's to Epon as an embedding medium.

C. On Section Labeling vs. Pre-Embedding Labeling

On-section labeling of resin sections is an effective method for many cells and tissues, although the labeling efficiency is relatively poor (Fig. 2; Kurth *et al.*, 1996; Schwarz and Humbel, 2007). Proteins that are easy to detect in adult tissues, however, might be below the detection limit in embryonic tissues with this method, whereas pre-embedding labeling at the light microscopical level leads to strong staining of the same proteins (Kurth, 2003). Staining of cryosections after Tokuyasu is more sensitive (Schwarz and Humbel, 2009; Slot and Geuze, 2007) and might offer an alternative approach for analyzing subcellular protein distribution even in the amphibian embryo. To our knowledge, the staining of ZO-1 (Fig. 3) is the first demonstration of this technique in a *Xenopus* embryo. Unfortunately, amphibian blastula or gastrula stages are extremely difficult to cut at low temperatures. In such cases, the whole-mount staining of thick vibratome sections in combination with nanogold-coupled markers and silver enhancement currently reveals the best compromise between preservation of ultrastructure and antigenicity (Fig. 4; Kurth, 2003).

Acknowledgments

The authors would like to thank Inge Zimmermann, and Metta Riebesell for excellent technical assistance, and E.M. Tanaka for the kind donation of reagents.

References

Akey, C. W., and Radermacher, M. (1993). Architecture of the *Xenopus* nuclear pore complex revealed by three-dimensional cryo-electron microscopy. *J. Cell Biol.* **122**, 1–19.

Allen, T. D., Rutherford, S. A., Murray, S., Sanderson, H. S., Gardiner, F., Kiseleva, E., Goldberg, M. W., and Drummond, S. P. (2007a). A protocol for isolating *Xenopus* oocyte nuclear envelope for visualization and characterization by scanning electron microscopy (SEM) or transmission electron microscopy (TEM. *Nat. Protoc.* **2**, 1166–1172.

Allen, T. D., Rutherford, S. A., Murray, S., Sanderson, H. S., Gardiner, F., Kiseleva, E., Goldberg, M. W., and Drummond, S. P. (2007b). Generation of cell-free extracts of *Xenopus* eggs and demembranated sperm chromatin for the assembly and isolation of in vitro-formed nuclei for western blotting and scanning electron microscopy (SEM). *Nat. Protoc.* **2**, 1173–1179.

Amaya, E., Offield, M. F., and Grainger, R. M. (1998). Frog genetics: *Xenopus tropicalis* jumps into the future. *Trends Genet.* **14**, 253–255.

Armstrong, J. B., and Graveson, A. C. (1988). Progressive patterning precedes somite segmentation in the Mexican axolotl (*Ambystoma mexicanum*). *Dev. Biol.* **126**, 1–6.

Bordzilovskaya, N. P., Dettlaff, T. A., Duhon, S. T., and Malacinski, G. M. (1989). Developmental-stage series of axolotl embryos. *In* "Developmental Biology of the Axolotl" (J. B. Armstrong, and G. M. Malacinski, eds.) pp. 201–219. Oxford University Press, NY.

Cardellini, P., Cirelli, A., and Citi, S. (2007). Tight junction formation in early *Xenopus laevis* embryos: Identification and ultrastructural characterization of junctional crests and junctional vesicles. *Cell Tissue Res.* **330**, 247–256.

Cerny, R., Meulemans, D., Berger, J., Wilsch-Bräuninger, M., Kurth, T., Bronner-Fraser, M., and Epperlein, H.-H. (2004). Combined intrinsic and extrinsic influences pattern cranial neural crest migration and pharyngeal arch morphogenesis in axolotl. *Dev. Biol.* **266**, 252–269.

Chen, W. Y., and Abatangelo, G. (1999). Functions of hyaluronan in wound repair. *Wound Repair Regen.* **7**, 79–89.

Cho, K. W.Y., Blumberg, B., Steinbeisser, H., and De Robertis, E. M. (1991). Molecular nature of Spemann's organizer: The role of the *Xenopus* homeobox gene goosecoid. *Cell* **67**, 1111–1120.

Coughlin, M., Brieher, W. M., and Ohi, R. (2007). Cell-free extract systems and the cytoskeleton: Preparation of biochemical experiments for transmission electron microscopy. *Methods Mol. Biol.* **369**, 199–212.

Crane, R. F., and Ruderman, J. V. (2006). Using *Xenopus* oocyte extracts to study signal transduction. *Methods Mol. Biol.* **322**, 435–443.

Cross, M. K., and Powers, M. A. (2009). Learning about cancer from frogs: Analysis of mitotic spindles in *Xenopus* egg extracts. *Dis. Model Mech.* **2**, 541–547.

Danilchik, M. V., Bedrick, S. D., Brown, E. E., and Ray, K. (2003). Furrow microtubules and localized exocytosis in cleaving *Xenopus laevis* embryos. *J. Cell. Sci.* **116**, 273–283.

Danilchik, M. V., and Brown, E. E. (2008). Membrane dynamics of cleavage furrow closure in *Xenopus laevis*. *Dev. Dyn* **237**, 565–579.

Danscher, G. (1981). Localization of gold in biological tissue. *Histochemistry* **71**, 1–16.

De Calisto, J., Araya, C., Marchant, L., Riaz, C. F., and Mayor, R. (2005). Essential role of non-canonical wnt signalling in neural crest migration. *Development* **132**, 2587–2597.

De Robertis, E. M. (2009). Spemann's organizer and the self-regulation of embryonic fields. *Mech. Dev.* **126**, 925–941.

Epperlein, H.-H., and Löfberg, J. (1990). The development of the larval pigment patterns in *Triturus alpestris* and *Ambystoma mexicanum*. *Adv. Anat. Embryol. Cell. Biol.* **118**, 1–99.

Epperlein, H. H., Meulemans, D., Bronner-Fraser, M., Steinbeisser, H., and Selleck, M. (2000b). Analysis of cranial neural crest migratory pathways in axolotl using cell markers and transplantation. *Development* **127**, 2751–2761.

Epperlein, H.-H., Radomski, N., Wonka, F., Walther, P., Wilsch, M., Müller, M., and Schwarz, H. (2000a). Immunohistochemical demonstration of hyaluronan and its possible involvement in axolotl neural crest cell migration. *J. Struct. Biol.* **132**, 19–32.

Epperlein, H.-H., Schwarz, H., Piendl, T., Löfberg, J., Studer, D., Spring, H., and Müller, M. (1997). Improved preservation of the subepidermal extracellular matrix in axolotl embryos using electron microscopical techniques based on cryoimmobilization. *J. Struct. Biol.* **118**, 43–61.

Epperlein, H.-H., Selleck, M. A., Meulemans, D., Mchedlishvili, L., Cerny, R., Sobkow, L., and Bronner-Fraser, M. (2007). Migratory patterns and developmental potential of trunk neural crest cells in the axolotl embryo. *Dev. Dyn.* **236**, 389–403.

Falck, P., Hanken, J., and Olsson, L. (2002). Cranial neural crest emergence and migration in the Mexican axolotl (*Ambystoma mexicanum*). *Zoology* **105**, 195–2002.

Fesenko, I., Kurth, T., Sheth, B., Fleming, T. P., Citi, S., and Hausen, P. (2000). Tight juntion biogenesis in the early *Xenopus* embryo. *Mech. Dev.* **96**, 51–65.

Fey, J., and Hausen, P. (1990). Appearance and distribution of laminin during development of *Xenopus laevis*. *Differentiation* **42**, 144–152.

Gajewski, A., and Krohne, G. (1999). Subcellular distribution of the *Xenopus* p58/lamin B receptor in oocytes and eggs. *J. Cell. Sci.* **112**, 2583–2596.

Giacomello, E., Vallin, J., Morali, O., Coulter, J. S., Boulekbache, H., Thiery, J. P., and Broders, F. (2002). Type I cadherins are required for differentiation and coordinated rotation in *Xenopus laevis* somitogenesis. *Int. J. Dev. Biol.* **46**, 785–792.

Gillespie, L. L., Armstrong, J. B., and Steinberg, M. S. (1985). Experimental evidence for a proteinaceous presegmental wave required for morphogenesis of axolotl mesoderm. *Dev. Biol.* **107**, 220–226.

Goldberg, M. W., Huttenlauch, I., Hutchison, C. J., and Stick, R. (2008). Filaments made from A- and B-type lamins differ in structure and organization. *J. Cell. Sci.* **121**, 215–225.

Griffith, J. M., and Posthuma, G. (2002). A reliable and convenient method to store ultrathin thawed cryosections prior to immunolabeling. *J. Histochem. Cytochem.* **50**, 57–62.

Griffiths, G., McDowall, A., Back, R., and Dubochet, J. (1984). On the preparation of cryosections for immunocytochemistry. *J Ultrastruct. Res.* **89**, 65–78.

Gurdon, J. B., Laskey, R. A., and Reeves, O. R. (1975). The developmental capacity of nuclei transplanted from keratinized cells of adult frogs. *J. Emryol. Exp. Morphol.* **34**, 93–112.

Habermann, B., Bebin, A. G., Herklotz, S., Volkmer, M., Eckelt, K., Pehlke, K., Epperlein, H.-H., Schackert, H. K., Wiebe, G., and Tanaka, E. M. (2004). An *Ambystoma mexicanum* EST sequencing project: Analysis of 17,352 expressed sequence tags from embryonic and regenerating blastema cDNA libraries. *Genome Biol.* **5**, R67.

Hall, B. K., and Hörstadius, S. (1988). "The Neural Crest." Oxford University Press, London.

Hamburger, V. (1988). "The Heritage of Experimental Embryology." Oxford University Press, London.

Hardin, J. D., and Keller, R. E. (1988). The behaviour and function of bottle cells during gastrulation of *Xenopus laevis*. *Development* **103**, 211–230.

Hascall, V. C., and Laurent, T. (1997). Hyaluronan: Structure and physical properties. Science of hyaluronan today (online), http://www.glycoforum.gr.jp/science/hyaluronan/HA01/HA01E.html (1997).

Hausen, P., and Riebesell, M. (1991). "The Early Development of *Xenopus Laevis*. An Atlas of the Histology." Springer-Verlag, Berlin.

Hayen, W., Goebeler, M., Kumar, S., Riessen, R., and Nehls, V. (1999). Hyaluronan stimulates tumor cell migration by modulating the fibrin fiber architecture. *J. Cell. Sci.* **112**, 2241–2251.

Heasman, J. (2006). Patterning the early *Xenopus* embryo. *Development* **133**, 1205–1217.
Heasman, J., Kofron, M., and Wylie, C. (2000). Beta-catenin signaling activity dissected in the early *Xenopus* embryo: A novel antisense approach. *Dev. Biol.* **222**, 124–134.
Hess, K., Steinbeisser, H., Kurth, T., and Epperlein, H.-H. (2008). Bone morphogenetic protein-4 and noggin signaling regulates pigment cell distribution in the axolotl trunk. *Differentiation* **76**, 206–218.
Kalt, M. R. (1971). The relationship between cleavge and blastocoel formation in *Xenopus laevis*. II. Electron microscopic observations. *J. Embryol. Exp. Morphol.* **26**, 51–66.
Kalt, M. R., and Tandler, B. (1971). A study of fixation of early amphibian embryos for electron microscopy. *J. Ultrastruct. Res.* **36**, 633–645.
Karnovsky, M. J. (1965). A formaldehyde–glutaraldehyde fixative of high osmolarity for use in electron microscopy. *J. Cell Biol.* **27**, 137A.
Keller, R. E. (1980). The cellular basis of epiboly: An SEM study of deep-cell rearrangement during gastrulation in *Xenopus laevis*. *J. Embryol. Exp. Morphol.* **60**, 201–234.
Keller, R. E. (1981). An experimental analysis of the role of bottle cells and the deep marginal zone in gastrulation of *Xenopus laevis*. *J. Exp. Zool.* **216**, 81–101.
Keller, R., Davidson, L. A., and Shook, D. R. (2003). How we are shaped: The biomechanics of gastrulation. *Differentiation* **71**, 171–205.
Kimelman, D. (2006). Mesoderm induction: From caps to chips. *Nat. Rev. Genet.* **7**, 360–372.
Kragl, M., Knapp, D., Nacu, E., Khattak, S., Maden, M., Epperlein, H. H., and Tanaka, E. M. (2009). Cells keep a memory of their tissue origin during axolotl limb regeneration. *Nature* **460**, 60–65.
Kurth, T. (2003). Immunocytochemistry of the amphibian embryo – from overview to ultrastructure. *Int. J. Dev. Biol.* **47**, 373–383.
Kurth, T., Fesenko, I. V., Schneider, S., Münchberg, F. E., Joos, T. O., Spieker, T. P., and Hausen, P. (1999). Immunocytochemical studies of the interactions of cadherins and catenins in the early *Xenopus* embryo. *Dev. Dyn.* **215**, 155–169.
Kurth, T., and Hausen, P. (2000). Bottle cell formation in relation to mesodermal patterning in the *Xenopus* embryo. *Mech. Dev.* **97**, 117–131.
Kurth, T., Schwarz, H., Schneider, S., and Hausen, P. (1996). Fine structural immunocytochemistry of catenins in amphibian and mammalian muscle. *Cell Tissue Res.* **286**, 1–12.
Lah, J. J., Hayes, D. M., and Burry, R. W. (1990). A neutral pH silver development method for the visualization of 1-nanometer gold particles in pre-embedding electron microscopic immunocytochemistry. *J. Histochem. Cytochem.* **38**, 503–508.
Laurent, T. C., Laurent, U. B., and Fraser, J. R. (1996). The structure and function of hyaluronan: An overview. *Immunol. Cell Biol.* **74**, 1–7.
Le Douarin, N. M., and Kalcheim, C. (1999). "The Neural Crest." 2nd edn, Cambridge University Press, Cambridge.
Lee, J.-Y., and Harland, R. M. (2007). Actomyosin contractility and microtubules drive apical constriction in *Xenopus* bottle cells. *Dev. Biol.* **311**, 40–52.
Löfberg, J., Nynäs-McCoy, A., Olsson, C., Jönsson, L., and Perris, R. (1985). Stimulation of initial neural crest cell migration in the axolotl embryo by tissue grafts and extracellular matrix transplanted on microcarriers. *Dev. Biol.* **107**, 442–459.
Löfberg, J., Perris, R., and Epperlein, H.-H. (1989a). Timing in the regulation of neural crest cell migration: Retarded "maturation" of regional extracellular matrix inhibits pigment cell migration in embryos of the white axolotl mutant. *Dev. Biol.* **131**, 168–181.
Löfberg, J., Epperlein, H. H., Perris, R., and Stigson, M. (1989b). Neural crest cell migration: A pictorial essay. In: Developmental biology of the axolotl (Armstrong, J. B. and Malacinsky, G. M., eds.), Oxford University Press, New York.
Lourim, D., and Krohne, G. (1993). Membrane-associated lamins in *Xenopus* egg extracts: Identification of two vesicle populations. *J. Cell Biol.* **123**, 501–512.
Luft, J. H. (1971a). Ruthenium red and violet. I. Chemistry, purification, methods of use for electron microscopy and mechanism of action. *Anat. Rec.* **171**, 347–368.
Luft, J. H., (1971b). Ruthenium red and violet. II. Fine structural localization in animal tissues. *Anat. Rec.* **171**, 369–415.

Maco, B., Fahrenkrog, B., Huang, N. P., and Aebi, U. (2006). Nuclear pore complex structure and plasticity revealed by electron and atomic force microscopy. *Methods Mol. Biol.* **322**, 273–288.

Mandato, C. A., Weber, K. L., Zandy, A. J., Keating, T. J., and Bement, W. M. (2001). *Xenopus* egg extract as a model system for analysis of microtubule, actin filament, and intermediate filament interactions. *Methods Mol. Biol.* **161**, 229–239.

Maresca, T. J., and Heald, R. (2006). Methods for studying spindle assembly and chromosome condensation in *Xenopus* egg extracts. *Methods Mol. Biol.* **322**, 459–474.

McDonald, K. (2007). Cryopreparation methods for electron microscopy of selected model systems. *Methods Cell Biol.* **79**, 23–56.

Mchedlishvili, L., Epperlein, H. H., Telzerow, A., and Tanaka, E. M. (2007). A clonal analysis of neural progenitors during axolotl spinal cord regeneration reveals evidence for both spatially restricted and multipotent progenitors. *Development* **134**, 2083–2093.

Merzdorf, C. S., Chen, Y. H., and Goodenough, D. A. (1998). Formation of functional tight junctions in *Xenopus* embryos. *Dev. Biol.* **195**, 187–203.

Miltner, M. J., and Armstrong, J. B. (1983). Spermatogeneis in the Mexican axolotl, *Ambystoma mexicanum*. *J. Exp. Zool.* **227**, 255–263.

Moor, H. (1987). Theory and practice of high-pressure freezing. *In* "Cryotechniques in Biological Electron Microscopy" (Steinbrecht, R. A., Zierold, K., eds.) 175–191. Springer, Berlin, Germany.

Moriya, N., Komazaki, S., Takahashi, S., Yokota, C., and Asashima, M. (2000). In vitro pancreas formation from *Xenopus* ectoderm treated with activin and retinoic acid. *Dev. Growth Differ.* **42**, 593–602.

Müller, H. A., and Hausen, P. (1995). Epithelial cell polarity in early *Xenopus* development. *Dev. Dyn.* **202**, 405–420.

Nakatsuji, N., Smolira, M. A., and Wylie, C. C. (1985). Fibronectin is visualized by scanning electron microscopy immnuocytochemistry on the substratum for cell migration in *Xenopus laevis* gastrulae. *Dev. Biol.* **107**, 264–268.

Nieuwkoop, P. D., and Faber, J. (1967). "Normal Table of *Xenopus laevis* (Daudin)." North-Holland, Amsterdam.

Nye, H. L., Cameron, J. A., Chernoff, E. A., and Stocum, D. L. (2003). Extending the table of stages of normal development of the axolotl: Limb development. *Dev. Dyn.* **226**, 555–560.

Olsson, L., Stigson, M., Perris, R., Sorrell, J. M., and Löfberg, J. (1996a). Distribution of keratan sulphate and chondroitin sulphate in wild type and white mutant axolotl embryos during neural crest cell migration. *Pigment Cell Res.* **9**, 5–17.

Olsson, L., Svensson, K., and Perris, R. (1996b). Effects of extracellular matrix molecules on subepidermal neural crest cell migration in wild type and white mutant (dd) axolotl embryos. *Pigment Cell Res.* **9**, 18–27.

Osafune, K., Nishinakamura, R., Komazaki, S., and Asashima, M. (2002). In vitro induction of the pronephric duct in *Xenopus* explants. *Dev. Growth Differ.* **44**, 161–167.

Perris, R., Löfberg, J. ,, Fällström, C., von Boxberg, Y., Olsson, L., and Newgreen, D. F. (1990). Structural and compositional divergencies in the extracellular matrix encountered by neural crest cells in the white mutant axolotl embryo. *Development* **109**, 533–551.

Putta, S., Smith, J. J., Walker, J. A., Rondet, M., Weisrock, D. W., Monaghan, J., Samuels, A. K., Kump, K., King, D. C., Maness, N. J., Habermann, B., Tanaka, E., et al., (2004). From biomedicine to natural history research: EST resources for ambystomatid salamanders. *BMC Genomics* **5**, 54.

Reichardt, L. F. (1993). Extracellular molecules and their receptors. *In* "Guide Book to the Extracellular Matrix and Adhesion Proteins" (Kreis, T., and Vale, R. (eds.)) pp. 3–11. Oxford University Press, Oxford, London.

Reintsch, W. E., and Hausen, P. (2001). Dorsoventral differences in cell–cell interactions modulate the motile behaviour of cells from the *Xenopus* gastrula. *Dev. Biol.* **240**, 387–403.

Sadaghiani, B., and Thiebaud, C. H. (1987). Neural crest development in the *Xenopus laevis* embryo, studied by interspecific transplantation and scanning electron microscopy. *Dev. Biol.* **124**, 91–110.

Schneider, S., Steinbeisser, H., Warga, R. M., and Hausen, P. (1996). Beta-Catenin translocation into nuclei demarcates the dorsalizing centers in frog and fish embryos. *Mech. Dev.* **57**, 191–198.

Schroeder, T. E. (1970). Neurulation in *Xenopus laevis*. An analysis and model based upon light and electron microscopy. *J. Embryol. Exp. Morphol.* **23**, 427–462.

Schroeder, J. A., Gelderblom, H. R., Hauroeder, B., Schmetz, C., Milios, J., and Hofstaedter, F. (2006). Microwave-assisted tissue processing for same-day-EM-diagnosis of potential bioterrorism and clinical samples. *Micron* **37**, 577–590.

Schwab, M. S., Gossweiler, U., and Dreyer, C. (1998). Subcellular distribution of distinct nucleolin subfractions recognized by two monoclonal antibodies. *Exp. Cell Res.* **239**, 226–234.

Schwarz, H., and Humbel, B. (2007). Correlative light and electron microscopy using immunolabeled resin sections. *Methods Mol. Biol.* **369**, 229–256.

Schwarz, H., and Humbel, B. M. (2009). Correlative light and electron microscopy. *In* "Hand Book of Cryo-Preparation Methods for Electron Microscopy" (Cavalier, A., Spehner, D., Humbel, B. M., eds.) pp 537–565. CRC Press, Boca Raton, FL.

Scott, J. E., Cummings, C., Brass, A., and Chen, Y. (1991). Secondary and tertiary structures of hyaluronan in aqueous solution, investigated by rotary shadowing-electron microscopy and computer simulation. *Biochem. J.* **274**, 699–705.

Shook, D. R., Majer, C., and Keller, R. (2004). Pattern and morphogenesis of presumptive superficial mesoderm in two closely related species, *Xenopus laevis* and *Xenopus tropicalis*. *Dev. Biol.* **270**, 163–185.

Sive, H. L., Grainger, R. M., and Harland, R. M. (2000). Early Development of *Xenopus Laevis*: A Laboratory Manual. Cold Spring Harbor Laboratory Press, New York.

Slack, J. M.W., Lin, G., and Chen, Y. (2008). The *Xenopus* tadpole: A new model for regeneration research. *Cell Mol. Life Sci.* **65**, 54–63.

Slot, J. W., and Geuze, H. J. (2007). Cryosectioning and immunolabeling. *Nat. Protoc.* **2**, 2480–2591.

Sobkow, L., Epperlein, H. H., Herklotz, S., Straube, W. L., and Tanaka, E. M. (2006). A germline GFP transgenic axolotl and its use to track cell fate: Dual origin of the fin mesenchyme during development and the fate of blood cells during regeneration. *Dev. Biol.* **290**, 386-397.

Spemann, H., and Mangold, H. (1924). Über die Induktion von Embryonalanlagen durch Implantation artfremder Organisatoren. *W. Roux' Arch. f. Entwicklungsmech.* **100**, 599–638.

Spurr, A. R. (1969). A low viscosity epoxy resin embedding medium for electron microscopy. *J. Ultrastr. Res.* **26**, 31–43.

Steinberg, M. S. (1957). A non-nutrient culture medium for amphibian embryonic tissue. *Carnegie Inst Wash. Year B* **56**, 347–348.

Stick, R., and Schwarz, H. (1983). Disappearance and reformation of the nuclear lamina structure during specific stages of meiosis in oocytes. *Cell* **33**, 949–958.

Stierhof, Y. D. (2009). Immunolabeling of ultrathin sections with enlarged 1 nm gold or Q-dots. *In* "Hand Book of Cryo-Preparation Methods for Electron Microscopy" (Cavalier, A., Spehner, D., Humbel, B. M., eds.) pp. 587–616. CRC Press, Boca Raton, FL.

Stoffler, D., Feja, B., Fahrenkrog, B., Walz, J., Typke, D., and Aebi, U. (2003). Cryo-electron tomography provides novel insights into nuclear pore architecture: Implications for nucleocytoplasmic transport. *J. Mol. Biol.* **328**, 119–130.

Studer, D., Graber, W., Al-Amoudi, A., and Eggli, P. (2001). A new approach for cryofixation by high-pressure freezing. *J. Microsc.* **2003**, 285–294.

Studer, D., Humbel, B. M., and Chiquet, M. (2008). Electron microscopy of high pressure frozen samples: Bridging the gap between cellular ultrastructure and atomic resolution. *Histochem. Cell Biol.* **130**, 877–889.

Tammaro, P., Shimomura, K., and Proks, P. (2008). *Xenopus* oocytes as a heterologous expression system for studying ion channels with the patch-clamp technique. *Methods Mol. Biol.* **491**, 127–139.

Tokuyasu, K. T. (1976). Membranes as observed in frozen sections. *J. Ultrastruct. Res.* **55**, 281–287.

Toole, B. P. (2004). Hyaluronan: From extracellular glue to pericellular cue. *Nat. Rev. Cancer* **4**, 528–539.

Van Herp, F., Coenen, T., Geurts, H. P., Janssen, G. J., and Martens, G. J. (2005). A fast method to study the secretory activity of neuroendocrine cells at the ultrastructural level. *J. Microsc.* **218**, 79–83.

Venable, J. H., and Coggeshall, R. (1965). A simplified lead citrate stain for use in electron microscopy. *J. Cell Biol.* **25**, 407–408.

Voss, S. R., Epperlein, H. H., and Tanaka, E. M. (2009). *Ambystoma mexicanum*, the axolotl: A versatile amphibian model for regeneration, development, and evolution studies.. *Cold Spring Harb. Protoc.*, doi: 10.1101/pdb.emo128.

Walther, P. (2008). High-resolution cryo-SEM allows direct identification of F-actin at the inner nuclear membrane of *xenopus* oocytes by virtue of its structural features. *J. Microsc.* **232**, 379–385.

Wang, L., Humbel, B. M., and Roubos, E. W. (2005). High-pressure freezing followed by cryosubstitution as a tool for preserving high quality ultrastructure and immunoreactivity in the *Xenopus laevis* pituitary gland. *Brain Res. Protoc.* **15**, 155–163.

Whight, T. N., Heinegard, D. K., and Hascall, V. C. (1991). Proteoglycans. *In* "Cell Biology of Extracellular Matrix" (Hay, E. D. ed.), 2nd edn., pp. 45–78. Plenum, NY.

Winklbauer, R., and Nagel, M. (1991). Directional mesoderm cell migration in the *Xenopus* gastrula. *Dev. Biol.* **148**, 573–589.

Winklbauer, R., Selchow, A., Nagel, M., and Angres, B. (1992). Cell interaction and its role in mesoderm cell migration during *Xenopus* gastrulation. *Dev. Dyn.* **195**, 290–302.

Youn, B. W., Keller, R. E., and Malacinski, G. M. (1980). An atlas of notochord and somite morphogenesis in several anuran and urodelean amphibians. *J. Embryol. Exp. Morphol.* **59**, 223–247.

Youn, B. W., and Malacinski, G. M. (1981). Somitogenesis in the amphibian *Xenopus laevis*: Scanning electron microscopic analysis of intrasomitic cellular arrangements during somite rotation. *J. Embryol. Exp. Morphol.* **64**, 23–43.

Zackson, S. L., and Steinberg, M. S. (1987). Chemotaxis or adhesion gradient? Pronephric duct elongation does not depend on distant sources of guidance information. *Dev. Biol.* **124**, 418–422.

CHAPTER 18

Modern Approaches for Ultrastructural Analysis of the Zebrafish Embryo

Nicole L. Schieber[*,†], **Susan J. Nixon**[†], **Richard I. Webb**[*], **Viola M.J. Oorschot**[‡], *and* **Robert G. Parton**[*,†]

[*]Centre for Microscopy and Microanalysis, The University of Queensland, Brisbane, 4072 Queensland, Australia

[†]Institute for Molecular Bioscience, The University of Queensland, Brisbane, 4072 Queensland, Australia

[‡]Department of Cell Biology, Cell Microscopy Center, University Medical Center Utrecht, 3584 CX Utrecht, The Netherlands

---

978-0-12-381007-6
DOI: 10.1016/S0091-679X(10)96018-4

Abstract

The zebrafish is a powerful vertebrate system with great advantages for both forward and reverse genetic screens and as a model for human disease conditions. Light microscopy has been used extensively to study zebrafish development but less frequently have these studies been combined with ultrastructural information. Zebrafish embryos are ideal for electron microscopy (EM) with a single transverse section containing many different cell types and tissues. However, conventional methods of EM do not provide optimal preservation of all tissues and are usually incompatible with immunolabelling and visualisation of expressed fluorescently tagged proteins. Here we examine methods that overcome these problems. We summarise a range of methods, applicable to the ultrastructural analysis of zebrafish embryos, including methods for fast freezing and processing of zebrafish embryos. These methods preserve antigenicity, ultrastructure and GFP fluorescence even after embedding in resin. In addition, they are compatible with electron tomography. These methods provide a new set of research tools that provide an additional level of information, complementing current methods for study of this widely used model system.

I. Introduction

Zebrafish embryos are a useful vertebrate model system for cell and developmental biology with great advantages for genetic and small molecule screening, and for understanding human disease conditions (Haffter and Nusslein-Volhard, 1996; Haffter *et al.*, 1996). New techniques for studying gene function, including TILLing, zinc finger nucleases and more established approaches such as morpholino-based knockdowns, have been effectively combined with light microscopic analyses of the transparent zebrafish embryos (Amsterdam and Hopkins, 2006; Doyon *et al.*, 2008; Meng *et al.*, 2008; Moens *et al.*, 2008; Pickart *et al.*, 2006). The zebrafish researchers therefore have a powerful range of techniques at their disposal. Less attention has been given to ultrastructural analysis; yet zebrafish embryos are also ideal for electron microscopic (EM) studies and EM provides additional levels of information. A single section of a zebrafish embryo can span the entire embryo and contain many different cell types and tissues. Conventional methods of EM may not provide optimal preservation of all tissues in an embryo and are usually incompatible with immunolabelling and visualisation of expressed fluorescently tagged proteins. The widespread use of the zebrafish system necessitates the development of additional methods to complement conventional EM processing methods.

A. Historical Perspective

With the increasing use of zebrafish embryos in developmental and cellular biology, the application of electron microscopic techniques that preserve native ultrastructure and allow immunolocalisation within the context of the whole organism has become increasingly important. Many of the earlier ultrastructural studies of the zebrafish embryo made use of routine preparation techniques involving chemical fixation protocols utilising fixation in glutaraldehyde followed by post-fixation in osmium tetroxide (Branchek and Bremiller, 1984; Metcalfe *et al.*, 1985). The size and complexity of the embryo have, however, presented particular challenges for the electron microscopist and it has often been difficult to obtain excellent preservation of all cell structures in all tissues. A recent study has shown that even variations of these standard protocols can give differing results (Chapman *et al.*, 2009). This study compared Millonig's fixative, which contains a high concentration of glutaraldehyde in a double-strength phosphate buffer, with 2.5% glutaraldehyde in cacodylate buffer and a similar fixative but followed by a tannic acid step. While none gave perfect fixation, with all tissues showing regions of better preservation than others, Millonig's fixative provided the best general fixation, particularly of myelin and nuclear membranes. Other studies have argued that the use of paraformaldehyde (PFA) in the primary fixative is paramount as glutaraldehyde alone is known to cause shrinkage (Akhtar *et al.*, 2008). Despite their small size, the penetration of fixatives has been a primary concern of researchers and hence this has supported the use of PFA as a component of the initial fixative. PFA is a fast-penetrating fixative but its cross-linking ability is poor in comparison to glutaraldehyde. However, a more complex fixative has been designed to provide even faster penetration of fixatives into the zebrafish embryo in the hope of achieving more rapid fixation and so better preservation of ultrastructure (Jontes *et al.*, 2000; Kljavin, 1987). This fixative that contains glutaraldehyde and PFA as well as dimethyl sulphoxide and acrolein has been used in the studies of retinal development and synaptogenesis with good results (Jontes *et al.*, 2000; Kljavin, 1987). Acrolein is an extremely fast-penetrating fixative with better cross-linking ability than PFA, and hence is a very useful fixative for embryos. However, safety issues still prevent its widespread use (Griffiths, 1993).

An extensive comparison of a range of fixatives and their effects on preservation of the zebrafish embryo has been carried out in earlier studies (Kalicharan *et al.*, 1998; Rawson *et al.*, 2000). In these studies the authors concentrated on achieving optimal structure for SEM studies and so their methods largely involved the use of non-coating techniques, which they claimed were necessary to achieve a stable sample while allowing observation of fine structural detail. They did, however, show that for TEM the use of a prefixative containing glutaraldehyde, PFA, acrolein and picric acid in conjunction with a post-fixation using osmium tetroxide/potassium ferrocyanide resulted in adequate preservation with good contrast. It should be noted however, that these studies mainly involved examination of the outer layers of the embryo, particularly the chorion, as they were interested in finding reasons for poor penetration of cryoprotectants prior to cryopreservation. In most studies zebrafish embryos are actually dechorionated prior to fixation so these regions do not normally have an impact on penetration of fixatives.

The transparency of the zebrafish embryo has made it amenable for light microscopy and many of the advanced imaging methods used in cultured cells are now being applied to the zebrafish (Huisken and Stainier, 2009; Scott *et al.*, 2007). Methods that combine the advantages of light microscopy with correlative EM studies would therefore be of immediate interest to zebrafish researchers. In this respect, the work of Luby-Phelps *et al.* (2003), in which it was shown that it is possible to achieve good correlative LM/EM results with preservation of the green fluorescent protein (GFP) signal during embryo processing so that the fluorescence could still be seen in resin sections, was a significant advance. Using mild fixation and subsequent embedding in LR White resin, they showed that GFP expression could be observed in unstained 1 μm sections and that the antigenicity was retained allowing labelling at both the LM and EM levels. While preservation was sufficient to allow subcellular localisation of the GFP, the fine structural detail was compromised by the light fixation involved.

II. Rationale

As summarised in the above section, chemical fixation followed by conventional embedding protocols in plastic resins have been widely used to study the ultrastructure of the zebrafish embryo and these methods have provided important insights into key developmental processes at the ultrastructural level (de Jong-Curtain *et al.*, 2009; Hall *et al.*, 2007; Nixon *et al.*, 2005; Parsons *et al.*, 2002). However, these techniques, originally developed for mammalian tissues and cells, can introduce artefacts due to the need for the chemical fixatives to penetrate the entire embryo to reach deep tissues and their inability to effectively fix all cellular constituents (Lucic *et al.*, 2005; McDonald, 2009; Studer *et al.*, 2008). In addition, these methods are not usually compatible with immunolabelling and do not take advantage of the range of methods available for fluorescently tagging proteins increasingly used to study proteins in real time in the zebrafish embryo (Amsterdam *et al.*, 1995; Liu *et al.*, 2008; Peters *et al.*, 1995). Tokuyasu protocols involving cryosectioning and subsequent immunolabelling (Tokuyasu, 1980) have been performed successfully on zebrafish embryos (Hill *et al.*, 2008; Nixon *et al.*, 2007) but do not give perfect ultrastructural preservation especially of structures such as the notochord (Nixon *et al.*, 2009). In recent years electron microscopy has been revolutionised by the increasing use of fast freezing techniques to avoid initial chemical fixation (McDonald *et al.*, 2007). In addition, electron microscopists are increasingly interested in gaining high-resolution 3D information by the use of electron tomography (Leis *et al.*, 2009). The ideal method would avoid chemical fixation and utilise the power of GFP (and other fluorescent proteins) as markers for light microscopy to correlate light microscopic observations with immunogold electron microscopy on the same sections. In addition, this technique should be compatible with 3D electron tomography.

In this article we summarise a number of methods for processing of zebrafish embryos, progressing from conventional methods of processing for ultrastructure to more technically demanding recently developed methods that allow correlative light and electron microscopy, immunogold labelling and electron tomography

(Nixon *et al.*, 2009). With this method GFP fluorescence is still retained after freeze substitution (FS) and Lowicryl embedding of high-pressure frozen (HPF) embryos.

III. Methods

A. Epon Embedding—Conventional

This is the standard method for the processing of zebrafish embryos for EM (Fig. 1A) and is applicable to all ages.

1. Fixation

Embryos were initially fixed with 2.5% EM grade glutaraldehyde in E3 media or phosphate buffered saline (PBS) (both giving similar results) on the bench for 2 h at

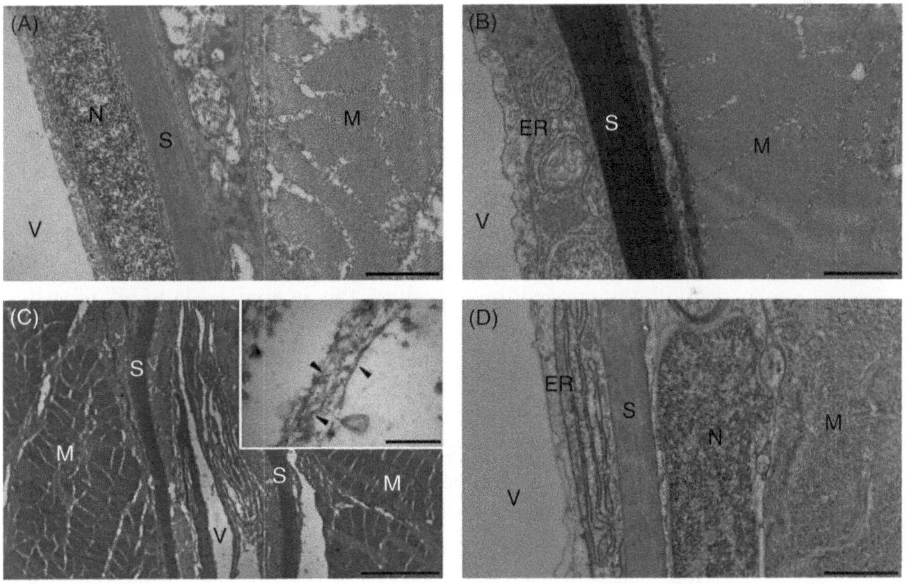

Fig. 1 Zebrafish embryos (5 days post-fertilisation, dpf) processed for EM using four different processing schemes. Images of the notochord are shown for comparison. (A) Embryo fixed and processed using the conventional method for Epon embedding. (B) Embryo fixed and processed using the microwave scheme before embedding in Epon. (C) Embryo fixed and processed using the Tokuyasu method, which ensures high antigenicity (see insert) but results in poorer morphological preservation of the notochord. (D) High-pressure frozen embryo processed by rapid FS and embedded in Lowicryl HM20. Note the excellent preservation of membranes and structure in tissue surrounding the notochord of the embryo. S, notochord sheath; M, muscle; ER, endoplasmic reticulum; V, notochord vacuole; N, nucleus. Scale bar in (A, B, D) = 1 μm, (C) = 5 μm, (C insert) = 500 nm.

room temperature. Following three 5-min rinses in PBS to remove the fixative, at which point they can be stored, the embryos were rinsed in 0.1 M cacodylate buffer twice for 10 min each. Embryos were post-fixed with 1% osmium tetroxide buffered in 0.1 M cacodylate for 1 h, then removed, rinsed once quickly with ultra high quality (UHQ) water, then twice more for 10 min each. An en bloc stain of 2% uranyl acetate in UHQ water for 1 h was then used to enhance sample contrast.

2. Dehydration

Samples were progressively dehydrated in a series of ethanol washes each 15 min in duration beginning with 2% uranyl acetate in 50% ethanol to further enhance sample contrast, followed by 50, 60, 70, 80, 90 and twice at 100%.

3. Embedding

A standard Epon 812 resin recipe was used for infiltration beginning with a 1:2 dilution of Epon resin to 100% ethanol, followed by 1:1 and 2:1, each for 30 min. Finally, samples were left overnight in 100% Epon before placing in labelled moulds and polymerisation in a sealed container with molecular sieve for 48 h at 60°C.

B. Epon Embedding—Microwave

This method takes advantage of a microwave-based processing scheme to more rapidly process embryos while producing improved morphological results (Fig. 1B). As with the conventional method, this is applicable to all zebrafish embryonic ages.

1. Fixation

A Pelco BioWave® microwave containing a ColdSpot™ maintains a temperature of 23°C throughout the processing steps for uniform microwave radiation. Embryos processed with the BioWave were fixed under vacuum with 2.5% EM grade glutaraldehyde in E3 or PBS on a cycle of 2 min on–off–on at 80 W, repeated twice. Vacuum has been suggested to improve fixative diffusion (Wendt *et al.*, 2004). Samples were then rinsed in 0.1 M cacodylate buffer, once in the fumehood and again in the microwave for 40 s at 250 W without vacuum. Under vacuum, the embryos were post-fixed with 1% osmium tetroxide buffered in 0.1 M cacodylate at 80 W on a cycle of 2 min on–off–on, which was repeated twice without replacing the solution. To remove the osmium solution, samples were rinsed in the fumehood with UHQ water, then at 250 W for 40 s with vacuum. Then to enhance sample contrast, an en bloc stain (2% aqueous uranyl acetate) was used in the microwave at 150 W with vacuum on 1 min on–off–on cycle.

2. Dehydration

En bloc staining was continued with 2% uranyl acetate in 50% ethanol in the microwave for a 1 min on–off–on cycle at 150 W with vacuum. Embryos were then progressively dehydrated in a series of ethanol washes (50, 60, 70, 80, 90 and twice at 100%), each for 40 s in the microwave without vacuum at 250 W.

3. Embedding

Resin infiltration was under vacuum beginning with a 1:2 dilution of standard recipe Epon 812 resin to 100% ethanol followed by 1:1 and 2:1, each for 3 min at 250 W. This was repeated twice more with 100% Epon before placing the embryos in moulds and polymerisation as above (Section A).

C. Frozen Sectioning

The Tokuyasu method (Tokuyasu, 1980) ensures the highest preservation of antigenicity but is less useful for morphological analyses and tomography. Embryos were processed for frozen sectioning (Fig. 1C) by modifications of methods described by Slot and Geuze (2007).

1. Fixation

To allow antigen detection, zebrafish embryos were fixed with either 4% EM grade PFA or 2% PFA/0.2% glutaraldehyde, both in E3. Immediately after removal from the incubator, an equal volume of the fixative (made at 2× concentration) to volume of E3 was added. This was left for 5 min at room temperature before being replaced with a 1× fixative solution, where it was left for 2 h at room temperature or overnight at 4°C. Once fixed, embryos were dissected while still in the fixative to leave only the area of interest and allow better gelatin infiltration. Fixed embryos were stored in 1% PFA and rinsed in PBS (3 × 5 min) when ready to embed.

2. Sample Preparation

Fixed embryos were first quenched at room temperature for 5 min (0.15% glycine in PBS). Then they were immersed in warmed gelatin in PBS at 37°C for 15–30 min, inverting periodically. For each embryo sample, a 7 μl drop of warmed gelatin was added to a parafilm-covered slide and kept on a heating block at 37°C. Following the 30 min infiltration, the embryo samples were placed on the gelatin drops, then a small piece of parafilm was rested on top to exert mild pressure and minimise the gelatin for easy trimming. The slide was placed on ice for 30 min or until the gelatin solidified and peeled off the slide. Excess gelatin could then be trimmed to leave a block suitable for mounting on cryo-stubs and then submerged in cryoprotectant, polyvinylpyrrolidone (PVP)-sucrose, rotating at 4°C overnight.

3. Mounting

Blocks were removed from the PVP-sucrose and placed onto cleaned cryo-stubs in the desired cutting orientation with a small amount of PVP-sucrose to adhere the embryo to the stub and immediately immersed in liquid nitrogen. From this point, the mounted samples were kept in liquid nitrogen or below −100°C to retain good morphology.

4. Sectioning

Zebrafish embryos were sectioned using a Leica UC6 ultramicrotome with FC6 cryoattachment. The blocks were initially trimmed at −100°C, to remove excess gelatin around the embryo. Once trimmed, a thick (100 nm) section was picked up with 2.3 M sucrose using a copper loop, and placed onto a glass slide. Sections were then stained for light microscopy examination with toludine blue that was added to the slide and tilted for 15–30 s to dissolve sucrose, then gently rinsed with water and dried. This was useful to check the embryo's orientation, allowing further trimming to obtain thin sections focussing on a particular area. Thin sections (Fig. 1C) were cut at −120°C and picked up with a 1:1 solution of 2.3 M sucrose in PBS to 2% methyl cellulose in distilled water, using a 3 mm diameter copper loop and placed onto a Formvar carbon-coated 100 mesh copper grid. The grids were kept on ice or at 4°C until labelled to protect antigenicity. The pickup solution was kept on ice and the copper loop kept in chilled water between pickups to wash.

5. Immunolabelling

Frozen thin sections (60–70 nm) placed on Formvar carbon-coated grids were immunolabelled by inverting onto a series of droplets as described by Slot and Geuze (2007). Grids were moved directly from one drop to the next without drying at any stage. Care must also be taken to ensure that the reverse side of the grid is kept dry. Grids were washed for 2 min in PBS to remove any remaining sucrose. The grids were then incubated on a drop of quench solution (20 mM glycine in PBS) for 10 min, and block solution (0.2% fish skin gelatin (FSG), 0.1% bovine serum albumin (BSA), 10 mM glycine in PBS) for a further 15 min before chosen specific primary antibody incubation (7 μl drop per grid) for 30 min. Between primary and secondary incubations, both of which were diluted in block solution, grids were washed twice in block solution and twice in PBS, each for 5 min. Gold-labelled secondary antibody or Protein A-gold incubation was for 30 min on 7 μl drops. Excess reagent was removed by washing with block solution (2 × 2 min) and PBS (3 × 2 min). The grids were then treated with 1% (w/v) glutaraldehyde in PBS for 5 min then rinsed with fresh UHQ water (5 × 2 min), using anti-capillary forceps that are washed and dried between water washes to ensure no phosphate remains.

Grids were contrasted with a 9:1 mixture of 2% methyl cellulose (MC) (in distilled water) to 4% aqueous uranyl acetate (UA). The grids were incubated briefly on two droplets of MC/UA, and then incubated for 10 min on a large droplet covered from light, all on a plastic plate on ice. A copper loop was used to pick up each grid and excess solution was gently drained away using filter paper. The loop with the grid was then left to dry upright for a minimum of 30 min. The grids were then removed from the loop and examined in a JEOL1011 transmission electron microscope (Fig. 1C).

D. High-Pressure Freezing—Epon

This method has been shown to be successful for zebrafish embryos from 6 h post-fertilisation (staged according to standard criteria; Kimmel *et al.*, 1995) to 6 days post-fertilisation (dpf; Fig. 2), providing high-quality preservation of ultrastructure in a near-native state.

1. High-Pressure Freezing

Embryos were anaesthetised and placed into 0.7% agarose in E3 media in 200 μm membrane carriers (Nixon *et al.*, 2009). To keep the sample at "living" temperature before freezing, they were positioned into the membrane while on the heating pad,

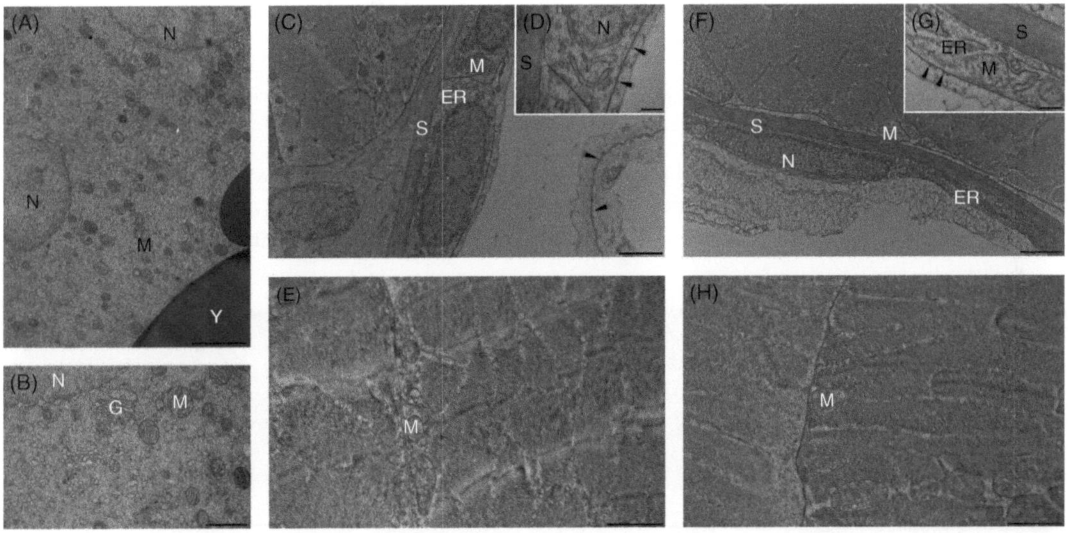

Fig. 2 Zebrafish embryos after HPF at a number of different developmental time points, rapid FS and embedding in Epon or Lowicryl HM20. (A–B) Six hour post-fertilisation (hpf) embedded in Epon. (C–E) Seventy-two hpf embedded in Lowicryl HM20; (D) High magnification of the notochord to illustrate preservation of the membrane and caveolae (arrow head); (E) Muscle. (F–H) Five dpf zebrafish embryos after embedding in Lowicryl HM20; (G) High magnification of notochord to illustrate preservation of the membrane and caveolae (arrowhead); (H) muscle. S, notochord sheath; Y, yolk; N, nucleus; M, mitochondria; G, Golgi; ER, endoplasmic reticulum. Scale bar in (A) = 2.5 μm, (B, E, H) = 1 μm, (C, F) = 2 μm, (D, G) = 500 nm.

curling the tail around so that the entire fish fits within the carrier. Samples were then cryofixed using the Leica EM PACT2 HPF.

2. Freeze Substitution

Frozen samples underwent freeze substitution in the Leica EM AFS2, which was cooled to −140°C. Membrane carriers were placed directly into cooled freeze substitution media (1% osmium tetroxide, 0.5% uranyl acetate, 5% water in acetone) at −140°C and warmed to −85°C over 2 h (27.5°C/h). Samples remained at −85°C for a further 73 h before being warmed to 20°C over 15 h (7°C/h) and maintained at that temperature for 2 h. Samples were then immediately washed twice in the microwave with acetone for 40 s each without vacuum at 250 W and removed from the membrane carriers.

3. Embedding

Infiltration was all under vacuum in the microwave at 250 W using a standard procure Epon 812 recipe. A 1:1 dilution of Epon to resin was added to the samples, followed by two changes of 100% Epon, each for 3 min. Embryos were then added to moulds and the resin polymerised by curing at 60°C for 48 h.

E. High-Pressure Freezing—HM20 Low-Temperature Embedding, Labelling and Tomography

HPF provides rapid preservation of the zebrafish embryo without chemical fixatives, thus maintaining antigenicity and GFP fluorescence (Figs. 3 and 4) when embedded in resins such as Lowicryl HM20, as well as excellent ultrastructure (Fig. 1D).

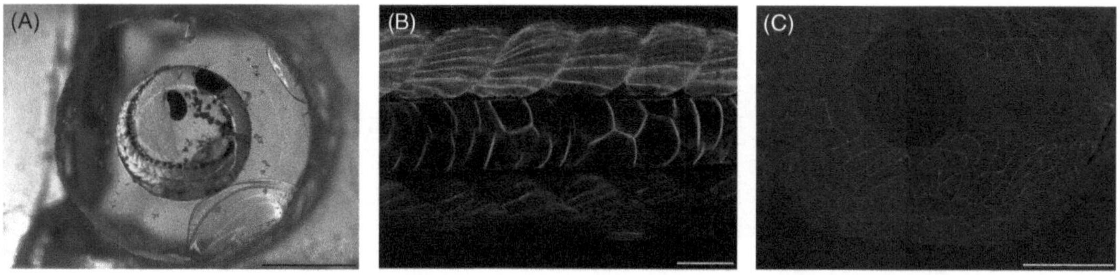

Fig. 3 High-pressure freezing (HPF) and rapid freeze substitution (FS) of zebrafish embryos retain GFP fluorescence when embedded in Lowicryl HM20. (A) Zebrafish embryos, 3 dpf, embedded in Lowicryl HM20 after HPF and rapid FS. The membrane carrier used for HPF has been removed. The embryo remains intact with no gross structural damage. (B) A 48 hpf zebrafish embryo expressing caveolin-3-GFP analysed live by confocal microscopy. Note the fluorescence in the muscle and notochord. (C) Transverse section of a 72 hpf zebrafish embryo expressing caveolin-3-GFP as above, after HPF, rapid FS and embedding in Lowicryl HM20. GFP expression is evident in the muscle and notochord. Scale bar in (A) = 1 mm, (B, C) = 50 μm. (See Plate no. 24 in the Color Plate Section.)

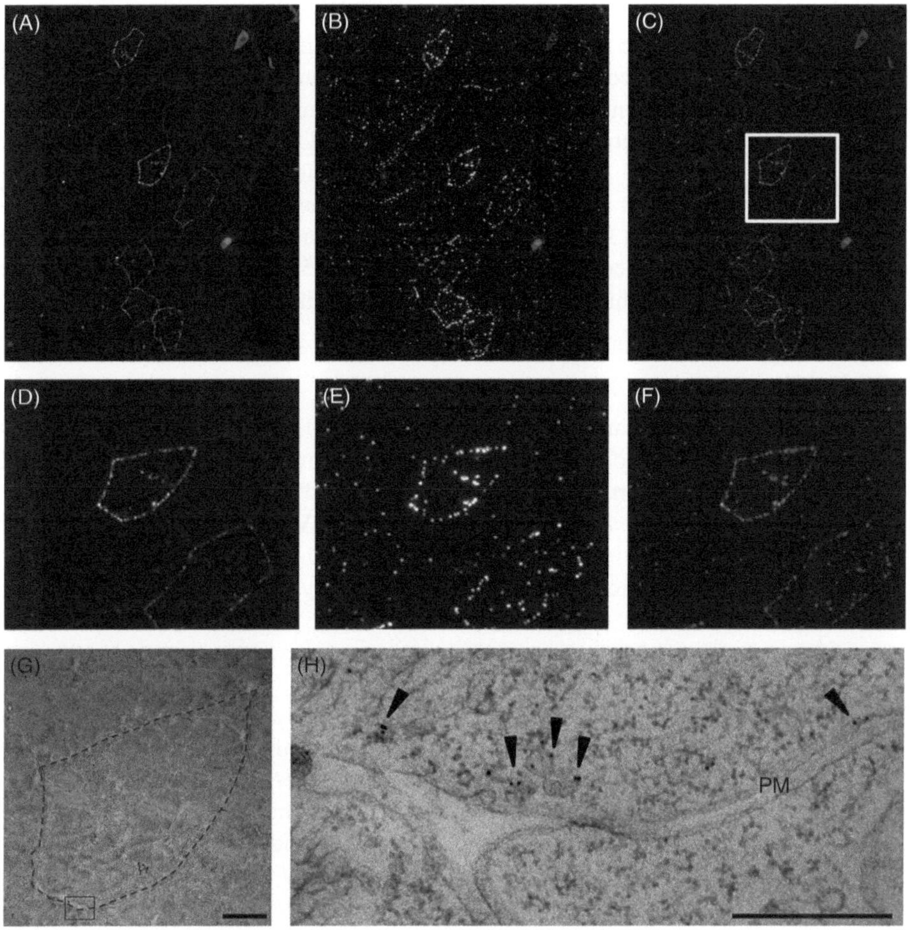

Fig. 4 Three dpf zebrafish embryos injected with Caveolin-3-GFP expressed under the *cav3* promotor at the 1–2 cell stage after HPF, rapid FS and embedding in Lowicryl HM20. Thick sections (200 nm) were cut and viewed for GFP expression by epifluorescence microscopy (A and D), labelled for GFP (B and E) and a merge of these images can be seen in panels (C and F). (D–F) Higher magnification of the same area boxed in (C). (G) Electron micrograph of the muscle fibre in (D–F). (H) Immunogold GFP labelling (arrow heads) in the membrane (PM) of the muscle (boxed area in G) demonstrates identical labelling as seen by epifluorescence allowing correlative light and electron microscopy. Scale bar in (G) = 2 μm, (H) = 500 nm. (See Plate no. 25 in the Color Plate Section.)

1. High-Pressure Freezing

Embryos were anaesthetised and cryofixed as described above (Section D).

2. Freeze Substitution

Freeze substitution in the Leica EM AFS2 was using FSP, which used pre-programming to automatically change solutions at the requested intervals. The machine was pre-cooled to −145°C and then membrane carriers were individually placed into compartments of the AFS2 buckets face up, keeping record of the samples and their corresponding position. Containers were also filled with the required freeze substitution solutions and cooled. Samples were fixed in acetone containing 0.2% uranyl acetate and 5% water at −85°C for 48 h. As large amounts of frost can be introduced during the loading of samples, an extra change of substitution solution is carried out after roughly 6 h. The temperature of the AFS was then brought to −50°C over 11 h (3°C/h) while still in the same solution. Samples were subsequently washed twice in acetone each for 20 min at −50°C. Fish were released from the membrane carriers while in acetone using a fine pin and the carrier removed from the FSP.

3. Embedding

A standard recipe of Lowicryl HM20 was prepared according to the manufacturer's recommendations. Infiltration was performed at −50°C first in a 1:1 dilution of acetone to resin for 1 h, followed by 1:2 for 1 h and 100% for 26 h (three changes of 100% Lowicryl: 1, 7, 18 h). Polymerisation was under UV light within the AFS2, with samples warmed from −50 to 20°C over 48 h, then polymerised for a further 48 h. After the blocks were removed from the buckets (Fig. 3A) they were stored away from light at −20°C.

4. Light Microscopy of HM20 Sections

Thick (200 nm) sections of zebrafish could be examined for GFP fluorescence immediately (Figs. 3C, 4A and D) or after labelling for immunofluorescence with fluorescently labelled secondary antibodies (Fig. 4B and E). Sections were cut with a 45° diamond knife on a Leica UC6 ultramicrotome and floated on a water bath before they were picked up with a copper loop and placed onto round (12 mm) coverslips. The sections were attached to the coverslips by removing excess water with filter paper and left covered to dry. To examine the sections without labelling, coverslips were mounted onto slides with Aqua Poly/Mount and sealed. Slides were kept covered and stored in the fridge if left for any period of time as this was shown to preserve the fluorescent signal.

For IF labelling, coverslips were incubated section side down in a series of drops protected from light. Initially, coverslips were placed in three block solution washes (1% w/v BSA in PBS) each for 10 min before incubation in a 50 μl drop of primary antibody for 30 min. This was repeated for the secondary antibody incubation (3 × 10 min block wash; 30 min antibody incubation). Final washes were in block solution (3 × 5 min) and water (3 × 2 min), with excess water gently removed by filter paper. Coverslips were then mounted onto slides with Mowiol and allowed to set overnight at room temperature. Sections were viewed with an Olympus BX51 light microscope (Fig. 4).

5. Immunolabelling of HM20 Sections

Thin Lowicryl sections (60–80 nm) can also be immunolabelled for electron microscopy. Sections picked up on Formvar carbon-coated grids were dried and inverted section side down on the following series of solutions. Throughout all stages, PBS, which is used in general immunolabelling protocols (see Section C), was replaced with 0.1 M PHEM + 137 mM NaCl, which resulted in reduced background. Grids were washed in PHEM for 2 min, quench solution (20 mM glycine in 0.1 M PHEM) for 10 min, and blocking buffer (0.5% FSG, 0.5% BSA, 10 mM glycine in PHEM) for 15 min. Then, grids were incubated for 30 min in specific primary antibody (7 μl), after which they were washed in block solution twice and PHEM twice (5 min each). Secondary antibody incubation (7 μl) was also for 30 min followed by two block washes and two PHEM washes (2 min each). The reaction was finally stabilised with 1% (w/v) glutaraldehyde in PHEM fixation for 5 min, then washed in water (3 × 2 min) and dried in forceps. Generally, grids were viewed unstained without further staining "on grid" (Fig. 4G and H).

6. Electron Tomography of Thick Sections

Thick (300 nm) sections were placed on Formvar-coated copper slot grids once an area of interest was found in thin sections. Colloidal gold particles (10 nm) were then affixed to both surfaces ensuring an even distribution, and lightly coated with carbon (Marsh, 2007). Using a Tecnai F30 (FEI Company), serial electron micrographs were taken from a tilt series over a ±60° range at 1° increments about two axis. These were later aligned with help of the fiducial gold markers and combined to calculate tomograms and produce 3D reconstructions and movies (for further details see Nixon *et al.*, 2007, 2009; Richter *et al.*, 2008).

IV. Materials

A. Epon Embedding—Conventional

Instrumentation: Ultra high quality water (UHQ, Elga Labwater, UK).

Materials: Petri dishes, transfer pipettes, silicon moulds, sealed plastic container.

Reagents: E3 media (5 mM NaCl, 0.17 mM KCl, 0.33 mM $CaCl_2$, 0.33 mM $MgSO_4$); cacodylate buffer (sodium cacodylate trihydrate, ProSciTech, Australia); EM grade glutaraldehyde, osmium tetroxide, Epon 812 resin, molecular sieve type 3A (ProSciTech, Australia); uranyl acetate (Electron Microscopy Sciences, Hatfield, PA, USA); ethanol; phosphate buffered saline.

B. Epon Embedding—Microwave

Instrumentation: Pelco BioWave® microwave containing a ColdSpot™ and vacuum chamber (Ted Pella, Inc., Redding, CA, USA), ultra high quality water (UHQ, Elga Labwater, UK).

Materials: Petri dishes, transfer pipettes, silicon moulds, sealed plastic container.

Reagents: E3 media (5 mM NaCl, 0.17 mM KCl, 0.33 mM $CaCl_2$, 0.33 mM $MgSO_4$), cacodylate buffer (sodium cacodylate trihydrate, ProSciTech, Australia);

EM grade glutaraldehyde, osmium tetroxide, Epon 812 resin, molecular sieve type 3A (ProSciTech, Australia); uranyl acetate (Electron Microscopy Sciences, Hatfield, PA, USA); ethanol; phosphate buffered saline.

C. Frozen Sectioning

Instrumentation: Ultra high quality water (UHQ, Elga Labwater, UK), heating block, rotator, Leica UC6 ultramicrotome with FC6 cryoattachment (Leica Microsystems, Wetzlar, Germany), Cressington carbon coater, JEOL1011 transmission electron microscope.

Materials: Parafilm, glass slides, fine-tipped forceps, anti-capillary forceps, cryostubs (ProSciTech, Australia), glass for glass knives or diamond knives (Diatome, Biel, Switzerland), 3 mm perfect copper loop, Protein-A gold (UMC-Utrecht, the Netherlands), 100 mesh copper grids (GCU100h, ProSciTech, Australia), filter paper.

Reagents: E3 media (5 mM NaCl, 0.17 mM KCl, 0.33 mM $CaCl_2$, 0.33 mM $MgSO_4$), cacodylate buffer (sodium cacodylate trihydrate, ProSciTech, Australia); EM grade glutaraldehyde and PFA (ProSciTech, Australia); glycine (Biotechnology grade, Amresco, Inc., OH, USA); gelatine (10% local food grade); phosphate buffered saline; 2.3 M sucrose, PVP-sucrose, fish skin gelatine, bovine serum albumin (Sigma-Aldrich, Inc., St. Louis, MO, USA); liquid nitrogen; toludine blue, Formvar (Merck, Darmstadt, Germany); methyl cellulose; uranyl acetate (Electron Microscopy Sciences, Hatfield, PA, USA).

D. High-Pressure Freezing—Epon

Instrumentation: Heating pad; Leica EM PACT2 HPF, Leica EM AFS2 (Leica Microsystems, Vienna, Austria); Pelco BioWave® microwave containing a ColdSpot™ and vacuum chamber (Ted Pella, Inc., Redding, CA, USA).

Materials: 200 μm membrane carriers, silicon moulds, sealed plastic container.

Reagents: Agarose; osmium tetroxide, Epon 812 resin, molecular sieve type 3A (ProSciTech, Australia); uranyl acetate (Electron Microscopy Sciences, Hatfield, PA, USA); acetone.

E. High-Pressure Freezing—HM20 Low-Temperature Embedding, Labelling and Tomography

Instrumentation: Heating pad; Leica EM PACT2 HPF, Leica EM AFS2 using Freeze Substitution Processor, Leica UC6 ultramicrotome (Leica Microsystems, Vienna, Austria); 45° diamond knife (Diatome, Biel, Switzerland); Olympus BX51 light microscope (Tokyo, Japan); Cressington carbon coater; Tecnai F30 (FEI Company).

Materials: 200 μm membrane carriers, copper loop, 12 mm round coverslips, copper slot grids (2 × 1 mm slot, GCu1M2M, ProSciTech, Australia), Protein-A gold (UMC-Utrecht, the Netherlands).

Reagents: Agarose; Lowicryl HM20, glutaraldehyde (ProSciTech, Australia); uranyl acetate (Electron Microscopy Sciences, Hatfield, PA, USA); acetone; Aqua Poly/Mount (Polysciences, Inc., USA); Mowiol, Formvar (Merck, Darmstadt, Germany); PHEM (PIPES, HEPES, EGTA, $MgCl_2$); sodium chloride, phosphate buffered saline.

V. Concluding Remarks and Perspectives

The HPF/FS method described here has a number of advantages over other methods providing a combination of good ultrastructural preservation, maintenance of antigenicity (for both immunogold labelling and direct GFP visualisation) and suitability for electron tomography. However, cryofixation can be time consuming and is not always practical. In addition, for larger specimens such as adult fish, cryofixation is impossible without tissue dissection, which itself causes deterioration of sensitive structures. In these cases, the use of chemical fixatives before HPF represents an important complementary technique (Sosinsky *et al.*, 2008). Ultrastructural preservation after chemical fixation and HPF has been shown to be superior to chemical fixation alone, particularly for labile structures, such as notochord and cytoskeleton (Sosinsky *et al.*, 2008). This hybrid method could easily be combined with the methods described here for immunolabelling and tomography after rapid FS and embedding in Lowicryl resins.

The methods described here correlate immunogold labelling and fluorescence on the same (or serial) sections. Quantum dots, although less electron dense than gold particles, have the advantage of allowing the visualisation of exactly the same probes by light and electron microscopy (Giepmans *et al.*, 2006) and our preliminary investigations suggest they will be readily applicable to labelling on the HM20 sections described here. Finally, these methods should be amenable to the elegant array tomography methods developed by Smith and coworkers (Micheva and Smith, 2007). This technique, which uses serial plastic sections and tomographic 3D reconstruction, combines large volume imaging of a specimen with 3D high-resolution analysis and depth-independent immunolabelling (Micheva and Smith, 2007). Recently, the new techniques of Focused Ion Beam/Scanning Electron Microscopy and Serial Block Face/Scanning Electron Microscopy have been used to look at blood vessel fusion events in zebrafish embryos (Armer *et al.*, 2009). Both these involve serial block face imaging with removal of thin layers of material between each image. They allow the ability to obtain 3D information at a multi-cell level. Application of these methods to the zebrafish embryo can provide large-scale information on cell–tissue organisation, as well as ultrastructural detail and molecular characterisation, in a system where proteins can be expressed, or downregulated, rapidly and efficiently. The full range of modern light and EM techniques now being developed in laboratories throughout the world (Gaietta *et al.*, 2006; Giepmans *et al.*, 2006) should be readily applicable to the zebrafish embryo in the coming years.

Acknowledgments

This work was supported by a grant from the National Health and Medical Research Council of Australia. The Institute for Molecular Bioscience is a Special Research Centre of the Australian Research Council. Confocal microscopy was performed at the Australian Cancer Research Foundation (ACRF)/Institute for Molecular Bioscience Dynamic Imaging Facility for Cancer Biology, which was established with the support of the ACRF. The authors thank Harriet Lo for transgenic fish and acknowledge the facilities as well as scientific and technical assistance from staff in the Australian Microscopy & Microanalysis Facility (AMMRF) at the Centre for Microscopy and Microanalysis at The University of Queensland.

References

Akhtar, S., Schonthaler, H. B., Bron, A. J., and Dahm, R. (2008). Formation of stromal collagen fibrils and proteoglycans in the developing zebrafish cornea. *Acta Ophthalmol.* **86**, 655–665.

Amsterdam, A., and Hopkins, N. (2006). Mutagenesis strategies in zebrafish for identifying genes involved in development and disease. *Trends Genet.* **22**, 473–478.

Amsterdam, A., Lin, S., and Hopkins, N. (1995). The *Aequorea victoria* green fluorescent protein can be used as a reporter in live zebrafish embryos. *Dev. Biol.* **171**, 123–129.

Armer, H. E., Mariggi, G., Png, K. M., Genoud, C., Monteith, A. G., Bushby, A. J., Gerhardt, H., and Collinson, L. M. (2009). Imaging transient blood vessel fusion events in zebrafish by correlative volume electron microscopy. *PLoS ONE* **4**, e7716.

Branchek, T., and Bremiller, R. (1984). The development of photoreceptors in the zebrafish, *Brachydanio rerio*. I. Structure. *J. Comp. Neurol.* **224**, 107–115.

Chapman, G. B., Tarboush, R., Eagles, D. A., and Connaughton, V. P. (2009). A light and transmission electron microscope study of the distribution and ultrastructural features of peripheral nerve processes in the extra-retinal layers of the zebrafish eye. *Tissue Cell* **41**, 286–298.

de Jong-Curtain, T. A., Parslow, A. C., Trotter, A. J., Hall, N. E., Verkade, H., Tabone, T., Christie, E. L., Crowhurst, M. O., Layton, J. E., Shepherd, I. T., Nixon, S. J., Parton, R. G., *et al.*, (2009). Abnormal nuclear pore formation triggers apoptosis in the intestinal epithelium of elys-deficient zebrafish. *Gastroenterology* **136**, 902–911.

Doyon, Y., McCammon, J. M., Miller, J. C., Faraji, F., Ngo, C., Katibah, G. E., Amora, R., Hocking, T. D., Zhang, L., Rebar, E. J., Gregory, P. D., Urnov, F. D., *et al.*, (2008). Heritable targeted gene disruption in zebrafish using designed zinc-finger nucleases. *Nat. Biotechnol.* **26**, 702–708.

Gaietta, G. M., Giepmans, B. N., Deerinck, T. J., Smith, W. B., Ngan, L., Llopis, J., Adams, S. R., Tsien, R. Y., and Ellisman, M. H. (2006). Golgi twins in late mitosis revealed by genetically encoded tags for live cell imaging and correlated electron microscopy. *Proc. Natl. Acad. Sci. U.S.A.* **103**, 17777–17782.

Giepmans, B. N., Adams, S. R., Ellisman, M. H., and Tsien, R. Y. (2006). The fluorescent toolbox for assessing protein location and function. *Science* **312**, 217–224.

Griffiths, G. (1993). "Fine Structure Immunocytochemistry." Springer-Verlag, Berlin-Heidelberg.

Haffter, P., Granato, M., Brand, M., Mullins, M. C., Hammerschmidt, M., Kane, D. A., Odenthal, J., van Eeden, F. J., Jiang, Y. J., Heisenberg, C. P., Kelsh, R. N., Furutani-Seiki, M., *et al.*, (1996). The identification of genes with unique and essential functions in the development of the zebrafish, *Danio rerio*. *Development* **123**, 1–36.

Haffter, P., and Nusslein-Volhard, C. (1996). Large scale genetics in a small vertebrate, the zebrafish. *Int. J. Dev. Biol.* **40**, 221–227.

Hall, T. E., Bryson-Richardson, R. J., Berger, S., Jacoby, A. S., Cole, N. J., Hollway, G. E., Berger, J., and Currie, P. D. (2007). The zebrafish candyfloss mutant implicates extracellular matrix adhesion failure in laminin alpha2-deficient congenital muscular dystrophy. *Proc. Natl. Acad. Sci. U.S.A.* **104**, 7092–7097.

Hill, M. M., Bastiani, M., Luetterforst, R., Kirkham, M., Kirkham, A., Nixon, S. J., Walser, P., Abankwa, D., Oorschot, V. M., Martin, S., Hancock, J. F., and Parton, R. G. (2008). PTRF-Cavin, a conserved cytoplasmic protein required for caveola formation and function. *Cell* **132**, 113–124.

Huisken, J., and Stainier, D. Y. (2009). Selective plane illumination microscopy techniques in developmental biology. *Development* **136**, 1963–1975.

Jontes, J. D., Buchanan, J., and Smith, S. J. (2000). Growth cone and dendrite dynamics in zebrafish embryos: Early events in synaptogenesis imaged in vivo. *Nat. Neurosci.* **3**, 231–237.

Kalicharan, D., Jongebloed, W. L., Rawson, D. M., and Zhang, T. T. (1998). Variations in fixation techniques for field emission SEM and TEM of zebrafish (*Branchydanio rerio*) embryo inner and outer membranes. *J. Electron Microsc.* **47**, 645–658.

Kimmel, C. B., Ballard, W. W., Kimmel, S. R., Ullmann, B., and Schilling, T. F. (1995). Stages of embryonic development of the zebrafish. *Dev. Dyn.* **203**, 253–310.

Kljavin, I. J. (1987). Early development of photoreceptors in the ventral retina of the zebrafish embryo. *J. Comp. Neurol.* **260**, 461–471.

Leis, A., Rockel, B., Andrees, L., and Baumeister, W. (2009). Visualizing cells at the nanoscale. *Trends Biochem. Sci.* **34**, 60–70.

Liu, N. A., Ren, M., Song, J., Rios, Y., Wawrowsky, K., Ben-Shlomo, A., Lin, S., and Melmed, S. (2008). In vivo time-lapse imaging delineates the zebrafish pituitary proopiomelanocortin lineage boundary regulated by FGF3 signal. *Dev. Biol.* **319**, 192–200.

Luby-Phelps, K., Ning, G., Fogerty, J., and Besharse, J. C. (2003). Visualization of identified GFP-expressing cells by light and electron microscopy. *J. Histochem. Cytochem.* **51**, 271–274.

Lucic, V., Forster, F., and Baumeister, W. (2005). Structural studies by electron tomography: From cells to molecules. *Annu. Rev. Biochem.* **74**, 833–865.

Marsh, B. J. (2007). Reconstructing mammalian membrane architecture by large area cellular tomography. *Methods Cell Biol.* **79**, 193–220.

McDonald, K. L. (2009). A review of high-pressure freezing preparation techniques for correlative light and electron microscopy of the same cells and tissues. *J. Microsc.* **235**, 273–281.

McDonald, K. L., Morphew, M., Verkade, P., and Muller-Reichert, T. (2007). Recent advances in high-pressure freezing: Equipment- and specimen-loading methods. *Methods Mol. Biol.* **369**, 143–173.

Meng, X., Noyes, M. B., Zhu, L. J., Lawson, N. D., and Wolfe, S. A. (2008). Targeted gene inactivation in zebrafish using engineered zinc-finger nucleases. *Nat. Biotechnol.* **26**, 695–701.

Metcalfe, W. K., Kimmel, C. B., and Schabtach, E. (1985). Anatomy of the posterior lateral line system in young larvae of the zebrafish. *J. Comp. Neurol.* **233**, 377–389.

Micheva, K. D., and Smith, S. J. (2007). Array tomography: A new tool for imaging the molecular architecture and ultrastructure of neural circuits. *Neuron* **55**, 25–36.

Moens, C. B., Donn, T. M., Wolf-Saxon, E. R., and Ma, T. P. (2008). Reverse genetics in zebrafish by TILLING. *Brief. Funct. Genomic. Proteomic.* **7**, 454–459.

Nixon, S. J., Carter, A., Wegner, J., Ferguson, C., Floetenmeyer, M., Riches, J., Key, B., Westerfield, M., and Parton, R. G. (2007). Caveolin-1 is required for lateral line neuromast and notochord development. *J. Cell. Sci.* **120**, 2151–2161.

Nixon, S. J., Webb, R. I., Floetenmeyer, M., Schieber, N., Lo, H. P., and Parton, R. G. (2009). A single method for cryofixation and correlative light, electron microscopy and tomography of zebrafish embryos. *Traffic* **10**, 131–136.

Nixon, S. J., Wegner, J., Ferguson, C., Mery, P. F., Hancock, J. F., Currie, P. D., Key, B., Westerfield, M., and Parton, R. G. (2005). Zebrafish as a model for caveolin-associated muscle disease; caveolin-3 is required for myofibril organization and muscle cell patterning. *Hum. Mol. Genet.* **14**, 1727–1743.

Parsons, M. J., Pollard, S. M., Saude, L., Feldman, B., Coutinho, P., Hirst, E. M., and Stemple, D. L. (2002). Zebrafish mutants identify an essential role for laminins in notochord formation. *Development* **129**, 3137–3146.

Peters, K. G., Rao, P. S., Bell, B. S., and Kindman, L. A. (1995). Green fluorescent fusion proteins: Powerful tools for monitoring protein expression in live zebrafish embryos. *Dev. Biol.* **171**, 252–257.

Pickart, M. A., Klee, E. W., Nielsen, A. L., Sivasubbu, S., Mendenhall, E. M., Bill, B. R., Chen, E., Eckfeldt, C. E., Knowlton, M., Robu, M. E., Larson, J. D., Deng, Y., *et al.*, (2006). Genome-wide reverse genetics framework to identify novel functions of the vertebrate secretome. *PLoS ONE* **1**, e104.

Rawson, D. M., Zhang, T., Kalicharan, D., and Jongebloed, W. L. (2000). Field emission scanning electron microscopy and transmission electron microscopy studies of the chorion, plasma membrane and syncytial layers of the gastrula-stage embryo of the zebrafish *Brachydanio rerio*: A consideration of the structural and functional relationships with respect to cryoprotectant penetration. *Aquac. Res.* **31**, 325–336.

Richter, T., Floetenmeyer, M., Ferguson, C., Galea, J., Goh, J., Lindsay, M. R., Morgan, G. P., Marsh, B. J., and Parton, R. G. (2008). High-resolution 3D quantitative analysis of caveolar ultrastructure and caveola-cytoskeleton interactions. *Traffic* **9**, 893–909.

Scott, E. K., Mason, L., Arrenberg, A. B., Ziv, L., Gosse, N. J., Xiao, T., Chi, N. C., Asakawa, K., Kawakami, K., and Baier, H. (2007). Targeting neural circuitry in zebrafish using GAL4 enhancer trapping. *Nat. Methods* **4**, 323–326.

Slot, J. W., and Geuze, H. J. (2007). Cryosectioning and immunolabeling. *Nat. Protoc.* **2**, 2480–2491.

Sosinsky, G. E., Crum, J., Jones, Y. Z., Lanman, J., Smarr, B., Terada, M., Martone, M. E., Deerinck, T. J., Johnson, J. E., and Ellisman, M. H. (2008). The combination of chemical fixation procedures with high pressure freezing and freeze substitution preserves highly labile tissue ultrastructure for electron tomography applications. *J. Struct. Biol.* **161**, 359–371.

Studer, D., Humbel, B. M., and Chiquet, M. (2008). Electron microscopy of high pressure frozen samples: Bridging the gap between cellular ultrastructure and atomic resolution. *Histochem. Cell Biol.* **130**, 877–889.

Tokuyasu, K. T. (1980). Immunochemistry on ultrathin frozen sections. *Histochem. J.* **12**, 381–403.

Wendt, K. D., Jensen, C. A., Tindall, R. and Katz, M. L. (2004). Comparison of conventional and microwave-assisted processing of mouse retinas for transmission electron microscopy. *J. Microsc.* **214**, 80–88.

CHAPTER 19

Transmission Electron Microscopy of Cartilage and Bone

Douglas R. Keene ***and*** **Sara F. Tufa**

Shriners Hospitals for Children, Micro-Imaging Center, Portland, Oregon 97239

Abstract

This review of cartilage microscopy is restricted to a brief description of the major molecular components in cartilage and bone and an in-depth discussion of methods for the preparation of these tissues for examination by transmission electron microscopy. Included within are simple methods for the isolation of cartilage-specific macromolecules, methods for immunoidentification of tissue components, bone decalcification protocols, and fixative recipes designed to stabilize cells and proteoglycan within the cartilage matrix. A discussion of cartilage prepared by high-pressure freezing (HPF)/freeze substitution (FS) is also presented.

METHODS IN CELL BIOLOGY, VOL. 96

978-0-12-381007-6
DOI: 10.1016/S0091-679X(10)96019-6

I. Introduction

In common with most other connective tissues, cartilage consists of relatively small numbers of cells distributed throughout a dense extracellular matrix. Cartilage lacks blood vessels and lymphatics, hence it is the matrix that regulates metabolites and nutrients to and from the cells. The composition and molecular structure of the matrix in large part determines the biological function of different cartilages; however, cartilages are grouped based on their morphologies. Hyaline cartilage includes articular, tracheal, nasal, and rib cartilage. The annulus fibrosis of the intervertebral disk is composed of fibrous cartilage. The ear contains elastic cartilage, and is the only cartilage matrix containing elastin (Heinegård and Paulsson, 1987). Chondrocytes create the matrix, primarily composed of fibrillar type II collagen and several populations of proteoglycans (Buckwalter and Rosenberg, 1988); however, the presence of minor collagens and noncollagenous proteins help to define specific cartilages. Sixty to eighty percent of the wet weight of cartilage is water that is organized and maintained by an unusually high concentration of glycosaminoglycans, which themselves contribute 30–35% of the dry weight of cartilage. These glycosaminoglycans include keratan sulfate, chondroitin-6-sulfate, chondroitin-4-sulfate, dermatan sulfate, and hyaluronan (Hardingham, 1986). With the exception of hyaluronan, glycosaminoglycans covalently bind to protein to form proteoglycans, which can extend or expand in solution depending on the relative proportions of keratan sulfate, chondroitin sulfate, and protein. Those containing chondroitin sulfate and keratan sulfate exist either as monomers or aggregates (Hardingham, 1986; Hascall, 1977). Monomers have molecular weights ranging from just over one million to about four million. In hyaline cartilages, proteoglycan monomers bind to hyaluronic acid (HA) via their HA-binding region to form large aggregates.

Proteoglycan concentration and distribution vary markedly among different tissues. In the less turgid tissues such as skin and tendon, cupromeronic blue staining (Scott and Haigh, 1985) reveals by transmission electron microscopy (TEM) that dermatan sulfate proteoglycans associate with collagen fibrils via the “d” and “e” bands (Scott and Haigh, 1985). In cornea (a more turgid tissue containing a higher concentration of proteoglycans), in addition to dermatan sulfate at the “d” and “e” bands, keratan sulfate proteoglycans interact with banded collagen fibrils at the “a” and “c” bands (Scott and Haigh, 1988). Cartilage has a greater concentration of proteoglycans than any other tissue and in the large cartilage proteoglycans, keratan sulfate and chondroitin sulfate share the same protein core, with the absolute numbers of each dependent on their position relative to cells, nutrients, and age (Scott and Stockwell, 2006). In highly aligned fibril bundles, such as those in tendon, proteoglycan is seen to span from one fibril to the next, locking adjacent fibrils in register. However in most areas of cartilage, proteoglycans do not appear to be regularly located on the fibril surface (Scott and Haigh, 1988).

It is likely that the proteoglycan–collagen fibril interaction is facilitated by the protein core of the proteoglycans, with their GAG chains free to interact with other components of the matrix, including other GAG chains (Lewis *et al.*, 2010; Scott, 1995). These

PG–PG interactions form relatively stable antiparallel connections, and importantly, *the collagen fibrils are oriented by the proteoglycans*, instead of the fibrils being merely embedded in a gel of proteoglycans. Cartilage elasticity is likely dependent on this dynamic supramolecular complex of proteoglycan bridges and collagen fibrils termed "shape modules" (Scott, 2003; Scott and Stockwell, 2006). The result of all these intermolecular bonds is that the matrix behaves very much as a single macromolecular complex (Heinegård and Paulsson, 1987).

Collagens contribute to approximately 50% of the dry weight, which give cartilage its form and tensile strength (Buckwalter and Rosenberg, 1988). The negatively charged proteoglycans provide a strong force to expand the tissue while the fibers of type II collagen prevent expansion beyond a certain level. When under load, the proteoglycans allow the cartilage to resist extensive compression and enable the tissue to expand back to its original shape after removal of the load (Kempson, 1979). Along with the aggregating proteoglycans, smaller dermatan sulfate proteoglycans have been identified (Rosenberg *et al.*, 1985), which have specific relationships with collagen fibrils and may mediate interaction between cartilage components and influence collagen fibril diameter.

A perpetual issue encountered in the preparation of cartilage for TEM is how to adequately cope with the enormously high content of proteoglycan within the cartilage matrix. In tissues prepared by standard EM fixation using glutaraldehyde followed by osmium tetroxide, proteoglycans appear as matrix granules, loosely defined as "small dense particles sometimes associated with collagen fibrils and usually linked together by delicate intergranular fibrils about 50Å wide" (Anderson and Sajdera, 1971), to which proteoglycan monomer and link protein can be localized using immunoelectron microscopy (Poole *et al.*, 1980; Ratcliffe *et al.*, 1984). For optimal cell preservation, all observations indicate that the proteoglycan matrix must be stabilized or the chondrocytes will not be well preserved. This is best accomplished by the addition of ruthenium hexammine trichloride (RHT) to the fixative or by cryopreservation methodology, such as high-pressure freezing (HPF)/freeze substitution (FS). However, following both methods, collagen fibrils are difficult to image among the high background of dense proteoglycan. The proteoglycan content of cartilage also imposes a formidable diffusion barrier to the infiltration of antibody into the matrix, necessitating treatment in proteoglycan-degrading enzymes such as hyaluronidase or chondroitinase as a prerequisite to "en bloc" (diffusion of primary and secondary antibody into unfixed tissue) techniques for immunoelectron microscopy. Therefore, one must compromise between inadequate display of collagen fibril profiles in favor of optimal cell morphology, or inadequate cell morphology in favor of well-visualized collagen fibrils. Proteoglycan filaments can be stained with cupromeronic blue and their binding positions mapped to the surface of the collagen fibrils at the resolution of the collagen banding pattern (Scott and Haigh, 1985). However, there is no methodology that allows resolution of finest structure of the proteoglycan matrix in ultrathin tissue sections. These issues are so formidable that many studies have depended on mechanical or chemical disruption of cartilage matrix to isolate and study the individual macromolecular complexes.

II. Rationale

No one TEM tissue preparation protocol succeeds in optimally visualizing all structural entities within cartilage or bone. One must choose among various fixation conditions based on which tissue entity one hopes to optimally visualize. This chapter gives insight into task-specific methodologies; therefore, it separately considers the isolation and preparation of individual cartilage macromolecules, the optimal visualization of collagen fibrils within a cartilage matrix, and the methodologies for specific stabilization of proteoglycans resulting in optimal preservation of chondrocytes. Immunocytochemical protocols for cartilage are also discussed, as are demineralization methods for calcified tissues.

III. Methods

A. Imaging Isolated Cartilage Macromolecules

Proteoglycan can be isolated from cartilage in 3 M $MgCl_2$, bound with cytochrome *c*, and spread onto a hypophase using a method nearly identical to the Kleinschmidt technique used for contrasting DNA. This technique produces images of well-spread highly resolved proteoglycan aggregates that are unrivaled by any other technique (Buckwalter and Rosenberg, 1988; Rosenberg *et al.*, 1975). Collagen fibrils and other lesser components of cartilage matrix may also be isolated by simply dicing the tissue (using a double-edge razor blade) into <0.5 mm cubes under a volatile buffer (0.1 M ammonium bicarbonate, pH 7.8), then shearing the tissue *while in ice-cold buffer* using a Omni 2000 tissue homogenizer until few visible tissue fragments remain. Larger tissue fragments settle over a few minutes and the supernatant can be rotary shadowed (Keene *et al.*, 1991). Though this method is "quick and dirty", excellent images of many species of macromolecules are easily obtained (Fig. 1). Proteoglycans are occasionally visualized (Fig. 2) but are rarely spread sufficiently to be resolved. Collagen fibrils are most easily imaged following this isolation technique by on-grid staining with 4% phosphotungstic acid (PTA) (Fig. 1E). Matrix components isolated by this crude homogenization technique can be identified using antibodies, either incubated in solution with the homogenate (Gregory *et al.*, 2001; Keene *et al.*, 1991) or on-grid as additional incubation steps prior to staining with PTA (Zhu *et al.*, 1999) using steps identical to those used for immunolabeling antigens exposed on the surface of sections (Fig. 1F). For isolation of pure pools of specific macromolecules, strong chaotrophic agents including 4 M guanidine-HCl (containing protease inhibitors) (Tang *et al.*, 1979) and *N*-ethylmaleimide are commonly used prior to purification by cesium chloride equilibrium density gradient centrifugation (Heinegård and Paulsson, 1987).

B. General Considerations for Tissue Preparation

As for any tissue prepared for TEM, cartilage samples should always be as fresh as possible. Wet the tissue during dissection in Dulbecco's Modification of Eagle's Medium

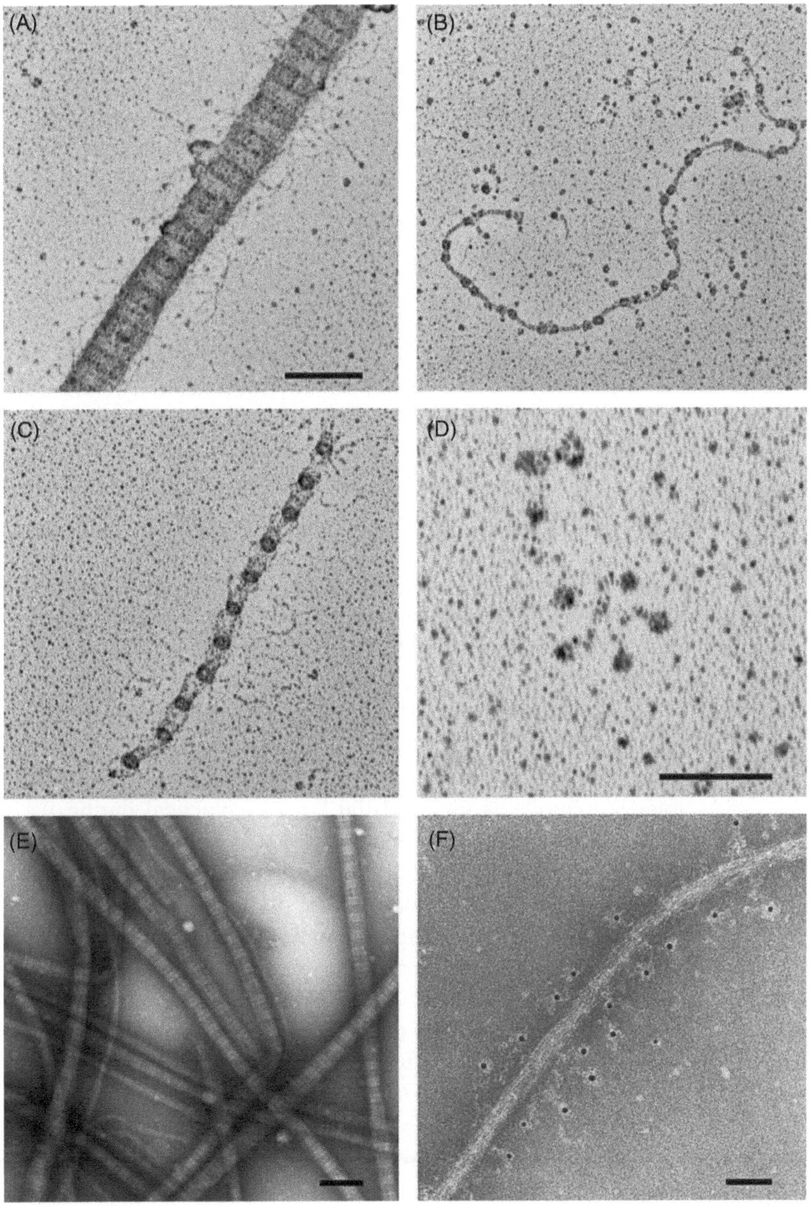

Fig. 1 Macromoleular complexes released from cartilage by shearing in a volatile buffer. The supernatant was either rotary shadowed (A–D) or negative stained (E, F; 4% PTA, pH 7.0). (A) Banded collagen fibril; (B) type VI collagen microfilament; (C) fibrillin microfibril; (D) cartilage oligomatrix matrix protein (COMP); (E) banded collagen fibrils, (F) a thin banded fibril immunolabeled *on grid* with type IIA collagen antibody followed by a −5 nm secondary gold conjugate. A–E, 8-year-old human articular cartilage; F, fetal human cartilage. Bars: A–C = 100 nm; D, F = 50 nm; E = 250 nm.

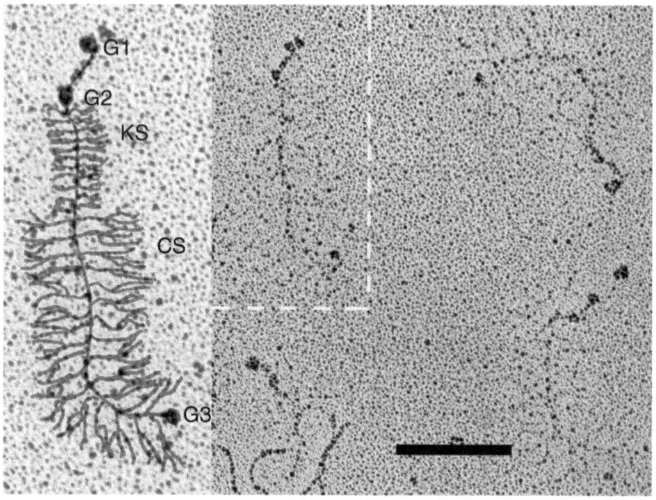

Fig. 2 Proteoglycan monomers isolated from fetal calf cartilage by shearing, then rotary shadowed. To the left is an artist's rendition outlining the approximate position of the G3 globular domain, the chondroitin sulfate (CS) and keratan sulfate (KS) chains, and the G2 and G1 globular domains. The G1 domain binds hyaluronic acid via link protein: many of hundreds of these monomers would link to HA to compose a proteoglycan aggregate. Bar = 200 nm.

(DMEM). If cutting bone, use either a single-edge razor blade or, if cutting calcified tissue, a disposable microtome blade made oil-free with acetone. Bone will cut (fracture) more easily if cut parallel to the long axis than in cross section. If cartilage is your only interest, remove as much of the bone as practical during the initial dissection. Use a double-edge razor blade for making the final cuts within the cartilage matrix. Double-edge razor blades vary greatly in strength and sharpness depending on manufacturer (see materials list). Slice, do not chop the tissue. Final dimensions of the tissue block should include a side 1 mm or slightly less. Expect a region approximately 50–100 μm along the free-cut edges of the tissue to contain poorly preserved chondrocytes, known as "edge effect" (Kashiwa *et al.*, 1975). If the tissue is calcified, orient the sample so that sections are cut parallel to the length of the bone shaft. The alternative cross section will result in more damage to glass and diamond knives.

We routinely examine 0.5–1.0-μm-thick sections in the light microscope to check sample orientation prior to cutting ultrathin sections from the same tissue block. Thick sections are removed from the water at the cutting edge of a glass or diamond knife with an eyelash stick, then transferred to a 50 μl drop of water on a clean glass slide, then dried on a hot surface at 60°C. While drying on the slide warmer, sections are spread by exposure to chloroform vapors affected by pipetting three drops of chloroform onto a Whatman #3 filter paper rectangle folded twice to form a tent suspended less than 1 cm over the sections. Ultrathin sections are spread similarly, using a filter paper rectangle folded so that a flat surface with three drops of chloroform is suspended over the diamond knife boat. Without effective spreading, expect significant compression artifacts and folds in both thick and ultrathin sections of cartilage and bone. We have not found a method as effective as chloroform. Chloroform is a proven carcinogen; we therefore perform this procedure only in a ventilated area.

We collect large (often 1 mm × 2 mm) ultrathin sections onto freshly glow-discharged 1 mm × 2 mm formvar-coated copper–palladium slot grids. We cast formvar films (with a silver interference color) from freshly cleaned glass microscope slides dipped in a solution of 0.35% formvar in ethylene dichloride and dry the slides in a nitrogen atmosphere within a "glove bag" (Keene, 1984). The importance of rendering the surface of the formvar hydrophilic via glow discharge cannot be overstated since otherwise it is very difficult to lay these large sections without puckering or folding. A home-made glow-discharge unit can be easily manufactured (Aebi and Pollard, 1987). Ultrathin sections are thoroughly dried prior to staining in saturated uranyl acetate in 50% ethanol for 15 min followed by Reynolds lead citrate for 1–3 min. The quality of cartilage staining is particularly sensitive to fresh lead citrate; stain solutions older than 3 weeks should be discarded.

C. Retention of Cartilage Matrix Proteoglycans in Tissues Using Cationic Dyes

The appearance of chondrocytes and their relationship to collagen and matrix granules within the cartilage matrix vary significantly depending on the fixation chemistry used to prepare the tissue. It was not until the introduction of osmium tetroxide (Palade, 1952) and methacrylate embedding media (Borysko, 1955) that chondrocytes and matrix components were clearly imaged in the TEM (Robinson and Cameron, 1956). In these early images, pericellular "moats" or halos surrounding chondrocytes were debated as potential artifacts of tissue preparation (Cameron and Robinson, 1958). We now know that standard fixation protocols including glutaraldehyde and osmium tetroxide are ineffective in retaining proteoglycans within cartilage. Indeed, Engfeldt and Hjertquist (1968) reported a 4–10% loss of proteoglycan during immersion in glutaraldehyde and up to a 60% loss following immersion in osmium tetroxide, for a combined loss of up to 70% proteoglycan during a process normally considered a stabilization protocol. Progressive extraction of matrix proteoglycan coincides with greater degrees of chondrocyte shrinkage within all levels of the cartilage growth plate. This can be quickly evaluated by observing 1-μm-thick epon sections of cartilage in the light microscope. Inadequate retention of proteoglycan is evidenced by light halos surrounding chondrocytes. In the TEM, chondrocytes in a proteoglycan extracted matrix will have a ruffled border, with only these extensions contacting the matrix proper; and in the case of mouse cartilage, there will be a crystalline deposit at the cell borders (Fig. 3). Poorly preserved chondrocytes are often heavily vesiculated (Fig. 3C and D). Chondrocytes have a specific set of molecules on their surfaces, including hyaluronate, that have high binding affinities to matrix proteoglycans (Hunziker *et al.*, 1983; Sommarin and Heinegård, 1986), evidence that the cell surface and matrix are intimately associated. Ruthenium Red, a hexavalent cation, can be used in simultaneous combination with glutaraldehyde and osmium tetroxide to *retain* proteoglycan (Thyberg, 1977). The reaction product appears as a dense, globular precipitate clustered on the surface of collagen fibrils (Luft, 1971).

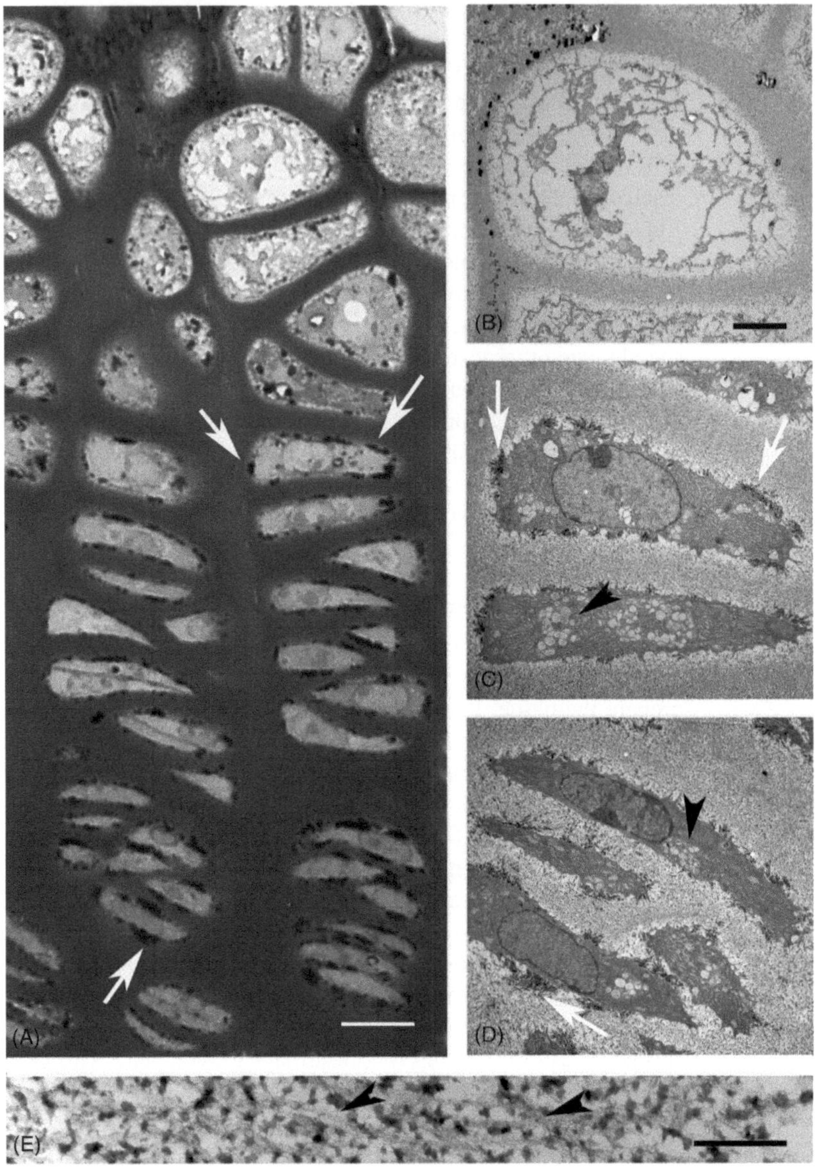

Fig. 3 Serial 1 μm and ultrathin sections of 10-day-old mouse cartilage fixed in 1.5% glutaraldehyde/1.5% paraformaldehyde including 0.05% tannic acid, 1% OsO_4 secondary fixation, dehydration in ethanol, followed by propylene oxide-assisted infiltration and embedding in Spurrs epoxy. (A) By light microscopy, chondrocytes throughout the growth plate appear shrunken within their lacunae, with dense crystalline deposits (white arrows) clustered at the cell periphery. At levels in the growth plate, corresponding to the adjacent LM, TEM shows (B) hypertrophic chondrocytes with discontinuous plasmalemmata, nuclear membranes, and RER. Prehypertrophic chondrocytes (C) contain numerous translucent vesicles (black arrows) within the cytoplasm and crystalline spicules at the cell periphery (white arrows), and also intact nuclei and RER. Few other cellular organelles are noted in these cells. Cells within the proliferative zone (D) are similar in appearance. Collagen fibrils (E) are somewhat resolved within the cartilage matrix (arrows). Bars: A, 10 μm; B–D, 4 μm; E, 100 nm. (See Plate no. 26 in the Color Plate Section.)

However, retention of proteoglycan is not sufficient for the preservation of chondrocytes. The use of 0.3–0.6% RHT included as an additive to glutaraldehyde, osmium tetroxide, and the buffer wash between leads to a finer, more uniform precipitation of proteoglycan throughout the matrix. With this greater stabilization of proteoglycans, chondrocytes are fully expanded within their lacunae and the "moat" artifact is eliminated. The chondrocyte plasmalemmata are intact and uniformly attached to the pericellular matrix. Significant improvements in the preservation of chondrocyte organelles and membranes are also evident (Fig. 4, mouse growth plate and Fig. 5, human growth plate) (Hunziker, 1982, 1983, and 1992). Glycogen within resting and proliferative chondrocytes is somewhat extracted and marginally contrasted following RHT exposure. Substituting osmium-potassium ferrocyanide for osmium tetroxide as the secondary fixative significantly improves glycogen retention (De Burijn, 1973) and also stabilizes cartilage matrix proteoglycan more so than does osmium alone (Shepard and Mitchell, 1989). In our experience, we find significant improvement in glycogen retention and in preservation of Golgi using 0.6% RHT to the primary fixative and buffer rinse and 0.6% RHT with 0.05 M potassium ferrocyanide in 1% OsO_4 as the secondary fixative (Fig. 6). This protocol does compromise stabilization of the proteoglycan matrix as compared to using RHT alone and as a result cells in the hypertrophic and prehypertrophic zone of the growth plate are not well stabilized (Suppl. Fig. A).

As eluded to earlier, for ultrastructural studies requiring the visualization of collagen fibrils within the matrix, RHT and/or potassium ferrocyanide used in concentrations necessary for optimal retention of proteoglycan are not appropriate for the visualization of collagen fibril profiles. Collagen fibrils are virtually invisible within a "sea" of densely stained proteoglycan (Figs. 4–6). Better definition of collagen fibrils may be obtained using less RHT (0.3%).

A significant impression of the supramolecular complex formed by proteoglycan may be gained with the use of cupromeronic blue (Fig. 7). 0.05% cupromeronic blue is added to the primary buffered glutaraldehyde fixative. The addition of magnesium chloride added to the fixative at the "critical electrolyte concentration" of 0.3 M (Scott and Haigh, 1985) insures that only sulfated proteoglycans stain; at lesser concentrations, undersulfated proteoglycans also stain. Fixation in glutaraldehyde/cupromeronic blue is for 17 h at 4°C, followed by a buffer wash, with no osmium postfixation step included. Cupromeronic blue staining can be intensified by incubation in aqueous 0.5% sodium tungstate followed by 0.5% sodium tungstate in 50% ethanol, each for 15 min. The tissue is then continued through ethanol dehydration to 100%, washed in propylene oxide, and embedded in epoxy. The protocols in much of the literature evaluating cupromeronic blue staining start with cryostat sections from unfixed frozen tissues. We prefer to expose fresh tissue to the glutaraldehyde/cupromeronic blue stain to avoid ice damage to chondrocytes; however, this method is inferior to glutaraldehyde-containing RHT for chondrocyte preservation. Cupromeronic blue appears to penetrate more slowly than does glutaraldehyde; therefore, the staining intensity in fresh tissue exposed to stain varies depending on how far into the edge of the tissue one examines. Regardless, this method of proteoglycan localization is ideal for three-dimensional electron tomography, as demonstrated recently in cornea (Lewis *et al.*, 2010).

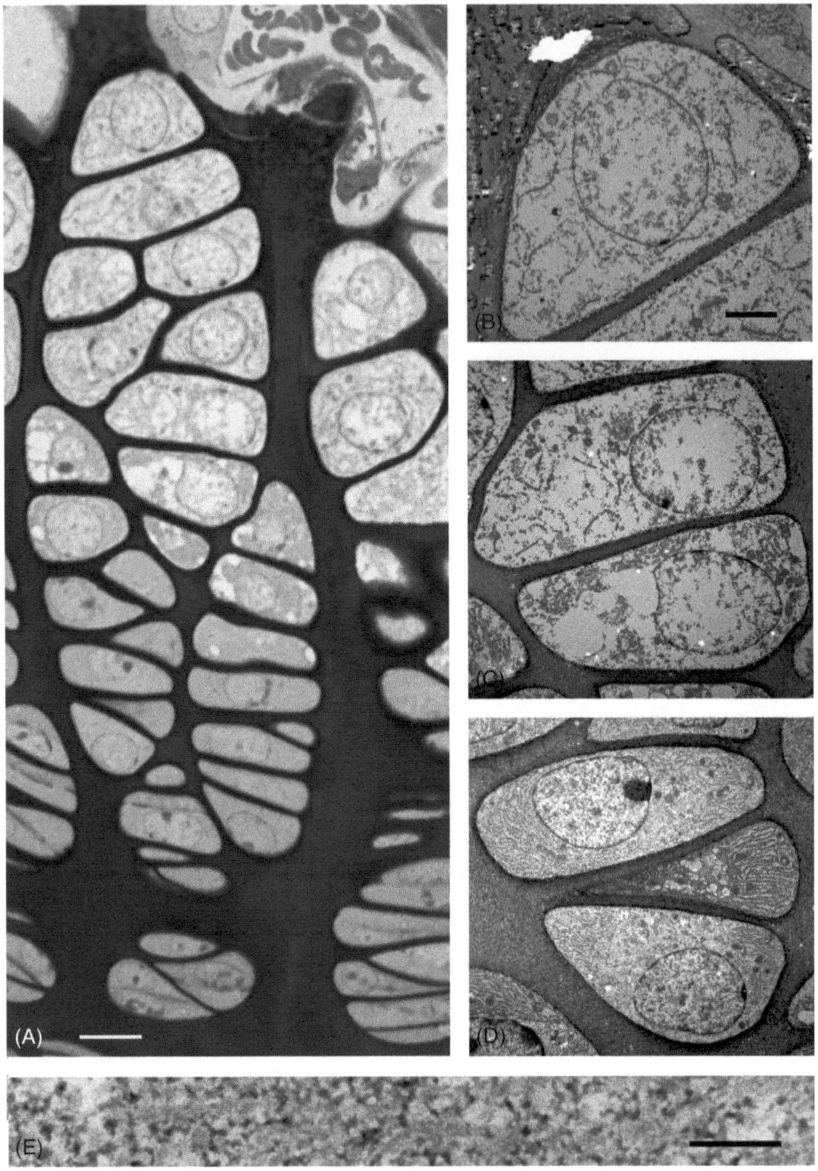

Fig. 4 Serial 1 μm and ultrathin sections of cartilage from the same 10-day-old mouse pictured in Fig. 3. This cartilage was fixed with primary and secondary fixatives containing 0.6% RHT and embedded in Spurrs epoxy. By light microscopy, chondrocytes at all levels of the growth plate fully occupy their lacunae, with a dense ring surrounding each chondrocyte. Hypertrophic chondrocytes (B) demonstrate mostly intact plasmalemmata and nuclear membranes, with the presence of a few membrane-bound organelles and RER within the cytoplasm. Prehypertrophic (C) and proliferative chondrocytes (D) display progressively more intracellular organelles with excellent membrane detail and a lack of open vesicles seen in Fig. 1C and D. No crystalline deposits are noted at the periphery of any chondrocytes within the growth plate. Collagen fibrils (E) are poorly resolved within the matrix. Bars: A, 10 μm; B–D, 4 μm; E, 100 nm. (See Plate no. 27 in the Color Plate Section.)

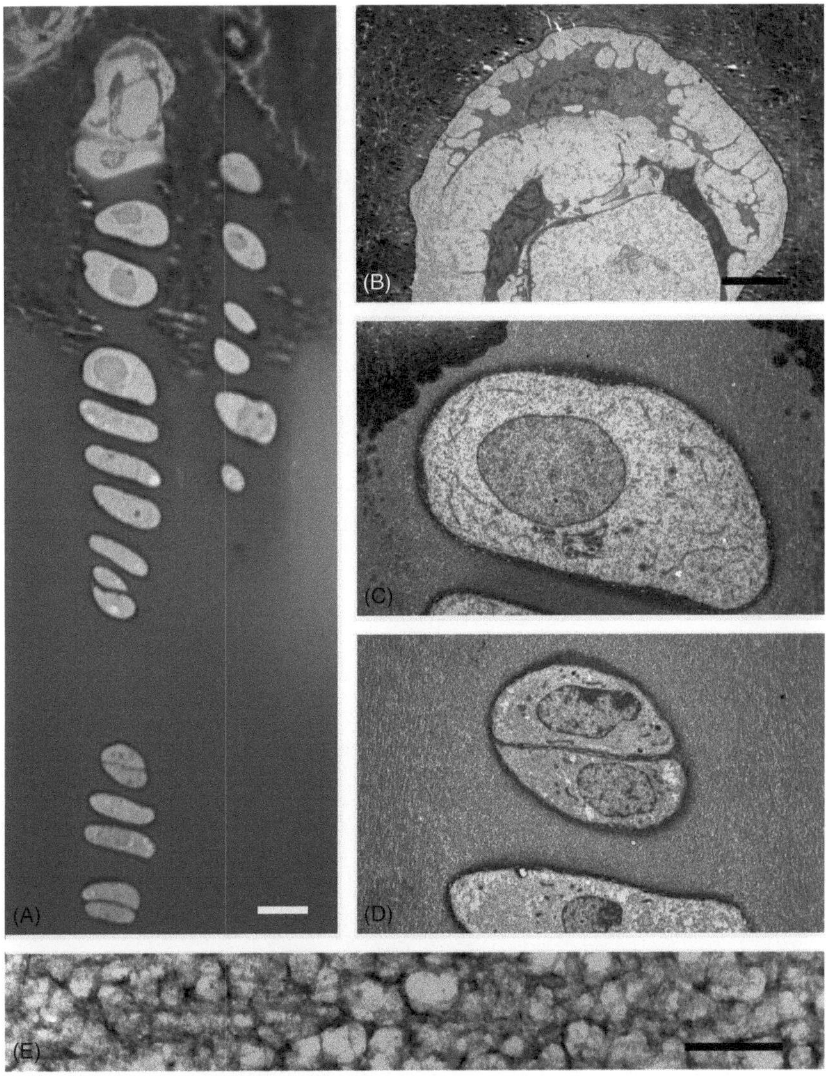

Fig. 5 Serial 1 μm and ultrathin sections of nondecalcified 15-month-old human cartilage/bone fixed in the presence of 0.6% RHT in both primary and secondary fixatives and embedded in Spurrs epon. As with similarly fixed mouse growth plate, chondrocytes throughout the growth plate fully occupy their lacunae, with a dense ring surrounding each chondrocyte (A). Within the bony matrix (B) an osteoblast is well preserved. Hypertrophic chondrocytes within the calcified cartilage (C) demonstrate mostly intact plasmalemmata; nuclear membranes and RER are well defined as are a few distinct organelles within the cytoplasm. Increasingly more organelles are present in chondrocytes positioned more toward the proliferative zone (D). Collagen fibrils (E) are poorly resolved within the matrix. Bars: A, 10 μm; B–D, 4 μm; E 100 nm. (See Plate no. 28 in the Color Plate Section.)

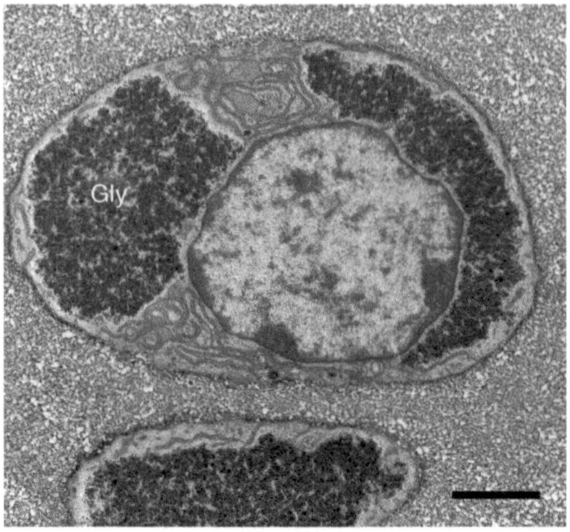

Fig. 6 Proliferative chondrocyte of 10-day-old mouse cartilage. Fixed in the presence of 0.6% RHT in the primary fixative and both 0.6% RHT and 0.05 M potassium ferrocyanide in the secondary fixative, glycogen (Gly) is well contrasted. Spurrs epoxy embedding. Bar = 2 μm.

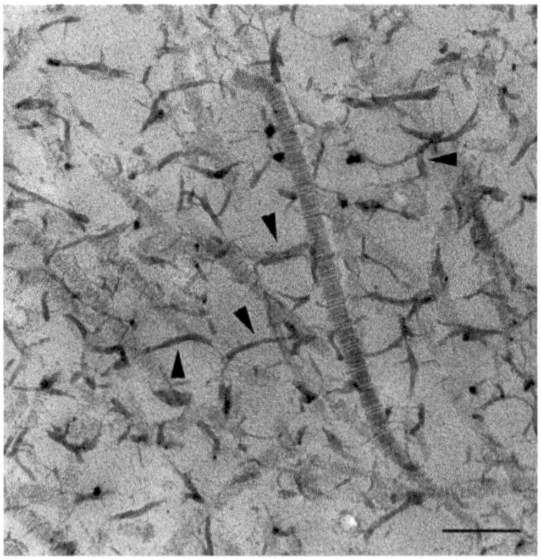

Fig. 7 Two-month-old human hyaline cartilage fixed with 0.025 M sodium acetate-buffered 2.5% glutaraldehyde containing 0.05% cupromeronic blue and 0.1 M magnesium chloride. Osmium fixation is omitted. Cupromeronic blue staining is enhanced with 0.5% sodium tungstate. Stained proteoglycans (arrow heads) are seen to occasionally intersect collagen fibrils and to span long distances between fibrils. Section stained in ethanolic uranyl acetate only. Bar = 200 nm.

As mentioned earlier, prior to initiating en bloc protocols designed to diffuse antibodies into the cartilage matrix, the proteoglycan component must first be destabilized. For these studies, pretreatment in 0.25 units/ml chondroitinase ABC for 2 h at 37°C (Keene *et al.*, 1995; Morris *et al.*, 2000; Zhu *et al.*, 1999) extracts the proteoglycan matrix sufficiently to allow en bloc immunolocalization using small immuno-gold conjugates (Fig. 8A and B). In our experience, even following digestion with chondroitinase, antibody diffuses into the dense collagen matrix of cartilage no more than

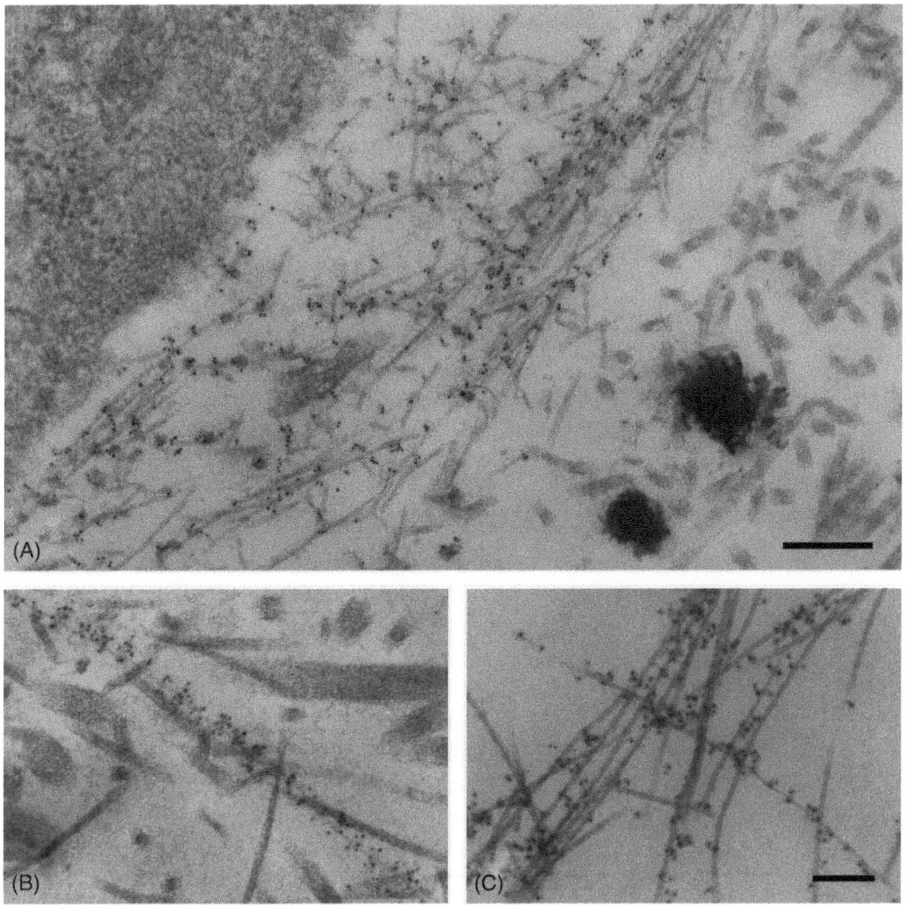

Fig. 8 En bloc immuno-gold labeling of cartilage predigested with 0.25 units/ml chondroitinase. (A) Antibody to type XI collagen exclusively labels thin collagen fibrils in fetal rat cartilage. (B) Antibody to fibrillin localizes to rare microfibrils within the matrix of 13-year-old human cartilage. (C) Chondroitinase in combination with several freeze–thaw cycles results in extensive cell damage, but improves penetration and immunolabeling density of type XI collagen in fetal bovine cartilage. Following penetration of primary and 5 nm gold secondary antibody conjugate, samples were fixed in 1.5% glutaraldehyde/1.5% paraformaldehyde with 0.05% tannic acid followed by 1% OsO_4 and embedded in Spurrs epon. Bars = 200 nm.

20 μm. Collagen fibrils will withstand many freeze–thaw cycles without significant change in ultrastructure, which may be a useful aid in opening the cartilage matrix considerably to the penetration of antibody (Fig. 8C).

D. Cryostabilization of Cartilage and Bone

As cryostabilization techniques have come of age, it is universally agreed that the most faithful representation of the native state of cells and tissues is gained by initially immobilizing the tissue in vitreous, noncrystalline ice. Many studies (reviewed by Gilke and Staehelin, 1986; see also Kellenberger, 1991) have shown that fast-freezing followed by freeze substitution dramatically improves the quality of ultrastructural preservation. Proper cryopreservation results in little or no detectable ice crystal damage. The lesser methods (plunge freezing, metal mirror freezing, propane jet freezing, etc.) are effectual only in freezing very thin tissue specimens (up to 10 μm). However, for *isolated* fibrils, plunge freezing has been very successful. A modification of the "bare grid method", solutions of macromolecular complexes are suspended across holey films and then plunge frozen in liquid ethane, allowing the observation of fine structure unhindered by stains (Meadows *et al.*, 2000). Careful control of atmospheric moisture, blotting, and control of the cryogen and microenvironment at its surface are all important in determining the quality of freezing and the rate of success. Successful freezing results in molecules embedded within a thin (10–100 nm) layer of amorphous ice, which can be examined in the cryo-TEM without further manipulation or staining (albeit contrast is very low). In our laboratory, we use a FEI Vitrobot (Iancu *et al.*, 2006) to facilitate freezing, which produces excellent results and is easy to use.

Herein it is sufficient to report that only high-pressure freezing is appropriate for the cryopreservation of cartilage and bone. HPF is a method relevant to biological samples (Moor *et al.*, 1980; Muller and Moor, 1984a, b) and is similar in concept to propane jet freezing, except at the moment just prior to exposing the tissue to liquid cryogen, the pressure surrounding the tissue is increased to about 2045 bar. High hydrostatic pressure inhibits the expansion of water as it crystallizes, suppressing the rate of ice crystal nucleation and growth, and also depresses the freezing point of water a few degrees (Moor, 1987; Moor *et al.*, 1980; Muller and Moor, 1984a, b). Good freezing can be expected up to a depth of about 200 μm. Following cryoimmobilization it is possible to section and examine the tissue while still in a frozen hydrated state (Al-Amoudi *et al.*, 2004), but due to technical difficulties, more commonly the samples are embedded in epoxy via a freeze substitution protocol and resin sections cut and viewed at room temperature. The components of the freeze substitution "cocktail" can vary, but will consist of either ethanol or acetone as the substitution media and likely also include some concentration of glutaraldehyde, osmium tetroxide, and/or uranyl acetate depending on the application (for a detailed review of methodology see McDonald, 2007; see also Chapters 4, 6, 15, 11, 25, and 26 of this volume). When compared to standard glutaraldehyde/OsO_4 fixation resulting in a 70% loss of proteoglycans (Engfeldt and Hjertquist, 1968), cryofixation followed by freeze substitution in methanol/glutaraldehyde results in a loss of only 2.2% proteoglycan (Hunziker and Herrmann, 1987). Furthermore, as compared to chemical fixation, proteoglycans are not precipitated; instead they contribute to a homogeneous gel-like substance in which collagen fibrils and cells are embedded (Keene and McDonald, 1993). Cartilage prepared by HPF followed by freeze substitution demonstrates a preservation quality of chondrocytes far superior to that of any chemical fixation routine, though preservation across large regions of matrix is rarely uniform. Vitrification of articular cartilage matrix, as evaluated by electron diffraction of frozen hydrated sections, has been shown to be zone-dependent and appears to correlate proportionally with proteoglycan concentration and inversely with water content (Hunziker *et al.*, 1996). When ice crystals begin to form and grow, matrix components become segregated into phases of water and concentrated biological residues, a phenomenon known as "segmentation". The importance of recognizing these ice crystal damage artifacts, particularly in cartilage matrix, cannot be overstated since the segregation pattern mimics that of native proteoglycan structure (Arsenault, 1994; Engfeldt *et al.*, 1994; Hunziker, 1993; Hunziker *et al.*, 1984; Keene and McDonald, 1993). In regions demonstrated to be free from ice crystal damage, collagen fibrils are seen to be embedded in a finely granular matrix with no filamentous structures (Fig. 9A). In progressively damaged regions, segregation artifacts can be recognized from mild to severe (Fig. 9B–D). With regard to chondrocyte structure, again the results obtained by cryostabilization are far superior to any chemical fixation protocol. Intimate contact between chondrocyte plasmalemmata and the matrix and elimination of halos surrounding chondrocytes is a universal feature among chondrocytes independent of their position within the growth plate. Nuclear membranes, even within hypertrophic chondrocytes are preserved intact. The fine structure of rough endoplasmic reticulum is exquisite, with ribosomes present as discrete, evenly spaced entities distributed along unbroken reticular membrane. Microtubules and microfilaments are linear and well contrasted. Polysomes are commonly imaged in tangential section. Cisternae membranes within Golgi organelles are clearly demarcated. Membrane-bound organelles including lysosomes, secretory granules, and vesicles can be differentiated (Figs. 9 and 10; also see Suppl. Fig. B) (Hunziker *et al.*, 1984). In addition to the characteristics of the tissue itself, there are many factors affecting the outcome of samples prepared by this methodology. In our experience utilizing either the Baltec HPM 010 or an early version of the Leica EMPact, our more successful freezing runs have been gained by (1) utilizing the smallest volume freezing planchet necessary to just barely contain the sample; (2) surrounding the tissue in a freezing media consisting of 50% fish gelatin/50% hexanes, (3) utilizing an acetone freeze substitution protocol including 2.0% OsO_4 and 0.1% uranyl acetate (UA); and (4) embedding in epon. The importance of a successful freeze substitution method cannot be overstated. After pursuing many variations in protocol, the method we have settled on is as follows: frozen samples are transferred into the automated freeze substitution (AFS) device under liquid nitrogen. Using LN2-cooled forceps, the samples are quickly dropped into freeze substitution media held at a temperature just above its freezing point (−90°C). The freeze substitution media we use for best preservation of ultrastructure consists of 12.5 ml 2% osmium

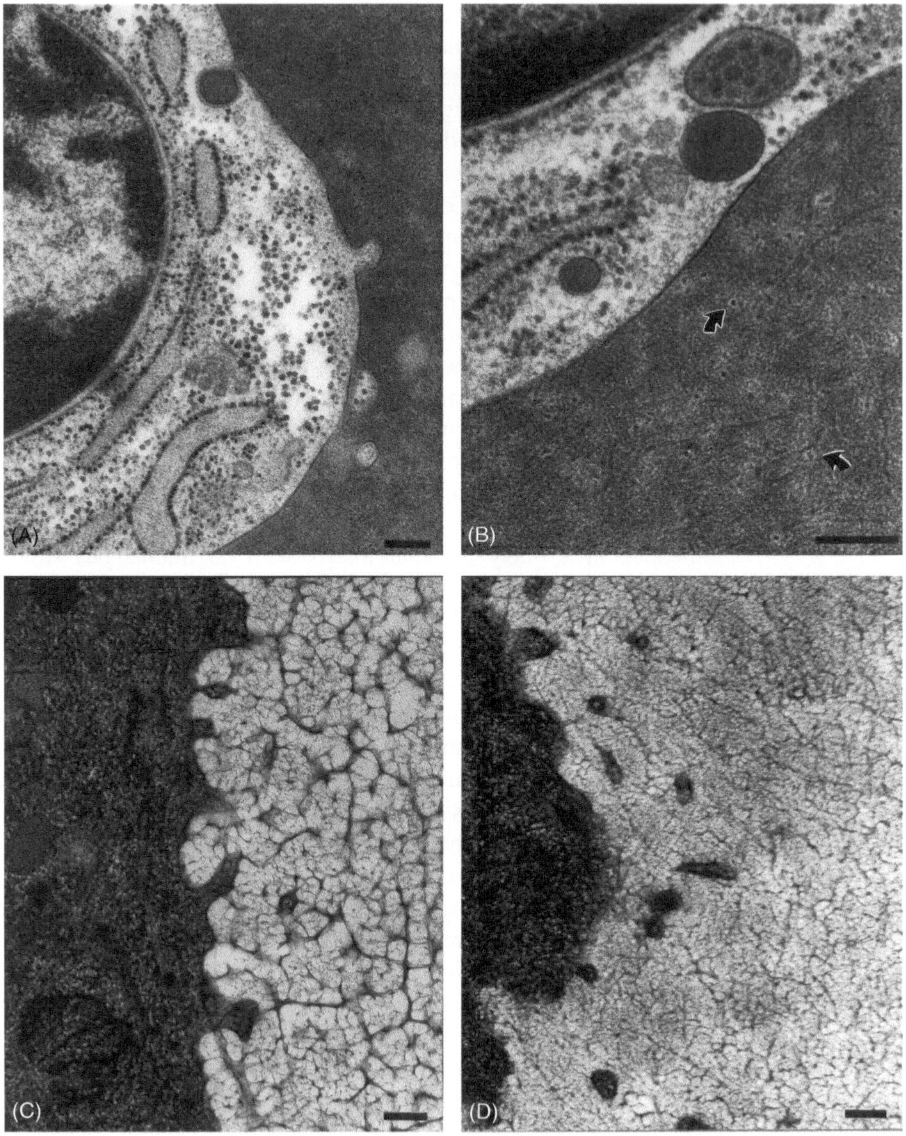

Fig. 9 High-pressure freezing/freeze substitution of 17-day fetal chick sterna: artifacts may be present in samples prepared by high-pressure freezing and freeze substitution. (A) Occasionally, extraction within the chondrocyte cytosol is evident. (B) "Halos" may surround collagen fibrils in the matrix in slightly damaged areas, and the matrix may also appear somewhat granular. (C) In severely damaged areas a meshwork of thick and thin filaments traverses the matrix, often accompanied by deterioration of chondrocyte morphology. (D) In moderately damaged areas a similar but finer network may be apparent. Bars = 200 nm. From Keene and McDonald, 1993. *J. Histochem. Cytochem. 41, 1121, 1141–1153*, with permission.

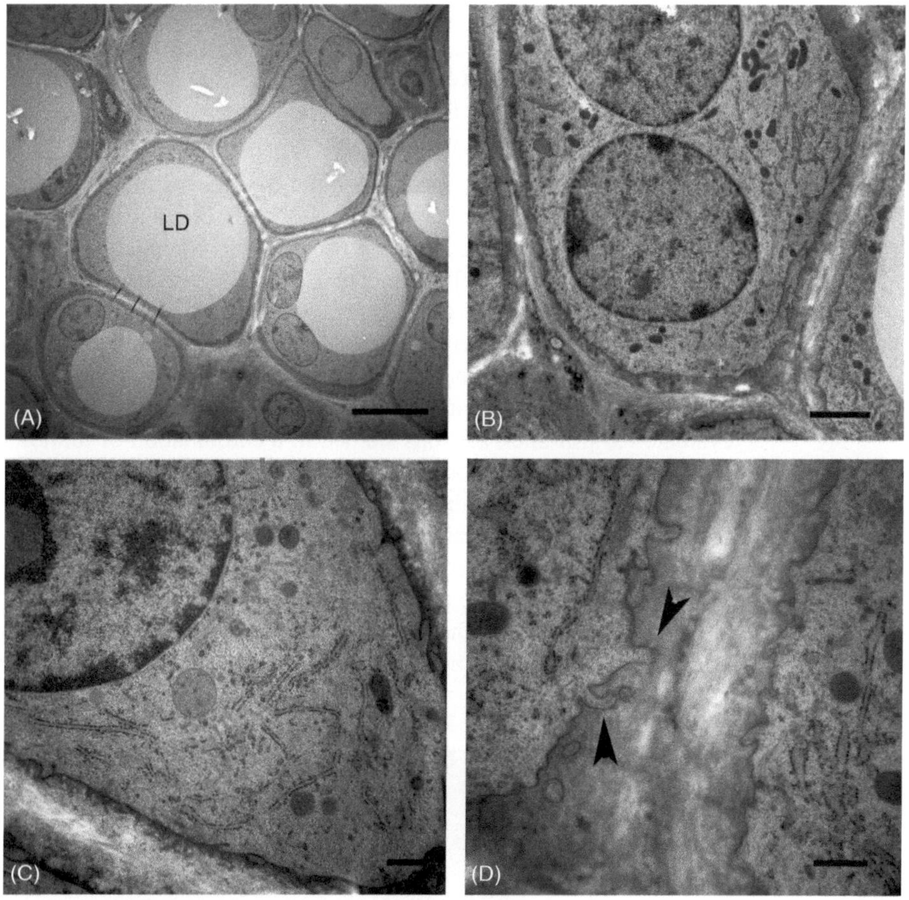

Fig. 10 Adult mouse ear high-pressure frozen and freeze substituted in acetone with 2% OsO_4 and 0.1% UA, then embedded in Spurrs epoxy. (A) Excellent preservation of spherical lipid droplet vesicles (LD) cell profiles is present in a large region. (B) Intracellular organelles including nuclear membranes, RER, and a variety of well-defined organelles (C, D) is unsurpassed by any chemical protocol. Cell borders (D) are in intimate contact with the elastin-rich matrix. Bars: A, 10 μm; B, 2 μm; C and D, 500 nm.

tetroxide in 100% acetone to which is added 250 μl of 5% UA in 100% methanol. Temperature and duration are as follows:

1. −90°C for 48 h
2. Warm to −80°C over 10 h
3. −80°C for 48 h
4. Warm to −50°C over 10 h
5. Replace freeze substitution media with 100% acetone at −50°C
6. Warm to −20°C over 3 h; rinse again in 100% acetone for 1 h
7. Replace 100% acetone with 1:1 acetone:Spurrs epon

8. Warm to 0°C over 3 h
9. Warm to +20°C over 1 h (in the AFS to prevent condensation)
10. 1:2 acetone:Spurrs. Remove from AFS and release the tissue from freezing hats (if necessary). Infiltrate for 1 h
11. Continue infiltration with epon over 4–5 h and embed

There are many potential variables in the freeze substitution protocol. Our protocol is conservative, but it does work for cartilage. For a comprehensive review of the practical aspects of tissue handling and solution composition, see McDonald (2007).

When evaluating HPF/FS samples in the TEM, even within the same tissue section there may be exceptionally well-preserved regions and also heavily damaged regions. Finding these well-preserved regions can be dependent on careful examination of several regions within individual blocks. Well-frozen tissue fragments of cartilage usually appear translucent when observed in epon using a stereoscope. Usually, we evaluate 1.0-μm-thick sections stained in a combination of toluidine blue O and basic fuchsin in the light microscope before collecting ultrathin sections from the same tissue block. In Koehler illumination, chondrocytes with minimal damage appear more rounded, larger, and considerably more translucent than damaged cells (Fig. 11). In our experience, elastic cartilage of mouse ear freezes exceptionally well (Fig. 10). This particular success may be due in part to the ease in which ear tissue can be handled, as it can be removed from the animal using a punch of the same diameter as the sample holder and inserted, full-thickness, into the holder. Likely due to its high water content,

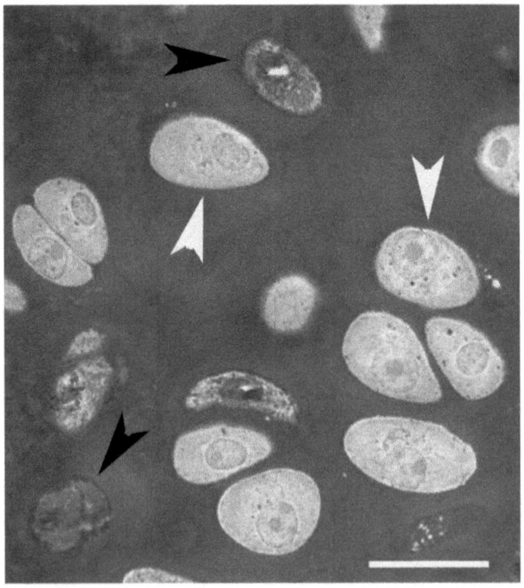

Fig. 11 A 1-μm-thick epoxy section of high-pressure frozen, freeze-substituted human cartilage as observed in the light microscope. This approach allows identification of potentially well-frozen chondrocytes (white arrows) and poorly preserved chondrocytes (black arrows). Bar = 10 μm. (See Plate no. 29 in the Color Plate Section.)

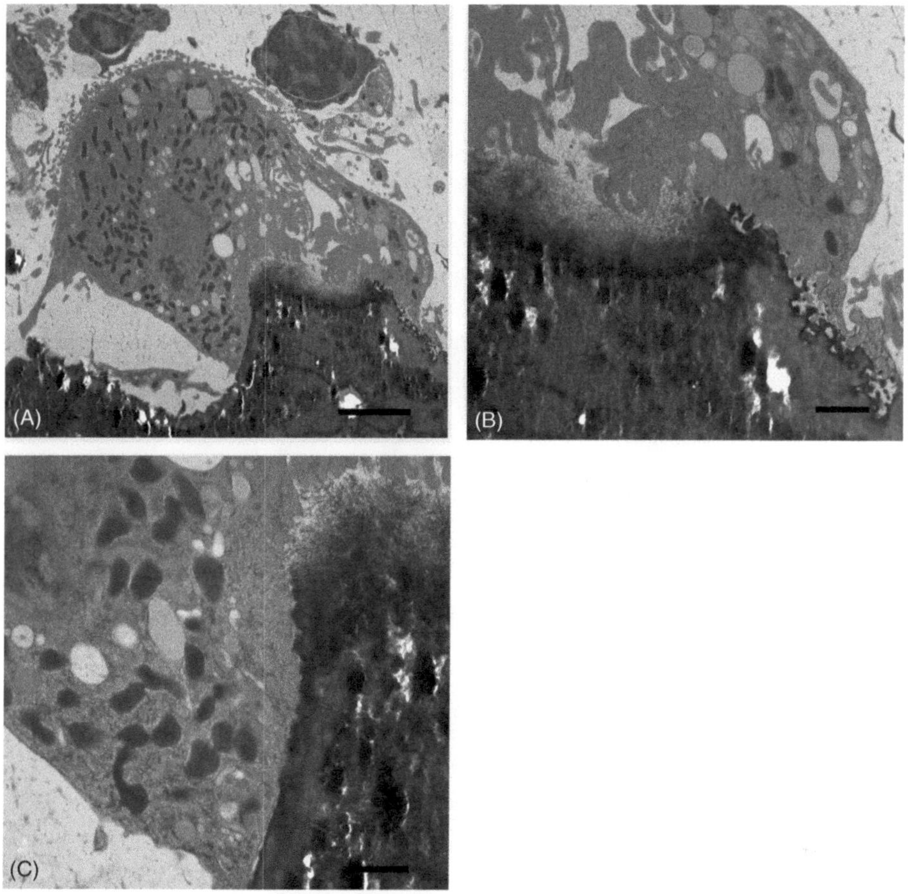

Fig. 12 Neonate mouse bone prepared by high-pressure freezing/freeze substitution in acetone containing 1% OsO_4 with 0.1% UA and embedding in Spurrs epoxy. Immature red blood cells (A, top) and an active osteoclast are well preserved. Higher magnification (B and C) reveals the sealing zones (at the leading edges of the cell/bone interface and the active resorptive zone). Bars: A, 4 μm; B, 1 μm; C, 500 nm.

cartilage matrix appears to be one of the more difficult tissues to successfully cryopreserve; however, adjacent regions within the bony matrix often demonstrate excellent preservation (Figs. 12 and 13).

E. Fixation in Glutaraldehyde-Based Solutions

Dilute 3% glutaraldehyde/3% paraformaldehyde supplied in 0.1 M cacodylate (Tousimis) with an equal part DMEM (Cellgro), hereafter designated as "primary fixative". Use DMEM for all rinses prior to dehydration. Additives to the primary fixative may include 0.05% w/v tannic acid (mixed well with significant agitation),

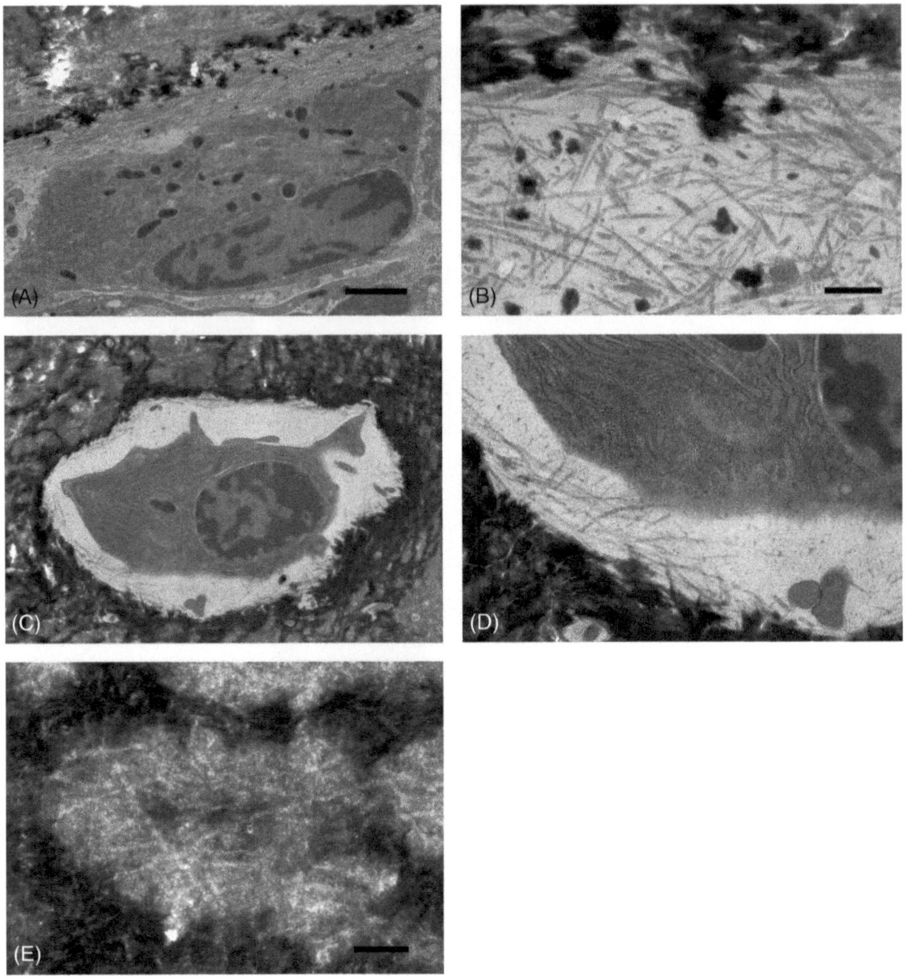

Fig. 13 Neonate mouse bone prepared by high-pressure freezing/freeze substitution in acetone containing 1% OsO_4 with 0.1% UA and embedded in Spurrs epon. (A) Osteoblast. (B) Osteoblast/bone interface. (C) Osteocyte. (D) Osteocyte/bone interface. (E) Bone matrix. Bars: A and C, 2 μm; B and D, 500 nm; E, 200 nm.

which will result in greater contrast of collagen fibrils. Use 0.05% cupromeronic blue and include up to 0.3 M magnesium chloride for localization of proteoglycans (fixation is for 17 h and omit osmium postfixation. See earlier discussion in Section C). For retention of proteoglycan facilitated by ruthenium, 0.6% RHT is added to the primary fixative, to the buffer wash following primary fixation, and also to the secondary fixative. The secondary fixative is 1% osmium tetroxide made by diluting 2% osmium tetroxide in water 1:1 with DMEM. For contrasting glycogen, add 0.05 M potassium ferrocyanide to the secondary fixative or to the secondary fixative containing 0.6% RHT (this will form a dense ugly black solution; do not worry).

1. Primary fixation is for 2 h at ambient temperature.
2. Rinse three times over 15 min in DMEM.
3. Fix for 1 h in 1% OsO_4 (potentially with additives).
4. Rinse 3× over 15 min in DMEM.
5. Dehydrate in a graded ethanol series (30, 50, 70, 90, 95%) each at 15 min; 100% ethanol 3× over 30 min; 100% propylene oxide 3× over 30 min.
6. Epon infiltration: 1:1 propylene oxide:epon 30 min; then 1:2 propylene oxide:epon, 1:3 propylene oxide:epon, then three changes of 100% epon for 1 h each.

We prefer Spurrs epoxy because of its low viscosity and reasonable cutting characteristics. A successful recipe with a good balance of hardness and cutting characteristics is ERL 4221 = 20.5 g; DER 736 = 7.15 g; NSA = 29.5 g; DMAE = 0.5 ml. Mix this volume for at least 30 min. On occasion we have used microwave energy to facilitate the steps inclusive from osmium fixation through epon infiltration; however, we experience slight extraction of chondrocytes with our protocol and have not yet determined if we might improve the results with some modification. Modern histological microwave ovens are significantly different from those present in the home: they have true variable power output (as apposed to cycling the magnetron on and off), they are ventilated so that noxious fumes are removed from the inside of the oven, hot-spots within the chamber are virtually eliminated, and sample cooling via an external water recirculator insures that the samples do overheat. The times and power levels used will depend on the manufacturer and model of the oven. Ours is a Pelco model 3450 microwave oven with a variable power accessory (MW power controller 3430), and we use a Pelco "cold spot" through which 10°C water is circulated by way of a Beckman closed loop water recirculator. The sample containers and solution volumes influence the outcome. With our microwave, we contain the samples in clear 2 ml eppendorf tubes with 600 μl of the following solutions, using times and power settings as follows:

1. 1% OsO_4: power setting 94 W. 2 min on, 2 min off, 4 min on, 5 min off, 2 min on, 2 min off, 4 min on
2. DMEM rinse: ×3 over 15 min on the bench (no MW)
3. 30, 50, 70, 95% EtOH: 94 W, 60 s each
4. 100% EtOH: 94 W, 60 s; replace EtOH. 94 W, 60 s
5. 100% propylene oxide. 94 W, 60 s; replace propylene oxide. 94 W, 60 s
6. 1:1 then 1:3 propylene oxide:epon. 250 W, 15 min each
7. 100% epon. Three changes. 250 W, 15 min each
8. Embed and polymerize (outside the microwave) at 70°C overnight.

F. Immunocytochemical Methods

An overall theme in this discussion of cartilage and bone is their inherent high density, not only due to closely packed collagen fibrils and mineralized matrix (in bone) but also due to proteoglycan. Successful immunoelectron microscopy therefore depends either on exposing epitopes on the surface of sections to antibody binding (on-section or surface labeling) or on selectively solubilizing the proteoglycan network, allowing the penetration of antibodies into the matrix (en bloc labeling). Exposing epitopes for on-section surface labeling may be accomplished by a variety of techniques. Poole *et al.* (1982) exposed proteoglycan, link protein, and collagens by working with 5 μm cryostat sections cut from articular cartilage frozen in Tissue-Tek O.C.T., which were mounted onto thermanox cover slips. These were then carried through an immuno-peroxidase procedure, fixed in glutaraldehyde and osmium tetroxide, and then embedded in epoxy. Although ice damage resulted in poor cellular structure, antibody labeling of the proteoglycan aggregates gave a pattern strikingly similar to that following glutaraldehyde and osmium fixation in the presence of RHT.

1. On-Section Labeling

More commonly, specimens are prepared for section surface labeling by lightly fixing the tissue and then embedding in resin. As in other organ systems, successful immunocytochemistry is dependent on many factors. Some antigens can still be localized even after exposure to concentrated fixatives and embedding in epon. For example, antibodies recognizing elastin (a component of perichondrium) successfully localize to sections cut from tissues fixed in 1.5% glutaraldehyde and 1% OsO_4, and embedded in Spurrs. Other antigens are unsuccessfully localized no matter how mild the preparation conditions. Among the more successful resins are Lowicryls (Armbruster *et al.*, 1982; Carlemalm *et al.*, 1985), which can be polymerized using UV light at temperatures at or below −20°C. In order to avoid (as much as possible) extraction of soluble tissue components during dehydration, tissues are exposed to increasing concentrations of dehydrates (ethanol or methanol) at progressively lower temperatures (PLT procedure), beginning with 30% alcohol at 0°C, then 50% alcohol at −10°C lowered to −20°C, and proceeding to more concentrated alcohols at temperatures as low as −50°C, taking care not to fall below the freezing point of residual solvent still residing in the tissue (Hobot, 1989; Robertson, 1992). In our experience, PLT dehydration and infiltration is very important in the preservation of both structure and antigenicity. This is true also for tissue embedded in LR White and polymerized in the oven at 60°C. Finding the best conditions for retaining antigenicity and good ultrastructure in preparation for section surface labeling is empirical, with several good choices for embedding media. McKee and coworkers localized osteopontin with colloidal gold on the surface of sections cut from bone fixed in 0.1% glutaraldehyde/4% paraformaldehyde embedded in Lowicryl K4M or LR White, and published many panels of micrographs depicting the quality of ultrastructural preservation and localization (McKee and Nanci, 1996). They also recognized that structures in thin sections originally containing mineral ultimately become decalcified after exposure to the aqueous solutions used during immunocytochemistry. There are also many potential fixatives that may be tried in an attempt to balance acceptable ultrastructure with good immunolabeling density. In our laboratory, we commonly use 0.1% glutaraldehyde/4% paraformaldehyde in DMEM, and our "go to" embedding media for section surface

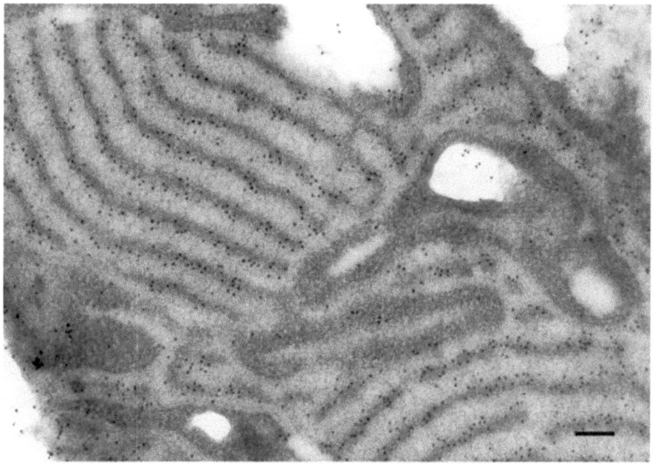

Fig. 14 Section surface immunolocalization of cartilage oligomeric protein (COMP) in pseudoachondroplasia cartilage chondrocyte. COMP localizes to finger print-like stacks of rough endoplasmic reticulum within chondrocytes in this dwarfing disorder (Maddox, 1997). Exposure of intracellular epitopes on the surface of sections cut from tissue fixed in 0.1% glutaraldehyde/4% paraformaldehyde and embedded in LR White after PLT dehydration is the most effective way to access this intracellular compartment. Bar = 100 nm.

immunolabeling is LR White (Fig. 14). We incorporate PLT dehydration facilitated by a Leica AFS device, with the protocol as follows:

1. Fix in 0.1% glutaraldehyde/4% paraformaldehyde in DMEM for 30 min on ice.
2. Rinse in ice cold DMEM 3× over 15 min.
3. Rinse in 0.15 M Tris-HCl, pH 7.5 with three changes for a minimum of 4 h at 4 °C.
4. 30% ethanol at 0°C and ending at −10°C for 15 min.
5. 50% ethanol beginning at −10°C and ending at −20°C for 15 min.
6. 70% EtOH at −20°C for 15 min.
7. 90% EtOH at −20°C for 15 min.
8. 90% EtOH (not 100% EtOH) is combined with LR White in a ratio of 1:1; 1:2, and 1:3 for 60 min each.
9. Two changes of 100% LR White at −20°C for 60 min each.
10. A last change of LR White beginning at −20°C and slowly warmed to ambient temperature over 30 min.

Care must be taken not to allow condensation within the specimen vials as solutions are changed. When polymerizing LR White, do not use silicon molds and do not include a paper label. We use polyethylene Wheaton "snap caps" as embedding chambers. These should be stored in the polymerization oven for several days prior to use to eliminate moisture trapped within the plastic. A disk cut from aclar film, placed at the bottom of the cap (with the samples and media above) will separate easily from polymerized LR White and facilitate crystal-clear viewing of the embedded samples. The presence of aclar also insures even polymerization of LR White, which may not otherwise be complete at the media/cap interface. *It is important to polymerize LR White in an oxygen-free atmosphere*, which may be facilitated by including a 1000-ml beaker freshly filled with dry ice, which sublimates to gaseous CO_2 in the 60°C oven during initial polymerization. Embedded samples in polymerized media are cut free from the surrounding media using a small vice and a spiral jeweler saw blade, which will cut in any direction.

Our second choice embedding media is Lowicryl K4M, with every step the same as for LR White embedding except the tissue is fully dehydrated to 100% EtOH and the media is polymerized in a nitrogen or CO_2 atmosphere at −20°C using UV light. A CO_2 atmosphere may be obtained in a closed but slightly vented chamber using dry ice. Alternatively, a nitrogen atmosphere is created in the AFS by slightly lowering the temperature of the chamber just after the last sample is added (this will bring fresh nitrogen into the chamber), then raising it again to reach the polymerization temperature.

Ultrathin sections are mounted onto formvar-coated 1 mm × 2 mm Ni slot grids and floated on 30 μl pools confined within the depressions of a multi-well Teflon plate (Ted Pella). The following solutions are all at ambient temperature:

1. 2.0% nonfat dry milk with 0.5% fish gelatin and 0.5% ovalbumin in 0.15 M Tris-HCl pH 7.5 for 30 min
2. Three quick changes in Tris-HCl
3. Primary antibody diluted 1:10 in Tris-HCl with 0.2% BSAc (Aurion) for 2 h
4. Three changes of Tris-HCL over 20 min
5. Appropriate 5 or 10 nm gold secondary conjugate diluted 1:9 in Tris-HCL with 0.2% BSAc for 60 min
6. Three rinses in Tris over 5 min
7. Three dH_2O rinses over 15 min, dry, then stain in uranyl acetate followed by Reynolds lead citrate

2. En bloc Immunoelectron Microscopy Protocol

Fresh tissue is kept moist with DMEM and trimmed with disposable double-edge razor blades and (if bony) disposable microtome blades so that one dimension of the tissue block is less than or equal to 1 mm. Rinse the sample briefly in DMEM, then incubate in a volume exceeding 10× that of the tissue in 0.25 units/ml chondroitinase ABC in DMEM for 2 h to overnight at ambient temperature. Samples are then rinsed 3× over 20 min in large volumes of DMEM, then incubated in 150 μl primary antibody diluted no more than 1:5 in DMEM overnight at 4°C. Importantly, we use ~ 4 ml borosilicate tubes (dimension O.D. 10 mm × L 75 mm) for all chondroitinase and antibody-labeling steps. Be sure that the tissues sink to the bottom of the tubes during all incubations. The tissues are gently agitated by keeping the tubes nearly vertical in a partitioned box on the platform of a small rotator (Sarstedt Sarmix GM1), which easily fits in even a small refrigerator. Following incubation in primary antibody, the samples are rinsed 3× over a minimum of 4 h in DMEM, then incubated in gold conjugated secondary antibody diluted 1:3 in DMEM, again on a rotator overnight at 4°C. The size of the colloidal gold conjugate is critical. 1 nm gold conjugates will diffuse into the tissue farthest and will also penetrate basement membranes; 5 nm will not penetrate basement membranes; 10 nm colloidal gold will penetrate chondroitinase-digested cartilage only a few tens of microns, and 20 nm conjugates are not useful at all. Gold conjugates with a size of 1 nm need to be "enhanced", a protocol designed to precipitate metal onto the 1-nm gold particulate, effectively making it visible in the TEM. We use a gold intensification kit (Nanoprobes, Inc.) as follows:

1. Tissues are confined within 4 ml glass borosilicate tubes and rinsed in DMEM.
2. Tissues and all gold enhancement components are cooled on ice.
3. Equal volumes of component A and B are combined and chilled on ice for 15 min, after which equal parts of C and D are added to solution AB.
4. Tissues are incubated in the ABCD solution for 15 min on ice, after which the tubes are transferred to a 25°C water bath for exactly 5 min. The idea here is that the gold enhancement solutions are allowed to penetrate the tissue at a low-enough temperature to inhibit any precipitation of gold. Warming the tissue quickly after low-temperature diffusion allows the enhancement to proceed uniformly within the tissue volume.
5. Samples are then rinsed in three quick changes of ice cold DMEM.
6. Samples may then be immersed in primary fixative, rinsed, and also immersed in secondary fixative including osmium tetroxide.

G. Specific Issues in Working with Calcified Cartilage and Bone

Decalcification (demineralization) of calcified cartilage and bone is often undertaken to soften the tissue for subsequent segmentation and ultramicrotomy. This is particularly important for densely mineralized tissues, such as mature long bones and teeth. However, for developmental studies of small animals (embryonic or neonatal mice and chicks), decalcification is often undesired and may not be necessary. In our laboratory, we cycle three 3.5 mm ultra-diamond knives so that one is away for re-sharpening or is new, one is mostly unscratched and used only for soft tissues, and a third has scratches to the point that it is unacceptable for soft tissues but sufficient for mineralized tissues. Our ultrathin sections of calcified tissues will often have knife marks, but we are willing to trade these defects for the ability to cut hard tissues. We cycle diamond histo-knives similarly. In our laboratory, it takes about 8 months for a knife to be fully cycled. Care is taken to trim bony regions away if we are focused only on the cartilage, and we orient our tissue so that we are cutting parallel to the axis of a long bone whenever possible. Cross sections of bones and growth plates will result in many more knife defects. If a project involves more mature long bones or teeth, demineralization following perfusion fixation is necessary. Acids, including formic, acetic, and citric, may be used; however, the chelating agent disodium EDTA seems the best choice for overall tissue structure and reliable staining (Callis and Sterchi, 1998). The time frame for decalcification can be measured in days, weeks, and months, depending on the type and size of the mineralized tissue being processed. First reported by Hellstrom and Nilsson (1992), one mechanism to significantly shorten the time for chelation by EDTA is to utilize microwave energy during demineralization. The use of a modern histological microwave, discussed earlier, is imperative to safety and to reproducible results. As reported by Tinling *et al.* (2004), a closed loop recirculator may be used to bathe bone samples in continually flowing EDTA at 20°C during microwave irradiation within the now current Pelco Laboratory Microwave (Pelco BioWave). The results demonstrated a microwave-assisted demineralization time (for 1 mm cross sections of adult gerbil femur) of 33 h as compared to 45 h for nonmicrowave demineralization *at the same temperature*. Although the mechanism of microwave-assisted decalcification is not fully determined, it is clearly not due to heating. In our own laboratory, we use a temperature probe (in an equal volume of water adjacent to the samples; do not put the probe into EDTA) to regulate temperature to an upper limit of 40°C. A closed loop recirculator set to 10°C circulates water through a Pelco "cold spot" placed on the floor of the microwave oven. Upon this surface, samples are confined in scintillation vials and submersed in 10 ml of (0.2 M EDTA in 50 mM Tris adjusted to pH 7.9 *with NaOH pellets*). Continual agitation of the samples during microwave decalcification is accomplished by bubbling air into the sample container. Perfusion fixed, intact adult mouse long bones can be decalcified within 48 h via the protocol that follows. The power setting on the microwave is maintained at 750 W throughout the entire procedure:

1. Four changes of EDTA, 90 min each
2. Fresh EDTA, overnight
3. Four changes of EDTA, 90 min each
4. Fresh EDTA, overnight

IV. Materials

Instrumentation

High-pressure freezer: Our Leica EM PACT has been replaced with the EMPACT2, Leica Microsystems

Automated freeze substitution device: Our Leica AFS has been replaced with the AFS2, Leica Microsystems

Controlled device for vitrification of aqueous samples suspended on grids: FEI Vitrobot Mark IV, FEI, Hillsboro, OR

Tissue homogenizer: Our Omni 2000 has been replaced by Omni TH, Omni International, Kennesaw, GA

Materials

Multi-well teflon plate: Ted Pella, Redding, CA # 10526-1

Disposable microtome blades: Accu-Edge 4685, Electron Microscopy Sciences, Hatfield, PA

Double-edge razor blades: Ted Pella, Redding, CA # 121-6

Single-hole copper–palladium 1×2 mm slot grids: Electron Microscopy Sciences, Hatfield, PA #G3010-CP (special order)

Glove bag inflatable glove chamber: Cole-Parmer, Vernon Hills, IL # EW-04408-38

Thermanox cover slips: Nalge Nunc International # 174950; Fisher Scientific # 12-565-88

Wheaton polyethylene "snap caps": Wheaton 242612; Fisher Scientific, 06450201

Aclar film: Ted Pella, Redding, CA # 10502

Spiral jewelers saw blade size 2/0 (cuts in any direction): Electron Microscopy Sciences, Hatfield, PA # 72015-02

4 ml borosilicate tubes (dimension O.D. 10 mm × L 75 mm): VWR, Batavia, IL # 89000-476

20 ml scintillation vials: VWR Scientific 66022-004

Specimen rotator (fits in refrigerator): Sarstedt (Newton, NC) Sarmix GM1 # 90.181.401

Reagents

Cupromeronic blue: Sigma # C1234

Sodium tungstate dihydrate: Sigma # 72069

Hexanes: $CH_3(CH_2)_4CH_3$, EMD Chemicals, Gibbstown, NJ # HX0299

Phosphotungstic acid (PTA): Electron Microscopy Sciences, Hatfield, PA # 19500

Ruthenium hexammine trichloride (RHT): Polysciences, Inc. Warrington, PA cat # 17253

Dulbecco's Modification of Eagle's Medium (DMEM): Cellgro, Manassas, VA # 10-013CV

Potassium ferrocyanide: Electron Microscopy Sciences, Hatfield, PA # 13746-66-2

Chondroitinase ABC: Sigma #C2905

3% glutaraldehyde/3% paraformaldehyde: Supplied in 0.1 M cacodylate buffer; 10 ml glass ampoules sealed with a nitrogen cap: Tousimis, Rockville, MD # 1010A

Tannic acid: Electron Microscopy Sciences, Hatfield, PA # 10502

Disodium ethylenediamine tetraacetate (EDTA): Fisher Scientific # S-311

Tris-HCl: Electron Microscopy Sciences, Hatfield, PA #11720

Epoxy tissue stain composed of toluidine blue O and basic fuchsin: EMS #14950

Spurr embedding kit: Ladd Research, Williston, VT # 21230

LR White embedding media: Ted Pella #18181

Immunocytochemicals

Nonfat dry milk: Carnation brand; your neighborhood grocery store

Cold water fish gelatine: Electron Microscopy Sciences, Hatfield, PA #25560

Ovalbumin: Sigma A5503-5G

Bovine serum albumin: Electron Microscopy Sciences, Hatfield, PA # 25554

Aurion 5 nm secondary gold conjugates: Electron Microscopy Sciences, Hatfield, PA

Gold intensification kit: Nanoprobes, Inc., Yaphank, NY GoldEnhance # EM 2113

V. Concluding Remarks

Cartilage is undeniably one of the more difficult tissues to adequately stabilize for TEM. Perfusion fixation is typically used as a reliable chemical fixation method for tissues having substantial vasculature; however, blood supply to cartilage is so limited that perfusion offers little advantage. When considering fixation options for cartilage, it is important to adequately stabilize proteoglycans if one hopes to image chondrocytes with reasonable resolution. This may be accomplished with the addition of RHT in both the primary and secondary fixatives or, better yet, by cryostabilization facilitated by HPF/FS. But neither of these methods allows the visualization of collagen fibrils within the cartilage matrix; for this evaluation the proteoglycan moiety must be extracted, which surprisingly is facilitated in by standard glutaraldehyde/osmium tetroxide fixation protocols. Tannic acid is added to glutaraldehyde, gaining additional contrast to collagen fibrils. The specialized chemical methods for cartilage stabilization are no more difficult or time consuming than standard fixation routines; however, your choice in initial chemical treatment will always result in poor visualization of one or more important tissue components. Being aware of the consequences of these fixation protocols is therefore critical to the goals sought in the final evaluation of tissue architecture.

References

Aebi, U., and Pollard, T. D. (1987). A glow-discharge unit to render electron microscope grids and other surfaces hydrophilic. *J. Electron Microsc. Tech.* **1**, 29–33.

Al-Amoudi, A., Chang, J. J., Leforestier, A., McDowall, A., Salamin, L. M., Norlen, L. P., Richter, K., Blanc, D., Studer, D., and Dubochet, J. (2004). Cryoelectron microscopy of vitreous sections. *EMBO J.* **23**, 3583–3588.

Anderson, H. C., and Sajdera, S. W. (1971). The fine structure of bovine nasal cartilage: Extraction as a technique to study proteoglycans and collagen in cartilage matrix. *J. Cell Biol.* **49**, 650–663.

Armbruster, B. L., Carlemalm, E., Chivovetti, R., Garrabito, R. M., Hobot, J. A., Kellenberger, E., and Villiger, W. (1982). Specimen preparation for electron microscopy using low temperature embedding resins. *J. Microsc.* **126**, 77–86.

Arsenault, A. L. (1994). Image analysis of the extracellular matrix. *Microscopy Res. Tech.* **5**, 409-421.

Borysko, E. (1955). Ph.D. Thesis, The Johns Hopkins University.

Buckwalter, J. A., and Rosenberg, L. C. (1988). Electron microscopic studies of cartilage proteoglycans. *Electron Micros. Rev.* **1**, 87–112.

Callis, G., and Sterchi, D. (1998). Decalcification of bone: Literature review and practical study of various decalcifying agents, methods, and their effects on bone histology. *J. Histochem.* **21**, 49–58.

Cameron, D. A., and Robinson, R. A. (1958). Electron microscopy of epiphyseal and articular cartilage matrix in the femur of the newborn infant. *J. Bone Joint Surg.* **40**, 163–170.

Carlemalm, E., Villiger, W., Hogot, J. A., Acetarin, J. D., and Kellenberger, E. (1985). Low temperature embedding with lowicryl resins: Two new formulations and some applications. *J. Microsc.* **140**, 55–63.

De Burijn, W. C. (1973). Glycogen: Its chemistry and morphologic appearance in the electron microscope. I. A modified OsO_4 fixative which selectively contrasts glycogen. *J. Ultrastruct. Res.* **42**, 29–50.

Engfeldt, B., and Hjertquist, S. O. (1968). Studies on the epiphysial growth zone I: The preservation of acid glycosaminoglycans in tissues in some histotechnical procedures for electron microscopy. *Virchows Arch. B. Cell Pathol.* **1**(3), 222–229.

Engfeldt, B., Reinholt, F. P., Hultenby, K., Widholm, S. M., and Muller, M. (1994). Ultrastructure of hypertrophic cartilage: Histochemical procedures compared with high pressure freezing and freeze substitution. *Calcif. Tissue Int.* **55**, 274–280.

Gilkey, J. C., and Staehelin, A. (1986). Advances in ultrarapid freezing for the preservation of cellular ultrastructure. *J. Electron Microsc. Tech.* **3**, 177–210.

Gregory, K. E., Keene, D. R., Tufa, S. F., Lunstrum, G. P., and Morris, N. P. (2001). Developmental distribution of collagen type XII in cartilage: Association with articular cartilage and the growth plate. *J. Bone Miner. Res.* **16**, 2005–2016.

Hardingham, T. E. (1986). Structure and biosynthesis of proteoglycan. *Thematology* **10**, 143–183.

Hascall, V. C. (1997). Interaction of cartilage proteoglycans with hyaluronic acid. *J. Supramol. Struct.* **7**, 101–120.

Heinegard, D., and Paulsson, M. (1987). Cartilage. *In* "Methods in Enzymology" (L. W. Cunningham, ed.) Vol. 145, pp. 336–363. Academic Press, San Diego.

Hellstrom, S., and Nilsson, M. (1992). The microwave oven in temporal bone research. *Acta Oto-Laryngoloica, Suppl.* **493**, 15–18.

Hobot, J. A. (1989). Lowicryls and low temperature embedding for colloidal gold methods. *In* "Colloidal Gold: Principles, Methods, and Applications" (M. A. Hyatt, ed.) Vol. 2, p. 75. Academic Press, San Diego.

Hunziker, E. B. (1993). Application of cryotechniques in cartilage tissue preservation and immunoelectron microscopy: Potentials and problems. *Microscopy Res. and Tech.* **24**, 457–464.

Hunziker, E. B., and Herrmann, W. (1987). In situ localization of cartilage extracellular matrix components by immunoelectron microscopy after cryotechnical tissue processing. *J. Histochem. Cytochem.* **35**, 647–655.

Hunziker, E. B., Herrmann, W., and Schenk, R. K. (1982). Improved cartilage fixation with ruthenium (III) hexammine trichloride (RHT). *J. Ultrastruct. Res.* **81**, 1–12.

Hunziker, E. B., Herrmann, W., and Schenk, R. K. (1983). Ruthenium hexammine trichloride (RHT)-mediated interaction between plasmalemmal components and pericellular matrix proteoglycans is responsible for the preservation of chondrocytic plasma membranes in situ during cartilage fixation. *J. Histochem. Cytochem.* **31**, 717–727.

Hunziker, E. B., Herrmann, W., Schenk, R. K., Mueller, M., and Moor, H. (1984). Cartilage ultrastructure after high pressure freezing, freeze substitution, and low temperature embedding. I. Chondrocyte ultrastructure—implications for the theories of mineralization and vascular invasion. *J. Cell Biol.* **98**, 267–276.

Hunziker, E. B., Ludi, A., and Herrmann, W. (1992). Preservation of cartilage matrix proteoglycans using cationic dyes chemically related to ruthenium hexammine trichloride. *J. Histochem. Cytochem.* **40**, 909–917.

Hunziker, E. B., Wagner, J., and Studer, D. (1996). Vitrified articular cartilage reveals novel ultra-structural features respecting extracellular matrix architecture. *Histochem. Cell Biol.* **106**, 375–382.

Iancu, C. V., Tivol, W. F., Schooler, J. B., Dias, D. P., Henderson, G. P., Murphy, G. E., Wright, E. R., Li, Z., Yu, Z., Briegel, A., Gan, L., He, Y., *et al.*, (2006). Electron cryotomography sample preparation using the vitrobot. *Nat. Protoc.* **6**, 2813–2819.

Kashiwa, H. K., Luchtel, D. L., and Park, H. Z. (1975). Chondroitin sulfate and electron lucent bodies in the pericellular rim about unshrunken hypertrophied chondrocytes of chick long bone. *Anat. Rec.* **183**, 359–372.

Keene, D. R. (1984). A method for producing dust-, streak- and hole-free formvar films in laboratories having high atmospheric humidity. *Stain Technol.* **59**, 56–58.

Keene, D. R., Maddox, B. K., Kuo, H.-J., Sakai, L. Y., and Glanville, R. W. (1991). Extraction of extendable beaded structures and their identification as fibrillin-containing extracellular matrix microfibrils. *J. Histochem. Cytochem.* **39**, 441–449.

Keene, D. R., and McDonald, K. (1993). The ultrastructure of the connective tissue matrix of skin and cartilage after high-pressure freezing and freeze-substitution. *J. Histochem. Cytochem.* **41**, 1141–1153.

Keene, D. R., Oxford, J. T., and Morris, N. P. (1995). Ultrastructural localization of collagens II, IX, and XI in the growth plate of human rib and fetal bovine epiphyseal cartilage: Type XI collagen is restricted to thin fibrils. *J. Histochem. Cytochem.* **43**, 967–979.

Kellenberger, G. E. (1991). The potential of cryofixation and freeze substitution: Observations and theoretical considerations. *J. Microsc.* **161**, 183–203.

Kempson, G. (1979). *In* "Adult Articular Cartilage" (M.A.R. Freeman, ed.) 2nd Ed., p. 333. Pitman, Tunbridge and Wells, England.

Lewis, P. N., Pinali, C., Young, R. D., Meek, K. M., Quantock, A. J., and Knupp, C. (2010). Structural interactions between collagen and proteoglycans are elucidated by three-dimensional electron tomography of bovine cornea. *Structure* **18**, 239–245.

Luft, J. H. (1971). Ruthenium red and violet I: Chemistry, purification, methods of use for electron microscopy and mechanism of action. *Anat. Rec.* **171**, 347–368.

Maddox, B. K., Keene, D. R., Sakai, L. Y., Charbonneau, N. L., Morris, N. P., Ridgway, C. C., Boswell, B. A., Sussman, M. D., Horton, W. A., Bächinger, H. P., and Hecht, J. T. (1997). The fate of cartilage oligomeric matrix protein is determined by the cell type in the case of a novel mutation in pseudoachondroplasia. *J. Biol. Chem.* **49**, 30993–30997.

McDonald, K. (2007). Cryopreparation methods for electron microscopy of selected model systems. *Methods Cell Biol.* **79**, 23–56.

McKee, M. D., and Nanci, A. (1996). Osteopontin at mineralized tissue interfaces in bone, teeth, and osseointegrated implants: Ultrastructural distribution and implications for mineralized tissue formation, turnover, and repair. *Microsc. Res. Tech.* **33**, 141–164.

Meadows, R. S., Holmes, D. F., Gilpin, C. J., and Kadler, K. E. (2000). Electron cryomicroscopy of fibrillar collagens. *Methods Mol. Biol.* **139**, 95–109.

Moor, H. (1987). Theory and practice of high pressure freezing. *In* "Cryotechniques in Biological Electron Microscopy" (R. A. Steinbrecht and K. Zierold, eds.) pp. 175–191. Springer-Verlag, Berlin.

Moor, H., Bellin, G., Sandri, C., and Akert, K. (1980). The influence of high pressure freezing on mammalian nerve tissue. *Cell Tissue Res.* **209**, 201–216.

Morris, N. P., Oxford, J. T., Davies, G.B.M., Smoody, B. F., and Keene, D. R. (2000). Developmentally regulated alternative splicing of the $\alpha1$(XI) collagen chain: Spatial and temporal segregation of isoforms in the cartilage of fetal rat long bones. *J. Histochem. Cytochem.* **48**, 725–741.

Muller, M., and Moor, H. (1984). Cryofixation of thick specimens by high pressure freezing. *Scan. Electron Microsc.* **1984**, 131–138.

Muller, M., and Moor, H. (1984). Cryofixation of suspensions and tissues by propane-jet freezing and high pressure freezing. *Proc. EMSA* **1**, 6–9.

Palade, G. E. (1952). The study of fixation for electron microscopy. *J. Exp. Med.* **95**, 285.

Poole, A. R., Pidoux, I., Reiner, A., and Rosenberg, L. (1982). An immunoelectron microscope study of the organization of proteoglycan monomer, link protein, and collagen in the matrix of articular cartilage. *J. Cell Biol.* **93**, 921–993.

Poole, A. R., Reiner, P. A., Tang, L.-H., Chio, H., and Rosenberg, L. (1980). Localization of proteoglycan monomer and link protein in the matrix of bovine articular cartilage: An immunohistochemical study. *J. Histochem. Cytochem.* **28**, 621–635.

Ratcliffe, A., Fryer, P. R., and Hardingam, T. E. (1984). The distribution of aggregating proteoglycan in articular cartilage: Comparison of quantitative immunoelectron microscopy with radioimmunoassay and biochemical analysis. *J. Histochem. Cytochem.* **32**, 192–201.

Robertson, D., Monaghan, P., Clarke, C., and Atherton, A. J. (1992). An appraisal of low-temperature embedding by progressive lowering of temperature into Lowicryl HM20 for immunocytochemical studies. *J. Microsc.* **168**, 85–100.

Robinson, R. A., and Cameron, D. A. (1956). Electron microscopy of cartilage and bone matrix at the distal epiphyseal line of the femur in the newborn infant. *J. Biophys. Biochem. Cytol.* **2**, 253–263.

Rosenberg, L. C., Choi, H. U., Tang, L. H., Johnson, T. L., Pal, S., Webber, C., Reiner, A., and Poole, A. R. (1985). Isolation of dermatan sulfate proteoglycans from mature bovine articular cartilages. *J. Biol. Chem.* **260**, 6304–6313.

Rosenberg, L., Hellmann, W., and Kleinschmidt, A. K. (1975). Electron microscopic studies of proteoglycan aggregates from bovine articular cartilage. *J. Biol. Chem.* **250**, 1877–1883.

Scott, J. E. (1995). Extracellular matrix, supramolecular organization and shape. *J. Anat.* **187**, 259–269.

Scott, J. E. (2003). Elasticity in extracellular matrix "shape modules" of tendon, cartilage, etc. A sliding proteoglycan-filament model. *J. Physiol.* **553**, 335–343.

Scott, J. E., and Haigh, M. (1985). Small-proteoglycan: Collagen interactions: Keratan sulphate proteoglycan associates with rabbit corneal collagen fibrils at the "a" and "c" bands. *Biosci. Rep.* **5**, 765–774.
Scott, J. E., and Haigh, M. (1988). Keratan sulphate and the ultrastructure of cornea and cartilage: A "stand-in" for chondroitin sulphate in conditions of oxygen lack? *J. Anat.* **158**, 95–108.
Scott, J. E., and Stockwell, R. A. (2006). Cartilage elasticity resides in shape module decoran and aggrecan sumps of damping fluid: Implications in osteoarthosis. *J. Physiol.* **574**, 643–650.
Shepard, N., and Mitchell, N. (1989). Improved chondrocyte morphology and glycogen retention in the secondary center of ossification following osmium-potassium ferrocyanide fixation. *J. Electron Microsc. Tech.* **11**, 83–89.
Sommarin, Y., and Heinegård, D. (1986). Four classes of cell-associated proteoglycans in suspension cultures of articular-cartilage chondrocytes. *Biochem. J.* **233**, 809–818.
Tang, L. H., Rosenberg, L., Reiner, A., and Poole, A. R. (1979). Proteoglycans from bovine nasal cartilage: Properties of a soluble form of link proteins. *J. Biol. Chem.* **254**, 10523–10536.
Thyberg, J. (1977). Electron microscopy of cartilage proteoglycans. *Histochem. J.* **9**, 259–266.
Tinling, S. P., Giberson, R. T., and Kullar, R. S. (2004). Microwave exposure increases bone demineralization rate independent of temperature. *J. Microsc.* **215**, 320–235.
Zhu, Y., Oganesian, A., Keene, D. R., and Sandell, L. J. (1999). Type IIA procollagen containing the cysteine-rich amino propeptide is deposited in the extracellular matrix of prechondrogenic tissue and binds to TGF-B1 and BMP-2. *J. Cell Biol.* **144**, 1069–1080.

CHAPTER 20

Electron Microscopy of the Mouse Central Nervous System

Wiebke Möbius[*], Benjamin Cooper[†], Walter A. Kaufmann[‡], Cordelia Imig[†], Torben Ruhwedel[*], Nicolas Snaidero[*], Aiman S. Saab[*], *and* Frédérique Varoqueaux[†]

[*]Department of Neurogenetics, Max-Planck-Institute of Experimental Medicine, Göttingen D-37075, Germany

[†]Department of Molecular Neurobiology, Max-Planck-Institute of Experimental Medicine, Göttingen D-37075, Germany

[‡]Department of Pharmacology, Innsbruck Medical University, Innsbruck A-6020, Austria

978-0-12-381007-6
DOI: 10.1016/S0091-679X(10)96020-2

Abstract

The high degree of similarity between mouse and human physiology and genomes makes mice the animal model of choice to study the functions and dysfunctions of the central nervous system (CNS). The considerable knowledge accumulated in the past decades and the steadily growing number of genetically modified mouse lines allow for the increasingly accurate understanding of biological circuits. Electron microscopy (EM) contributes to unravel the biology of neuronal networks and the myelinating glia by allowing a fine morphological scrutiny of the nervous tissue. We provide detailed descriptions of the conventional processing as well as cryopreparation methods such as high-pressure freezing (HPF), freeze-substitution (FS), and SDS-digested freeze-fracture replica labeling (SDS-FRL) on selected CNS regions such as the retina, optic nerve, and cerebellum. By taking example of the ribbon synapse in the retina and myelinated retinal ganglion cell axons of the optic nerve, we discuss the advantages and drawbacks of these methods in a comparative way.

I. Introduction

The vertebrate CNS comprises the retina, the brain, and the spinal cord, and contains trillions of neurons of great variety aiming to integrate, propagate, and respond to sensory or cellular signals. Offering a huge diversity of shapes, neurons have developed specialized subcellular compartments allowing them to optimize their ability to process and transmit information, and communicate with each other in the process of synaptic transmission. Chemical synapses are neuronal hallmarks, asymmetric junctions at which a presynaptic neuron passes on information to a receiving postsynaptic neuron. The presynaptic compartment is devoted to regulated neurotransmitter release, while the postsynaptic compartment transduces this signal. Both compartments are complex structures with specific scaffold, adhesion, and receptor proteins. It is becoming increasingly clear that many disorders, ranging from autism spectrum disorders and mental retardation to Parkinson disease are due to or are contributed by synaptic dysfunctions (Belmonte and Bourgeron, 2006; Nikolaus *et al.*, 2009). Hence, the unraveling of synaptic mechanisms remains a major goal of today's cellular neuroscience, and relies on multidisciplinary research. Morphological and in particular ultrastructural observations are essential to substantiate physiological findings. There exists a great diversity of synapses, likely reflecting their functional requirements, such as transient versus sustained release, high speed of release, graded transmission etc. Here, we take the retinal ribbon synapse as an example to illustrate the outcome of different protocols described herein. This type of synapse is present also in the pineal gland and in the cochlea (Lagnado, 2003); for a review see Sterling and Matthews (2005).

The retina is a highly organized network of deceiving morphological simplicity allowing nevertheless for complex information coding (Gollisch and Meister, 2010). It contains five major neuronal cell types (for about 50 distinct subtypes) whose cell bodies distribute in three nuclear layers, and whose processes transfer and modulate information at two synaptic plexiform layers (Fig. 1; Masland, 2001a,b; Wässle, 2004). Excitation is conveyed through graded transmission by the ribbon synapse-bearing photoreceptor and bipolar cells, while inhibitory modulation of this main excitatory pathway is provided by horizontal and amacrine cells (Fig. 1). At the innermost layer of the retina, ganglion cells are the final recipients of integrated visual information. They emit axonal processes, which exit the eye at the optic disc and fasciculate to form the optic nerve, conveying visual information to higher brain areas (Fig. 1).

The insulating sheath of myelin wrapping axons is another archetypical feature of long-range projecting neurons. The myelin sheath is required for the efficient action potential propagation at relatively low energy costs. By providing this advantage, the evolutionary innovation of myelin is considered a prerequisite for the development of compact and efficient vertebrate brains (Möbius *et al.*, 2008). Myelin is a structurally and molecularly specialized membrane structure enriched in galactocerebroside and sulfatide (Nonaka and Kishimoto, 1979; Norton and Cammer, 1984; Norton and Poduslo, 1973). Because of its lipid-rich nature, it is particularly delicate to preserve for ultrastructural investigation and will be also focused upon in this chapter.

Considering the history of myelin research, the development of concepts and the pace of discovery were tightly connected to the technical developments in imaging, including electron microscopy (EM). After Virchow introduced the term "myelin" (myelos = marrow) in 1854, it took another 100 years and the invention of EM to resolve its ultrastructure and to prove that myelin in the peripheral nervous system (PNS) consists of a spirally wrapped and tightly condensed multilayered membrane sheath produced by Schwann cells (Fernandez-Moran, 1950; Geren, 1954; Robertson, 1955; Sjöstrand, 1949). More effort is required to understand the structure of myelin formed by oligodendrocytes in the central nervous system (CNS). This was partially due to preservation difficulties, but also due to the fact that in adult tissue continuities between the myelin sheath and the oligodendroglial cell body are formed by small processes and therefore difficult to observe. Studies on re-myelination in cat spinal cord clearly showed for the first time that the mechanism of myelin formation in the CNS resembles largely that in the PNS (Bunge *et al.*, 1961, 1962). Here it was demonstrated that oligodendroglial processes contact axons and form the myelin sheath by spiral wrapping around the axon, as already proposed (Peters, 1960a,b). As demonstrated later (Bunge and Glass, 1965; Hirano and Dembitzer, 1967; Kruger and Maxwell, 1966; Peters, 1964), oligodendrocytes significantly differ from Schwann cells since they are able to myelinate several axons and axonal segments.

The function of oligodendrocytes and the myelin sheath in axonal support exceeds its role in saltatory action potential propagation. This is illustrated by the occurrence of severe disorders, in which neurodegeneration is caused by genetic defects primarily affecting oligodendrocytes, such as the hypomyelinating leukodystrophies Pelizaeus–Merzbacher disease (PMD) and spastic paraplegia type 2 (SPG2) (Bonnet-Dupeyron *et al.*, 2008; Griffiths *et al.*, 1995, 1998; Inoue, 2005; Saugier-Veber *et al.*, 1994). Another example for the contribution of myelin to a neurodegenerative disease is multiple sclerosis (MS), which is described as an inflammatory-mediated demyelinating disorder (Trapp and Nave, 2008). Apart from these examples there is a large number of other myelin-related disorders (Barkovich, 2007; Boespflug-Tanguy *et al.*, 2008).

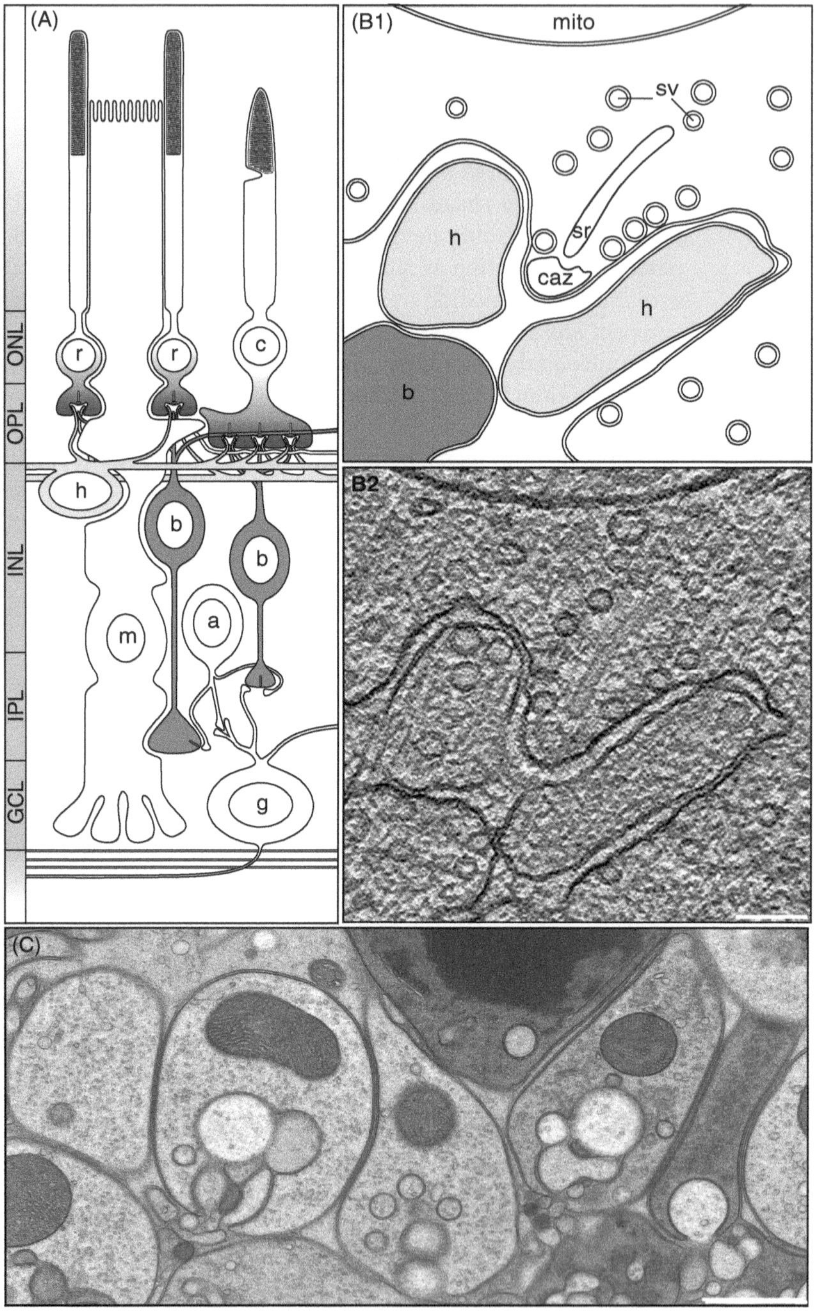

Fig. 1 (*Continued*)

Therefore, mouse mutants and disease models are valuable tools to study the biology of myelinating glial cells and the peculiar nature of axon–glia interactions (Edgar and Nave, 2009).

Here, we refer mostly to the optic nerve as a model tissue to study CNS myelin, since it is well accessible and can be removed quickly and intact for fixation by conventional immersion fixation or high-pressure freezing (Moor, 1987; Moor *et al*., 1980). Moreover, the optic nerve is a relatively simple system with a limited number of glial cell types, and a parallel orientation of all axons, making it suitable for the systematic morphological analysis of myelin in normal as well as pathological conditions.

II. Rationale

Imaging of the CNS is crucial to understand its functions and dysfunctions. The morphology of conventionally prepared samples, obtained by aldehyde-based perfusion and epoxy-resin embedding, has shaped our view of the CNS for decades, and likely depicts most of CNS ultrastructure accurately. However, cryotechniques, such as high-pressure freezing/freeze-substitution (HPF/FS) and freeze-fracture replica, allow for a more accurate and insightful appraisal of ultrastructure and composition of biological samples.

We refer to the retina, a well-described finely layered neuronal network, as a model system to study conventional and ribbon synapses, and to the well-accessible, relatively simple optic nerve as a model to study CNS myelin. In detail, we describe both the conventional preparation and embedding of aldehyde-perfused samples, and the HPF/FS of native and fixed samples for ultrastructural observation after Epon embedding.

Toward the end of this chapter, we illustrate SDS-digested freeze-fracture replica labeling (SDS-FRL) of cerebellar tissue. This technique, performed on fixed or unfixed, high-pressure-frozen samples, represents a powerful tool to address questions on the structural organization of biomembranes, and allows immunodetection and quantitative analysis of integral membrane proteins beyond limitations of thin-section EM (Fujimoto, 1995, 1997; Masugi-Tokita and Shigemoto, 2007).

Fig. 1 (A) Schematic representation of retinal circuitry. A main transmission pathway conveys an excitatory signal from rod (r) and cone (c) photoreceptors via bipolar cells (b) onto ganglion cells (g); populations of inhibitory horizontal (h) and amacrine (a) cell modulate this signal. ONL, INL: Outer and inner nuclear layers, respectively; OPL, IPL: Outer and inner plexiform layers, respectively; GCL: Ganglion cell layer. (B1) Schematic representation and (B2) single tomographic slice of synaptic elements visible at a photoreceptor ribbon triad upon HPF, freeze-substitution, and Epon embedding of fresh mouse retinal tissue. Caz: cytomatrix active zone; sr: synaptic ribbon; sv: synaptic vesicle. (C) Overview of mouse retinal tissue at the OPL prepared by HPF and freeze-substitution. All processes show an excellent ultrastructural preservation. Inner and outer membranes are smooth and intact; mitochondrial cristae are regular, neither swollen nor disrupted. Synaptic vesicles are nicely distinguishable. There is no reticulation within the nucleus and little extracellular space. Scale bar = 100 nm in B, 1 μm in C. (See Plate no. 30 in the Color Plate Section.)

III. Methods

A. Conventional Fixation and Embedding

For the fixation of mouse CNS tissue by perfusion, glutaraldehyde is the most frequently used fixative (Schultz and Karlsson, 1965), alone or in combination with formaldehyde (Karnovsky, 1965). Cacodylate buffers have a long history in EM (Glauert, 1974), since they are easy to prepare, stable during storage, and allow the addition of low concentrations of calcium (1–3 mM) without precipitation. However, cacodylate buffers are toxic and must be handled with care. In general, phosphate buffers can be used instead, which are non-toxic and also easy to prepare.

Commonly, two fixatives are used, both yielding a well-preserved ultrastructure of the neuropile, including synapses and myelin; the first one based on phosphate buffer according to Karlsson and Schultz (1965) and Schultz and Karlsson (1965) and the second one based on cacodylate buffer ("Perfixol," Griffiths *et al.*, 1981). As the more widely used, we discuss only results obtained by using the first protocol (see Section V.A). Of note, immersion fixation of freshly dissected nerves or retina samples yields an equally well-preserved ultrastructure as perfusion fixation. Aldehyde-fixed samples are classically being further dehydrated and embedded in Epon (see below). Alternatively, they can be further processed by HPF/FS (see Section V.B). In such a mixed approach, a rapid fixation may circumvent a possible anoxia during preparation of brain tissue for HPF, and protect against mechanical damage during handling of fragile structures (Sosinsky *et al.*, 2008); on the other hand, it allows avoiding dehydration artifacts frequently observed upon conventional infiltration and embedding procedures at room temperature.

1. Fixatives

a. "Karlsson–Schultz": 2.5% Glutaraldehyde/4% Formaldehyde in Phosphate Buffer pH 7.3 According to Karlsson and Schultz (1965)

Paraformaldehyde (16%) stock solution:

- Mix 80 g paraformaldehyde with 450 ml H_2O.
- Stir at 60°C for 15–20 min.
- Add droplets of 1 M NaOH until solution turns clear.
- Add H_2O to a final volume of 500 ml.
- Filter with a Nalgene™ filter unit and freeze aliquots for storage (−20°C).
- Thaw aliquots at 60°C for use, the solution should be clear.

Fixative working solution (200 ml):

- 0.36 g $NaH_2PO_4 \cdot H_2O$
- 3.1 g $Na_2HPO_4 \cdot 2H_2O$
- 1 g NaCl
- 130 ml H_2O

- 20 ml glutaraldehyde 25% in H_2O, EM-grade (end 2.5%)
- 50 ml paraformaldehyde 16% in H_2O (end 4%)
- Filter with a Nalgene™ filter unit, adjust pH to 7.3 if necessary

The amount of glutaraldehyde in this fixing solution can be reduced to 0.2% when samples should be processed for immunoelectron microscopy. In case of sensitive antigens, glutaraldehyde can also be omitted, but this will result in a less-preserved ultrastructure. An excellent and detailed protocol for cryosectioning according to Tokuyasu (1973) and immunolabeling of thawed cryosections is provided by Peters and Pierson (2008) and therefore not explained further here.

b. "Perfixol": 5% Glutaraldehyde/4% Formaldehyde in Cacodylate Buffer pH 7.2 According to Griffiths et al. (1981)

Buffer (0.08 M Na-cacodylate, pH 7.2):

- Dissolve 8.56 g Na-cacodylate in 450 ml H_2O.
- Adjust pH to 7.2 with 1 M HCl.
- Add H_2O to a final volume of 500 ml.

Fixative working solution (500 ml):

- 150 ml 0.08 M cacodylate buffer
- 250 ml 8% paraformaldehyde in H_2O (final: 4%)
- 100 ml glutaraldehyde 25% in H_2O, EM-grade (final: 5%)
- 0.33 g $CaCl_2 \cdot 2H_2O$
- Filter with a Nalgene™ filter, adjust pH to 7.2 if necessary.

2. Perfusion Fixation

a. Required Items

- Anesthetics: Avertin
- Styrofoam board
- Peristaltic pump, tubing, and adaptor
- Hypodermic injection syringe and needles
- 27-gauge winged infusion set
- Dissection tools (scissors, forceps, clamps, needles)
- Flushing solution: PBS or HBSS.

b. Anesthetics: Avertin

- Take 2 g Avertin
- Add 2 ml tertiary amylalcohol
- Warm up 96.5 ml H_2O to maximum 40°C prior to adding it to the Avertin/amylalcohol solution
- Stir for 30 min
- Filter through a 0.2 μm filter

- Make 5 ml aliquots in dark tubes and store at −20°C; Avertin is sensitive to light!
- After thawing the solution is ready to use. Store in the dark at 4°C for maximum 4 weeks.

*c. **Perfusion***

- Prepare the perfusion tubing: it should be filled with the flushing solution without air bubbles.
- Deeply anesthetize the mouse by injecting intraperitoneally 0.2 ml Avertin per 10 g body weight. The anesthesia usually lasts approximately 1 h. Check anesthesia by pinching a paw. The mouse should not twitch.
- Stretch out the mouse on a Styrofoam board with needles (Suppl. Fig. 1A).
- Remove the skin by making a transversal cut just below the diaphragm; cut through the diaphragm just at the very edge of the chest, open the chest with two lateral cuts; bend the chest backward and fix it with needles to allow a good access to the heart (Suppl. Fig. 1B–F).
- Insert a needle (27-gauge winged infusion set) into the left ventricle. Control the position of the needle so that it should not injure the aorta. Make a small cut into the right atrium (Suppl. Fig. 1G and H).
- Perfuse with warm PBS or HBSS (approximately 10 ml) to clear the blood vessels.
- Replace the flushing solution with the fixative without introducing any air bubbles into the system; perfuse with approximately 70 ml of the fixative solution at room temperature. The mouse will jerk briefly and then become stiff.
- Dissect out the eyes, optic nerves, and brain under the stereomicroscope (see below).
- Place the tissue samples immediately into vials that contain cold fixative.
- Store the tissue at 4°C until needed. For prolonged storage, the fixative should be replaced by 1% formaldehyde; otherwise, the tissue will get very hard and brittle over time.

3. Dissection

- The retina is dissected out from the eyecup and flattened out: The eye is gently taken out by pulling the skin around the eyelid taut, causing the eyeball to protrude slightly and thus permitting to grip the optic nerve beneath the eye with curved forceps. The optic nerve is then cut with spring scissors behind the forceps to avoid damaging the retina; the eye is transferred to PBS, stabilized with tweezers at the root of the optic nerve, and punctured with a sharp needle at the center of the sclera; then while holding the eye at the sclera, the eye cup is opened up with small spring scissors by a circular cut below the base of the ciliary muscle, at the level of the ora serrata; after the lens and vitreous humor are removed, the sclera is carefully flipped over—the retina can then be isolated by a cut of the optic nerve. For embedding, the retina is flattened out by making four incisions, and can be handled with a brush.
- For the dissection of optic nerves and brain, the fur is removed carefully from the skull. The procedures for dissecting the optic nerve after perfusion fixation is very similar to the dissection of fresh optic nerves (see Section III.B.3.b). Shortly, the

connective tissue around the eye is cut to gain access to the eye socket; the optic nerve is cut behind the eyeball and the eye is removed. The skull is then cut open laterally on both sides in a caudo-rostral direction, without injuring the brain. The bone is flipped to the side and the brain is carefully taken out. The optic nerves will stay attached to the brain via the chiasm. If they do not slip out easily from the skull, they should be cut as close to the bone as possible.

- After removal of the optic nerves by cutting them at the chiasm, the brain is placed into a mouse brain slicer matrix and sliced into 1 mm coronal or sagittal slices with razor blades. Alternatively, thinner brain slices can be made using a vibratome.

4. Embedding in Epon

Brain slices, retina pieces, or optic nerves are placed into small-capped glass vials. Alternatively, thin vibratome sections or retina pieces, which often curl up during dehydration, can be placed into a 24-well plate and covered with small round glass cover slips until dehydration has reached the 90% ethanol step.

- 3×10 min wash at 4°C with buffer (e.g., phosphate or cacodylate buffer)
- 4 h incubation in 2% OsO_4 at 4°C in the same buffer
- 3×10 min wash at 4°C with the same buffer
- 2×10 min incubation at 4°C in 30% ethanol
- 2×10 min incubation at 4°C in 50% ethanol
- 2×10 min incubation at 4°C in 70% ethanol
- 2×10 min incubation at 4°C in 90% ethanol
- 4×10 min incubation at 4°C in 100% ethanol
- 3×10 min incubation at RT in propylenoxide
- 1 h infiltration at RT in propylenoxide:Epon mix 2:1
- 1 h infiltration at RT in propylenoxide:Epon mix 1:1
- Overnight infiltration at RT in propylenoxide:Epon mix 2:1
- 4 h infiltration at RT in pure Epon (Table I)

Table I
Epon Recipe

Glycidether 100	21.4 g
DDSA	14.4 g
MNA	11.3 g
total:	47.1 g
Stir for 10 min	
DMP-30	0.84 ml
Stir for 20 min, avoid air bubbles	

Pour the first three Epon components into a beaker placed on a balance. The accelerator (DMP-30) is added by using a 1 ml disposable syringe.

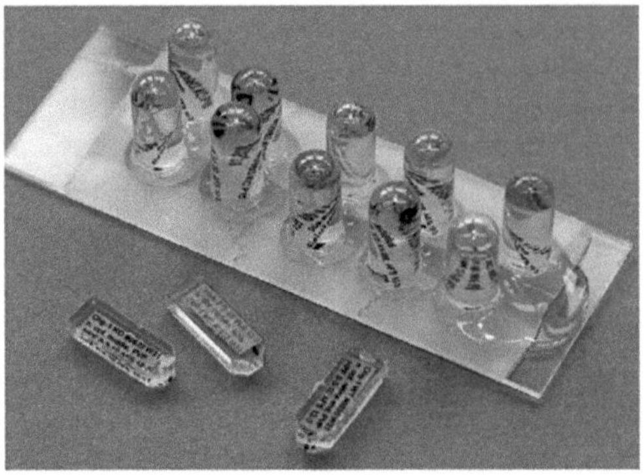

Fig. 2 Epon embedding: For flat or longitudinal embedding, samples are placed on an objective slide which was wrapped with Parafilm^{TM}. This helps to detach the blocks later. An Epon-filled gelatine capsule is placed upside down on the sample. For transverse sectioning, nerves are placed into embedding molds. Small labels are included in the blocks to identify the sample.

- Transfer into embedding molds or Epon-filled gelatine capsules. For flat embedding or longitudinal embedding of nerves, place the samples on a glass slide wrapped with Parafilm^{TM}, which greatly helps to detach the blocks later on. Then an Epon-filled gelatine capsule size 0 is placed upside-down on top of the tissue (Fig. 2). Small labels (e.g., 3 lines 3 pt size fond) with sample details can be included to help identify the blocks
- Polymerize for 24 h at 60°C
- Remove gelatine capsules by cooking the blocks 15 min in water

5. Preparation and Contrasting of Sections

The sample or the area of choice is trimmed; ultrathin sections (70–90 nm) are placed on either copper slot or 100 mesh hexagonal Formvar-coated EM grids. The making of Formvar films is described in detail in Peters and Pierson (2008) and Slot and Geuze (2007). The air-dried sections are contrasted as follows.

a. Required Solutions

4% uranyl acetate in H_2O, filtered with a 0.22 μm Millipore^{TM} filter

Lead citrate according to (Reynolds, 1963):

- Add 1.33 g lead (II) nitrate to 30 ml boiled/degassed H_2O.
- Add 1.76 g Na_3 citrate dihydrate.
- Shake vigorously and stir for 30 min (lead nitrate will convert to lead citrate).
- Add droplets of 1 M NaOH until solution turns clear.
- Fill to a final volume of 50 ml, store at 4°C.

b. Procedure Place a piece of Parafilm™ firmly onto a 96-well plate, press and move an Eppendorf Cap over the parafilm to form "indentations" before removing the protecting sheet. Droplets of staining solutions are then placed into the indentations. Incubate the grids with the sections facing down on the droplets as follows:

- 30 min incubation in 4% uranyl acetate in H_2O
- 3×1 min wash in droplets of H_2O
- 6 min incubation in lead citrate (place KOH pastilles near the droplets to gather CO_2 from ambient air; do not breathe on the droplets, cover them with a lid during the incubation)
- 4×1 min wash in droplets of H_2O
- Dry the grids with a pointed tip of a filter paper

Alternatively, grids can be washed by several dippings in three beakers filled with H_2O. When dried, the sections can be observed in the transmission electron microscope. Depending on the result, incubation times have to be optimized for different samples.

B. High-Pressure Freezing

There are three commercially available types of high-pressure freezers (McDonald, 2007). We use a BAL-TEC HPM 010 with specimen carriers of 2 mm inner diameter and indentations of 0.1 and 0.2 mm and the flat side of the 0.3 mm hat as a lid (Kaech, 2009).

1. Materials

The following requisites should be at hand before dissecting the living tissue:

- Binocular
- Tissue chopper
- Specimen carriers
- Space filler of choice
- Fine-tipped paint brush, if yeast paste is used as filler
- Forceps: Dumont #5 fine tipped and mirror finish for handling fresh tissue
- Scissors for cutting skin and bones
- Fine spring scissor for cutting the optic nerve
- Scalpel
- Perfect loop™
- Plastic petri dishes
- Nitrocellulose membrane
- Blotting paper
- Ice box with ice-cold PBS
- Disposable syringe (1 ml)
- Perforated cryotubes

2. Space Fillers

The space filler is required to avoid air inclusion in the specimen carrier around the sample, since this reduces the quality of freezing. Ideally, space fillers should transfer heat effectively during cooling and also act as cryoprotectants (Hess, 2007; McDonald, 2007; McDonald *et al.*, 2007). Apart from the widely used 1 -hexadecene, solutions with a high concentration of cells (yeast paste) or protein (BSA), or high molecular weight compounds like dextran or polyvinylpyrrolidone (PVP) are examples of useful space fillers (McDonald *et al.*, 2007).

We make yeast paste by mixing fresh yeast with ice-cold PBS until the paste has a creamy consistence. We have tested several other space fillers including 20% dextran or 20% BSA in artificial cerebrospinal fluid (ACSF) prepared according to Tekkök *et al.* (2007), 0.4% low melting point (LMP) agarose in PBS and 20% PVP (MW = 10,000) in PBS. Depending on the freeze-substitution "cocktail," most of the fillers turn black during freeze-substitution thereby obscuring the sample. This can be very inconvenient if the sample has to be embedded and cut in a defined orientation. Using BSA as filler results in a black/brownish disk in which the sample is hardly visible, and which tends to crack easily. This problem is less severe when using 20% PVP in PBS as space filler. Dextran is only insufficiently infiltrated by Epon making the blocks very difficult to cut. LMP agarose has to be kept at around 37°C and does not turn black. In some cases, this can be an advantage, but we prefer to keep the freshly dissected nerves at ice temperature since this seems to be beneficial for the integrity of the traumatized axons. Yeast paste can be removed from the sample after freeze-substitution and Epon infiltration by gently tapping with forceps. Therefore, we prefer to use yeast paste or 20% PVP in PBS as space fillers.

3. Sample Preparation

a. Retina The eye is dissected as described in Section III.A.3, with the following modifications: after the lens is removed, the eye cup is flattened out after making four equidistant incisions, placed retina-side up in a petri dish, and covered with a wet nitrocellulose membrane; the retina adheres to the membrane, which can be flipped over; the sclera can be gently lifted up and excised by severing the optic nerve at the level of the optic disc (Fig. 3).

b. Optic Nerve The preparation of the fresh mouse optic nerve should be performed with greatest care. The application of any force to the nerve, like pulling, has to be avoided. If possible, a person experienced in the preparation of optic nerves for electrophysiology measurements should be asked for assistance. After decapitation, the fur is removed from the skull following a midline skin incision and then pulling the skin toward the nose (Suppl. Fig. 2A). Behind the eye, incisions are made with a scalpel to gain access to the eye socket without injuring the eyeball (Suppl. Fig. 2B). Scissors are inserted behind the eyeball to cut blindly the optic nerve (Suppl. Fig. 2C). A white spot should be visible at the eyeball, indicating the remainder of

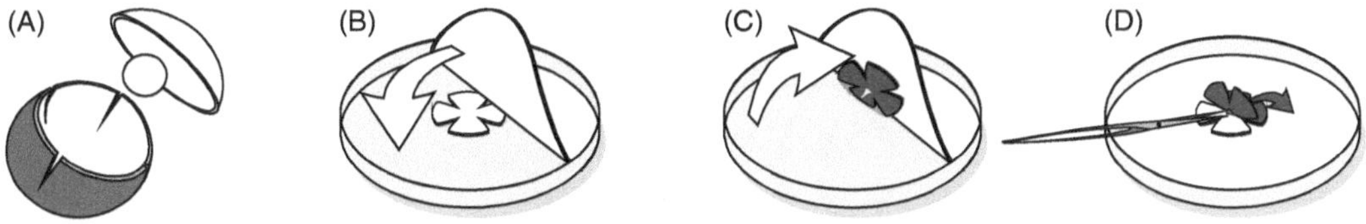

Fig. 3 Dissection of the fresh mouse retina: The eye is rapidly enucleated by cutting the optic nerve with spring scissors. In PBS, the eyecup is opened up at the level of the ora serrata; after the lens and vitreous humor are removed, four equidistant incisions are made (A). The flattened eyecup is placed in a petri dish and covered with a wet nitrocellulose filter (B). The filter is then flipped over (C), the sclera can be lifted up and the retina separated by severing the optic nerve at the level of the optic disc (D).

the optic nerve (Suppl. Fig. 2D). The skull is cut laterally with a scissor and the bone flipped sideward to expose the brain (Suppl. Fig. 2E). Then forceps are inserted in a 45° angle into the brain to lift it carefully from the skull. The optic nerves should slip easily out of the skull and stay attached to the brain (Suppl. Fig. 2F). By pouring ice-cold PBS with a syringe over the brain, the optic nerves are straightened without being touched (Suppl. Fig. 2G). The optic nerves are released by cutting with the fine spring scissor between chiasm and brain (Suppl. Fig. 2H), picked up gently at the chiasm with forceps (Suppl. Fig. 2I), and transferred into ice-cold PBS in a petri dish as quickly as possible to protect the severed axons from metabolic stress. Now the sample is ready for loading into the specimen carriers (Suppl. Fig. 2J).

4. Sample Loading

a. Retina The nitrocellulose membrane holding the retina is placed on three wet blotting papers on the stage of a tissue chopper (Fig. 4A). The retina is kept permanently wet with ice-cold, oxygenated Ames medium. Slices of 200 μm are made and trimmed at their edges with a scalpel blade (Fig. 4B). Drops of filler (usually 20% BSA in Ames medium) and specimen carriers (0.1/0.2 mm) are placed on a cooled metal slab covered with Parafilm^{TM} (Fig. 4C–E). The retina stubs are transferred in drops through filler and turned on their side with a fine brush (Fig. 4C). A selected sample is picked up with a Perfect loop^{TM} and transferred into a filled carrier (Fig. 4D). If necessary, excess of solution can be removed with a piece of blotting paper. A second sample carrier (0/0.3 mm) is placed as a lid (Fig. 4E). The sandwich is then ready to be inserted into the holder of the HPF machine for freezing. The entire procedure (from dissection to freezing) should not take more than 5 min.

b. Optic Nerve Since the optic nerve is too long to be inserted intact into the specimen carrier, it has to be cut in the middle and a part of the chiasm has to be removed. For cutting, the optic nerve is placed under the binocular in ice-cold PBS on the back side of a plastic petri dish, if yeast paste is used as filler. By holding the

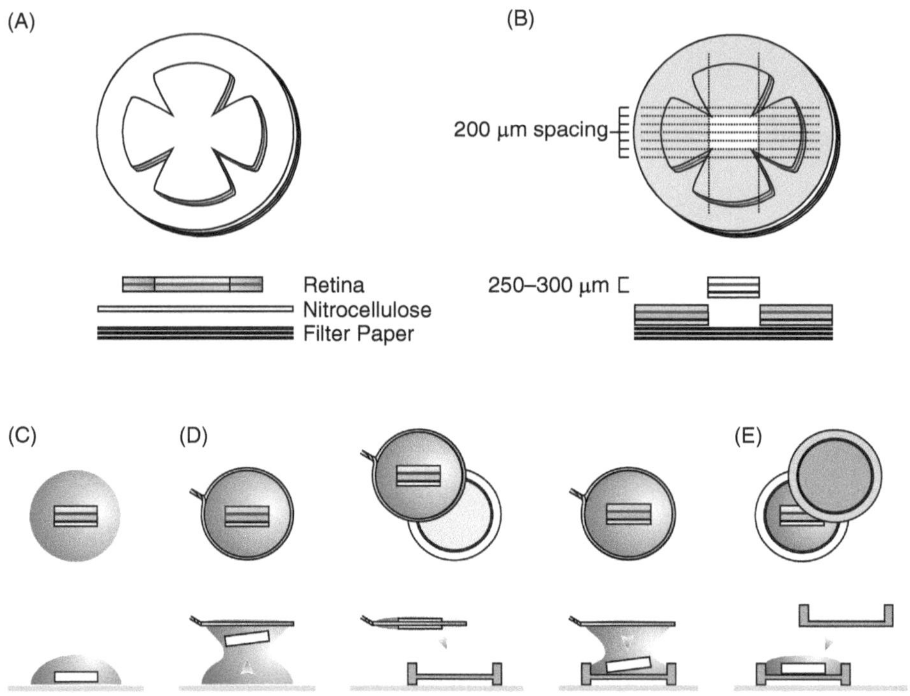

Fig. 4 Loading of the specimen carriers with retina slices: The nitrocellulose membrane holding the retina is placed on three wet blotting papers on the stage of a tissue chopper (A). The retina is held permanently wet with ice-cold Ames medium. Slices of 200 μm are made and trimmed at their edges with a scalpel blade (B). Drops of filler and specimen carriers (0.1 / 0.2 mm) are placed on a cooled metal slab covered with Parafilm™ (C–E). The retina stubs are transferred in filler and turned on their side with a fine brush (C). A selected sample is picked up with a Perfect loop™ and transferred into a filled carrier (D). Excess of solution can be removed with a piece of blotting paper. A second sample carrier (0/0.3 mm) is placed as a lid on top of the loaded carrier (E). (See Plate no. 31 in the Color Plate Section.)

nerve at the chiasm, it can be gently chopped with a scalpel. This procedure results in three pieces: one with the chiasm and two attached stumps of the optic nerves, and two free pieces of optic nerves. The specimen carriers are submerged as well in the droplet of PBS so that the pieces of tissue "swim" into the carriers, guided by forceps (Fig. 5A). We always place the chiasm with attached nerves in one carrier and the other two pieces of the optic nerves in another carrier. Then the carriers are lifted from the droplet and the PBS is removed by blotting with a tip of blotting paper (Fig. 5B). Yeast paste is applied by wetting a pointed paint brush with the paste and then taking a small amount of it with forceps by squeezing the paint brush gently. The yeast paste is then released from the forceps directly into the loaded carrier (Fig. 5C). With a little experience, the exact required quantity of paste can be applied in this way to slightly overfill the carrier. Finally, a second sample carrier (0.3 mm) is placed flat side down as a lid on top of the loaded carrier after wetting the flat side

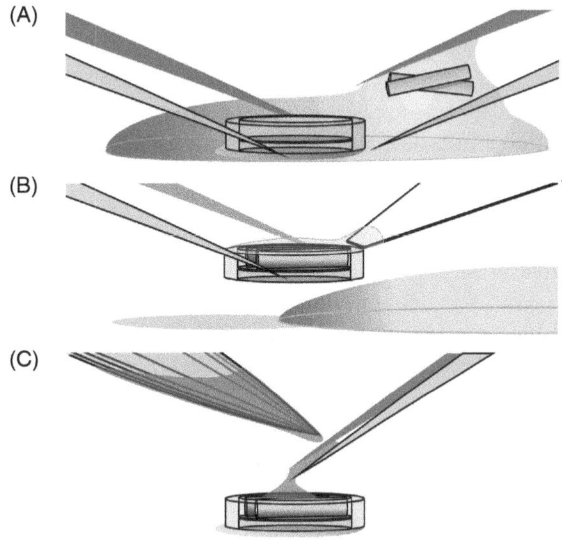

Fig. 5 Loading of the specimen carriers with pieces of the optic nerve: The sample is guided into the carrier with forceps, under PBS immersion (A). The carriers are then lifted up and the PBS is removed with a tip of blotting paper (B). Yeast paste is applied by wetting a pointed paint brush with the paste and then taking a small amount with forceps by squeezing the paint brush gently. The yeast paste is then released from the forceps directly into the loaded carrier (C). (See Plate no. 32 in the Color Plate Section.)

with 1-hexadecene as described (Kaech, 2009). The whole sandwich is then inserted into the holder and frozen.

Alternatively, 20% PVP in PBS is used as filler; the freshly dissected nerve and the specimen carriers are submerged in an ice-cold droplet of this filler. After placing the sample into the carrier as described above, the level of liquid is adjusted with a pointed piece of blotting paper, before adding the lid carrier and freezing the sample.

5. Storage

After freezing, the carrier is immersed in liquid nitrogen and the lid of the carrier sandwich is removed. Samples are transferred into cryotubes and stored in liquid nitrogen. Cryotubes should be perforated beforehand at their upper half and in the lid with a hot needle to ensure that they do not float in liquid nitrogen. Of note, a hole in the bottom part of the tubes would make the nitrogen leak out quickly during transfer, which can lead to a rapid loss of the sample integrity.

C. Freeze-Substitution and Epon Embedding

After cryoimmobilization, the high-pressure frozen samples are subjected to freeze-substitution for dehydration and chemical stabilization of their fine structure. For this purpose fixatives are added to an organic solvent, such as ethanol or acetone.

Freeze-substitution is carried out at −90°C, since acetone will freeze at about −94°C. We perform the freeze-substitution in the AFS-2 unit from Leica (Vienna, Austria) by using the Leica aluminum tins as containers equipped with disposable insets (flow-through rings). This allows us to place samples into 10 separated wells, a very useful feature when a large number of different samples are processed. Each well can hold 2–3 specimen carriers. It is wise to keep in mind that at higher temperatures the "cocktail" might evaporate, so the tins should always be covered with the corresponding "covers for flat embedding" (Leica). Moreover, one should monitor the level of liquid throughout the whole procedure (particularly during extended 4°C periods) or perform the freeze-substitution in screw-cap vials.

We have chosen two protocols yielding different membrane contrast. The first one, which we termed the "classical OsO_4 protocol," is based on a protocol according to Van Harreveld and Crowell (1964), which was also applied to cerebellar tissue (Van Harreveld *et al.*, 1965) and is widely used (Humbel and Müller, 1986; Studer *et al.*, 1989; Vanhecke *et al.*, 2003). The second protocol was published by the team of S. Marty (INSERM U1024, ENS, Paris; (Siksou *et al.*, 2007)) for the analysis of synapses from organotypic hippocampal slices; we called it the "tannic acid–OsO_4 protocol," because here tannic acid is used as a mordant preceding the OsO_4 incubation to increase contrast. Crystalline OsO_4 is dissolved in acetone at room temperature; tannic acid is soluble in acetone and uranyl acetate is added to the substitution "cocktail" from a 5% stock solution in methanol which is kept at −20°C.

1. The "Classical" OsO_4 Protocol

- 48 h incubation in 2% OsO_4 in acetone at −90°C
- Warm up to −60°C by increments of 2°C/h (15 h)
- 8 h incubation at −60°C
- Warm up to −30°C by increments of 2°C/h (15 h)
- 8 h at −30°C
- 3×20 min wash in acetone
- Warm up to 4°C (1 h on ice)
- 3 h on ice: infiltration in Epon:acetone mix 1:2
- Overnight incubation in Epon:acetone mix 2:1 at RT
- Overnight incubation in pure Epon (freshly made) at RT
- Removal of the samples from the specimen carrier, transfer into embedding molds, and polymerization for 24 h at 60°C

This procedure yields a very fine and light contrast of the membranes, while ribosomes stand out nicely (see Section V.B).

2. The Tannic Acid–OsO_4 Protocol

- 100 h incubation in 0.1% tannic acid in acetone at −90°C
- 4×30 min wash in acetone at −90°C (2 h)
- 7 h incubation in 2% OsO_4 in acetone at −90°C

- Warm up to −20°C by increments of 5°C/h (14 h)
- 16 h incubation at −20°C
- Warm up to 4°C by increments of 10°C/h (2.4 h)
- 1 h incubation at 4°C
- 3×20 min washing with acetone at 4°C (1 h)
- 3 h infiltration in Epon:acetone mix 1:1
- Overnight incubation in 90% Epon in acetone at RT
- 6 h incubation in pure Epon (freshly made) at RT
- Removal of the samples from the specimen carrier, transfer in embedding molds, and polymerization for 24 h at 60°C

This protocol gives an excellent membrane contrast, which can be further enhanced by complementing the 2% OsO_4 incubation cocktail with 0.1% uranyl acetate (see Section V.B). This increases the intrinsic contrast within the sample making it more suitable for EM-tomography and 3D reconstruction. If needed, ultrathin sections can be further contrasted as described and analyzed with the electron microscope.

D. SDS-Digested Freeze-Fracture Replica Labeling (SDS-FRL)

Although unfixed tissue can be processed, brains are preferably fixed by transcardial perfusion using low concentrations of formaldehyde (0.5–2%; phosphate- or cacodylate-buffered) rendering tissue handling easier and improving reproducibility of immunolabeling (Fukazawa *et al.*, 2009). Glutaraldehyde fixation should be avoided to preserve antigenicity of epitopes and to increase solubilization efficiency of cleaning agents in subsequent steps (Fukazawa *et al.*, 2009; Severs and Robenek, 2008). Fixed brains are cut with a Vibroslicer at a thickness of < 200 μm, and regions of interest are dissected from these slices and cryoprotected by immersion in a glycerol solution. For tensile fracture (Severs, 2007), samples are then mounted between two specimen carriers by means of double-sided adhesive tape forming a ring-shaped chamber (Fig. 6). That way sandwiched samples are rapidly frozen by use of a high-pressure freezing machine. We used the HPM 010 for our purposes. Once being frozen, samples can be stored in liquid nitrogen for several months. For replication, samples are then fractured by a double replica method in a freeze-fracture replica machine like the BAF 060 (Leica Microsystems) using a double replica specimen table (Fukazawa *et al.*, 2009; Severs, 2007). Fractured faces are then replicated by deposition of a first thin carbon layer (2–5 nm) to ensure immobilization of macromolecules on the replica, a second platinum/carbon layer (2 nm) for shadowing the membrane structure, and a third carbon layer (15–20 nm) for strengthening the material. Replicas are removed from the freeze-fracture machine, brought to atmospheric pressure, and allowed to thaw to room temperature in buffered saline. Subsequently, replicas are cleaned (dissociation of biological material not immobilized in the replica membrane) by means of a solubilization medium. The choice of the appropriate medium depends on further usage of the replicas. In case of immunolabeling, a sodium dodecyl sulfate (SDS) containing solubilization medium is required (Fujimoto, 1995; Fukazawa *et al.*, 2009), and in case of ultrastructural analysis of biomembranes

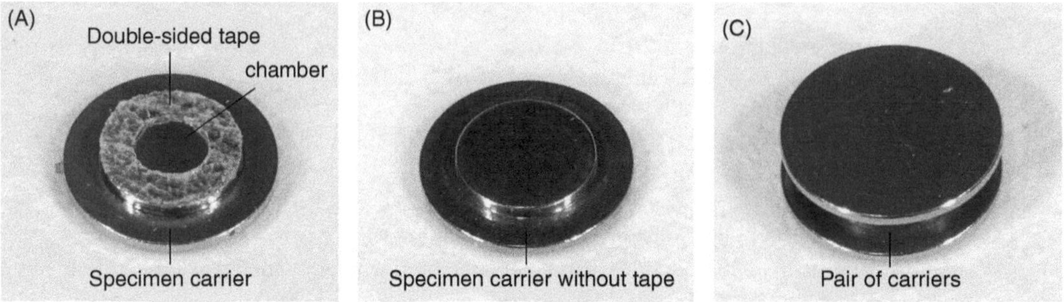

Fig. 6 Specimen carriers for tensile fracture. (A) A ring of double-sided adhesive tape is attached to a specimen carrier with a flat platform (diameter, 3 mm) thereby building a chamber where the sample can be mounted. (B) Another specimen carrier without tape is used as a lid thus forming a sandwich with the carrier holding the sample. (C) A pair of carriers is ready to be frozen in the high-pressure freezing machine. The sandwiched sample can be fractured, separating the pair of carriers by opening the double replica specimen table in the freeze-fracture device.

without immunolabeling other cleaning agents like sodium hypochlorite, sulphuric acid, or chromic acid containing media are preferred (Severs, 2007; Walther, 2003). Replicas can be kept in the respective solubilization medium for several months at room temperature until further processing. For immunodetection of intramembrane macromolecules, replicas are subjected to an indirect immunogold labeling procedure (Hagiwara *et al.*, 2005; Kaufmann *et al.*, 2009; Masugi-Tokita *et al.*, 2007; Tarusawa *et al.*, 2009). Finally, replicas are washed in distilled water, mounted on Formvar-coated EM grids, and analyzed in a conventional transmission electron microscope.

1. Materials

The following materials should be at hand for preparation, freeze-fracture replication, and immunolabeling of samples:

- Vibroslicer
- Binocular
- Scalpel with peaked blade
- Fine-haired paint brush and fine forceps
- Fine spring scissors
- Specimen carriers for tensile fracture
- Acetone and metal polish
- Double-sided adhesive tape (150 μm thickness or less)
- Hollow punch (1–2 mm diameter hole)
- High-pressure freezing machine
- Perforated cryotubes
- Liquid nitrogen
- Liquid nitrogen storage dewar

- Freeze-fracture replica machine
- Double replica specimen table
- Carbon rods with and without platinum inserts
- Filaments for electron beam gun
- Hybridization oven or autoclave
- Sealable glass vials
- Glass rods with platinum wire and fine-tipped glass rods
- Orbital shaker
- Porcelain spotting tray
- Filter paper and parafilm
- Primary antibodies and gold-conjugated secondary antibodies
- EM line-grids coated with Formvar film

Required solutions:

- Saline: 0.9% NaCl in distilled water
- Fixative: 1% formaldehyde plus 15% of a saturated solution of picric acid in phosphate buffer (PB; 0.1M, pH 7.4)
- Cryoprotection medium: 30% glycerol in 0.1M PB (pH 7.4)
- Solubilization medium: 2.5% SDS plus 20% sucrose in 15mM Tris-HCl (pH 8.3)
- Washing buffer for immunolabeling: 0.05% bovine serum albumin (BSA) plus 0.05% NaN_3 in TBS, pH 7.4
- Blocking buffer for immunolabeling: 5% BSA plus 0.05% NaN_3 in TBS, pH 7.4
- Antibody diluent: 2% BSA plus 0.05% NaN_3 in TBS, pH 7.4

2. Sample Preparation

Animals are perfusion-fixed transcardially, and tissue is removed and sliced with a Vibroslicer. The area of interest is trimmed, cryoprotected, and rapidly frozen in a high-pressure freezing machine. The following procedure can be followed:

- Anesthetize the mouse
- Perfuse with warm saline to clear the blood vessels
- Perfuse with chilled fixation solution (perfusion rate, 7 ml per min; duration, 10 min)
- Remove the organ/tissue of interest immediately after fixation
- Wash tissue in phosphate buffer (0.1 M, pH 7.4) containing 0.05% NaN_3
- Slice tissue with a Vibroslicer at 140 μm thickness
- Trim area of interest from obtained slices with a scalpel or a razor blade
- Immerse samples in cryoprotection medium (overnight at 4°C)
- Mount samples onto specimen carriers for tensile fracture: the sample is placed in a hole of double-sided adhesive tape attached to a specimen carrier and covered with cryoprotection medium; another carrier without tape is placed on top so that the sample is sandwiched in between them (Fig. 6)
- Freeze the samples: the pair of carriers with the sandwiched sample is rapidly frozen in the high-pressure freezing machine and stored in liquid nitrogen until further processing

3. Freeze-Fracture Replication

Frozen samples are fractured in a freeze-fracture machine and replicated with deposition of carbon and platinum/carbon. Replicas are removed from the carriers and unexposed tissue is solubilized.

- Place frozen samples in a double replica specimen table (under liquid nitrogen)
- Insert the replica table in the freeze-fracture replica machine
- Fracture the samples by physical force opening of the double replica specimen table, thereby separating the carriers (fracture temperature, −110 to −120°C)
- Replicate exposed tissue surfaces by evaporation of carbon and platinum/carbon: first layer of carbon (rotating) from a 90° angle (thickness 5 nm), second layer of platinum/carbon (unidirectional) from a 60° angle (thickness 2 nm), and third layer of carbon (rotating) from a 90° angle (thickness 15–20 nm)
- Remove the specimen table from the freeze-fracture machine and bring the samples to atmospheric pressure
- Remove replicas from specimen carriers immediately upon retrieval and allow them to thaw to room temperature by immersing them in phosphate buffer (0.1 M, pH 7.4)
- Transfer replicas into solubilization medium and heat to 80°C (18 h on a shaking platform in a hybridization oven) thereby dissolving unexposed tissue from replica membranes

4. Immunolabeling

Replicas are subjected to an immunogold labeling procedure:

- Wash replicas in fresh solubilization medium (10 min at RT)
- Wash in washing buffer (3×10 min at RT)
- Block nonspecific binding sites in blocking buffer (30 min at RT)
- Apply primary antibodies in antibody diluent (overnight at 7°C on a rocker plate)
- Wash in washing buffer (3×10 min at RT)
- Block nonspecific binding sites in blocking buffer (30 min at RT)
- Apply secondary antibodies in antibody diluent (overnight at 7°C on a rocker plate)
- Wash in washing buffer (3×10 min at RT)
- Wash in double-distilled water (2×3 min at RT)
- Mount replicas on Formvar-coated EM grids (100-line copper grids)
- Observe replicas under a stereomicroscope to obtain information about tissue architecture and orientation (optional)
- Analyze replicas under a transmission electron microscope (80–100 kV)

IV. Materials

A. Reagents

For the preparation of fixatives for the conventional sample preparation, we used paraformaldehyde from Serva (# 31628), glutaraldehyde from EMS (25% in

H_2O, # 16216), and Na-cacodylate from Sigma (# C-0250). To make the anesthetics, we used Avertin (2,2,2-Tribromoethanol) from Sigma-Aldrich(# T 48402), tertiary amyl alcohol (2-Methyl-2-butanol) from Merck (# 806193). HBSS was obtained from Gibco (# 14170120) and Ames medium from Sigma (#A1420). The Epon mixture was composed of components from Serva: glycidether 100 (# 21045), DDSA (# 20755), MNA (# 29452), and DMP-30 from Plano (#R1065). OsO_4 was obtained from EMS (ampoules with crystals, # 19110), uranyl acetate from SPI-CHEM (# 02624-AB), and lead nitrate from Merck (# 7398). PVP (MW = 10,000, # P2307) and tannic acid (A.C. S. reagent # 403040) were obtained from Sigma. We used glass-distilled acetone from EMS (100 ml bottles, #10015) as freeze-substitution solvent, because it does not turn black after adding OsO_4. SDS was obtained from Serva (#20760), glycerol and BSA from Sigma (# G5516 and # A9647, respectively), and gold-conjugated antibodies from British BioCell Int.

B. Materials

EM grids (copper slot # G2500C and 100 mesh hexagonal copper # G2410C) and gelatine capsules size 0 (# G292-10) were obtained from Plano (Wetzlar, Germany). AFS-2 consumables were purchased from Leica, Vienna, Austria: universal container for plastic capsules (# 16702735), flow-through rings (# 16707157), and cover for flat embedding mold (# 16702747) as well as materials for replica preparation like carbon rods, platinum inserts, and tungsten cathodes. The specimen carriers for HPM 010 were available at Microscopy Services, Flintbek, Germany (#241 and 242) or Wohlwend GmbH (Sennwald, Switzerland). Porcelain spotting trays were obtained from Groepl C. Elektronenmikroskopie (# L4102).

C. Instrumentation

Dumont # 5 forceps with mirror finish (# 11251-23) and fine spring scissors (# 15396-00) were purchased from FST. The Perfect loop™ was from EMS (# 70944). We also used a mouse brain slicer matrix from Zivic Instruments (# BSM001.1 and # BSM001.2), a McIlwain tissue chopper (Ted Pella Inc, # 10180), Hybridization oven BFED53 (Binder), an Orbital shaker from IKA® (# 01.472801), and a stereomicroscope (SZX12) from Olympus (#6L02262). The Vibroslicer (VT1000S, # 047235612), the ultramicrotome (Ultracut S), and the freeze-fracture replica machine (BAF-060) were from Leica Microsystems (Vienna, Austria). The high-pressure freezing machine (HPM-010) was obtained from BAL-TEC (Balzers, Liechtenstein).

V. Results and Discussion

Rodents have served for decades as vertebrate models to understand the structural and functional complexity of the CNS. The recent availability of genetically engineered mouse lines provides a further incentive and means to unravel the morphological basis of CNS functions and dysfunctions.

Obviously, the highest quality of tissue preservation should be strived for, approaching a native-like state. Currently, the most life-like rendering at the ultrastructural level is achieved by HPF followed by cryo-electron microscopy of vitrified sections (CEMOVIS) (Al-Amoudi *et al.*, 2004; Dubochet *et al.*, 2007). However, this technique is not widely available, and experimental purposes might require alternative methods.

When standard features (e.g., synapse numbers, cell density) need to be analyzed in large cohorts of samples or animals, perfusion fixation and conventional embedding remain valid, as they are easy and highly reproducible methods (see Section III.A); however, in some cases like for the retina or myelin, typical ultrastructural features are difficult to preserve by conventional techniques. Then, for qualitative and quantitative descriptions, cryopreparation methods have to be applied to meet the highest standards of morphological preservation (Studer *et al.*, 2008). We have reported here the use of two well-established cryo-methods: First, we have described HPF/FS for the routine preparation of retinal tissues and optic nerves. We have highlighted the numerous advantages of this method. Second, we have described the application of a less commonly used cryo-method, freeze-fracture replica labeling and its great potential for neurobiological samples.

A. Chemical Fixation

After its introduction as a fixation reagent for EM (Sabatini *et al.*, 1963), glutaraldehyde became a universal routine fixative. By cross-linking proteins rapidly and irreversibly, it allows the preservation of fine ultrastructural details. Yet, it also leads to typical alterations of morphology, such as volume changes (Lee, 1984), undulating membranes, and shape changes of the endosomal organelles (Murk *et al.*, 2003). These artifacts are probably partially caused by osmotic effects of the fixing solution/vehicle buffer, since its osmolarity cannot match the different osmolarities of all subcellular compartments.

Whereas acceptable preservation of neuronal ultrastructure of brain tissue can be achieved upon perfusion fixation using aldehyde solutions, it is rarely the case in the retina. This may be due to its vascularization pattern, as blood vessels do not reach within all layers, and its sensitivity to hypoxia and excitotoxicity (Izumi *et al.*, 2000). Therefore, the retina is often fixed by immersion, after being dissected out—since the protective sclera would not allow for a fast penetration of the fixative. While this preparation yields decent overall preservation of the retina (Eldred *et al.*, 1983), it remains plagued by numerous artifacts: undulating and disrupted membranes, washed-out cytoplasm, large extracellular spaces, and damaged organelles (golgi cisternae, mitochondria etc.; Fig. 7A). The osmolarity of the fixative is also responsible for retinal alteration (Margo and Lee, 1995). Altogether, this limits the morphological examination of key features, such as the photoreceptor and bipolar ribbon synapses, and renders the outcome of any structure–function study arguable.

The preservation of myelin is another "notorious problem" in ultrastructure research, since the multilamellar structure tends to split upon chemical fixation (Feirabend *et al.*, 1994, 1998). This artifact could veil relevant pathological structural changes in mutant mice. Glutaraldehyde seems to minimize this effect, therefore some protocols apply up to 5% glutaraldehyde, such as "perfixol" (Griffiths *et al.*, 1981). A hypertonic buffer system like the phosphate buffer according to Karlsson and Schultz (1965) and Schultz and Karlsson (1965) with an osmolarity of 440 mOsmols proved beneficial for myelin preservation, but the membrane splitting is still frequently observed (Fig. 8A–C).

We experienced that the optic nerve myelin is also evenly well preserved by either chemical immersion fixation (Fig. 8B) or perfusion fixation (Fig. 8C), since the optic nerve is quite resilient if dissected carefully. However, since the brain is sensitive to the oxygen deprivation, HPF of inner brain regions involves the danger of artifacts caused by anoxia, when the time-span between tissue extraction and cryoimmobilization is extended. To circumvent this problem, tissue extraction can be achieved using a biopsy needle system (Hohenberg *et al.*, 1996; Shimoni and Müller, 1998; Vanhecke *et al.*, 2003). Alternatively, the brain tissue can be dissected after perfusion fixation and then processed for HPF/FS (Fig. 8D).

B. Cryopreparation

The use of HPF/FS has proven to be greatly beneficial for the preservation of retinal ultrastructure (Figs. 1 and 7). As for the optic nerve, it is essential to carry out the dissection and freezing of the retinal samples with the greatest care and minimal delay. When preparation time is kept to 5–10 min, all processes show an excellent ultrastructural preservation. Within the thickness of the sample, however, gradual artifacts due to the formation of ice crystals can be identified. Close to the surface, the neuropile is dense but smooth in appearance; identifiable specializations (i.e., terminals, dendrites, glial processes) have a distinguishable cytoplasmic content, and all membranes are smooth and run parallel to each other (Figs. 1 and 7). Mitochondria harbor regular cristae, and no sign of nuclear reticulation is observed (Fig. 7D1). In this case, synaptic vesicles, for example, are nicely distinguishable within and at the membrane of presynaptic ribbon terminals (Fig. 7F), a feature that will allow unprecedented accuracy in the structure–function analysis of numerous mutant mice. Deeper in the tissue, typical artifacts progressively appear and can serve as landmarks to assess the quality of a sample regardless of its origin (Fig. 7C–E). Even in very well-preserved areas, limited "blebbing" of the nuclear envelope can be observed and may be disregarded. In addition, limited mitochondrial alterations can be found in otherwise well-preserved regions. Straight cristae are often observed in some cell types as well as a "blotchy" aspect (Fig. 7D2), likely reflecting a suboptimal resin infiltration. In contrast, when mitochondria exhibit swollen or heavily disrupted cristae, and membranes become increasingly irregular and "wobbly," the ultrastructure should be analyzed with hesitation (Fig. 7C3, D3). Finally, reticulation of the chromatin (Fig. 7E3) or within the cytoplasm is definitely a sign of severe ice crystal damage. Suboptimal freezing can also result in the artifactual formation of a multitude of small, rounded elements across large, undefined extracellular spaces. Such areas should obviously not be considered for analysis.

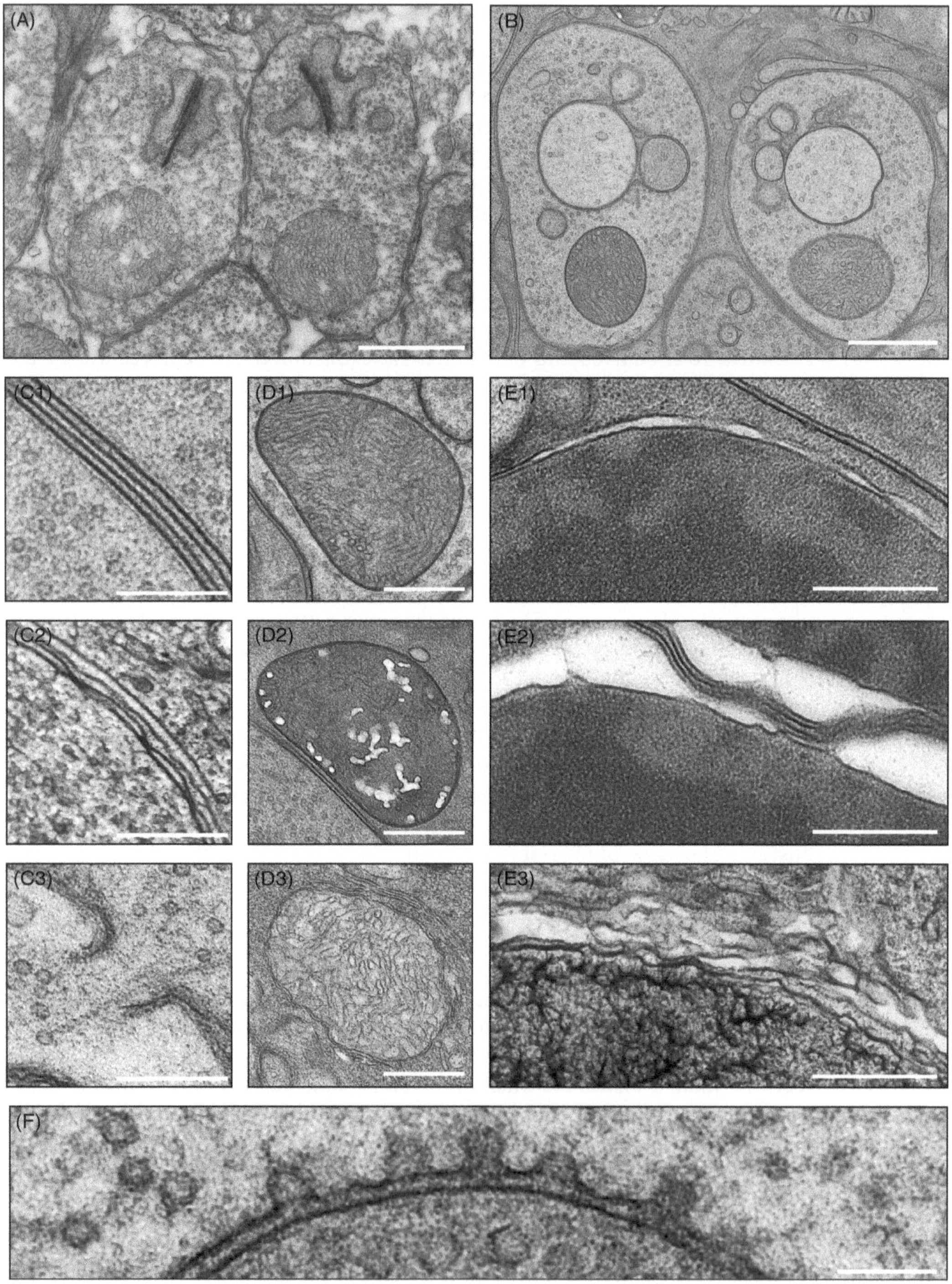

Fig. 7 (*Continued*)

By comparing samples of conventionally processed aldehyde-fixed tissue with samples from the same fixed specimen after HPF/FS, the effect of the dehydration procedure on the morphology can be studied. In the prefixed and freeze-substituted optic nerve, we have observed that myelin membrane splitting was still detectable (Fig. 8D). This implies that membrane splitting is an effect of fixation and not induced by the dehydration procedure. However, the visibility and preservation of membranes was superior compared to the conventionally processed samples. Moreover, the myelin thickness was increased, probably due to less shrinkage during freeze-substitution compared to conventional dehydration. The inner tongue is larger and displays more detail. These changes in morphology are obvious. Therefore, for quantitative studies, for example, on myelin thickness, one has to keep in mind that data obtained from high-pressure frozen and freeze-substituted specimen are not directly comparable with data extracted from conventional samples.

The optic nerve is a quite robust and a small CNS structure, which is ideally suited for cryopreparation techniques, after one has acquired the necessary skills for the dissection and sample loading steps (Suppl. Figs. S2 and 4). Systematic testing of different space fillers revealed that both fresh yeast paste and 20% PVP (MW = 10.000) are useful fillers with quite different properties: Yeast paste from fresh yeast is preferred compared to dry yeast, since the former is more homogenous and contains less air bubbles after mixing with ice-cold PBS. With some experience it is possible to apply the exact required quantity of yeast paste into the specimen carrier without overfilling it. After freeze-substitution and Epon infiltration the paste can be removed by scraping carefully before orienting and placing the samples into the embedding molds. In contrast, 20% PVP behaves very differently. It is a yellowish and slightly viscous liquid, in which we submerge the specimen carrier and the fresh optic nerve pieces during sample loading. By lifting the carrier while holding the optic nerve pieces with forceps in place, the nerves can be guided into the indentation of the specimen carrier. Then, before adding the lid to close the sandwich, only the level of liquid has to be adjusted with a fine tip of blotting paper. The

Fig. 7 Ultrastructural preservation of tissue and organelles upon HPF/FS of fresh retina (A, C–F) versus Epon embedding of perfusion-fixed retina (B). (A) Upon HPF, the tissue appears well contrasted, smooth, and intact at the OPL. (B) Upon perfusion-fixation, numerous artifacts are recurrently observed, such as tissue disruption, undulated and discontinuous membranes, damaged organelles (mitochondria), washed-out cytoplasm, and large extracellular spaces. (C–E) Upon HPF also, ultrastructural preservation varies within a sample; several features can be chosen to judge the degree of integrity of the sample, among which aspect of the membranes (C1–C3), mitochondria (D1–D3), and nucleus (E1–E3) are determinant. In the best-preserved areas, membranes are smooth, parallel, and continuous (C1); mitochondrial cristae are flat, regular, and readily recognizable (D1); and the nucleus harbors homogenous chromatin and a regularly spaced double membrane (E1). Wherever freezing was suboptimal, membranes were undulated (C2) or worse, segmented (C3); mitochondrial cristae were swollen (D3) or disrupted; the nuclear envelope was disjointed (E2) or worse, the chromatin was reticulated (E3). Of note, a blotchy aspect of mitochondria (D2) reflects a poor resin infiltration. (F) In an intact sample, synaptic or endocytic vesicles can be clearly observed in the cytoplasm and at the plasma membrane of photoreceptor ribbon terminals. Scale bars = 1 μm in (A) and (B), 250 nm in (C), 500 nm in (D) and (E), and 100 nm in (E).

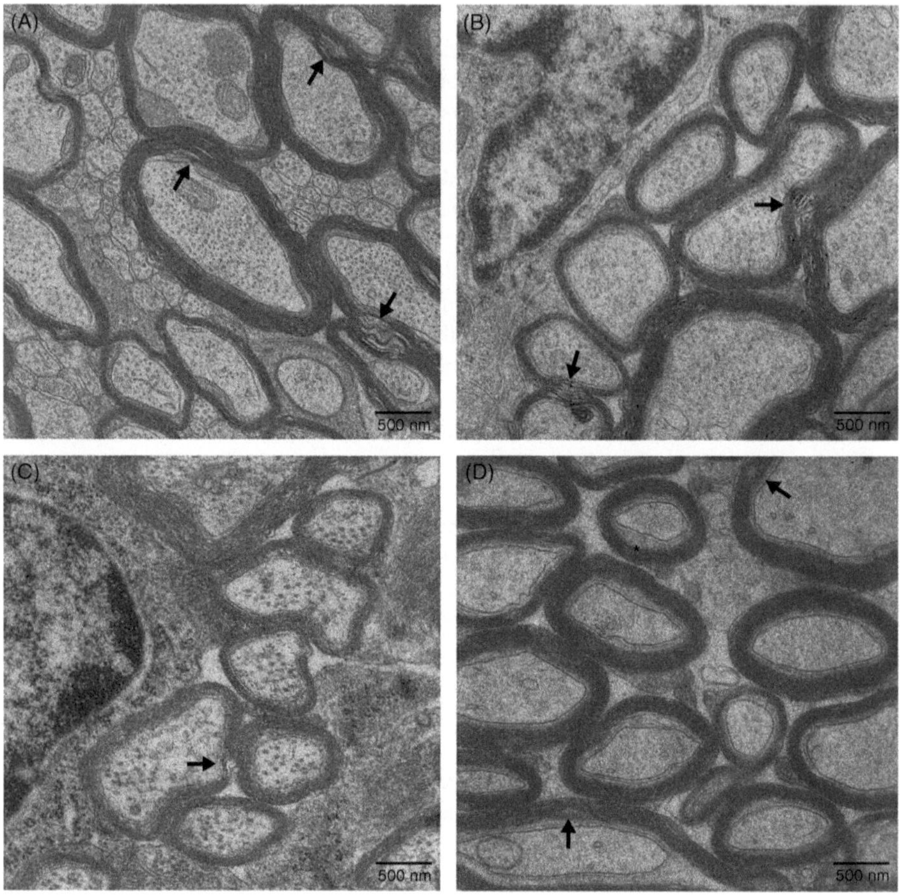

Fig. 8 Chemical fixation: Aspect of aldehyde-fixed tissue (Karlsson–Schultz fixation solution) from the mouse corpus callosum (A) and the optic nerve (B–D) upon immersion (B), perfusion (A, C), or perfusion-fixation followed by HPF/FS (D). (A) Upon perfusion-fixation, cytoskeleton and mitochondria are well preserved in axons of the corpus callosum, yet myelin suffers from membrane splitting (arrows). (B) After immersion fixation, the ultrastructure of the optic nerve is well preserved except for the splitting of myelin membranes (arrows). (C) Upon perfusion-fixation, membrane visibility and contrast are poor and myelin membrane splitting is obvious in a piece of mouse optic nerve. (D) Upon HPF/FS of another piece of the same optic nerve with the tannic acid–OsO_4 protocol, membranes are smooth and their contrast visibly enhanced. The inner tongue is enlarged and the myelin sheath is thicker. Yet, myelin membrane splitting is still detectable (arrows), implying that this artifact is induced by chemical fixation. For a comparison, see the ultrastructure of freshly prepared tissue in Fig. 9. Scale bars = 500 nm.

PVP filler hardly ever contains air bubbles. Some attention has to be paid to remove air bubbles that stick in the specimen carrier indentation after submerging it into the filler. After freeze-substitution and resin infiltration, the sample tends to stick to the specimen carrier. By gently circumscribing the sample with an injection needle, the whole piece containing the nerves can be removed without damage. Although PVP turns dark brown

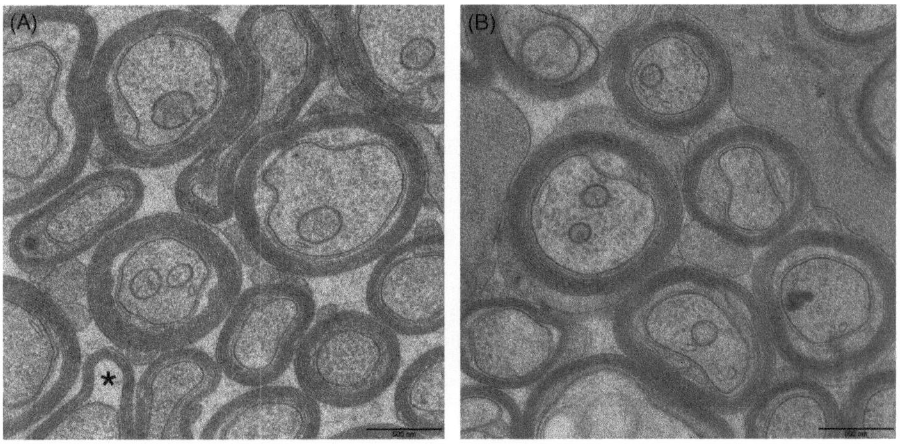

Fig. 9 The effect of different space fillers for HPF: Ultrastructure of fresh mouse optic nerve after HPF using yeast paste (A) versus 20% PVP in PBS (B) as fillers and freeze-substitution according to the tannic acid–OsO_4 protocol. (A) Samples frozen with yeast paste display mostly more contrast and large extracellular spaces. Distortions of the axon–glia interface are often observed (asterisk in (A)). (B) PVP-filled samples show a little bit less contrast and mostly less ice crystal damage. Compared to chemically fixed samples, the inner tongue of the myelin sheath is larger and shows more detail. Scale bars = 500 nm.

during freeze-substitution, the shape of the nerves can still be seen by using a light source from below. This makes it possible to orient the nerves for transverse sectioning. We experienced that samples frozen with yeast paste are slightly different from samples frozen in 20% PVP (Fig. 9). Samples with yeast paste tend to have more contrast and larger extracellular space (Fig. 9A) compared to PVP samples, which suffer less from freezing damage (Fig. 9B).

The most astonishing difference between cryopreserved and conventionally processed samples is the superior preservation of myelin and the fine detail in the inner tongue as well as in other glial processes. Artifacts associated with the cryopreparation procedure can be caused either by handling the fresh tissue or by ice crystal damage. A typical handling artifact concerning the preservation of myelin is the detachment of myelin lamellae as shown in Fig. 10A. This is probably caused by stretching the nerve. Another frequent artifact of the same category is the separation of the axon and the myelin sheath (Fig. 10B and C). Since the integrity of the axon–myelin interface is easily disturbed, great attention has to be paid to the proper handling of fresh nerves. Also the space filler could play a role here, since we observed less axon–myelin separation when using PVP, possibly due to less air inclusions in the filler. We observed the same artifact after plunge freezing (data not shown), so it is not solely caused by pressure. We also observed cracks in the myelin sheath (Fig. 10C). Whether this is a pressure artifact or caused by "freeze fracture" during removal of the lid carrier is not clear to us. Apart from that, typical signs of ice crystal damage are clustered neurofilaments in the axon, undulating axonal membrane, artificially fused membranes, clustered appearance of GFAP (glial fibrillary acidic protein) in the

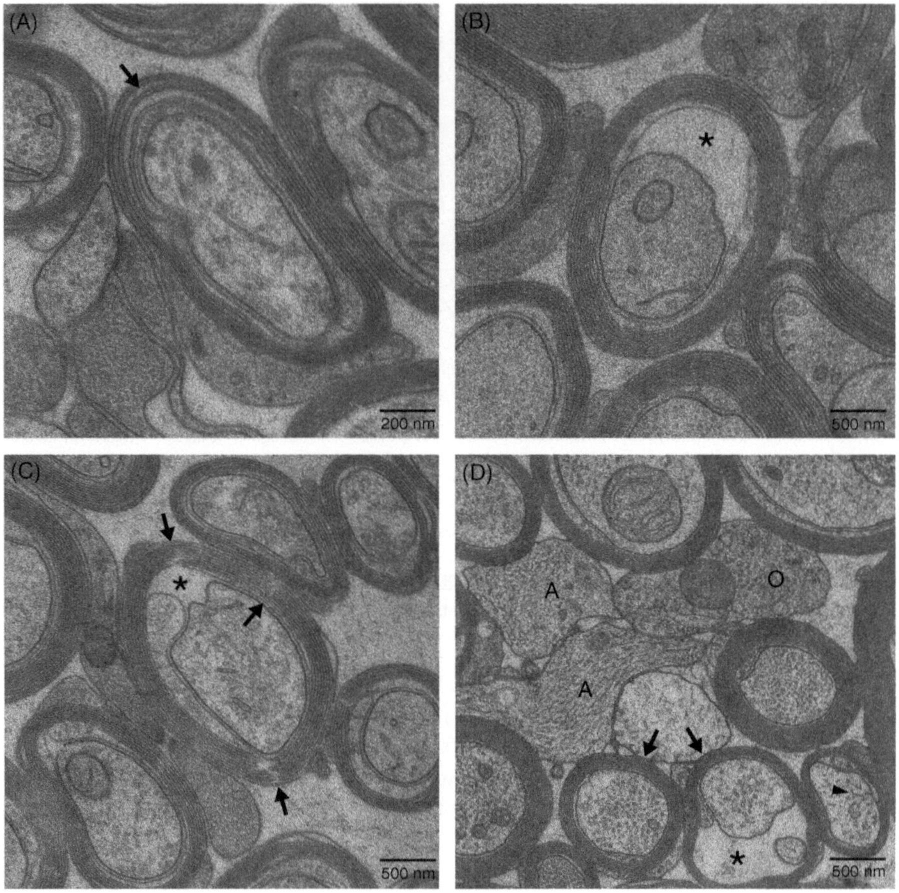

Fig. 10 Artifacts of cryopreparation: Mouse optic nerve after HPF using yeast paste as space filler and after freeze-substitution following the tannic acid–OsO_4 protocol. (A) Detached myelin lamellae (arrow) are probably caused by handling, for example, pulling the nerve. (B) Distortions of the axon–myelin coherence are also frequently observed (asterisk, also in C and D). They are not caused by pressure, since they also occur during plunge freezing at ambient pressure (data not shown). Probably the separation of myelin and axon is caused by handling the fragile fresh nerve. The use of PVP as space filler seems to be beneficial for the preservation of the axon–myelin coherence may be due to less air inclusions in the filler. Sometimes in addition to myelin–axon separation (asterisk) the myelin sheath shows cracks. Because this occurs in otherwise well-frozen areas, this artifact is not related to ice crystal damage, but rather caused by fracture during opening of the specimen carrier sandwich or other unknown effect. (D) Signs of typical ice crystal damage: Two astrocytic processes (marked "A") next to an oligodendrocytic process (O) show clustered GFAP (glial fibrillary acidic protein) filaments. Clustering of neurofilaments is visible within the axons (arrowhead). A frequent freezing damage is the artificial fusion of membranes (arrows) here visible between a naked axon showing also clustered microtubule and neurofilaments and neighboring myelin sheaths. Moreover, the membrane of this axon is not smooth. Scale bars = 200 nm in (A), 500 nm in (B), (C), and (D).

astrocytic processes (Fig. 10D), and the typical aggregation pattern in the nucleus (data not shown). To our experience it is important to keep the sample size as small as possible. Therefore, best results were obtained with optic nerves of young mice which fit into the 0.1 mm specimen carrier. However, myelin seems to be totally resilient to freezing damage.

The choice of chemical stabilizing agents in the freeze-substitution defines the final contrast and appearance of the tissue in the electron microscope. We tested two different protocols, one traditional using only 2% OsO_4 in acetone as fixing agent (Van Harreveld and Crowell, 1964; Van Harreveld *et al.*, 1965) in a 3-step procedure with 8 h at −90, −60, and −30°C (Hohenberg *et al.*, 1994; Humbel and Müller, 1986; Studer *et al.*, 1989; Vanhecke *et al.*, 2003), the other one using tannic acid in acetone at −90°C as a mordant followed by a OsO_4 alone or a combination of OsO_4 and uranyl acetate (Siksou *et al.*, 2007). The first protocol results in a weak and fine membrane contrast, but ribosomes and chromatin appear well contrasted (Fig. 11A). The second protocol is designed for increased contrast. Using tannic acid during freeze-substitution was already described for plant samples (discussed in Hess (2007)), embryonic tissue (Allenspach, 1993), cell culture (Bridgman and Reese, 1984; Giddings, 2003), frog retina (Wilson *et al.*, 1998), and prefixed rat brain (Sosinsky *et al.*, 2008). It can also be applied after freeze-substitution (Jimenez *et al.*, 2009) and at the stage of post-staining of sections (McDonald and Müller-Reichert, 2002). Tannic acid effects membrane contrast by incubation at −90°C, since it is removed before continued substitution with OsO_4 at −90°C. This protocol for freeze-substitution of optic nerves results in clearly visible membranes, ribosomes, and chromatin (Fig. 11B). Complementation of the 2% OsO_4 freeze-substitution "cocktail" with 0.1% uranyl acetate increases the intrinsic contrast of the sample. This becomes obvious only after omitting the post-staining of the sections. In noncontrasted longitudinal sections of a node of Ranvier the contrast is already quite good (Fig. 11C and E), but after addition of uranyl acetate the contrast of unstained sections is greatly enhanced (Fig. 11D and F). This effect turned out to be very helpful for the application of 3D methods, such as EM-tomography or serial block-face imaging (data not shown). In Fig. 11F the septate-like junctions (*arrow*) linking the paranodal loops (*asterisk*) of the terminating myelin sheath to the axonal membrane are clearly visible without further staining. However, they do not have the appearance of the so-called transverse bands (Hirano and Dembitzer, 1969) described as periodic paranodal structures in Kruger and Maxwell (1966) which the same junctions are visualized more clearly after conventional processing.

C. Freeze-Fracture Replica Labeling

First attempts to apply freeze-fracture EM to the analysis of biological material were made in the 1950s and 1960s (Moor and Mühlethaler, 1963; Steere, 1957). This technique boomed then in the 1970s and 1980s (Afzelius and Maunsbach, 2004) when it was used to analyze the structural organization of various biological membranes and organelles. An example for such investigations is illustrated in Fig. 12 showing a cross-fractured axon of the rat cerebellum. At low magnification the axoplasm with fractured microfilaments and

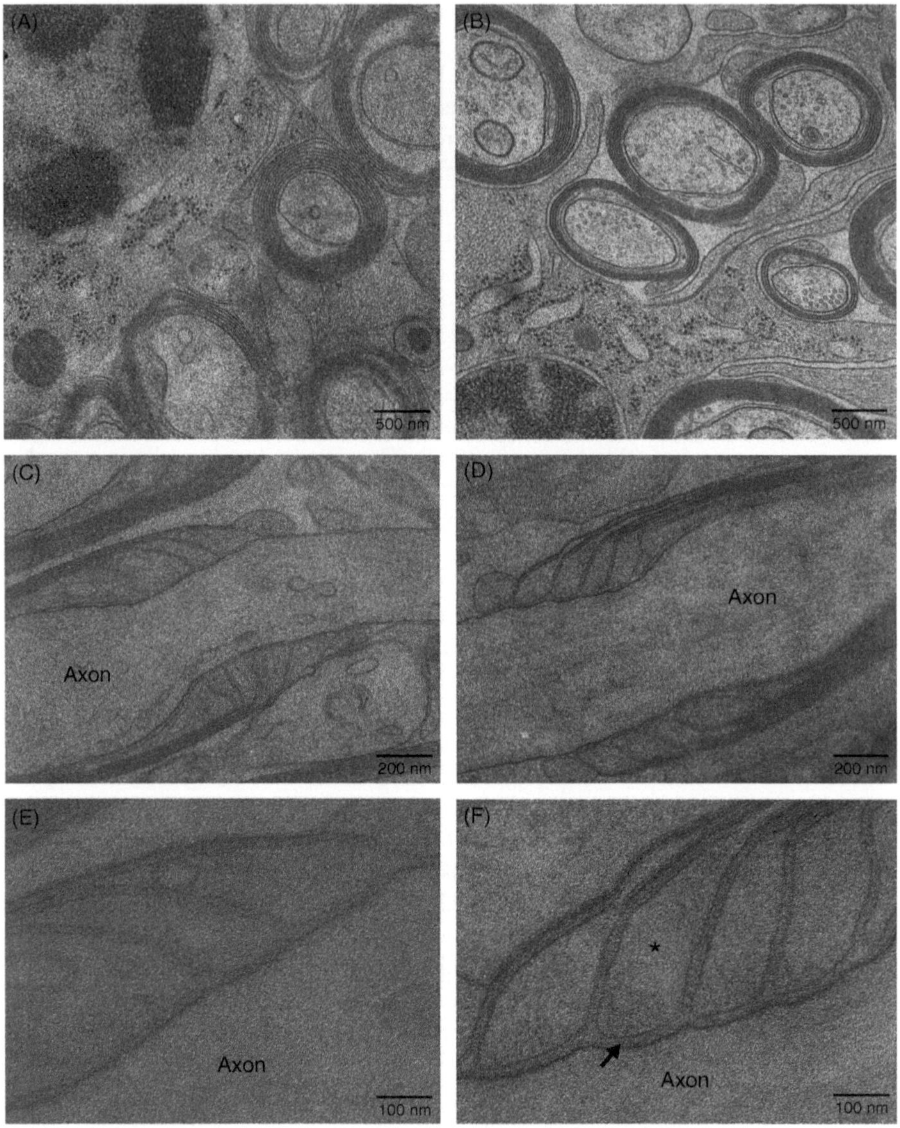

Fig. 11 Effect of different freeze-substitution cocktails: (A) Mouse optic nerve substituted after HPF/FS following the "classical" OsO_4 protocol. Membranes show a weak contrast, but ribosomes and chromatin appear clearly contrasted. The lack of contrast makes it difficult to identify the glial cell on the left. Bar: 500 nm. (B) Freeze-substitution using the tannic acid–OsO_4 protocol. Membranes, ribosomes, chromatin, and cytoskeletal elements are well contrasted. The glial cell in the lower left corner extending slender processes between axons shows characteristics of an astrocyte. (C) and (E) Node of Ranvier in a noncontrasted longitudinal section from a sample substituted following the tannic acid–OsO_4 protocol. Membranes are well visible but the contrast is too low to see more details. (D) and (F) Noncontrasted longitudinal section of a node in a sample substituted according to tannic acid–OsO_4 protocol with addition of 0.1% uranyl acetate to the 2% OsO_4 cocktail. Complementation with 0.1% uranyl acetate greatly enhances the contrast making this kind of sample ideal for three-dimensional reconstruction by EM-tomography or serial block-face imaging. Paranodal loop (asterisk), septate-like junction (arrow). Scale bars = 500 nm in (A) and (B), 200 nm (C) and (D), and 100 nm (E) and (F).

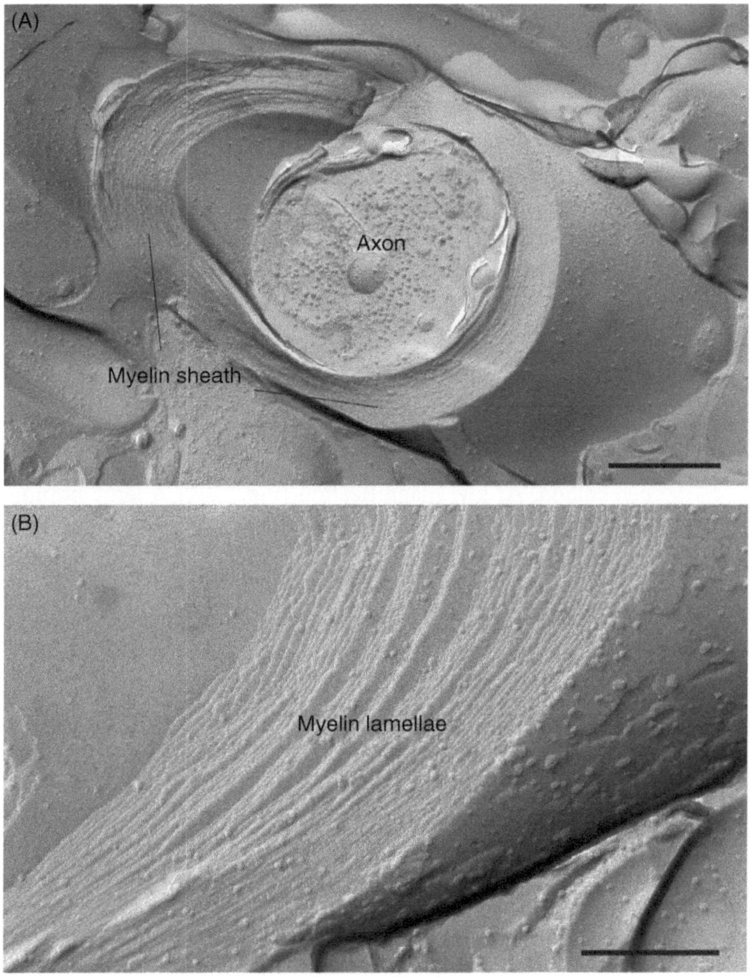

Fig. 12 Freeze-fractured and replicated axons of the rat cerebellum. (A) Low-power micrograph of a cross-fractured axon surrounded by a myelin sheath. Scattered microfilaments and subcellular organelles can be differentiated in the axoplasm. (B) In an aspect of the myelin sheath at higher magnification, several layers of myelin lamellae stubbed with intramembrane particles are visible. Scale bars = 600 nm in (A), 350 nm in (B).

subcellular organelles and the surrounding myelin sheath are visible (Fig. 12A). At higher magnification (Fig. 12B), several layers of myelin lamellae stubbed with scattered intramembrane particles can be differentiated within the myelin sheath. Such intramembrane particles represent protein aggregations in the plasma membrane. However, the molecular composition of these aggregations remains obscure when performing conventional freeze-fracture techniques—a likely reason why interest in this technique faded out. Freeze-fracture replication received a substantial boost again when Fujimoto

(Fujimoto, 1995) combined classical freeze-fracture EM with SDS digestion for immunocytochemical labeling of membrane proteins (SDS-FRL). This innovative method suddenly allowed the identification of proteins on replicas and their precise localization in particular cell types at a submicron level. It was originally applied to liver, small intestine, and kidney samples (Fujimoto, 1995, 1997), and later modified and used for the analysis of brain tissue (Fukazawa *et al.*, 2009). It has several advantages compared to conventional immunogold electron microscopy, such as pre-embedding or post-embedding immunogold labeling with respect to sensitivity and quantifiability (Masugi-Tokita and Shigemoto, 2007). Reliable quantification of immunolabeling has often been hampered by the limited accessibility of antigens buried in the tissue. In SDS-FRL, antibodies can target molecules exposed in the biomembrane without diffusion restrictions resulting in high sensitivity and reproducibility. An example of such a replicated and immunolabeled sample is illustrated in Fig. 13. Antibodies conjugated to 10 nm gold particles and directed against the alpha principal subunit of large-conductance calcium-activated potassium (BK) channels were applied on replicas of the rat cerebellum. At low magnification (Fig. 13A), a large portion of a neuron soma and the very proximal portions of primary dendrites are visible. Immunogold particles labeling BK channels are unevenly distributed over the plasma membrane forming two distinct pools. One pool is scattered and the other pool is clustered in the extrasynaptic plasma membrane (Fig. 13B). In distal portions of dendrites, only scattered BK channels can be detected. They are localized to dendritic trunks and spines although at low abundance (Fig. 13C; for details see Kaufmann *et al.* (2009).

In freeze-fracture, the fracture plane tends to follow the central hydrophobic core of a biomembrane thereby splitting the membrane into two leaflets, an outer extracellular E-face and an inner protoplasmic P-face (Fukazawa *et al.*, 2009; Severs, 2007). Integral membrane proteins are thereby allocated unpredictably to either of the membrane faces or to both depending on the nature of the molecule. Proteins in the E-face can then be visualized using antibodies directed against epitopes at their extracellular domains, whereas proteins allocated to the P-face can be visualized using antibodies directed against intracellular domains. In case of BK channel localization, antibodies directed against an epitope at the intracellular C-terminal domain of the alpha subunit were applied (Kaufmann *et al.*, 2009). Thus, the immunolabeling is found at the P-face of the plasma membrane. The E-face is free of any immunolabeling (Fig. 13B). For quantitative analysis, both pools need to be considered. In case of knife fracture, material above the blade path is discarded while both sides of the fracture plane are retained and replicated in tensile fracture. This allows the localization of integral membrane proteins on both faces of the membrane and quantification of the whole pool of a specific protein. Tensile fracture is thus preferred for quantitative immunolabeling studies. In some instances, the fracture plane does not follow the core of a biomembrane at all and the specimen is cross-fractured allowing an *en face* view on the specimen. As exemplified, an axon terminal synapsing with a convex-fractured dendritic spine is cross-fractured and clear synaptic vesicles devoid of any immunolabeling are visible in the bouton (Fig. 13C). In general, the plane of fracture is unpredictable but influenced by the nature of the tissue, degree of chemical treatment

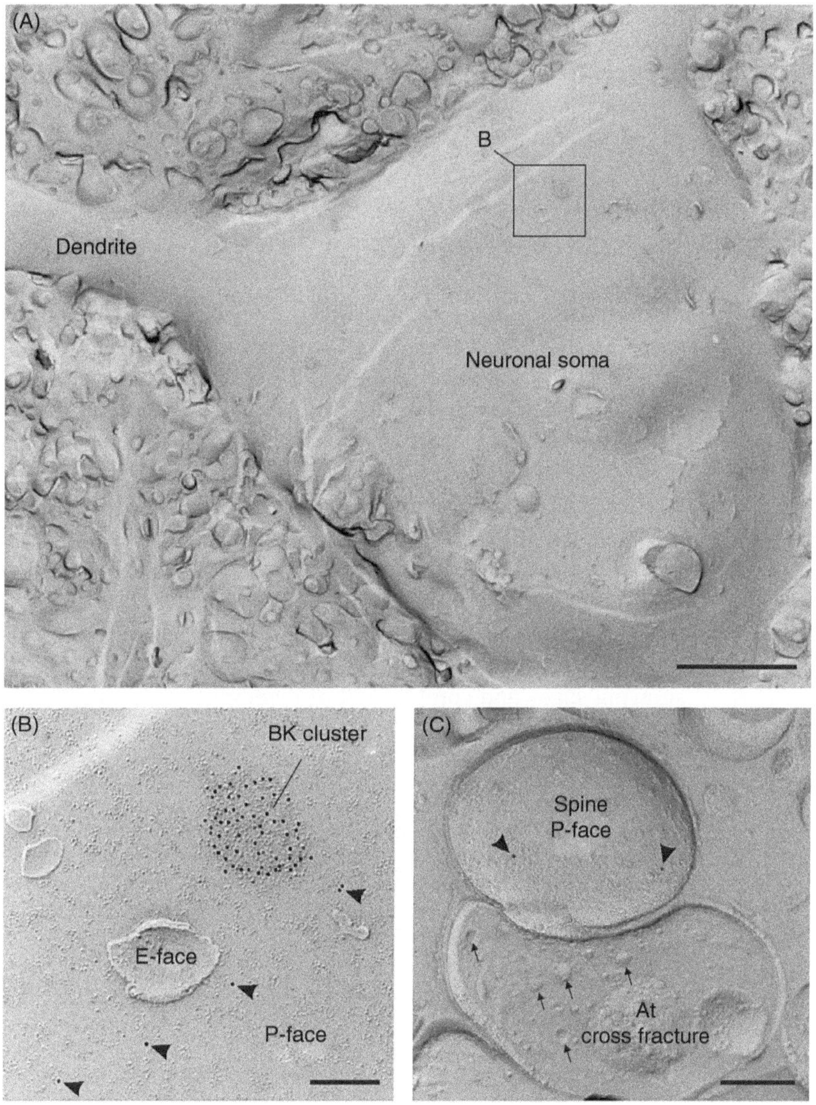

Fig. 13 Immunolocalization of large-conductance calcium-activated potassium (BK) channels in a neuron of the rat cerebellum with SDS-digested freeze-fracture replica labeling (SDS-FRL). (A) Low-power micrograph of a neuronal soma immunolabeled with antibodies conjugated to 10-nm gold particles and directed against epitopes at an intracellular C-terminal domain of the BK channel alpha subunit. Hence, channels can be detected at the P-face of the plasma membrane (antibodies kindly provided by H.G. Knaus, Innsbruck Medical University, Austria). (B) In an aspect of the plasma membrane at higher magnification (boxed area in A), a cluster of BK channels associated with a cluster of intramembrane particles, next to some scattered BK channels can be seen (indicated with arrowheads). (C) Scattered BK channels (arrowheads) can also be observed postsynaptic in the plasma membrane P-face of a dendritic spine. The presynaptic axon terminal (At) is cross-fractured and small synaptic vesicles are detectable (indicated with arrows). Scale bars = 10 μm in (A), 300 nm in (B), and 100 nm in (C).

(e.g., concentration of fixative used) as well as physical settings like fracture temperature. These parameters need to be established for each sampling area, and may be varied to meet specific goals of a study.

VI. Concluding Remarks

EM plays an important role in the analysis of the CNS in mouse mutants that often serve as disease models. Most published data are based on conventional sample preparation by chemical fixation and Epon embedding. For quantitative studies involving large cohorts of samples this is a valid approach, but for the analysis of fine structural details cryopreparation methods are offering the best sample preservation. This is especially the case for retina and myelin which are rarely well preserved by conventional fixation and embedding procedures. HPF/FS are well-established cryopreparation methods, but for every sample of choice the procedures have to be adjusted and associated artifacts revealed. Here, a comparative approach by also using chemically prefixed samples for HPF/FS seems useful. Prefixed samples are also used for freeze-fracture cytochemistry after HPF. This technique is the only method for transmission electron microscopy (TEM) which reveals a planar view on membranes showing shape and composition of membrane domains.

Acknowledgments

We are grateful to Klaus Armin Nave (Department of Neurogenetics, MPI of Experimental Medicine, Göttingen) and Nils Brose (Department of Molecular Neurobiology, MPI of Experimental Medicine, Göttingen) for their continuous support. Furthermore, we like to acknowledge Martin Doerre for his excellent technical assistance and wish to thank Michael W. Hess (Innsbruck Medical University, Austria) for reading and commenting on our manuscript.

References

Afzelius, B. A., and Maunsbach, A. B. (2004). Biological ultrastructure research; the first 50 years. *Tissue Cell* **36**, 83–94.

Al-Amoudi, A., Chang, J. J., Leforestier, A., McDowall, A., Salamin, L. M., Norlen, L. P., Richter, K., Blanc, N. S., Studer, D., and Dubochet, J. (2004). Cryo-electron microscopy of vitreous sections. *EMBO J.* **23**, 3583–3588.

Allenspach, A. (1993). Ultrastructure of early chick embryo tissues after high pressure freezing and freeze substitution. *Microsc. Res. Tech.* **24**, 369–384.

Barkovich, A. J. (2007). Myelin mishaps. *Ann. Neurol.* **62**, 107–109.

Belmonte, M. K., and Bourgeron, T. (2006). Fragile X syndrome and autism at the intersection of genetic and neural networks. *Nat. Neurosci.* **9**, 1221–1225.

Boespflug-Tanguy, O., Labauge, P., Fogli, A., and Vaurs-Barriere, C. (2008). Genes involved in leukodystrophies: A glance at glial functions. *Curr. Neurol. Neurosci. Rep.* **8**, 217–229.

Bonnet-Dupeyron, M. N., Combes, P., Santander, P., Cailloux, F., Boespflug-Tanguy, O., and Vaurs-Barriere, C. (2008). PLP1 splicing abnormalities identified in pelizaeus-merzbacher disease and SPG2 fibroblasts are associated with different types of mutations. *Hum. Mutat.* **29**, 1028–1036.

Bridgman, P. C., and Reese, T. S. (1984). The structure of cytoplasm in directly frozen cultured cells. I. Filamentous meshworks and the cytoplasmic ground substance. *J. Cell Biol.* **99**, 1655–1668.

Bunge, M. B., Bunge, R. P., and Pappas, G. D. (1962). Electron microscopic demonstration of connections between glia and myelin sheaths in the developing mammalian central nervous system. *J. Cell Biol.* **12**, 448–453.

Bunge, M. B., Bunge, R. P., and Ris, H. (1961). Ultrastructural study of remyelination in an experimental lesion in adult cat spinal cord. *J. Biophys. Biochem. Cytol.* **10**, 67–94.

Bunge, R. P., and Glass, P. M. (1965). Some observations on myelin-glial relationships and on the etiology of the cerebrospinal fluid exchange lesion. *Ann. N. Y. Acad. Sci.* **122**, 15–28.

Dubochet, J., Zuber, B., Eltsov, M., Bouchet-Marquis, C., Al-Amoudi, A., and Livolant, F. (2007). How to "read" a vitreous section. *Methods Cell Biol.* **79**, 385–406.

Edgar, J. M., and Nave, K. A. (2009). The role of CNS glia in preserving axon function. *Curr. Opin. Neurobiol.* **19**, 498–504.

Eldred, W. D., Zucker, C., Karten, H. J., and Yazulla, S. (1983). Comparison of fixation and penetration enhancement techniques for use in ultrastructural immunocytochemistry. *J. Histochem. Cytochem.* **31**, 285–292.

Feirabend, H. K., Choufoer, H., and Ploeger, S. (1998). Preservation and staining of myelinated nerve fibers. *Methods* **15**, 123–131.

Feirabend, H. K., Kok, P., Choufoer, H., and Ploeger, S. (1994). Preservation of myelinated fibers for electron microscopy: A qualitative comparison of aldehyde fixation, microwave stabilisation and other procedures all completed by osmication. *J. Neurosci. Methods* **55**, 137–153.

Fernandez-Moran, H. (1950). Electron microscope observations on the structure of the myelinated nerve fiber sheath. *Exp. Cell Res.* **1**, 143–149.

Fujimoto, K. (1995). Freeze-fracture replica electron microscopy combined with SDS digestion for cytochemical labeling of integral membrane proteins. Application to the immunogold labeling of intercellular junctional complexes. *J. Cell. Sci.* **108**(Pt 11), 3443–3449.

Fujimoto, K. (1997). SDS-digested freeze-fracture replica labeling electron microscopy to study the two-dimensional distribution of integral membrane proteins and phospholipids in biomembranes: Practical procedure, interpretation and application. *Histochem. Cell Biol.* **107**, 87–96.

Fukazawa, Y., Masugi-Tokita, M., Tarusawa, E., Hagiwara, A., and Shigemoto, R. (2009). SDS-digested freeze-fracture replica labeling (SDS-FRL). *In* "Handbook of Cryopreparation Methods for Electron Microscopy" (A. Cavalier, D. Spehner, and B. M. Humbel, eds.), CRC Press Inc, Boca Raton, FL.

Geren, B. B. (1954). The formation from the schwann cell surface of myelin in the peripheral nerves of chick embryos. *Exp. Cell Res.* **7**, 558–562.

Giddings, T. H. (2003). Freeze-substitution protocols for improved visualization of membranes in high-pressure frozen samples. *J. Microsc.* **212**, 53–61.

Glauert, A. M. (1974). Part I: Fixation, dehydration and embedding of biological specimens. *In* "Practical Methods in Electron Microscopy" (A. M. Glauert, ed.) Vol. 3, pp. 1–207. North-Holland Publishing Company/American Elsevier Publishing Co., Amsterdam/New York.

Gollisch, T., and Meister, M. (2010). Eye smarter than scientists believed: Neural computations in circuits of the retina. *Neuron* **65**, 150–164.

Griffiths, I. R., Duncan, I. D., and McCulloch, M. (1981). Shaking pups: A disorder of central myelination in the spaniel dog. II. Ultrastructural observations on the white matter of the cervical spinal cord. *J. Neurocytol.* **10**, 847–858.

Griffiths, I., Klugmann, M., Anderson, T., Yool, D., Thomson, C., Schwab, M. H., Schneider, A., Zimmermann, F., McCulloch, M., Nadon, N., and Nave, K. A. (1998). Axonal swellings and degeneration in mice lacking the major proteolipid of myelin. *Science* **280**, 1610–1613.

Griffiths, I. R., Schneider, A., Anderson, J., and Nave, K. A. (1995). Transgenic and natural mouse models of proteolipid protein (PLP)-related dysmyelination and demyelination. *Brain Pathol.* **5**, 275–281.

Hagiwara, A., Fukazawa, Y., Deguchi-Tawarada, M., Ohtsuka, T., and Shigemoto, R. (2005). Differential distribution of release-related proteins in the hippocampal CA3 area as revealed by freeze-fracture replica labeling. *J. Comp. Neurol.* **489**, 195–216.

Hess, M. W. (2007). Cryopreparation methodology for plant cell biology. Methods Cell Biol. **79**, 57–100.

Hirano, A., and Dembitzer, H. M. (1967). A structural analysis of the myelin sheath in the central nervous system. *J. Cell Biol.* **34**, 555–567.

Hirano, A., and Dembitzer, H. M. (1969). The transverse bands as a means of access to the periaxonal space of the central myelinated nerve fiber. *J. Ultrastruct. Res.* **28**, 141–149.

Hohenberg, H., Mannweiler, K., and Müller, M. (1994). High-pressure freezing of cell suspensions in cellulose capillary tubes. *J. Microsc.* **175**, 34–43.

Hohenberg, H., Tobler, M., and Müller, M. (1996). High-pressure freezing of tissue obtained by fine-needle biopsy. *J. Microsc.* **183**, 133–139.

Humbel, B. M., and Müller, M. (1986). "Freeze Substitution and Low Temperature Embedding. *In* Science of Biological Specimen Preparation 1985." pp. 175–183. SEM Inc, AMF O'Hare, Chicago, IL.

Inoue, K. (2005). PLP1-related inherited dysmyelinating disorders: Pelizaeus-merzbacher disease and spastic paraplegia type 2. *Neurogenetics* **6**, 1–16.

Izumi, Y., Hammerman, S. B., Benz, A. M., Labruyere, J., Zorumski, C. F., and Olney, J. W. (2000). Comparison of rat retinal fixation techniques: Chemical fixation and microwave irradiation. *Exp. Eye Res.* **70**, 191–198.

Jimenez, N., Vocking, K., van Donselaar, E. G., Humbel, B. M., Post, J. A., and Verkleij, A. J. (2009). Tannic acid-mediated osmium impregnation after freeze-substitution: A strategy to enhance membrane contrast for electron tomography. *J. Struct. Biol.* **166**, 103–106.

Kaech, A. (2009). Bal-tec HPM 010 high-pressure freezing machine. *In* "Handbook of Cryo-Preparation Methods for Electron Microscopy" (A. Cavalier, D. Spehner103–128. CRC Press, Boca Raton, FL.

Karlsson, U., and Schultz, R. L. (1965). Fixation of the central nervous system from electron microscopy by aldehyde perfusion. I. Preservation with aldehyde perfusates versus direct perfusion with osmium tetroxide with special reference to membranes and the extracellular space. *J. Ultrastruct. Res.* **12**, 160–186.

Karnovsky, M. J. (1965). A formaldehyde-glutaraldehyde fixative of high osmolality for use in electron microscopy. *J. Cell Biol.* **27**, 137A.

Kaufmann, W. A., Ferraguti, F., Fukazawa, Y., Kasugai, Y., Shigemoto, R., Laake, P., Sexton, J. A., Ruth, P., Wietzorrek, G., Knaus, H. G., Storm, J. F., and Ottersen, O. P. (2009). Large-conductance calcium-activated potassium channels in purkinje cell plasma membranes are clustered at sites of hypolemmal microdomains. *J. Comp. Neurol.* **515**, 215–230.

Kruger, L., and Maxwell, D. S. (1966). Electron microscopy of oligodendrocytes in normal rat cerebrum. *Am. J. Anat.* **118**, 411–435.

Lagnado, L. (2003). Ribbon synapses. *Curr. Biol.* **13**, R631.

Lee, R. M. (1984). A critical appraisal of the effects of fixation, dehydration and embedding on cell volume. *In* "The Science of Biological Specimen Preparation" (J. P. Revel, T. Barnard61–70. SEM Inc, AMF O'Hare, Chicago, IL.

Margo, C. E., and Lee, A. (1995). Fixation of whole eyes: The role of fixative osmolarity in the production of tissue artifact. *Graefes Arch. Clin. Exp. Ophthalmol.* **233**, 366–370.

Masland, R. H. (2001). The fundamental plan of the retina. *Nat. Neurosci.* **4**, 877–886.

Masland, R. H. (2001). Neuronal diversity in the retina. *Curr. Opin. Neurobiol.* **11**, 431–436.

Masugi-Tokita, M., and Shigemoto, R. (2007). High-resolution quantitative visualization of glutamate and GABA receptors at central synapses. *Curr. Opin. Neurobiol.* **17**, 387–393.

Masugi-Tokita, M., Tarusawa, E., Watanabe, M., Molnar, E., Fujimoto, K., and Shigemoto, R. (2007). Number and density of AMPA receptors in individual synapses in the rat cerebellum as revealed by SDS-digested freeze-fracture replica labeling. *J. Neurosci.* **27**, 2135–2144.

McDonald, K. (2007). Cryopreparation methods for electron microscopy of selected model systems. *Methods Cell Biol.* **79**, 23–56.

McDonald, K. L., Morphew, M., Verkade, P., and Müller-Reichert, T. (2007). Recent advances in high-pressure freezing: Equipment- and specimen-loading methods. *Methods Mol. Biol.* **369**, 143–173.

McDonald, K., and Müller-Reichert, T. (2002). Cryomethods for thin section electron microscopy. *Methods Enzymol.* **351**, 96–123.

Möbius, W., Patzig, J., Nave, K. A., and Werner, H. B. (2008). Phylogeny of proteolipid proteins: Divergence, constraints, and the evolution of novel functions in myelination and neuroprotection. *Neuron Glia Biol.* **4**, 111–127.

Moor, H. (1987). Theory and practice of high pressure freezing. *In* "Cryotechniques in Biological Electron Microscopy" (R. A. Steinbrecht, K. Zierold, ed.) Vol. 17, pp. 175–191. Springer-Verlag, Berlin.

Moor, H., Bellin, G., Sandri, C., and Akert, K. (1980). The influence of high pressure freezing on mammalian nerve tissue. *Cell Tissue Res.* **209**, 201–216.

Moor, H., and Mühlethaler, K. (1963). Fine structure in frozen-etched yeast cells. *J. Cell Biol.* **17**, 609–628.

Murk, J. L., Posthuma, G., Koster, A. J., Geuze, H. J., Verkleij, A. J., Kleijmeer, M. J., and Humbel, B. M. (2003). Influence of aldehyde fixation on the morphology of endosomes and lysosomes: Quantitative analysis and electron tomography. *J. Microsc.* **212**, 81–90.

Nikolaus, S., Antke, C., and Muller, H. W. (2009). *In vivo* imaging of synaptic function in the central nervous system: I. Movement disorders and dementia. *Behav. Brain Res.* **204**, 1–31.

Nonaka, G., and Kishimoto, Y. (1979). Levels of cerebrosides, sulfatides, and galactosyl diglycerides in different regions of rat brain. Change during maturation and distribution in subcellular fractions of gray and white matter of sheep brain. *Biochim. Biophys. Acta* **572**, 432–441.

Norton, W. T., and Cammer, W. (1984). Isolation and characterization of myelin. *In* "Myelin" (P. Morell, ed.) pp. 147–195. Plenum Press, New York.

Norton, W. T., and Poduslo, S. E. (1973). Myelination in rat brain: Changes in myelin composition during brain maturation. *J. Neurochem.* **21**, 759–773.

Peters, A. (1960). The formation and structure of myelin sheaths in the central nervous system. *J. Biophys. Biochem. Cytol.* **8**, 431–446.

Peters, A. (1960). The structure of myelin sheaths in the central nervous system of xenopus laevis (daudin). *J. Biophys. Biochem. Cytol.* **7**, 121–126.

Peters, A. (1964). Observations on the connexions between myelin sheaths and glial cells in the optic nerves of young rats. *J. Anat.* **98**, 125–134.

Peters, P. J., and Pierson, J. (2008). Immunogold labeling of thawed cryosections. *Methods Cell Biol.* **88**, 131–149.

Reynolds, E. S. (1963). The use of lead citrate at high pH as an electron-opaque stain in electron microscopy. *J. Cell Biol.* **17**, 208–212.

Robertson, J. D. (1955). The ultrastructure of adult vertebrate peripheral myelinated nerve fibers in relation to myelinogenesis. *J. Biophys. Biochem. Cytol.* **1**, 271–278.

Sabatini, D. D., Bensch, K., and Barrnett, R. J. (1963). Cytochemistry and electron microscopy. The preservation of cellular ultrastructure and enzymatic activity by aldehyde fixation. *J. Cell Biol.* **17**, 19–58.

Saugier-Veber, P., Munnich, A., Bonneau, D., Rozet, J. M., Le Merrer, M., Gil, R., and Boespflug-Tanguy, O. (1994). X-linked spastic paraplegia and pelizaeus-merzbacher disease are allelic disorders at the proteolipid protein locus. *Nat. Genet.* **6**, 257–262.

Schultz, R. L., and Karlsson, U. (1965). Fixation of the central nervous system for electron microscopy by aldehyde perfusion. Ii. Effect of osmolarity, ph of perfusate, and fixative concentration. *J. Ultrastruct. Res.* **12**, 187–206.

Severs, N. J. (2007). Freeze-fracture electron microscopy. *Nat. Protoc.* **2**, 547–576.

Severs, N. J., and Robenek, H. (2008). Freeze-fracture cytochemistry in cell biology. *Methods Cell Biol.* **88**, 181–204.

Shimoni, E., and Müller, M. (1998). On optimizing high-pressure freezing: From heat transfer theory to a new microbiopsy device. *J. Microsc.* **192**, 236–247.

Siksou, L., Rostaing, P., Lechaire, J. P., Boudier, T., Ohtsuka, T., Fejtová, A., Kao, H. T., Greengard, P., Gundelfinger, E. D., Triller, A., and Marty, S. (2007). Three-dimensional architecture of presynaptic terminal cytomatrix. *J. Neurosci.* **27**, 6868–6877.

Sjöstrand, F. S. (1949). An electron microscope study of the retinal rods of the guinea pig eye. *J. Cell. Physiol.* **33**, 383–403.

Slot, J. W., and Geuze, H. J. (2007). Cryosectioning and immunolabeling. *Nat. Protoc.* **2**, 2480–2491.

Sosinsky, G. E., Crum, J., Jones, Y. Z., Lanman, J., Smarr, B., Terada, M., Martone, M. E., Deerinck, T. J., Johnson, J. E., and Ellisman, M. H. (2008). The combination of chemical fixation procedures with high pressure freezing and freeze substitution preserves highly labile tissue ultrastructure for electron tomography applications. *J. Struct. Biol.* **161**, 359–371.

Steere, R. L. (1957). Electron microscopy of structural detail in frozen biological specimens. *J. Biophys. Biochem. Cytol.* **3**, 45–60.

Sterling, P., and Matthews, G. (2005). Structure and function of ribbon synapses. *Trends Neurosci.* **28**, 20–29.

Studer, D., Humbel, B. M., and Chiquet, M. (2008). Electron microscopy of high pressure frozen samples: Bridging the gap between cellular ultrastructure and atomic resolution. *Histochem. Cell Biol.* **130**, 877–889.

Studer, D., Michel, M., and Muller, M. (1989). High pressure freezing comes of age. *Scanning Microsc. Suppl.* **3**, 253–268, discussion 268–9

Tarusawa, E., Matsui, K., Budisantoso, T., Molnar, E., Watanabe, M., Matsui, M., Fukazawa, Y., and Shigemoto, R. (2009). Input-specific intrasynaptic arrangements of ionotropic glutamate receptors and their impact on postsynaptic responses. *J. Neurosci.* **29**, 12896–12908.

Tekkök, S. B., Ye, Z., and Ransom, B. R. (2007). Excitotoxic mechanisms of ischemic injury in myelinated white matter. *J. Cereb. Blood Flow Metab.* **27**, 1540–1552.

Tokuyasu, K. T. (1973). A technique for ultracryotomy of cell suspensions and tissues. *J. Cell Biol.* **57**, 551–565.

Trapp, B. D., and Nave, K. A. (2008). Multiple sclerosis: An immune or neurodegenerative disorder? *Annu. Rev. Neurosci.* **31**, 247–269.

Van Harreveld, A., and Crowell, J. (1964). Electron microscopy after rapid freezing on a metal surface and substitution fixation. *Anat. Rec.* **149**, 381–385.

Van Harreveld, A., Crowell, J., and Malhotra, S. K. (1965). A study of extracellular space in central nervous tissue by freeze-substitution. *J. Cell Biol.* **25**, 117–137.

Vanhecke, D., Graber, W., Herrmann, G., Al-Amoudi, A., Eggli, P., and Studer, D. (2003). A rapid microbiopsy system to improve the preservation of biological samples Prior To high-pressure freezing. *J. Microsc.* **212**, 3–12.

Walther, P. (2003). Recent progress in freeze-fracturing of high-pressure frozen samples. *J. Microsc.* **212**, 34–43.

Wässle, H. (2004). Parallel processing in the mammalian retina. *Nat. Rev. Neurosci.* **5**, 747–757.

Wilson, M. T., Farmer, M. A., and Karwoski, C. J. (1998). Ultrastructure of the frog retina after high-pressure freezing and freeze substitution. *J. Microsc.* **189**, 219–235.

CHAPTER 21

Rapidly Excised and Cryofixed Rat Tissue

Dimitri Vanhecke, Werner Graber, ***and*** **Daniel Studer**

Institute of Anatomy, University of Berne, CH-3000 Bern 9, Switzerland

Abstract

All preparation efforts of biological samples in electron microscopy are focused to preserve structures as close as possible to the native state. To achieve this goal with tissues, it is of advantage to have a very short time between excision and fixation. The most common approach is chemical fixation: cross-linking of the tissue samples with aldehydes followed by postfixation with osmium tetroxide. Here, the fastest approach for tissue samples is perfusion. However, the diffusion of the fixation solution from

METHODS IN CELL BIOLOGY, VOL. 96

978-0-12-381007-6
DOI: 10.1016/10.1016/S0091-679X(10)96021-4

blood vessels into the depth of the tissue is still slow and does not allow an overall instant fixation of a single cell. As a result, osmotic effects become evident (swelling or shrinkage of cell organelles). Another possibility is to take a tissue sample from the experimental animal. Excision of tissue can last quite some time, which results in even more pronounced autolytic induced osmotic effects. Furthermore, the animal does not survive the procedure in most cases. Alternatively, microbiopsies are an elegant technique to rapidly excise small quantities of tissue. Some tissues, such as liver and muscle, may be obtained using a non-lethal approach. To avoid the artifacts introduced by chemical fixation, high-pressure freezing of microbiopsies (brain, liver, kidney, and muscle) is a powerful alternative to chemical fixation. Here, we describe the microbiopsy method, and high-pressure freezing/freeze-substitution (HPF/FS) as a follow-up procedure. Cryosectioning of high-pressure frozen samples is optimally preserving the ultrastructure; however, it is not considered to be a routine approach yet.

I. Introduction

A. General Aspects

Cryofixation aims for the vitrification of the sample, i.e., the solidification of the water content into vitreous water without the formation of ice crystals (Shimoni and Muller, 1998; Studer *et al.*, 1995). Successful cryofixation of biological samples results in superior ultrastructural morphology compared to chemically fixed samples (Moor and Mühlethaler, 1963). Such well-frozen biological samples yield an ultrastructure close to the living state at the distinct time point of immobilization (Adrian *et al.*, 1984; Eggli and Graber, 1994; Kiss *et al.*, 1990; Royer and Kinnamon, 1996; Studer *et al.*, 1989, 1995; Von Schack *et al.*, 1993). Furthermore, improved preservation of antigens was reported and hence improved probe binding is achieved (Hernandez-Verdun *et al.*, 1991; Monaghan *et al.*, 1998) and a significant decrease in loss and shifts of ions was observed (Somlyo *et al.*, 1985; Zierold, 1991), which is also reflected in the reduction of osmotic effects (Studer *et al.*, 1992).

B. Freezing Performance

Achieving vitrification of biological materials can be challenging. The success of vitrification mainly depends on the concentration of solutes in the sample and on its size. The solute concentrations in aqueous samples define the cooling rate necessary for vitrification; the size defines the absolute achievable cooling rate at a distinct position in the sample. Characteristic for the melting point of water is a minimum of about −22°C at a pressure of 2045 bar (Kanno *et al.*, 1975). Applying pressure in this range during freezing reduces the critical cooling rate for vitrification by a factor of 100 (Sartori *et al.*, 1993). The effects of the pressure are minimal, owing to the very low compressibility of water (Suppes *et al.*, 2003), yet they are existent (loss of liquid crystal structure of DNA: Leforestier *et al.*, 1996; change of lipidic phase: Semmler *et al.*, 1998).

Depending on the rate of cooling, crystallized (ice) or vitreous water is formed. Ice induces segregation by phase separation. These segregation patterns can be wrongly interpreted as real structures (compare Hunziker and Schenk, 1984 with Studer *et al.*, 1995). Vitreous water is an amorphous solid and induces no segregation in the samples. Consummate cryofixation by high-pressure freezing leads to vitreous samples that are optimally preserved (Al-Amoudi *et al.*, 2007; Studer *et al.*, 1995). Suboptimal cryofixation of tissues, yet without detectable segregation patterns (e.g., after freeze-substitution or freeze-fracturing) is considered as good cryofixation (well-frozen samples). The possible ice crystal ramifications are very fine, leading to segregation patterns below the information limit in plastic embedded sections (a few nm). Bad freezing induces perceptible ice crystal ramifications that segregate the sample noticeably. Note that vitrification can only be ascertained—most commonly by electron diffraction—at temperatures below −150°C. Under ambient conditions (e.g., after freeze-substitution), no discrimination can be made between vitrified and well-frozen samples.

Whereas untreated biological samples frozen using ambient pressure methods (such as plunge freezing and metal mirror freezing) produce only superficial vitreous layers (<20 μm, Escaig, 1982), biological samples of a thickness of 200 μm can be completely vitrified by high-pressure freezing (Sartori *et al.*, 1993; Shimoni and Muller, 1998; Studer *et al.*, 1995). Vitrification of apple leaf tissue (Michel *et al.*, 1992) and bulk animal tissue using the nematode *Caenorhabditis elegans* were shown (Hohenberg *et al.*, 1994). Vitrification of mammalian cartilage was demonstrated. It was shown that formation of vitreous water could only be achieved in regions with high proteoglycan and low water concentrations (Studer *et al.*, 1995). For liver, only selective vitrification occurs (Richter, 1994): electron diffraction patterns showed that the tissue contained crystalline frozen regions as well as vitreous areas. Vitreous human skin samples were obtained and desmosomes therein described at a resolution of 3.4 nm (Al-Amoudi *et al.*, 2007). In summary, high-pressure freezing to date is the only freezing approach that is successful in vitrifying bulk tissues (up to 200 μm in thickness).

C. Sample Preparation Prior to High-Pressure Freezing

The advantages of high-pressure cryofixation are restricted by the time needed for sample extraction before freezing. In order to avoid autolytic processes, which result in the deterioration of the ultrastructure (Muller and Moor, 1984), fast sample excision and careful preparation prior to cryofixation are a prerequisite. Detrimental effects became noticeable in mouse brain tissue when the handling time between cardiac arrest and cryofixation exceeded 30 s (Van Harreveld *et al.*, 1965). This estimation was based on morphological criteria.

Several approaches for fast tissue excision have been proposed. Excision and cryofixation—in parallel—of tissue samples by using a chilled hollow needle (Von Zglinicki *et al.*, 1986) is, from a theoretical point of view, optimal since there is no faster approach possible and no high pressure is applied to the sample; however, only a thin superficial layer of 20 μm may be preserved of the excised tissue. Less fast, but better sample handling times in the order of minutes, can be realized with the use of microbiopsy systems. These tools, originally developed for clinical use, were first applied in the collection of samples for high-pressure freezing by Hohenberg *et al.* (1996). However, the speed of sampling was highly dependent on the skill of the operator. First, because the specimen had to be trimmed further to the size of the sample holders with basic cutting tools, and second, because the small tissue blocks had to be transferred to the holders of the cryofixation device. Even more dependent on personal skills is the method proposed by Shimoni and Muller (1998). A gold capillary sharpened at one end was injected at high speed into the tissue, retrieved, and high-pressure frozen. The exact location of the excised tissue could not be determined before injection. Furthermore, laborious searching for the capillary tube in the tissue before cryofixation was time consuming.

The systems described depend mainly on the skills of the experimenter and, moreover, are not accurate enough (Shimoni and Muller, 1998) or produce too large biopsies (Hohenberg *et al.*, 1996). For routine tissue preparation prior to high-pressure freezing, an easy-to-handle microbiopsy system was established (Vanhecke *et al.*, 2003).

II. Rationale

With the possibility to describe macromolecular structures in tissues by electron microscopy (Ortiz *et al.*, 2006), any improvement of structural preservation during sample preparation is very important. The handling time prior to fixation is a crucial issue: reduced handling times will inevitably result in improved tissue quality and, therefore, improved capture of the living state. The objective was to accomplish short handling times, but still allow for a rapid transfer to a high-pressure freezing machine without depending too heavily on technical skills or the experience of the experimenter. This paradigm naturally has a direct positive impact on the reproducibility of the experiments and indirectly on the accuracy of potential quantification procedures. The microbiopsy system described here accomplishes the excision of a sufficiently small sample, transfer, and high-pressure freezing in less than 30 s by inexperienced, but well-instructed test persons. We explain here the use of the commercially available microbiopsy system of Leica Microsystems (Vienna, Austria), which was commercialized on the basis of our developments (Vanhecke *et al.*, 2003).

III. Methods

A. Microbiopsy System

The microbiopsy system described here (Fig. 1) was developed for optimal compatibility with a commercial high-pressure freezing machine, the Leica EMPact

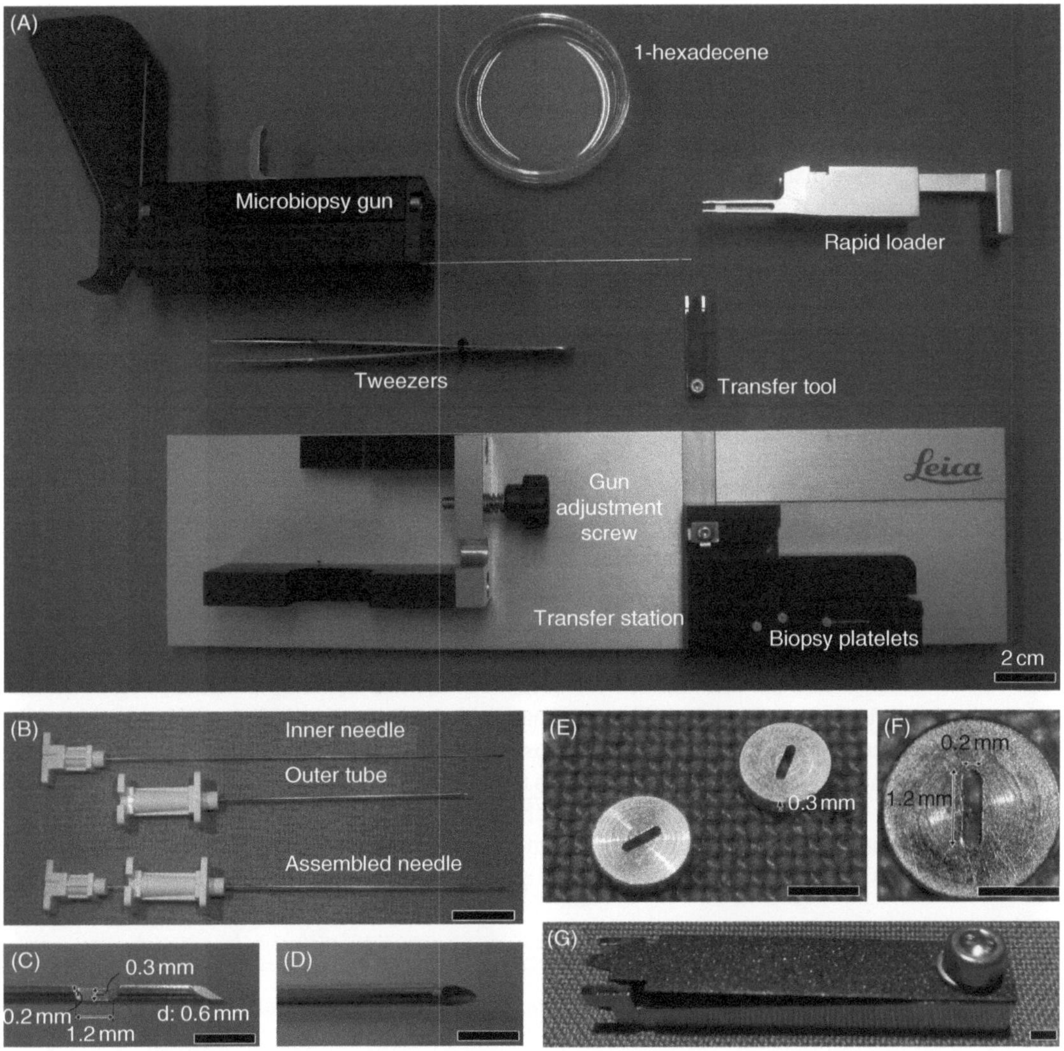

Fig. 1 The microbiopsy system. (A) An overview of the tools involved. The microbiopsy gun excises the tissue and transfers it, with the help of the transfer tool, to the cavity of a biopsy platelet, which is loaded onto the rapid loader tool of the Leica EMPact II RTS. (B) The excision mechanism dissected: a needle, equipped with a cavity, is encased by a tube, equipped with a sharpened end. A larger magnification of the inner needle's cavity (C) and the outer tube's sharpened end (D) are provided. (E) The biopsy platelets, the link between the excision and the high-pressure freezing in more detail. The dimensions, which correspond to the dimensions of the cavity of the inner needle, are indicated in (F). (G) The transfer tool, the device that facilitates the transfer of the sample from the needle cavity to the platelet. Grid bars: 2 cm (A–B) and 2 mm (C–G).

(Leica Microsystems, Vienna, Austria); both with the original EMPact (Studer *et al.*, 2001) and the updated EMPact II RTS version (McDonald *et al.*, 2007; Verkade, 2008). Essentially, the tools described below are two different systems: a procedure to dissect a small sample of tissue from the donor organ (microbiopsy gun) and a high-pressure cryofixation approach (rapid loader). The link between both systems is the biopsy platelet (Fig. 1E and F). The dimensions of the cavity of these platelets are, on one hand, sufficiently small to allow the formation of amorphous ice in biological samples, but on the other hand, large enough to take up samples of tissue that are not too affected by dissection edge effects.

The transfer is carried out on the transfer station (Fig. 1A), which provides the setup for alignment of the two aforementioned systems.

The microbiopsy gun (Fig. 1A) contains a dissection needle (inner needle: Fig. 1B and C) equipped with a cavity of dimensions matching those of the biopsy platelet (compare Fig. 1C with 1F). Upon insertion of the needle (with retracted outer tube) into the tissue, the cavity fills through elasticity of the tissue. Tissue is excised by the outer tube, which is armed with a sharpened, slanted edge (Fig. 1D). The shot of the biopsy gun is simply the movement of this outer tube over the inner needle; the tissue in the cavity becomes dissected. An excised microbiopsy is very prone to dehydration; therefore, the needle is opened (retraction of surrounding tube) in 1-hexadecene. This manipulation guarantees that the tissue does not dry during the following transfer.

The microbiopsy needle does not penetrate skin easily, and impossibly penetrates bone. Rats are anesthetized for brain biopsies and the skin is detached from the skull. After having fixed the rat in a stereotactic frame, holes are drilled into the skull in order to reach the organ with the microbiopsy needle. For all other organs, the skin is opened or removed first with small scissors. In the case of liver, small cuts were sufficient. However, for such a small organ as kidney, the rat had to be dissected to such an extent that the animal had to be killed afterward. In all tissues, a slight bending of the needle, so that the cavity is on the concave side, greatly helps to fill completely the cavity with tissue.

B. Preparation of the Transfer Station

To decrease the transfer time, a biopsy platelet (Fig. 2A) must be loaded into the clamp of the rapid loader (Fig. 2B and C) prior to the actual biopsy. Attention should be paid to the direction of the slot: it must be arranged parallel to the stalk of the biopsy loader (Fig. 2C). Proper functioning of the entire system is based on a well-aligned situation of platelet slot (Fig. 2D), the transfer tool (Figs. 1G and 2E), and the needle cavity (Fig. 2F and G). The position of the needle cavity can be adjusted using the gun adjustment screw (Fig. 1A). Poor alignment will result in time-consuming, contra-productive handling.

C. Uptake of the Excised Tissue in the Biopsy Platelet

The outer needle is retracted by the gun handle while dipped in 1-hexadecene (Fig. 1A) and the content of the inner needle cavity is exposed (Fig. 3B). The microbiopsy gun is placed in the gun bay of the transfer station so that the needle

Fig. 2 Preparation and alignment of the microbiopsy system. (A) The loading station for the biopsy platelets. As demonstrated in (B), the platelet inserted in the rapid loader is orientated with the platelet slot parallel to the loader's longitudinal axis. Correct assembling yields an aligned platelet slot as shown in (C) and at larger magnification in (D). The design of the transfer station, and the transfer tool, assure a position of this tool's tip directly above the platelet's slot as can be seen in (E). Finally, accurate alignment of the needle tip—with retracted outer tube—produces a correctly configured system as shown in (F) and in top view in (G). The position of the needle tip is adjusted using the screw of the gun adjustment screw (see Fig. 1A). All bars: 2 mm.

cavity is located precisely over the platelet slot (Fig. 3A), followed by the transfer of the biopsy into the biopsy platelet slot by means of the transfer tool (Fig. 3C). The result is a biopsy platelet filled with the biopsy sample introduced in the rapid loading holder of EMPact II RTS (Fig. 3D).

D. High-Pressure Freezing

Once the tissue is in the biopsy platelet fixed in the rapid loading holder, the sample can then be further processed as described for the Leica EMPact II RTS (Vanhecke and Studer, 2009). In summary, the rapid loading holder is pushed into the primed rapid loader device (RTS). The next steps are automated: the platelet is fixed into the pod of the rapid loading device, introduced by the same device into the high-pressure freezing part of EMPact II and high-pressure frozen. With this approach the samples are frozen

Fig. 3 Excision and transfer of a biopsy. (A) Depiction of an orthogonal view of the biopsy platelet aligned with the loaded cavity of the inner needle of the microbiopsy gun. The biopsy shown in (B) immediately after taking and immersion into 1-hexadecene is transferred from the needle cavity into the platelet slot by the transfer tool (C). Finally, as shown in (D), the biopsy fits the slot of the biopsy platelet well and is ready for high-pressure freezing. Bar: 2 mm.

within 20–30 s after the initial microbiopsy needle introduction into the organ. In the case of an EMPact I, the platelets have to be fixed and transferred manually into the machine for high-pressure freezing. Even with the EMPact I, handling times of 30 s or less can be achieved.

E. Freeze-Substitution and Plastic Embedding

The frozen specimens are freeze-substituted in acetone containing 2% osmium tetroxide, in an AFS cryostat (Leica Microsystems) and embedded in Epon/Araldite. The samples are substituted at −90°C for 27 h and for 8 h at −60°C and −30°C. Samples are warmed up to 4°C, washed three times for 10 min in 100% acetone and embedded in Epon/Araldite by increasing the resin concentration (resin/acetone 30%/70%, 70%/30%, and 100%/0%). Incubation time is 3 h at 4°C for the first two solutions and 12 h for 100% resin at room temperature. Finally, fresh resin is added to the sample prior to polymerization at 60°C for 5 days.

The embedded samples are ultrathin sectioned and stained with uranyl acetate and lead citrate.

IV. Materials

A. Animals

Wistar rats obtained from the Central Animal Facility (University of Bern) were treated according to Swiss animal protection laws. The rats were anesthetized by injecting 0.12 ml per 100 g body weight of the following narcotics mixture: 1 part

Prequillan, 5 parts Xylapan, 10 parts Narketan (all supplied by Chassot, Belp, Switzerland). After having taken the biopsies the rats were killed under anaesthesia by neck dislocation.

B. Reagents and Tools

Suppliers are: Plano, Wetzlar, Germany (Plano)
Merck, Darmstadt, Germany (Merck)
Fluka, Sigma-Aldrich, Buchs, Switzerland (Fluka)
Leica Microsystems, Vienna, Austria (Leica)
Roth, Arlesheim, Switzerland (Roth)
Migros do it, Switzerland (Migros)
Oxkem Ltd, Reading, Berkshire (Oxkem)

We used needles (Plano, order nr: N5013) to pin down anesthetized rats on cork plates 30 × 30 cm (Roth, order nr: 1391.1) before removing the skin with surgical scissors (Plano, order nr: T554) and forceps (Dumont Pin24 and SS; Plano order nr: T520A and 5264, respectively). For the biopsies, the biopsy needles (Leica order nr: 16706887) and biopsy carriers (Leica order nr: 16706892) were used. DREMEL 400-4/80 (Migros) is a cordless electric drill we used to make 2-mm-diameter holes into the rat skull. High-pressure freezing consumables used were 1-hexadecene (Fluka order nr: 52280) to avoid dehydration of the sample and liquid nitrogen (produced by the Chemistry Department, University of Bern). For the substitution process, acetone (Merck order nr: 100014) and osmium tetroxide (Oxkem) were used. Finally, the samples were embedded in Epon/Araldite (Fluka): a mixture of 475 ml Epon (order nr: 45345), 245 ml DDSA (order nr: 45346), and 280 ml MNA (order nr: 45347). Prior to use 1.5% of BDMA (order nr: 13365) was added to the mixture. The staining was performed according to Reynolds (1963), with 2% uranyl acetate (Fluka, order nr: 73943), lead citrate (= lead nitrate (Merck, order nr: 107398) and sodium citrate (Merck, order nr: 106448)).

The stereotactic frame was self-made (commercially available frames can be purchased from Kopf Instruments, Tujunga, California). The microbiopsy tool is available as a preparation tool with EMPact II plus RTS (Leica). The programmable cryostat for freeze-substitution (EM AFS) and the ultramicrotome (Ultracut UCT) are both obtained from Leica. The sections were observed in an electron microscope EM400 (Philips (FEI), Eindhoven, The Netherlands) and recorded on a Morada digital camera (OlympusSIS, Münster, Germany).

V. Results and Discussion

Images of tissue biopsies were published in Vanhecke *et al.* (2003, 2006). Electron micrographs of brain, kidney, liver, and muscle biopsies are shown in Figs. 4–6. The samples show excellent structural preservation after microbiopsy excision and high-pressure freezing within 30 s. An example of segregation effects is

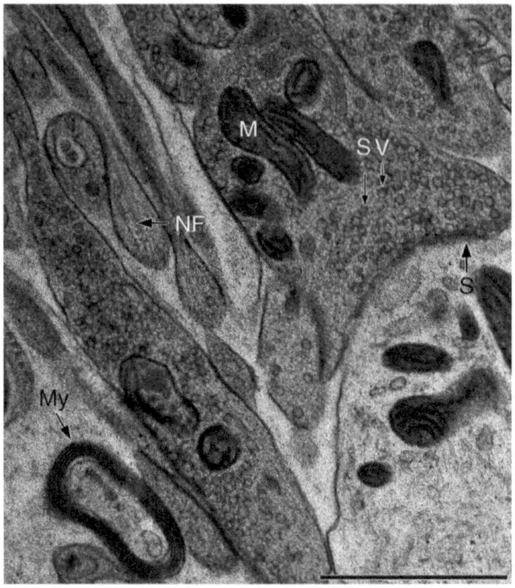

Fig. 4 Electron micrograph of a rat brain. From the cortex of the brain a biopsy was taken. A bouton with synaptic vesicles (SV) is forming a synapse (S) with a dendrite. The tube-like structure filled with vesicles at the lower left is likely a bouton en passant located close to a myelinated axon (My). Neurofilaments (NF) are nicely depicted. Bar: 1 μm.

shown in Fig. 5B. Such a sample is characterized by reticulate patterns, caused by the exclusion of all solutes (ions, proteins, etc.) from the ice crystal. During the growth of the crystal (depending on the cooling rate), the solutes are concentrated at the periphery of the crystal, yielding the reticulate structure. This structure is definitively ice crystal induced. The state of the water in well-frozen parts of the tissue (ice or vitreous water) can be objectively ascertained in a cryo-electron microscope at temperatures below −150°C by electron diffraction. The information on the state of water is lost when follow-up methods involve warming-up phases (freeze-substitution, freeze-drying, freeze-fracturing). The state of the ice after freezing in such samples cannot be retrieved, but parameters based on the interpretation of the ultrastructure can provide hints that the tissue is well frozen: the sample was either vitrified or the ice crystals are below the resolution limit. In Fig. 4, considerable differences in uniformity of staining can be seen between different compartments (compare the mitochondria with the surrounding cytoplasm). This may be explained in the following way: mitochondria were maybe vitreous after freezing (based on the fact that staining is uniform); however, the cell cytoplasm probably was affected by ice ramifications (less uniform staining), which are below the resolution limit and hence, invisible. In well-frozen samples, intact membranes have a smooth appearance caused by the cytoplasmic turgor (see, e.g., mitochondria, endoplasmatic reticulum, and nuclear envelope in Fig. 5A) and do not form

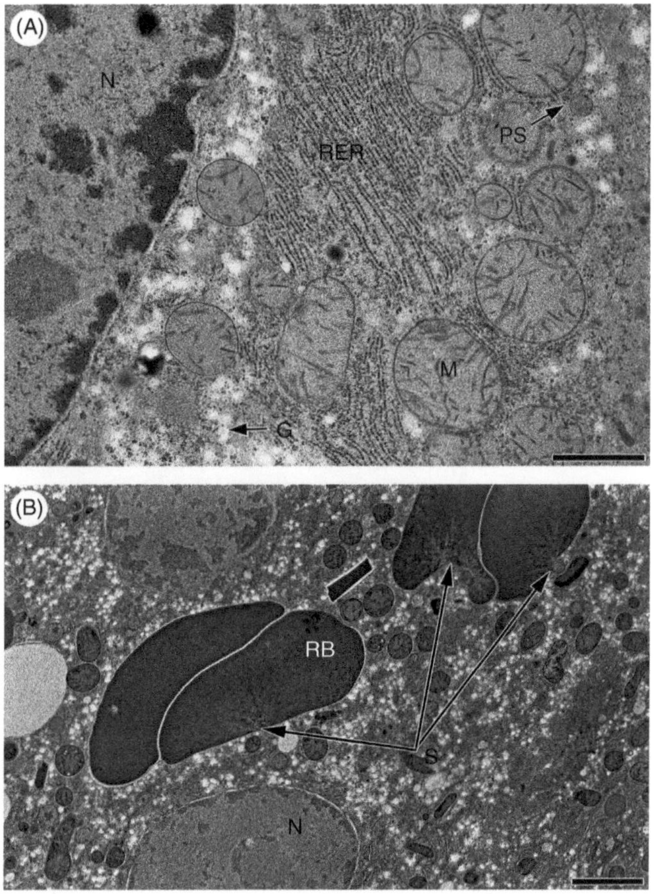

Fig. 5 (A) A liver biopsy was taken to characterize hepatocytes. Well-preserved organelles are shown. Nucleus (N), mitochondria (M), rough endoplasmatic reticulum (RER), peroxisome (PS), and glycogen (G) are marked. Bar: 1 μm. (B) A micrograph of a liver biopsy showing two heavy artifacts that occasionally occur. The erythrocytes (RB) were squeezed between hepatocytes during excision with the microbiopsy needle and the erythrocytes are showing heavy segregation patterns (S). Bar: 2 μm.

irregularly corrugated or wavy borders. This indicates that there were no osmotic shocks during preparation. Regular arranged structures, such as the myofilaments of muscle in cross section (Fig. 6A), will become disordered when disrupted by growing ice crystal ramifications. As a matter of fact, erythrocytes are exceptionally prone to ice crystal formation and show segregation patterns (Fig. 5B) whereas the surrounding tissue is well frozen. Tissue regions showing intact erythrocytes, such as in the muscle shown in Fig. 6A, are typically free of ice crystal damage. The kidney shown in Fig. 6B summarizes a number of the discussed features of a well-frozen sample: smooth membranes, running parallel to their counterparts of the neighboring cell. The cytosol does not appear empty, or shows non-uniformly distribution contrast,

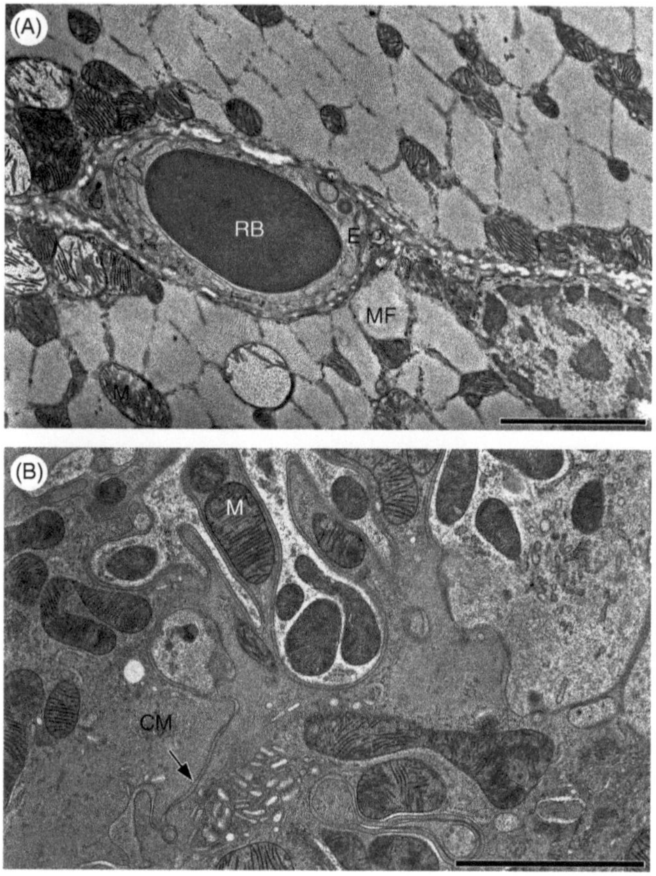

Fig. 6 (A). A cross section of a muscle biopsy: a capillary delineated by an endothelial cell (E) is completely filled with an erythrocyte (RB). Cross-sectioned myofibrils (MF) and mitochondria (M) are beautifully preserved. (B) Biopsy of a well-preserved piece of kidney. Membranes (CM) and mitochondria (M) show perfect structures. Bars: 2 μm.

but is evenly stained, which points to little or no loss of small molecules. Reticulate structures as seen in Fig. 5B are not observed in Figs. 4, 5A, and 6. In essence good criteria for well-frozen samples are straight membranes and an evenly stained cytoplasm.

A. Positive Aspects

Samples could be prepared in a very short time period (<30 s). This is the time limit introduced by early cryofixation experiments (Van Harreveld *et al.*, 1965). Based on our knowledge, we judge the ultrastructural preservation of the samples to be excellent.

Concerning handling, inexperienced but well-trained persons were able to perform all steps (from excising until the tissue is cryofixed by high-pressure freezing) in the time period below 30 s. However, to perform serial biopsies at different time points, it is recommended to engage three people: one person to take care of the animal (anesthesia) and to perform the biopsy, the second person to perform the biopsy transfer into the rapid loader, and the third person to handle the high-pressure freezing machine and the short-term storage of frozen samples. The high-pressure frozen samples can be directly mounted in a cryoultramicrotome and vitreous sections can be produced (Vanhecke *et al.*, 2003).

B. Possible Improvements

Some aspects of the current approach need further improvements. First, the biopsy needle is still too large. For rat organs, the size of the needle is appropriate. For applications involving mouse tissue (especially brain tissue), however, the large needle size is problematic. The tip of the inner needle is too long for the mouse brain and the animal is sacrificed upon the first biopsy. The fact that erythrocytes in the liver, as shown in Fig. 5B, are squeezed into the gall capillaries shows that this piece of tissue was suffering heavy distortion during excision.

Second, a processing time of 30 s is fast for most tissues, but to our understanding, should be even faster for nervous tissue preparation. When structure–function correlations on a molecular level have to be achieved, preparation time has to be further reduced. A fully automated biopsy needle, specimen transfer, and freezing process should allow sample preparation in less than 5 s, as it is the case for the rapid freezing of cell cultures (Verkade, 2008). Ideally, with such a technical improvement long-lasting experiments (i.e., collection of samples at different time points) should also be possible for a single person.

Acknowledgments

B. Krieger, Institute of Anatomy, Bern was accomplishing photographic paper work. D. Studer was supported by the Swiss National Foundation (grant no.: 3100A0-118394).

References

Adrian, M., Dubochet, J., Lepault, J., and McDowall, A. W. (1984). Cryo-electron microscopy of viruses. *Nature* **308**, 32–36.

Al-Amoudi, A., Diez, D. C., Betts, M. J., and Frangakis, A. S. (2007). The molecular architecture of cadherins in native epidermal desmosomes. *Nature* **450**, U832–U838.

Eggli, P., and Graber, W. (1994). Improved ultrastructural preservation of rat ciliary body after high pressure freezing and freeze substitution: A perspective view based upon comparison with tissue processed according to a conventional protocol or by osmium tetroxide/microwave fixation. *Microsc. Res. Tech.* **29**, 11–22.

Escaig, J. (1982). New instruments which facilitate rapid freezing at 83 K and 6 K. *J. Microsc.* **126**, 221–229.

Hernandez-Verdun, D., Quintana, C., Masson, C., Gautier, T., and Arnoult, J. (1991). Cryofixation, cryosubstitution, cryoembedding for visualizing of nuclear ultrastructure and for immunodetection in HeLA-cells. *Biol. Cell* **72**, 121–132.

Hohenberg, H., Mannweiler, K., and Muller, M. (1994). High-pressure freezing of cell-suspensions in cellulose capillary tubes. *J. Microsc.* **175**, 34–43.

Hohenberg, H., Tobler, M., and Muller, M. (1996). High-pressure freezing of tissue obtained by fine-needle biopsy. *J. Microsc.* **183**, 133–139.

Hunziker, E. B., and Schenk, R. K. (1984). Cartilage ultrastructure after high pressure freezing, freeze-substitution, and low temperature embedding. II. Intercellular matrix ultrastructure—preservation of proteoglycans in their native state. *J. Cell Biol.* **98**, 277–282.

Kanno, H., Speedy, R., and Angell, C. (1975). Supercooling of water to −92°C under pressure. *Science* **189**, 880–881.

Kiss, J. Z., Giddings, T. H., Staehelin, L. A., and Sack, F. D. (1990). Comparison of the ultrastructure of conventionally fixed and high-pressure frozen freeze substituted root-tips of *Nicotiana* and *Arabidopsis*. *Protoplasma* **157**, 64–74.

Leforestier, A., Richter, K., Livolant, F., and Dubochet, J. (1996). Comparison of slam-freezing and high-pressure freezing effects on the DNA cholesteric liquid crystalline structure. *J. Microsc.* **184**, 4–13.

McDonald, K. L., Morphew, M., Verkade, P., and Muller-Reichert, T. (2007). Recent advances in high-pressure freezing: Equipment- and specimen-loading methods. *Methods Mol. Biol.* **369**, 143–173.

Michel, M., Gnagi, H., and Muller, M. (1992). Diamonds are a cryosectioners best friend. *J. Microsc.* **166**, 43–56.

Monaghan, P., Perusinghe, N., and Muller, M. (1998). High-pressure freezing for immunocytochemistry. *J. Microsc.* **192**, 248–258.

Moor, H., and Mühlethaler, K. (1963). Fine structure in frozen-etched yeast cells. *J. Cell. Biol.* **17**, 609–628.

Muller, M., and Moor, H. (1984). Cryofixation of thick specimens by high pressure freezing. *In* "The Science of Biological Specimen Preparation for Microscopy and Microanalysis" (J. Revel, T. Barnard, and G. Haggis, eds.). Scanning Microscopy International, Chicago.

Ortiz, J. O., Forster, F., Kurner, J., Linaroudis, A. A., and Baumeister, W. (2006). Mapping 70s ribosomes in intact cells by cryoelectron tomography and pattern recognition. *J. Struct. Biol.* **156**, 334–341.

Reynolds, E. (1963). The use of lead citrate at high pH as an electron-opaque stain in electron microscopy. *J. Cell Biol.* 17, 208–212.

Richter, K. (1994). High-density morphologies of ice in high-pressure frozen biological specimens. *Ultramicroscopy.* **53**, 237–249.

Royer, S. M., and Kinnamon, J. C. (1996). Comparison of high-pressure freezing/freeze substitution and chemical fixation of catfish barbel taste buds. *Microsc. Res. Tech.* **35**, 385–412.

Sartori, N., Richter, K., and Dubochet, J. (1993). Vitrification depth can be increased more than 10-fold by high-pressure freezing. *J. Microsc.* **172**, 55–61.

Semmler, K., Wunderlich, J., Richter, W., and Meyer, H. W. (1998). High-pressure freezing causes structural alterations in phospholipid model membranes. *J. Microsc.* **190**, 317–327.

Shimoni, E., and Muller, M. (1998). On optimizing high-pressure freezing: From heat transfer theory to a new microbiopsy device. *J. Microsc.* **192**, 236–247.

Somlyo, A. P., Bond, M., and Somlyo, A. V. (1985). Calcium content of mitochondria and endoplasmic-reticulum in liver frozen rapidly in-vivo. *Nature* **314**, 622–625.

Studer, D., Graber, W., Al-Amoudi, A., and Eggli, P. (2001). A new approach for cryofixation by high-pressure freezing. *J. Microsc.* **203**, 285–294.

Studer, D., Hennecke, H., and Muller, M. (1992). High-pressure freezing of soybean nodules leads to an improved preservation of ultrastructure. *Planta* **188**, 155–163.

Studer, D., Michel, M., and Muller, M. (1989). High pressure freezing comes of age. *Scanning Microsc. Suppl.* **3**, 253–268, discussion 268–269.

Studer, D., Michel, M., Wohlwend, M., Hunziker, E. B., and Buschmann, M. D. (1995). Vitrification of articular-cartilage by high-pressure freezing. *J. Microsc.* **179**, 321–332.

Suppes, G. J., Egan, S., Casillan, A. J., Chan, K. W., and Seckar, B. (2003). Impact of high pressure freezing on DH5 alpha *Escherichia coli* and red blood cells. *Cryobiology* **47**, 93–101.

Van Harreveld, A., Crowell, J., and Malhotra, K. (1965). A study of extracellular space in central nervous tissue by freeze-substitution. *J. Cell Biol.* **25**, 117–137.

Vanhecke, D., Eggli, P., Graber, W., and Studer, D. (2006). A new microbiopsy system enables rapid preparation of tissue for high pressure freezing. *In* "Methods in Molecular Biology" (D. J. Taatjes and B. T. Moosman, eds.) Humana Press, Totowa, New Jersey, Vol. 319, pp. 463–477.

Vanhecke, D., Graber, W., Herrmann, G., Al-Amoudi, A., Eggli, P., and Studer, D. (2003). A rapid microbiopsy system to improve the preservation of biological samples prior to high-pressure freezing. *J. Microsc.* **212**, 3–12.

Vanhecke, D., and Studer, D. (2009). High pressure freezing with the Leica EMPact. *In* "Handbook of Cryo-Preparation Methods for Electron Microscopy" (A. Cavalier, D. Spehner, and B. Humbel, eds.). Taylor & Francis Group, Boca Raton, FL.

Verkade, P. (2008). Moving EM: The rapid transfer system as a new tool for correlative light and electron microscopy and high throughput for high-pressure freezing. *J. Microsc.* **230**, 317–328.

Von Schack, M. L., Fakan, S., Villiger, W., and Muller, M. (1993). Cryofixation and cryosubstitution—a useful alternative in the analyses of cellular fine-structure. *Eur. J. Histochem.* **37**, 5–18.

Von Zglinicki, T., Rimmler, M., and Purz, H. J. (1986). Fast cryofixation technique for x-ray-microanalysis. *J. Microsc.* **141**, 79–90.

Zierold, K. (1991). Cryofixation methods for ion localization in cells by electron-probe microanalysis—a review. *J. Microsc.* **161**, 357–366.

CHAPTER 22

The Actin Cytoskeleton in Whole Mount Preparations and Sections

Guenter P. Resch*, Edit Urban†, *and* Sonja Jacob†

*IMP-IMBA-GMI Electron Microscopy Facility, Institute of Molecular Biotechnology of the Austrian Academy of Sciences, 1030 Vienna, Austria

†Institute of Molecular Biotechnology of the Austrian Academy of Sciences, 1030 Vienna, Austria

Abstract

In non-muscle cells, the actin cytoskeleton plays a key role by providing a scaffold contributing to the definition of cell shape, force for driving cell motility, cytokinesis, endocytosis, and propulsion of pathogens, as well as tracks for intracellular transport. A thorough understanding of these processes requires insight into the spatial and temporal organisation of actin filaments into diverse higher-order structures, such as networks, parallel bundles, and contractile arrays. Transmission and scanning electron microscopy can be used to visualise the actin cytoskeleton, but due to the delicate nature of actin filaments, they are easily affected by standard preparation protocols, yielding variable degrees of ultrastructural preservation.

978-0-12-381007-6
DOI: 10.1016/S0091-679X(10)96022-6

In this chapter, we describe different conventional and cryo-approaches to visualise the actin cytoskeleton using transmission electron microscopy and discuss their specific advantages and drawbacks. In the first part, we present three different whole mount techniques, which allow visualisation of actin in the peripheral, thinly spread parts of cells grown in monolayers. In the second part, we describe specific issues concerning the visualisation of actin in thin sections. Techniques for three-dimensional visualisation of actin, protein localisation, and correlative light and electron microscopy are also included.

I. Introduction and Rationale

Actin is a major protein in eukaryotic cells forming polarised filamentous polymers (F-actin, microfilaments) with an average diameter of 7 nm and a helical arrangement of the subunits (Steinmetz *et al.*, 1997). Apart from its well-known function in muscle contractility, F-actin participates in many important cellular processes. Together with other proteins it determines the shape of a cell by adding stability to the cortex and by providing the structural core for protrusions like microvilli or synapses (reviewed in Cingolani and Goda, 2008; Revenu *et al.*, 2004). Furthermore, actin drives dynamic processes such as cell motility, initiated by lamellipodial protrusion (Bray, 2001), cytokinesis via a contractile ring of actin and myosin (reviewed in Glotzer, 2005), and endocytosis (reviewed in Kaksonen *et al.*, 2006; Merrifield, 2004). Actin filaments can also serve as tracks for vesicle and organelle transport via myosin motors (reviewed in Ross *et al.*, 2008). Besides these functions in physiology, there are pathogens that hijack the cellular actin machinery to drive their own propulsion, such as *Listeria monocytogenes, Shigella flexneri*, and vaccinia virus (reviewed in Gouin *et al.*, 2005). This actin-driven movement can be reconstituted *in vitro*, using biomimetic assays (Loisel *et al.*, 1999; Plastino *et al.*, 2004), which are expected to provide an additional insight into the architecture and dynamics of the actin cytoskeleton. To regulate the polymerisation and organisation of actin filaments into higher-order architectures fulfilling these diverse roles, a vast number of actin-associated proteins exist. They serve either a regulatory function in controlling polymerisation and depolymerisation or a structural role (reviewed in Winder and Ayscough, 2005).

The distribution of actin in cells and tissues can be visualised in the fluorescence microscope, using fluorescent actin fusion proteins or fluorescently labelled phalloidin derivatives that bind F-actin (Prast *et al.*, 2006). Analysis of the detailed architecture of the filament networks and bundles requires the resolution of the electron microscope. Hence, preparation techniques to visualise the actin cytoskeleton by electron microscopy (EM) are of great interest for cell biology in general, and of special importance for studies of cell motility, infection biology, and endocytosis. Using the appropriate methodology is particularly important as the actin cytoskeleton is a very delicate structure, and artefacts can be introduced by many preparative steps (Maupin-Szamier and Pollard, 1978; Resch *et al.*, 2002a; Ris, 1985; Small, 1981).

The techniques we describe and discuss in this chapter centre around the visualisation of the actin cytoskeleton in non-muscle cells. We cover whole mount techniques, where cultured cells are examined as a whole at their periphery without sectioning (Section II), as well as plastic- and cryo-sectioning approaches suitable for inner parts of cells and tissue (Section III). We compare different methods focussing on their advantages and disadvantages. The visualisation of synthetic actin filaments, the actin cytoskeleton in plants and fungi, and actin in skeletal muscle are not discussed. We also have to omit approaches using scanning electron microscopy (SEM; Walther, 2008; Wepf, personal communication), or freeze drying/replica transmission electron microscopy (TEM; DiNardo *et al.*, 2005; Flanagan *et al.*, 2001; Heuser and Kirschner, 1980).

In addition to standard preparation protocols, we cover correlative light and electron microscopy (CLEM), protein localisation, and three-dimensional reconstruction of the actin cytoskeleton. CLEM, an increasingly popular approach (Braet and Geerts, 2009), combines light microscopy (LM) and EM on the very same sample and feature of interest. For example, dynamic information or data about the distribution of endogenous or microinjected fluorescent proteins can be complemented with high-resolution data from EM, linking function and structure. To be suitable for CLEM of the actin cytoskeleton, a preparation method has to be compatible with high-resolution LM, be highly reproducible, and yield high-resolution data that represent the natural state of the specimen (Plitzko *et al.*, 2009; Svitkina *et al.*, 1995).

To understand the role of the numerous actin-associated proteins in shaping and regulating the architecture of the cytoskeleton, it is important to establish their spatial distribution within cells. Hence, another focus of each approach discussed will be the applicability of immuno(gold)labelling. As an alternative to immunolabelling, the binding of myosin II fragments to actin was used in early work on the cytoskeleton to distinguish actin from other cytoskeletal filaments (Heuser and Kirschner, 1980; Ishikawa *et al.*, 1969). Both heavy meromyosin (HMM), a dimer composed of two head (S1) and neck (S2) regions, and myosin heads alone (subfragment S1) were employed for labelling. Nowadays, this technique is mainly used to determine the polarity of actin filaments by giving them a distinct arrow- or rope-like pattern upon decoration. This allows identification of the functionally very distinct plus and minus ends of the filaments (Small *et al.*, 1978; Tilney and Portnoy, 1989; Vignjevic *et al.*, 2003).

Since the actin cytoskeleton is a three-dimensional structure we finally discuss approaches, using electron tomography (ET; reviewed in Steven and Aebi, 2003), to resolve the spatial architecture of actin networks in cells. Here, we focus on recent studies of lamellipodia, the thin protruding zones at the front of moving cells. The importance of detailed structural information on this "organelle" is illustrated by recent controversies about branching of actin filaments (Small *et al.*, 2008; Svitkina and Borisy, 1999) or about the existence of two distinct filament networks at the leading edge (Ponti *et al.*, 2004; Vallotton and Small, 2009).

II. Whole Mount Techniques

On flat surfaces, i.e., cell culture dishes or on microscope slides, many cell types spread out with very thin parts at the periphery and a thicker region around the nucleus.

At a time when the instrumentation to routinely produce ultrathin sections was not yet available, Porter *et al.* (1945) introduced an approach in which cultured cells are examined as a whole in the TEM. Today, this approach referred to as "whole mount technique" is still being used frequently in cytoskeleton research, mostly for studying cell motility. First, this approach obviates the need for embedding the sample in resin and the problems associated with this procedure (Section III). Second, the finest cell extensions of adherent cells, such as lamellipodia and filopodia, are not ideally accessible by thin section EM because their thickness (Koestler *et al.*, 2009) is just in the range of one or two thin sections. However, in most whole mount techniques, structural studies are restricted to the cell periphery where the specimen is thin enough to be penetrated by the electron beam. Even with intermediate or high-voltage TEM, satisfactory visualisation of structures is becoming increasingly difficult towards the nucleus due to the dense cytoplasm. Hence, whole mounts are mainly being used when the processes of interest, such as protrusion events in cell motility, occur close to the cell periphery. To uncover actin networks for improved visualisation, most whole mount protocols use detergent extraction, removing the membrane and cytosolic protein. The insoluble remainder is often referred to as "cytoskeleton", a term also used with this meaning throughout this text.

Popular model systems include the slime mould *Dictyostelium discoideum* (Claviez *et al.*, 1986; Chapter 9 by Koonce and Gräf, this volume; Medalia *et al.*, 2002) because of the ease of cultivation and the simple structure of its genome, keratocytes from fish and amphibia because of their fast locomotion and regular, characteristic shape (Small *et al.*, 1995; Svitkina *et al.*, 1997; Urban *et al.*, 2010), fibroblasts (Nemethova *et al.*, 2008; Verkhovsky *et al.*, 1995) for historic reasons, melanoma cells (Koestler *et al.*, 2008; Resch *et al.*, 2002b; Svitkina *et al.*, 2003), and growth cones of neurons (Burnette *et al.*, 2007; Korobova and Svitkina, 2008; Lewis and Bridgman, 1992). With modifications, the techniques outlined in this section are also applicable to subcellular structures containing actin such as actin-driven vesicles or intracellular pathogens, and biomimetic assays.

A. Negative Staining

Negative staining is a well-established method for visualisation of macromolecular complexes and organelles in TEM. A solution of heavy metal salt is used to rinse the specimen. Following air drying, the electron-transparent specimen is embedded and supported in a thin amorphous layer of stain that generates an electron-dense surrounding, providing negative contrast in the TEM. Commonly used negative stains include uranyl acetate (UA), uranyl formate, ammonium molybdate, and phosphotungstic acid. The technique was invented by Brenner and Horne (1959) and first applied on mosaic viruses mixed with phosphotungstic acid. Ever since, it was used for visualisation of countless biological specimens. Brown *et al.* (1976) introduced it as a method for ultrastructural visualisation of the actin cytoskeleton in whole mounts: Their approach was based on Triton X-100 extraction followed by glutaraldehyde (GA) fixation and UA negative staining. Subsequent studies by Small and Celis (1978) and Small *et al.*

(1978) showed, with optimised buffer conditions, that the actin cytoskeleton could be well visualised by this type of approach.

Preservation of the actin cytoskeleton was further improved by the introduction of simultaneous extraction and fixation in a Triton/GA mixture followed by phalloidin treatment and negative staining with neutral sodium silicotungstate stain (SST; Höglund *et al.*, 1980; Small, 1981). These modifications allowed a more detailed analysis of the structural organisation of dense and highly ordered actin filaments and to observe even the substructure of the filaments *in situ*. In addition, stereo-image pairs indicated retention of three-dimensionality in negatively stained actin cytoskeletons (Small *et al.*, 1982). The negative stain method for cytoskeletons has been successfully applied in studies on numerous cell types (Claviez *et al.*, 1986; Höglund, 1985; Koestler *et al.*, 2008; Rottner *et al.*, 1999; Small *et al.*, 1995), *Listeria* actin comet tails (Sechi *et al.*, 1997), and biomimetic assays (Wiesner, Resch, Small, and Carlier, unpublished data).

1. Method

For negative staining, cells are directly grown on EM grids coated either with a Formvar (Auinger and Small, 2008) or a carbon film (Resch *et al.*, unpublished data). In the latter case a brief glow discharge treatment of the grids is advisable to increase the hydrophilic properties to promote matrix coating or cell attachment to the surface. Au or gilded grids are the most adequate for direct cell culture. Cu grids should be avoided because of the cell toxicity of Cu released from the grids and the magnetism of Ni grids can influence the alignment of the microscope. The mesh size of the grid can vary but below 150 mesh, some loss in film stability can be expected.

The procedure to plate cells onto grids is principally the same as for glass coverslips, including UV sterilisation of grids. To encourage cell spreading on the film, the grids are incubated in a drop of matrix proteins or poly-L-lysine in a humid chamber prior to plating. Detached cells, in an appropriate dilution and a suitable medium, are prepared in multiwell dishes. Subsequently, one grid only is placed into each well, with the film side facing up. The required cell density—one or two cells per grid square depending on cell and mesh size—has to be checked in the phase contrast light microscope before further processing. Once the cells are adequately spread, they can be washed briefly in pre-warmed phosphate buffered saline (PBS) or serum-free medium, then extracted and fixed at room temperature. For this procedure, droplets of approximately 100 μl extraction, fixation, and post-fixation solutions are prepared on a piece of Parafilm in a moist chamber.

The commonly used extraction solution consists of Triton and GA in cytoskeleton buffer (CB; 10 mM MES, 150 mM NaCl, 5 mM EGTA, 5 mM glucose, 5 mM $MgCl_2$; pH 6.1) that is designed to minimise extraction of cytoskeletal proteins. According to cell type, the concentration of Triton (0.05–1.00%) and GA (0.1–1.0%) in the initial fixation mixture, the pH of the buffer, as well as the time of initial fixation have to be optimised: For melanoma cells and fibroblasts, the actin cytoskeleton is well visualised with a mixture of 0.50% Triton and 0.25% GA in CB, pH 6.5, while for thicker cells, such as fish keratocytes and *Dictyostelium*, a higher detergent concentration (0.75–1.00% Triton and 0.25% GA in CB, pH 6.1) is suggested. After 1 min incubation in the extraction/fixation solution, the grid is transferred gently, without drying, with a pair of forceps into a fixative of 2% GA in CB (pH 7.0) supplemented with phalloidin (1–10 μg/ml) to stabilise the actin filaments. This second fixation step lasts at least 20–30 min. If timing requires, the grids can be stored in another post-fixative solution containing 2% GA and 10 μg/ml phalloidin in CB, pH 7.0, for several days in a moist chamber at 4°C.

Prior to staining, the grid is carefully transferred from the fixative into CB for a brief wash to remove the fixative. For negative staining, the grid is clamped in a pair of forceps and held vertically. Before the sample has time to dry out the cell side of the grid is rinsed with a few drops of negative stain (in total approximately 100 μl). Excess liquid is drained with the torn edge of a filter paper from the back side of the grid, from the edge of the grid and from between the tips of the forceps. Care should be taken not to touch the front (cell) side of the grid. The sample is allowed to air dry and is immediately ready for inspection in a TEM (Fig. 1). To avoid re-crystallisation of the stain the air-dried sample can be stored in a desiccator up to a few days. Longer storage is not recommended.

Different negative stains can be applied but neutral SST (pH 7.5 adjusted with NaOH) has so far yielded in the best-preserved actin filament order at the leading edge. Another conventionally applied stain is UA (pH 4, unadjusted) that produces higher contrast than SST and better visualisation of microtubules and intermediate filaments (IF), but less well-preserved cytoskeletal actin (Small and Celis, 1978). Negative stains are generally used as a 2 or 3% aqueous solution for conventional TEM (up to 120 kV). The addition of carbohydrate as trehalose (up to 0.5%) to the negative stain solution has the advantage of creating a thicker protective layer of stain without excessively high contrast (Harris *et al.*, 1995).

2. Instrumentation and Materials

Instrumentation: Inverted phase contrast light microscope, desiccator, routine transmission electron microscope.

Materials: Formvar-coated 200 mesh Au grids, stainless-steel forceps with fine tips, multiwell dishes for cell culture, moist chamber, Whatman No. 1 filter paper.

Reagents: Cell culture medium, PBS (8.0 g/l NaCl, 0.2 g/l KCl, 1.2 g/l Na_2HPO_4, 0.2 g/l KH_2PO_4; pH 7.3), CB (10 mM MES, 150 mM NaCl, 5 mM EGTA, 5 mM glucose, 5 mM $MgCl_2$; pH 6.1), extraction solution (0.1–1.0% GA, 0.05–1.00% Triton 100-X, 1 μg/ml phalloidin in CB; pH 6.5), fixation solution (2% GA, 1 μg/ml phalloidin in CB; pH 7.0), post-fixation solution (2% GA, 10 μg/ml phalloidin in CB; pH 7), 1–8% SST, pH 7.5, or 2% UA, pH 3.5. The negative staining solutions should be filtered through a 0.2 μm filter prior to use. During pH adjustment of SST, the pH is checked on a daily basis until it is stable.

3. Correlative Microscopy

Auinger and Small (2008) describe a very reproducible approach to correlate data from LM with negative staining EM. For this approach, cells are grown on coverslips

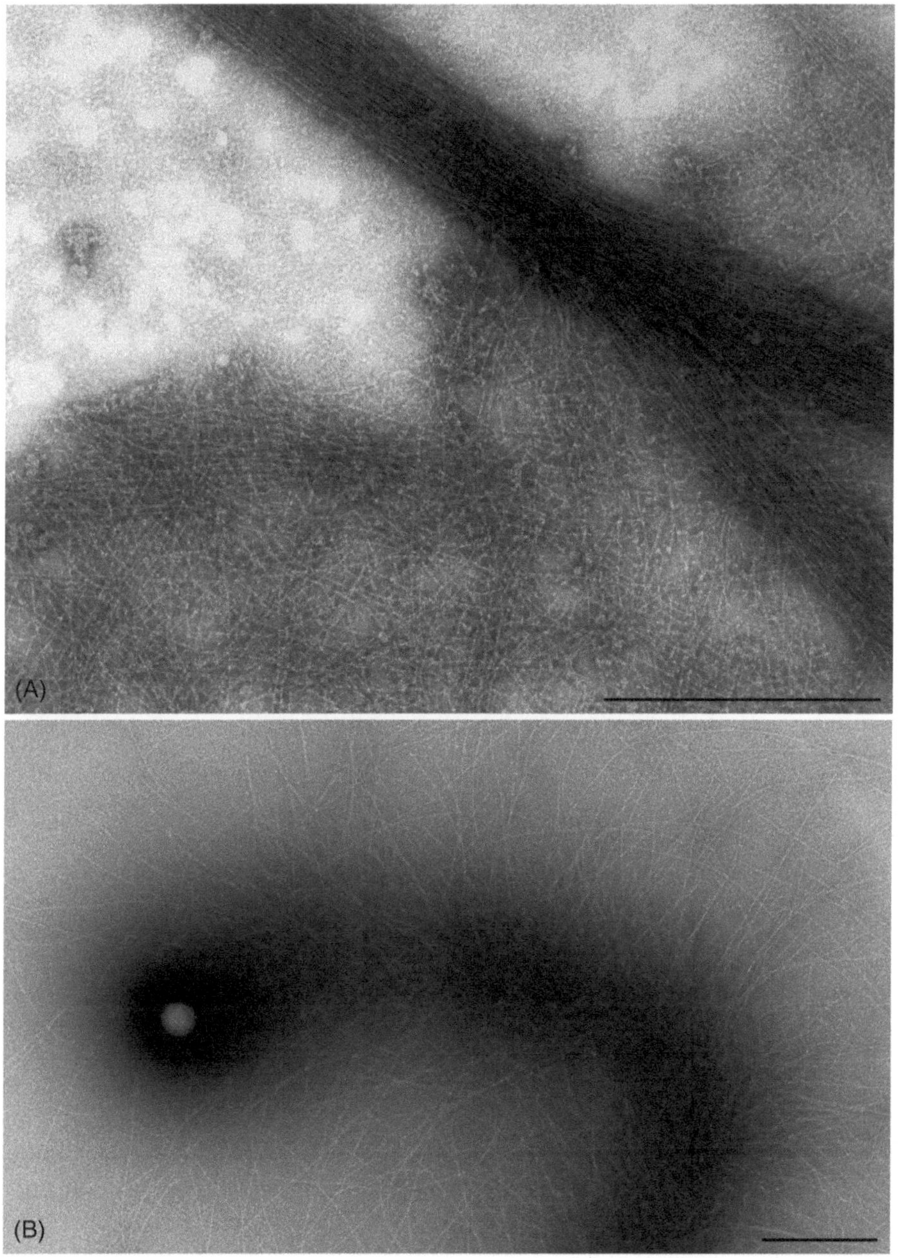

Fig. 1 Negative staining of the actin cytoskeleton. (A) Electron micrograph of the actin cytoskeleton of a CAR fish fibroblast, extracted and described as in the text, and negatively stained with 2% sodium silicotungstate (SST), showing a filopodium embedded in the actin network of the lamellipodium. (B) Negatively stained actin comet tail formed in an *in vitro* motility assay on an N-WASP-coated bead, negatively stained with 2% SST (Wiesner *et al.* unpublished data). Bars: 0.5 μm (A), 0.5 μm (B).

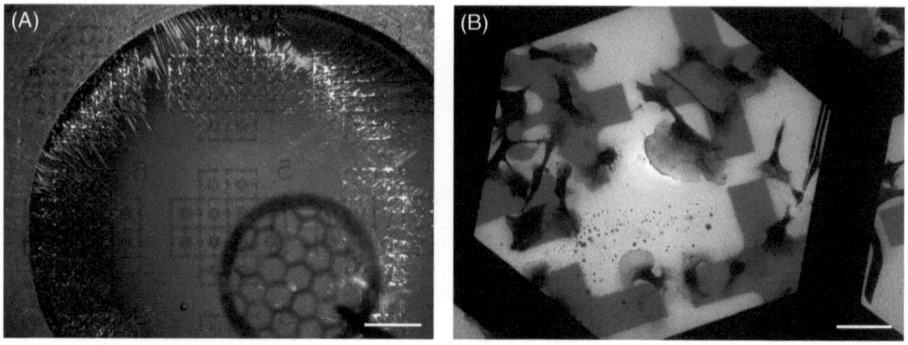

Fig. 2 Relocation of cells for correlative light and negative staining electron microscopy. (A) Positioning of a grid with the Formvar film immobilised on a support ring. (B) Low-magnification overview of the relocation pattern and negatively stained cells in the electron microscope. Bars: 1 mm (A), 0.1 mm (B).

coated with a Formvar film. To facilitate later handling of the film, a rim of concentrated Formvar solution (≈6%) is applied onto the edge of the coverslips. For cell relocation a finder grid pattern is deposited onto the film by evaporating Au using custom-made "negative grid masks" (Small, 1984) from Pyser (www.pyser-sgi.com) so that the pattern is easily visible under a dissecting microscope (Fig. 2A). The coverslip/film combination is sterilised under UV light for at least 30 min for further cell culture use. The cells are transfected for expression of fluorescent proteins of interest and plated on the coverslips. The imaging conditions and modes in the LM are dependent on the experimental requirements. At a chosen time of imaging, the growth medium is exchanged for extraction/fixation solutions as described above. After a final LM image of the fixed cell was taken, the coverslip is removed from the microscope chamber and transferred into post-fixative where it can be kept at 4°C for several days. Grease used for sealing the microscope chamber still attached to the coverslip does not pose a problem at this stage.

The next step is the film removal from the coverslips: the coverslip is placed film side up into a petri dish containing CB. With the help of two pairs of forceps, the thin Formvar film is gently detached from the coverslip at the thick Formvar rim. During detachment of the whole film, it is inverted and floated on the buffer surface. After the film is immobilised using a metal ring resting just below the buffer surface, a 50 mesh Cu grid is placed onto the film so that the feature of interest lies in the central hole. This is achieved under a dissecting microscope with gentle operation of a micromanipulator holding forceps with a grid (Fig. 2A). A piece of Parafilm is used to pick up the grid attached to the Formvar film, negative staining is carried out as described above, and the feature of interest can be relocated in the TEM using the electron-dense pattern (Fig. 2B). This method was used to correlate migration velocity and ultrastructural organisation of actin cytoskeleton in melanoma cells (Koestler *et al.*, 2008) and to investigate the contribution of filopodia to the construction of the cytoskeleton (Nemethova *et al.*, 2008).

4. Protein Localisation

Immunolabelling can be combined with the negative staining technique to localise actin-associated proteins. After a fixation using a slightly lower or equal concentration of GA as compared to the protocol outlined above, the fixed cytoskeleton is labelled with primary antibodies (pAb) against the protein of interest followed by washing steps in PBS and an incubation with a secondary antibody (sAb) carrying a 5 or 10 nm Au tag that is easily visible after negative staining. After washing, we recommend a post-fixation in 2% GA, 10 μg/ml phalloidin in CB for at least 20 min at room temperature or at 4°C overnight to stabilise cytoskeletal structures and antibody binding prior to negative staining. The localisation of filamin (Small, 1988), VASP (Rottner *et al.*, 1999), or Abi1 (Koestler *et al.*, 2009) were shown by immunogold labelling.

For the identification of actin filaments, or the determination of their polarity, S1 labelling can be used in conjunction with negative staining. Small *et al.* (1978, 1995) used this approach to demonstrate the uniform polarity of actin filaments in lamellipodia.

5. Three-Dimensional Reconstruction

Recently, ET was applied to cytoskeletons embedded in a thick layer of negative stain (Urban *et al.*, 2010). Using this technique, it could be shown that a high degree of three-dimensionality is retained after negative staining of the lamellipodium and the double helical structure of actin filaments can be resolved throughout the tomograms (Fig. 3 and Movie S1 at http://www.elsevierdirect.com/companions/9780123810076). The sample

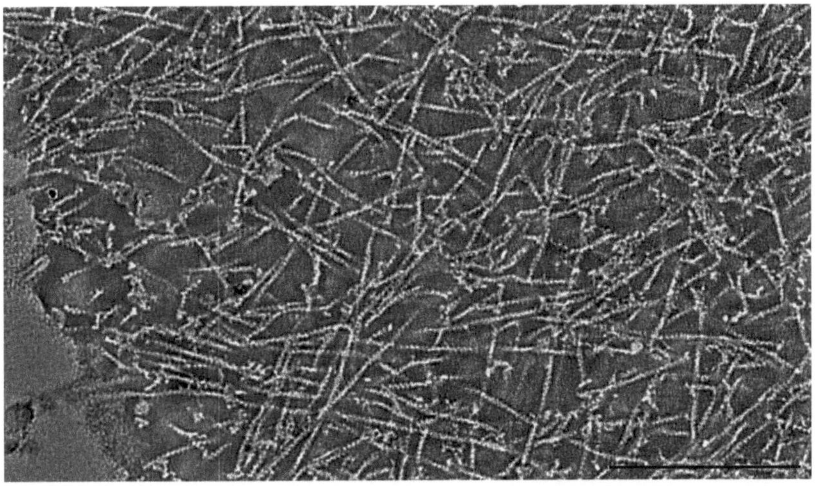

Fig. 3 Actin filament organisation in a (computational) tomogram section of a negatively stained fish keratocyte lamellipodium. The cells were plated on Formvar-coated finder grids, extracted and fixed as described in the text, and negatively stained in 4% sodium silicotungstate. The reprojection was calculated from two tilt series around orthogonal axes (Urban *et al.*, 2010). The actin substructure is evident on all of the filaments; note that the continuity and the true length of the filaments only becomes visible when inspecting the data in 3D, not in a single 0.75 nm section as shown here. The scan through the entire tomogram is shown in Supplementary Movie S1 at http://www.elsevierdirect.com/companions/9780123810076. Bar: 200 nm.

preparation is the same as described above except that cells are plated onto Formvar-coated NHF-A 200 mesh Au finder grids (Maxtaform) to allow the relocation of the cells for double tilt series. To improve contrast in intermediate voltage TEM, and for increased support of the specimen during air drying, a higher concentration of stain (4–8% SST) mixed with protein-coated 10 nm Au fiducials (see Section II.C.1) is used.

For a detailed overview on the acquisition and reconstruction of 3D data by ET, see the following section on cryo-ET. In the case of the negatively stained cytoskeletons presented by Urban *et al.* (2010), the tilt series were acquired on a 300 kV FEG TEM at –190°C using a tilt range of typically ±60° with the Saxton tilt scheme based on 1° increments. A defocus value of –5 μm and an on-screen magnification of 27,500× were used for image acquisition at normal dose. The tomograms were generated from two tilt series obtained around orthogonal axes.

6. Discussion

The advantage of negative staining is that the method is relatively fast, simple, and does not require special equipment. It gives very high contrast and the susceptibility to radiation damage is insignificant. It is generally accepted that the air drying step included in negative staining can cause partial collapse of the specimen (Adrian *et al.*, 1998; Ohi *et al.*, 2004). However, as shown recently by ET (Urban *et al.*, 2010), in well-preserved actin networks of the thinner peripheral regions of cultured cells, three-dimensionality is retained up to a certain extent, and the organisation of actin filaments can be investigated in three dimensions, contrary to earlier claims (Svitkina, 2007). A disadvantage is the common variability in the spreading of the stain, leading to differences in thickness and contrast within and in-between grids, limiting the yield of successful CLEM experiments (Auinger and Small, 2008).

As all "cytoskeleton" methods, negative staining uses extraction, a procedure that is meant to remove the plasma membrane and cytosolic proteins completely, but leave the cytoskeleton untouched. Factors that decide about the reliable preservation of the cytoskeleton upon extraction are the nature of the detergent, the presence of a stabiliser (e.g., aldehyde fixative), and the composition of a buffer vehicle (Lindroth *et al.*, 1992). Even though a lot of time was spent optimising these conditions for negative staining, the results have to be assessed critically by monitoring the extraction of proteins or the loss of fluorescent signal by extraction.

B. Critical Point Drying/Platinum Replica

Critical point drying (CPD) was introduced by Anderson (1951) and is frequently used in SEM of biological specimens. The samples are dried via a phase transition from liquid to gas using elevated pressure and temperature beyond the liquid–vapor critical point of the surrounding liquid. In this way, destructive surface tension effects at the liquid–gas phase boundary as encountered with air drying can be circumvented, resulting in specimens with considerably less compression and collapse. For practical reasons—the critical point of water is at 373°C and 22 MPa—biological specimens are dehydrated in a graded series of ethanols or acetone and subsequently transferred to liquid CO_2 (critical point at 31°C and 7.4 MPa) in which they are dried (Allen *et al.*, 2008).

For drying whole mount specimens for TEM, this technique was applied first on erythrocytes (Anderson, 1951). In the following decades, it became a very popular tool to characterise the cryptic "cytomatrix" (Porter, 1984): Among visualising numerous organelles and cytoskeletal elements as actin bundles and microtubules, Buckley and Porter (1967, 1975) identified a loose meshwork of cytoplasmic filaments. However, individual actin filaments were difficult to resolve and hard to distinguish from IF. In later studies, based mainly on CPD preparations of extracted and unextracted cells, the existence of a novel network of cytoplasmic filaments, named the "microtrabecular lattice", was proposed (Schliwa and van Blerkom, 1981; Wolosewick and Porter, 1976, 1979). Subsequent work using freeze drying (Heuser and Kirschner, 1980) as well as negative staining (Small, 1981) suggested that the microtrabecular lattice was an artefact of the CPD technique. A systematic study by Ris (1985) attributed the appearance of the "microtrabecular lattice" to an artefact of CPD with impure CO_2, emphasising the importance of using completely water- and ethanol-free CO_2 to prevent distortions.

Later, CPD of cytoskeletons was combined with heavy metal shadowing of the dried specimens to produce enhanced contrast (Svitkina *et al.*, 1984). Pt was chosen because of a reasonably low melting temperature and acceptable grain size at the magnifications used. In 1995, Tatyana Svitkina introduced an improved CPD/Pt shadowing approach for visualisation of the cytoskeleton in whole mount specimens (Svitkina, 2007, 2009; Svitkina and Borisy, 1998; Svitkina *et al.*, 1995). In contrast to the protocols employed earlier, this adaptation of the technique uses a rapid detergent extraction supplemented with high-molecular-weight polyethylene glycol (PEG) to stabilise the cytoskeleton, triple fixation for improved preservation of filaments, and UA treatment during dehydration aiming for increased protein stability.

The CPD/Pt replica technique has gained a lot of popularity for studies of cytoskeletal architecture, and has been used with numerous cell types (Bear *et al.*, 2002; Burnette *et al.*, 2007; Mejillano *et al.*, 2004; Svitkina and Borisy, 1999; Svitkina *et al.*, 1997; Yang *et al.*, 2007). Cameron *et al.* (2001), Vignjevic *et al.* (2003), Plastino *et al.* (2004), and others used this approach for the visualisation of actin in biomimetic systems and in the tail of *Listeria*. Based on results obtained by this method, a model for dendritic (branched) organisation of the lamellipodium via the Arp2/3 complex has been proposed (Carlier *et al.*, 2003; Pollard and Borisy, 2003; Svitkina and Borisy, 1999), and models for the formation of filopodia have been revisited (Svitkina *et al.*, 2003).

1. Method

For the CPD/Pt shadowing technique, cells are grown either on standard glass coverslips or on small pieces of glass coverslips, about the size of EM grids. An asymmetric shape of the coverslips (cut with a diamond pencil) can aide unambiguous identification of the cell-containing side throughout the procedure. Biomimetic assays or pathogens can be adsorbed to glass surfaces, with some of the specimen being washed out during preparation (Cameron *et al.*, 2001; Vignjevic *et al.*, 2003). Coating of the coverslip with poly-L-lysine can improve adherence, but will lead to increased background actin.

As soon as the cells are ready for preparation, culture medium is removed from the dish and the cells are quickly rinsed with PBS, or a serum-free physiological solution such as DMEM (Dulbecco's Modified Eagle's Medium), pre-warmed to cell culture temperature. In all steps, the exchange of media is performed quickly; the specimens must not dry! As outlined before, cytoskeletal structures in cells have to be made accessible for contrasting and freed of the membrane and cytosolic proteins by chemical extraction. As these procedures call for a delicate balance between sufficient extraction and the risk of loss or collapse of cytoskeletal structures, Svitkina (2007) proposed a number of different extraction procedures, depending on the aim of the experiment. For visualisation of the cytoskeleton in general, 1.0% Triton X-100 and 1.0–4.0% PEG (35 kDa) as a stabiliser in PIPES-EGTA-Mg buffer (PEM; 100 mmol/l PIPES, 1 mmol/l EGTA, 1 mmol/l $MgCl_2$; pH 6.9) is used (Fig. 4A and A′). For visualisation of lamellipodia, a relatively strong extraction using 0.2–1.0% Triton and 2 μmol/l phalloidin as the sole stabiliser in PEM buffer is employed (Fig. 4B and B′). For visualisation of very labile, easily extracted components of the cytoskeleton, a mixture of 0.5% Triton and 0.25% GA in PEM for simultaneous extraction/fixation can be used, a protocol originally developed for negative staining (Höglund *et al.*, 1980). The extraction time is less critical than the ratio of Triton and GA, a wide range of 3–10 min is suggested. All extraction procedures are carried out at room temperature, not at 4 or 37°C. Following extraction, the specimens are washed in PEM a few times. When using PEG in the extraction buffer, these washes should be prolonged to facilitate exchange of the viscous solution. For procedures that use phalloidin as a stabiliser during extraction, the washing buffer is supplemented with 1 μmol/l phalloidin.

Subsequently, specimens undergo a three-step fixation procedure at room temperature according to Svitkina *et al.* (1995): First, specimens are fixed for 20 min in 2% GA in 100 mmol/l cacodylate buffer, pH 7.3. Second, they are fixed for 20 min in a freshly prepared solution of aqueous 0.1% low-molecular-weight tannic acid (TA), followed by brief washes and a 5 min incubation in water. In a third step of fixation, 0.1% aqueous UA is applied for another 20 min. Proper washing is essential before this step to remove free TA, which forms electron-dense precipitates with UA. The order in which the three fixatives are applied is essential, as TA is thought to bind to both (fixed) protein (Seagull and Heath, 1979) and heavy metal (U), hence mediating the formation of a shell that protects filaments against further processing steps.

For dehydration and CPD, the specimens are loaded horizontally into a wire basket fitting into the CPD's chamber (Fig. 5A). The coverslips are transferred individually from the UA into the wire basket submerged in water, with the cells facing up and pieces of lens paper (Fig. 5B) as a spacer between two coverslips. No more than 12

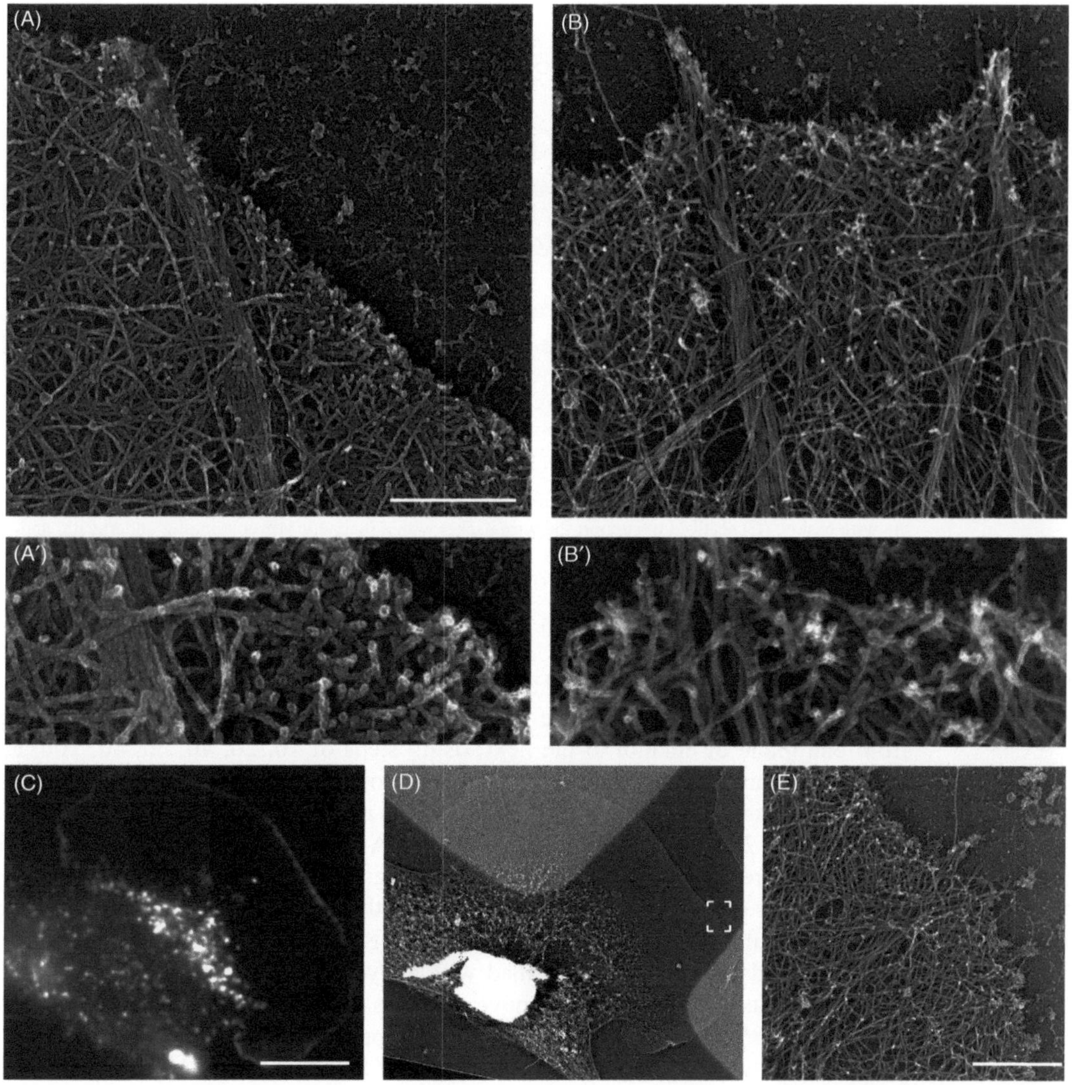

Fig. 4 Visualisation of the actin cytoskeleton by the critical point drying/Pt replica approach. (A, A′) Cell edge of B16F1 mouse melanoma cells extracted for 10 min with 1.0% Triton, 4% PEG, and 2 μmol/l phalloidin in PEM. (B, B′) The same cell line extracted for 5 min with 1.0% Triton and 2 μmol/l phalloidin in PEM. This procedure, excluding PEG or GA, is designed to remove all soluble protein; the structures appear clearer but also more loosened up than in the standard protocol. (C) Fluorescence micrograph of a mouse melanoma cell expressing EGFP-tagged β-subunit of capping protein, which localises to the cell front. (D) Low-magnification electron micrograph showing the same cell and the Au patterns evaporated onto the coverslip used for relocation (E) High-magnification view of the region highlighted in (D). Bars: 0.5 μm (A and B), 10 μm (C and D), and 0.5 μm (E). Modified from *Methods Cell Biol.* **79,** 295–319 with permission.

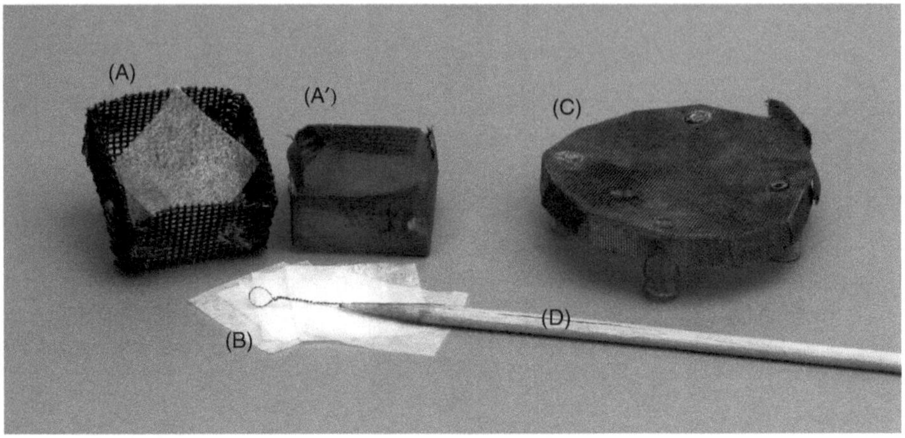

Fig. 5 Set of special tools used for CPD/Pt replica technique. (A) Bottom and (A′) lid of the wire basket used for dehydration and CPD, (B) square pieces of lint-free lens paper used as spacers between coverslips in the wire basket, (C) scaffold to support the wire basket with a magnetic stirrer spinning underneath, and (D) 3 mm Au loop mounted on a wooden stick for transferring replicas.

small coverslips should be loaded at once. Finally, the wire basket is closed with a lid made out of wire (Fig. 5A′). The basket is then transferred quickly without drying from one grade of ethanol to the next, using steps of 10, 20, 40, 60, 80, 100, and 100%, for 5 min each. To improve the exchange of ethanols, the wire basket is placed on top of a wire scaffold (Fig. 5C) that allows a magnetic stirrer to spin underneath. After another 20 min incubation in 0.1–0.2% UA in anhydrous EtOH (which should be prepared fresh, but a bit in advance as it takes time to dissolve) without stirring, the specimens are washed briefly three times with stirring in anhydrous EtOH, dried over a molecular sieve, preferably with indicator pearls.

After dehydration, the CPD is filled with anhydrous EtOH and the wire basket is transferred into the chamber. To remove all traces of water from CO_2 (Lindroth *et al.*, 1988; Ris, 1985), the CO_2 bottle should be equipped with a water-absorbing filter. The chamber is washed with at least 10 exchanges of liquid CO_2 at 5–10°C to remove all traces of EtOH. Following the instructions for the CPD, temperature and pressure are raised, the specimens are dried and retrieved from the instrument.

After CPD, the coverslips are removed from the wire basket and transferred immediately to a high-vacuum evaporator with a rotary table equipped for evaporation of Pt and C. The coverslips are mounted on the adhesive part of memo notes, which in turn are fixed with double-sided tape on the rotary table (Svitkina, 2009). Once the required vacuum ($\approx$1 mPa) is reached, $\approx$2.7 nm Pt are evaporated onto the rotating specimen at an angle of 45°. As the thickness of the Pt layer is critical for the quality of the replica, use of a quartz film thickness monitor is recommended. Following Pt evaporation, the replica is backed with a 3–5 nm layer of C, evaporated onto the rotating sample at 90°.

The replicas retrieved from the high-vacuum evaporator are mounted on EM grids: First, the replica on the coverslip is scratched with a needle to yield fragments that comfortably fit onto EM grids. To release replicas, they are floated on top of diluted (10%) hydrofluoric acid in plastic (!) dishes, e.g., multiwell dishes as used for cell culture. Within a short time, the coverslip will fall off, leaving the floating replica behind. Using a 3-mm-diameter Pt or Au loop (Fig. 5D), the replicas are then transferred to dishes filled with water containing traces of household detergent or Triton (a few μl of 0.5% detergent per well). After two washes on distilled water to remove the detergent, the floating replicas are picked up from below with Formvar-coated 50 mesh Cu grids. An eyelash as used for manipulating ultrathin sections can be used to position the replica correctly. Excess water is drained from the side of the grid and after drying, specimens can be analysed in a routine TEM (Fig. 4A and B).

2. Instrumentation and Materials

Instrumentation: Critical point dryer (Leica EM CPD, Tousimis Samdri, etc.) supplied with CO_2 dried through a water-absorbing filter (e.g., Tousimis LCO2 Filter Assembly; www.tousimis.com), high-vacuum evaporator set up for evaporation of Pt and C onto a rotating target, quartz film thickness monitor for high-vacuum evaporator, routine transmission electron microscope.

Materials: 7–10 mm glass coverslips for cell culture, diamond pencil, cell culture dishes, wire basket for CPD (Figs. 5A and A′), lens paper (Fig. 5B), wire scaffold to support the wire basket with a magnetic stirrer spinning underneath (Fig. 5C), molecular sieve, memo notes, Pt and C sources for evaporation (e.g., Pt wire of 0.2 mm and carbon rods), fine needle, fine stainless-steel forceps, multiwell dishes, 3 mm Pt or Au loop (Fig. 5D), Formvar-coated 50 mesh grids, filter paper.

Reagents: PBS (8.0 g/l NaCl, 0.2 g/l KCl, 1.2 g/l Na_2HPO_4, 0.2 g/l KH_2PO_4; pH 7.3), PEM buffer (100 mmol/l PIPES, 1 mmol/l EGTA, 1 mmol/l $MgCl_2$; pH 6.9), extraction buffer (different formulations used according to sensitivity of specimen; see text), 2% GA in 100 mmol/l cacodylate buffer, pH 7.3, aqueous 0.1% low-molecular-weight TA (Mallinckrodt Baker, KY, USA), aqueous 0.1% UA, 10, 20, 40, 60, 80, 100% and anhydrous EtOH, 0.1–0.2% UA in 100% EtOH, hydrofluoric acid, household detergent or Triton X-100.

3. Correlative Microscopy

The CPD/Pt replica protocol introduced by Svitkina *et al.* (1995) is tailored for a high success rate with CLEM (Fig. 4C–E). To obtain correlative data, only a number of minor changes (Svitkina and Borisy, 1998) have to be incorporated into the protocol outlined above.

As cells are directly grown on coverslips, correlation of data from LM and EM preparations is rather straightforward. To relocate cells, coverslips with a finder pattern are used that can be visualised using both LM and EM. This finder pattern is produced by evaporating Au through a 400 mesh finder grid onto a glass coverslip. The Au layer should be stabilised at 160°C overnight. Cells that grow in the parts free of Au are used. The setup for LM should be optimised to allow rapid exchange of the cell culture medium for the extraction buffer. This can be done by mounting the coverslips with the finder pattern into petri dishes with a hole in the bottom, using silicone grease or silicone (as used for fish tanks). Coverslips are detached from these petri dishes after chemical fixation. Immersion oil left on the bottom of the coverslip should be removed beforehand, using a cotton swab with an appropriate solvent. If you use silicone grease to seal the petri dish for LM, cut off the greased edges with a diamond pen to avoid contaminating the specimen.

Following dehydration, CPD, and shadowing, but while the replica is still attached to the coverslip, the area of interest has to be identified under a stereo microscope. The replica should be scratched to free the part of interest. Special care has to be taken when the floating replica is picked up with the grid, not to cover the area of interest with a grid bar. This can be done under a dissecting microscope as well, optionally with the help of a micromanipulator holding the forceps with the grid (Auinger and Small, 2008). Alternatively, Formvar-coated slot grids can be used: GS2/3X1 Increased Support Slot Grids from Gilder Grids (www.gildergrids.co.uk) have proven particularly useful for all kinds of applications in our hands.

The correlative LM/EM approach has been used in a number of studies to characterise cell motility by phase contrast microscopy before EM fixation (Svitkina and Borisy, 1999), to identify myosin patches (Svitkina *et al.*, 1995), to observe dynamics and localisation of fluorescently labelled actin and fascin (Svitkina *et al.*, 2003) and constitutively active mDia2-GFP (Yang *et al.*, 2007), and to elucidate their role in filopodia formation.

4. Protein Localisation

Indirect immunolabelling can be combined with CPD/Pt shadowing (for a protocol, see Svitkina, 2007). For optimal structural preservation, it is advisable to use a pAb that reacts with GA-fixed material and to do the labelling between GA and TA fixation. If antigenicity is lost by GA fixation, Svitkina and Borisy (1998) recommend not to use weaker fixatives such as formaldehyde or even precipitating ones like MeOH. Instead, they suggest carrying out the immunolabelling after extraction, before chemical fixation (Verkhovsky *et al.*, 1995). A drawback of this approach is that with weak or no fixation, the cytoskeletal structure might deteriorate and conclusions about localisation should be made with these limitations in mind. To obtain good visualisation of the Au particles on the replica, a size of at least 10 nm is required, but larger particles are preferable. This approach has been used to localise myosin II (Svitkina *et al.*, 1997; Verkhovsky *et al.*, 1995), subunits of the Arp2/3 complex (Korobova and Svitkina, 2008; Svitkina and Borisy, 1999), ADF/cofilin (Svitkina and Borisy, 1999), and other actin-related proteins in cytoskeletons.

Myosin S1 labelling can be combined with the CPD/Pt replica technique as well. In contrast to the arrowhead pattern obtained in negatively stained samples, decorated actin filaments have a rope-like appearance. As only the surface of the filament directed to the Pt source is shadowed, the directionality can be deduced from the turn of the helix. S1 decoration is carried out preceding chemical fixation; for lab protocols, see Svitkina and Borisy (1998) and Svitkina (2007). This method has been applied to cytoskeletons (Semenova *et al.*, 2008; Svitkina *et al.*, 1995) and to synthetic systems (Vignjevic *et al.*, 2003).

5. Three-Dimensional Reconstruction

One aim of CPD is to minimise specimen distortion and collapse, to avoid artefacts, and to preserve a high degree of three-dimensionality. Stereoscopic images in Yang *et al.* (2007), Korobova and Svitkina (2008), and Semenova *et al.* (2008) show that the three-dimensional organisation of the actin in CPD/Pt shadowed preparations of cytoskeletons is well retained. However, Pt shadowing creates a somewhat unilateral view of the cytoskeleton due to contrasting only from the top and decreases with the depth of the specimen. Hence, such specimens are unlikely to be amenable to three-dimensional reconstruction as offered by ET.

6. Discussion

The CPD/Pt replica approach produces appealing images of the actin cytoskeleton with high contrast, clear visualisation of filaments, and with a very apparent three-dimensional organisation. It has become the standard for visualisation of the actin cytoskeleton in text books (Alberts *et al.*, 2008; Bray, 2001). This technique is, however, only being used by very few labs worldwide.

Enhancement of contrast by Pt shadowing is limited to the superficial layers of the specimen. Hence, structural information is lost, for example, for three-dimensional reconstruction, despite good three-dimensional preservation of the cytoskeleton. On the other hand, very thick and compact structures are better visualised than in negative staining, where only electron-dense masses of stain without resolvable detail can be seen. This was shown by Gimona *et al.* (2003) on podosomes and by Cameron *et al.* (2001) using motility assays, where regions close to the bead can be visualised (compare with Fig. 1B).

The most important criterion, however, is that the preparation of a specimen reliably represents the situation in the living cell. As already discussed in the previous section on negative staining, considerations concern the formation of artefacts due to the chemical extraction procedure, which is required to make cytoskeletal structures accessible for visualisation. Also, an issue related to the CPD/Pt replica technique that deserves critical attention is the triple fixation, including prolonged treatment with solutions at a very acidic pH (2% aqueous UA has a pH of $\approx$4, 0.2% aqueous TA has a pH of $\approx$3). In addition, dehydration and CPD can potentially introduce artefacts by shrinkage and distortion (Boyde, 1978; Lindroth *et al.*, 1992). These considerations are particularly important as evidenced by differences in actin organisation seen by the application of alternative techniques using the same cell types: after CPD, filaments in lamellipodia appear strongly branched, whereas after negative staining they do not (Koestler *et al.*, 2008; Small *et al.*, 2008; Urban *et al.*, 2010). To test the CPD/Pt replica technique for the possibility that it induces branch-like artefacts, Resch *et al.* (2002a) subjected F-actin *in vitro* to the CPD/Pt replica procedure and obtained highly branched networks, not seen in the same F-actin preparations observed by cryo-EM. This needs evaluation to ascertain if the frequent branches seen in CPD/Pt replica specimens are actual structural features or preparation artefacts.

C. Cryo-Electron Tomography

In cryo-transmission electron microscopy, specimens embedded in amorphous (i.e., vitrified) ice are imaged at liquid nitrogen (LN_2) temperature or lower in a TEM. The low temperature is required both to prevent the water from recrystallisation and to avoid its sublimation in the microscope's vacuum. This technique, introduced by Dubochet and McDowall (1981) and Adrian *et al.* (1984), is today the only approach in TEM used routinely to image biological specimens in their natural, aqueous environment. In contrast to conventional EM procedures, no heavy metal stains need to be added, though the addition of negative stain is an option (Adrian *et al.*, 1998). Contrast in such unstained frozen-hydrated specimens is predominantly generated by phase contrast (Erickson and Klug, 1971; Lepault and Pitt, 1984). To account for the susceptibility of cryo-specimens to beam damage, a low-dose mode has to be used for data acquisition. Low-dose imaging minimises exposure of the region of interest, e.g., by using a displaced area for focussing. The use of phase contrast and low-dose exposures contributes, however, to a rather low signal-to-noise ratio characteristic for cryo-images. Because of the lack of destructive treatments for specimen dehydration and staining, it has become the standard technique for reconstruction of macromolecules by TEM at unprecedented resolution in the subnanometer range (reviewed in Jiang and Ludtke, 2005).

Following the initial work on isolated macromolecular complexes and sectioned material, this technique was used to resolve the internal organisation of small eukaryotic cells: Frederik *et al.* (1991) and O'Toole *et al.* (1993) produced the first images of frozen hydrated blood platelets. In the two-dimensional projection images, F-actin superimposed with the crowded cytoplasmic environment of whole, unextracted cells was difficult to trace. Chemical extraction and fixation before freezing (Resch *et al.*, 2002b) was used to improve visualisation of actin meshworks by cryo-EM, but at the risk of inducing artefacts by the more invasive specimen preparation.

In the 1990s, advances in microscope hardware and computing power made three-dimensional reconstructions by (cryo-)ET practically feasible (reviewed by Steven and Aebi, 2003). This opened the way to study actin filaments in unextracted cells, as superimposing material is removed in the computational sections of the tomograms. One of the first technical studies on cryo-ET by Grimm *et al.* (1997) used F-actin-filled vesicles as test specimens and the filaments were resolved reasonably well. Shortly after, whole prokaryotic cells were reconstructed three-dimensionally using cryo-ET (Grimm *et al.*, 1998). Subsequently, Medalia *et al.* (2002) succeeded in producing the first reconstruction of a part of a whole frozen eukaryotic cell: At the cell edge of a *Dictyostelium* amoeba, actin filaments were resolved among other features, at a resolution of 5–6 nm. This study set the scene for further work on the actin cytoskeleton in filopodia (Lucic *et al.*, 2007; Medalia *et al.*, 2007), on the structure of the actin cytoskeleton upon virus entry (Maurer *et al.*, 2008) on the organisation of actin in actively protruding cell edges (Urban *et al.*, 2010), and on actin-like proteins in prokaryotes (Kürner *et al.*, 2005).

In this section, we describe the visualisation of the actin cytoskeleton in unextracted, unfixed whole mounts by cryo-ET. Despite technical challenges, this approach is considered optimal to obtain direct information about the organisation of actin networks *in vivo*. For a protocol on cryo-EM of chemically extracted and fixed cytoskeletons (Fig. 6D), see Resch *et al.* (2006).

1. Method

As for negative staining (Section II.A.1), cells for visualisation in cryo-EM are grown directly on coated EM grids. 200 mesh Au grids are a suitable compromise for stability of the support film and obstruction of the beam at high tilt. Most studies use grids with perforated C films with few and small holes (i.e., holey-carbon grids), either home-made (de Carlo, 2009; Grassucci *et al.*, 2007) or commercially available products such as Quantifoil R 1/4 (www.quantifoil.de). Alternatively, C-coated Formvar films have been used for cryo-ET (Schwartz *et al.*, 2007). Cells are plated on the grids as described previously in the section on negative staining. Attachment and spreading of the cells is repeatedly monitored with phase contrast LM. Once the desired state of cells is reached, the specimens can be frozen.

For plunge (immersion) freezing, a number of different devices can be used: In basic home-made devices, an EM grid is held by tweezers, excess liquid is removed manually with a piece of filter paper and the device guides the immersion of the grid into a cup of secondary cryogen—liquefied ethane, propane, or a mixture of both (Tivol *et al.*, 2008)—sitting in a bath of LN_2 (shown in de Carlo, 2009). More sophisticated custom-made versions include provisions to control humidity and temperature in the moments before and after blotting, double-sided blotting, or automated plunge timers (Cyrklaff *et al.*, 1990). Meanwhile, there are also a number of instruments commercially available, such as the FEI Vitrobot (Frederik and Hubert, 2005; www.fei.com/vitrobot), the Leica EM GP (www.leica-microsystems.com), and the Gatan Cp3 (www.gatan.com). These instruments all aim to simplify the preparation procedure and to provide standardised, highly reproducible freezing conditions.

No matter which type of instrument is used for freezing, the key problems remain the same: A water layer with a thickness suitable for reliable vitrification, imaging, and preservation of the specimen must be produced by blotting most of the liquid from the grid with filter paper. At all times after freezing, the specimen must be kept below devitrification temperature and ice contamination on the surface of the vitrified specimen must be avoided. Therefore, only pre-cooled and frost-free tools are to be used, the atmosphere in the room used for freezing should be dry, and the grid should be

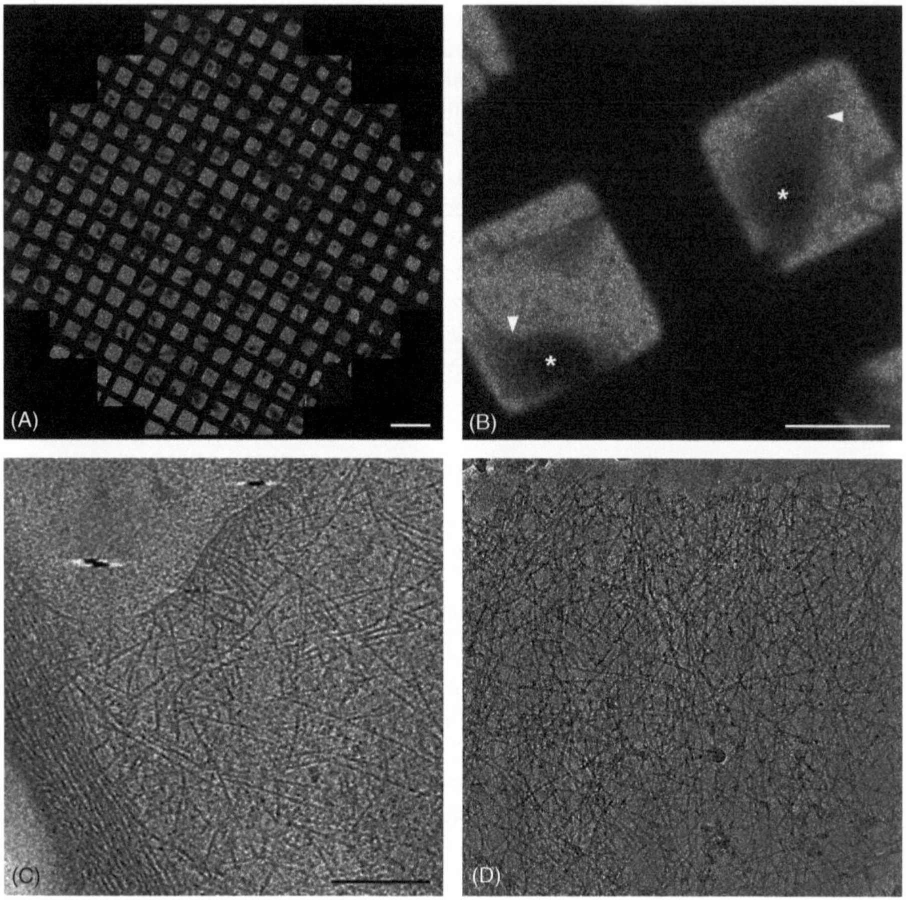

Fig. 6 Cryo-electron microscopy and tomography of cells directly grown on grids. (A) Low-magnification montage of a whole grid with cells in a thin layer of evenly thick ice. The specimen on a Quantifoil R1/4 film was frozen with the Leica EM GP; the montage was prepared using SerialEM. (B) Cells at intermediate magnification in low-dose search mode. Cell bodies are marked with asterisks; thin regions potentially suitable for electron tomography of the lamellipodium are indicated with arrowheads. (C) Computational 6.0 nm section through a tomogram of a lamellipodium region of a vitreously frozen CAR fish fibroblast, showing actin filaments arranged in a meshwork and a bundle as well as the plasma membrane. The complete scan through the tomogram is shown in Supplementary Movie S2 at http://www.elsevierdirect.com/companions/9780123810076. (D) Lamellipodium region of a mouse melanoma cell extracted with a Triton/GA mix, chemically fixed in GA, and plunge frozen (Resch *et al.*, 2002b). Note the clear visibility of filaments and actin substructure even in a projection image. Bars: 200 μm (A), 50 μm (B), 250 nm (C and D).

stored under LN_2 at all times and only transferred through a dry N_2 atmosphere but not through moist air. For in-depth protocols on the art of plunge freezing specimens, see Grassucci *et al.* (2007) and de Carlo (2009), both using home-made instruments, or Iancu *et al.* (2007) and Frederik *et al.* (2009), on the Vitrobot.

Prior to freezing the cells, we keep the grids in an incubator nearby to avoid unnecessary temperature shifts. Each grid is picked up individually from the cell culture dish with the forceps fitting the plunge freezer and mounted in the instrument set up with LN_2 and liquid ethane, ready to freeze. Since only a very small volume of liquid adheres to the grid, we immediately add a defined volume (e.g., 4 μl) of pre-warmed cell culture medium or buffer to prevent drying of the specimen and to standardise conditions. This also allows the delivery of 10 nm Au colloid particles used as fiducials for the reconstruction of the tilt series. To avoid osmotic shock and aggregation of the Au particles, we do not add just Au colloid (which comes suspended in water) or colloid mixed with medium or buffer. Instead, we use protein-coated Au particles (Garvalov *et al.*, 2006; Lucic *et al.*, 2007; Urban *et al.*, 2010) suspended in fully concentrated cell culture medium or buffer. Subsequently, the grid is blotted from the back side and frozen.

With each new cell line, medium, or grid type, different blotting times should be tested to define optimal conditions. Using the Leica EM GP for freezing of either melanoma cells or fibroblasts grown on Quantifoil R1/4 films, good results were obtained with the chamber at 37°C and 90% relative humidity, blotting for 1–3 s from the back side of the film with Whatman No. 1 filter paper and immediate freezing. Blotting from the back side is essential as it reduces (yet does not eliminate!) the risk of damaging cells and support film by avoiding direct contact with the filter paper. After freezing, the grid is transferred quickly to a cryo grid box sitting in LN_2 (liquid phase!). Once all grids are frozen, the grid box is closed and kept in LN_2 until further use.

For cryo-TEM, a microscope equipped with a cooled specimen chamber, a cryo-transfer system (holder or cartridge system), and low-dose software is required. In addition, tomography requires a CCD camera, a goniometer and a specimen holder capable of high tilt, and acquisition software (see below). An intermediate voltage instrument equipped with an energy filter is beneficial.

With the same precautions as outlined above (temperature, humidity), the grid is loaded into the cryo-transfer system. Once the grid is inserted into the microscope and has reached stability, we prepare a montage of the whole grid from images acquired at low magnification (100–200×) and minimal dose. This allows ready identification of regions with an intact support film, suitable ice thickness, and also cell profiles (Fig. 6A). Software that supports montaging includes EM-TOOLS (www.tvips.com), Leginon (Suloway *et al.*, 2005; www.leginon.org), SerialEM (Mastronarde, 2005; bio3d.colorado.edu/SerialEM/), and the TOM toolbox (Sartori *et al.*, 2007; www.biochem.mpg.de/baumeister/tom_e/). Suitable areas on the grid are screened at an intermediate magnification with minimal dose on a binned CCD live image to localise the features of interest. A strong underfocus will significantly improve the visualisation of unstained cellular features. Once a feature of interest is found (Fig. 6B), the thickness and quality of the ice, as well as the number of fiducial markers, is checked by a low-dose exposure of the area. At this stage, it is crucial to critically assess the integrity of the frozen cell, as some cell types have a very strong tendency to burst upon blotting.

Recording the tilt series in low dose involves a number of tasks: autofocussing, tracking of the area of interest or application of prediction algorithms to compensate for mechanical inaccuracies of the stage, adjustment of illumination/exposure time to account for increasing thickness of the specimen at high tilt, recentering of the slit of an energy filter, etc. A number of programs with different levels of automation are available for these tasks, both freeware and commercial ones: EM-TOMO (www.tvips.com), SerialEM (Mastronarde, 2005; bio3d.colorado.edu/SerialEM), the TOM Toolbox (Nickell *et al.*, 2005; www.biochem.mpg.de/baumeister/tom_e/), UCSF Tomography/Leginon (Suloway *et al.*, 2009; Zheng *et al.*, 2007; www.leginon.org), and Xplore3D (www.fei.com).

For visualisation of the actin cytoskeleton in the periphery of whole mounts, we acquire single tilt series on a 300 kV FEG TEM at LN_2 temperature. We use a tilt range higher than ±55° with a 2° increment to limit the dose, yielding about 60 images per series with a nominal total dose of 6000 electrons/nm^2, and a defocus of −10 μm at a nominal magnification of 22,000×. Before acquisition of the tilt series, eucentricity is refined in the same grid square. To reduce degradation of the contrast and blurring of the image due to increased inelastic scattering of electrons with increased thickness of the specimen (Grimm *et al.*, 1997), we use zero-loss filtering via a post-column energy filter with a slit width of 20 eV. For a discussion on how the individual parameters influence the quality of the resulting tomogram, see Koster and Bárcena (2006). Tilting the specimen not just around one but also around a second orthogonal axis (i.e., double tilt tomography; Penczek *et al.*, 1995) reduces the amount of data lost and reconstruction artefacts created by the inability to acquire projection images from all angles: The "missing wedge" is reduced to a "missing pyramid". Due to restrictions of electron dose, this double tilt approach is not used very frequently in cryo-ET, but it is feasible (Henderson *et al.*, 2007).

Once the tilt series has been acquired, a volume is reconstructed from the projection images. This is a multistep procedure, including alignment of the tilted views into a common framework of coordinates based on cross-correlation and/or the fiducial markers and backprojection of the tilted views into a volume. Program packages, such as IMOD (Kremer *et al.*, 1996; bio3d.colorado.edu/imod), Inspect3D (www.fei.com), and the TOM Toolbox (Nickell *et al.*, 2005; www.biochem.mpg.de/baumeister/tom_e), provide the required functionality. Apart from the clear visualisation of the features of interest and the signal-to-noise ratio, the quality of a tomographic reprojection can be assessed by examining the Au fiducials: In *x*/*y*, they should be evenly round, and elongated but not "banana"-shaped in *z*. This elongation affecting all structures arises from anisotropic resolution due to the incomplete sampling of the volume in the tilt series. A further artefact in reprojections arising from the missing data is the weaker representation of actin filaments in certain orientations relative to the tilt axis and the optical axis. Both effects can be reduced using double-tilt tomography (Mastronarde, 1997).

Even though the signal-to-noise ratio in the tomogram is significantly improved as compared to the individual tilted views (Fig. 6C and Movie S2 at http://www.elsevierdirect.com/companions/9780123810076), denoising the data is still an important issue, especially for further automated analysis. A number of different advanced 3D filtering methods suitable for cryo-electron tomographic data are discussed in Hegerl and Frangakis (2006). Some of these algorithms are already implemented in the software packages for reconstruction mentioned above. Data from ET can be presented as electronic sections (tomographic slices) at any angle through the reconstructed volume. Software such as IMOD (Kremer *et al.*, 1996) allows manual modelling of three-dimensional structures based on the electronic sections (for an example, see Urban *et al.*, 2010). Alternatively, individual voxels can be assigned automatically either to the structure or the background based on their brightness and represented as a surface-rendered three-dimensional object (for an example, see Medalia *et al.*, 2002).

2. Instrumentation and Materials

Instrumentation: Inverted phase contrast light microscope, freeze plunging device (manual or automatic: FEI Vitrobot, Leica EM GP, Gatan Cryoplunge 3), liquid nitrogen specimen storage tank, intermediate voltage transmission electron microscope equipped for cryo-EM (see text) and software for acquisition of grid maps and tilt series (see text), software for tomographic reconstruction, filtering and modelling (see text).

Materials: Stainless-steel forceps with fine tips, 200 mesh Au grids coated with a perforated carbon film with few small holes for cell culture, multiwell dishes, liquid nitrogen, ethane (quality 4.5 or better), Whatman No. 1 filter paper.

Reagents: Cell culture medium, protein-coated 10 nm colloidal Au, commercially available (Garvalov *et al.*, 2006; Lucic *et al.*, 2007) or self-made (Urban *et al.*, 2010).

3. Correlative Microscopy

For correlating data from LM with cryo-EM/ET there are two strategies reviewed in Plitzko *et al.* (2009): Either to perform LM on the frozen specimens, using a special cryo-light microscope, or to use LM to observe living specimens and to freeze samples just after the occurrence of specific cellular events and stages.

In a cryo-LM suitable for CLEM, the specimen has to be kept below −140°C and free of ice contamination, and it has to support reasonably high resolution and preferably also epifluorescence microscopy. Prototypes were introduced by Sartori *et al.* (2007) and Schwartz *et al.* (2007). Frozen cells grown on grids were imaged by phase contrast and epifluorescence microscopy. The same cells or fluorescently labelled structures were relocated in the cryo-EM, undamaged and still vitrified. While this approach does not allow the visualisation of cell dynamics, the specimen in the LM corresponds exactly to what will be examined in the EM. It also allows assessment of the damage to support film and cells by blotting and freezing, an estimate of the ice thickness (Sartori *et al.*, 2007) and studies of weak fluorescent signals by long-time exposure due to reduced photobleaching under cryo conditions (Schwartz *et al.*, 2007). This approach will be simplified significantly once integration of a light microscope into a cryo-TEM has been achieved, as it has already been realised in a TEM for room temperature specimens (Agronskaia *et al.*, 2008).

The second strategy is to observe living cells cultured on grids on the LM, to transfer them quickly to an immersion freezing device, and to freeze them. In contrast to the first approach, dynamic events before freezing can be investigated and potential artefacts by blotting and freezing (see discussion below) can be identified more easily. However, cells can be lost due to breakages of the support film, displaced by blotting, if not strongly attached, or be obscured in a too thick ice layer. Due to the time needed for transferring and freezing the specimen, the time resolution of this approach is rather low.

4. Protein Localisation

Most conventional immunolabelling strategies cannot be combined with cryo-EM (reviewed in Leis *et al.*, 2009). First, chemical fixation, permeabilisation as required for intracellular epitopes, and repeated incubation and washing steps are incompatible with the idea of minimally invasive specimen preparation. Second, labelling cannot be carried out after cryofixation at below devitrification temperature and with the specimen embedded in solidified water.

Hence, alternative strategies have to be employed to localise proteins in cells imaged by cryo-EM/ET. One way is to correlate data from LM and EM, as outlined in the previous section, with the drawback of reduced resolution of the label. Future developments might open the way to protein localisation in cryo-EM by pattern recognition techniques (reviewed in Frangakis and Förster, 2004), expression of tags that can be visualised directly in EM (Mercogliano and DeRosier, 2007), or microinjection of tagged proteins or antibodies as used for LM (Rottner *et al.*, 2001) with 1.4 nm nanogold as a potential marker (He *et al.*, 2009).

5. Discussion

Cryo-ET of whole cells (reviewed in Leis *et al.*, 2009; Steven and Aebi, 2003) is receiving significant attention since this technology is perceived as a key advance to resolve cellular ultrastructure beyond the reach of conventional methods of specimen preparation. Undoubtedly, this technique allows structural preservation at the nanoscale hitherto unseen and is most likely the only way to produce a true "snapshot" of a single moment in the life of a cell. Despite a lot of enthusiasm for this technique it is still very young and very demanding in terms of hardware and expertise in acquisition, processing, and interpretation of data. Importantly, the application of cryo-ET on whole mounts, as described above, is restricted to peripheral regions of the cell, with a thickness of up to a few hundred nanometers. If regions deeper inside the cells have to be visualised, cryo-sectioning (Section III.B), focussed ion beam thinning of the specimen (Marko *et al.*, 2007), or cryo-planing (Rigort *et al.*, 2010) is required.

One particular difficulty of the cryo-ET of unextracted cells is to localise features of interest, or to identify a feature in the functional context of the cell. This is because of the weak contrast that unstained specimens exhibit and the limited possibility to judge the region to be imaged before acquisition of the tomographic tilt series, due to susceptibility to beam damage. To readily locate features, and to reliably correlate the structure of the actin cytoskeleton with cellular function, correlative microscopy (Section II.C.3), or a homogenous population of cells, e.g. showing protrusion, or treadmilling all over the cell edge (Small *et al.*, 2008; Urban *et al.*, 2010) can be used.

Finally, it is dangerous to think that plunge freezing of cells is a preparation technique entirely free of artefacts: living cells are more easily perturbed than a suspension of molecules. Frequently, we have observed disintegrated cells, ruptured plasma membranes or organelles, and cell protrusions being drawn into holes or cracks in the carbon film (Wonesch and Resch, unpublished data). We attribute these preparation artefacts to the flow of liquid during blotting and the surface tension of the receding liquid layer—and, most likely, there are numerous far more subtle changes occurring. Bearing this in mind is particularly important as the low contrast inherent to cryo-images makes it difficult to assess the condition of a cell as a whole at low magnification.

III. Sectioning Techniques

For electron microscopy of structures located deeper inside cells, or in tissues, sectioning in a matrix suitable for microtomy is the most frequently used approach. Embedding media for soft biological specimens used in TEM are plastics such as epoxy or acrylic resins, ice or—much less frequently used—re-extractable media including PEG (Wolosewick, 1980) or poly-methylmethacrylate (Gorbsky and Borisy, 1985). In this chapter, we discuss the pitfalls of visualising the actin cytoskeleton in plastic sections after chemical or physical fixation, as well as in vitrified cryo-sections.

A. Plastic Sections

In a pioneering study using thin sections of epoxy embedded cells, Ishikawa *et al.* (1969) revealed the presence of F-actin in extracted non-muscle cells by labelling the filaments with HMM. Subsequent progress in extending the knowledge on the actin arrangements in non-muscle cells was rather slow due to the difficulties of filament preservation at the ultrastructural level using standard plastic embedding: While densely packed actin bundles, as in skeletal muscle or stress fibres, are relatively stable structures, this is not the case for loose filament meshworks. As already seen in early studies, e.g., on nerve growth cones (Yamada *et al.*, 1971), cortical actin was observed as a relatively unstructured flocculent material. In order to improve preservation of F-actin networks, the effect of several preparative steps on their integrity was investigated: Maupin-Szamier and Pollard (1978) showed that post-fixation with OsO_4, as routinely used in TEM preparations can lead to fragmentation of the actin filaments. A study by Small (1981) confirmed these findings and, furthermore, highlighted the susceptibility of actin networks to destruction by dehydration in organic solvents. However, TA, used at low concentrations during or after aldehyde fixation offered some protection from degradation to F-actin (Begg *et al.*, 1978; Maupin and Pollard, 1983; Seagull and Heath, 1979; Svitkina *et al.*, 1995).

Using chemical fixation, dehydration, and embedding with a number of modifications, the actin cytoskeleton could be visualised in varying clarities in yeast (Kamasaki *et al.*, 2005, 2007), stress fibres, microvilli (Begg *et al.*, 1978), and tendon (Canty *et al.*, 2006). Furthermore, the actin tails of pathogenic bacteria (Gouin *et al.*, 1999; Tilney and Portnoy, 1989; van Kirk *et al.*, 2000), vaccinia virus (Cudmore *et al.*, 1995) *in situ*, and biomimetic assays (Noireaux *et al.*, 2000) were analysed using this method. Some of these studies use decoration of actin with myosin heads (HMM or S1), which also stabilises and aids visualisation of the filaments besides allowing determination of their polarity. In direct comparison, the differences in preservation between native and labelled filaments become obvious (Gouin *et al.*, 1999; Ishikawa *et al.*, 1969; Tilney and Portnoy, 1989; van Kirk *et al.*, 2000). Importantly, this comes at the expense of having to first extract cells with detergent to infiltrate the myosin fragments.

With the introduction of high-pressure freezing (HPF; Moor and Riehle, 1968; Studer *et al.*, 2001) and freeze substitution (FS, reviewed in Humbel and Schwarz, 1989) as routine techniques for plastic embedding of cells and tissues, alternative possibilities for the preservation of ultrastructure in tissue emerged. In a number of studies employing HPF and FS, filamentous structures were designated by the authors as actin filaments: in *Dictyostelium* spores (Sameshima *et al.*, 2002), cultured cells (Jiménez *et al.*, 2006; Sawaguchi *et al.*, 2002), and tissues like nervous tissue (Siksou *et al.*, 2007, 2009) or taste buds (Royer and Kinnamon, 1996). Standard FS "recipes", including normal concentrations of OsO_4 in the substitution cocktail, were employed. The contribution of the individual fixatives to the preservation of actin and the benefits of including tannic acid (Siksou *et al.*, 2007) have yet to be investigated.

1. Method

For conventional ultrathin section preparations of cells and tissues aiming to visualise actin networks, we suggest some modifications to the standard protocols (Hayat, 2000). First, we include 0.2% TA in the primary aldehyde fixative for improved fixation of actin. OsO_4 post-fixation should not be omitted, but the concentration is lowered to 0.2% and the incubation is carried out on ice.

In the TEM, F-actin appears only slightly thinner than the more robust IF (10 nm) with which they should not be confused. IF typically follow curved trajectories whereas actin filaments are closely linear and in more concentrated arrays (Small, 1988; Yamada *et al.*, 1971). A straightforward way to judge the preservation of actin filaments in a conventional preparation is to examine cell protrusions such as filopodia and microvilli: Both the presence and the organisation of actin in these structures are well characterised.

For immunolabelling of actin and its associated proteins, the usual pre- and post-embedding techniques (Griffiths, 1993) are applicable (Langanger *et al.*, 1984;

Schmitz and Drenckhahn, 1993; Sekiya-Kawasaki *et al.*, 2003). Due to a lack of clearly visible filaments, this approach was also used to localise actin. Structural features of the filament networks cannot be visualised using immunolabelling, however.

2. Instrumentation and Materials

Instrumentation: Desiccator (optional, for long-time resin infiltration of specimens), polymerisation oven, mill for trimming (optional), ultramicrotome, routine TEM.

Material: Plastic tubes and dishes, embedding moulds, fine degreased razor blades for trimming, diamond knife, eyelash mounted on a stick for the manipulation of sections, Formvar-coated 50 mesh grids.

Reagents: Isotonic buffer (e.g., 100 mmol/l phosphate buffer), 1–2% GA and 0.2% TA in isotonic buffer, 0.2% OsO_4 in isotonic buffer, 40, 60, 80, 95% and anhydrous EtOH, propylene oxide (should be avoided for safety reasons; optional intermedium for larger specimens), monomeric epoxy resin (Agar 100, EMbed 812, Spurr's resin, etc.), lead citrate, UA.

3. Discussion

Thin-section EM is by no means an ideal technique to visualise the actin cytoskeleton in non-muscle cells: First, using standard techniques for chemical fixation, dehydration, and embedding without special precautions, loose meshworks of actin filaments are poorly preserved and only visible as a "spongy material" or "polygonal meshwork" (Kadurugamuwa *et al.*, 1991; Prévost *et al.*, 1992; Sabaneyeva *et al.*, 2009; Taunton *et al.*, 2000; Yamada *et al.*, 1971). Second, the thin filaments barely set themselves apart from the cytoplasmic background. Third, the continuity of the filaments and the three-dimensional organisation of the cytoskeleton are lost by the sectioning process, unless tomography is used. The modifications outlined above also come at the cost of insufficient fixation of other cellular components (lowered OsO_4) or permeabilisation of the specimen (myosin head decoration). While there are promising results using HPF and FS, there is no systematic study investigating the effects of different FS reagents on the preservation of actin.

B. Cryo-EM of Vitrified Sections

The aim of this technique is to study the ultrastructure of bulk biological material in its native fully hydrated state in the absence of chemical fixation, dehydration, and heavy metal staining (Al-Amoudi *et al.*, 2004). After rapid freezing and microtomy below −140°C, the vitrified specimens are observed in a cryo-EM (see also Section II.C on cryo-ET). The method already emerged in the 1980s (McDowall *et al.*, 1983) but in order to obtain reproducible results the freezing and sectioning procedures had to be optimised. First, HPF became a routine procedure to vitrify biological samples in the range of up to 200 μm thickness (Studer *et al.*, 2001). Second, the cryo-sectioning process for thin vitreous sections was investigated in detail and optimised (Al-Amoudi *et al.*, 2002). In 2004, Dubochet and co-workers coined the term "CEMOVIS" and showed the first results on the ultrastructure of bacterial membranes and on epidermal desmosomes in human skin tissue in their native unstained state in the cryo-EM (Al-Amoudi *et al.*, 2004).

Successful high-resolution imaging by CEMOVIS is dependent on the different preparative steps. The biological sample (cell suspension, monolayer, or tissue biopsy) is exposed to extracellular cryoprotectant—commonly 20% dextran—prior to freezing. It is then vitrified by HPF and then kept constantly below the devitrification temperature. In the cryo-microtome below −140°C, it is cut into vitreous sections with a dry diamond knife. The 30–80 nm ultrathin sections are subsequently mounted onto pre-cooled grids for transfer to the cryo-EM for imaging under low-dose conditions. As cryo-EM of whole mounts, CEMOVIS also benefits from a combination with cryo-ET for 3D reconstruction of the supramolecular organisation in cells (Al-Amoudi *et al.*, 2007; Castano-Diez *et al.*, 2007).

Problems in CEMOVIS arise from the cutting of the vitreous sections. There are, in particular, the clearly visible knife marks, the cutting-induced compression, crevasses or fractures, and chatter, i.e., periodic variation in sample thickness (Al-Amoudi *et al.*, 2005). These limitations can only be minimised by precisely adjusting cutting conditions and the environment. Another difficulty of the method is that artefacts induced by HPF cannot be completely excluded. The freezing quality is variable and depends on the physical properties of the sample, the water content, and on the amount of cryoprotectant present. The effect of pressure on fine biological structures is poorly understood but cannot be neglected. For example, vitrification by HPF was proposed to lead to rearrangement of the molecular organisation in DNA structure and to changes in membrane structure (Leforestier *et al.*, 1996; Semmler *et al.*, 1998). Furthermore, this approach also suffers from the difficulties of localising features of interest, as already outlined in the section on cryo-ET. While cytoskeletal structures, such as microtubules and actin filaments, can be visualised in ultrathin cryo-sections of cultured cells as shown in recent studies (Castano-Diez *et al.*, 2007; Gruska *et al.*, 2008), this approach still poses too many technical challenges to be used as a routine technique for visualisation of the actin cytoskeleton.

IV. Summary

Difficulties in visualising the actin cytoskeleton by EM arise mainly from the susceptibility of actin filaments to standard specimen preparation procedures. Further challenges result from the arrangement of actin filaments in three-dimensional arrays, whose organisation is difficult to capture in projection images and sections. The contrast of individual filaments relative to the cytoplasmic background is also very weak and they are easily confused with the much more stable IF.

If the biological question at hand permits, the use of a whole mount approach (Section II) is strongly recommended. Whole mount techniques using conventional

TEM (negative staining and critical point drying/Pt replica) obviate preparative steps such OsO_4 post-fixation, embedding, and sectioning. Three-dimensionality is also less of an issue when working with thin sheets of cytoplasm. While the structures visualised with these two approaches appear similar at first glance, minor distortions can influence the interpretation of filament organisation and interaction. Conflicting results (reviewed in Small *et al.*, 2008) call for careful evaluation of the techniques and the results obtained. Of all whole mount methods discussed, cryo-ET uses the least invasive and, hence, preferable way to prepare specimens, but it comes with specific technical difficulties and limitations. The use of this technique in visualisation of the actin cytoskeleton will increase, provided that techniques to freeze cells reliably, and to investigate ultrastructure in the context of biological function will be further developed and implemented.

If plastic embedding of specimens cannot be obviated, attention should be paid to the precautions outlined in Section III. The structural preservation of actin can be evaluated using known structures such as microvilli and actin networks in lamellipodia as references. Care has to be taken to distinguish IF from actin.

Acknowledgments

The authors would like to thank Dr. J. Victor Small for his feedback on this manuscript; Dr. Daniela Nicastro, Dr. Andres Kaech, and Dr. Roger Wepf for the inspiring discussion; and Marlene Brandstetter, Nicole Fellner, Karlo Pavlovic, and Hannes Tkadletz for their assistance with preparing this chapter. The work of G.P.R. is funded by the City of Vienna/Zentrum für Innovation und Technologie via the Spot of Excellence Grant "Center of Molecular and Cellular Nanostructure", and E.U. and S.J. are supported by the Austrian Science Fund (FWF project P21292-B09 to J. V. Small) and the Vienna Science and Technology Fund (WWTF).

References

Adrian, M., Dubochet, J., Fuller, S. D., and Harris, J. R. (1998). Cryo-negative staining. *Micron* **29**, 145–160.

Adrian, M., Dubochet, J., Lepault, J., and McDowall, A. W. (1984). Cryo-electron microscopy of viruses. *Nature* **308**, 32–36.

Agronskaia, A. V., Valentijn, J. A., van Driel, L. F., Schneijdenberg, C. T., Humbel, B. M., van Bergen En Henegouwen, P. M., Verkleij, A. J., Koster, A. J.,, and Gerritsen, H. C. (2008). Integrated fluorescence and transmission electron microscopy. *J. Struct. Biol.* **164**(2), 183–189.

Al-Amoudi, A., Diez, D. C., Betts, M. J., and Frangakis, A. S. (2007). The molecular architecture of cadherins in native epidermal desmosomes. *Nature* **450**, 832–837.

Al-Amoudi, A., Dubochet, J., and Studer, D. (2002). Amorphous solid water produced by cryosectioning of crystalline ice at 113 K. *J. Microsc.* **207**, 146–153.

Al-Amoudi, A., Norlen, L. P., and Dubochet, J. (2004). Cryo-electron microscopy of vitreous sections of native biological cells and tissues. *J. Struct. Biol.* **148**, 131–135.

Al-Amoudi, A., Studer, D., and Dubochet, J. (2005). Cutting artefacts and cutting process in vitreous sections for cryo-electron microscopy. *J. Struct. Biol.* **150**, 109–121.

Alberts, B., Johnson, A., Lewis, J., Raff, M., Roberts, K., and Walter, P. (2008). "Molecular Biology of the Cell." 5th edn., Garland Science, New York, Milton Park.

Allen, T. D., Rutherford, S. A., Murray, S., Drummond, S. P., Goldberg, M. W., and Kiseleva, E. (2008). Scanning electron microscopy of nuclear structure. *Methods Cell Biol.* **88**, 389–409.

Anderson, T. (1951). Techniques for the preservation of three-dimensional structure in preparing specimens for the electron microscope . *Trans. N.Y. Acad. Sci.* **13**, 130–134.

Auinger, S., and Small, J. V. (2008). Correlated light and electron microscopy of the cytoskeleton. *Methods Cell Biol.* **88**, 257–272.

Bear, J. E., Svitkina, T. M., Krause, M., Schafer, D. A., Loureiro, J. J., Strasser, G. A., Maly, I. V., Chaga, O. Y., Cooper, J. A., Borisy, G. A., and Gertler, F. B. (2002). Antagonism between ena/VASP proteins and actin filament capping regulates fibroblast motility. *Cell* **109**, 509–521.

Begg, D. A., Rodewald, R., and Rebhun, L. I. (1978). The visualization of actin filament polarity in thin sections. Evidence for the uniform polarity of membrane-associated filaments. *J. Cell Biol.* **79**, 846–852.

Boyde, A. (1978). Pros and cons of critical point drying and freeze drying for SEM. *Scan. Electron Microsc.* **2**, 303–314.

Braet, F., and Geerts, W. J. (2009). Foreword to the themed issue on correlative microscopy. *J. Microsc.* **235**, 239–240.

Bray, D. (2001). "Cell Movements: From Molecules to Motility." Garland, New York.

Brenner, S., and Horne, R. W. (1959). A negative staining method for high resolution electron microscopy of viruses. *Biochim. Biophys. Acta* **34**, 103–110.

Brown, S., Levinson, W., and Spudich, J. A. (1976). Cytoskeletal elements of chick embryo fibroblasts revealed by detergent extraction. *J. Supramol. Struct.* **5**, 119–130.

Buckley, I. K., and Porter, K. R. (1967). Cytoplasmic fibrils in living cultured cells. A light and electron microscope study. *Protoplasma* **64**, 349–380.

Buckley, I. K., and Porter, K. R. (1975). Electron microscopy of critical point dried whole cultured cells. *J. Microsc.* **104**, 107–120.

Burnette, D. T., Schaefer, A. W., Ji, L., Danuser, G., and Forscher, P. (2007). Filopodial actin bundles are not necessary for microtubule advance into the peripheral domain of aplysia neuronal growth cones. *Nat. Cell Biol.* **9**, 1360–1369.

Cameron, L. A., Svitkina, T. M., Vignjevic, D., Theriot, J. A., and Borisy, G. G. (2001). Dendritic organization of actin comet tails. *Curr. Biol.* **11**, 130–135.

Canty, E. G., Starborg, T., Lu, Y., Humphries, S. M., Holmes, D. F., Meadows, R. S., Huffman, A., O'Toole, E. T., and Kadler, K. E. (2006). Actin filaments are required for fibripositor-mediated collagen fibril alignment in tendon. *J. Biol. Chem.* **281**, 38592–38598.

Carlier, M. F., Le Clainche, C., Wiesner, S., and Pantaloni, D. (2003). Actin-based motility: From molecules to movement. *Bioessays* **25**, 336–345.

Castano-Diez, D., Al-Amoudi, A., Glynn, A. M., Seybert, A., and Frangakis, A. S. (2007). Fiducial-less alignment of cryo-sections. *J. Struct. Biol.* **159**, 413–423.

Cingolani, L. A., and Goda, Y. (2008). Actin in action: The interplay between the actin cytoskeleton and synaptic efficacy. *Nat. Rev. Neurosci.* **9**, 344–356.

Claviez, M., Brink, M., and Gerisch, G. (1986). Cytoskeletons from a mutant of *Dictyostelium discoideum* with flattened cells. *J. Cell. Sci.* **86**, 69–82.

Cudmore, S., Cossart, P., Griffiths, G., and Way, M. (1995). Actin-based motility of vaccinia virus. *Nature* **378**, 636–638.

Cyrklaff, M., Adrian, M., and Dubochet, J. (1990). Evaporation during preparation of unsupported thin vitrified aqueous layers for cryo-electron microscopy. *J. Electron Microsc. Tech.* **16**, 351–355.

de Carlo, S. (2009). Plunge freezing (holey carbon method). *In* "Handbook of Cryo-Preparation Methods for Electron Microscopy" (A. Cavalier, D. Spehner, and B. M. Humbel eds.) CRC Press, Boca Raton, FL.

DiNardo, A., Cicchetti, G., Falet, H., Hartwig, J. H., Stossel, T. P., and Kwiatkowski, D. J. (2005). Arp2/3 complex-deficient mouse fibroblasts are viable and have normal leading-edge actin structure and function. *Proc. Natl. Acad. Sci. U.S.A.* **102**, 16263–16268.

Dubochet, J., and McDowall, A. W. (1981). Vitrification of pure water for electron microscopy. *J. Microsc.* **124**, RP3–RP4.

Erickson, H. P., and Klug, A. (1971). Measurement and compensation of defocusing and aberrations by Fourier processing of electron micrographs. *Philos. Trans. R. Soc. Lond. Ser. B* **261**, 105–118.

Flanagan, L. A., Chou, J., Falet, H., Neujahr, R., Hartwig, J. H., and Stossel, T. P. (2001). Filamin A, the arp2/3 complex, and the morphology and function of cortical actin filaments in human melanoma cells. *J. Cell Biol.* **155**, 511–517.

Frangakis, A. S., and Förster, F. (2004). Computational exploration of structural information from cryo-electron tomograms. *Curr. Opin. Struct. Biol.* **14**, 325–331.

Frederik, P. M., and Hubert, D. H. (2005). Cryoelectron microscopy of liposomes. *Methods Enzymol.* **391**, 431–448.

Frederik, P. M., Stuart, M. C., Bomans, P. H., Busing, W. M., Burger, K. N., and Verkleij, A. J. (1991). Perspective and limitations of cryo-electron microscopy. From model systems to biological specimens. *J. Microsc.* **161**, 253–262.

Frederik, P. M., de Haas, F., and Storms, M.M.H. (2009). Controlled vitrification. *In* "Handbook of Cryo-Preparation Methods for Electron Microscopy" (A. Cavalier, D. Spehner, and B. M. Humbel eds.) CRC Press, Boca Raton, FL.

Garvalov, B. K., Zuber, B., Bouchet-Marquis, C., Kudryashev, M., Gruska, M., Beck, M., Leis, A., Frischknecht, F., Bradke, F., Baumeister, W., Dubochet, J., and Cyrklaff, M. (2006). Luminal particles within cellular microtubules. *J. Cell Biol.* **174**, 759–765.

Gimona, M., Kaverina, I., Resch, G. P., Vignal, E., and Burgstaller, G. (2003). Calponin repeats regulate actin filament stability and formation of podosomes in smooth muscle cells. *Mol. Biol. Cell* **14**, 2482–2491.

Glotzer, M. (2005). The molecular requirements for cytokinesis. *Science* **307**, 1735–1739.

Gorbsky, G., and Borisy, G. G. (1985). Microtubule distribution in cultured cells and intact tissues: Improved immunolabeling resolution through the use of reversible embedment cytochemistry. *Proc. Natl. Acad. Sci. U.S.A.* **82**, 6889–6893.

Gouin, E., Gantelet, H., Egile, C., Lasa, I., Ohayon, H., Villiers, V., Gounon, P., Sansonetti, P. J., and Cossart, P. (1999). A comparative study of the actin-based motilities of the pathogenic bacteria *Listeria monocytogenes, Shigella flexneri* and *Rickettsia conorii. J. Cell. Sci.* **112**(Pt 11), 1697–1708.

Gouin, E., Welch, M. D., and Cossart, P. (2005). Actin-based motility of intracellular pathogens. *Curr. Opin. Microbiol.* **8**, 35–45.

Grassucci, R. A., Taylor, D. J., and Frank, J. (2007). Preparation of macromolecular complexes for cryo-electron microscopy. *Nat. Protoc.* **2**, 3239–3246.

Griffiths, G. (1993). "Fine Structure Immunocytochemistry." Springer-Verlag, Berlin, Heidelberg, New York, London, Paris, Tokyo, Hong Kong, Barcelona, Budapest.

Grimm, R., Barmann, M., Hackl, W., Typke, D., Sackmann, E., and Baumeister, W. (1997). Energy filtered electron tomography of ice-embedded actin and vesicles. *Biophys. J.* **72**, 482–489.

Grimm, R., Singh, H., Rachel, R., Typke, D., Zillig, W., and Baumeister, W. (1998). Electron tomography of ice-embedded prokaryotic cells. *Biophys. J.* **74**, 1031–1042.

Gruska, M., Medalia, O., Baumeister, W., and Leis, A. (2008). Electron tomography of vitreous sections from cultured mammalian cells. *J. Struct. Biol.* **161**, 384–392.

Harris, J. R., Gebauer, W., and Markl, J. (1995). Keyhole limpet haemocyanin: Negative staining in the presence of trehalose. *Micron* **26**, 25–33.

Hayat, M. (2000). "Principles and Techniques of Electron Microscopy: Biological Applications." Cambridge University Press. Cambridge, New York, Melbourne, Madrid.

He, Y., Jensen, G. J., and Bjorkman, P. J. (2009). Nanogold as a specific marker for electron cryotomography. *Microsc. Microanal.* **15**, 183–188.

Hegerl, R., and Frangakis, A. S. (2006). Denoising of electron tomograms. *In* "Electron Tomography—Methods for Three-Dimensional Visualisation of Structures in the Cell" (J. Frank, ed.) Springer, New York.

Henderson, G. P., Gan, L., and Jensen, G. J. (2007). 3-D ultrastructure of O. Tauri: Electron cryotomography of an entire eukaryotic cell. *PLoS One* **2**, e749.

Heuser, J. E., and Kirschner, M. W. (1980). Filament organization revealed in platinum replicas of freeze-dried cytoskeletons. *J. Cell Biol.* **86**, 212–234.

Höglund, A. S. (1985). The arrangement of microfilaments and microtubules in the periphery of spreading fibroblasts and glial cells. *Tissue Cell* **17**, 649–666.

Höglund, A. S., Karlsson, R., Arro, E., Fredriksson, B. A., and Lindberg, U. (1980). Visualization of the peripheral weave of microfilaments in glia cells. *J. Muscle Res. Cell. Motil.* **1**, 127–146.

Humbel, B., and Schwarz, H. (1989). Freeze-substitution for immunocytochemistry. *In* "Immuno-Gold Labeling in Cell Biology" (A. Verkleij and J. Leunissen, eds.) CRC Press, Boca Raton, FL.

Iancu, C. V., Tivol, W. F., Schooler, J. B., Dias, D. P., Henderson, G. P., Murphy, G. E., Wright, E. R., Li, Z., Yu, Z., Briegel, A., Gan, L., He, Y., *et al.*, (2007). Electron cryotomography sample preparation using the vitrobot. *Nat. Protoc.* **1**, 2813–2819.

Ishikawa, H., Bischoff, R., and Holtzer, H. (1969). Formation of arrowhead complexes with heavy meromyosin in a variety of cell types. *J. Cell Biol.* **43**, 312–328.

Jiang, W., and Ludtke, S. J. (2005). Electron cryomicroscopy of single particles at subnanometer resolution. *Curr. Opin. Struct. Biol.* **15**, 571–577.

Jiménez, N., Humbel, B. M., van Donselaar, E., Verkleij, A. J., and Burger, K.N.J. (2006). Aclar discs: A versatile substrate for routine high-pressure freezing of mammalian cell monolayers. *J. Microsc.* **221**, 216–223.

Kadurugamuwa, J. L., Rohde, M., Wehland, J., and Timmis, K. N. (1991). Intercellular spread of *Shigella flexneri* through a monolayer mediated by membranous protrusions and associated with reorganization of the cytoskeletal protein vinculin. *Infect. Immun.* **59**, 3463–3471.

Kaksonen, M., Toret, C. P., and Drubin, D. G. (2006). Harnessing actin dynamics for clathrin-mediated endocytosis. *Nat. Rev. Mol. Cell Biol.* **7**, 404–414.

Kamasaki, T., Arai, R., Osumi, M., and Mabuchi, I. (2005). Directionality of F-actin cables changes during the fission yeast cell cycle. *Nat. Cell Biol.* **7**, 916–917.

Kamasaki, T., Osumi, M., and Mabuchi, I. (2007). Three-dimensional arrangement of F-actin in the contractile ring of fission yeast. *J. Cell Biol.* **178**, 765–771.

Koestler, S. A., Auinger, S., Vinzenz, M., Rottner, K., and Small, J. V. (2008). Differentially oriented populations of actin filaments generated in lamellipodia collaborate in pushing and pausing at the cell front. *Nat. Cell Biol.* **10**, 306–313.

Koestler, S. A., Rottner, K., Lai, F., Block, J., Vinzenz, M., and Small, J. V. (2009). F- and G-actin concentrations in lamellipodia of moving cells. *PLoS One* **4**, e4810.

Korobova, F., and Svitkina, T. (2008). Arp2/3 complex is important for filopodia formation, growth cone motility, and neuritogenesis in neuronal cells. *Mol. Biol. Cell* **19**, 1561–1574.

Koster, A. J., and Bárcena, M. (2006). Cryotomography: Low-dose automated tomography of frozen-hydrated specimens. *In* "Electron Tomography—Methods for Three-Dimensional Visualisation of Structures in the Cell" (J. Frank, ed.) Springer, New York.

Kremer, J. R., Mastronarde, D. N., and McIntosh, J. R. (1996). Computer visualization of three-dimensional image data using IMOD. *J. Struct. Biol.* **116**, 71–76.

Kürner, J., Frangakis, A. S., and Baumeister, W. (2005). Cryo-electron tomography reveals the cytoskeletal structure of *Spiroplasma melliferum*. *Science* **307**, 436–438.

Langanger, G., de Mey, J., Moeremans, M., Daneels, G., de Brabander, M., and Small, J. V. (1984). Ultrastructural localization of alpha-actinin and filamin in cultured cells with the immunogold staining (IGS) method. *J. Cell Biol.* **99**, 1324–1334.

Leforestier, A., Richter, K., Livolant, F., and Dubochet, J. (1996). Comparison of slam-freezing and high-pressure freezing effects on the DNA cholesteric liquid crystalline structure. *J. Microsc.* **184**, 4–13.

Leis, A., Rockel, B., Andrees, L., and Baumeister, W. (2009). Visualizing cells at the nanoscale. *Trends Biochem. Sci.* **34**, 60–70.

Lepault, J., and Pitt, T. (1984). Projected structure of unstained, frozen-hydrated T-layer of *Bacillus brevis*. *EMBO J.* **3**, 101–105.

Lewis, A. K., and Bridgman, P. C. (1992). Nerve growth cone lamellipodia contain two populations of actin filaments that differ in organization and polarity. *J. Cell Biol.* **119**, 1219–1243.

Lindroth, M., Bell, P. B., and Fredriksson, B. A. (1988). Comparison of the effects of critical point-drying and freeze-drying on cytoskeletons and microtubules. *J. Microsc.* **151**, 103–114.

Lindroth, M., Bell, P. B., Fredriksson, B. A., and Liu, X. D. (1992). Preservation and visualization of molecular structure in detergent-extracted whole mounts of cultured cells. *Microsc. Res. Tech.* **22**, 130–150.

Loisel, T. P., Boujemaa, R., Pantaloni, D., and Carlier, M. F. (1999). Reconstitution of actin-based motility of *Listeria* and *Shigella* using pure proteins. *Nature* **401**, 613–616.
Lucic, V., Kossel, A. H., Yang, T., Bonhoeffer, T., Baumeister, W., and Sartori, A. (2007). Multiscale imaging of neurons grown in culture: From light microscopy to cryo-electron tomography. *J. Struct. Biol.* **160**, 146–156.
Marko, M., Hsieh, C., Schalek, R., Frank, J., and Mannella, C. (2007). Focused-ion-beam thinning of frozen-hydrated biological specimens for cryo-electron microscopy. *Nat. Methods* **4**, 215–217.
Mastronarde, D. N. (1997). Dual-axis tomography: An approach with alignment methods that preserve resolution. *J. Struct. Biol.* **120**, 343–352.
Mastronarde, D. N. (2005). Automated electron microscope tomography using robust prediction of specimen movements. *J. Struct. Biol.* **152**, 36–51.
Maupin, P., and Pollard, T. D. (1983). Improved preservation and staining of HeLA cell actin filaments, clathrin-coated membranes, and other cytoplasmic structures by tannic acid-glutaraldehyde-saponin fixation. *J. Cell Biol.* **96**, 51–62.
Maupin-Szamier, P., and Pollard, T. D. (1978). Actin filament destruction by osmium tetroxide. *J. Cell Biol.* **77**, 837–852.
Maurer, U. E., Sodeik, B., and Gruenewald, K. (2008). Native 3D intermediates of membrane fusion in herpes simplex virus 1 entry. *Proc. Natl. Acad. Sci.* **105**(30), 10559–10564.
McDowall, A. W., Chang, J. J., Freeman, R., Lepault, J., Walter, C. A., and Dubochet, J. (1983). Electron microscopy of frozen hydrated sections of vitreous ice and vitrified biological samples. *J. Microsc.* **131**, 1–9.
Medalia, O., Beck, M., Ecke, M., Weber, I., Neujahr, R., Baumeister, W., and Gerisch, G. (2007). Organization of actin networks in intact filopodia. *Curr. Biol.* **17**, 79–84.
Medalia, O., Weber, I., Frangakis, A. S., Nicastro, D., Gerisch, G., and Baumeister, W. (2002). Macromolecular architecture in eukaryotic cells visualized by cryoelectron tomography. *Science* **298**, 1209–1213.
Mejillano, M. R., Kojima, S., Applewhite, D. A., Gertler, F. B., Svitkina, T. M., and Borisy, G. G. (2004). Lamellipodial versus filopodial mode of the actin nanomachinery: Pivotal role of the filament barbed end. *Cell* **118**, 363–373.
Mercogliano, C. P., and DeRosier, D. J. (2007). Concatenated metallothionein as a clonable gold label for electron microscopy. *J. Struct. Biol.* **160**, 70–82.
Merrifield, C. J. (2004). Seeing is believing: Imaging actin dynamics at single sites of endocytosis. *Trends Cell Biol.* **14**, 352–358.
Moor, H., and Riehle, U. (1968). Snap-freezing under high pressure: A new fixation technique for freeze-etching. *Proc. Fourth Eur. Reg. Conf. Elect. Microsc.* **2**, 33.
Nemethova, M., Auinger, S., and Small, J. V. (2008). Building the actin cytoskeleton: Filopodia contribute to the construction of contractile bundles in the lamella. *J. Cell Biol.* **180**, 1233–1244.
Nickell, S., Forster, F., Linaroudis, A., Net, W. D., Beck, F., Hegerl, R., Baumeister, W., and Plitzko, J. M. (2005). TOM software toolbox: Acquisition and analysis for electron tomography. *J. Struct. Biol.* **149**, 227–234.
Noireaux, V., Golsteyn, R. M., Friederich, E., Prost, J., Antony, C., Louvard, D., and Sykes, C. (2000). Growing an actin gel on spherical surfaces. *Biophys. J.* **78**, 1643–1654.
Ohi, M., Li, Y., Cheng, Y., and Walz, T. (2004). Negative staining and image classification—powerful tools in modern electron microscopy. *Biol. Proced. Online* **6**, 23–34.
O'Toole, E., Wray, G., Kremer, J., and McIntosh, J. R. (1993). High voltage cryomicroscopy of human blood platelets. *J. Struct. Biol.* **110**, 55–66.
Penczek, P., Marko, M., Buttle, K., and Frank, J. (1995). Double-tilt electron tomography. *Ultramicroscopy* **60**, 393–410.
Plastino, J., Olivier, S., and Sykes, C. (2004). Actin filaments align into hollow comets for rapid VASP-mediated propulsion. *Curr. Biol.* **14**, 1766–1771.
Plitzko, J. M., Rigort, A., and Leis, A. (2009). Correlative cryo-light microscopy and cryo-electron tomography: From cellular territories to molecular landscapes. *Curr. Opin. Biotechnol.* **20**, 83–89.
Pollard, T. D., and Borisy, G. G. (2003). Cellular motility driven by assembly and disassembly of actin filaments. *Cell* **112**, 453–465.

Ponti, A., Machacek, M., Gupton, S. L., Waterman-Storer, C. M., and Danuser, G. (2004). Two distinct actin networks drive the protrusion of migrating cells. *Science* **305**, 1782–1786.

Porter, K. R. (1984). The cytomatrix: A short history of its study. *J. Cell Biol.* **99**, 3s–12s.

Porter, K. R., Claude, A., and Fullam, E. F. (1945). A study of tissue culture cells by electron microscopy. *J. Exp. Med.* **81**, 233–246.

Prast, J., Gimona, M., and Small, J. V. (2006). Immunofluorescence microscopy of the cytoskeleton: Combination with green fluorescent protein tags. *In* "Cell Biology" (J. E. Celis, ed.) Elsevier Science, Amsterdam, Boston, Heidelberg, London, New York, Oxford, Paris, San Diego, San Francisco, Singapore, Sydney, Tokyo.

Prévost, M. C., Lesourd, M., Arpin, M., Vernel, F., Mounier, J., Hellio, R., and Sansonetti, P. J. (1992). Unipolar reorganization of F-actin layer at bacterial division and bundling of actin filaments by plastin correlate with movement of *Shigella flexneri* within HeLa cells. *Infect. Immun.* **60**, 4088–4099.

Resch, G. P., Goldie, K. N., Hoenger, A., and Small, J. V. (2002a). Pure F-actin networks are distorted and branched by steps in the critical-point drying method. *J. Struct. Biol.* **137**, 305–312.

Resch, G. P., Goldie, K. N., Krebs, A., Hoenger, A., and Small, J. V. (2002b). Visualisation of the actin cytoskeleton by cryo-electron microscopy. *J. Cell. Sci.* **115**, 1877–1882.

Resch, G. P., Small, J. V., and Goldie, K. N. (2006). Electron microscopy of extracted cytoskeletons: Negative staining, cryoelectron microscopy, and correlation with light microscopy. *In* "Cell Biology—A Laboratory Handbook" (J. E. Celis, ed.) Elsevier Science. Amsterdam, Boston, Heidelberg, London, New York, Oxford, Paris, San Diego, San Francisco, Singapore, Sydney, Tokyo.

Revenu, C., Athman, R., Robine, S., and Louvard, D. (2004). The co-workers of actin filaments: From cell structures to signals. *Nat. Rev. Mol. Cell Biol.* **5**, 635–646.

Rigort, A., Bäuerlein, F. J. B., Leis, A., Gruska, M., Hoffmann, C., Laugks, T., Böhm, U., Eibauer, M., Gnaegi, H., Baumeister, W. and Plitzko, J. M. (2010). Micromachining tools and correlative approaches for cellular cryo-electron tomography. *J. Struct. Biol.* (electronically published).

Ris, H. (1985). The cytoplasmic filament system in critical point-dried whole mounts and plastic-embedded sections. *J. Cell Biol.* **100**, 1474–1487.

Ross, J. L., Ali, M. Y., and Warshaw, D. M. (2008). Cargo transport: Molecular motors navigate a complex cytoskeleton. *Curr. Opin. Cell Biol.* **20**, 41–47.

Rottner, K., Behrendt, B., Small, J. V., and Wehland, J. (1999). VASP dynamics during lamellipodia protrusion. *Nat. Cell Biol.* **1**, 321–322.

Rottner, K., Krause, M., Gimona, M., Small, J. V., and Wehland, J. (2001). Zyxin is not colocalized with vasodilator-stimulated phosphoprotein (VASP) at lamellipodial tips and exhibits different dynamics to vinculin, paxillin, and VASP in focal adhesions. *Mol. Biol. Cell* **12**, 3103–3113.

Royer, S. M., and Kinnamon, J. C. (1996). Comparison of high-pressure freezing/freeze substitution and chemical fixation of catfish barbel taste buds. *Microsc. Res. Tech.* **35**, 385–412.

Sabaneyeva, E. V., Derkacheva, M. E., Benken, K. A., Fokin, S. I., Vainio, S., and Skovorodkin, I. N. (2009). Actin-based mechanism of *Holospora obtusa* trafficking in *Paramecium caudatum*. *Protist* **160**, 205–219.

Sameshima, M., Kishi, Y., Osumi, M., Mahadeo, D., and Cotter, D. A. (2002). Electron microscopy of actin rods and bundles in *Dictyostelium discoideum* by high-pressure freezing. *J. Electron Microsc. (Tokyo)* **51**, 337–340.

Sartori, A., Gatz, R., Beck, F., Rigort, A., Baumeister, W., and Plitzko, J. M. (2007). Correlative microscopy: Bridging the gap between fluorescence light microscopy and cryo-electron tomography. *J. Struct. Biol.* **160**, 135–145.

Sawaguchi, A., McDonald, K. L., Karvar, S., and Forte, J. G. (2002). A new approach for high-pressure freezing of primary culture cells: The fine structure and stimulation-associated transformation of cultured rabbit gastric parietal cells. *J. Microsc.* **208**, 158–166.

Schliwa, M., and van Blerkom, J. (1981). Structural interaction of cytoskeletal components. *J. Cell Biol.* **90**, 222–235.

Schmitz, F., and Drenckhahn, D. (1993). Distribution of actin in cone photoreceptor synapses. *Histochemistry* **100**, 35–40.

Schwartz, C. L., Sarbash, V. I., Ataullakhanov, F. I., McIntosh, J. R., and Nicastro, D. (2007). Cryo-fluorescence microscopy facilitates correlations between light and cryo-electron microscopy and reduces the rate of photobleaching. *J. Microsc.* **227**, 98–109.

Seagull, R. W., and Heath, I. B. (1979). The effects of tannic acid on the *in vivo* preservation of microfilaments. *Eur. J. Cell Biol.* **20**, 184–188.

Sechi, A. S., Wehland, J., and Small, J. V. (1997). The isolated comet tail pseudopodium of *Listeria monocytogenes*: A tail of two actin filament populations, long and axial and short and random. *J. Cell Biol.* **137**, 155–167.

Sekiya-Kawasaki, M., Groen, A. C., Cope, M.J.T.V., Kaksonen, M., Watson, H. A., Zhang, C., Shokat, K. M., Wendland, B., McDonald, K. L., McCaffery, J. M., and Drubin, D. G. (2003). Dynamic phosphoregulation of the cortical actin cytoskeleton and endocytic machinery revealed by real-time chemical genetic analysis. *J. Cell Biol.* **162**, 765–772.

Semenova, I., Burakov, A., Berardone, N., Zaliapin, I., Slepchenko, B., Svitkina, T., Kashina, A., and Rodionov, V. (2008). Actin dynamics is essential for myosin-based transport of membrane organelles. *Curr. Biol.* **18**, 1581–1586.

Semmler, K., Wunderlich, J., Richter, W., and Meyer, H. W. (1998). High-pressure freezing causes structural alterations in phospholipid model membranes. *J. Microsc.* **190**, 317–327.

Siksou, L., Rostaing, P., Lechaire, J.-P., Boudier, T., Ohtsuka, T., Fejtová, A., Kao, H.-T., Greengard, P., Gundelfinger, E. D., Triller, A., and Marty, S. (2007). Three-dimensional architecture of presynaptic terminal cytomatrix. *J. Neurosci.* **27**, 6868–6877.

Siksou, L., Triller, A., and Marty, S. (2009). An emerging view of presynaptic structure from electron microscopic studies. *J. Neurochem.* **108**, 1336–1342.

Small, J. V. (1981). Organization of actin in the leading edge of cultured cells: Influence of osmium tetroxide and dehydration on the ultrastructure of actin meshworks. *J. Cell Biol.* **91**, 695–705.

Small, J. V. (1984). Simple procedures for the transfer of grid images onto glass coverslips for the rapid relocation of cultured cells. *J. Microsc.* **137**, 171–175.

Small, J. V. (1988). The actin cytoskeleton. *Electron Microsc. Rev.* **1**, 155–174.

Small, J. V., Auinger, S., Nemethova, M., Koestler, S., Goldie, K. N., Hoenger, A., and Resch, G. P. (2008). Unravelling the structure of the lamellipodium. *J. Microsc.* **231**, 479–485.

Small, J. V., and Celis, J. E. (1978). Filament arrangements in negatively stained cultured cells: The organization of actin. *Cytobiologie* **16**, 308–325.

Small, J. V., Herzog, M., and Anderson, K. (1995). Actin filament organization in the fish keratocyte lamellipodium. *J. Cell Biol.* **129**, 1275–1286.

Small, J. V., Isenberg, G., and Celis, J. E. (1978). Polarity of actin at the leading edge of cultured cells. *Nature* **272**, 638–639.

Small, J. V., Rinnerthaler, G., and Hinssen, H. (1982). Organization of actin meshworks in cultured cells: The leading edge. *Cold Spring Harb. Symp. Quant. Biol.* **46**(Pt 2), 599–611.

Steinmetz, M. O., Stoffler, D., Hoenger, A., Bremer, A., and Aebi, U. (1997). Actin: From cell biology to atomic detail. *J. Struct. Biol.* **119**, 295–320.

Steven, A. C., and Aebi, U. (2003). The next ice age: Cryo-electron tomography of intact cells. *Trends Cell Biol.* **13**, 107–110.

Studer, D., Graber, W., Al-Amoudi, A., and Eggli, P. (2001). A new approach for cryofixation by high-pressure freezing. *J. Microsc.* **203**, 285–294.

Suloway, C., Pulokas, J., Fellmann, D., Cheng, A., Guerra, F., Quispe, J., Stagg, S., Potter, C. S., and Carragher, B. (2005). Automated molecular microscopy: The new leginon system. *J. Struct. Biol.* **151**, 41–60.

Suloway, C., Shi, J., Cheng, A., Pulokas, J., Carragher, B., Potter, C. S., Zheng, S. Q., Agard, D. A., and Jensen, G. J. (2009). Fully automated, sequential tilt-series acquisition with leginon. *J. Struct. Biol.* **167**, 11–18.

Svitkina, T. (2007). Electron microscopic analysis of the leading edge in migrating cells. *Methods Cell Biol.* **79**, 295–319.

Svitkina, T. (2009). Imaging cytoskeleton components by electron microscopy. *Methods Mol. Biol.* **586**, 187–206.

Svitkina, T. M., and Borisy, G. G. (1998). Correlative light and electron microscopy of the cytoskeleton of cultured cells. *Methods Enzymol.* **298**, 570–592.

Svitkina, T. M., and Borisy, G. G. (1999). Arp2/3 complex and actin depolymerizing factor/cofilin in dendritic organization and treadmilling of actin filament array in lamellipodia. *J. Cell Biol.* **145**, 1009–1026.

Svitkina, T. M., Bulanova, E. A., Chaga, O. Y., Vignjevic, D. M., Kojima, S., Vasiliev, J. M., and Borisy, G. G. (2003). Mechanism of filopodia initiation by reorganization of a dendritic network. *J. Cell Biol.* **160**, 409–421.

Svitkina, T. M., Shevelev, A. A., Bershadsky, A. D., and Gelfand, V. I. (1984). Cytoskeleton of mouse embryo fibroblasts. Electron microscopy of platinum replicas. *Eur. J. Cell Biol.* **34**, 64–74.

Svitkina, T. M., Verkhovsky, A. B., and Borisy, G. G. (1995). Improved procedures for electron microscopic visualization of the cytoskeleton of cultured cells. *J. Struct. Biol.* **115**, 290–303.

Svitkina, T. M., Verkhovsky, A. B., McQuade, K. M., and Borisy, G. G. (1997). Analysis of the actin-myosin II system in fish epidermal keratocytes: Mechanism of cell body translocation. *J. Cell Biol.* **139**, 397–415.

Taunton, J., Rowning, B. A., Coughlin, M. L., Wu, M., Moon, R. T., Mitchison, T. J., and Larabell, C. A. (2000). Actin-dependent propulsion of endosomes and lysosomes by recruitment of N-WASP. *J. Cell Biol.* **148**, 519–530.

Tilney, L. G., and Portnoy, D. A. (1989). Actin filaments and the growth, movement, and spread of the intracellular bacterial parasite, listeria monocytogenes. *J. Cell Biol.* **109**, 1597–1608.

Tivol, W. F., Briegel, A., and Jensen, G. J. (2008). An improved cryogen for plunge freezing. *Microsc. Microanal.* **14**, 375–379.

Urban, E., Jacob, S., Nemethova, M., Resch, G. P., and Small, J. V. (2010). Electron tomography reveals unbranched networks of actin filaments in lamellipodia. *Nat Cell Biol.* **12**, 429–435.

Vallotton, P., and Small, J. V. (2009). Shifting views on the leading role of the lamellipodium in cell migration: Speckle tracking revisited. *J. Cell. Sci.* **122**, 1955–1958.

van Kirk, L., Hayes, S. F., and Heinzen, R. A. (2000). Ultrastructure of *Rickettsia rickettsii* actin tails and localization of cytoskeletal proteins. *Infect. Immun.* **68**, 4706–4713.

Verkhovsky, A. B., Svitkina, T. M., and Borisy, G. G. (1995). Myosin II filament assemblies in the active lamella of fibroblasts: Their morphogenesis and role in the formation of actin filament bundles. *J. Cell Biol.* **131**, 989–1002.

Vignjevic, D., Yarar, D., Welch, M. D., Peloquin, J., Svitkina, T., and Borisy, G. G. (2003). Formation of filopodia-like bundles *in vitro* from a dendritic network. *J. Cell Biol.* **160**, 951–962.

Walther, P. (2008). High-resolution cryo-SEM allows direct identification of F-actin at the inner nuclear membrane of *Xenopus* oocytes by virtue of its structural features. *J. Microsc.* **232**, 379–385.

Winder, S. J., and Ayscough, K. R. (2005). Actin-binding proteins. *J. Cell. Sci.* **118**, 651–654.

Wolosewick, J. J. (1980). The application of polyethylene glycol (PEG) to electron microscopy. *J. Cell Biol.* **86**, 661–675.

Wolosewick, J. J., and Porter, K. R. (1976). Stereo high-voltage electron microscopy of whole cells of the human diploid line, WI-38. *Am. J. Anat.* **147**, 303–323.

Wolosewick, J. J., and Porter, K. R. (1979). Microtrabecular lattice of the cytoplasmic ground substance. Artifact or reality. *J. Cell Biol.* **82**, 114–139.

Yamada, K. M., Spooner, B. S., and Wessells, N. K. (1971). Ultrastructure and function of growth cones and axons of cultured nerve cells. *J. Cell Biol.* **49**, 614–635.

Yang, C., Czech, L., Gerboth, S., Kojima, S., Scita, G., and Svitkina, T. (2007). Novel roles of formin mDia2 in lamellipodia and filopodia formation in motile cells. *PLoS Biol.* **5**, e317.

Zheng, S. Q., Keszthelyi, B., Branlund, E., Lyle, J. M., Braunfeld, M. B., Sedat, J. W., and Agard, D. A. (2007). UCSF tomography: An integrated software suite for real-time electron microscopic tomographic data collection, alignment, and reconstruction. *J. Struct. Biol.* **157**, 138–147.

CHAPTER 23

Three-Dimensional Cryo-Electron Microscopy on Intermediate Filaments

Robert Kirmse, Cédric Bouchet-Marquis, Cynthia Page, ***and*** **Andreas Hoenger**

The Boulder Laboratory for 3-D Microscopy of Cells, University of Colorado at Boulder, Boulder, Colorado 80309-0347

Abstract

Together with microtubules and actin filaments (F-actin), intermediate filaments (IFs) form the cytoskeleton of metazoan cells. However, unlike the other two entities that are extremely conserved, IFs are much more diverse and are grouped into five different families. In contrast to microtubules and F-actin, IFs do not exhibit a polarity,

978-0-12-381007-6
DOI: 10.1016/S0091-679X(10)96023-8

which may be the reason that no molecular motors travel along them. The molecular structure of IFs is less well resolved than that of the other cytoskeletal systems. This is partially due to their functional variability, tissue-specific expression, and their intrinsic structural properties. IFs are composed mostly of relatively smooth protofibrils formed by antiparallel arranged α-helical coiled–coil bundles flanked by small globular domains at either end. These features make them difficult to study by various electron microscopy methods or atomic force microscopy (AFM). Furthermore, the elongated shape of monomeric or dimeric IF units interferes with the formation of highly ordered three-dimensional (3-D) crystals suitable for atomic resolution crystallography. So far, most of the data we currently have on IF macromolecular structures come from electron microscopy of negatively stained samples, and fragmented α-helical coiled–coil units solved by X-ray diffraction. In addition, AFM allows the observation of the dynamic states of IFs in solution and delivers a new view into the assembly properties of IFs. Here, we discuss the applicability of cryo-electron microscopy (cryo-EM) and cryo-electron tomography (cryo-ET) for the field. Both methods are strongly related and have only recently been applied to IFs. However, cryo-EM revealed distinct new features within IFs that have not been seen before, and cryo-ET adds a 3-D view of IFs revealing the path and number of protofilaments within the various IF assemblies.

I. Introduction

Intermediate filaments (IFs) constitute one of the three major cytoskeletal filament assemblies found in metazoan cells, along with microtubules and microfilaments (F-actin). Together, these three filament systems operate as an integrated, dynamic network (Franke *et al.*, 1978; Fuchs and Karakesisoglou, 2001; Herrmann *et al.*, 2000; Parry and Steinert, 1999). However, while the majority of mutations in microtubules or actin are lethal, and their structures are highly conserved, IF proteins differ considerably in sequence and are grouped into five different families based on their structure and function (De Pianto and Coulombe, 2004; Herrmann and Aebi, 2004). Numerous mutations in the various types of IFs contribute to a broad set of pathological conditions (Magin *et al.*, 2004; Omary *et al.*, 2004). The best understood function of IFs is that of a scaffold for maintaining cell and tissue integrity (Omary *et al.*, 2004; 2006). For example, IFs play a crucial part in tissue repair by binding and modulating the activity of several types of effector proteins, such as receptors, kinases, adaptors, and heat shock proteins (De Pianto and Coulombe, 2004). Currently, more than 70 different IF proteins identified so far have been grouped into five sequence homology classes: desmins, keratins, vimentins, lamins, and neurofilaments (De Pianto and Coulombe, 2004; Herrmann *et al.*, 2004; Parry, 2007). They can be further subdivided into three independent groups according to their mode of assembly (Herrmann and Aebi, 2004). However, despite the large diversity among IF proteins, they share a basic structural building plan exhibiting a 45 nm-long central α-helical "rod" domain that is flanked by non-α-helical N- and C-terminal end domains, termed "head" and "tail," respectively

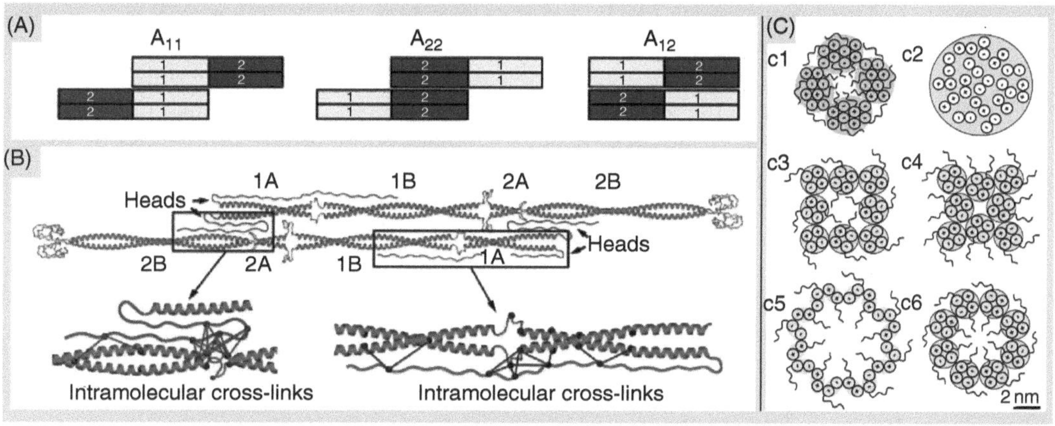

Fig. 1 Structural models for the assembly of IFs: (A) Models of possible vimentin tetramer arrangements as elucidated by cross-linking experiments illustrated in (B). (C) Packing models of protofibrils, composed of tetramers. Currently, our tomographic data obtained on vimentin filaments *in vitro*, and shown in Fig. 2, suggests a packing arrangement similar to that seen in (C (c1)). However, the cross sections of unidentified IFs shown in Fig. 3B of the preliminary data section indicate that other options may occur as well (see also the cross sections of keratins in Norlén and Al-Amoudi (2004)). Illustrations adapted from Herrmann and Aebi (2004). (See Plate no. 33 in the Color Plate Section.)

(see Fig. 1). Two of these molecules dimerize in parallel and in register according to so-called hydrophobic A- and D-type interactions (from a repetitive pattern of heptads) to form an α-helical coiled–coil structure. These dimers constitute the basic building blocks of all IF assemblies (Fuchs and Hanukoglu, 1983; Mücke *et al.*, 2004b; Steinert and Parry, 1985).

IFs have been extensively studied for their roles in cell biology and pathophysiology for numerous diseases, but their supramolecular structure is still only partially resolved and, therefore, are the subject of many current studies. Successful studies include many electron microscopy investigations, mostly carried out using negative staining or metal surface shadowing (e.g., see: Fowler and Aebi, 1983, 1986; reviewed in: Herrmann and Aebi, 2004; Herrmann *et al.*, 2007; Kreplak *et al.*, 2008). To further advance our knowledge on the very dynamic assembly process of IFs, atomic force microscopy (AFM) has been applied to IFs imaged in physiological buffer solution (Kirmse *et al.*, 2007; Kreplak *et al.*, 2008; Wickert *et al.*, 2005). Furthermore, systematic mass-per-length measurements have been carried out by scanning transmission electron microscopy and this allowed estimating the assembly geometry of intact filaments from protofibrils (Engel *et al.*, 1985; see also Fig. 1). Finally, X-ray studies have been performed on some parts of the coiled–coil segments (Strelkov *et al.*, 2002, 2003), but the crystallization of a full IF monomer or dimer has not been achieved yet and the only X-ray studies on intact filaments and assembly intermediates have come from small-angle X-ray scattering (Sokolova *et al.*, 2006).

Three-dimensional (3-D) cryo-electron microscopy (cryo-EM) is a relatively recent approach to study IF ultrastructure and function. The apparent reluctance to solve the 3-D building plan of IFs may partially have to do with the structural diversity among the different IF families. However, probably much more important, in this respect, is the lack of large globular building blocks that can be used for averaging approaches as utilized for microtubules (e.g., see Beuron and Hoenger, 2001) or F-actin (e.g., see Bremer and Aebi, 1992). With the exception of lamins (Ben-Harush *et al.*, 2009; Dhe-Paganon *et al.*, 2002), which show quite large globular tail domains, most IFs are too smooth to reveal an axial repeat that can be visualized directly by cryo-EM (i.e., a sequence of strong blobs as found in F-actin and tubulin protofilaments). IFs also lack the polarity of microtubules and F-actin. Hence, this makes them difficult to study by EM and hard to analyze them by established image-averaging procedures such as helical reconstruction (Beuron and Hoenger, 2001; DeRosier and Moore, 1970) or 3-D volume averaging of particles in subvolumes extracted from tomograms (Nicastro *et al.*, 2006; Cope *et al.*, 2010).

The first cryo-EM study was published by Watts *et al.* (2002) on isolated trichocyte (hard α-keratin) IFs. More recently, the studies by Goldie *et al.* (2007), carried out on assembled vimentin filaments, were also in solution. The 3-D analysis in this work revealed a clear four-protofibril building plan for intact vimentin filaments (see Fig. 1C (c1)). Further studies followed and IFs were analyzed *in situ* by a recently revived vitrified sectioning approach (Al-Amoudi *et al.*, 2004; Dubochet *et al.*, 2007; Hsieh *et al.*, 2002; McDowall *et al.*, 1983). In contrast to freeze-substitution and plastic-embedding, vitrified sectioning preserves the specimens in a frozen-hydrated state and typically precludes any kinds of chemical fixations, dehydration, or any embedding of the cell in polymerizable resin. Instead, the frozen specimens are directly mounted in a microtome and sectioned (Ladinsky *et al.*, 2006). The resulting sections reveal an inherently low contrast and can be hard to interpret (Dubochet *et al.*, 2007). Due to the brittleness of frozen sections there are crevasses and sometimes problems with compression. However, molecular detail and physiological conformations are much better preserved (for IFs see Norlén, 2008; Norlén and Al-Amoudi, 2004; for microtubules Bouchet-Marquis *et al.*, 2007). The combination of vitrified cryosectioning with the newest progress in data acquisition for cryo-electron tomography (cryo-ET) (Mastronarde, 2005; Zhang *et al.*, 2003; Zheng *et al.*, 2004) now provides an extremely powerful tool with which to elucidate the 3-D architecture of cells, organelles, and cytoskeletal structures such as the IF meshwork at molecular resolution (e.g., see:Jensen and Briegel, 2007; Kürner *et al.*, 2005; Medalia *et al.*, 2002).

II. Rationale

Since this "Methods in Cell Biology" series already published an extensive electron microscopy chapter on IF structure in 2008 (Kreplak *et al.*, 2008), we will focus here primarily on the progresses that have been made in this field by introducing cryo-EM methods (Dubochet *et al.*, 1988; Taylor and Glaeser, 1976) and cryo-ET (reviewed in: Lucic *et al.*, 2005; McIntosh *et al.*, 2005; Norlén *et al.*, 2007). In addition, we compare the results from cryo-EM and cryo-ET with that of AFM. Besides AFM that works in solution, the rapid freezing process of IFs directly in physiologic buffer conditions renders these two methods the most suitable ones for dynamic studies into IF assembly and maintenance. The preservation warrants molecular detail to atomic resolution, and it is mostly the image-recording technology available today combined with 3-D reconstruction approaches that limits the interpretations of structural data to a few nanometers. Currently the best electron microscopy data produce between 1.5 and 2 nm detail. This can be combined by computational docking procedures with higher resolution data such as X-ray or NMR maps, as far as available.

III. Methods

A. Cryo-Electron Microscopy and Cryo-Electron Tomography

1. Introduction to Cryo-EM and Cryo-ET

Currently, the most promising method to produce physiologically relevant 3-D information of IFs in molecular detail is cryo-EM (Figs. 2C,E, 3A, and 5C,D; Dubochet *et al.*, 1988) in combination with tomographic 3-D reconstruction, called cryo-ET (Figs. 2, 3, and 7; Lucic *et al.*, 2005). Fig. 2D shows a 0.9 nm thick *xy*-slice through a cryo-tomogram of isolated vimentin filaments. Fig. 2F shows the area marked in Fig. 2D as a 1.8 nm thick *xz*-slice, which is also shown in Figs. 3D and 7C. While the images in Figs. 2F and 7C show some internal density, which might be an artifact of the projection along slightly bent filaments and the missing cone of data (see Fig. 7 sketch), image 3B shows a 2 nm *xz*-slice that reveals four protofibrils forming a full vimentin filament with remarkable detail. Cryo-ET uses a back-projection procedure of two-dimensional (2-D) projections recorded from a 3-D object at small angular increments (e.g., 1°) to restore the 3-D information of an object. The projections are typically collected with a goniometer stage that allows tilting of the specimen in the electron beam between ±70° (in an ideal situation). Hence, tomography is also a very powerful tool for the reconstruction of vitrified sections obtained from intact cells (Fig. 3). The major plus of tomographic 3-D reconstruction is that it can be applied to virtually anything that can be imaged by electron microscopy. Unlike most conventional 3-D reconstruction methods (e.g., helical or icosahedral analysis, single-particle averaging tilt-series reconstruction of 2-D crystalline assemblies) tomography can be applied to large, flexible, and dynamic structures because no *a priori* assumptions of symmetry constraints are made. In addition, tomography does not rely on the structural integrity of multiple particles to be averaged. Unfortunately, this convenience comes at a price: tomograms require recording large tilt series of 100–150 individual tilted images. While this is less of a problem for fixed and resin-embedded specimens that are typically recorded at room temperature (RT), it is an extremely critical factor for cryo-ET as frozen-hydrated samples suffer substantially from exposure to the electron beam. Typically, to obtain 3-D data at about 2–3 nm resolution the specimens may be exposed to a cumulative dose of no more than

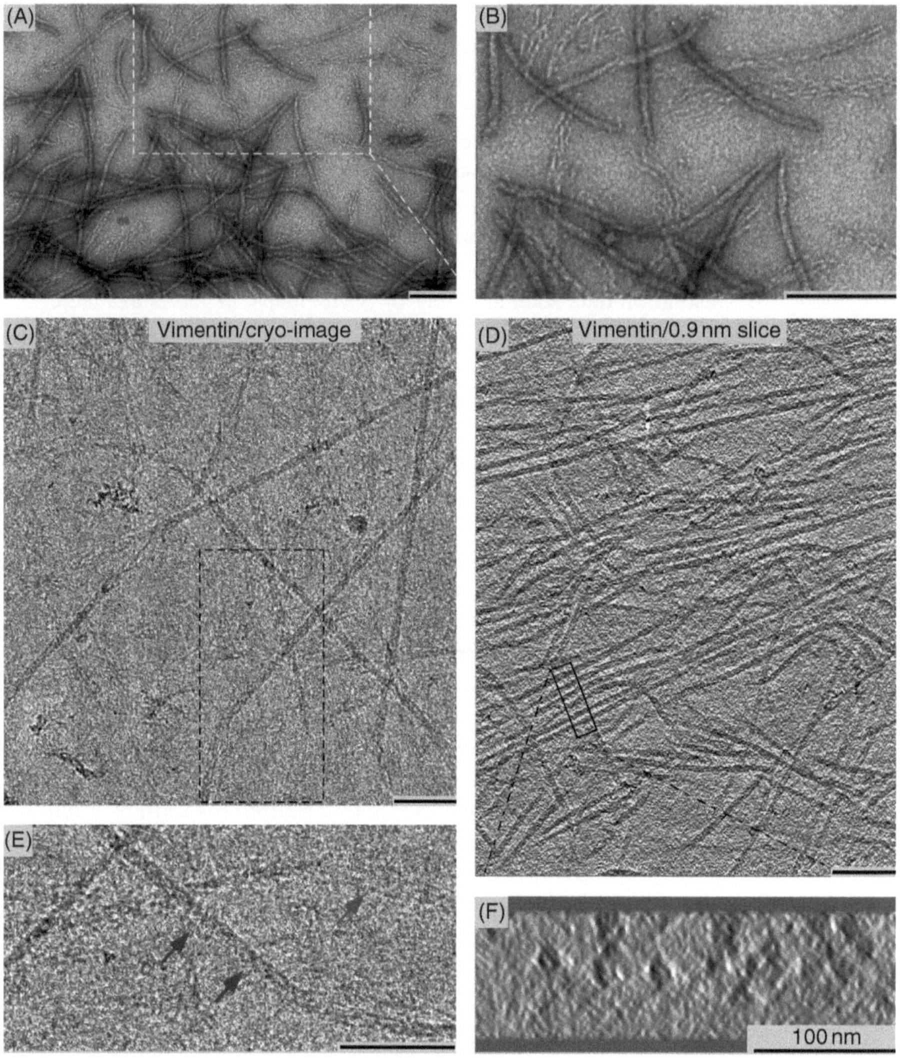

Fig. 2 Negative staining versus cryo-EM of vimentin filaments. Images (A) and (B) show in detail a partially unraveled filament arrangement and evidence of a substructural organization by protofibrils. (C), (D), (E), and (F) show cryo-EM of vimentin. Cryo-micrographs and tomograms of unfixed and unstained vimentin IFs. With both imaging techniques a protofibril sub-arrangement is revealed in the open or partially unraveled sections of filaments. The major difference seen when comparing results from negative stain- or cryo-microscopy is that under vitrified conditions vimentin IFs are predominantly straight, whereas with negative stain filaments are often curved. In cryo-EM the filaments are not adsorbed on a support instead they are embedded in ice. This is in contrast to negative staining EM or AFM where the filament is imaged after absorption to a surface. It is also important to notice that shorter filaments that do not span the entire width of a hole in the holey carbon film are also straight (E), suggesting that this attribute is a result of the filament structure rather than support proximity. (F) A cross section from the tomogram in (D) suggests that a filament is composed of four smaller fibers. (See Plate no. 34 in the Color Plate Section.)

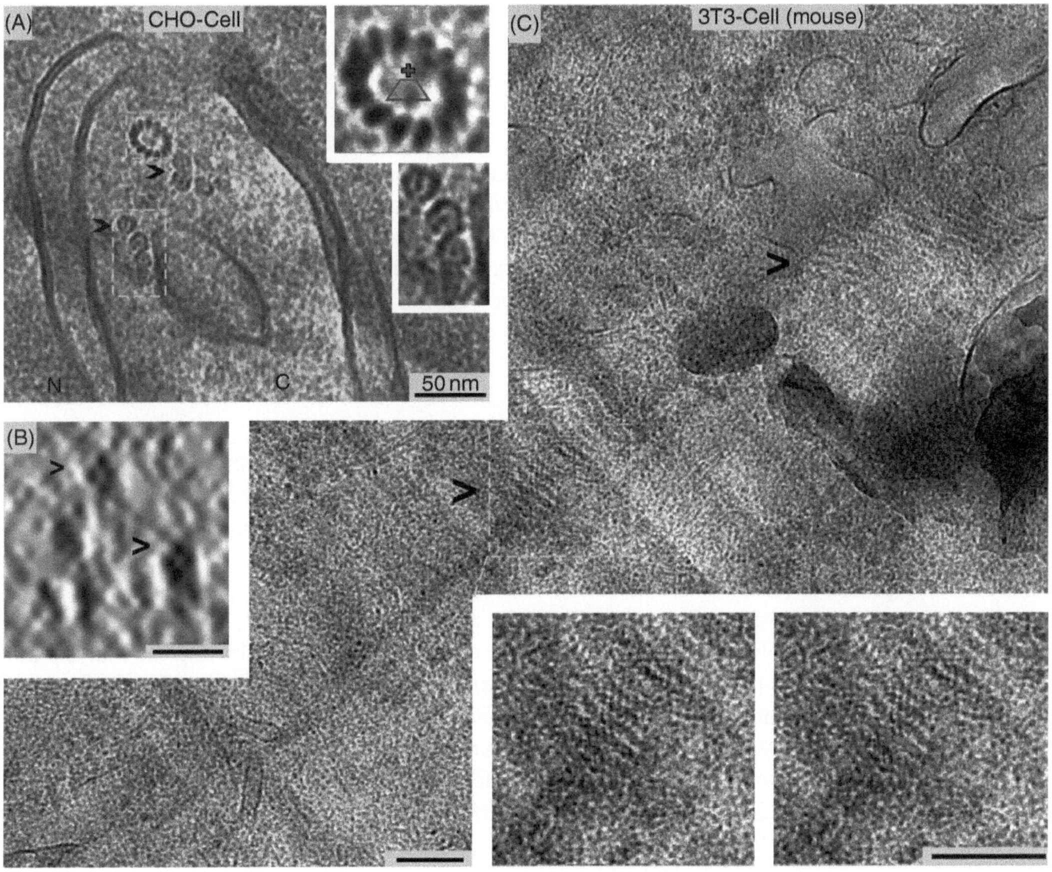

Fig. 3 IFs observed in their cellular context by vitrified sectioning. (A) 80 nm thick cryosection of a CHO (Chinese Ovarian Hamster) cell showing IF bundles and a microtubule in the vicinity of the nuclear envelope. Insets in (A): the polarity of the 13 protofilaments microtubule can be derived directly from the characteristic, slightly triangular shape of the protofilaments, and the cross-sectioned IFs show a round structure and an inner high-density core. (B) *In vitro* assembled vimentin IFs (after 60 min of assembly) embedded in vitreous ice and observed by cryo-ET. A 10 nm thick slice was extracted from the tomogram to reveal diamond-shaped cross-sectioned vimentin filaments dominated by four centers of mass according to four protofibrils. (C) Overview of an A31 3T3 cell cytoplasm showing longitudinally oriented IF bundles. Inset shows the rope-like structure of the IFs along their axes. (See Plate no. 35 in the Color Plate Section.)

100–200 $e^-/Å^2$. This means that for a tilt series of 120 images, each image has to be recorded at a dose of ~1 $e^-/Å^2$ that produces very noisy, low-contrast images. A second major drawback of tomography, in general, comes from limited tilt-capabilities of microscope stages. Ideally, one would like to tilt a full 180° range. However, this is currently not possible due to the geometrical limitations of the specimen holders and grid supports. Hence, a single-tilt data set suffers from a so-called missing wedge of data and reveals nonisotropic resolution with limited detail in the *z*-direction and image details appear smeared out along the *z*-axis (see Figs. 2F, 3B, and 7). Double-tilt tomograms only feature a missing pyramid of data, but the beam sensitivity of frozen-hydrated samples usually prevents the recording of two perpendicular tilt series by cryo-ET. Although the key feature of tomography is its independence from averaging 2-D image elements, so-called 3-D volume averaging may be a powerful tool to further analyze tomographic 3-D data and enhance the signal : noise ratio of some image detail (e.g., see the results of our PEET software in Cope *et al.* (2010) and Nicastro *et al.* (2006)). Obviously, any averaged data have to be structurally reproducible; so far we did not succeed in defining structurally identical image elements from cryo-ET data of IFs that would have produced a clearer view of the molecular arrangements of dimeric and tetrameric subunits.

2. Specimen Preparation Methods for IFs

Recombinantly expressed and purified vimentin protein is prepared for assembly by removing the urea and increasing the pH of the buffer. Thawed aliquots of the purified protein, stored in 8 M urea and 5 mM Tris buffer containing 1 mM EDTA, 0.1 mM EGTA, and 1 mM DTT pH 7.5, are dialyzed into the same buffer without urea (pH 8.4).

The assembly of vimentin from the tetrameric building blocks is initiated with the addition of NaCl. The dialyzed protein is diluted to 0.4 mg/ml using the dialysis buffer at pH 8.4 as described above. To set up a single assembly reaction 20 μl of protein is placed in one microcentrifuge tube and 20 μl of two times concentrated assembly buffer (45 mM Tris pH 7.5, 320 mM NaCl, 2 mM $MgCl_2$) is placed in a second tube. These tubes are preincubated at 37°C for 3–5 min. The reaction begins by diluting the protein solution with an equal volume of assembly buffer and letting the reaction proceed for the allotted time. The final assembly conditions are 0.2 mg/ml protein in 25 mM Tris pH 7.5, 160 mM NaCl, 1 mM $MgCl_2$.

After the appropriate incubation time, 3 μl of the reaction mix is pipetted on a glow-discharged holey carbon grid. The solution is immediately blotted off the grid and the grid is plunge frozen in liquid ethane. Blotting and plunging is done either by hand, using a custom-made gravity-operated plunge-freezing device, with an approximate blot time (3 s) and ambient temperature and humidity levels, or using an FEI Vitrobot Mark 1V. Those grids prepared with the Vitrobot are blotted for 2.5–3.5 s in a chamber equilibrated to 4°C at 95% humidity. After freezing the grids are stored under liquid nitrogen (LN_2) until they are viewed in a FEI Tecnai TF20 transmission electron microscope operated at 200 kV.

2-D projections were acquired using a dose of 30–70 $e^-/Å^2$ and at a nominal magnification of 50 k. Tomograms were imaged at 29 k and 50 k using a total dose for the tilt series of 200 electrons. The tomograms were set up so that the beam intensity varied as 1/cosine of the tilt angle. The tilt angle increments (set at 1.5° or 2.0°) were proportional to the cosine of the tilt angle (Saxton scheme). The tilt range of the tomograms was approximately ±60° for 29 k magnification and from +65° to −65° for the 50 k magnification.

B. Observation of IFs in their Cellular Context in Vitrified Sections

In order to gain fine structural information on IFs, cryo-electron microscopy of vitreous sections (CEMOVIS: Al Amoudi *et al.*, 2004; Hsieh *et al.*, 2002; McDowall *et al.*, 1983) was performed on Chinese Ovarian Hamster (CHO) cells (Fig. 3A) and A31 3T3 mouse cells (Fig. 3C). This technique consists of cutting ultrathin (~50 nm) vitrified cryosections from a high-pressure frozen (Studer *et al.*, 2004) pellet of cells without being exposed to freeze-substitution or any type of chemical fixation. Those sections have to be treated like any frozen-hydrated specimen and are imaged at LN_2 temperature in the electron microscope, in our case with a GATAN-626 cryo-holder. Compared to other more invasive electron microscopy methods, CEMOVIS precludes the use of contrasting agents applied to the sections. Therefore, frozen-hydrated cryosections only exhibit image contrast defined by the biological material's own density, analogous to cryo-EM of isolated protein or protein–RNA complexes. 2-D as well as 3-D acquisitions can be performed on cryosections (see Fig. 3).

Fig. 3A shows a bundle of IFs together with a microtubule located within a protuberance of the cytoplasm (marked as C in Fig. 3A) into the nuclear space (marked as N in Fig. 3A). Both compartments are separated by the nuclear envelope. The visibility of the double lipid layer and the microtubule protofilaments demonstrates the remarkable preservation of biological structures in vitrified cryosections. The strong contrast of the two small bundles of IFs (Fig. 3A, black arrow heads) comes from their perpendicular orientation to the section plane, and the resulting superposition of their mass along a 50–60 nm-long stretch. One of them is located in the close vicinity of a cross-sectioned microtubule exhibiting the typical 13 protofilament arrangement of microtubules (for similar cryosection images of microtubules see Bouchet-Marquis *et al.* (2007)). Images of cross-sectioned cellular IFs do not suffer from the missing wedge effects seen in tomographically reconstructed end-on views (Figs. 3B and 7C). They have an overall round shape when oriented perpendicular to the section's surface. Moreover, it shows a fine density attached to the inner side of the filament pointing in. The presence of an inner density seems to be directly related to the IF's 3-D organization. It shows different packing than the four-protofibril arrangement (see Fig. 1C(c1)) compared with tomographically reconstructed end-on views of the vimentin IFs shown in Fig. 3B. Hence, the IFs in Fig. 3A, though not unambiguously identified, strongly resemble the keratin structures previously observed by Norlén and Al-Amoudi (2004). The presence of keratins would not be unexpected for that location in CHO cells. IFs arranged along the section plane (Fig. 3C) are much harder to see.

Bundles of IFs can also be followed through the volume of the cytoplasm by performing cryo-ET on cryosections. However, the filaments are much harder to see and are virtually invisible in single 2-D cryo-images. Fig. 3 is extracted from a tomographic reconstruction of a 70 nm thick cryosection using the IMOD software package (Kremer *et al.*, 1996). The tilt series ranges from ±60° at a magnification of 29 k (0.7 nm pixel size). Even if suffering from an inherent low contrast-to-noise ratio, many IFs (black arrowheads) can be found and followed throughout the volume of the section. Those filaments show a diameter of roughly 12.5–13.5 nm on average. These dimensions are relatively high compared to values found in most of the textbooks (Herrmann and Aebi, 2004) (~7–11 nm) but they are not surprising; CEMOVIS allows the observation of hydrated structures, avoiding dehydration artifacts (Dubochet and Sartori Blanc, 2001). The better preservation allows a more realistic measurement of structures dimensions observed in their native environment.

C. Atomic Force Microscopy

1. Introduction to the Method

The basic principle for today's AFM was developed in 1985 with the invention of the scanning tunneling microscope (STM; Baró *et al.*, 1985) that was honored by the Nobel Prize to Profs. Gerd Binnig and Heinrich Rohrer (together with Ernst Ruska for his invention of the electron microscope back in 1931; Robinson, 1986). The STM was the first instrument that was able to produce real-space images of a conducting surface with atomic resolution. Since then AFM became widely used in physics, chemistry, and life sciences for its ability to produce high-resolution images of a wide variety of samples, such as polymers, DNA, or even cell surfaces. The AFM's ability to process native samples under physiological conditions (e.g., buffered media at 37°C) is extremely useful. In contrast to other imaging techniques, AFM is a mechano-optical device that produces morphometric data, meaning each acquired data point has a defined *x*-, *y*-, and *z*-coordinate. Images built from AFM data are topographic images with the contrast based on the sample structure. Image color is arbitrary, since any color could be assigned to a specific height on the *z*-coordinate (Fig. 6G and H).

The parts of a modern AFM are a stylus attached to a flat spring, termed cantilever, a piezo-driven *xyz*-scanner, a laser, and a four-quadrant photodiode (Fig. 6F). The laser beam is focused on the back of the cantilever and reflected on the photodiode. Movement of the stylus over the sample area is facilitated by the *xyz*-piezo scanner. In general, two different imaging modes (either tapping or contact mode) can be distinguished (for recent reviews see Ludwig *et al.*, 2008; Puchner and Gaub, 2009). In the first static mode, the movement of the cantilever caused by weak attractive or repulsive forces results in an altered deflection of the laser beam. During the movement of the cantilever a constant deflection (e.g., constant force) is maintained by moving the cantilever vertically by the *z*-element of the piezo scanner in response to surface features of the sample. The second mode is generally referred to as resonance mode. Here the cantilever is oscillated at its resonance frequency. Utilizing a feedback loop the oscillation is kept constant during scanning and again the cantilever is moved vertically in response to the sample features. Both modes offer different advantages for a distinct range of applications. In resonance mode, for example, AFM can be operated in noncontact, intermittent contact, phase imaging, and force modulation scanning mode. Examples for static mode are contact scanning and force spectroscopy. Resolution in AFM critically depends on sample properties and the shape of the tip at the end of the cantilever. Subnanometer-scale features can be resolved on proteins reconstituted *in vitro*.

2. AFM on IFs

a. Preparation of Vimentin and Desmin Human vimentin and desmin were recombinantly expressed and purified as described elsewhere (Hofmann *et al.*, 1992). The purified proteins can be stored at −80°C in 8 M urea, 5 mM Tris-HCl (pH 7.5), 1 mM DTT, 1 mM EDTA, and 10 mM methyl ammonium chloride. The day before use the protein was dialyzed into a 2 mM sodium phosphate buffer (pH 7.5) containing 1 mM DTT. Dialysis was performed at RT in a stepwise fashion (8, 4, 2, and 0 M Urea) and continued overnight at 4°C into fresh buffer without urea. The next day the protein was dialyzed for an additional hour in 2 mM sodium phosphate buffer (pH 7.5) without DTT at RT. To start filament polymerization, assembly buffer (2× AP) containing 200 mM KCl in 2 mM sodium phosphate (pH 7.5) was added in equal volumes to a 0.2 or 0.4 mg/ml protein solution. Assembly was allowed to proceed for the desired time at 37°C. After a chosen assembly time, ranging from seconds to 1 h, the assembly was terminated by the addition of an equal amount of stop buffer (0.2% glutaraldehyde (w/v) in 1× AP) for 30 s and then diluting into 1× AP to a final protein concentration between 1 and 10 μg/ml. Stopping the assembly without utilizing glutaraldehyde was done by diluting the protein solution directly into 1× AP. Aliquots (30 μl) were allowed to adsorb on a solid support (see below) for 5–20 min prior to AFM imaging. Adsorption to a solid support (muscovite mica, highly ordered pyrolytic graphite (HOPG), or hydrophilic glass) essentially stopped filament assembly for the unfixed samples.

b. Data Acquisition by AFM Mica and HOPG were freshly cleaved immediately before use with standard Scotch or duct tape removing the top layers of the material glued to a steel disc, thus exposing a fresh clean surface. The hydrophilic glass was prepared as described previously (Mücke *et al.*, 2005; 2009). In brief, standard glass coverslips were treated for 1 h in "piranha" solution (1:3 solution of H_2SO_{4conc}: 30% H_2O_2), rinsed with distilled water, exposed to ultrasound (50 kHz) for 15 min, and dried under a stream of nitrogen. Such treatment left the glass surface covered with Si-OH groups rendering it hydrophilic for some hours. The solid supports were glued to steel discs either directly or on a piece of Teflon, which in turn was glued to the steel disc. The steel discs are magnetically mounted on the piezoelectric scanner of the AFM.

All samples were scanned in liquid using oxide-sharpened 100 μm long cantilevers with a nominal spring constant of 0.38 N/m and tip radii between 5 and 40 nm (NP-S, Veeco, Santa Barbara, CA, USA). In addition, we utilized Super Sharp Silicon tips (SSS-FM, Nanosensors, Nano and More GmbH, Wetzlar, Germany), which are built from monolithic silicon typically with a tip radius of 2 nm and a spring constant of 4.2 N/m. All images were recorded in tapping mode on a Nanoscope IIIa controller (Veeco) at 512 × 512 pixels. The scanning speed was in the range of 1–2 Hz.

The main difference between AFM and EM is the possibility of investigating hydrated and unfixed IFs at RT with AFM, while EM always requires solidification of the observed material (drying, freezing) due to the exposure to high vacuum. However, attention has to be paid to the influence of the solid support on the architecture of the sample as well as the geometry of the tip. Fig. 4B and C shows IFs adsorbed to mica and imaged in phosphate or reassembly buffer (see Section IV) using tapping mode. Vimentin filaments (Fig. 4B, left panel) typically exhibited a height of 3–5 nm and a full width at half maximum (FWHM) of 40–70 nm. At the ultrastructural level individual filaments were observed, which appear rough with small breaks or gaps along their length. In contrast desmin filaments adsorbed to mica were generally 8–10 nm in height with a FWHM of about 30 nm. At lower magnifications they appear as dense spatially localized networks

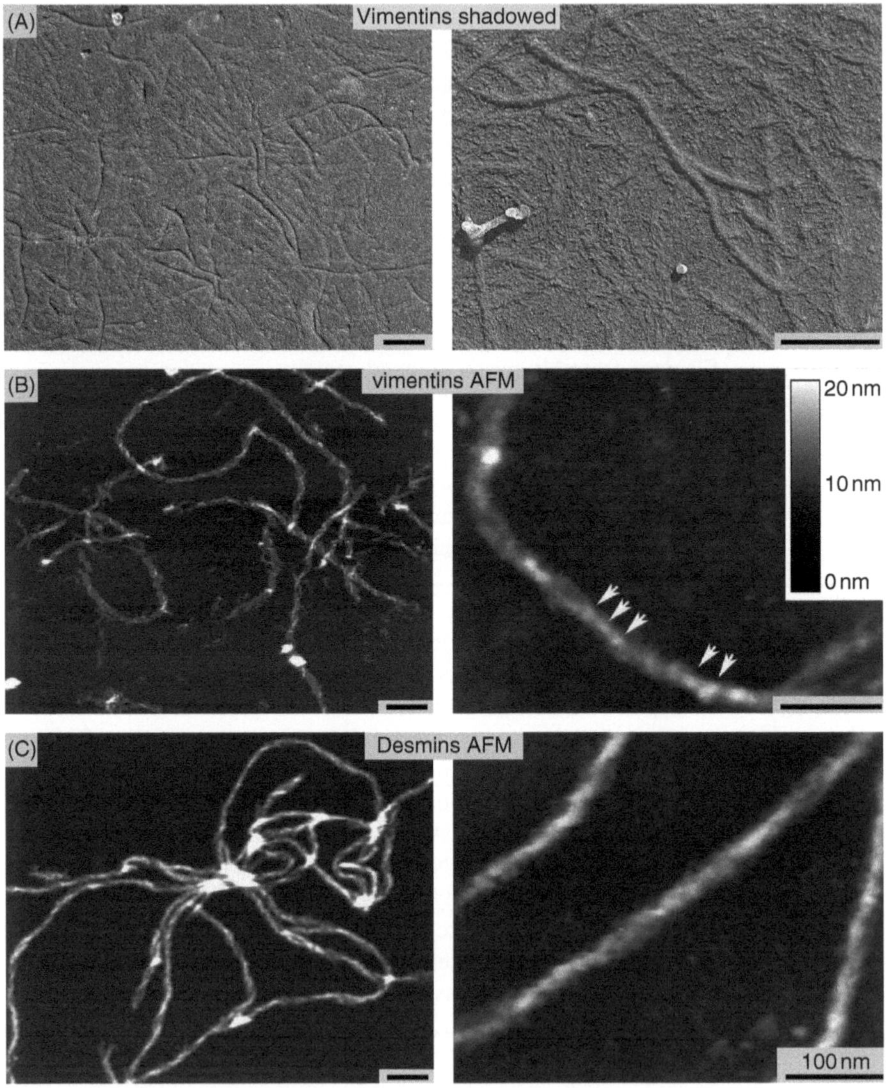

Fig. 4 (*Continued*)

(not shown). Higher magnifications (Fig. 4C, left panel) of such networks reveal compact filaments with a right-handed twist.

Sample preparation for conventional EM of IFs involves fixation with glutaraldehyde and adsorption to a glow-discharged carbon-coated EM-grid followed by staining and dehydration (Herrmann *et al.*, 1996). These conditions can be mimicked for AFM imaging. Following assembly, the IF samples are fixed with 0.1% glutaraldehyde and allowed to adsorb to HOPG, which is similar to the surface properties of a carbon-coated EM-grid. After fixation mainly compact filaments with a height of 5–8 nm and a FWHM of 25–30 nm can be observed (Fig. 4B and C, right panels). After glutaraldehyde treatment vimentin filaments appear compact and reveal subfilamentous structures. In some cases a clear 20 nm beading can be seen (Fig. 3B, right panel, arrows). Desmin appears as compact as in the unfixed samples, but with either pronounced 15–18 nm thick stripes along the filaments or a large variability in the angle between the repeat and the filament axis. All attempts to process the helical features so far had a blurring effect, which is mainly due to the slightly different morphology of one filament to its neighbor. A cause for the differences might be induced by the surface adsorption or tip–sample interactions of the AFM.

Besides observing mature IFs on solid support, visualization of IF precursor and the so-called unit-length filament (ULF) is also possible (Fig. 5A). Here we blocked the assembly with glutaraldhyde after 3 s. The resulting structures were adsorbed to hydrophilic glass and scanned with SSS-FM cantilevers. Short ULF-like particles could be observed with a length of 60–80 nm, a height of 10 nm, and a FWHM of 30 nm. These dimensions are compatible with previously published observations of ULFs utilizing negative staining and EM (Herrmann *et al.*, 1999).

Fig. 4 Surface imaging of IFs by unidirectional metal shadowing and tapping-mode AFM images. Both methods produce surface topography images, but with a different spatial resolution. (A) Overview and larger magnification of unidirectionally shadowed vimentin filaments polymerized for 1 h at 37°C. Shadowed images are obtained from plunge-frozen/vitrified specimens that were subsequently freeze-dried and metal shadowed. The images reveal a good view of the helical properties and some 3-D aspects of the filament structures and their subfibrils. Sub-protofibrils are well resolved. Both the length and width of filaments and subfibrils are accurately reproduced. However, in particular, the loosely packed filaments appear collapsed on the carbon support, which might have caused some distortions (arrows). (B) Overview and larger magnification of vimentin filaments assembled for 1 h at 37°C and (C) of desmin filaments assembled for only 10 min at 37°C. Note the more rapid assembly of desmin filaments, which produces the same length and more compacted filaments after 10 min versus 1 h that it takes to assemble vimentin into similar length filaments and which are still not as compact as desmin IFs. AFM is currently the only method that allows imaging of supramolecular structures in solution at RT with molecular detail. The images shown here were taken in 2 mM phosphate buffer, 100 mM KCl (pH 7.5) at RT with NP-S tips on mica (B left, C left) and HOPG (B right, C right). Hence, AFM is the method of choice to observe dynamic processes, such as IF assembly. While the length of these filaments is correctly reproduced, the width is distorted due to the geometry of the AFM tip. As a consequence, subfibers and loosely packed filaments are typically less well resolved and may appear with a different shape according to their orientation to the tip-scanning direction (images adapted from Mücke *et al.* (2005)). (See Plate no. 36 in the Color Plate Section.)

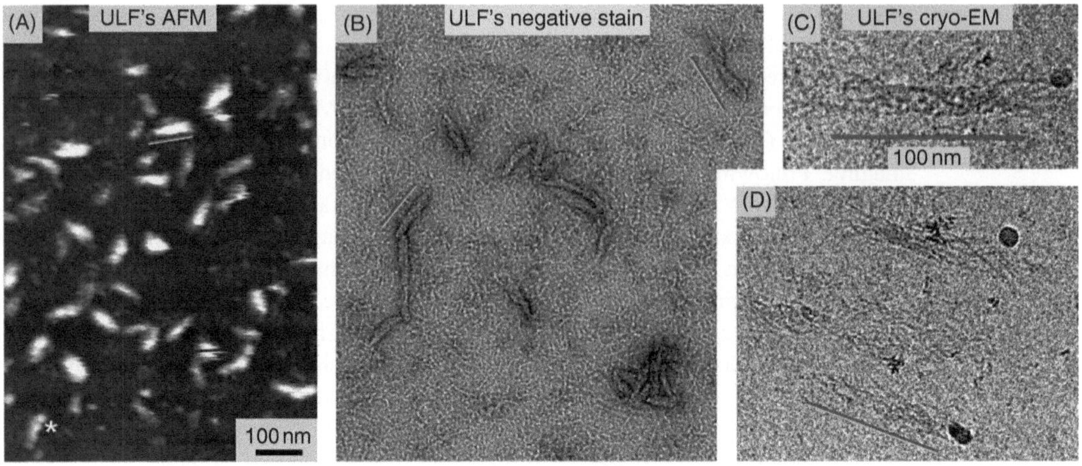

Fig. 5 Comparison of AFM negative staining and cryo-EM images. AFM (A) negative staining (B) and cryo-EM images (C, D) of short assembly units called unit-length filaments (ULFs) of vimentin as they occur during the early onset phase of polymerization. In these images assembly of vimentin filaments was stopped almost immediately after initiation using chemical fixation (A), staining (B), or plunge-freezing into ethane (C, D.) AFM of glutaraldehyde-fixed ULFs (A), using SSS-FM tips. Negatively stained unfixed vimentin (B), and Cryo-EM (C) and (D). In the case of AFM and negative stain, ULFs of about 100 nm lengths are evident. Cryo-EM micrographs reveal a substructural arrangement similar to that seen with full-length filaments that are unraveling or only loosely packed with 3–4 nm spacings inbetween protofibrils (e.g., see also Fig. 2A and C). (See Plate no. 37 in the Color Plate Section.)

IV. Materials

A. AFM (Instrument/equipment)

- Multimode with Nanoscope IIIa controller (Veeco): Multimode AFM with up to 20×20 μm scan area operated by a Nanoscope IIIa controller.
- Muscovite mica (Plano, Wetzlar, Germany): naturally occurring mineral that can be easily cleaved to yield nearly atomically flat fresh and polar surfaces.
- HOPG, Mica (Plano): carbon-based material that can be easily cleaved to yield nearly atomically flat fresh nonpolar surfaces similar to mica but with properties closer to a carbon-coated grid.
- AFM NP-S tips, Veeco: one of the standard tips and cantilevers used in AFM to image biological samples in solution with tip radii from 5 nm or larger.
- AFM SSS-FM (Nanosensors, Nano and More GmbH): specialized AFM tips with finer end radii (~2 nm) and stiffer cantilevers.

B. IF Biochemistry

- Purified vimentin or desmin was stored in 8 M Urea (plus 5 mM Tris, 1 mM EDTA, 0.1 mM EGTA, 1 mM DTT, 10 mM methyl ammonium chloride pH 7.5) at −80°C.

- 5 mM Tris pH 8.4: buffer used for vimentin and desmin after dialysis; when dialyzed into this buffer, vimentin and desmin will mainly be in tetrameric form.
- 2 mM Sodium phosphate buffer pH 7.5: buffer used for vimentin and desmin after dialysis; when dialyzed into this buffer, vimentin and desmin will mainly be in tetrameric form. This buffer resembles the physiological conditions more closely than the Tris buffers.
- Reassembly Buffer: 2× Tris assembly buffer (45 mM Tris pH 7.5, 320 mM NaCl): when using the Tris buffer, assembly of the IFs is started by raising the ionic strength using NaCl. The assembly buffer is mixed 1:1 with the dialyzed protein solution to yield the final assembly conditions (22.5 mM Tris, 160 mM NaCl, pH 7.5).
- 2× Phosphate assembly buffer (2 mM sodium phosphate pH 7.5, 200 mM KCl): when using the phosphate buffer, assembly of the IFs is started by raising the ionic strength using KCl. The assembly buffer is mixed 1:1 with the dialyzed protein solution to yield the final assembly conditions (2 mM phosphate, 100 mM KCl, pH 7.5).
- Glutaraldehyde 0.2% in Tris or phosphate buffer: used to chemically stop filament assembly. The glutaraldehyde solution is mixed 1:1 with the assembled protein to a final concentration of 0.1% glutaraldehyde.
- Piranha solution (1:3 H_2SO_4, concentration 30% H_2O_2): the solution was prepared fresh before use (careful it gets really hot!) to clean cover slips for 1 h and generate hydrophilic surface properties.

C. Cryo-Electron Microscopy (Plunge Freezing)

- Vitrobot Mark IV (FEI Company, Hillsboro, OR, and Eindhoven, the Netherlands): Fully automated plunge-freezing device that we used for humidity-controlled plunge-freezing experiments. The Vitrobot allows the maintainance of any desired humid environment up to 99% humidity.
- Custom-made gravity-operated plunge-freezing device: used for plunge-freezing for experiments where humidity was not a critical issue.
- Commercially produced holey carbon grids (4.0 μm holes, 2.0 μm spaces, C-Flat, Protochips Inc., Raleigh, NC, USA).
- Colloidal gold (5 or 10 nm, British-Biocell International, Cardiff, UK): Colloidal gold particles are used as fiducial markers for the alignment of individual projections from tomographic tilt series to generate a back-projected 3-D volume.

D. Cryo-Electron Microscopy on Vitrified Sections

- Cell lines: CHO (used as epidermal model for keratins), A31 3T3 (fibroblasts used for vimentins). However, any type of mammalian cell type would serve the purpose we attempted here.
- Culture media: Dubelco's modified eagle medium, 2% L-glutamine, 10% FCS, 1 mg glucose per ml in 5% CO_2. Cells cultures are kept at 37°C all the time.

- High-pressure freezer: BalTec HPM010 (Leica, Vienna, Austria): Used to produce frozen-hydrated and vitrified blocks of cell pellets and tissues that can either be used directly for thin sectioning as shown here or be subjected to freeze-substitution and plastic-embedding.
- Cryo-ultramicrotome Ultracut UCT microtome from Leica Microsystems. This instrument uses an ultrasharp diamond knife to cut thin sections (50–100 nm) of frozen-hydrated material (see Figs. 3A,C and 6A–E). The microtome used here was slightly modified from the commercial version to accommodate the production of vitrified cryo-sections.
- Cryo-diamond knives: 45° (DIATOME, Biel, Switzerland), specifically designed for cryo-ultramicrotomy on vitrified samples (Fig. 6A–E). Although there are knifes with a smaller cutting angle such as 35° and 25° usually producing less compression, we used the 45° knife because it produces less crevasses that we prefer over compression artifacts.

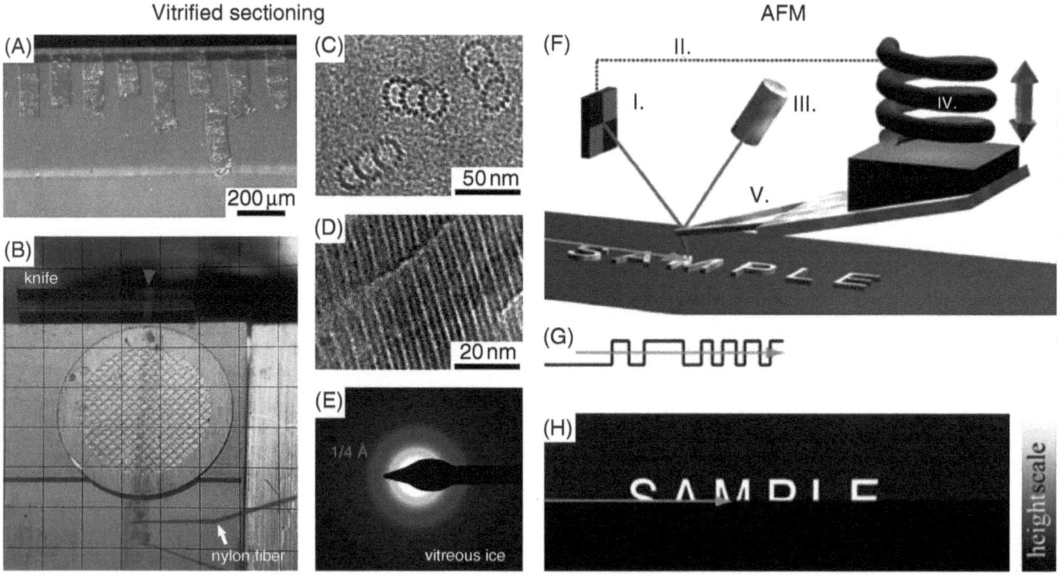

Fig. 6 Preparation of vitrified sections and image recording by AFM. (A) Ribbons of vitrified sections lying at the surface of a 45° cryo-diamond knife. 50 nm thick sections show a distinctive transparency. (B) Placing of ribbons of vitrified sections on the surface of a 200 mesh, carbon-coated EM-grid using a nylon fiber. (C, D) Examples of well-preserved cellular assemblies observed in vitreous sections. (C) Part of the centrosome composed of three triplets of cross-sectioned microtubules. (D) Membranous array showing a well-preserved spacing between each other. (E) The diffraction pattern consisted of the two diffuse rings characteristic of vitreous ice, presenting a spacing of 0.317 and 0.214 nm, respectively. (F) A modern AFM is composed of a stylus that is attached to a flat spring, termed cantilever. This cantilever is controlled by a piezo-driven *xyz*-scanner, and its movement when scanning a surface (G, H) is recorded by a laser, and a four-quadrant photodiode. The laser beam is focused on the back of the cantilever and reflected on the photodiode that is used directly for a feedback loop controlling the positioning of the tip. (See Plate no. 38 in the Color Plate Section.)

E. Cryo-Electron Microscopy and Electron Tomography, Equipment

- Cryo-electron microscope: Tecnai-F20 (FEI Company). The microscope features a ±70° tilt-goniometer.
- Cryo-specimen holder Gatan-626 and GATAN-614 (GATAN Inc., Pleasanton, CA, USA): Used for imaging frozen-hydrated specimens at LN_2 temperature (see Figs. 2C–F, 3, 5C, D, 6C, D, and 7).
- SerialEM data acquisition software (Mastronarde, 2005; for a free download of that software see http://bio3d.colorado.edu/SerialEM/index.html): This is our own product designed for automated acquisition of tomographic tilt series from RT and cryo-samples.
- IMOD software package for tomographic 3-D reconstruction and analysis of tilt series: This again is our own product developed and constantly improved by David Mastronarde at the Boulder 3-D laboratory (Kremer *et al.*, 1996; for a free download of that software see http://bio3d.colorado.edu/imod).

V. Discussion

A. *In vitro* Studies on Isolated Frozen-Hydrated IFs

In this chapter we compare cryo-EM and cryo-ET on isolated IFs with data obtained by other EM methods as well as AFM. Here we would like to evaluate the technical

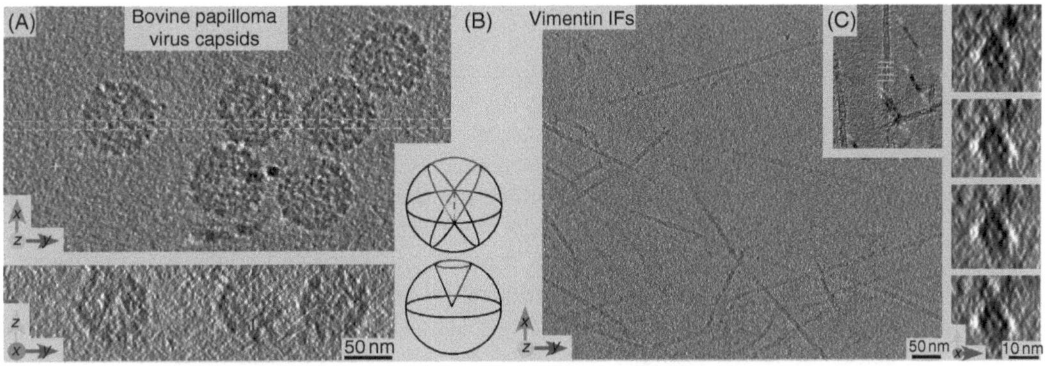

Fig. 7 Effect of the missing wedge in cryo-tomogramphy data as shown for bovine papilloma virus capsids and vimentin IFs. (A) This effect is particularly obvious within the viral structure due to the icosahedral symmetry in the virus. With isotropic resolution very similar projections should be produced in all directions. While the particles in *xy*-projection (top panel) appear isotropically resolved, the loss in *z*-direction is substantial (bottom panel). Reducing this information loss to a missing cone (lower sketch) or missing pyramid would not fully eliminate that loss, but it would add valuable data and substantially improve the quality of the tomogram. (B, C) 10 nm tomographic slice of a cryo-electron tomogram of vimentin IFs. The missing wedge of resolution is not seen in this orientation. A stretch of a filament is shown and the cross sections are marked with red bars (inset). The missing wedge effect is now visible as a "stretching" effect along the *z*-axis, lowering the resolution in that direction. (See Plate no. 39 in the Color Plate Section.)

difficulties of data acquisition and interpretation from cryo-EM over the preexisting data obtained by negative stain microscopy, metal shadowing, or AFM. What exactly is now the unique contribution that cryo-EM and cryo-ET did, or will do, to the field of IF structure? Ideally, this chapter should be read together with the chapter of Kreplak *et al.* (2008) published in a previous "Methods in Cell Biology" book series. Here, we discuss the power of EM and AFM for IF structure and function by critically comparing cryo-EM (Fig. 2C–F) with (1) negative staining EM with or without chemical fixation (Fig. 2A and B), (2) metal shadowing (Fig. 4A), and (3) AFM (Figs. 4B,C, and 5A). AFM, metal shadowing, and negative staining EM all produce rather bent filaments that may be very compact, or partially unraveled, depending on the state of assembly and buffer (salt) conditions. Interestingly, the state of unraveling (see Fig. 2B) appears to be induced by a highly cooperative process. Unraveled filaments look identically unraveled over long stretches. This indicates that the filaments unravel upon a trigger signal that rapidly spreads along an entire filament. This rapid, cooperative unraveling is also observed by cryo-EM; however, there is one significant difference seen only in cryo-EM pictures: the compacted filaments may appear much stiffer compared with any other preparation (Fig. 2C–F). The reason for this may be that filaments prepared for cryo-EM are embedded in ice and imaged over holes without carbon support to recover a maximum of image contrast. All other microscopy methods work with a support structure of some sort. Hence, one could speculate that the adsobtion to a carbon or Formvar film renders the filaments more bent and kinkier (Mücke *et al.*, 2009). Cryo-EM, AFM (Fig. 4B–C), and freeze-drying/metal shadowing (Fig. 4A) all share the possibility of keeping filaments in physiological buffer. Today AFM is often carried out directly in solution. For cryo-EM and shadowing, samples need to be ultrafast-frozen, but the buffer conditions do not change during preparation. In contrast, buffer conditions change when preparing samples for negative staining (Fig. 2A and B).

However, a comparison of vimentin filaments prepared by negative staining (Fig. 2A and B), freeze-drying/metal shadowing (Fig. 4A), or imaged in solution by AFM (Fig. 4B and C) reveals similar conformational state. Compacted and partially unraveled filaments can be seen at about the same rate, depending on the state of assembly, with compact filaments mostly bent. Partially unraveled filaments expose individual protofibrils that can be seen by all three methods, as well as cryo-EM. Regarding image resolution, negative staining and metal shadowing show similar detail, while the resolution achieved by AFM is somewhat lower with the typical widening effect due to the tip geometry, which is most pronounced along the scanning direction (see Figs. 4B,C, 5A, and 6F–H). However, the lack of resolution is compensated by other unique properties of AFM. AFM is the only method where IFs can be observed at molecular resolution directly in solution and in a fully dynamic state. The lack of an apparent helical or 2-D crystalline symmetry makes it difficult to assess the 3-D structure of the filaments. Negative staining is typically used to generate 2-D projections. AFM produces an accurate 3-D topograph with precise measurements along the *z*-axis. Metal shadowing produces a pseudo-3-D appearance with an inaccurate measure for the *z*-axis. However both methods are incapable of imaging internally hidden structures (e.g., the inner cores of a virus) and the surfaces of complexes in contact with the support film.

1. Conclusions

The suitability of electron microscopy or AFM to study the ultrastructure of IFs can be summarized as follows:

- Negative staining EM can be compared to a "Volkswagen": It is a cheap, practical, and reliable method to assess a large array of aspects of IF structure and function such as assembly and disassembly dynamics via ULFs (see Fig. 5), as well as stability. However, negative staining often has some serious limitations when it comes to physiological accuracy and gaining dynamic information: "It gets you to most places, but not everywhere." Factors to be considered are: exposure of specimens to nonphysiological pH values (e.g., the pH of Uranyl acetate is ~4.5), salt conditions, as well as the potentially disruptive effect of adsorption to a substrate during air-drying.
- AFM and unidirectional metal shadowing on IFs can both be compared to a high-performing "sports car": Unmatched performance in some aspects, e.g., when it comes to imaging in solution, and on highly regular surfaces. But for most applications AFM may be a bit unpractical or provide only limited information and therefore not an everyday choice. AFM is the only method that works in solution on IF under assembly conditions revealing their dynamic properties, but is accompanied by low interpretable resolution on most biological samples (> 5 nm, better on 2-D crystalline assemblies), tip artifacts, adsorption to a substrate, and limited 3-D capabilities. In principle, the same applies for shadowing except that the power of shadowing lies in producing high-resolution surface images with a pseudo-3-D effect. The use of AFM and EM of specimens on different solid supports demonstrates the ability to access ultrastructural features of the observed filaments both in a dried-fixed versus a hydrated state and the ability of the IFs to modify their internal 3-D structure according to the surrounding environment (i.e., to the chemical properties of mica, HOPG, and hydrophilic glass). The ability for rearrangements, explaining the assembly polymorphism of IFs or the low bending stiffness, was already proposed in Troncoso *et al.* (1988) and Mücke *et al.* (2004a). However, the stiff appearance of IFs in cryo-EM (Figs. 2C–E and 7B) suggests a possible sensitivity of the structural integrity to surface adsorption. The ability to generate real-space topographic images of samples on a surface makes AFM an excellent tool to complement EM and 3-D EM studies when addressing questions about the ultrastructure and possible structural arrangements of IFs.
- Cryo-EM of isolated/*in vitro* assembled IFs in ice (in particular in combination with cryo-ET) compares to a luxury "sports utility vehicle" (SUV): it stands for high-end performance, best quality, and practicality, however, with some limitations. Sometimes it may be simply a costly overkill. Cryo-EM offers great versatility and high performance regarding achievable molecular resolution and the

maintenance of physiologically relevant conditions. Cryo-EM also offers full 3-D reconstruction capabilities including the possibility of imaging internal structures, but the equipment is expensive, image contrast is low, selective staining of individual components is usually not possible, and frozen-hydrated specimens are very sensitive to beam-induced damage. Hence, cryo-EM, while essential for the acquisition of some data, may not always be the best choice (i.e., in cases where high molecular resolution is not critical and where negative staining or AFM could perform equally well such as for imaging time-resolved dynamic conformations).

B. *In Situ* Studies on IFs in Vitrified Cryosections

As a major part of the cytoskeleton, IFs produce a large 3-D network within the cell. Similar to the other two networks (for an intracellular study on actin see: Medalia *et al.*, 2002; Chapter 22 by Resch *et al.*, this volume; for microtubules see: Höög *et al.*, 2007; Melloy *et al.*, 2007) the 3-D architecture of IFs can only be studied within an intact cellular context. Therefore we also tested the applicability of CEMOVIS (Al-Amoudi *et al.*, 2004) and cryo-ET on IFs maintained within their cellular context (Fig. 3). So far most investigations on the 3-D arrangement of cytoskeletal assemblies and other structures, such as organelles, vesicles, have been performed with either plastic-section tomography (see Hoenger and McIntosh, 2009; O'Toole *et al.*, 2007) or, in exceptional cases, Tokuyasu cryo-sections in combination with immunolabeling (Tokuyasu, 1986; see also: Ladinsky and Howell, 2007). This is in contrast to vitrified sectioning (Al-Amoudi *et al.*, 2004; Dubochet *et al.*, 2007; Hsieh *et al.*, 2002; McDowall *et al.*, 1983) where neither staining nor immunolabeling can be applied. Once a sample is vitrified it has to be kept at a low temperature to maintain its vitreous state (temperatures below approximately −140°C or 133 K). On the downside, vitrified sections often show cutting artifacts such as compression and formation of crevasses (Dubochet *et al.*, 2007).

The advantages, but also drawbacks, of vitrified sectioning for structural investigations of IFs *in situ* are illustrated in Fig. 3 (see also Norlén (2008); Norlén and Amoudi (2004); Norlén *et al.* (2009)). Vitrified sectioning of cells or tissues revealed spectacular images of microtubules (e.g., see Fig. 6C; Bouchet-Marquis *et al.*, 2007) with remarkable structural preservations despite the intrinsic compression problems of vitrified sectioning. Actin or actin bundles usually can be identified based on the helical parameters of F-actin. However, the smooth surface and the dynamic properties of IFs often make it very difficult to properly image them within the packed content of an intact cell. In addition, there are five major families of IFs, and since they are structurally relatively similar and do not "wear nametags," depending on the cell type and location within a cell it might be difficult to identify them properly. Fig. 3A and B shows end-on projections of IFs in cells (3A: possibly keratin) and *in vitro* (3B: vimentin). In both cases the end-on projections superimpose sufficient density to count or at least estimate the protofibril numbers that form the particular filament. However, the end-on views of filaments (Fig. 3A) are a plain 2-D projection with isotropic resolution, while the ones in Figs. 3C and 7C are an *xz*-view within a tomographic reconstruction and appear stretched out due to the missing cone of data in the *z*-direction. Fig. 3B clearly shows four density domains for vimentin, yet the reconstruction shown in Fig. 7C also revealed a high-density core in vimentins, questioning the relevance of an all-exclusive four-protofibril type packing for vimentins as modeled in Fig. 1C(c1). Is it possible that vimentins alternate between models such as the ones suggested in Fig. 1C(c1 and c2)? At least for the cases shown in Figs. 3B and 7C we know for sure that we are looking at vimentins *in vitro*. The nature of the filaments recorded *in situ* (Fig. 3A and C) is less certain. The end-on views shown in Fig. 3A reveal a similar pattern as the vimentins in Fig. 7C with 6–7 surrounding domains with a high-density core. This pattern, however, is also highly similar to keratin filaments as shown in Norlén and Amoudi (2004) and their location near the nucleus of a CHO cell suggests that they could indeed be keratins.

Obviously, vitrified sectioning not only produces filaments in end-on views, but also may cut through bundles in all directions (see Fig. 3C). These orientations often produce so little contrast that they are only discovered when visualized within thin (10–15 nm) tomographic sections after a successful reconstruction. Thin tomographic sections omit the density contributions from protein structures above and below and, thereby, strongly enhance the visibility of fine structures somewhat similar to confocal fluorescence light microscopy where fluorochromes are selectively excited within a thin focal plane only. Otherwise the superimposed densities through a fully vitrified cryosection, even if it is as thin as 50–80 nm, when imaged at an extremely low-dose condition (1–2 $e^-/Å^2$), impairs the signal so much that single IFs remain almost invisible.

1. Conclusions

In summary, the application of vitrified cryosectioning for IF structure and function studies appears very promising, as it is for other cellular 3-D studies, but the method is far from being routinely established for the IF field, similar to cryo-EM and cryo-ET in general. The rise of cryo-tomography, however, for the last decade or so, initiated a boom in cell biological applications that also affected the field of IFs. The possibility of generating 3-D structures without the need for molecular symmetry and/or image element averaging is too appealing to be ignored for IF-related structural studies. Currently, tomography is, in fact, the only way to obtain complete 3-D data from IFs, both *in vitro* and *in situ*. However, the typical issues associated with cryo-EM in general also apply to IFs, if not even more than for other fields such as microtubules and actin structure (see also Chapter 22 by Resch *et al.*, this volume). Accordingly, cryo-EM data of IFs are still rare, progress is slow, and the kinds of data shown here are currently produced by a few groups only. New developments are needed to boost the field, such as (1) site-specific high-density labels that allow mapping of the internal molecular detail of IFs without disrupting filament formation; (2) 3-D reconstruction software explicitly designed for IF 3-D structure applications; and (3) improved microscope hardware, such as more sensitive detectors, image filters, and possibly corrector devices for chromatic (Cc) and spherical (Cs) distortions. These filters are now commercially available on the most recent types of FEI-Company (Titan-Krios) and Zeiss electron microscopes.

Acknowledgments

This work has been supported by grant P41-RR000592 from NCRR (to Andreas Hoenger) and by a postdoctoral fellowship of the German Research Foundation (DFG) to Robert Kirmse. We thank Mary Morphew (University of Colorado at Boulder, Boulder, Colorado) for the images displayed in fig. 7A, and Kenneth N. Goldie (University of Basel, Switzerland) for images displayed in figs. 2C, E, and 3B.

References

Aebi, U., Cohn, j, Buhle, L., and Gerace, L. (1986). The nuclear lamina is a meshwork of intermediate-type filaments. *Nature* **323**, 560–564.

Aebi, U., Fowler, W. E., Rew, P., and Sun, T. T. (1983). The fibrillar substructure of keratin filaments unraveled. *J. Cell Biol.* **97**, 1131–1143.

Al-Amoudi, A., Norlén, L., and Dubochet, J. (2004). Cryo-electron microscopy of vitreous sections of native biological cells and tissues. *J. Struct. Biol.* **148**, 131–135.

Baró, A. M., Miranda, R., Alamán, J., García, N., Binnig, G., Rohrer, H., Gerber, C., and Carrascosa, J. L. (1985). Determination of surface topography of biological specimens at high resolution by scanning tunneling microscopy. *Nature* **315**, 253–254.

Ben-Harush, K., Wiesel, N., Frenkiel-Krispin, D., Moeller, D., Soreq, E., Aebi, U., Herrmann, H., Gruenbaum, Y., and Medalia, O. (2009). The supramolecular organization of the *C. Elegans* nuclear lamin filament. *J. Mol. Biol.* **386**, 1392–1402.

Beuron, F., and Hoenger, A. (2001). Structural analysis of the microtubule-kinesin complex by cryo-electron microscopy. *Methods Mol. Biol.* **164**, 235–254.

Bouchet-Marquis, C., Zuber, B., Glynn, A. M., Eltsov, M., Grabenbauer, M., Goldie, K. N., Thomas, D., Frangakis, A. S., Dubochet, J., and Chrétien, D. (2007). Visualization of cell microtubules in their native state. *Biol. Cell.* **99**, 45–53.

Bremer, A., and Aebi, U. (1992). The structure of the F-actin filament and the actin molecule. *Curr. Opin. Cell Biol.* **4**, 20–26.

Cope, J., Gilbert, S., Rayment, I., Mastronarde, D., and Hoenger, A. (2010). Cryo-electron tomography of microtubule-kinesin motor complexes. *J. Struct. Biol.* **170**(2), 257–265.

De Pianto, C., and Coulombe, P. A. (2004). Intermediate filaments and tissue repair. *Exp. Cell Res.* **301**, 68–76.

DeRosier, D., and Moore, P. B. (1970). Reconstruction of three-dimensional images from electron micrographs of structures with helical symmetry. *J. Mol. Biol.* **52**, 355–369.

Dhe-Paganon, S., Werner, E. D., Chi, Y. I., and Shoelson, S. E. (2002). Structure of the globular tail of nuclear lamin. *J. Biol. Chem.* **277**, 17381–17384.

Dubochet, J., Adrian, M., Chang, J. J., Homo, J. C., Lepault, J., McDowall, A. W., and Schultz, P. (1988). Cryo-electron microscopy of vitrified specimens. *Q. Rev. Biophys.* **21**, 129–228.

Dubochet, J., and Sartori Blanc, N. (2001). The cell in absence of aggregation artifacts. *Micron* **32**, 91–99.

Dubochet, J., Zuber, B., Eltsov, M., Bouchet-Marquis, C., Al-Amoudi, A., and Livolant, F. (2007). How to "read" a vitreous section. *Methods Cell Biol.* **79**, 385–406.

Engel, A., Eichner, R., and Aebi, U. (1985). Polymorphism of reconstituted human epidermal keratin filaments: Determination of their mass-per-length and width by scanning transmission electron microscopy (STEM). *J. Ultrastruct. Res.* **90**, 323–335.

Fowler, W. E., and Aebi, U. (1983). Preparation of single molecules and supra molecular complexes for high resolution metal shadowing. *J. Ultrastruct. Res.* **83**, 319–334.

Franke, W. W., Schmid, E., Osborn, M., and Weber, K. (1978). Different intermediate-sized filaments distinguished by immune fluorescence microscopy. *Proc. Natl. Acad. Sci. U.S.A.* **75**, 5034–5038.

Fuchs, E., and Hanukoglu, I. (1983). Unraveling the structure of the intermediate filaments. *Cell* **34**, 332–334.

Fuchs, E., and Karakesisoglou, I. (2001). Bridging cytoskeletal intersections. *Genes Dev.* **15**, 1–14.

Goldie, K. N., Wedig, T., Mitra, A. K., Aebi, U., Herrmann, H., and Hoenger, A. (2007). Dissecting the 3-D structure of vimentin intermediate filaments by cryo-electron tomography. *J. Struct. Biol.* **158**, 378–385.

Herrmann, H., and Aebi, U. (2004). Intermediate filaments: molecular structure, assembly mechanism, and integration into functionally distinct intracellular scaffolds. *Annu. Rev. Biochem.* **73**, 749–789.
Herrmann, H., Bär, H., Kreplak, L., Strelkov, S. V., and Aebi, U. (2007). Intermediate filaments: From cell architecture to nanomechanics. *Nat. Rev. Mol. Cell. Biol.* **8**, 562–573.
Herrmann, H., Haner, M., Brettel, M., Ku, N. O., and Aebi, U. (1999). Characterization of distinct early assembly units of different intermediate filament proteins. *J. Mol. Biol.* **286**, 1403–1420.
Herrmann, H., Haner, M., Brettel, M., Muller, S. A., Goldie, K. N., Fedtke, B., Lustig, A., Franke, W. W., and Aebi, U. (1996). Structure and assembly properties of the intermediate filament protein vimentin: The role of its head, rod and tail domains. *J. Mol. Biol.* **264**, 933–953.
Herrmann, H., Strelkov, S. V., Feja, B., Rogers, K. R., Brettel, M., Lustig, A., Häner, M., Parry, D. A., Steinert, P. M., Burkhard, P., and Aebi, U. (2000). The intermediate filament protein consensus motif of helix 2b: Its atomic structure and contribution to assembly. *J. Mol. Biol.* **289**, 817–832.
Hoenger, A., and McIntosh, J. R. (2009). Probing the macromolecular organization of cells by electron tomography. *Curr. Opin. Cell Biol.* **21**, 89–96.
Hofmann, I., Franke, W. W., and Herrmann, H. (1992). Identification of a nonapeptide motif in the vimentin head domain involved in intermediate filament assembly. *J. Mol. Biol.* **223**, 637–650.
Höög, J. L., Schwartz, C., Noon, A. T., O'Toole, E. T., Mastronarde, D. N., McIntosh, J. R., and Antony, C. (2007). Organization of interphase microtubules in fission yeast analyzed by electron tomography. *Dev. Cell.* **12**, 349–361.
Hsieh, C. E., Marko, M., Frank, J., and Mannella, C. A. (2002). Electron tomographic analysis of frozen-hydrated tissue sections. *J. Struct. Biol.* **138**, 63–73.
Jensen, G. J., and Briegel, A. (2007). How electron cryotomography is opening a new window onto prokaryotic ultrastructure. *Curr. Opin. Struct. Biol.* **17**, 260–267.
Kirmse, R., Portet, S., Mücke, N., Aebi, U., Herrmann, H., and Langowski, J. (2007). A quantitative kinetic model for the in vitro assembly of intermediate filaments from tetrameric vimentin. *J. Biol. Chem.* **282**, 18563–18572.
Kremer, J. R., Mastronarde, D. N., and McIntosh, J. R. (1996). Computer visualization of three-dimensional image data using IMOD. *J. Struct. Biol.* **116**, 71–76.
Kreplak, L., Richter, K., Aebi, U., and Herrmann, H. (2008). Electron microscopy of intermediate filaments: Teaming up with atomic force and confocal laser scanning microscopy. *Methods Cell Biol.* **88**, 273–297.
Kürner, J., Frangakis, A., and Baumeister, W. (2005). Cryo-electron tomography reveals the cytoskeletal structure of spiroplasma melliferum. *Science* **307**, 436–438.
Ladinsky, M. S., and Howell, K. E. (2007). Electron tomography of immunolabeled cryosections. *Methods Cell Biol.* **79**, 543–558.
Ladinsky, M. S., Pierson, J. M., and McIntosh, J. R. (2006). Vitreous cryo-sectioning of cells facilitated by a micromanipulator. *J. Microsc.* **224**, 129–134.
Lucic, V., Forster, F., and Baumeister, W. (2005). Structural studies by electron tomography: From cells to molecules. *Annu. Rev. Biochem.* **74**, 833–865.
Ludwig, T., Kirmse, R., Poole, K., and Schwarz, U. S. (2008). Probing cellular microenvironments and tissue remodeling by atomic force microscopy. *Pflugers Arch.* **456**(1), 127–137.
Magin, T. M., Reichelt, J., and Hatzfeld, M. (2004). Emerging functions: Diseases and animal models reshape our view of the cytoskeleton. *Exp. Cell Res.* **301**, 91–102.
Mastronarde, D. N. (2005). Automated electron microscope tomography using robust prediction of specimen movements. *J. Struct. Biol.* **152**, 36–51.
McDowall, A. W., Chang, J. J., Freeman, R., Lepault, J., Walter, C. A., and Dubochet, J. (1983). Electron microscopy of frozen hydrated sections of vitreous ice and vitrified biological samples. *J. Microsc.* **131**, 1–9.
McIntosh, R., Nicastro, D., and Mastronarde, D. N. (2005). New views of cells in 3D: An introduction to electron tomography. *Trends Cell Biol.* **15**, 43–51.
Medalia, O., Weber, I., Frangakis, A., Nicastro, D., Gerisch, G., and Baumeister, W. (2002). Macromolecular architecture in eukaryotic cells visualized by cryoelectron tomography. *Science* **298**, 1209–1213.
Melloy, P., Shen, S., White, E., McIntosh, J. R., and Rose, M. D. (2007). Nuclear fusion during yeast mating occurs by a three-step pathway. *J. Cell Biol.* **179**, 659–670.

Mücke, N., Kirmse, R., Wedig, T., LeterrierKreplak, J. F., and Kreplak, L. (2005). Investigation of the morphology of intermediate filaments adsorbed to different solid supports. *J. Struct. Biol.* **150**, 268–276.
Mücke, N., Klenin, K., Kirmse, R., Bussiek, M., Herrmann, H., Hafner, M., and Langowski, J. (2009). Filamentous biopolymers on surfaces: Atomic force microscopy images compared with brownian dynamics simulation of filament deposition. *PLoS One* **4**, e7756.
Mücke, N., Kreplak, L., Kirmse, R., Wedig, T., Herrmann, H., Aebi, U., and Langowski, J. (2004a). Assessing the flexibility of intermediate filaments by atomic force microscopy. *J. Mol. Biol.* **335**, 1241–1250.
Mücke, N., Wedig, T., Bürer, A., Marekov, L. N., Steinert, P. M., Langowski, J., Aebi, U., and Herrmann, H. (2004b). Molecular and biophysical characterization of assembly-starter units of human vimentin. *J. Mol. Biol.* **340**, 97–114.
Nicastro, D., Schwartz, C., Pierson, J., Gaudette, R., Porter, M. E., and McIntosh, J. R. (2006). The molecular architecture of axonemes revealed by cryoelectron tomography. *Science* **313**, 944–948.
Norlén, L. (2008). Exploring skin structure using cryo-electron microscopy and tomography. *Eur. J. Dermatol.* **18**, 279–284.
Norlén, L., and Al-Amoudi, A. (2004). Stratum corneum keratin structure, function, and formation: The cubic rod-packing and membrane templating model. *J. Invest. Dermatol.* **123**, 15–32.
Norlén, L., Masich, S., Goldie, K., and Hoenger, A. (2007). Structural analysis of vimentin and keratin intermediate filaments by cryo-electron tomography. *Exp. Cell Res.* **313**, 2217–2227.
Norlén, L., Oktem, O., and Skoglund, U. (2009). Molecular cryo-electron tomography of vitreous tissue sections: Current challenges. *J. Microsc.* **235**, 293–307.
Omary, M. B., Coulombe, P. A., and McLean, W. H. (2004). Intermediate filament proteins and their associated diseases. *N. Engl. J. Med.* **351**, 2087–2100.
Omary, M. B., Ku, N.-O., Tao, G.-Z., Toivola, D. M., and Liao, J. (2006). "Heads and tails" of intermediate filament phosphorylation: Multiple sites and functional insights. *Trends Biochem. Sci.* **31**, 383–394.
O'Toole, E. T., Giddings, T. H. Jr., and Dutcher, S. K. (2007). Understanding microtubule organizing centers by comparing mutant and wild-type structures with electron tomography. *Methods Cell Biol.* **79**, 125–143.
Parry, D. A., and Steinert, P. M. (1999). Intermediate filaments: Molecular architecture, assembly, dynamics and polymorphism. *Q. Rev. Biophys.* **32**, 99–18.
Parry, D.A.D., Strelkov, S. V., Burkhard, P., Aebi, U., and Herrmann, H. (2007). Towards a molecular description of intermediate filament structure and assembly. *Exp. Cell Res.* **313**, 2204–2216.
Puchner, E. M., and Gaub, H. E. (2009). Force and function: Probing proteins with AFM-based force spectroscopy. *Curr. Opin. Struct. Biol.* **19**, 605–614.
Robinson, A. L. (1986). Electron microscope inventors share Nobel physics Prize: Ernst Ruska built the first electron microscope in 1931; Gerd Binnig and Heinrich Rohrer developed the scanning tunneling microscope 50 years later. *Science* **23**, 821–822.
Sokolova, A. V., Kreplak, L., Wedig, T., Mücke, N., Svergun, D. I., Herrmann, H., Aebi, U., and Strelkov, S. V. (2006). Monitoring intermediate filament assembly by small-angle X-ray scattering reveals the molecular architecture of assembly intermediates. *Proc. Natl. Acad. Sci.* **103**, 16206–16211.
Steinert, P. M., and Parry, D. A. (1985). Intermediate filaments: Conformity and diversity of expression and structure. *Annu. Rev. Cell Biol.* **1**, 41–65.
Strelkov, S. V., Herrmann, H., Geisler, N., Wedig, T., Zimbelmann, R., Aebi, U., and Burkhard, P. (2002). Conserved segments 1a and 2b of the intermediate filament dimer: Their atomic structures and role in filament assembly. *EMBO J.* **21**, 1255–1266.
Studer, D., Graber, W., Al-Amoudi, A., and Eggli, P. (2001). A new approach for cryofixation by high-pressure freezing. *J. Microsc.* **203**, 285–294.
Taylor, K. A., and Glaeser, R. M. (1976). Electron microscopy of frozen hydrated biological specimens. *J. Ultrastruct. Res.* **55**, 448–456.
Tokuyasu, K. T. (1986). Application of cryoultramicrotomy to immunocytochemistry. *J. Microsc.* **143**, 139–149.
Troncoso, J., Eichner, R., Engel, A., Aebi, U., and Häner, M. (1988). Unifying principles in intermediate filament structure and assembly. *Protoplasma* **145**, 73–78.

Watts, N. R., Jones, L. N., Cheng, N., Wall, J. S., Parry, D. A., and Steven, A. C. (2002). Cryo-electron microscopy of trichocyte (hard alpha-keratin) intermediate filaments reveals a low-density core. *J. Struct. Biol.* **137**, 109–118.

Wickert, U., Mücke, N., Wedig, T., Mueller, S. A., Aebi, U., and Herrmann, H. (2005). Characterization of the in vitro co-assembly process of the intermediate filament proteins vimentin and desmin: Mixed polymers at all stages of assembly. *Eur. J. Cell Biol.* **84**, 379–391.

Zhang, P., Borgnia, M. J., Mooney, P., Shi, D., Pan, M., O'Herron, P., Mao, A., Brogan, D., Milne, J. L., and Subramaniam, S. (2003). Automated image acquisition and processing using a new generation of 4k × 4k CCD cameras for cryo electron microscopic studies of macromolecular assemblies. *J. Struct. Biol.* **143**, 135–144.

Zheng, Q. S., Braunfeld, M. B., Sedat, J. W., and Agard, D. A. (2004). An improved strategy for automated electron microscopic tomography. *J. Struct. Biol.* **147**, 91–101.

CHAPTER 24

Correlative Time-Lapse Imaging and Electron Microscopy to Study Abscission in HeLa Cells

Julien Guizetti[*], Jana Mäntler[†], Thomas Müller-Reichert[‡], *and* Daniel W. Gerlich[*]

[*]Institute of Biochemistry, Swiss Federal Institute of Technology Zurich (ETHZ), CH-8093 Zurich, Switzerland

[†]Max Planck Institute of Molecular Cell Biology and Genetics (MPI-CBG), 01307 Dresden, Germany

[‡]Medical Theoretical Center, TU Dresden, 01307 Dresden, Germany

978-0-12-381007-6
DOI: 10.1016/S0091-679X(10)96024-X

Abstract

HeLa cells are widely used as a model system to study cell division. The last step of cell division, abscission, occurs at an about 1 μm wide intercellular bridge that connects the post-mitotic sister cells. Abscission often occurs long after ingression of the cleavage furrow, and no efficient methods to synchronize cells to this stage are available. Here, we have developed a correlative fluorescence time-lapse imaging and electron microscopic approach using Aclar sheets with engraved grid patterns. This grid pattern, leaving a negative imprint on thin-layer embedded samples, allows identification of cells selected from the time-lapse imaging for serial-section electron microscopy. This method facilitates the ultrastructural analysis of specific stages of abscission.

I. Introduction

Abscission is the essential last step of the cell division process, which irreversibly splits post-mitotic sister cells (reviewed in Steigemann and Gerlich, 2009). Prior to abscission, cells partition most of their cytoplasmic contents by the ingression of a cleavage furrow. This initial phase of cytokinesis leads to the formation of an intercellular bridge containing dense microtubule bundles (reviewed in Glotzer, 2009). Previous electron microscopy studies showed that these microtubule bundles overlap in the central midbody zone in an antiparallel manner (Euteneuer and McIntosh, 1980; Mullins and Biesele, 1977). At the overlap zone, microtubules are coated with a highly electron dense matter, and appear as a characteristic structure termed midbody.

Post-mitotic sister cells remain connected by the intercellular bridge for up to several hours before abscission takes place. The mechanism of abscission is not understood in detail, but it involves the disassembly of midbody-associated microtubules and fission of the plasma membrane. An abscission assay measuring diffusion of cytoplasmic components between the separating cells showed that these two events are tightly correlated (Steigemann *et al.*, 2009). Midbody microtubule bundles disassemble sequentially on both sides of the midbody, likely mediated by the microtubule severing protein Spastin (Connell *et al.*, 2009). Furthermore, a large number of membrane trafficking factors have been found essential for abscission, including endosomal and recycling vesicle proteins, like the small GTPases Rab11, Rab35, or Arf6 (Echard *et al.*, 2004; Kouranti *et al.*, 2006), and both v- and t-SNARES (Gromley *et al.*, 2005; Low *et al.*, 2003). A long list of additional abscission factors have been identified through proteomics of purified midbodies and loss-of-function screening (Echard *et al.*, 2004; Skop *et al.*, 2004).

How cytoskeletal and membrane dynamics coordinately promote cell fission is not known, but different models for the abscission mechanism have been proposed (reviewed in Steigemann and Gerlich, 2009). One model suggests that mechanical traction forces between sister cells could rupture the intercellular bridge once it has been sufficiently destabilized by microtubule remodeling and constriction. The rupture of the intercellular bridge would need subsequent sealing by cellular wound healing. An alternative model proposes that Golgi-derived and endocytic vesicles fill the intercellular bridge adjacent to the midbody to promote final cell splitting. This model proposes that vesicles fuse with each other and with the plasma membrane, but does not explain how this could topologically resolve plasma membranes by fission. A third possible mechanism for abscission involves plasma membrane ingression lateral to the midbody during microtubule disassembly.

Experimental data that would allow to discriminate between these three models is scarce. This is due to the rapid and unsynchronized progression through abscission, which takes only few minutes, compared with the stable intercellular bridge stage that can vary substantially in length. The diameter of the intercellular bridge (in HeLa cells about 1 μm) is too small to resolve all structural dynamics with sufficient detail by live cell imaging using light microscopy.

To overcome these limitations, correlative light and electron microscopy therefore enables to gather information about cellular dynamics obtained from time-resolved light microscopy, as well as ultrastructural detail from electron microscopy. Light microscopy can also provide molecular specificity by labeling proteins of interest with fluorescent marker proteins (reviewed in Tsien, 1998). In combination with electron microscopy (Muller-Reichert *et al.*, 2007), this can reveal details underlying the transient and small-scale process of abscission.

GFP-tagged α-tubulin allows to stage mitosis, cytokinetic furrow ingression, as well as abscission, based on midbody-associated microtubule disassembly (Steigemann *et al.*, 2009). Time-lapse imaging of cells expressing this marker therefore enables to identify intermediate stages of abscission. Here, we present methodology for time-resolved imaging of cells progressing through cytokinesis, followed by ultrastructural analysis of the same cell by transmission electron microscopy. For this approach we use the HeLa "Kyoto" strain of human cervix cancer cells. These cells stably attach to the culture surface and show very limited cellular migration, which makes particularly suitable for long-term live-cell imaging (Steigemann *et al.*, 2009). These cells are also particularly suitable to efficiently generate cell lines stably expressing GFP-tagged markers (Poser *et al.*, 2008; Steigemann *et al.*, 2009).

II. Rationale

The described technique combines time-resolved live imaging of cells during cytokinesis with an ultrastructural analysis of intercellular bridges to dissect the mechanism of abscission. The methods presented here include time-resolved fluorescence light microscopy, chemical fixation, retrieving of live-imaged cells for serial thin sectioning, and ultrastructural analysis by transmission electron microscopy. In the following paragraphs, we provide a step-by-step description of these methods.

III. Methods

A. Cell Culture, Synchronization and Plating of Cells

In this part we describe the preparation of HeLa cells for live-cell imaging suitable for their subsequent identification within the plastic-embedded electron microscopy sample.

HeLa cells stably expressing α-tubulin-EGFP were grown in 75 cm^2 culture flasks with tilted neck. The flasks were filled with 13 ml of DMEM medium and cells were split in a way that cell cultures were around 80% confluent on the day of the experiment.

Aclar® was used as a support film on which cells were grown (Jimenez *et al.*, 2006). Prior to cell seeding, a quadratic grid pattern of 5×5 lines with 1 mm spacing was scratched onto the Aclar surface, using a syringe needle. The Aclar was then cut down to a piece of about 1×1 cm to fit tightly into a chamber of an 8-well Labtek® (Fig. 1A). One corner of the Aclar support was cut off to allow removal of the Aclar sheet once imaging and fixation of the sample was completed, and one square of the grid pattern was marked by a scratch to unambiguously label the side of cell attachment and identify correct orientation. The Aclar was washed in pure ethanol, plasma cleaned (25 mA, negative polarization, 1 min), and placed into a Labtek chamber. Hereafter, the film was coated with poly-L-lysine to further improve the adherence of the HeLa cells to the surface. The diluted poly-L-lysine solution (1:10 in water) was applied for 5 min, then removed and the Aclar sheet was dried under the sterile bench for 30 min. Finally, the Aclar surface was washed twice with PBS. After this treatment, surfaces were ready for the seeding of the synchronized HeLa cells (Fig. 1B–C).

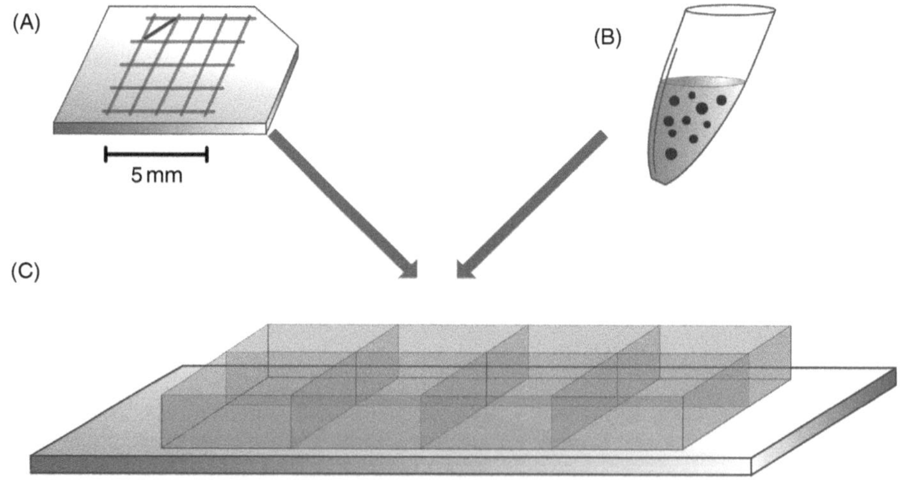

Fig. 1 Preparation of HeLa cells for live-cell imaging. (A) Aclar sheet with scratched grid pattern. (B) HeLa cells after mitotic shake off. (C) Chambered coverslip with eight wells (LabTek®). (See Plate no. 40 in the Color Plate Section.)

To synchronize the cells, we performed a mitotic shake off. This method collects mitotic cells that are less attached to the culture flask, by detaching them from the surface using mechanical shearing force. To achieve this, the flask containing medium and cells were tapped on a flat surface about 5× and then about 20× with the corner of the flask against an edge. Thereafter the detached cells float in the medium. To first remove dead cells and cells that are persistently arrested in mitosis, a primary shake off was performed and discarded. After the primary shake off, fresh and pre-warmed medium was added to the cells. Incubation for 2–3 h allowed enough new cells to enter mitosis. Then, the secondary shake off was carried out to collect a cell population highly enriched in mitosis. The supernatant of the secondary shake off was transferred to a 15 ml Falcon tube. (The flask with the remaining interphase cells could be re-used for subsequent replica experiments i.e. new rounds of shake offs.) The cells were spun down for 3 min at 1000 rpm. After carefully removing the supernatant, leaving the cell pellet intact, the tube was flicked to facilitate the re-suspension of the cells. Such a cell pellet usually contained around $1.5–3 \times 10^5$ cells. The cells were then resuspended in 500 μl of pre-warmed medium and the cell concentration determined by a cell counting chamber. We consider a cell density of $8–10 \times 10^4$ cells per well for an 8-well Labtek chambered coverslip as optimal to be seeded (Fig. 1C).

B. Live-Cell Imaging and Annotation of Selected Cell Pairs

The Labtek chambered coverslip was brought to the light microscope as quickly as possible to capture cells still in anaphase. Our laser scanning confocal microscope, equipped with a humidified incubation chamber providing 5% CO_2 atmosphere, and an automated stage, allows multi-position time-lapse imaging with autofocus (Rabut and Ellenberg, 2004). mEGFP-α-tubulin as well as DIC images, in which the scratched grid was nicely visible, were recorded using a 20× objective 0.8 NA. (Due to the thickness of the Aclar film, we needed a sufficient focal depth to cover the distance between cells and bottom of Labtek chamber.) It is not essential to use a confocal microscope, as we obtained sufficient image quality for staging midbody microtubule disassembly also on a widefield epifluorescence system (with a 20× 0.8 NA short-distance air objective). However, robust autofocusing and reproducible stage positioning are key to the experiment (Conrad and Gerlich, 2010).

In any case, it is essential to first optimize the imaging and incubation conditions such that the cells divide without potential perturbations due to photodamage or inappropriate culture conditions on the microscope. A detailed discussion on how to optimize long-term live imaging conditions can be found in Schmitz and Gerlich (2009).

We first searched for the intersection of the central gridlines of the 5 × 5 grid on the Aclar sheets. At this position the X and Y values of the stage were set to zero and the position was marked by the software. Thereafter 19 additional locations evenly distributed over the Aclar grid were picked (Fig. 2A–B). Screenshots of the selected stage positions to document the X,Y values of all positions were saved for later

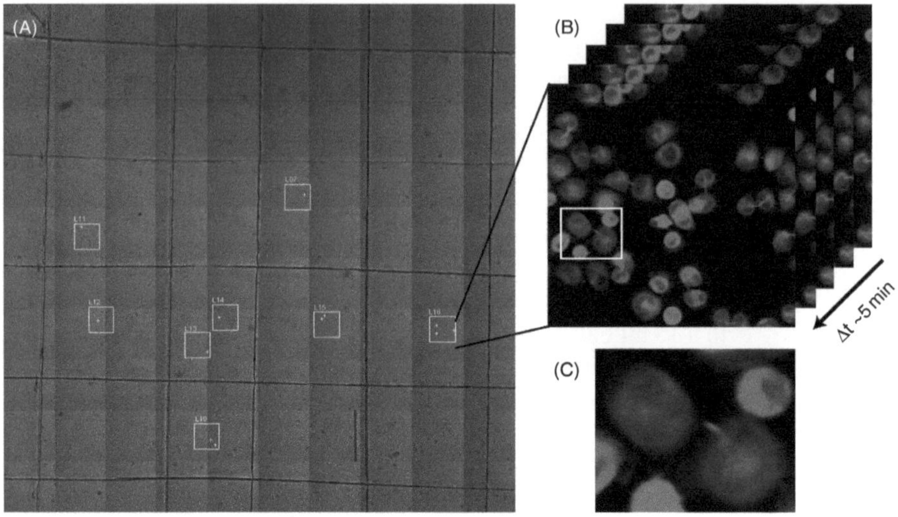

Fig. 2 Live-cell imaging and annotation of cell pairs. (A) Aclar sheet with grid pattern and annotation of selected regions. (B) Live-cell imaging of selected area. (C) Imaging of a selected cell pair after chemical fixation. (See Plate no. 41 in the Color Plate Section.)

analysis. Then 3-D movies of z-stacks of 8 μm thickness in three slices with a time lapse of 5.5 min for a total duration of about 100 min after plating were acquired (Fig. 2B). This time point had been determined beforehand containing the peak of intercellular bridges in the process of microtubule disassembly. The medium in the well was then removed and 600 μl of PBS containing 2.5% glutaraldehyde were added to fix the cells. The chemical fixation was carried out directly in the Labtek within the microscope holder. The cells were fixed for 1.5 h.

After addition of the fixative a tile image (10 × 10 stitched images with a total size of 5 × 5 mm) of the entire field of cells on the grid was acquired (Fig. 2A). This overview picture later served as a map to identify specific cells in the plastic embedded samples. Next, the 20 acquired time-lapse movies were visually inspected to identify those locations containing a cell fixed at stage of interest (Fig. 2B). Using the microscope control software we returned to the selected locations and acquired higher resolution images of the respective cells (Fig. 2C). The cells within the movie images were marked using the annotation tool of the LSM510 software.

Then the movie frames on the tile image were highlighted using ImageJ. Quadratic regions on the tile image were labeled according to the X,Y position values relative to the center of the grid (0,0) (Fig. 2A). However, during the fixation step the Labtek can drift. If necessary we corrected for that drift by determining how many μm the intersection of the central grid lines shifted away from its original position (0,0). This correction value was subtracted from the original X,Y position values using an Excel table. The conversion from μm to pixels was carried out using Excel software.

Furthermore, the position of the selected cell pairs within the movie fields were manually labeled. Taken together, this yielded an overview map of the whole Aclar grid with labeling of all cells of interest (Fig. 2A).

C. Sample Preparation for Electron Microscopy

Next the samples for transmission electron microscopy were prepared. All further steps were carried out at room temperature, within the Labtek chamber using volumes of about 700 μl. Special care was necessary at all steps to avoid preparation artifacts due to drying out.

The glutaraldehyde was removed after 90 min of fixation and the cells were washed 3× with PBS. For contrasting the sample, a 2% OsO_4 solution was added. After 60 min of incubation, the cells were washed again 3× with PBS. At this point the sample could be stored at 4°C overnight. After washing 3× with bi-distilled water we sequentially incubated the cells for 5 min in 35, 50, 70, 85, and 95% ethanol. Finally, we incubated 3× for 5 min in 100% ethanol from which the water was extracted beforehand using a molecular sieve.

For the infiltration steps we used gradually increasing Epon/Araldite concentrations in 100% ethanol: first 20 min in 25% Epon/Araldite, then 20 min in 50% Epon/Araldite, followed by 30 min in 75% Epon/Araldite, and finally 2× 45 min in pure Epon/Araldite.

Thin-layer embedding was accomplished on glass slides as previously described (Muller-Reichert *et al.*, 2003). Briefly, folded pieces of Parafilm were used as spacers. These pieces of Parafilm were placed on Teflon-coated glass slides at such a distance that an Aclar piece of 1 × 1 cm size could be placed on top, just bridging the two spacers. Between the Parafilm spacers we placed a droplet of pure Epon/Araldite, the Aclar piece was removed from the Labtek chamber and placed upside down on the droplet with the edges of the Aclar piece resting upon the spacers (Fig. 3A). Then the resin polymerized for 48 h at 60°C. After resin polymerization, the Aclar sheet was peeled off leaving the cells in the thin layer of resin with the mirror-image grid pattern behind (Fig. 3B).

Using a scalpel, areas containing selected cells were cut out and remounted on "dummy" blocks for ultramicrotomy as previously described (Fig. 3C; Muller-Reichert *et al.*, 2003). Samples were then trimmed for serial sectioning. We collected thin sections (70 nm) on Formvar-coated copper slot grids (Fig. 3D). Sections were post-stained with 2% uranyl acetate in 70% methanol followed by Reynold's lead citrate and imaged by transmission electron microscopy (Fig. 3E).

D. Retrieving Cells in Epoxy Resin, Data Collection, and Analysis

To retrieve cells of interest that were labeled based on the movie data, the pattern of the cells on the section at low magnification was compared with the one in the last frame of the movie. The relative positions and shapes of the cells allowed us to identify post-mitotic sister cells of interest. Then those thin sections containing the midbody of

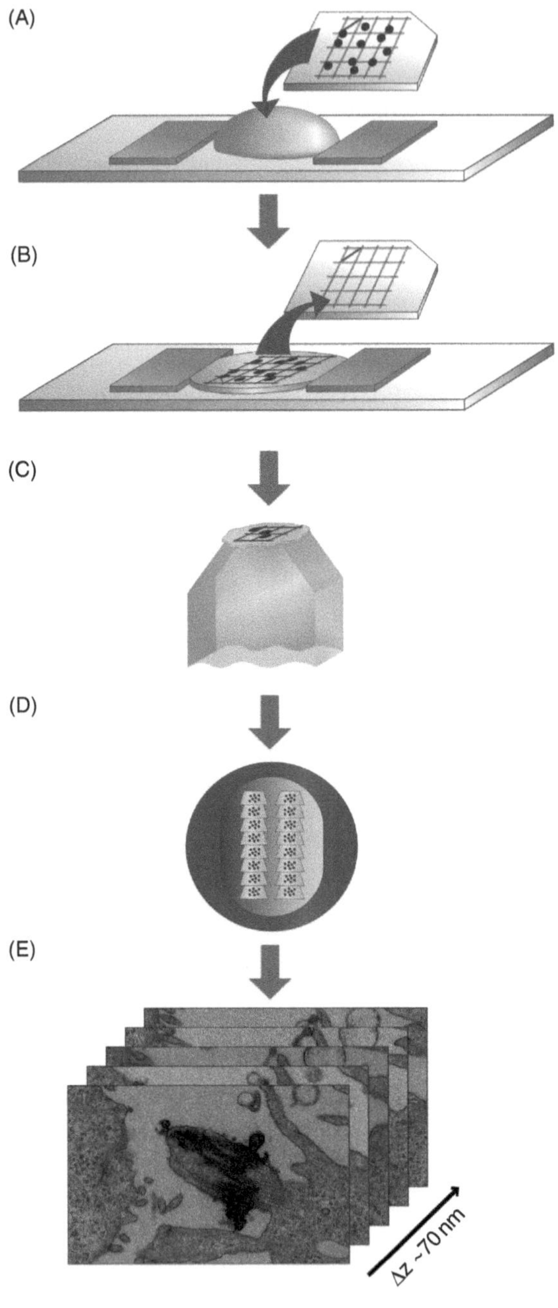

Fig. 3 Processing of samples for transmission electron microscopy. (A) Thin-layer embedding of HeLa cells on glass slide. (B) Removal of Aclar sheet after resin polymerization. (C) Remounting of sample on "dummy" block. (D) Serial sectioning of selected area. (E) Electron microscopy of serial sections. (See Plate no. 42 in the Color Plate Section.)

selected sister cells were identified and series of sections were imaged in the correct order. The electron microscopy images were annotated as such they could be clearly and easily correlated with the light microscopic images (Experiment, Location, Cell, Grid, Section; e.g. E05_L11_C02_G13_01). Eventually, this yielded: (1) a time lapse movie showing progression through telophase, (2) a fixed-cell image showing the tubulin-GFP pattern of selected midbodies at higher resolution, and (3) a stack of electron microscopy images of the intercellular bridge of the same cell. This allowed us to correlate the stage of telophase with the ultrastructural features of the intercellular bridges.

IV. Instrumentation and Materials

A. Cell Culture, Synchronization, and Plating

Instrumentation: Plasma cleaner, centrifuge.

Materials: Aclar® embedding film (Electron Microscopy Sciences, EMS #50425-10), syringe and ruler, HeLa cell line (Kyoto strain) stably expressing tubulin-EGFP, EasyFlasks® (Nunc #156472), cell counter, 8-well Labtek® II (Nunc #154534), falcons, scissors, and tweezers.

Reagents: Poly-L-Lysine (Sigma #P4707), PBS (Gibco #10010-023), ethanol, and GlutaMAX DMEM medium (Gibco #**31966**).

B. Live-Cell Imaging and Annotation of Selected Cell Pairs

Instrumentation: Confocal laser scanning imaging system (Zeiss LSM510Screening) with stable automated multi-position stage and EMBL incubation chamber.

Materials: Labtek® holder (custom made), LSM510 imaging software, Excel, and ImageJ.

Reagents: Glutaraldehyde 2.5% (EMS #16220) in PBS (Gibco #10010).

C. Sample Preparation for Electron Microscopy

Instrumentation: Objective marker with a diamond tip (Leica Microsystems, Vienna, Austria), ultramicrotome (Leica UCT), oven at 60°C for resin polymerization.

Materials: Glass slides, Parafilm®, and "dummy" blocks.

Reagents: Teflon® solution, PBS (Gibco #10010), osmium tetroxide 2% aqueous (EMS #19152), ethanol, molecular sieve (Merck #1.05704), and Epon/Araldite (EMS; EMbed-812 #14900, Araldite-502 #10900, DDSA #13710, DMP-30 #13600)

D. Retrieving Cells in Epoxy Resin, Data Collection and Analysis

Instrumentation: Transmission electron microscope (TECNAI 12, FEI, Eindhoven, The Netherlands; Zeiss 10CA, Oberkochen, Germany).

Materials: LSM510 Image examiner software (Zeiss).

V. Discussion

We have developed a correlative fluorescence time-lapse imaging and transmission electron microscopy approach to study abscission in HeLa cells. This allowed us (1) to record multiple time-lapse movies at selected stage locations, (2) to determine the post-mitotic stage of intercellular bridges, (3) to identify the selected cells after thin-layer embedding in Epon/Araldite, and (4) to analyze the structure of selected intercellular bridges in serial thin sections by electron microscopy.

The protocol presented here is easily extendable to other experimental conditions, for example RNAi depletion of candidate abscission regulators, or their inhibition by chemical compounds. Loss-of-function studies with the methodology presented here will reveal phenotypes with an unprecedented level of detail. Using cells expressing other fluorescent markers will enable to dissect the exact localization of specific components, for example secretory vesicles or recycling endosomes. Moreover, immuno-electron microscopy could be applied to precisely analyze the location of specific gene products. This, however, would require a different fixation protocol without the use of osmium tetroxide, infiltration with a different resin, such as LR White, and a modified protocol for thin-layer embedding (Muller-Reichert *et al.*, 2003).

As an alternative to chemical fixation, high-pressure freezing/freeze-substitution can be used for sample preparation. This would require to culture mitotic shake off cells on sapphire discs for cryoimmobilization. However, sapphire discs for high-pressure freezing are much smaller than Aclar sheets used here for chemical fixation. Therefore, the number of cells per sapphire disc that can be analyzed may be considerably smaller. In addition, it is not possible to scratch a grid pattern onto their surface. To facilitate the identification of selected cells on sapphire discs, however, an interesting approach has been developed recently (Chapter 28 by McDonald *et al.*, this volume). By placing a finder grid onto a sapphire disc prior to carbon evaporation, it is possible to obtain a pattern that is sufficient to mark the position of specific cell pairs. This way, a similar correlative microscopy approach as presented for chemically fixed cells should be possible.

High-pressure freezing would be particularly suitable for 3-D analysis by electron tomography. Cryoimmobilization within milliseconds and a subsequent fixation and dehydration during freeze-substitution at deep temperatures preserves details like the spacing and structure of individual microtubules and their ends.

The correlative fluorescence time-lapse and electron microscopic approach described here can be applied to all cell types that attach efficiently to Aclar, and it is possible to use it for the study of other biological processes that require a combined readout of cellular dynamics and ultrastructural detail. We anticipate that correlative live and electron microscopy will be key to uncover the mechanisms underlying abscission, as well as other biological processes that occur transiently and on small scales.

Acknowledgments

The authors would like to thank Louis Kerr (MBL Woods Hole) for providing access to the transmission electron microscope and Michael H. A. Schmitz (ETHZ) for generating the mEGFP-α-tubulin/H2B-mCherry expressing HeLa "Kyoto" cell line. The original HeLa "Kyoto" cell line was obtained from S. Narumiya, Kyoto University. The authors would also like to thank the staff of the LMC Facility (ETHZ) for support and Dr. Roberta Mancini (ETHZ) for providing reagents. DWG was funded by Marine Biological Laboratory (MBL) summer research fellowships of the Evelyn and Melvin Spiegel Fellowship Fund. TMR was supported by the MBL Summer Research Fellowship (2007) and Research Award (2009).

References

Connell, J. W., Lindon, C., Luzio, J. P., and Reid, E. (2009). Spastin couples microtubule severing to membrane traffic in completion of cytokinesis and secretion. *Traffic* **10**(1), 42–56.

Conrad, C., and Gerlich, D. W. (2010). Automated microscopy for high-content RNAi screening. *J. Cell Biol.* **188**(4), 453–461.

Echard, A., Hickson, G. R., Foley, E., and O'Farrell, P. H. (2004). Terminal cytokinesis events uncovered after an RNAi screen. *Curr. Biol.* **14**(18), 1685–1693.

Euteneuer, U., and McIntosh, J. R. (1980). Polarity of midbody and phragmoplast microtubules. *J. Cell Biol.* **87**(2 Pt 1), 509–515.

Glotzer, M. (2009). The 3Ms of central spindle assembly: Microtubules, motors and MAPs. *Nat. Rev. Mol. Cell. Biol.* **10**(1), 9–20.

Gromley, A., Yeaman, C., Rosa, J., Redick, S., Chen, C. T., Mirabelle, S., Guha, M., Sillibourne, J., and Doxsey, S. J. (2005). Centriolin anchoring of exocyst and SNARE complexes at the midbody is required for secretory-vesicle-mediated abscission. *Cell* **123**(1), 75–87.

Jimenez, N., Humbel, B. M., van Donselaar, E., Verkleij, A. J., and Burger, K. N. (2006). Aclar discs: a versatile substrate for routine high-pressure freezing of mammalian cell monolayers. *J. Microsc.* **221**(Pt 3), 216–223.

Kouranti, I., Sachse, M., Arouche, N., Goud, B., and Echard, A. (2006). Rab35 regulates an endocytic recycling pathway essential for the terminal steps of cytokinesis. *Curr. Biol.* **16**(17), 1719–1725.

Low, S. H., Li, X., Miura, M., Kudo, N., Quinones, B., and Weimbs, T. (2003). Syntaxin 2 and endobrevin are required for the terminal step of cytokinesis in mammalian cells. *Dev. Cell* **4**(5), 753–759.

Muller-Reichert, T., Hohenberg, H., O'Toole, E. T., and McDonald, K. (2003). Cryoimmobilization and three-dimensional visualization of C. Elegans ultrastructure. *J. Microsc.* **212**(Pt 1), 71–80.

Muller-Reichert, T., Srayko, M., Hyman, A., O'Toole, E. T., and McDonald, K. (2007). Correlative light and electron microscopy of early caenorhabditis elegans embryos in mitosis. *Methods Cell Biol.* **79**, 101–119.

Mullins, J. M., and Biesele, J. J. (1977). Terminal phase of cytokinesis in D-98s cells. *J. Cell Biol.* **73**(3), 672–684.

Poser, I., Sarov, M., Hutchins, J. R., Heriche, J. K., Toyoda, Y., Pozniakovsky, A., Weigl, D., Nitzsche, A., Hegemann, B., Bird, A. W., Pelletier, L., Kittler, R., *et al.* (2008). BAC TransgeneOmics: A high-throughput method for exploration of protein function in mammals. *Nat. Methods* **5**(5), 409–415.

Rabut, G., and Ellenberg, J. (2004). Automatic real-time three-dimensional cell tracking by fluorescence microscopy. *J. Microsc.* **216**(Pt 2), 131–137.

Schmitz, M. H., and Gerlich, D. W. (2009). Automated live microscopy to study mitotic gene function in fluorescent reporter cell lines. *Methods Mol. Biol.* **545**, 113–134.

Skop, A. R., Liu, H., Yates, J., Meyer, B. J., and Heald, R. (2004). Dissection of the mammalian midbody proteome reveals conserved cytokinesis mechanisms. *Science* **305**(5680), 61–66.

Steigemann, P., and Gerlich, D. W. (2009). Cytokinetic abscission: Cellular dynamics at the midbody. *Trends Cell Biol.* **19**(11), 606–616.

Steigemann, P., Wurzenberger, C., Schmitz, M. H., Held, M., Guizetti, J., Maar, S., and Gerlich, D. W. (2009). Aurora B-mediated abscission checkpoint protects against tetraploidization. *Cell* **136**(3), 473–484.

Tsien, R. Y. (1998). The green fluorescent protein. *Annu. Rev. Biochem.* **67**, 509–544.

CHAPTER 25

Viral Infection of Cells in Culture—Approaches for Electron Microscopy

Paul Walther*, **Li Wang**[†], **Sandra Ließem**[‡], ***and*** **Giada Frascaroli**[†]

*Central Electron Microscopy Facility, Ulm University, Ulm D-89081, Germany

[†]Institute of Virology, University Hospital Ulm, Ulm D-89081, Germany

[‡]Institute of Pathology, University Hospital Hannover, Hannover D-30625, Germany

Abstract

In this article we present and discuss three electron microscopical approaches to investigate the pathway of human cytomegalovirus (Herpesviridae) infection in cultivated human macrophages and fibroblasts. These methods include surface scanning

978-0-12-381007-6
DOI: 10.1016/S0091-679X(10)96025-1

electron microscopy and high-pressure freezing combined with freeze-substitution and electron tomography. With this set of methods, basically all steps of the viral cycle can be investigated and documented. These methods are especially helpful to investigate membrane structures, since viral particles of a diameter of 150 nm fit well into a semi-thin section of about 500 nm that can be imaged in three dimensions with STEM tomography.

I. Introduction

We investigated entry and egress of cytomegalovirus (CMV) in human macrophages and fibroblasts cultivated *in vitro*. CMV is a DNA virus of the Herpesviridae family coding for approximately 200 open reading frames (Murphy *et al.*, 2003). The virion consists of an icosahedral capsid surrounded by the tegument protein layer and lipid envelope containing viral glycoproteins. The steps of entry and egress of the viral cycle of CMV are schematically outlined in Fig. 1 (throughout this chapter we refer to stages 1–11 as outlined). The viral particles first attach to the cell surface (stage 1, Lehmann *et al.*, 2005). It is proposed that they then enter the cell by membrane fusion (stage 2) leading to a naked capsid in the cytosol (stage 3) that travels and associates to the nuclear pore where the DNA is released into the nucleus (stage 4). The second described entering mechanism is endocytosis (stage 5), most likely macropinocytosis, resulting in viral particles retained in a vesicle (stage 6). The capsids can possibly escape these vesicles by fusion of the viral membrane with the vesicle (stage 7). In the nuclei of infected cells, mature C-capsids are formed by loading the viral DNA (vDNA) into B-capsids (Church and Wilson, 1997; Newcomb *et al.*, 1994; Sheaffer *et al.*, 2001). The C-capsids then exit the nucleus by a primary envelopment at infoldings of the inner nuclear membrane (stage 8, Buser *et al.*, 2007). According to this generally accepted model, the primary membrane envelope fuses with the outer nuclear membrane and naked capsids are released into the cytosol (stage 9, Mettenleiter, 2002). These capsids acquire tegument proteins and a secondary envelope by budding into cytoplasmic membranes (stage 10) and are then secreted by exocytosis (stage 11).

During the past two decades, two approaches for sample preparation and imaging by electron microscopy have emerged. The first is direct imaging of the biological structures using the contrast formed by the low-mass biological structures (e.g., Dubochet *et al.*, 2007; see also Chapter 2 by Jensen *et al.*, this volume). This method requires imaging of the sample in a frozen-hydrated state, where contrast is low and beam sensitivity is high. Searching for rare events, such as a viral particle entering a cell, is, therefore, difficult. The second method, which was used for this study, is to enhance contrast and reduce beam sensitivity by staining the sample structures with heavy metals and fixing the sample and replacing the water with an embedding medium such as Epon. These samples can stand the electron beam without obvious beam damage for much longer time periods and the operator has the possibility to search for rare events. The obvious limitation of this approach is that heavy

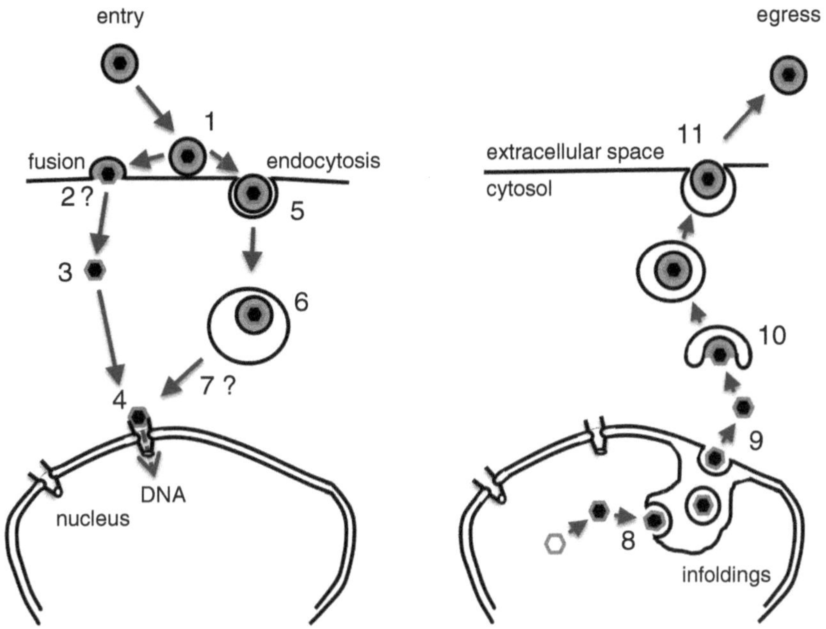

Fig. 1 Schematic diagram of entry and egress of cytomegalovirus (Herpesviridae) as observed in the electron microscope. The viral particles touch the cell surface (1) and can enter the cell via two pathways: The first is membrane fusion (2) leading to a naked capsid in the cytosol (3) that travels and associates to the nuclear pore, where the DNA is released into the nucleus (4). The second described entering mechanism is endocytosis (5), most likely macropinocytosis, resulting in viral particles retained in a vesicle (6). The capsids can possibly escape these vesicles by fusion of the viral membrane with the vesicle (7). The fusion steps, labelled with question marks, are difficult (in our hands actually impossible) to observe in the electron microscope, possibly because the events are rare and fast and, therefore, difficult to capture. Egress starts with budding of the DNA-filled capsids (primary envelopment), mainly into infoldings of the inner nuclear membrane (8, Buser *et al.*, 2007). According to the most accepted model, the membrane of the particle fuses with the outer nuclear membrane (9), leaving a naked viral particle to the cytosol. There it acquires tegument proteins and buds into cytoplasmic membranes (10; secondary envelopment). The result is a viral particle in a vesicle and it leaves the cell by exocytosis (11). Modified after *J. Virol.* **76,** 1537–1547. (See Plate no. 43 in the Color Plate Section.)

metals can label the biological structures only with a limited accuracy and one can never take full advantage of the resolution power of an electron microscope (in the range of the size of atoms) when applying this second method.

II. Rationale

The purpose of this article is to study the events of viral egress and infection in the model system of cultivated cells. This can be done with a variety of different preparation methods for electron microscopy. Electron microscopy is a very powerful tool to investigate viral infections of cultivated cells. Depending on the method of sample preparation it allows investigation of cellular events at a macromolecular level. Electron microscopy, therefore, can bridge the gap between light microscopical methods, especially fluorescence microscopy with subcellular resolution, and biochemical and genetic methods. Only the electron microscope is able to visualize and thereby exactly localize membrane fusion and fission events during viral entry and egress (see Mettenleiter, 2002). With biochemical or genetical approaches, an exact localization of the observed processes is not possible. Electron microscopy is, therefore, an indispensable tool for investigation of viral infection in cells.

III. Methods

As mentioned in Section I, this article focuses on indirect imaging methods, where samples are heavy metal labelled and dehydrated. The classical method to do this is chemical fixation, resin embedding and thin sectioning (see Mettenleiter, 2002). In recent years a number of new developments in biological electron microscopy have been introduced:

- high-resolution scanning electron microscopy (SEM) suitable to investigate attachment of virions to the cell surface (Walther, 2008),
- high-pressure freezing allowing a rapid arrest of cellular processes thus preserving a momentary situation of a dynamic event, similar to a flashlight photo (McDonald *et al.*, 2007),
- freeze-substitution serving to dehydrate the samples at cold temperatures with accurate preservation of membranes and macromolecular compounds (Buser and Walther, 2008), and
- electron tomography allowing for three-dimensional reconstruction of subcellular structures (Yakushevska *et al.*, 2007).

These new developments have been applied and combined in this study and will be presented in the following section.

A. Imaging the Surface of Infected Cells by SEM

The distribution of viral particles on the cell surface can give information about the uptake and release mechanisms of a virus. Modern field emission SEMs provide a resolution comparable to a conventional transmission electron microscope (TEM), with the advantage that bulk samples can be investigated.

1. Preparation Protocol

Human macrophages infected with human CMV were chemically fixed with 2.5% glutaraldehyde (in PBS and 1% saccharose) for 1 h at room temperature.

After washing with PBS, the cells were post-fixed with OsO_4 (2% in PBS) for 1 h at room temperature. After washing with PBS, the samples were gradually dehydrated in 30, 50, 70, 90 and 100% propanol (5 min at each step). The cells were critical point dried using carbon dioxide as translation medium (Critical Point Dryer CPD 030, Bal-Tec, Principality of Liechtenstein). The samples were double-layer coated as described in a BAF 300 freeze etching device (Bal-Tec, Principality of Liechtenstein) by electron beam evaporation with 3 nm of platinum–carbon from an angle of 45° (unidirectional) and with about 6 nm of carbon perpendicularly (90° rotary shadowing) in order to enhance electrical conductivity and mechanical strength (Walther, 2008; Walther and Hentschel, 1989). Samples were then investigated in a Hitachi S-5200 in-lens field emission SEM at an accelerating voltage of 10 kV. We used the backscattered electron image in order to visualize the underlying platinum layer that is in close contact to the biological structure of interest (Walther, 2008).

2. Results

A macrophage 30 min post-infection with human CMV is shown in Fig. 2. Many viral particles are attached to the cell surface. Identification of viral particles, however, is not always unambiguous, since dense bodies, small cell protrusions such as microvilli or lipid vesicles might look very similar. The double-layer coating is, thereby, helpful since the spherical nature of the particle becomes evident, when looking at the round shadow formed by each particle. We plan to solve this problem in future experiments by either immunogold labelling the viral particles or correlative microscopy with GFP-labelled viral particles. Using such prepared samples, we hope to gain some information of the uptake mechanism by investigating the distribution of the particles on the cell surface. It is challenging to capture the very moment of entry of a viral particle into the cell, which occurs either by membrane fusion or by endocytosis. We have been unable, so far, to clearly identify these events by SEM. This is possible since these events are relatively rare and fast, and, therefore, difficult to trap.

B. High-Pressure Freezing of Infected Cells

Cryo-fixation has been established as a method for fast immobilization of cells and small tissue pieces within a few milliseconds. Thereby, dynamic physiological processes, such as viral infection and egress, can be immobilized at different stages. High-pressure freezing (Moor and Riehle, 1968) enables us to freeze native samples up to a thickness of about 200 μm without the damaging formation of ice crystals. It has been demonstrated that high-pressure freezing is a very suitable method for the study of viral egress in host cells (Buser *et al.*, 2007; Chevillotte *et al.*, 2009; Rupp *et al.*, 2007; Wild *et al.*, 2002). Importantly, ultrarapid immobilization makes it possible to detect intermediate steps of viral egress with short lifetimes.

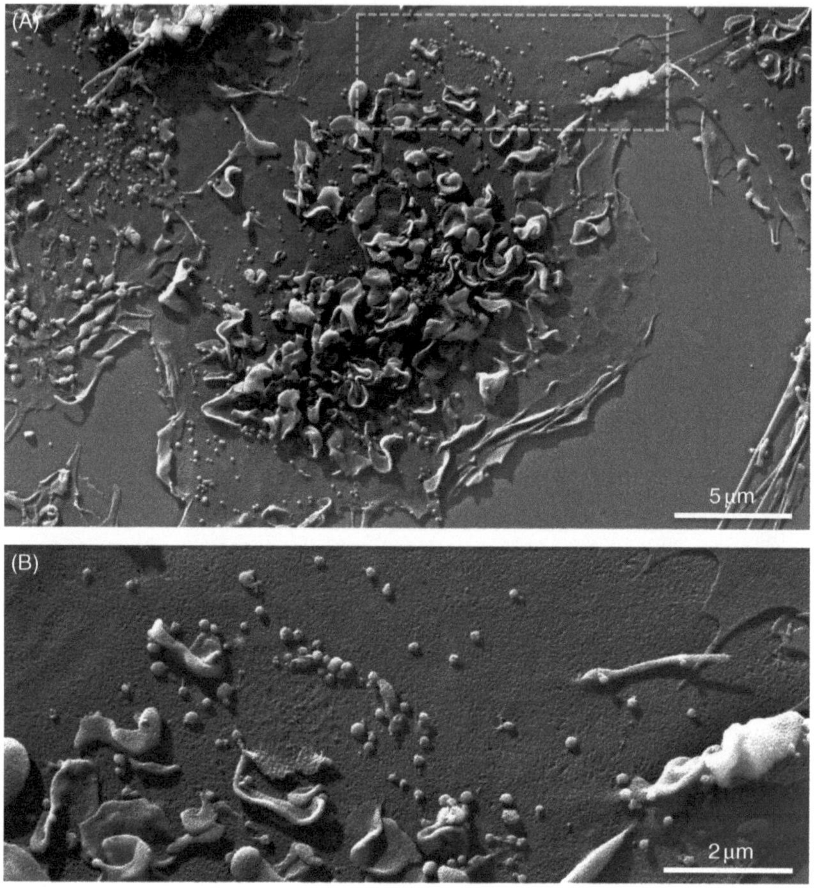

Fig. 2 Surface of a macrophage infected with cytomegalovirus 30 min post-infection and prepared by conventional methods. (A) The cell surface consists of numerous surface extensions and some viral particles are attached to the plasma membrane. The unidirectional coating helps to judge the shape of the particles. (B) An area of the cell (indicated in A) at higher magnification showing round structures of 150 nm diameter, that most likely represent viral particles (stage 1 of Fig. 1). These particles are not observed on the sapphire disc between the cells.

Sapphire discs with a diameter of 3 mm and a thickness of 0.05 mm were introduced as supporting material for cells in culture (Eppenberger-Eberhardt *et al.*, 1997; see also Chapter 24 by Guizetti *et al.* and Chapters 15 and 28 by McDonald *et al.*, this volume). These discs are well suited for light microscope investigation prior to immobilization by high-pressure freezing. We adapted this method in our laboratory. For this purpose, a high-pressure freezer (the Compact HPF 01) has, therefore, been placed in an S2 virological laboratory. This allows us to freeze cells infected with human CMV without pre-fixation or any other pre-treatment.

1. Preparation Protocol

The sapphire discs were cleaned by sonication for 15 min each in 60% sulphuric acid, soap water and twice in absolute ethanol, and allowed to dry. The discs were then coated with approximately 20 nm carbon by electron beam evaporation in a BAF 300 and dried for 8 h or overnight in an oven at 120°C to increase the stability of the carbon film. The coated sapphire discs were then carefully immersed in complete medium and the cells were seeded on top. After the specified amount of time post-infection the samples were high-pressure frozen with an HPM01. Two aluminium planchettes (3 mm diameter; Engineering Office M. Wohlwend, Switzerland) were immersed in hexadecene (i.e., 1-hexadecene), the lower one being flat, while the upper one had a cavity of 100 µm. The lower planchette was inserted into the sample holder, then the sapphire disc with the adherent cells was removed from the culture dish (usually a 24-well plate) with tweezers, quickly deposited on a filter paper to remove excess medium, immersed in hexadecene and placed in the sample holder on top of the flat side of the planchette. The "sandwich" construction was then completed with the upper planchette with the 100-µm cavity pointing down toward the cells and high-pressure frozen in a Compact HPF 01 (Engineering Office M. Wohlwend GmbH, Sennwald, Switzerland). Since hexadecene is not miscible with water it does not serve as a cryoprotectant. Instead it functions in proper transfer pressure and heat transfer during the freezing process by providing a good contact between the planchettes and the sample. The entire sample loading/mounting process takes approximately 10–20 s for a trained person. After the freezing process the samples are quickly transferred to the cryo-box filled with liquid nitrogen where the samples can be stored until freeze-substitution (Buser *et al.*, 2007).

C. Freeze-Substitution and Conventional TEM

Freeze-substitution is a method to gently dehydrate the high-pressure frozen samples by using an organic solvent at a temperature of –90°C. At this temperature the organic solvent starts to dissolve the ice and the sample is slowly warmed up afterwards, so that the added fixatives, such as osmium tetroxide and uranyl acetate can react and fix the sample (Schwarz *et al.*, 1993). The fully dehydrated samples are then embedded into Epon/Araldite or into methacrylates, such as Lowicryl or LR-Gold (reviewed by Müller and Moor, 1984). These plastic embedded samples are cut and analyzed in a TEM at ambient temperatures. It has been demonstrated in numerous studies that freeze-substitution leads to a more life-like structural preservation than conventional fixation and dehydration at room temperature (Ebersold *et al.*, 1981; McDonald *et al.*, 2007; Szczesny *et al.*, 1996).

Our previous observation that membrane visibility is greatly affected by the freeze-substitution protocol (Walther and Ziegler, 2002) has been used as a base for further investigation of contrast formation during freeze-substitution. The protocols have been improved and adapted to the cell monolayer—CMV system (Buser and Walther, 2008). As a major point we demonstrated that the visibility of membranes can be improved by adding water to the substitution medium.

1. Preparation Protocol

The high-pressure frozen samples were carefully removed from the aluminium planchettes in liquid nitrogen and immersed in 1 ml substitution medium re-cooled to –90°C. The freeze-substitution medium contained 0.2% osmium tetroxide, 0.1% uranyl acetate and 5% water in acetone. We used a specially designed home-made computer-controlled substitution apparatus (A. Ziegler and W. Fritz, Central Electron Microscopy Unit, University of Ulm, unpublished) to slowly warm the samples from –90 to 0°C over a period of 16 h in an exponential warming protocol (Walther and Ziegler, 2002). The samples were kept at 0°C for 1 h and at room temperature for another hour in order to increase sample contrast. Samples were washed with water-free acetone or propanol and embedded in a two-step Epon series: 50% Epon in acetone or propanol for 1 h, then 100% Epon for 6 h. The Epon was then polymerized for 48 h at 60°C. Ultrathin sections (60 nm) were cut using a Leica Ultracut UCT ultramicrotome and a diamond knife (Diatome, Switzerland) and collected on bare copper grids. The ultrathin sections were imaged with a Zeiss EM 10, or a Philips 400 transmission electron microscope, at an acceleration voltage of 80 kV. The images were recorded on Agfa negative film, the negatives were scanned and the resulting digital images processed with Adobe Photoshop.

The following preparation protocol was used to prepare the freeze-substitution medium:

1. Dissolve osmium tetroxide in water to obtain a 4% solution (w/v) at room temperature.
2. Dissolve 6 mg of uranyl acetate in 300 μl of this 4% osmium tetroxide solution at room temperature.
3. Add 5.7 ml of acetone to this solution at room temperature, fill the solution into Eppendorf cups (1 ml each) and put them to the pre-cooled freeze-substitution device (i.e., –90°C or colder).
4. Wait to add the samples until the temperature is at least –90°C.

2. Results of High-Pressure Frozen and Freeze-Substituted Samples Imaged by Conventional TEM

Fig. 3 shows a macrophage infected with CMV. The sample has been prepared by high-pressure freezing and freeze-substitution and imaged in a conventional TEM. Freeze-substitution particularly retains the membranes that play a key role during the viral pathway. In Fig. 3A, three different stages of virus entry can be observed (stages 1, 3 and 6). Many viral particles sit on the cell surface (arrows; stage 1). In contrast to surface scanning microscopy, the dark viral core can be identified in the thin section, and the viral particles (arrows) can be unambiguously discerned from dense bodies or other vesicles on the cell surface. Two viral particles are seen within a vesicle (arrowheads; stage 6). It is difficult to identify viral particles or capsids that enter the cell, since they are small compared to the size of the cell and it is rather unlikely to trap them in an ultrathin section. Fig. 3B shows another viral particle in a vesicle,

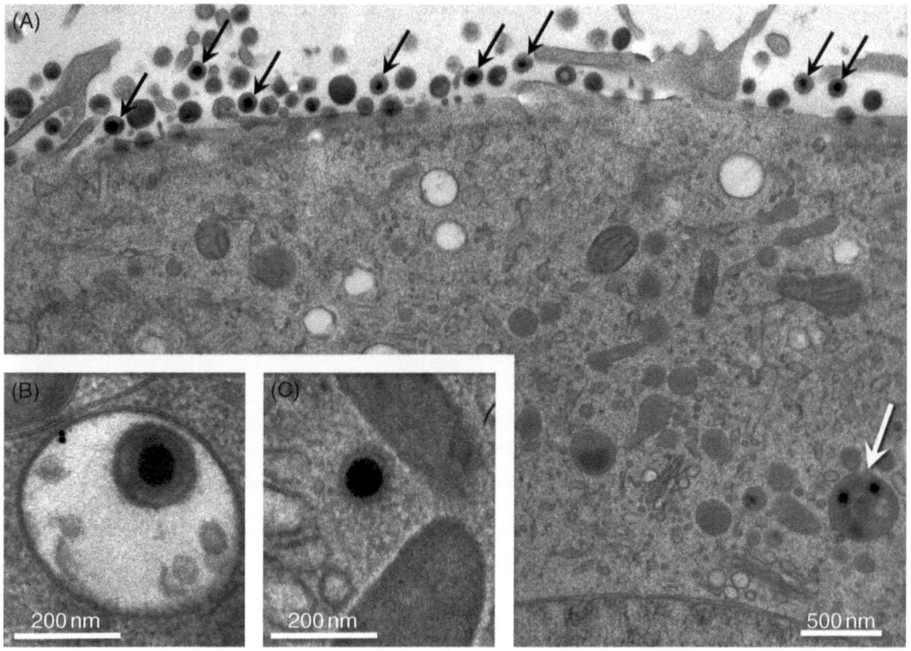

Fig. 3 Macrophages infected with cytomegalovirus. The sample has been prepared by high-pressure freezing and freeze-substitution and was imaged in a conventional TEM. This method especially retains the membranes that play a key role during the viral life cycle. (A) Two stages of viral entry can be observed. Many viral particles sit on the cell surface (black arrows; stage 1). Two viral particles stay within a vesicle (white arrow; stage 6). (B) Image representing another viral particle in a vesicle, which, according to stage 6 is the result of an endocytotic event. (C) A naked capsid, which, according stage 3 must be the result of a fusion event. Such a fusion event has not been visualized yet (stage 2), most likely because this stage is short lived.

representing an endocytotic event. A naked capsid, most likely the result of a membrane fusion event (stage 3), taking place either at the cell surface or within the cell through vesicle fusion with the viral membrane (Wang *et al.*, 2009) is shown in Fig. 3C. Fig. 4 exhibits another situation at the surface of a macrophage. The large infolding possibly represents an endocytotic event. Filamentous structures (arrows) may represent F-actin, a marker for macropinocytosis (similar structures can also be observed in supplemental material, movie 1). It is, however, difficult to discern F-actin from intermediate filaments (Sailer *et al.*, 2010).

D. STEM Tomography

Electron tomography allows full three-dimensional data acquisition of, e.g., a semi-thin section (about 500 nm thick). Practically, this is realized by tilting the semi-thin section in the TEM over a range of about 140° and acquiring about 100

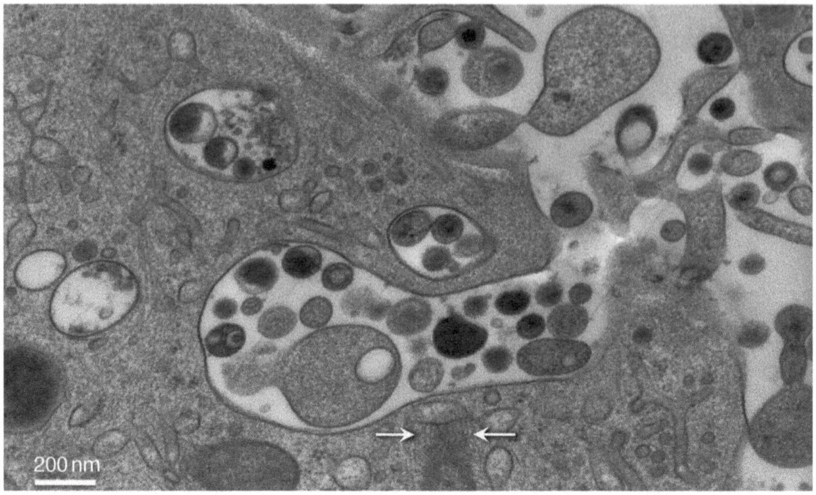

Fig. 4 Macrophage infected with human cytomegalovirus. The inclusion may represent an endocytotic event. Filamentous structures (arrows) may represent F-actin, a marker for macropinocytosis.

images at different angles. From this large amount of data the tomographic model is calculated by backprojection (reviewed by Baumeister, 2005; see also Chapter 2 by Jensen, Chapter 4 by O'Toole and Chapter 6 by Giddings *et al.*, this volume).

It results that electron tomography of high-pressure frozen samples is a suitable method to investigate the dynamic events of viral envelopment and the related cell biological effects. For this project we found that *scanning* transmission electron microscopy (STEM; Yakushevska *et al.*, 2007) is especially useful since it enables tomographic reconstruction of relatively thick sections of up to 600 nm, since in contrast to conventional TEM the STEM image formation is not limited by inelastic scattering but by widening of the primary beam (Höhn *et al.*, 2009; see also Chapter 5 by Hanssen *et al.*, this volume). Experiments with sections up to 1 μm thick have been basically successful; however, the resolution decreased remarkably due to widening of the primary beam caused by scattering events in the sample (data not shown). So, using sections thicker than 600 nm works well at low magnification; at higher magnifications, however, it is most likely not a useful method.

1. Preparation Protocol

For STEM tomography, samples were high-pressure frozen, freeze-substituted and embedded in Epon as described above. Semi-thin sections with a thickness of 400–600 nm were cut with a Leica Ultracut UCT ultramicrotome using a diamond knife (Diatome, Switzerland). Samples were collected on bare copper grids with grid bars in one direction to prevent the grid bars from hiding the biological structures at high tilt angles. For tomography, it is essential that the sections are mounted as flat as possible.

It turned out to be rather difficult to attach the relatively thick sections on the grid bars. Therefore, the copper grids were coated with poly-L lysine (10% in water) before the sections were attached. The grids were then warmed on a heating table to a temperature of 60°C to flatten the sections. Afterwards, the sections were again coated with poly-L-lysine to attach the 15 nm colloidal gold particles (Aurion, The Netherlands) on both sample sides. These gold particles serve as fiducial markers for the calculation of the tomograms. Finally, the mounted sections were carefully coated with 5 nm carbon from both sides by electron beam evaporation to increase electrical conductivity and mechanical stability. Improving electrical conductivity helps to reduce mass loss caused by ionization due to inelastic scattering (Walther *et al.*, 1995). Also, drift is reduced when electrical conductivity is reduced by carbon coating. The tilt series (–72° to +72°) were recorded with a 300 kV field emission STEM (Titan 80-300 TEM, FEI, Eindhoven) with a high-angle annular dark field detector (Fischione, Export, PA, USA). Tomograms were reconstructed by weighted backprojection using the IMOD software package (Kremer *et al.*, 1996; Mastronarde, 2008). The Amira software was used for segmentation and data display (Höhn *et al.*, 2009).

2. Results of High-Pressure Freezing, Freeze-Substitution and STEM Tomography

Different stages of secondary envelopment of viral particles in the cytoplasm (stage 10) are shown in Fig. 5.) The sample is prepared as described above. Fig. 5A is a computer-generated section through the reconstructed tomogram (i.e., a tomographic slice) and Fig. 5B is a 3-D model of this dataset. Membranes occur with high contrast and the two membrane leaflets can be well discerned. This technique allows observation of the different stages of envelopment. Capsids (red) with attached tegument (yellow) that are not yet enveloped with a membrane (1a–c) can be seen in Fig. 5B, but not in Fig. 5A, since they are not in the section plane. An envelope that is almost closed (2), two capsids (3) that share an envelope and a vesicle membrane are visible in both images. Some virions have completed their secondary envelopment and sit in vesicles (4a–c) as a consequence of the budding event.

IV. Materials

A. Imaging the Surface of Infected Cells by SEM

Instrumentation: Critical point dryer (Bal-Tec, Principality of Liechtenstein; actual distributor: Leica Scientific Instruments, Vienna, Austria), Balzers BAF 300 freeze etching device (Bal-Tec, Principality of Liechtenstein; actual distributor: Leica Scientific Instruments, Vienna, Austria), Hitachi S-5200 in-lens field emission scanning electron microscope (Hitachi, Tokyo, Japan)

Materials: Specimen stubs (Hitachi, Tokyo, Japan)

Reagents: Osmium tetroxide (ChemPur, Karlsruhe, Germany), glutaraldehyde (Merck KG, Darmstadt, Germany)

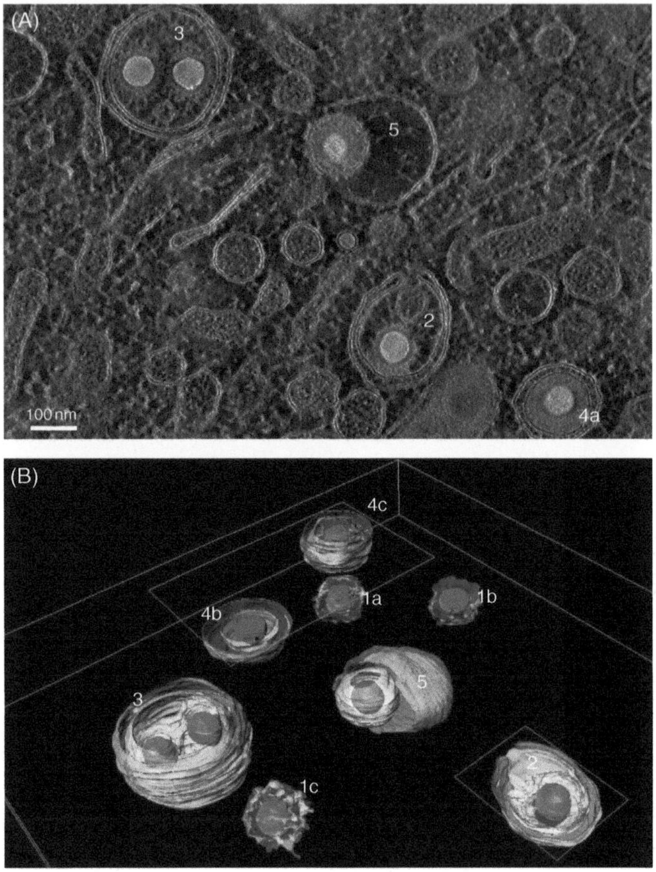

Fig. 5 Different stages of secondary envelopment in the cytoplasm (according to Fig. 1, stage 10). The sample was high-pressure frozen, freeze-substituted and the tilt series for tomographic reconstruction was taken in a 300 kV STEM using the dark field mode. (A) Tomographic slice. Membranes occur with high contrast and the two membrane leaflets can be well discerned. (B) Three-dimensional model of the same tomogram. Modelling allows observation of the different stages of envelopment. Capsids (red) with attached tegument (yellow) that are not yet enveloped with a membrane (1a–c; not seen in the tomographic slice in (A). An envelope that is almost closed (2) and two capsids (3) that share an envelope and a vesicle membrane are visible in both (A) and (B). Some virions have completed their secondary envelopment and sit in vesicles as a consequence of the budding event (4a–c). We also observed virions in vesicles that are cracked (5). It is not clear yet whether this is an effect of viral envelopment or represents a preparation artefact (Landwehr *et al.*, 2008). (See Plate no. 44 in the Color Plate Section.)

B. High-Pressure Freezing of Infected Cells

Instrumentation: High-pressure freezer Wohlwend Compact 01 (Engineering Office M. Wohlwend GmbH, Sennwald, Switzerland)

Materials: Sapphire discs (3 mm diameter, thickness 0.05 mm; Engineering Office M. Wohlwend GmbH, Sennwald, Switzerland), aluminium planchettes as covers for high-pressure freezing (3 mm diameter; flat and with 0.1 mm depression; Engineering Office M. Wohlwend GmbH, Sennwald, Switzerland)

Reagents: 1-Hexadecene for synthesis (Merck KG, Darmstadt, Germany)

C. Freeze-Substitution and Conventional TEM

Instrumentation: Freeze-substitution apparatus (home-made construction by A. Ziegler and W. Fritz, Ulm University), ultramicrotome (Leica Scientific Instruments, Vienna, Austria), diamond knife (Diatome AG, Switzerland), Zeiss EM 10 transmission electron microscope (Carl Zeiss, Oberkochen, Germany)

Materials: 3.05 mm copper grids (Plano GmbH, Wetzlar, Germany)

Reagents: Acetone p.a. (Merck KG, Darmstadt, Germany), osmium tetroxide (ChemPur, Karlsruhe, Germany), uranyl acetate (Merck KG, Darmstadt, Germany), Epon (Fluka, Buchs, Switzerland)

D. STEM Tomography

Instrumentation: Titan 80–300 field emission transmission electron microscope (FEI Corporation, Eindhoven, The Netherlands) with STEM unit and STEM detector (Fischione, Export, PA, USA), Single Tilt Tomography Holder (Fischione, Export, PA, USA), Plasma Cleaner TP218 (Binder Labortechnik, Hebertshausen, Germany)

Materials: 3 mm grids for tomography (Plano GmbH, Wetzlar, Germany)

Reagents: 15 nm colloidal gold as fiducial markers (Aurion, Wageningen, The Netherlands)

V. Discussion

The result of a scientific experiment depends heavily on the methods used. This rule is especially obvious in electron microscopy, since direct observation of a living system is not possible at a subcellular level. Fixation of the sample by cross-linking chemical fixation, or by fast freezing, can at best preserve the situation at a certain time point and, therefore, dynamic events can at best be observed as a series of "snapshots". This represents probably the most fundamental limitation of electron microscopy in life science, as explained in the following problem. The model presented in Fig. 1 works with the assumption that step 9 represents a fusion of the primary envelope with the outer nuclear membrane. An alternative interpretation, however, has been proposed (Leuzinger *et al.*, 2005; Wild, 2008; Wild *et al.*, 2002). It has been suggested that step 9, which is observed in the electron microscope without doubt, represents a budding event from a capsid coming from the cytosol and budding into the perinuclear space. The authors explain the existence of naked capsids in the cytosol by a breakdown of the nuclear pore complexes leading to a leak of capsids from the nucleus to the cytoplasm. They conclude further that de-envelopment does not exist and that, as a consequence, all envelopment events represent primary envelopments. This alternative model raises the interesting question as to what extent dynamic processes can be interpreted and understood with static electron microscopical images: Whereas the partially enveloped capsids at the outer nuclear membranes are generally interpreted as membrane fusion leaving capsids to the perinuclear space, Wild *et al.* (2002) interpret similar images as a budding event in the opposite direction. Future experiments with mutants showing defects in envelopment may shed light onto this unsolved problem.

The methods described in this article, based on labelling of the biological structures with heavy metals, bear the advantage that they can be applied to a wide variety of cells and viruses. The prepared samples are not very beam sensitive and, therefore, one can spend as much time as needed on the microscope to search for infected cells and for the area of interest. This is especially important when viral entry has to be investigated (stages 1–6), since viral particles entering cells are small and do not frequently occur and, therefore, are difficult to find. In our opinion, currently, it would be impossible to capture the right area using low-dose techniques.

Our high-pressure freezing device is located in an S2 safety laboratory to prevent infection of the staff during the experiment. Human CMV requires this S2 safety standard. In case higher safety standards are necessary, pre-fixation with glutaraldehyde prior to bringing the samples to our laboratories is inevitable. A study on dengue virus (Welsch *et al.*, 2009) has shown that samples, high-pressure frozen after prefixation with glutaraldehyde and freeze-substituted, still show superior ultrastructure compared to conventionally fixed and dehydrated samples. However, clear ultrastructural changes compared to samples high-pressure frozen from the living stage could be observed (see also Chapter 2 by Jensen *et al.*, this volume). One of these ultrastructural changes concerns the size of the mitochondrial intermembrane space. When high-pressure frozen from the physiologically active state, inner and outer membranes are in very close apposition, indicating a metabolically active state (Lang and Bronk, 1978; Walther *et al.*, 2009). On the other hand, samples that have been chemically fixed before high-pressure freezing show an intermembrane space of about 40–100 nm, indicating that the membrane potential is no longer fully established.

This article covers only a small portion of the different electron microscope approaches that allow the study of viral infections of cells. A large potential for the future lies in correlative techniques, i.e., the combination of fluorescence light with electron microscopy.

Acknowledgments

We thank Eberhard Schmid for many improvements of the freeze-substitution protocol and of the mounting of the semi-thin sections on the copper grids and Renate Kunz and Reinhard Weih for excellent technical assistance. We thank Sarah Bek for help with scanning electron microscopy. Special thanks go to David Mastronarde for making the great “IMOD” software available. This work was supported by the DFG Schwerpunktprogramm 1175.

References

Baumeister, W. (2005). A voyage to the inner space of cells. *Protein Sci.* **14**, 257–269.

Buser, C., and Walther, P. (2008). Freeze-substitution: The addition of water to polar solvents enhances the retention of structure and acts at temperatures around –60°C. *J. Microsc.* **230**, 268–277.

Buser, C., Walther, P., Mertens, T., and Michel, D. (2007). Cytomegalovirus primary envelopment occurs at large infoldings of the inner nuclear membrane. *J. Virol.* **81**, 3042–3048.

Chevillotte, M., Buser, C., Linta, L., Frascaroli, G., Lüske, A., Landwehr, S., Mertens, T., and von Einem, J. (2009). The major tegument protein pp65 of human cytomegalovirus is required for the incorporation of pUL69 and pUL97 into the virus particle and for viral growth in macrophages. *J. Virol.* **83**, 2480–2490.

Church, G. A., and Wilson, D. W. (1997). Study of herpes simplex virus maturation during a synchronous wave of assembly. *J. Virol.* **71**, 3603–3612.

Dubochet, J., Zuber, B., Eltsov, M., Bouchet-Marquis, C., Al-Amoudi, A., and Livolant, F. (2007). How to "read" a vitreous section. *Methods Cell Biol.* **79**, 385–406.

Ebersold, H. R., Cordier, J. L., and Lüthy, P. (1981). Bacterial mesosomes: Method dependent artifacts. *Arch. Microbiol.* **130**, 19–22.

Eppenberger-Eberhardt, M., Aigner, S., Donath, M. Y., Kurer, V., Walther, P., Zuppinger, C., Schaub, M. C., and Eppenberger, H. M. (1997). IGF-I and bFGF differentially influence atrial natriuretic factor and alpha-smooth muscle actin expression in cultured atrial compared to ventricular adult rat cardiomyocytes. *J. Mol. Cell. Cardiol.* **29**, 2027–2039.

Höhn, K., Pusapati, G. V., Seufferlein, T., Adler, G., and Walther, P. (2009). 3D reconstruction of Golgi stacks in brefeldin A treated cells. First results using STEM tomography. *In* "Life Sciences" (M. A. Pabst and G. Zellnig eds.) MC2009, Vol. 2, pp. 43–44, DOI: 10.3217/978-3-85125-062-6-248, Verlag der TU Graz, Graz (Austria) 2009.

Kremer, J. R., Mastronarde, D. N., and McIntosh, J. R. (1996). Computer visualization of three-dimensional image data using IMOD. *J. Struct. Biol.* **116**, 71–76.

Landwehr, S., von Einem, J., Mertens, T., and Walther, P. (2008). Cytomegalovirus membrane envelopment revisited—a STEM-tomography study based on high-pressure freezing and freeze substitution. *In* "Life Science" (A. Aretz, B. Hermanns-Sachweh, and J. Mayer eds.) EMC 2008, Vol. 3, pp. 81–82, Springer-Verlag, Berlin, Heidelberg, DOI: 10.1007/978-3-540-85228-5_41.

Lang, R.D.A., and Bronk, J. R. (1978). A study of rapid mitochondrial structural changes *in vitro* by spray-freeze-etching. *J. Cell Biol.* **77**, 134–147.

Lehmann, M. J., Sherer, N. M., Marks, C. B., Pyparet, M., and Mothes, W. (2005). Actin- and myosin-driven movement of viruses along filopodia precedes their entry into cells. *J. Cell Biol.* **170**, 317–325.

Leuzinger, H., Ziegler, U., Fraefel, C., Schraner, E. M., Glauser, D., Held, I., Ackermann, M., Müller, M., and Wild, P. (2005). Herpes simplex virus 1 envelopment follows two diverse pathways. *J. Virol.* **79**, 13047–13059.

Mastronarde, D. N. (2008). Correction for non-perpendicularity of beam and tilt axis in tomographic reconstructions with the IMOD package. *J. Microsc.* **230**, 212–217.

McDonald, K. L., Morphew, M., and Müller-Reichert, T. (2007). Recent advances in high pressure freezing. From: Methods in molecular biology. *In* "Electron Microscopy: Methods and Protocols" (J. Kuo, ed.) 2nd edn., Vol. 369, pp. 143–173. Humana Press Inc., Totowa, NJ.

Mettenleiter, T. C. (2002). Herpesvirus assembly and egress. *J. Virol.* **76**, 1537–1547.

Moor, H., and Riehle, U. (1968). Snap-freezing under high pressure: A new fixation technique for freeze-etching. *Proc. 4th Eur. Reg. Conf. Electron Microsc.* **2**, 33–34.

Müller, M., and Moor, H. (1984). Cryofixation of thick specimens by high pressure freezing. *In* "Science of Biological Specimen Preparation 1983" (J. P. Revel, T. B. Barnard, and G. H. Haggis, eds.) pp. 131–138. SEM Inc., AMF O'Hare (Chicago).

Murphy, E., Yu, D., Grimwood, J., Schmutz, J., Dickson, M., Jarvis, M. A., Hahn, G., Nelson, J. A., Myers, R. M., and Shenk, T. E. (2003). Coding potential of laboratory and clinical strains of human cytomegalovirus. *Proc. Natl. Acad. Sci. U.S.A.* **100**, 14976–14981.

Newcomb, W. W., Homa, F. L., Thomsen, D. R., Ye, Z., and Brown, J. C. (1994). Cell-free assembly of the herpes simplex virus capsid. *J. Virol.* **68**, 6059–6063.

Rupp, B., Ruzsics, Z., Buser, C., Adler, B., Walther, P., and Koszinowski, U. H. (2007). A random screen for dominant-negative mutants of the cytomegalovirus nuclear egress protein M50. *J. Virol.* **81**, 5508–5517.

Sailer, M., Höhn, K., Lück, S., Schmidt, V., Beil, M., Walther, P. (2010). Novel electron tomographic methods to study the morphology of keratin filament networks. *Microsc. Microanal.* **16**, 1–10.

Schwarz, H., Hohenberg, H., and Humbel, B. M. (1993). Freeze-substitution in virus research: A preview. *In* "Immunoelectron Microscopy in Virus Diagnosis and Research" (A. D. Hyatt and B. T. Eaton, eds.) pp. 97–118. CRC Press Inc, Boca Raton, FL.

Sheaffer, A. K., Newcomb, W. W., Gao, M., Yu, D., Weller, S. K., Brown, J. C., and Tenney, D. J. (2001). Herpes simplex virus DNA cleavage and packaging proteins associate with the procapsid prior to its maturation. *J. Virol.* **75**, 687–698.

Szczesny, P. J., Walther, P., and Müller, M. (1996). Light damage in rod outer segments: The effects of fixation on ultrastructural alterations. *Curr. Eye Res.* **15**, 807–814.

Walther, P., Wehrli, E., Hermann, R., Müller, M. (1995). Double layer coating for high resolution low temperature SEM. *J Microsc.* **179**, 229–237.

Walther, P. (2008). High resolution cryoscanning electron microscopy of biological samples. *In* "Biological Low-Voltage Scanning Electron Microscopy" (H. Schatten and J. B. Pawley eds.) pp. 245–261. Springer, New York.

Walther, P., and Hentschel, J. (1989). Improved representation of cell surface structures by freeze substitution and backscattered electron imaging. *Scanning Microsc.* **3**(Suppl. 3), 201–211.

Walther, P., Höhn, K., and Krisp, H. (2009). What is the true size of the mitochondrial intermembrane space? A study using high-pressure freezing and STEM tomography. *In* "Life Sciences" (M. A. Pabst and G. Zellnig, eds.) MC2009, Vol. 2, pp. 49–50, DOI: 10.3217/978-3-85125-062-6-248, Verlag der TU Graz, Graz (Austria) 2009.

Walther, P., and Ziegler, A. (2002). Freeze substitution of high-pressure frozen samples: The visibility of biological membranes is improved when the substitution medium contains water. *J. Microsc.* **208**, 3–10.

Wang, L., Walther, P., Mertens, T., and Frascaroli, G. (2009). Entry mechanisms of the human cytomegalovirus (HCMV) into GM- and M-monocyte derived macrophages. *In* "Life Sciences" (M. A. Pabst and G. Zellnig, eds.) MC2009, Vol. 2 pp. 204–205, DOI: 10.3217/978-3-85125-062-6-248, Verlag der TU Graz, Graz (Austria) 2009.

Welsch, S., Miller, S., Romero-Brey, I., Merz, A., Bleck, C. K., Walther, P., Fuller, S. D., Antony, C., Krijnse-Locker, J., and Bartenschlager, R. (2009). Composition and three-dimensional architecture of the dengue virus replication and assembly sites. *Cell Host Microbe.* **5**, 365–375.

Wild, P. (2008). Electron microscopy of viruses and virus-cell interactions. *Methods Cell Biol.* **88**, 497–524.

Wild, P., Schraner, E. M., Cantieni, D., Loepfe, E., Walther, P., Müller, M., and Engels, M. (2002). The significance of the Golgi complex in envelopment of bovine herpesvirus 1 (BHV-1) as revealed by cryobased electron microscopy. *Micron.* **33**, 327–337.

Yakushevska, A. E., Lebbink, M. N., Geerts, W. J., Spek, L., van Donselaar, E. G., Jansen, K. A., Humbel, B. M., Post, J. A., Verkleij, A. J., and Koster, A. J. (2007). STEM tomography in cell biology. *J. Struct. Biol.* **159**, 381–391.

CHAPTER 26

Intracellular Membrane Traffic at High Resolution

Jan R.T. van Weering[*], Edward Brown[*], Thomas H. Sharp[*,†], Judith Mantell[*,‡], Peter J. Cullen[*], *and* Paul Verkade[*,‡,$]

[*]Department of Biochemistry, School of Medical Sciences, University of Bristol, University Walk, Bristol, BS8 1TD, United Kingdom

[†]School of Chemistry, University of Bristol, Cantock's Close, Bristol, BS8 1TS, United Kingdom

[‡]Wolfson Bioimaging Facility, School of Medical Sciences, University Walk, Bristol, BS8 1TD, United Kingdom

[$]Department of Physiology and Pharmacology, School of Medical Sciences, University Walk, Bristol, BS8 1TD, United Kingdom

METHODS IN CELL BIOLOGY, VOL. 96

978-0-12-381007-6
DOI: 10.1016/S0091-679X(10)96026-3

I. Abstract

Membrane traffic between organelles is essential for a multitude of processes that maintain cell homeostasis. Many steps in these tightly regulated trafficking pathways take place in microdomains on the membranes of organelles, which require analysis at nanometer resolution. Electron microscopy (EM) can visualize these processes in detail and is mainly responsible for our current view of morphology on the subcellular level. This review discusses how EM can be applied to solve many questions of intracellular membrane traffic, with a focus on the endosomal system. We describe the expansion of the technique from purely morphological analysis to cryo-immuno-EM, correlative light electron microscopy (CLEM), and 3D electron tomography. In this review we go into some technical details of these various techniques. Furthermore, we provide a full protocol for immunolabeling on Lowicryl sections of high-pressure frozen cells as well as a detailed description of a simple CLEM method that can be applied to answer many membrane trafficking questions. We believe that these EM-based techniques are important tools to expand our understanding of the molecular details of endosomal sorting and intracellular membrane traffic in general.

II. Introduction

A. Intracellular Membrane Traffic

Membrane-mediated transport is one of the major fields of focus in cell biology research. It is essential for delivering newly synthesized proteins to the location where they can perform their function, maintaining lipid bilayers, storing signal molecules, recycling or degrading used proteins, and providing membrane expansion in cell division. Membrane carriers are tubular- or vesicular-shaped structures that are derived from the limiting membrane of the donor organelle where the cargo is initially located. To produce such a carrier, the donor bilayer is deformed into a budding profile (Fig. 1A). This process is regulated by proteins that can reshape membranes by forming a scaffold on the membrane of a specific curvature and/or inserting domains into the lipid bilayer that introduce membrane curvature (reviewed by McMahon and Gallop, 2005). Budding of the donor membrane coincides with sorting of specific cargo. Once the carrier is formed and loaded with cargo, the tubular/vesicular carrier detaches from the donor membrane in the fission step. This process is regulated by several proteins that narrow the carrier neck, e.g., dynamin, in combination with longitudinal force that is applied by motor proteins that pull the carrier along the cytoskeleton. After transport to the target organelle, the carrier is recognized by tethering proteins. The transport vesicle subsequently docks to the target membrane, involving proteins from the SNARE family. Finally, the carrier fuses with the target organelle to deliver its cargo.

The cell contains many different trafficking routes (Fig. 1B). Most of the steps in the trafficking processes are therefore highly regulated to obtain the required specificity. In

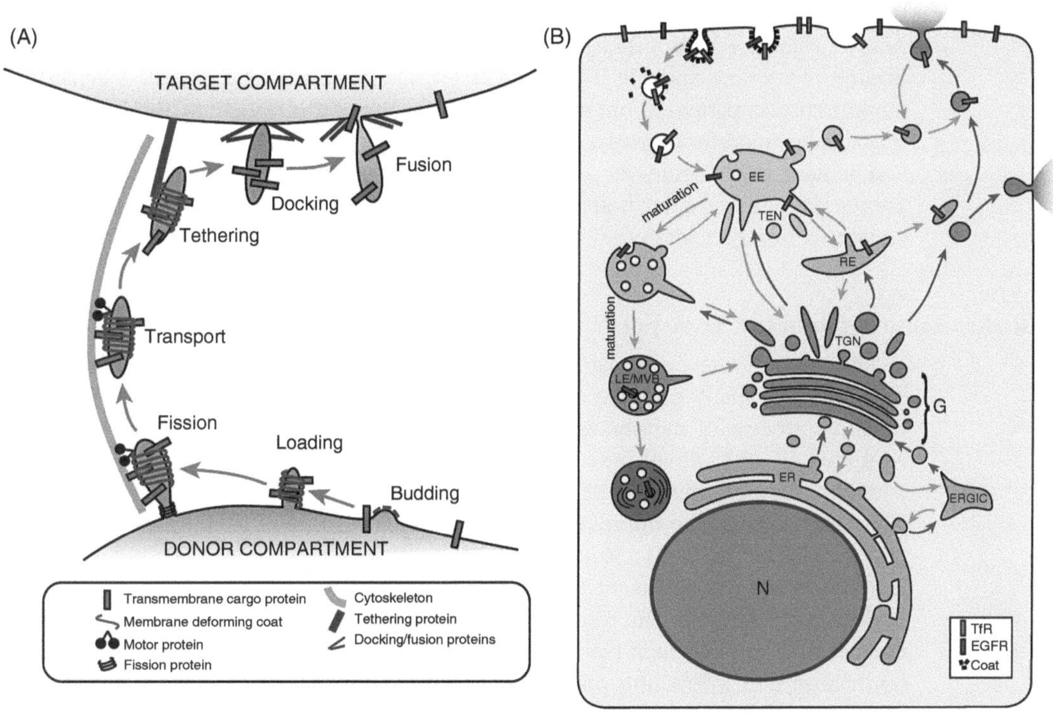

Fig. 1 Steps and routes of intracellular membrane transport. (A) Schematic diagram of the different steps in which membrane transport is generally divided. A bud is formed on the donor membrane, which is further shaped into a tubular/vesicular membrane carrier and loaded with the cargo. The membrane carrier pinches off from the donor compartment (fission) and is transported to the target compartment along the cytoskeleton. At the target compartment, the membrane carrier is tethered and docked before it fuses with the target membrane to deliver the cargo. (B) Cartoon of a cell depicting the major membrane trafficking routes. Blue arrows mark the biosynthetic pathway, orange arrows show the degradative route, and gray arrows mark the many recycling routes. See Section I.a. for details. ER, endoplasmic reticulum; G, Golgi complex; ERGIC, ER–Golgi intermediate compartment; TGN, *trans*-Golgi network; EE, early endosome; TEN, tubular endosomal network; RE, recycling endosome; LE, late endosome; MVB, multivesicular body; L, lysosome; TfR, transferrin receptor; EGFR, epidermal growth factor receptor. (See Plate no. 45 in the Color Plate Section.)

general, there are two main routes: the biosynthetic pathway and the degradative route. The biosynthetic pathway (Fig. 1B, blue arrows) starts with protein production in the endoplasmic reticulum (ER). Proteins are transported to the Golgi complex for modification either directly or through an intermediate compartment. In the *trans*-Golgi network (TGN) the proteins are sorted for transport to different locations in the cell where each protein performs its function, for instance, in the plasma membrane or in the endosomal system.

The degradative route (Fig. 1B, orange arrows) starts with the internalization or endocytosis of material from the plasma membrane. The endocytosed vesicles fuse with the early endosome where the cargo is sorted in a tubular endosomal network. Many proteins are recycled back to their steady-state compartments, including the plasma membrane and the TGN. There are many points of crosstalk between both major transport pathways but it is beyond the scope of this review to highlight them all. Proteins destined for degradation are sorted into intraluminal vesicles of the endosome and remain there while the endosomal vacuole matures into a late endosome also known as multivesicular body (MVB). The MVB fuses with the lysosome where its content is finally degraded.

B. Visualization of Membrane Transport Pathways

The use of fluorescently labeled proteins and advanced live-cell imaging techniques has identified many proteins and protein complexes that fulfill essential roles in the dynamic process of membrane traffic. However, the resolution of any microscope is restricted by the wavelength of the source of imaging radiation; in the case of conventional light microscopes this results in a limit in resolution of around 200 nm. This resolution can be improved by various super-resolution techniques (Betzig *et al.*, 2006; Gustafsson, 2005; Rust *et al.*, 2006; Schönle *et al.*, 2000), but even these techniques are not sufficient to visualize processes that take place at the nanometer scale. Furthermore, in fluorescence microscopy only labeled proteins are visible, not their cellular environment (so-called reference space). Finally, it is difficult to reliably co-localize non-interacting proteins to specific subdomains on membrane organelles using light microscopy due to insufficient resolution and the lack of reference space information (compare Rink *et al.*, 2005 and Hughes *et al.*, 2009). The only technique so far that can localize proteins at the nanometer resolution is EM. Immuno-EM methods have proven to be very powerful tools for cell biology and have made many important contributions to the field. Immuno-EM as we know it today started with the presentation of the first electron-dense marker, Ferritin, to label antibodies for EM (Singer, 1959). Over the years, the technique has been improved and expanded to a range of EM-based methods, such as CLEM and electron tomography. These techniques can allow cell biologists to gain novel insights in the process of membrane traffic.

C. Development of Ultrastructural Labeling

Membrane carriers of different trafficking routes often appear quite similar in transmission EM (TEM) and can only be identified by the presence or absence of specific marker proteins. Several labeling techniques can be applied in TEM to visualize the different trafficking routes and localize specific proteins at the ultrastructural level. We will discuss a couple of these labeling techniques as applied in the endosomal system.

The first method of ultrastructural labeling was performed by internalization of a fluid-phase EM marker by living cells and following the fate of that marker by fixing the cells at different time intervals. Using this method, Heuser and Reese (1973) analyzed synaptic vesicle endocytosis in the neuromuscular junction during different stimulation protocols, by which different numbers of synaptic vesicles fuse with the membrane to release their neurotransmitter. The nerve terminal was incubated with medium containing horseradish peroxidase (HRP). HRP itself is not visible in the EM but it catalyzes the precipitation of diaminobenzidine (DAB) in the presence of hydrogen peroxide, which appears as electron-dense material after osmium staining (Seligman *et al.*, 1968). Heuser and Reese (1973) observed that synaptic vesicles were being recycled locally in the nerve terminal by fusing with endosomes and proposed a model for the presynaptic vesicle cycle. In the same issue of that journal, Ceccarelli and co-workers loaded the nerve terminals with HRP or dextran and, combined with electrophysiological experiments, came to the conclusion that synaptic vesicles are not fully retrieved to an endosomal compartment as suggested by Heuser and Reese, but were refilled at the site of release (Ceccarelli *et al.*, 1973). The relative importance of these different types of synaptic vesicle recycling is one of the most studied processes in neuronal cell biology and is still actively debated (e.g., Balaji and Ryan, 2007; Granseth *et al.*, 2006; Zhang *et al.*, 2009).

As an alternative to fluid phase uptake, endocytosis of specific receptors can be followed by labeling their substrate. This technique revealed different sorting routes in the endocytic network and characterized the different organelles in the pleiomorphic endosomal system. The iron transporter transferrin (Tf) is still one of the most used endosomal markers. The receptor for transferrin (TfR) is constitutively internalized and recycled back to the plasma membrane. The Tf–iron complex binds the TfR at the cell surface and is transported into the cell by endocytosis (Fig. 1B). In the early endosome, the Tf–TfR complex releases the iron ions and is recycled back to the cell surface. This recycling pathway can take either a direct route from the early endosome to the plasma membrane or a slow recycling route through the recycling endosome compartment (Hopkins *et al.*, 1994) (Fig. 1B). Tf coupled to a reporter molecule (e.g., fluorophores, gold particles, or HRP) is applied to study processes of endocytosis and early endosomal sorting. Due to the efficient recycling of the Tf–TfR complex, late endosomes no longer contain Tf. Therefore, Tf can be used as a tool to study the maturation process from early to late endosomes. Mari and co-authors showed in an elegant quantitative TEM study that the maturation of early to late endosomes correlates with an increase of intraluminal vesicles in the endosomal vacuole (Mari *et al.*, 2008). The accumulation of intraluminal vesicles in late endosomes gave them the name of MVBs. However, the function of this organelle was unclear until labeled epidermal growth factor (EGF) identified MVBs as part of the endosomal–lysosomal degradation pathway (Gorden *et al.*, 1978; Haigler *et al.*, 1979). When EGF associates with its receptor, it stimulates cell proliferation upon which the ligand–receptor complex is internalized. The complex is then targeted to the intraluminal vesicles of MVBs for breakdown to prevent uncontrolled growth (Stoscheck and Carpenter, 1984). This makes EGF and the EGF receptor (EGFR) a popular tool to study the degradative pathway.

Lysosomes were characterized at the ultrastructural level by a slightly different approach. Essner and Novikoff (1961) used the activity of an endogenous acid phosphatase present in lysosomes to form a lead-phosphate precipitate on ultrathin sections. They found that the electron-dense precipitate was contained in "dense bodies"; organelles with high protein content and multilaminar internal membranes. These ultrastructural hallmarks are still used to recognize lysosomes in TEM.

Although the above-mentioned labeling techniques provided many new insights and are still used today, they are limited by the accessibility of the compartment from the outside of the cell or the presence of unique enzymes that catalyze a suitable reaction that is detectable in the EM. The application of immuno-histochemistry, i.e., labeling proteins using specific antibodies, should have overcome these limitations but it has proven to be complicated to combine with good ultrastructural preservation until the introduction of the Tokuyasu cryosectioning technique in 1973 (Tokuyasu, 1973). This simple method made immuno-EM widely accessible and still sets the standard for protein localization in the membrane traffic field at the nanometer resolution.

III. Methods

A. The Tokuyasu Cryosectioning Technique

1. Principle of the Tokuyasu Method

The key to a successful immuno-EM experiment is to retain both ultrastructural detail and antigenicity of the molecules in the ultrathin section. As biological material is too soft to be used for ultrathin sectioning straight away, the material is generally embedded in a resin to provide sufficient support. However, these resins cover the epitopes in the section and prevent/decrease antibody recognition. Some resins preserve the antigenicity of the proteins just at the section surface (methacrylic resins such as Lowicryl HM20, further described in Section -III.B.2.). Alternatively, biological material can be frozen to gain sufficient rigidity for ultrathin sectioning and still retain antigenicity. However, simple freezing of biological material will produce ice of a crystalline nature. These ice crystals take up more space than liquid water, thereby severely damaging the ultrastructure of the cell (Brown *et al.*, 2009; McDonald, 2007). Tokuyasu (1973) introduced a method to prevent ice crystal formation by infusing chemically fixed cells with high concentrations of sucrose. Sucrose is highly hydrophilic, which at high concentrations binds sufficient water molecules to prevent water from adopting its crystalline organization when cooled down. The non-crystalline ice (so-called vitreous ice) that is formed occupies the same amount of space as water and has minimal impact on the ultrastructure. This development allowed ultrathin sectioning of frozen cells that have preserved most of their ultrastructural information as well as antigenicity of their molecules.

2. The Tokuyasu Approach

Since its introduction, the Tokuyasu method has been improved and perfected (for a detailed protocol see, e.g., Slot and Geuze, 2007). Also, the introduction of diamond

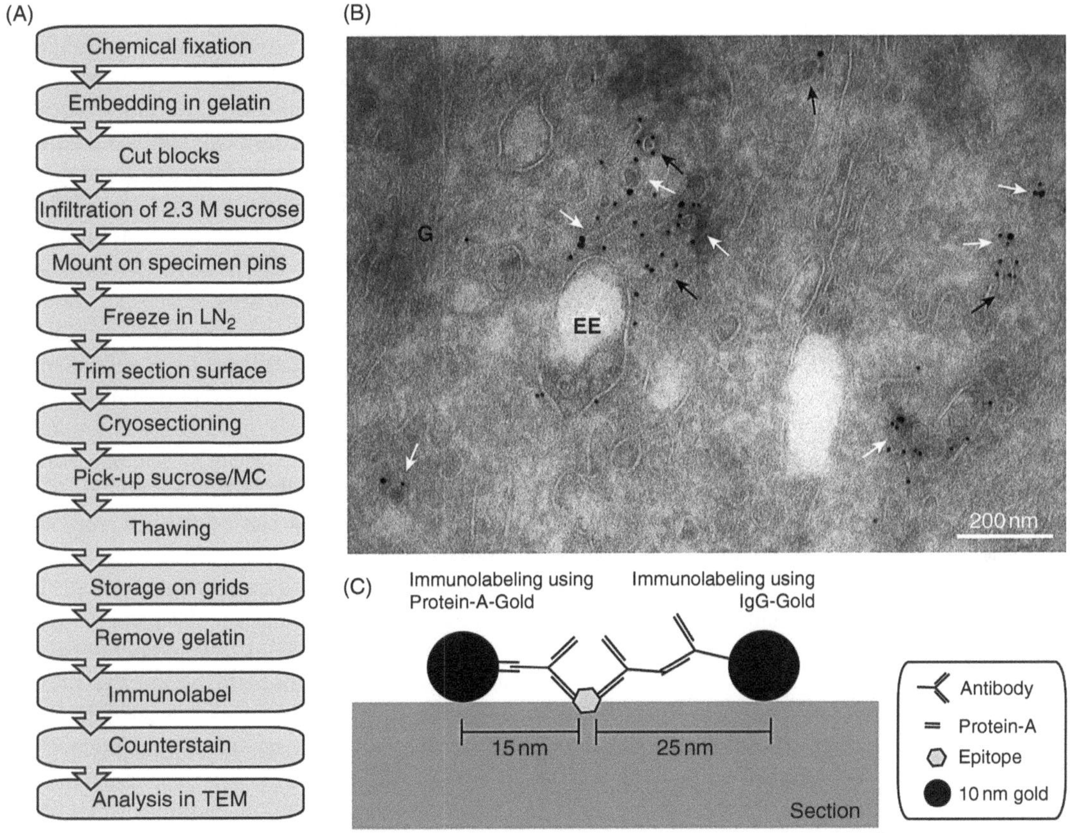

Fig. 2 Immunolabeling the endosomal compartment using the Tokuyasu method. (A) Flow diagram of a typical immuno-EM experiment using the Tokuyasu method (see Section "The Tokuyasu Approach" for details). LN_2, liquid nitrogen; MC, methylcellulose; TEM, transmission electron microscope. (B) Electron micrograph of immunogold labeling for GFP and RFP of HeLa cells, expressing GFP-SNX1 and RFP-Rab5, processed according to the Tokuyasu method. Ultrathin sections were labeled by GFP and RFP antibodies and PAG (10 and 15 nm, respectively). White arrows indicate tubular/vesicular profiles that are labeled by both sizes of gold; black arrows indicate membrane structures only labeled for one protein. G, Golgi complex; EE, early endosomal vacuole. Scale bar represents 200 nm. (C) Cartoon of immunogold labeling of an epitope on the surface of the ultrathin section. Protein-A-Gold markers have a higher resolution (within 15 nm distance of the epitope) compared to IgG-Gold (within 25 nm distance of the epitope) due to the larger size of the secondary antibody. Adapted from Verkade *et al.*, 1997.

cryo-knives as well as major improvements in the stability of the ultramicrotomes have made the technique much easier and accessible as a "standard" EM technique. The flowchart (Fig. 2A) shows the workflow of a Tokuyasu cryo-immuno-EM experiment. In brief, cells and tissues can be fixed by conventional fixatives, such as paraformaldehyde (PFA) and glutaraldehyde (GA), in a sufficiently strong buffer solution (e.g., Sorensen's phosphate buffer, buffering capacity of PBS is not strong enough) to scavenge the free protons produced during the fixation process. GA is preferable to PFA for preservation of ultrastructural detail, but its disadvantage is that more epitopes are distorted, thereby reducing the immunolabeling efficiency. Therefore, combinations of PFA and GA fixative are a popular choice for biological samples in order to obtain the optimal balance between ultrastructural preservation and antigenicity of the material. It is advisable to test different concentrations and/or combinations of fixative when starting an immuno-EM experiment on new cell-types/tissues or using new antibodies. Generally, successful labeling with an antibody/fixative combination in immuno-fluorescence light microscopy is a good indicator for a successful immuno-EM experiment.

After the material is fixed, the extracellular space is stabilized by a gelatin matrix. Monolayers of cells can be scraped, pelleted, and embedded in gelatin or embedded as an intact layer ("flat embedding"; for details see Oorschot *et al.*, 2002). These gelatin-encapsulated blocks of cells or tissue are infiltrated with a 2.3 M sucrose solution, mounted on specimen holders and frozen in liquid nitrogen (LN_2) for cryosectioning. As the material is now cryo-protected by sucrose, blocks can be thawed, remounted, and refrozen without damaging the ultrastructure.

The cryo-ultramicrotome is the only piece of specialized equipment required for the Tokuyasu technique, but most conventional ultramicrotomes can be used as such by attaching a cryo-chamber. Ultrathin sections are collected at low temperatures, generally −100 to −120°C. Other temperatures can be chosen to obtain the preferred rigidity of the block, for instance, filter-grown cells are sectioned at −90°C in our lab. Ultrathin cryosections are captured in sucrose solution, usually supplemented with methylcellulose to provide additional support (Liou *et al.*, 1996). The sections are thawed and stored, still covered by the pick-up solution, on carbon/formvar-coated grids at 4°C.

After removal of the gelatin, the ultrathin sections can be immuno-stained in a manner that is very similar to widely used immuno-fluorescence labeling of cells for confocal microscopy. The quality of the antibody (in combination with the applied fixative) is the main determinant of a successful immunolabeling experiment. New antibodies should be tested in different concentrations for optimal labeling. The antibodies can be visualized using electron-dense markers, such as colloidal gold particles of different sizes, which allows co-localization studies of multiple proteins in the same section (Fig. 2B). Gold particles can be directly linked to the primary antibody, attached to a secondary antibody raised against a specific species, or coated in Protein-A/G that binds antibodies with high affinity (Griffiths, 1993). Finally, the tissue in the section can be stained with uranyl acetate to improve contrast in TEM. It should be noted that in the Tokuyasu procedure uranyl acetate acts as a negative stain and deposits along the membranes rather than staining the membranes itself (Fig. 2B). The full procedure from fixing the cells up to analysis in the TEM can be completed in two to three days, which makes the Tokuyasu technique a popular technique to determine the nature of the many different membrane transport carriers in the cell and localize their proteins of interest at the ultrastructural level.

3. Remarks

As mentioned above, antibodies can be labeled indirectly by either secondary antibodies (IgG) coupled to gold (IgG-G) or by Protein-A-Gold (PAG). PAG binds several species of antibodies (not all, to check compatibility see, e.g., Harlow and Lane, 1988) at the Fc part of the primary antibody. This results in a one-to-one labeling pattern, i.e., each gold particle labels one primary antibody. IgG-G on the other hand recognizes multiple epitopes on a single primary antibody. This results in a labeling pattern by which multiple (generally 2–3) gold particles label a single primary antibody. This difference makes PAG-based labeling more suitable for quantification, while the IgG-G technique can amplify labeling when the number of epitopes is low. Another advantage of the PAG method is that it improves the resolution of the immunolabeling. The length of an IgG molecule can be estimated to be approximately 16 nm in its crystal structure. A secondary IgG antibody and gold particles will bind to the middle of the Fc region resulting in a distance of 8–12 nm for one antibody (Verkade *et al.*, 1997). Therefore, the total distance from the detected molecule to the center of a 10 nm gold particle linked to a secondary antibody can span up to 25 nm (Fig. 2C). With a molecular weight of 40–60 kDa Protein-A is significantly smaller than the 150 kDa of a complete IgG molecule. Therefore, the gold particles will be located closer to the epitope in PAG-labeled sections (within ~15 nm) compared to IgG-G labeling. Other strategies for improving the labeling resolution are using Fab fragments (~50 kDa) of antibodies instead of the complete antibody or directly linking the primary antibody to gold particles.

Both the PAG and IgG-G labeling technique can be used to study multiple proteins at the same time. For IgG-G, this can be achieved by using primary antibodies of different species that can be selectively recognized by IgG-G linked to different sizes of gold. For PAG, epitopes can be labeled by different sizes of gold consecutively by including a 5-min GA fixation step after each labeling step. This fixation step distorts the Fc part of the primary antibody; therefore, the same species of primary antibody can be used in multiple rounds. It is important to keep in mind that double labeling can introduce two artifacts: co-labeling, the different-sized gold particles label the same primary antibody, and steric hindrance, gold particles obscuring epitopes in the consecutive labeling reaction. Therefore, it is advisable to always include single labeled control conditions for each primary antibody to check that the labeling pattern is not affected in the double labeling condition.

Although the Tokuyasu cryo-immuno method currently is the best method for high antibody labeling in EM, it should be noted that the labeling efficiency of such experiments is estimated to be around 10–15% (D'Amico and Skarmoutsou, 2008; Griffiths and Hoppeler, 1986).

B. Cryofixation and High-Pressure Freezing/Freeze-Substitution

1. Principle of Cryofixation

Although the Tokuyasu cryosectioning method was a great step forward in the applications of EM, the method does have disadvantages. The major disadvantage is that the material has to be chemically fixed due to the osmotic pressure of the 2.3 M sucrose solution. Chemical fixation introduces several artifacts, which can have a great impact on the data (Brown *et al.*, 2009; McDonald, 2007; Murk *et al.*, 2003). First, chemical fixation is relatively slow (in the order of seconds to minutes depending on the tissue and fixative), which can be too slow if you are studying a process that is significantly faster (see, e.g., Rosenmund and Stevens, 1997). Second, chemical fixation of membrane proteins creates pores in the plasma membrane, which allows exchange of ions and small molecules between the extracellular medium and the cytosol during the fixation (Penttila *et al.*, 1974). Third, many cytoskeletal elements are destabilized during the chemical fixation (Small, 1981) and highly dynamic structures like sorting tubules can collapse upon chemical fixation (Jan R.T. van Weering and Peter J. Cullen, unpublished observation). Fourth, chemical fixation can dehydrate the specimen due to osmotic difference between the fixative and the cell (Kellenberger *et al.*, 1992; Murk *et al.*, 2003). Finally, chemical fixation can distort the epitope on the protein of interest, thereby reducing efficient immunolabeling. Although these artifacts of chemical fixation may have no major consequences on the outcome of the experiment, in some cases it might be crucial to circumvent them by using cryofixation.

The objective of cryofixation is to physically immobilize the specimen by freezing without ice crystal formation. As discussed above, vitreous ice can be obtained by using high concentrations of sucrose. Alternatively, water in living specimens can be vitrified by cooling down the samples so rapidly that water is not organized in ice crystals (Steinbrecht and Zierold, 1984). The cooling rate required for such vitreous freezing of water is about 10^5 K/s. This cooling rate can be achieved in very thin specimens (such as liposomes or platelets) by "plunge freezing" these samples in liquid ethane. Thicker specimens can be vitrified by high-pressure freezing (HPF). At 2000 bar the cooling rate for vitreous ice is reduced to 10^3 K/s that allows vitreous freezing of specimens up to a thickness of 200 μm (Moor, 1987; Studer *et al.*, 2008). HPF of biological specimens introduces minimal artifacts when well frozen and allows analysis of the ultrastructure of cells in their most natural state.

2. HPF Post-Processing

The optimal method to observe an HPF specimen is directly in its frozen-hydrated state at temperatures close to that of liquid nitrogen or below (cryo-EM). Plunge-frozen, isolated material only contains a very thin layer of frozen water and can be directly observed in a cryo-EM without any additional fixation or staining. Thicker specimens can be cryosectioned directly after HPF and observed in cryo-EM. This technique is called CEMOVIS (cryo-EM of vitreous sections) and is discussed by Resch *et al.*, Chapter 22 in this volume. Since no post-freezing labeling and/or staining is used in CEMOVIS, proteins can only be labeled by live labeling before the specimen is frozen. When labeling and/or staining of the material is required, the cryofixation needs to be replaced by chemical fixation to allow the sample to be handled at higher temperatures. This process is called freeze-substitution (FS).

FS is done by warming up the sample in a very controlled manner while fixing and dehydrating the sample at minimal temperatures. The dehydration step is done with an organic solvent, such as acetone, that is liquid at the starting temperature for an FS experiment (often −90°C). The dehydration and chemical fixation processes at these temperatures are slow, but are thought not to introduce as many artifacts as fixation and dehydration at room temperature. However, shrinkage and reduced antigenicity can still occur. This critically depends on the organic solvent in which the FS is performed. Methanol, for instance, extracts more water and cytoplasmic material compared to acetone (Weibull *et al.*, 1984) and is therefore hardly used in FS anymore. Addition of chemical compounds such as osmium tetroxide, uranyl acetate, and glutaraldehyde mainly determine whether antigenicity is retained. Ultimately, it is the resin that further determines both the ultrastructure and immuno-reactivity. In Fig. 3A we provide an

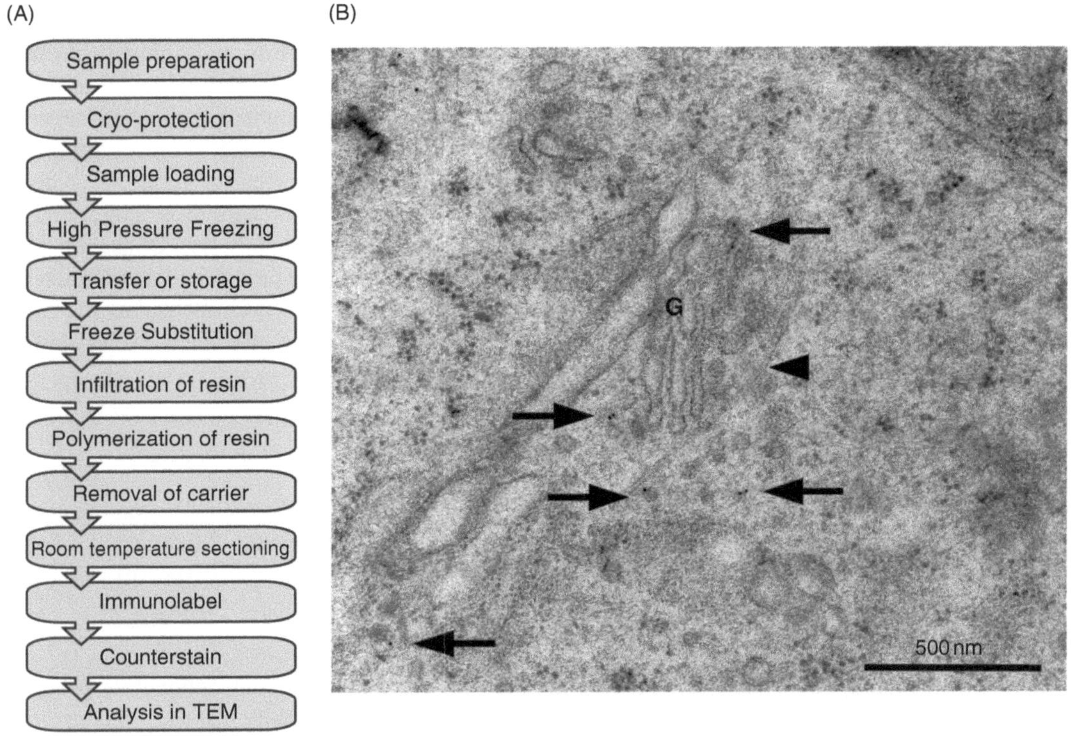

Fig. 3 Immunolabeling of VSVG using HPF/FS and Lowicryl embedding. (A) Flow diagram of a typical HPF/FS experiment using Lowicryl HM20 embedding and immunolabeling (see Section "Protocol for Immunolabeling Using HPF and Lowicryl Embedding" for details). (B) Electron micrograph of immunogold labeling for GFP in a HeLa cell after HPF/FS and Lowicryl HM20 embedding. HeLa cells grown on sapphire disks were transfected with a construct expressing a temperature-sensitive VSVG-GFP. With the temperature block present, the protein accumulates in the Golgi complex (G). 30 min after release of the temperature block the cells were high-pressure frozen and processed. Most of the protein has reached the cell surface but there is still protein present inside the Golgi and in vesicular profiles near the Golgi (arrows). This procedure preserves antigenicity as well as structural elements such as microtubules (arrowhead), which are very sensitive to chemical fixation. Scale bar represents 500 nm.

overview of the process. For the best ultrastructural preservation, we use a relatively short FS protocol that only lasts 23 h after which the sample is infiltrated with Epon and hardened in an oven. The FS cocktail contains 1% osmium tetroxide and 0.1% uranyl acetate in acetone (see, e.g., Brown *et al.*, 2009; McDonald *et al.*, 2007; Verkade, 2008). This protocol is unsuitable for subsequent immunolabeling because the osmium has destroyed most antigenic sites and Epon sections expose very few antigens. It is thought that the surface of Epon is much smoother compared to other resins such as Lowicryls (Hayat, 2000). A rougher surface results in a larger surface area of the section and therefore more epitopes will be exposed. In addition, Lowicryls are liquid at relatively low temperatures (from −35 to −80°C) and they can be polymerized at these temperatures using UV light. This makes Lowicryls very suitable as embedding medium after HPF. Lowicryls come in a polar/hydrophilic (K4M and K11M) and non-polar/hydrophobic (HM20 and HM23) form. We prefer Lowicryl HM20 for our studies as it gives both a good ultrastructure and antigenicity, albeit that the labeling efficiency is still lower compared to the Tokuyasu method (see Section III.A.1 and compare Figs. 2B and 3B). To utilize the higher labeling efficiency of Tokuyasu sections, HPF samples can be processed with an FS protocol compatible with the Tokuyasu technique (Ripper *et al.*, 2008; van Donselaar *et al.*, 2007). After the fixation and dehydration have been performed in FS, the sample is rehydrated and further processed using the Tokuyasu method (see for details van Donselaar *et al.*, 2007). This procedure is very time consuming and has not been applied to intracellular membrane studies although it could offer a method to study sensitive membrane structures such as sorting tubules with the immunolabeling efficiency of the Tokuyasu method.

3. Protocol for Immunolabeling Using HPF and Lowicryl Embedding

a. High-Pressure Freezing For our high-pressure freezing experiments we use the Leica EM PACT2 + Rapid Transfer System (RTS) (see Section "Materials" for details; Verkade, 2008). For detailed protocols of sample preparation for HPF of cell culture grown on sapphire disks, we refer to previous publications (McDonald *et al.*, 2007; Verkade, 2008). A few notes on the cell culture and HPF:

1. Sapphire disks are used as opposed to glass due to better conductive properties necessary for the rapid freezing process. However, sapphire disks are expensive and can be replaced by relatively cheap aclar film as cell support (Jiménez *et al.*, 2006). Aclar films are bought as sheets and 1.5 mm circles have to be punched out. In our hands cell growth is similar on aclar compared to glass or plastic and the HPF procedure works equally well.
2. Grow cells to full confluency. This is important because then they will detach easily from the growth support.
3. Cells in suspension are also very suitable for HPF. We leave the cell suspension to sediment (do not spin), and pipette 1 μl of the thick cell suspension into the well of

a 100-μm-deep HPF membrane carrier (Leica Microsystems), which is frozen in the Leica EM PACT2.

4. We generally add a cryo-protectant to the growth medium, but this is not always required (Verkade, 2008). We prefer BSA as it improves the sectioning properties of the block.

b. Freeze-Substitution A short FS protocol for pure ultrastructural analysis (i.e., embedding in Epon/Araldite) has been described previously (Verkade, 2008). Here, we will focus on an FS protocol for embedding in Lowicryl (HM20) that provides both good ultrastructure and immunolabeling. We make use of the Automated Freeze Substitution device (AFS2, Leica Microsystems) equipped with the attachment for automated exchange of the reagents (Freeze Substitution Processor, FSP, Leica Microsystems). We particularly value the use of the FSP in these embedding experiments as it brings the researcher in minimal contact with the toxic Lowicryl resin. The FSP exchanges the media through a syringe that is connected with the sample chambers via flow-through baskets.

1. Bring the AFS2 to −90°C and fill the flow-through baskets with the FS "cocktail". In the example shown (Fig. 3B), we have used 0.1% uranyl acetate in acetone but we have also had good experience with adding an additional 0.01% osmium tetroxide or just incubation in pure acetone (see Brown *et al.*, 2009).
   - Uranyl acetate hardly dissolves in acetone. A 10% stock solution is made in methanol and frozen in LN_2. The stock solution is diluted 1:100 in acetone.
2. Perform FS using the FSP.
   - 24 h at −90°C in the FS cocktail.
   - Raise the temperature to −50°C with steps of 5°C/h.
   - 24 h at −50°C in the FS cocktail.
   - Three washes with pure acetone, 1 h each.
   - Gradual infiltration with Lowicryl HM20, mixed with acetone, 25, 50, 75, 100%, 2 h each step.
   - 24 h at −50°C with UV treatment.
   - Raise the temperature to 20°C, 5°C /h with UV light switched on.
   - 24–48 h at 20°C with UV light switched on. The length of this step is not critical, but ascertain that the Lowicryl is fully polymerized. With the current use of LEDs instead of the older UV light sources the polymerization has become more reliable.

We have recently shortened this Lowicryl embedding procedure based on successful experiments described elsewhere (Hawes *et al.*, 2007). The initial step at −90°C is shortened to only 5 h and all warm-up steps are done with an increase of 20°C/h. The ultrastructure of these samples looks very similar to samples processed using the longer procedure (not shown).

3. After the polymerization, the Lowicryl blocks are removed from the flow-through baskets before sectioning (see Verkade, 2008). The sapphire disks are removed from the resin by repeated submergence in boiling water followed by liquid nitrogen. Lowicryl is harder than Epon, but removal of the sapphire disk is actually much easier. This procedure leaves the cells directly exposed at the surface of the Lowicryl.

c. Sectioning and Immunolabeling

4. Sectioning of Lowicryl HM20 blocks is a bit different from Epon since the hydrophobic resin sections do not easily float on water. To reduce compression or even complete loss of the section, the water level is lowered in the knife trough so the initial gliding is done over the diamond surface.
5. Immunolabeling the Lowicryl sections is similar to labeling Tokuyasu sections; however, the quenching step using glycine or sodium borohydrate can be skipped since no aldehydes were used in the procedure. The sections are taken through the following steps at room temperature:
6. Incubate 15 min 0.1% BSA in PBS (blocking solution).
7. Label 1 h with the primary antibody in blocking solution (10 μl single grid or 20 μl for two grids/drop and so on). Note that antibodies need to be tested for Lowicryl labeling. For example, we use a goat anti-GFP (Rockland, 600-101-215) that works well on Lowicryl HM20 embedded samples processed following the above-described method (see Fig. 3B).
8. Wash three times with blocking solution.
9. Label 30–60 min with secondary antibody in blocking solution.
10. Wash three times with blocking solution.
11. Wash six times with distilled water.
12. Contrast with uranyl acetate and lead citrate.

4. Remarks

In general, immunolabeling efficiency after HPF/FS and plastic embedding in Lowicryl is considered to be lower than with the Tokuyasu immuno-EM method, although a direct comparison has not been described. One of the main advantages of using the HPF method is that cytoskeletal elements, which are an integral part of the membrane trafficking pathways, are well preserved (compare Figs. 2B and 3B).

C. Correlative Light Electron Microscopy

1. Introduction to CLEM

EM has given us many insights into cell biology, but its static nature makes it difficult if not impossible to reconstruct a highly dynamic process such as membrane trafficking. Combining the advantages of ultrafast fluorescent live-cell imaging with the potential of EM for high-resolution imaging is one of the major objectives of

CLEM (Brown *et al.*, 2009; McDonald, 2009; Verkade, 2008). Only advanced CLEM techniques can allow us to reliably visualize and characterize intracellular trafficking events at high resolution, e.g., only with the accompanying live-cell data is it possible, for example, to discriminate between fusion and budding of a membrane carrier in EM (Verkade, 2008). Trying to achieve this, however, introduces many technical challenges that have led to a wide variety of CLEM procedures. Some approaches focus on the retracing aspect where light microscopy is mainly used to locate a region of interest, e.g., the fluorescence indicates the presence of a cell (Plitzko *et al.*, 2009). Others highlight the live-cell imaging (Polishchuk *et al.*, 2000) or fixation aspect (Verkade, 2008). The choice of the CLEM method ultimately depends on the research question. For intracellular membrane transport studies, the LM part of a CLEM experiment should include a live-cell imaging component to reconstruct the dynamic process.

The need for CLEM can be illustrated by the fact that there still remains confusion over issues, such as the co-localization of Tf and EGF receptors throughout the endocytic pathway (Leonard *et al.*, 2008). Light microscopy alone (even when performed with a higher-resolution LM technique, such as total internal reflection fluorescence microscopy (TIRF-M)) lacks sufficient resolution to conclude whether these receptors are found on the same organelle. Furthermore, the reference space provides essential structural information on the subdomains within one organelle where these receptors are present (Fig. 4B).

In a landmark paper, introducing the term CLEM, the morphology of carriers operating between the Golgi apparatus and the plasma membrane was established (Polishchuk *et al.*, 2000). These carriers were followed using VSVG-GFP as a biosynthetic reporter protein and chemically fixed at time-specific time points. Detection of previously followed carriers in the EM was performed by immunolabeling of GFP with secondary antibodies attached to either gold or HRP to precipitate DAB. This approach allowed the authors to study the fine ultrastructure of the organelles involved in Golgi-to-plasma-membrane traffic and relate these structures back to their stage in this membrane trafficking pathway.

2. CLEM Markers

One of the most important decisions to make when performing a CLEM experiment is the choice of the marker or probe. A CLEM probe must be fluorescent in the LM and electron-dense in the EM. Different experimental aims also require different probes. Thus far, no genetically encoded CLEM probe has been described that can be expressed in the cytoplasm of living cells. The current CLEM markers can only enter the cell by live labeling through endocytosis or microinjection (Verkade, 2008).

FluoroNanoGold is composed of a fluorescent moiety, such as fluorescein or rhodamine, and an ultrasmall gold particle (0.8–1.4 nm) attached to either an antibody or to streptavidin (Robinson and Takizawa, 2009). Since ultrasmall gold is not directly visible in a standard EM, it has to be silver-enhanced for

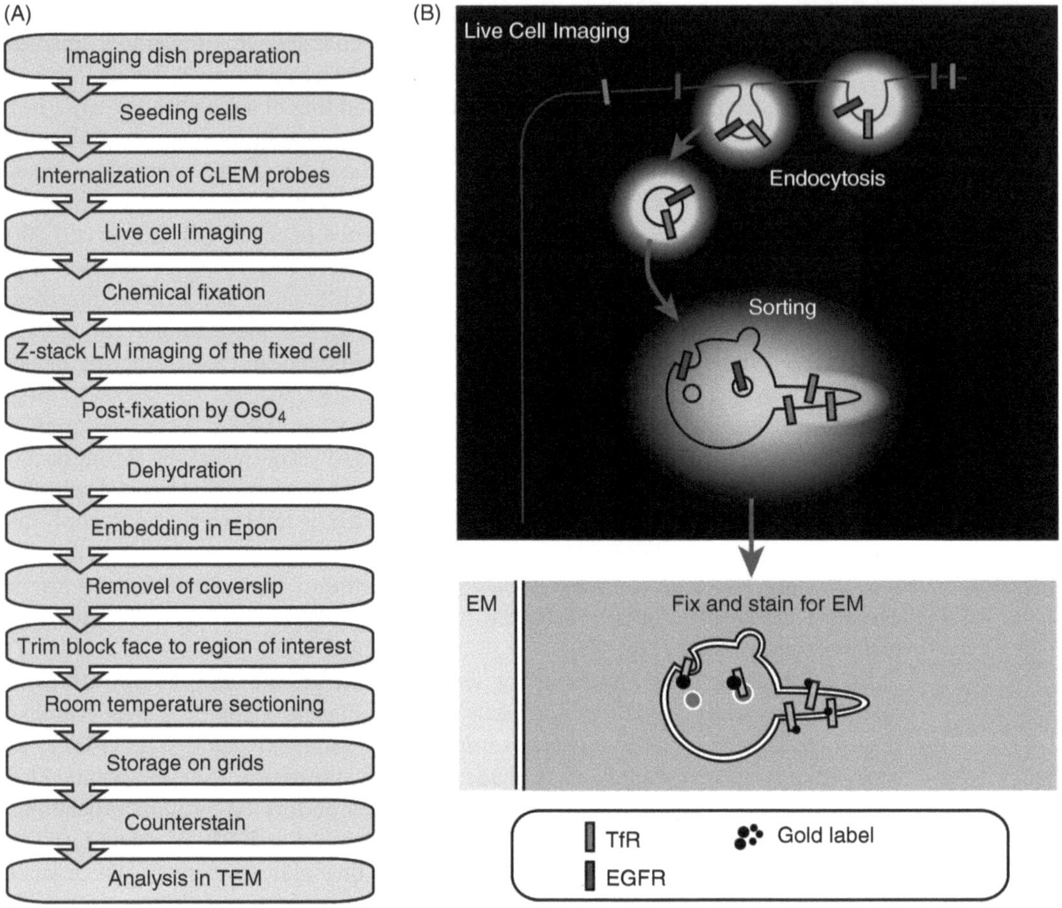

Fig. 4 The principle of CLEM. (A) Flow diagram of a typical CLEM experiment using the "simple" approach described in Section III.3.c consisting of live labeling, chemical fixation, and Epon embedding. OsO_4, osmium tetroxide. (B) Cartoon of a CLEM experiment on RFP-EGFR and GFP-TfR trafficking. Both types of cargo are internalized by endocytosis, showing up as yellow fluorescent compartments in live-cell imaging due to the co-localization of GFP and RFP. In the early endosome, GFP-TfR is sorted into tubular/vesicular carriers that retrieve the receptor to plasma membrane. In the fluorescence microscope, these carriers appear green due to the high concentration of GFP-TfR. The RFP-EGFR is collected in intraluminal vesicles in the endosomal vacuole, which shows up as predominately red. Although this sorting process is visible at the LM level, the endosome itself or its subdomains are not visible. By capturing the event by HPF and analyzing the sample by EM, it is possible to define the localization of GFP-TfR and RFP-EGFR in their different subdomains within the early endosome compartment. (See Plate no. 46 in the Color Plate Section.)

reliable imaging (Tchelidze *et al.*, 2006). The range of colors and sizes available as well as the high contrast achieved with gold particles have resulted in widespread use of these fluoro-gold probes throughout the CLEM community. Similarly, quantum dots (QDs) can be used to study endocytic events. QDs are brightly fluorescent nanocrystals with an electron-dense core usually comprising of either cadmium selenide or cadmium telluride (Giepmans *et al.*, 2005; Michalet *et al.*, 2001). QDs can be conjugated to antibodies or other targeting molecules, such as Protein-A or streptavidin (Giepmans *et al.*, 2006). Although QDs are available in colors that span the whole visible spectrum (Jaiswal and Simon, 2004), there are only three that have been reported to be reliably differentiated in the EM; spherical-blue QD525, spherical-green QD565, and cylindrical-red QD655 (Giepmans *et al.*, 2005). Because the cadmium in the quantum dot is less electron-dense than colloidal gold particles these particles can appear gray in the TEM (Giepmans *et al.*, 2005, but see Fig. 5K and N). As a probe for CLEM, QDs suffer from the same drawbacks as other exogenous labels: they can only be utilized for live-cell imaging after live labeling.

Live labeling can be circumvented by genetic manipulation of the protein of interest to adjoin either a fluorescent moiety such as GFP, or a targeting motif for the bi-arsenical molecules FlAsH and ReAsH. Fluorescein bi-arsenical hairpin-binding (FlAsH) and the resorufin red variant (ReAsH) are small, membrane-permeable fluorescent molecules that selectively label a tetracysteine motif (CCxxCC) appended to the protein of interest (Griffin *et al.*, 1998). Although well suited to imaging in the LM, as well as for more specialized techniques such as pulse-chase experiments (Gaietta *et al.*, 2002), neither FlAsH nor ReAsH are directly visible in the EM. However, ReAsH can photo-polymerize DAB in the same manner as HRP to form an osmiophilic precipitate (Adams *et al.*, 2002; Gaietta *et al.*, 2002). The drawbacks to using DAB include a diffuse staining and the possibility that DAB will drift from the site of production and fill the organelle/region surrounding the protein of interest. DAB is therefore a relatively low-resolution technique that should be used when the protein is encapsulated within an organelle.

Genetically encoded fluorescent proteins (Tsien, 2009), such as GFP and mRFP, are the most reliable targeting method for LM due to the one-to-one labeling ratio; if the protein is there it will be fluorescent and vice versa. Advantages include fluorescence, a wide variety of colors, and with the advent of photo-switchable and photo-convertible proteins the ability for super-resolution and pulse-chase experiments. The major limitation of GFP as a probe for CLEM is the total lack of visibility in the EM; proteins are not electron-dense and fluorescence is usually quenched by the fixation and embedding methods. Commercial antibodies to GFP are available but can only be applied after fixation and do not show one-to-one labeling efficiency. One method of visualization involves photo-conversion of DAB by production of singlet oxygen from fluorescing GFP (Monosov *et al.*, 1996). Known as GFP recognition after photobleaching (GRAB) the disadvantages of using DAB remain, although successful electron tomography has been demonstrated (Grabenbauer *et al.*, 2005; Meiblitzer-Ruppitsch *et al.*, 2008).

We have summarized and commented on the properties of the probes discussed above (Table I). In the next sections we will present applications of these probes for three different CLEM approaches: (1) a simple CLEM approach, (2) CLEM using Tokuyasu, and (3) CLEM using HPF.

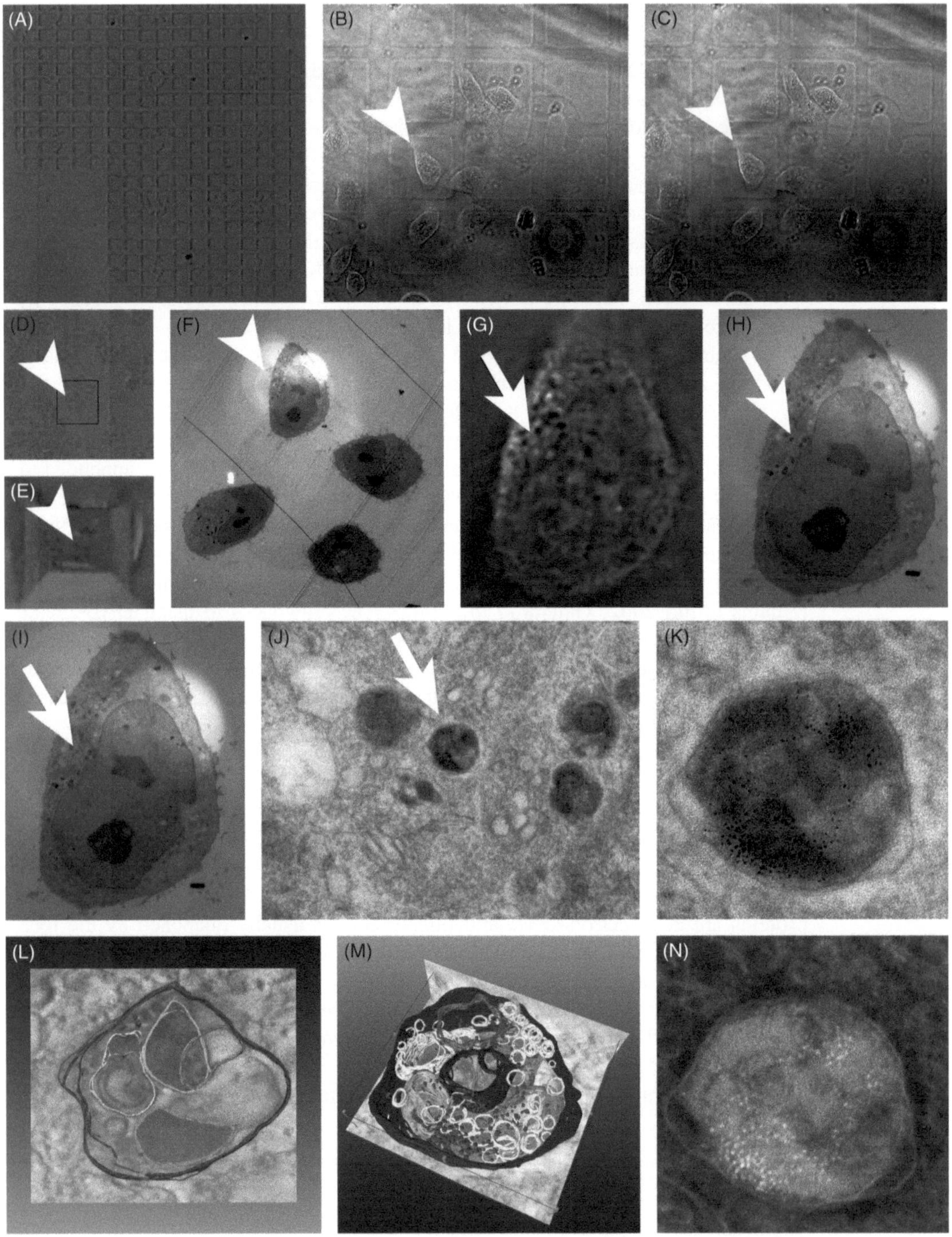

Fig. 5 (*Continued*)

Table I
Summary of Various Techniques and Probes Used for CLEM

	FPs	FlAsH/ReAsH	QDs	FluoroNanoGold	Immunogold	HRP
Live-cell LM	++	++	++[a]	+[a]	–	–
EM	–	–	+	++	++	+[b]
CLEM	–/+[b]	–/+[b]	++	+++	++	+[b]
Membrane penetration in live cell	NA	+	–	–	–	–
Penetration after fixation for EM	NA	NA	+	+	+	+
Multiprotein labeling	+	–	+++	++	+	–
Endogenous target protein	–	–	+	+	+	+
Pulse-chase	+[c]	++	–/+[a]	–/+[a]	–	–
Specificity	+++	++	+	+	+	+
Time delay for fluorescence	Minutes to hours	<20 min	0[a]	0[a]	NA	NA

Different techniques are scored from – (not suitable) to +++ (highly suited) based on the research reviewed above and on our own judgments. "Membrane penetration" indicates the ability of the probe to cross the plasma membrane non-invasively to label the intracellular target. "Penetration after fixation" refers to the depth of labeling in the sample that the probe can achieve after fixation with glutaraldehyde; HRP can be genetically attached to an intracellular target, in which case DAB is the penetrating reagent. "Multiprotein labeling" relies on different colors and sizes of the probe being distinguished in both LM and EM; for example, while there are two colors of FlAsH/ReAsH, they both bind to the same motif and therefore label the same protein. "Endogenous target proteins" are those with no genetic modification, so although labeling with antibodies enables endogenous protein targeting, both fluorescent proteins (FPs) and FlAsH/ReAsH require genetic modification. "Pulse-chase" is a live-cell imaging technique that allows the researcher to image the same protein at different time-points in different colors, i.e., newer proteins appear green whereas old ones are red. "Specificity" is a measurement of the fidelity of labeling; FPs label one to one due to the nature of genetic covalent modifications, whereas antibodies can label one protein more than once or not at all; FlAsH/ReAsH are comparable to antibodies. Longevity is the resistance of the probe to photobleaching, whereas phototoxicity is the ability of the probe to perturb the cell. NA = not applicable.

[a] Specifically for endocytosis events or direct microinjection of probes

[b] Requires photo-polymerization of DAB and osmium staining

[c] Only for photoconvertible FPs, such as Kaede

Fig. 5 A simple CLEM experiment including 3-D reconstruction by electron tomography. Example images of a CLEM experiment using the simple method described in Section "Introduction to CLEM". (A) The finder pattern of the embossed coverslip. (B) DIC image of HeLa cells grown on the embossed coverslip. Arrowhead points to the cell of interest (images B–F). (C) Overlay of B and a fluorescence image of the cells that have internalized EGF-biotin coupled to streptavidin-QD 655. (D) Stereomicroscopic image of the region of the cells of interest after chemical fixation and embedding in the Epon resin. (E) As D after trimming the block face. (F) EM overview image highlighting the area containing the cell of interest. (G) Zoomed image of the overlay of the DIC and fluorescence of the cell of interest. Arrow points to the organelle containing EGF-biotin coupled to streptavidin-QD 655 (image G–J). (H) EM image of the same region as G. (I) Overlay of the fluorescence image and the electron micrograph of the same region as G. (J) Electron micrograph of the subcellular area containing the organelle of interest. (K) Standard high-magnification bright field EM image of the organelle of interest. (L) Screen shot of the modeling part of an electron tomography experiment where structures of interest have to be manually segmented. (M) Screen shot of the modeling of the structure of interest. (N) HAADF image of the same region as K. (See Plate no. 47 in the Color Plate Section.)

3. A Simple CLEM Approach

CLEM techniques can be very technically challenging, aiming to combine ultrafast live-cell imaging with high immunolabeling efficiency and/or HPF for maximal ultrastructural preservation in EM (see Sections III.C.4 and III.C.5). For many questions, it might not be necessary to use such high-end techniques. When optimal preservation of antigenicity, instantaneous fixation, and/or maximal ultrastructural detail are not essential for the experiment, live-cell imaging followed by a simple chemical fixation of cells on an embossed coverslip may be sufficient (Fig. 4A).

Such a technique is significantly easier and has a much higher success rate than the other two methods discussed below (Sections III.C.4 and III.C.5). Here, we describe a detailed protocol for a simple CLEM approach, where we illustrate the degradative pathway taken by EGF receptors after internalization (Fig. 5).

a. Preparing Live-Cell Imaging Dish

1. Drill a 5 mm hole in the center of a standard 35 mm cell culture dish.
2. Attach an embossed coverslip using vacuum grease to the underside of the dish so that the raised finder pattern (Fig. 5A) is facing upwards and positioned within the hole.

b. Internalization of Endocytic Markers

3. Seed cells onto the coverslip and leave to grow to 30–70% confluence.
4. Prepare the CLEM marker. In our experiment, we pre-bind the EGF-biotin with Streptavidin-QD 655, 30 min at 37°C.
5. Remove growth media and replace with the CLEM marker solution (EGF-QD-655 in Fig. 5).
6. After a suitable incubation period remove the excess CLEM marker.
7. Wash the cells with warm PBS to remove any unbound ligands and add approximately 100 μl of imaging medium (phenol red-free imaging medium results in better quality LM images).

c. Live-Cell Imaging

8. Record the location of the cell of interest in relation to its surrounding by taking low-magnification bright field images (Fig. 5B).
9. Image fluorescence marker. Live-cell imaging through the embossed coverslip is compatible with almost any fluorescence microscope setup, with the exception of TIRF-M due to the varying thickness of glass coverslip.

d. Fixation and Embedding

10. Fix the cells by adding 1 ml of 2.5% glutaraldehyde in 0.1 M cacodylate buffer (pH 7.4) to the dish.
11. Record a z-stack of fluorescence images with ~0.2 μm steps before autofluorescence from fixation masks the fluorescent signal.

12. Remove the glutaraldehyde after 30 min of fixation and replace with cacodylate buffer.
13. Stain the cell by adding 1% osmium tetroxide in cacodylate buffer for 30 min.
14. Remove the osmium and wash with cacodylate buffer.
15. Replace cacodylate buffer with deionized water.
16. Dehydrate the sample in a graded series of ethanol going from 70 to 100% ethanol.
17. Embed in Epon and polymerize resin at 60°C for 24 h.

e. Relocation and Sectioning

18. Remove the coverslip by submerging the dish in LN_2 and immediate subsequent plunging into boiling water.
19. Carefully lift off the coverslip from the Epon block and cut away the cell culture dish. If the entire coverslip does not detach from the Epon block, repeat step 18.
20. Trace back the region of interest using the bright field images and a stereomicroscope. Trim region down to a suitable size (Fig. 5D and E).
21. Acquire serial sections of the trimmed block at the desired thickness (~70 nm for standard TEM, ~300 nm for tomography).
22. Contrast cells with uranyl acetate and lead citrate following standard protocols.
23. Relocate cells of interest (Fig. 5F). The structure of interest within a selected cell can be found with the aid of an overlay of the fluorescence onto the EM image (Fig. 5H–J). The structure of interest is now available for further detailed analysis, using standard bright field TEM (Fig. 5K), electron tomography (Fig. 5L and M), and scanning TEM (STEM) imaging (Fig. 5N, Section "Electron Tomography").

4. CLEM Using Tokuyasu

CLEM using the Tokuyasu method aims to combine dynamic live-cell imaging data with the immunolabeling efficiency of Tokuyasu sections (see Section III.A.1.). This method is technically challenging; a detailed description of the procedure is published elsewhere (van Rijnsoever *et al.*, 2008). Here, we will give a brief overview of the procedure.

An embossed coverslip is coated with Formvar followed by gelatin. This coverslip is mounted onto a culture dish in which cells expressing the fluorescent protein of interest are seeded. After live imaging, a PFA-based chemical fixation (e.g., 4% PFA and 0.05% GA) is applied to preserve antigenicity. Chemical fixation is intrinsically slow, therefore this CLEM approach is not recommended for studying highly dynamic processes or studying structures that are not well preserved during the fixation process. An advantage of this fixation method is that one can record a z-stack of fluorescent images of the fixed material for later relocation of the same structure in EM, while this is not possible when using cryofixation (see Section III.3.e.). DIC images are also taken to map the location of the cell on the grid. After post-fixation overnight in 2% PFA and 0.2% GA, the cell is relocated and marked with nail varnish. Cells are subsequently embedded in 6% gelatin (not 12%, which is used in standard Tokuyasu cryo-immuno-EM protocol, since the gelatin will be fixed later on). A layer of erythrocytes in 12% gelatin is added and solidified to facilitate orientation of the block. The sample is further fixed with 0.5% PFA and incubated with cold sucrose for 3 days on a rocker. A block containing the region of interest is cut out and mounted on a specimen pin. The specimen is then frozen in liquid nitrogen, serial cryosectioned at −100°C, and sections are collected in droplets of methylcellulose/sucrose. Every fifth grid is immunolabeled and viewed in EM. Then, using the z-stack from the LM, the sections containing the structures of interest can be narrowed down and immunolabeled for detailed analysis.

5. CLEM Using HPF

For optimal preservation of ultrastructure, the most suitable CLEM experiment is HPF with FS. This technique usually focuses on the trafficking of internalized CLEM markers rather then immunolabeling of fluorescent proteins. Surface immunolabeling of the section for endogenous proteins is possible when combined with FS to Lowicryl HM20, although this remains technically challenging. HPF-CLEM can combine live-cell imaging with the superior fixation quality of high-pressure freezing (Verkade, 2008). A disadvantage compared with Tokuyasu CLEM is that the reference z-stack for serial sectioning cannot be taken after cryofixation since there is currently no device that can integrate cryo-immobilization with light microscopy (see also Buser, Chapter 10 in this volume). Therefore, it is necessary to record the z-stack before physically transferring the sample between the light microscope and the HPF apparatus. When studying processes such as membrane trafficking, the speed of this transfer is critical. To maximally reduce the time interval between imaging and HPF, we perform live-cell imaging with the sample already held in the HPF rapid loader (Verkade, 2008). An unavoidable consequence of this setup is that it creates a "distance" between the cell and objective. With the modifications described previously (Brown *et al.*, 2009) this distance has been reduced and the system of cell retracing improved. Briefly, the technique involves imaging cells grown on sapphire disks with a carbon-coated finder pattern. The sapphire disks are held in modified HPF carriers and imaging is performed in an imaging dish containing medium with 20% BSA to act as a filler and cryoprotectant during HPF. Another modified rapid loader has recently been described that allows imaging in an intact imaging dish (McDonald, 2009). The time between recording the last image on the light microscope and having a fixed specimen can be as little as 4 s (Verkade, 2008).

Frozen samples have to undergo an FS protocol and resin embedding with either Epon or Lowicryl as described above (Section "HPF Post-Processing"). Cells are relocated with the aid of the carbon finder pattern, which transfers to the surface of the polymerized resin block. After trimming down to the area of interest, the block is serially sectioned with a diamond knife using a standard room temperature ultramicrotome. Achieving serial sections under room temperature conditions is significantly less challenging as compared to serial cryosectioning for the Tokuyasu approach.

6. Remarks

CLEM techniques are valuable tools that allow the cell biologist to meet the two main criteria for answering intracellular membrane trafficking questions: combining high spatial resolution with dynamic information of the trafficking pathway of interest. In Table II, we compare the three CLEM techniques discussed above and highlight the strengths and weaknesses of each. Since CLEM can be time consuming and technically challenging, it is important to choose the appropriate method that can provide an answer to a specific research question.

D. Electron Tomography

1. Principles and Use of Electron Tomography

The geometry and size of the membrane carriers are thought to play an important part in efficient sorting and transport of specific cargo (Carlton and Cullen, 2005; Traer *et al.*, 2007). TEM usually produces a two-dimensional projection of a three-dimensional sample, which makes the geometry of organelles in the sample hard to interpret. For example, cellular features that are seen as a single entity may actually be small features in different planes within the section. Furthermore, organelles and other cellular structures may only be partially present in the standard 70 nm sections, which can make interpretation difficult. For instance, vesicles and perpendicular-sectioned tubules both appear as circular projections in TEM. In line with this, structures might

Table II
Comparison of Three Different CLEM Techniques, Highlighting Some Properties of Each Technique and the Advantages and Disadvantages

	Tokuyasu	HPF	Simple
Quality of live-cell imaging	Good	Not optimal due to distance between cells and objective	Good
Labels	Internalized markers and high-efficiency immunolabeling	Internalized markers and surface immunolabeling with HM20	Only internalized markers
Time resolution	Seconds to minutes	4 s	Seconds to minutes
Processing	Technically challenging	Automated if using an AFS with FSP	Simple
Retracing cells	Embossed coverslip	Finder grids or carbon-coated finder pattern	Embossed coverslip
Sectioning	Cryosectioning; difficult to achieve serial sections	Standard diamond knife sectioning of resin block	Standard diamond knife sectioning of resin block
Preservation of ultrastructure	Good	Excellent	Good
Time to complete experiment	4–5 days	4–7 days	2–3 days
Pitfalls	Formvar film too thick, cryosectioning, immunolabeling	Making the live cell carrier, achieving a well-frozen sample	Removal of the coverslip

For further comments see Section III.3.

be bent, coming in and out of the section plane, and appear as two separate structures in the TEM projection. We can solve these difficulties by either combining serial thin sections or even better by viewing a semi-thick sample from different tilt angles (Frey *et al.*, 2006; McIntosh *et al.*, 2005). If a whole series of tilted images is collected, it is possible to computationally reconstruct a three-dimensional view or "tomogram". Unless samples are specially shaped into a "rod" using a focused ion beam (FIB) instrument, we are unable to collect data through a full 360° rotation, which would yield perfect reconstructions. Resolution in electron tomography is limited by the sample holder geometry. Since the FIB preparation of samples for tomography is not routinely established yet (Marko *et al.*, 2007), reconstructions usually contain a "missing wedge" of information.

To create a tomogram, the sample is generally tilted from −65° to +65°, typically with 1° increments (see also Chapters 4, 6, and 23 by O'Toole, Giddings *et al.*, and Hoenger *et al.*, this volume). Reconstructions can be improved by collecting so-called dual axes series, where a second set of data is collected from the same sample area at 90° to the first and the resulting tomograms are merged. Better results are achieved with smaller tilt increments, but a balance must then be found between acquisition time and electron dose to the sample to obtain the optimal image quality.

The images in a tilt series need to be further aligned and then the tomogram can be mathematically reconstructed using Fourier transform backprojection algorithms (see e.g. Frank, 2006). The alignment can be improved by the use of "fiducials" or large gold markers applied to the top and bottom surfaces of the sample. These are then used in feature tracking software to better align images in the series. Many different software packages and procedures are available for performing reconstructions and helping with visualizations, including freeware packages such as IMOD (Kremer *et al.*, 1996; Mastronarde, 1997). Finally, these structures can be rendered as three-dimensional models (Fig. 5L and M and supplementary movie 1 at http://www.elsevierdirect.com/companions/9780123810076). It should be noted that modeling of the tomogram is time consuming and can introduce subjectivity.

The thickness of a specimen that can be imaged depends on the microscope performance (mostly the acceleration voltage). Examples where electron tomography has been used include visualization of the complex structure of the Golgi apparatus (Ladinsky *et al.*, 1994, 1999), structural studies of mitochondria (Perkins *et al.*, 1997, 2009), and the microtubule cytoskeleton and centrosomes (Moritz *et al.*, 1995; O'Toole *et al.*, 2003). Ladinsky and co-workers combined serial tomographic reconstructions to create a 1 μm^3 volume containing the Golgi complex from several 300 nm tomographic sections (Ladinsky *et al.*, 1999). In our own lab, we have recently combined five serial 300 nm sections to reconstruct a multivesicular body labeled with different markers (Brown *et al.*, 2009).

Nowadays, sections with thicknesses of 200–300 nm can also be analyzed routinely by tomography using conventional 120 kV TEM mainly due to the implementation of computer-assisted acquisition available on modern microscopes. With the advance of intermediate-voltage electron microscopes with field emission guns and the combination with zero-loss electron energy loss spectroscopy (EELS) or STEM techniques, single sections of up to 1–2 μm thickness have been reconstructed (Beorchia *et al.*, 1993; Sousa *et al.*, 2009) (see also Chapter 25 by Walther *et al.*, this volume). The so-called high-angle annular dark field (HAADF) STEM detector collects electrons scattered to high diffraction angles by the sample. The signal is proportional to the square of the atomic number of the atoms in the sample. This opens up the possibility of discriminating similar-sized particles of different atomic compositions (e.g., silver vs. gold) (Sousa *et al.*, 2007). HAADF images can be superior for reconstructions due to a much better signal-to-noise ratio (Yakushevska *et al.*, 2007). As an example, whereas QDs are not easily visualized in standard bright field EM mode (Fig. 5K), the quantum dots "light up" using HAADF STEM (Fig. 5N; see also Brown *et al.*, 2009). However, it takes significantly longer to collect the data in the STEM mode compared to bright field TEM.

In electron tomography we should be aware of sample shrinkage. Due to locally produced heat leading to evaporation, the section will shrink during repetitive beam exposure. In order to facilitate alignment of the collected images the sample has to be pre-irradiated (i.e., pre-illuminated) before starting the collection of tilted views. This shrinkage of the section is especially an issue when tomography is applied to Tokuyasu and native cryosections, which are considerably more "vulnerable" to beam damage than resin sections. Electron tomography has been successfully applied on Tokuyasu cryosections, especially using nanogold markers to improve the labeling efficiency of the thick section (Tchelidze *et al.*, 2006; Zeuschner *et al.*, 2006). In our own lab, we recently used thick immunogold-labeled Tokuyasu cryosections to demonstrate the localization of different proteins of the COPII complex on budding vesicles of the transitional endoplasmic reticulum (Hughes *et al.*, 2009). Observing the sample at low temperature using a cryo-specimen holder, where the sample is held at liquid nitrogen temperatures, may help to reduce shrinkage but adds an extra degree of complexity to the experiment. This holder can also be used to perform tomography on frozen samples, cryo-electron tomography (cryo-ET), which is described in more depth by Kirmse *et al.*, Chapter 23 in this volume. A beautiful example of cryo-ET applied in the field of membrane traffic is the imaging of clathrin-coated transport vesicles (Cheng *et al.*, 2007). The technique can, in principle, be used to freeze whole cells and analyze the structure of protein complexes in their natural environment in nanometer resolution (reviewed by Baumeister *et al.*, 1999 and demonstrated by Medalia *et al.*, 2002). This detail is essential to answer some of the key questions on the molecular mechanism of membrane traffic and cargo sorting.

IV. Outlook

Although intracellular membrane transport is a highly dynamic process, the importance of "still" images at high resolution as provided by EM is still invaluable. The introduction of new EM techniques, such as CLEM and electron tomography, especially in combination with immunolabeling techniques, will open up new avenues of research to gain even deeper insights into the mechanisms of transport. In this paper we have presented an overview of the most important techniques currently available for transport studies. We are convinced the use of live-cell imaging techniques in combination with high-end EM techniques, such as CLEM and tomography, will soon set the standard for membrane transport studies.

V. Materials and Equipment

A. Equipment

AFS2 + FSP (Leica Microsystems, Vienna, Austria)
Cryo-ultramicrotome EM-UC6 + FC6 (Leica Microsystems, Vienna, Austria)
Diamond knife (Diatome, Biel, Switzerland)
EM PACT2 + RTS (Leica Microsystems, Vienna, Austria)
Forceps (Manufactures des Outils Dumont, Switzerland)

B. Materials

Acetone (EMS, Hatfield, USA)
BSA, fraction V (Sigma-Aldrich Co., Poole, UK)
Cell culture dish, 35 mm (Corning Inc., Corning, USA)
CO_2-independent medium (GIBCO/Invitrogen Life Technologies, Paisley, UK)
EGF-Biotin (Invitrogen Life Technologies, Paisley, UK)
Embossed coverslip (MatTek Inc., Ashland, USA)
Epon812 (TAAB, Aldermaston, UK)
Ethanol (Fisher, Loughborough, England)
GFP antibody, 600-101-215 Goat (Rockland Inc., Gilbertsville, USA)
GFP antibody, A11122 Rabbit (Invitrogen Life Technologies, Paisley, UK)
Glutaraldehyde, 25% EM grade (Agar Scientific, Stansted, UK)
Lowicryl HM20 Mono Step (Agar Scientific, Stansted, UK)
Membrane carrier, 100 μm deep (Leica Microsystems, Vienna, Austria)
Methanol (Fisher, Loughborough, UK)
Osmium tetroxide (EMS, Hatfield, USA)
Phenol red-free medium (GIBCO/Invitrogen Life Technologies, Paisley, UK)
Sapphire disks, 1.5 mm (Leica Microsystems, Vienna, Austria)
Sodium cacodylate buffer (EMS, Hatfield, USA)
Streptavidin-Quatum dot 655 (Invitrogen Life Technologies, Paisley, UK)
Uranyl acetate (Agar Scientific, Stansted, UK)

Acknowledgments

The Wellcome Trust Fund supports JRTvW and JM; EB is supported by BBSRC and THS is supported by the EPSRC.

References

Adams, S. R., Campbell, R. E., Gross, L. A., Martin, B. R., Walkup, G. K., Yao, Y., Llopis, J., and Tsien, R. Y. (2002). New biarsenical ligands and tetracysteine motifs for protein labeling *in vitro* and *in vivo*: Synthesis and biological applications. *J. Am. Chem. Soc.* **124**, 6063–6076.

Balaji, J., and Ryan, T. A. (2007). Single-vesicle imaging reveals that synaptic vesicle exocytosis and endocytosis are coupled by a single stochastic mode. *Proc. Natl. Acad. Sci. U.S.A.* **104**, 20576–20581.

Baumeister, W., Grimm, R., and Walz, J. (1999). Electron tomography of molecules and cells. *Trends Cell Biol.* **9**, 81–85.

Beorchia, A., Heliot, L., Menager, M., Kaplan, H., and Ploton, D. (1993). Applications of medium-voltage STEM for the 3-D study of organelles within very thick sections. *J. Microsc.* **170**, 247–258.

Betzig, E., Patterson, G. H., Sougrat, R., Lindwasser, O. W., Olenych, S., Bonifacino, J. S., Davidson, M. W., Lippincott-Schwartz, J., and Hess, H. F. (2006). Imaging intracellular fluorescent proteins at nanometer resolution. *Science* **313**, 1642–1645.

Brown, E., Mantell, J., Carter, D. A., Tilly, G., and Verkade, P. (2009). Studying intracellular transport using high-pressure freezing and correlative light electron microscopy. *Semin. Cell Dev. Biol.* **20**, 910–919.

Carlton, J. G., and Cullen, P. J. (2005). Coincidence detection in phosphoinositide signaling. *Trends Cell Biol.* **15**, 540–547.

Ceccarelli, B., Hurlbut, W. P., and Mauro, A. (1973). Turnover of transmitter and synaptic vesicles at the frog neuromuscular junction. *J. Cell Biol.* **57**, 499–524.

Cheng, Y., Boll, W., Kirchhausen, T., Harrison, S. C., and Walz, T. (2007). Cryo-electron tomography of clathrin-coated vesicles: Structural implications for coat assembly. *J. Mol. Biol.* **365**, 892–899.

D'Amico, F., and Skarmoutsou, E. (2008). Quantifying immunogold labelling in transmission electron microscopy. *J. Microsc.* **230**, 9–15.

Essner, E., and Novikoff, A. B. (1961). Localization of acid phosphatase activity in hepatic lysosomes by means of electron microscopy. *J. Biophys. Biochem. Cytol.* **9**, 773–784.

Frank, J. 2006. Electron Tomography: Methods for Three-Dimensional Visualization of Structures in the Cell. Springer Science + Business Media, New York, NY, USA.

Frey, T. G., Perkins, G. A., and Ellisman, M. H. (2006). Electron tomography of membrane-bound cellular organelles. *Annu. Rev. Biophys. Biomol. Struct.* **35**, 199–224.

Gaietta, G., Deerinck, T. J., Adams, S. R., Bouwer, J., Tour, O., Laird, D. W., Sosinsky, G. E., Tsien, R. Y., and Ellisman, M. H. (2002). Multicolor and electron microscopic imaging of connexin trafficking. *Science* **296**, 503–507.

Giepmans, B. N., Adams, S. R., Ellisman, M. H., and Tsien, R. Y. (2006). The fluorescent toolbox for assessing protein location and function. *Science* **312**, 217–224.

Giepmans, B. N., Deerinck, T. J., Smarr, B. L., Jones, Y. Z., and Ellisman, M. H. (2005). Correlated light and electron microscopic imaging of multiple endogenous proteins using quantum dots. *Nat. Methods* **2**, 743–749.

Gorden, P., Carpentier, J. L., Cohen, S., and Orci, L. (1978). Epidermal growth factor: Morphological demonstration of binding, internalization, and lysosomal association in human fibroblasts. *Proc. Natl. Acad. Sci. U.S.A.* **75**, 5025–5029.

Grabenbauer, M., Geerts, W. J., Fernadez-Rodriguez, J., Hoenger, A., Koster, A. J., and Nilsson, T. (2005). Correlative microscopy and electron tomography of GFP through photooxidation. *Nat. Methods* **2**, 857–862.

Granseth, B., Odermatt, B., Royle, S. J., and Lagnado, L. (2006). Clathrin-mediated endocytosis is the dominant mechanism of vesicle retrieval at hippocampal synapses. *Neuron* **51**, 773–786.

Griffin, B. A., Adams, S. R., and Tsien, R. Y. (1998). Specific covalent labeling of recombinant protein molecules inside live cells. *Science* **281**, 269–272.

Griffiths, G. (1993). "Fine Structure Immunocytochemistry." Springer-Verlag, Berlin Heidelberg.

Griffiths, G., and Hoppeler, H. (1986). Quantitation in immunocytochemistry: Correlation of immunogold labeling to absolute number of membrane antigens. *J. Histochem. Cytochem.* **34**, 1389–1398.

Gustafsson, M. G. (2005). Nonlinear structured-illumination microscopy: Wide-field fluorescence imaging with theoretically unlimited resolution. *Proc. Natl. Acad. Sci. U.S.A.* **102**, 13081–13086.

Haigler, H. T., McKanna, J. A., and Cohen, S. (1979). Direct visualization of the binding and internalization of a ferritin conjugate of epidermal growth factor in human carcinoma cells A-431. *J. Cell Biol.* **81**, 382–395.

Harlow, E., and Lane, D. (1988). "Antibodies: A Laboratory Manual." Cold Spring Harbor Laboratory Press, Cold Spring Habor, NY, USA.

Hawes, P., Netherton, C. L., Mueller, M., Wileman, T., and Monaghan, P. (2007). Rapid freeze-substitution preserves membranes in high-pressure frozen tissue culture cells. *J. Microsc.* **226**, 182–189.

Hayat, M. A. (2000). "Principles and Techniques of Electron Microscopy: Biological Applications." Cambridge University Press, Cambridge, UK.

Heuser, J. E., and Reese, T. S. (1973). Evidence for recycling of synaptic vesicle membrane during transmitter release at the frog neuromuscular junction. *J. Cell Biol.* **57**, 315–344.

Hopkins, C. R., Gibson, A., Shipman, M., Strickland, D. K., and Trowbridge, I. S. (1994). In migrating fibroblasts, recycling receptors are concentrated in narrow tubules in the pericentriolar area, and then routed to the plasma membrane of the leading lamella. *J. Cell Biol.* **125**, 1265–1274.

Hughes, H., Budnik, A., Schmidt, K., Palmer, K. J., Mantell, J., Noakes, C., Johnson, A., Carter, D. A., Verkade, P., Watson, P., and Stephens, D. J. (2009). Organisation of human ER-exit sites: Requirements for the localisation of sec16 to transitional ER. *J. Cell Sci.* **122**, 2924–2934.

Jaiswal, J. K., and Simon, S. M. (2004). Potentials and pitfalls of fluorescent quantum dots for biological imaging. *Trends Cell Biol.* **14**, 497–504.

Jiménez, N., Humbel, B. M., van Donselaar, E., Verkleij, A. J., and Burger, K. N. (2006). Aclar discs: A versatile substrate for routine high-pressure freezing of mammalian cell monolayers. *J. Microsc.* **221**, 216–223.

Kellenberger, E., Johansen, R., Maeder, M., Bohrmann, B., Stauffer, E., and Villiger, W. (1992). Artefacts and morphological changes during chemical fixation. *J. Microsc.* **168**, 181–201.

Kremer, J. R., Mastronarde, D. N., and McIntosh, J. R. (1996). Computer visualization of three-dimensional image data using IMOD. *J. Struct. Biol.* **116**, 71–76.

Ladinsky, M. S., Kremer, J. R., Furcinitti, P. S., McIntosh, J. R., and Howell, K. E. (1994). HVEM tomography of the trans-Golgi network: Structural insights and identification of a lace-like vesicle coat. *J. Cell Biol.* **127**, 29–38.

Ladinsky, M. S., Mastronarde, D. N., McIntosh, J. R., Howell, K. E., and Staehelin, L. A. (1999). Golgi structure in three dimensions: Functional insights from the normal rat kidney cell. *J. Cell Biol.* **144**, 1135–1149.

Leonard, D., Hayakawa, A., Lawe, D., Lambright, D., Bellve, K. D., Standley, C., Lifshitz, L. M., Fogarty, K. E., and Corvera, S. (2008). Sorting of EGF and transferrin at the plasma membrane and by cargo-specific signaling to EEA1-enriched endosomes. *J. Cell Sci.* **121**, 3445–3458.

Liou, W., Geuze, H. J., and Slot, J. W. (1996). Improving structural integrity of cryosections for immunogold labeling. *Histochem. Cell Biol.* **106**, 41–58.

Mari, M., Bujny, M., Zeuschner, D., Geerts, W., Griffith, J., Petersen, C., Cullen, P. J., Klumperman, J., and Geuze, H. (2008). SNX1 defines an early endosomal recycling exit for sortilin and mannose 6-phosphate receptors. *Traffic* **9**, 380–393.

Marko, M., Hsieh, C., Schalek, R., Frank, J., and Mannella, C. (2007). Focused-ion-beam thinning of frozen-hydrated biological specimens for cryo-electron microscopy. *Nat. Methods* **4**, 215–217.

Mastronarde, D. N. (1997). Dual-axis tomography: An approach with alignment methods that preserve resolution. *J. Struct. Biol.* **120**, 343–352.

McDonald, K. L. (2007). Cryopreparation methods for electron microscopy of selected model systems. *Methods Cell Biol.* **79**, 23–56.

McDonald, K. L. (2009). A review of high-pressure freezing preparation techniques for correlative light and electron microscopy of the same cells and tissues. *J. Microsc.* **235**, 273–281.

McDonald, K. L., Morphew, M., Verkade, P., and Muller-Reichert, T. (2007). Recent advances in high-pressure freezing: Equipment- and specimen-loading methods. *Methods Mol. Biol.* **369**, 143–173.

McIntosh, R., Nicastro, D., and Mastronarde, D. (2005). New views of cells in 3d: An introduction to electron tomography. *Trends Cell Biol.* **15**, 43–51.

McMahon, H. T., and Gallop, J. L. (2005). Membrane curvature and mechanisms of dynamic cell membrane remodelling. *Nature* **438**, 590–596.

Medalia, O., Weber, I., Frangakis, A. S., Nicastro, D., Gerisch, G., and Baumeister, W. (2002). Macromolecular architecture in eukaryotic cells visualized by cryoelectron tomography. *Science* **298**, 1209–1213.

Meiblitzer-Ruppitsch, C., Vetterlein, M., Stangl, H., Maier, S., Neumüller, J., Freissmuth, M., Pavelka, M., and Ellinger, A. (2008). Electron microscopic visualization of fluorescent signals in cellular compartments and organelles by means of DAB-photoconversion. *Histochem. Cell Biol.* **130**, 407–419.

Michalet, X., Pinaud, F., Lacoste, T. D., and Dahan, M. (2001). Properties of fluorescent semiconductor nanocrystals and their application to biological labeling. *Single Mol.* **2**, 261–276.

Monosov, E. Z., Wenzel, T. J., Luers, G. H., Heyman, J. A., and Subramani, S. (1996). Labeling of peroxisomes with green fluorescent protein in living *P. pastoris* cells. *J. Histochem. Cytochem.* **44**, 581–589.

Moor, H. (1987). Theory and practice of high pressure freezing. 175–191. *In* "Cryo-Techniques in Biological Electron Microscopy" (R. A. Steinbrecht and K. Zierold, eds.). Springer-Verlag, Berlin Heidelberg.

Moritz, M., Braunfeld, M. B., Sedat, J. W., Alberts, B., and Agard, D. A. (1995). Microtubule nucleation by gamma-tubulin-containing rings in the centrosome. *Nature* **378**, 638–640.

Murk, J.L.A.N., Posthuma, G., Koster, A. J., Geuze, H. J., Verkleij, A. J., Kleijmeer, M. J., and Humbel, B. M. (2003). Influence of aldehyde fixation on the morphology of endosomes and lysosomes: Quantitative analysis and electron tomography. *J. Microsc.* **212**, 81–90.

Oorschot, V., de Wit, H., Annaert, W., and Klumperman, J. (2002). A novel flat-embedding method to prepare ultrathin cryosections from cultured cells in their *in situ* orientation. *J. Histochem. Cytochem.* **50**, 1067–1080.

O'Toole, E. T., McDonald, K. L., Mäntler, J., McIntosh, J. R., Hyman, A. A., and Müller-Reichert, T. (2003). Morphologically distinct microtubule ends in the mitotic centrosome of *Caenorhabditis elegans*. *J. Cell Biol.* **163**, 451–456.

Penttila, A., Kalimo, H., and Trump, B. F. (1974). Influence of glutaraldehyde and-or osmium tetroxide on cell volume, ion content, mechanical stability, and membrane permeability of ehrlich ascites tumor cells. *J. Cell Biol.* **63**, 197–214.

Perkins, G. A., Renken, C., Martone, M. E., Young, S. J., Ellisman, M. H., and Frey, T. G. (1997). Electron tomography of neuronal mitochondria: Three-dimensional structure and organization of cristae and membrane contacts. *J. Struct. Biol.* **119**, 260–272.

Perkins, G. A., Sun, M. G., and Frey, T. G. (2009). Chapter 2 correlated light and electron microscopy/electron tomography of mitochondria *in situ*. *Meth. Enzymol.* **456**, 29–52.

Plitzko, J. M., Rigort, A., and Leis, A. (2009). Correlative cryo-light microscopy and cryo-electron tomography: From cellular territories to molecular landscapes. *Curr. Opin. Biotechnol.* **20**, 83–89.

Polishchuk, R. S., Polishchuk, E. V., Marra, P., Alberti, S., Buccione, R., Luini, A., and Mironov, A. A. (2000). Correlative light-electron microscopy reveals the tubular-saccular ultrastructure of carriers operating between Golgi apparatus and plasma membrane. *J. Cell Biol.* **148**, 45–58.

Rink, J., Ghigo, E., Kalaidzidis, Y., and Zerial, M. (2005). Rab conversion as a mechanism of progression from early to late endosomes. *Cell* **122**, 735–749.

Ripper, D., Schwarz, H., and Stierhof, Y. D. (2008). Cryo-section immunolabelling of difficult to preserve specimens: Advantages of cryofixation, freeze-substitution and rehydration. *Biol. Cell* **100**, 109–123.

Robinson, J. M., and Takizawa, T. (2009). Correlative fluorescence and electron microscopy in tissues: Immunocytochemistry. *J. Microsc.* **235**, 259–272.

Rosenmund, C., and Stevens, C. F. (1997). The rate of aldehyde fixation of the exocytotic machinery in cultured hippocampal synapses. *J. Neurosci. Methods* **76**, 1–5.

Rust, M. J., Bates, M., and Zhuang, X. (2006). Sub-diffraction-limit imaging by stochastic optical reconstruction microscopy (STORM). *Nat. Methods* **3**, 793–795.

Schönle, A., Glatz, M., and Hell, S. W. (2000). Four-dimensional multiphoton microscopy with time-correlated single-photon counting. *Appl. Opt.* **39**, 6306–6311.

Seligman, A. M., Karnovsky, M. J., Wasserkrug, H. L., and Hanker, J. S. (1968). Nondroplet ultrastructural demonstration of cytochrome oxidase activity with a polymerizing osmiophilic reagent, diaminobenzidine (DAB). *J. Cell Biol.* **38**, 1–14.

Singer, S. J. (1959). Preparation of an electron-dense antibody conjugate. *Nature* **183**, 1523–1524.
Slot, J. W., and Geuze, H. J. (2007). Cryosectioning and immunolabeling. *Nat. Protoc.* **2**, 2480–2491.
Small, J. V. (1981). Organization of actin in the leading edge of cultured cells: Influence of osmium tetroxide and dehydration on the ultrastructure of actin meshworks. *J. Cell Biol.* **91**, 695–705.
Sousa, A. A., Aronova, M. A., Kim, Y. C., Dorward, L. M., Zhang, G., and Leapman, R. D. (2007). On the feasibility of visualizing ultrasmall gold labels in biological specimens by STEM tomography. *J. Struct. Biol.* **159**, 507–522.
Sousa, A. A., Hohmann-Marriott, M. F., Zhang, G., and Leapman, R. D. (2009). Monte Carlo electron-trajectory simulations in bright-field and dark-field STEM: Implications for tomography of thick biological sections. *Ultramicroscopy* **109**, 213–221.
Steinbrecht, R. A., and Zierold, K. (1984). A cryoembedding method for cutting ultrathin cryosections from small frozen specimens. *J. Microsc.* **136**, 69–75.
Stoscheck, C. M., and Carpenter, G. (1984). Down regulation of epidermal growth factor receptors: Direct demonstration of receptor degradation in human fibroblasts. *J. Cell Biol.* **98**, 1048–1053.
Studer, D., Humbel, B. M., and Chiquet, M. (2008). Electron microscopy of high pressure frozen samples: Bridging the gap between cellular ultrastructure and atomic resolution. *Histochem. Cell Biol.* **130**, 877–889.
Tchelidze, P., Sauvage, C., Bonnet, N., Kilian, L., Beorchia, A., O'Donohue, M. F., Ploton, D., and Kaplan, H. (2006). Electron tomography of amplified nanogold immunolabelling: Improvement of quality based on alignment of projections with sinograms and use of post-reconstruction deconvolution. *J. Struct. Biol.* **156**, 421–431.
Tokuyasu, K. T. (1973). A technique for ultracryotomy of cell suspensions and tissues. *J. Cell Biol.* **57**, 551–565.
Traer, C., Rutherford, A., Palmer, K., Wassmer, T., Oakley, J., Attar, N., Carlton, J., Kremerskothen, J., Stephens, D., and Cullen, P. J. (2007). SNX4 coordinates endosomal sorting of TfnR with dynein-mediated transport into the endocytic recycling compartment. *Nat. Cell Biol.* **9**, 1370–1380.
Tsien, R. Y. (2009). Constructing and exploiting the fluorescent protein paintbox (Nobel lecture). *Angew. Chem. Int. Ed. Engl.* **48**, 5612–5626.
van Donselaar, E., Posthuma, G., Zeuschner, D., Humbel, B., and Slot, J. (2007). Immunogold labeling of cryosections from high-pressure frozen cells. *Traffic* **8**, 471–485.
van Rijnsoever, C., Oorschot, V., and Klumperman, J. (2008). Correlative light-electron microscopy (CLEM) combining live-cell imaging and immunolabeling of ultrathin cryosections. *Nat. Methods* **5**, 973–980.
Verkade, P. (2008). Moving EM: The rapid transfer system as a new tool for correlative light and electron microscopy and high throughput for high-pressure freezing. *J. Microsc.* **230**, 317–328.
Verkade, P., Schrama, L. H., Verkleij, A. J., Gispen, W. H., and Oestreicher, A. B. (1997). Ultrastructural co-localization of calmodulin and B-50/growth-associated protein-43 at the plasma membrane of proximal unmyelinated axon shafts studied in the model of the regenerating rat sciatic nerve. *Neuroscience* **79**, 1207–1218.
Weibull, C., Villiger, W., and Carlemalm, E. (1984). Extraction of lipids during freeze-substitution of *Acholeplasma laidlawii*-cells for electron microscopy. *J. Microsc.* **134**, 213–216.
Yakushevska, A. E., Lebbink, M. N., Geerts, W. J., Spek, L., van Donselaar, E. G., Jansen, K. A., Humbel, B. M., Post, J. A., Verkleij, A. J., and Koster, A. J. (2007). STEM tomography in cell biology. *J. Struct. Biol.* **159**, 381–391.
Zeuschner, D., Geerts, W., van Donselaar, E., Humbel, B., Slot, J., Koster, A. J., and Klumperman, J. (2006). Immuno-electron tomography of ER exit sites reveals the existence of free COPII-coated transport carriers. *Nat. Cell Biol.* **8**, 377–383.
Zhang, Q., Li, Y., and Tsien, R. W. (2009). The dynamic control of kiss-and-run and vesicular reuse probed with single nanoparticles. *Science* **323**, 1448–1453.

CHAPTER 27

3D Versus 2D Cell Culture: Implications for Electron Microscopy

Michael W. Hess[*], Kristian Pfaller[*], Hannes L. Ebner[*], Beate Beer[†,a], Daniel Hekl[†], *and* Thomas Seppi[†]

[*]Division of Histology and Embryology, Innsbruck Medical University, Innsbruck A-6020, Austria

[†]Department of Therapeutic Radiology and Oncology, Innsbruck Medical University, Innsbruck A-6020, Austria

[a] Present affiliation: Institute of Legal Medicine, Innsbruck Medical University, Innsbruck A-6020, Austria

978-0-12-381007-6
DOI: 10.1016/10.1016/S0091-679X(10)96027-5

Abstract

Cell culture systems are indispensable tools for basic research and a wide range of clinical *in vitro* studies. However, conventional 2D cell cultures poorly mimic the conditions in the living organism. This limitation may seriously compromise the reliability and significance of data obtained from such approaches. Therefore, we present here a comparative study on selected 3D and 2D cell cultures of U87-MG human glioblastoma cells that were processed by means of high-pressure freezing and freeze-substitution as well as by conventional chemical fixation and Tokuyasu cryo-section immuno-labeling. Three-dimensional cultures comprised pseudo-vascularized cultures, fiber and bead scaffold cultures, and spheroid cultures. Cell cultures in dishes and on coverslips were the static 2D culture systems used as reference models. We will discuss morphological and immunocytochemical observations with respect to the feasibility of the cell culture systems investigated for the state-of-the-art electron microscopy.

Abbreviations

2D	two-dimensional
3D	three-dimensional
EM	electron microscopy
FA	formaldehyde
FS	freeze-substitution/freeze-substituted
GA	glutaraldehyde
HPF	high-pressure freezing/high-pressure frozen
PVT	pseudo-vascularized tumor
RT	room temperature
SEM	scanning electron microscopy
TEM	transmission electron microscopy
UA	uranyl acetate

I. Introduction

Cell culture systems are indispensable tools for a wide range of clinical *in vitro* studies and for basic research. The classically preferred model is a static dish culture. This approach mainly generates adherent two-dimensional (2D) cell monolayers, which do not reflect the situation *in vivo*, where cells grow within a complex three-dimensional (3D) microenvironment (Lee *et al.*, 2008; Vunjak-Novakovic and Freed, 1998). In addition, static dish cultures involve feeding of the cells by replacing the overlaid culture medium every 2–3 days, instead of continuously supplying and removing metabolites and catabolites, respectively (Smalley *et al.*, 2006), as it is provided by vascular perfusion *in vivo* (Simionescu and Simionescu, 1988). This discrepancy in growth conditions leads to severe restrictions concerning the suitability of 2D culture methods as reliable *in vitro* tools (Indovina *et al.*, 2007; Meads *et al.*, 2009; Weigelt and Bissell, 2008). The morphology and physiology of cells grown as dish cultures differ substantially from their counterparts *in vivo*, compromising not only fundamental investigations in cell and developmental biology but also in clinical research.

Three-dimensional cultures are supposed better to mimic *in vivo* conditions. They exhibit a higher degree of structural complexity and homeostasis, analogous to tissues and organs. Certain aspects of morphogenesis were recently elucidated only with the aid of 3D culture models, such as the formation and malformation of glandular acini (Bissell *et al.*, 2003; Debnath *et al.*, 2003) or axon outgrowth and myelination (Gingras *et al.*, 2008). Furthermore, it was shown that migration of dendritic cells differs substantially in 2D and 3D environments, as a function of specific architecture and coordination of the actin cytoskeleton (Lammermann *et al.*, 2009).

The most widely used 3D models are spheroid cell aggregates and scaffold culture systems, since they can easily be generated (Friedrich *et al.*, 2009; Kunz-Schughart *et al.*, 1998; Provin *et al.*, 2009; Sutherland, 1988). Scaffolds exhibit quite different properties and chemical compositions but serve all the same purpose, namely to provide an extracellular mechanical support and/or to facilitate adhesion and growth of cells in 3D (Lee *et al.*, 2008). Widely used systems include a fiber meshwork made of biocompatible polymers (Monaghan *et al.*, 2003), solid or porous (gelatin) beads such as Cytodex^{TM} (Hagen and Grunewald, 2008; Zeuschner *et al.*, 2006) or Culti-Spher^{TM} (Rasey *et al.*, 1996), as well as extracellular matrix or its substitutes (Lee *et al.*, 2007). Spheroid and scaffold 3D models have been propagated especially for investigations of the behavior of spatially arranged cell communities at the scale of avascular micro-metastases, or of intervascular microregions of larger tumors (Sutherland, 1988). In addition, simple artificially vascularized 3D tissues have been established and characterized for *in vitro* and xenograft drug screening purposes (e.g., Decker *et al.*, 2004; Funatsu and Nakazawa, 2002; Griffith and Swartz, 2006, see also Knazek *et al.*, 1972).

The increasing use of 3D culture techniques in several research areas is also accompanied by technical challenges for microscopy. Whereas 2D cultures allocate a convenient and well-established basis for almost any kind of imaging (as well as for physiological and biochemical analyses), the suitability of novel 3D culture systems has still to be evaluated and optimized for most experimental approaches. In particular, the following major technical challenges seem worth mentioning with regard to specimen preparation of 3D-cell cultures for electron microscopy (EM). Conventional chemical fixation at ambient temperatures is usually slower for 3D cell communities than for 2D cell cultures. Delayed fixation is a major source of fixation artifacts (Hayat, 2000; Maunsbach and Afzelius, 1999) and becomes even more pronounced when the bulky 3D cell communities are growing within scaffold gels, such as gelatin or Matrigel^{TM}. Furthermore, 3D cultures subjected to immersion fixation frequently show gradients of structural preservation quality, similar to bulk tissue. Other technical problems associated with EM preparation of certain 3D cell culture systems relate to the cell density of the sample. Harvesting small, barely visible cell clusters in their native, unfixed state may be difficult, though mandatory for cryo-fixation techniques. Moreover, efficient and unbiased analysis of ultra-thin EM sections requires a fair amount of cells packed within a relatively small volume. This prerequisite is not easily met if scaffolds are only scarcely populated, or if scaffolds themselves occupy most of the sample volume. Section profiles of three neighboring Cytodex gelatin beads, for example, cover most of the area of an average 0.5×0.5 mm Tokuyasu cryo-section, not leaving much space for the cells under investigation. Examples for problems associated with light microscopy of 3D cultures are mentioned in footnote[1].

II. Rationale

Here we report our experience with selected 3D and 2D cell culture systems regarding their feasibility for the state-of-the-art EM (Suppl. Fig. 1). In particular, we present morphological and immuno-cytochemical observations on U87-MG glioblastoma cells cultured in 3D arrays as pseudo-vascularized cultures, fiber and bead scaffolds, and as spheroid cultures[2]. Conventional static 2D cultures were used as reference models (i.e., culture dish and coverslip). The preparation techniques dealt with include cryo-fixation by means of high-pressure freezing (HPF) and freeze-substitution (FS) followed by resin embedding, as well as standard chemical fixation by aldehydes for resin embedding and Tokuyasu cryo-section immuno-EM.

III. Methods

A. Cell Culture

1. Standard 2D Cell Culture

Two- dimensional batch cultures of the human glioblastoma cell line U87-MG were maintained under standard conditions in cell culture dishes at 36.8°C in a humidified atmosphere of 5% CO_2 in air. Incubation medium was renewed every 48 h. Cells were subcultured after detachment by incubation in 1 ml Trypsin–EDTA (ethylenediaminetetraacetic acid) solution. For TEM and SEM different kinds of coverslips were placed in

---

[1] Technical problems encountered with light microscopy of 3D-cultures concern, for example, adverse optical properties of scaffolds as well as the enlarged *z*-axis of the 3D specimens. Certain scaffolds (e.g., fiber meshwork, bead-like carriers) are, at best, semi-transparent; others, such as amorphous or nano-structured scaffolds, show a suboptimal refractive index. Both conditions may interfere with (high-resolution) imaging. Furthermore, certain high-end approaches such as TIRF microscopy are not easily applied to 3D cultures, if possible at all. For a recent methodological discussion of light microscopical and biochemical approaches for 3D cultures, see (Debnath *et al.*, 2003).

[2] The present paper is not meant as a vademecum for successfully generating 3D cell cultures. This must be tested empirically for any relevant cell type under investigation. Even concerning suitable EM preparation techniques, the cited literature is not exhaustive. Therefore, we apologize to all whose work—though related to the topic—is not discussed.

the culture dishes. Special coverslips made of synthetic sapphire (Ø 2 or 3 mm, 50 μm thick) were introduced for HPF by Martin Müller and colleagues to enable correlative (live cell) light microscopy and cryo-based EM (Eppenberger-Eberhardt *et al.*, 1991; Hess *et al.*, 2000; Schwarb, 1990). Sapphire coverslips/disks were cleaned with 96% ethanol and bidistilled water (alternatively, concentrated H_2SO_4 can be used instead of ethanol), coated with a ~4 to 10-nm carbon layer and autoclaved. Optionally, coverslips were additionally coated with sterile poly-L-lysine, laminin, or collagen to enhance cell attachment[3].

2. Conventional Static 3D Cell Culture

Fibre scaffolds. In-house fabricated polymeric scaffold supports were placed into six-well cell culture retainers (Cellstar®, Greiner Bio-One GmbH, Austria) in order to allow apical, as well as basal, medium supply during cell growth and tissue culture maintenance. U-87 MG single-cell suspensions of 5×10^6 cells per approach, derived from confluent cell monolayer cultures, were carefully transferred onto polycarbonate scaffolds. After 14 days of culturing under static condition, the inter-fiber space of the inoculated scaffolds was colonized by U-87 MG cells forming a compact three-dimensionally grown tumor-cell tissue.

Gelatin bead scaffolds (CultiSpher™). Single-cell suspensions of 5×10^5 cells derived from confluent U-87 MG monolayer cultures were transferred into bacterial-grade polystyrene culture dishes backed with a 2-mm layer of high-density agar (2.5% (w/v) in RPMI-1640) and filled with 20 ml of culture medium. CultiSpher macroporous gelatin beads (Suppl. Fig. 6A) were soaked with phosphate-buffered saline (PBS) and sterilized at 120°C for 20 min. After addition of 5000 CultiSpher per dish, colonization of the gelatin beads took place by U-87 MG cells under standard conditions in the incubator. The overgrown beads were harvested after 14 days of continuous static culturing.

Three-Dimensional Spheroids. Single-cell suspensions of 5×10^5 cells derived from confluent U-87 MG monolayer cultures were transferred into bacterial-grade polystyrene culture dishes supplemented with 20 ml of culture medium. Small aggregates of tumor cells spontaneously formed on the uncharged dish surface within 24 h. The medium was replaced every 2 days by transferring the spheroid suspension into a 50-ml conical centrifuge tube (BD Biosciences, Austria) allowing larger cell aggregates to settle for 1 min without centrifugation. The supernatant, including single cells and debris, was thoroughly removed; fresh medium was added and transferred to a new culture dish following the resuspension of spheroids. After 10 days of culturing, spheroids of > 500 μm were sorted out by sieving (stainless steel mesh: BSS 30) and the harvested spheroids were ready for EM processing.

---

[3] Another suitable substrate worth briefly mentioning are semi-permeable filters (e.g., Costar® Transwell, from CORNING, NY), optionally coated with poly-L-lysine or BD Matrigel™ (from BD BIOSCIENCES, USA). They are mandatory for achieving apicobasal polarization of 2D cultured epithelial cells (Felder *et al.*, 2002), but—admittedly—not for culturing U87 glioblastoma (for examples of EM applications of such filters, see Ruemmele *et al.*, 2010).

3. Pseudo-Vascularized 3D Cultures (PVT: Pseudo-Vascularized Tumor)

Spatially arranged and pseudo-vascularized U87-MG tumor cell cultures were grown and maintained with the aid of a patent-pending incubation container (ArtifiCell^{TM}, Inventive GmbH, Austria: Seppi *et al.*, unpublished data). The construction is based on similar principles as the cylindrical "lotus-root organoid module" described by others (see, for example Figs. 4 and 5 in Funatsu and Nakazawa, 2002). In the ArtifiCell incubation chamber, 500-μm-wide Midgee HoopTM polysulfone hollow fibers with pore sizes of 10^5 NMWCO (from Amersham Biosciences, UK) are hexagonally arranged at an intercapillary distance of 400 μm. Following inoculation with 5×10^6 U-87 MG cells, each tissue container was continuously perfused with fresh culture medium at optimized flow rates. After 5 days of culturing, the intercapillary space was entirely filled with a pseudo-vascularized tumor cell-tissue ready for EM processing.

B. Cryo-Processing for EM

1. High-Pressure Freezing (Moor, 1987; Moor and Riehle, 1968; Müller and Moor, 1984)

Sapphire coverslips/disks were mounted in the following order in-between two HPF-aluminum carriers that have been prewetted with culture medium: sapphire disks with cells facing up were placed onto the flat side of a 0.3-mm-deep carrier and covered with a 0.05- to 0.1-mm-deep carrier. The sandwiches obtained are subjected to HPF. Frozen sandwiches are stored in LN_2 for later use or transferred under LN_2 into cryo-vials filled with frozen FS media using precooled tweezers.

Plastic fiber scaffolds were punched using an ophthalmological or dermatological punch or cut using a sharp scalpel to make them fit in a 0.2- to 0.3-mm-deep specimen carrier (prefilled with culture medium). An additional carrier served as a lid with the flat or the hollow side facing down, depending on the sample thickness. (Filter membranes are treated similarly: filters are cut into triangles (Ø<2 mm), fitting into a 0.1-mm-deep, prewetted HPF-carrier, and covered with an additional carrier, flat side facing down.)

CultiSphers or spheroids were sandwiched together with culture medium between two cup-shaped aluminum specimen carriers (0.1–0.3 mm depth) and subjected to HPF (if pipetting is not feasible, a microaspiration device may be used (Triffo *et al.*, 2008)).

Pseudo-vascularized 3D cultures must be handled like soft tissue and, thus, represent a challenge during the sampling process. So far, we have not tested the HPF of such specimens. Therefore, we refer to published literature on the HPF of fragile soft tissues with the emphasis on fine-needle biopsies (Hohenberg *et al.*, 1996; Shimoni and Muller, 1998; Vanhecke *et al.*, 2003).

2. Freeze-Substitution and Resin Embedding for Morphology

Frozen sandwiches were transferred under LN_2 into cryo-vials filled with frozen FS media. For morphological studies, we used acetone containing 1% (w/v) osmium tetroxide (OsO_4) (Van Harreveld and Crowell, 1964) plus 0.1–0.2 % (w/v) uranyl acetate (UA). If necessary, this cocktail was supplemented with 4% (v/v) water (Buser and

Walther, 2008; Walther and Ziegler, 2002) to enhance specimen contrast (for suggestions how to prepare FS cocktails, see Hess, 2007; McDonald and Muller-Reichert, 2002). FS was performed at –90° for ~12–36 h (Humbel and Müller, 1986; Studer *et al.*, 1989) with the aid of an FS device, followed by warming up of the samples for another ~12–36 h (the use of short or extended time schedules was depending on purely pragmatic reasons, e.g., whether data were urgently needed or FS could be conveniently performed over the weekend). After thoroughly rinsing the specimens with acetone at temperatures between +4 and +20°C the samples were embedded in epoxy resins based on Epon and/or Araldite (Mollenhauer, 1964). Note that FS samples should not be embedded in Spurr's low-viscosity epoxy mixture (Spurr, 1969), because this resin may harm FS specimens (Hess, 1990). FS with pure, anhydrous acetone, optionally supplemented with UA, followed by embedment in (meth)acrylate resins can be used for immuno- and lectin-cytochemistry, but this aspect is not discussed here (for reviews, see Hess, 2007; Humbel and Müller, 1986; Schwarz and Humbel, 2007).

C. Conventional Processing at Ambient Temperature for EM

1. Chemical Fixation

Petri dish cultures of adherent cells were fixed by replacing cell culture medium with excess formaldehyde (FA) and/or glutaraldehyde (GA) fixative at RT (~20 ml for Ø 10 cm dishes). Fixation with 4% FA, optionally supplemented with 0.1–0.2% ultra-pure GA, is a standard fixation method for immuno-EM sample processing (Liou *et al.*, 1996; Tokuyasu, 1973, 1998). The recommended minimum fixation time is ~2 h at RT, but in the case of pure FA it is advisable to extend fixation up to ~3 days at +4°C (E. Jokitalo, pers. comm., see also Peters and Pierson, 2008; Webster, 1999). After fixation, samples were optionally stored in 1% FA at +4°C even for several months.

Coverslips and plastic scaffolds (or filter membranes) were immersed at RT into GA and/or FA (see above). For morphology, primary fixation with 2.5% GA (~2 h, RT) was applied according to standard protocols. GA-fixed samples were stored in buffer at +4°C (Sabatini *et al.*, 1963).

CultiSpher beads and spheroids were treated similarly to suspension cultures. Briefly, cell culture medium containing spheroids in suspension was mixed with an equal volume of double-strength fixative at RT and gently shaken. Following 15 min incubation, the solution was replaced by fresh, standard strength fixative and the fixation was continued as described above.

Pseudo-vascularized 3D cultures were first perfused with serum-free culture medium (10 min, +37°C) for rinsing the capillaries, followed by perfusion fixation with FA for 15 min at RT. After perfusion fixation, the PTFE tube housing the 3D cell cultures around the artificial capillaries was cut at both ends with a razor blade. Next, the tube was sealed with 10% gelatin to prevent loss of material during the subsequent tissue processing. This gelatin plug was then crosslinked with 4% FA for another 2 h at RT and finally the 3D cell community was removed from the Teflon tube (Fig. 1), cut into pieces, and stored.

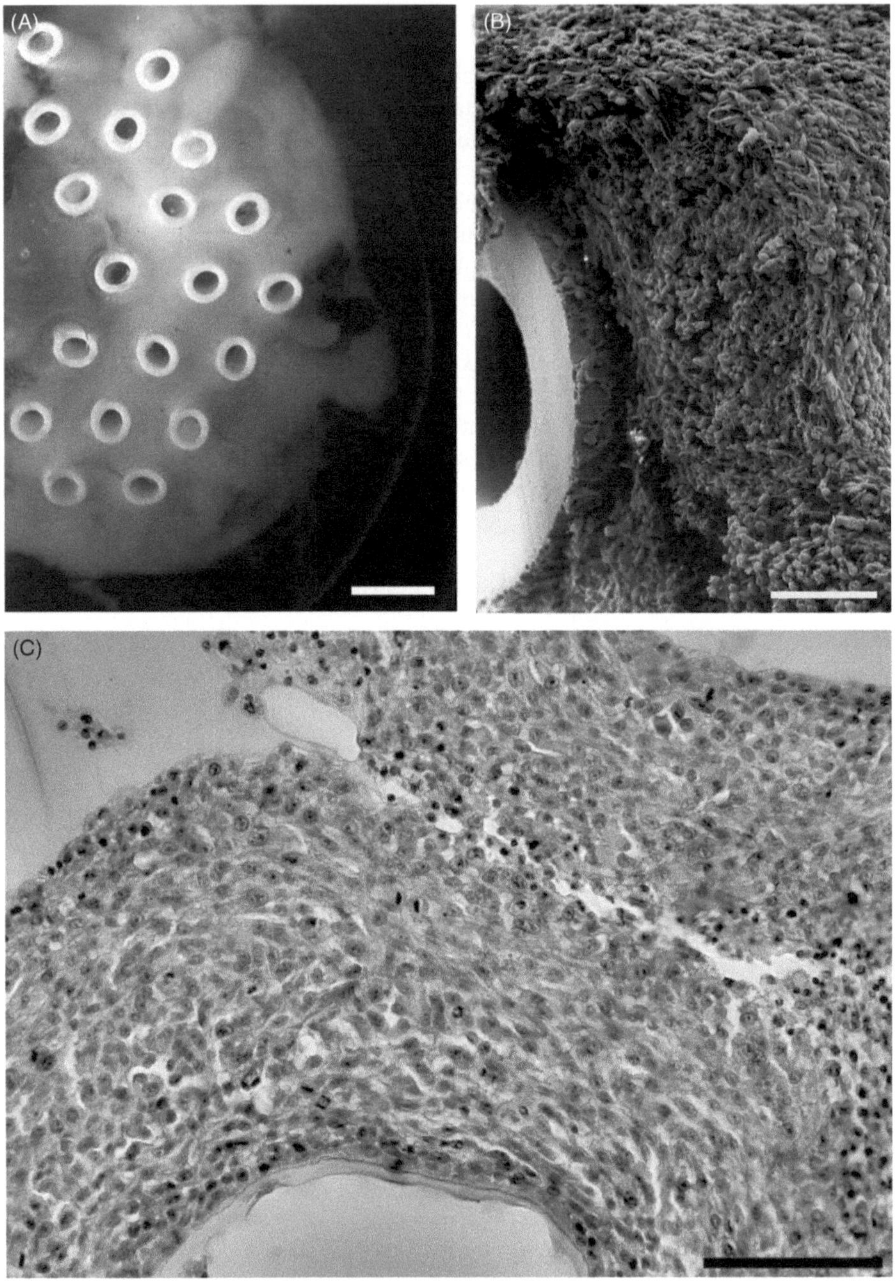

Fig. 1 PVT of U87 glioblastoma cells. (A) Macroscopical view of U87 cells growing in a 3D arrangement in the intercapillary space of an ArtifiCell© cell culture container. This image shows the PVT cell community after perfusion fixation with FA and embedding in gelatin before removal of the artificial capillaries. Scale bar=1 mm. (B) SEM overview of densely packed, 3D-arranged U87 cells surrounding their supplying artificial capillary. Scale bar=100 μm. (C) Paraffin section (H & E staining) of U87 PVT (capillaries removed after fixation). Scale bar=100 μm.

2. Resin Embedding for Morphology

FA-fixed 3D cultures were postfixed with 2.5% buffered GA (2 h, RT) and rinsed before postfixation and *en bloc* staining in 1% aqueous, unbuffered OsO_4. Following several rinses with water, a final postfixation/*en bloc* staining step was carried out with 0.1% aqueous UA (overnight, +4°C). Specimen dehydration and embedding in epoxy resin mixtures were performed according to standard procedures (Mollenhauer, 1964).

3. Tokuyasu Cryo-Section Immuno-EM

For indirect immunogold labeling of thawed cryo-sections according to Tokuyasu (Tokuyasu, 1973) samples were processed as previously described (Liou *et al.*, 1996). Briefly, FA-fixed pieces of 3D cultures, or cells scraped from the dish, were embedded in 10–12% (w/v) gelatin and solidified on ice. Small cubes (side length <1 mm) were cut from the gelatinized sample, cryo-protected with 2.3 M sucrose (in 0.1 M buffer; overnight at 4°C), mounted on aluminum stubs of a cryo-ultramicrotome, and frozen in LN_2 or in the microtome's cryo-chamber (at –90°C). Thin sections (50–80 nm) cut at ~ −110°C were picked up in a 1:1 mixture of methyl cellulose (2% v/v) and 2.3 M sucrose, immuno-labeled, and embedded in UA-methylcellulose.

4. Scanning EM

Cell cultures were fixed with FA (4% w/v) or GA (2.5% v/v). Subsequently, the samples were processed for SEM according to standard procedures, as previously described (Haller *et al.*, 2001). Briefly, sample processing comprised of postfixation with OsO_4, dehydration with ethanol, critical-point drying, and sputter-coating.

IV. Materials

A. Cell Culture

- The human glioblastoma cell line U87-MG (ATCC, LGC Standards GmbH; Wesel, Germany) was used for all examples shown here.
- Culturing medium was Eagle's minimum essential medium (Sigma-Aldrich, Austria) supplemented with 10% fetal bovine serum, 2 mM glutamine, 100 U/ml penicillin, and 100 μG/ml streptomycin (all Gibco, Invitrogen, Austria).
- Trypsin–EDTA solution (10×; Sigma-Aldrich, Austria) was used for detaching and separating 2D batch cultures.
- Culture dishes were used for 2D batch cultures (Ø 84 mm; Cellstar®, Greiner Bio-One GmbH, Austria) designated for Tokuyasu cryo-section immuno-EM.
- Bacterial-grade polystyrene culture dishes (Greiner Bio-One GmbH, Austria), which prevent cell attachment, were used for spheroid cultures.

- Coverslips made of glass or plastic (e.g., Thermanox®, NUNC, Nalge Nunc International, Rochester, NY) were used for conventional TEM and SEM of standard 2D cultures.
- Sapphire coverslips/disks used for HPF were obtained from Engineering Office Martin Wohlwend (CH-9466 Sennwald, Switzerland; martin-wohlwend@bluewin.ch) and are also available from Leica-Microsystems (Vienna, Austria). When using sapphire disks for TEM, one should consider that handling sapphires with tweezers may produce fine splinters. Minute, almost invisible splinters may become deposited in or near the cell layer, which apparently shortens the life of diamond knives; the same is obviously true for scratched sapphires (Reipert and Wiche, 2008). For avoiding this risk, polymer filter membranes are an alternative 2D culture substrate (Jimenez *et al.*, 2006; Morphew and McIntosh, 2003). Unlike resin sectioning, however, cryo-sectioning of filters possibly harms diamonds as well (W. Möbius, personal communication).
- Polymer scaffolds made of a network of polycarbonate fibers were used for static 3D cultures (scaffold diameter: 19 mm; mean scaffold thickness: 250±30 μm; mixed interfiber mesh range: BSS 440–150).
- CultiSpher^{TM} macroporous gelatin beads (Percell Biolytics AB, Sweden, or Sigma) were another kind of scaffold employed here.
- Standard cell culture equipment is not specified here.

B. Electron Microscopy

- 0.1–0.2 M buffer solution, pH~7.4 (e.g., phosphate, cacodylate, and PHEM).
- 4–8% (w/v) FA solution in 0.1 M buffer, which has been depolymerized from paraformaldehyde (Sigma); solutions were used freshly or made from frozen stocks.
- 0.1–3% (v/v) GA in 0.1 M buffer; using concentrations of 2–3% is a standard procedure to prepare morphological samples (Maunsbach and Afzelius, 1999) and EM-grade quality is sufficient for this purpose; 0.05–0.2% GA (as a supplement to 2–4% FA) is advised for immuno-EM (Liou *et al.*, 1996): in this case highest quality (double-distilled GA) is required for preserving antigenicity.
- 1% (w/v) OsO_4 in distilled water or acetone, dissolved from crystals.
- 0.1% (w/v) UA in water or acetone, diluted from a 10–20% (w/v) UA stock in methanol.
- 10–12% (w/v) gelatin in 0.1 M phosphate or PHEM buffer.
- Epoxy resins based on Epon and/or Araldite (including its substitutes); for detailed recipes, see Glauert and Lewis, 1998; Matsko and Mueller, 2005; Mollenhauer, 1964.
- GA, OsO_4, UA, and epoxy resins from Agar (Stansted, U.K.), EMS (Hatfield, PA), Polysciences (Warrington, PE, U.S.A.), or Ted Pella (Redding, CA).
- Methyl cellulose (2% v/v) in distilled water.
- 2.3 M sucrose in 0.1 M phosphate buffer.
- 2-ml cryo-vials which permit a one-hand operation (i.e., vials with screw-caps plus the appropriate cryo-racks with locking bottoms that match the cryo-vial; e.g., Nalgene, from Nalge Nunc International, Rochester, NY; note that these vials are not explicitly LN_2-certified!).

- Standard aluminum carriers for HPF from Engineering Office Martin Wohlwend or from Leica; in addition, certain special carriers for HPF were from M. Wohlwend: i.e., with wells only being 0.05-mm deep, or 0.3-mm-deep ones with polished backs—both especially suitable for HPF of sapphire disks.
- Tweezers, scalpels, slot, or mesh grids (copper) coated with Formvar and carbon from Agar, EMS, Polysciences or Ted Pella, aluminum pins for ultra-cryotomy from Leica (but rivets from local hardware stores serve the same purpose).
- Further EM instruments used in this study: dissecting microscope (i.e., Leica EM workstation), high-pressure freezing apparatus HPM-010 (BAL-TEC, Balzers, Liechtenstein), automated freeze-substitution device AFS (LEICA), critical-point dryer CPD-030 (BAL-TEC), specimen-coating unit MED-020 (BAL-TEC), ultramicrotome Ultracut S optionally equipped with FCS cryo-chamber (LEICA), diamond knives for resin section microtomy and ultra-cryo-tomy (DIATOME, Biel, Switzerland), field-emission SEM DSM982-Gemini (ZEISS, Oberkochen, Germany), transmission EM Philips CM 120 (from F.E.I., Eindhoven, the Netherlands), digital camera Morada (Olympus-SIS, Münster, Germany).

V. Results

A. Pseudo-Vascularized 3D Cultures (PVT: Figs. 1–4)

Micrographs from PVT of U87 glioblastoma cells are reflecting a cellular organization that is well mimicking histiotypic patterns (our unpublished data) and excellent ultrastructure preservation. Both compact and more loosely connected tumor cell communities proved morphologically well preserved as visualized by TEM and SEM of conventionally fixed specimens (Figs. 2A, 3, 4, Suppl. Fig. 2). Two subpopulations of U87 cells were present within the same PVT culture (Suppl. Fig. 2B-C). They possibly represent two stages of differentiation, according to a recent interpretation of similar subcellular features of this cell line (Yu *et al.*, 2008). Tokuyasu cryo-sections showed strong, distinct immunogold-labeling that could be unequivocally localized to clearly identifiable, well-structured subcellular constituents (Fig. 4; Suppl. Fig. 3C). For all these morphological and immuno-EM studies, specimens were fixed with 4% FA as primary fixative (optionally followed by postfixation with GA, OsO4, and UA). Interestingly, the preservation quality achieved with 4% FA was not distinguishable from that obtained with GA. HPF/FS has not yet been performed with fresh specimens derived from PVT 3D culture because of technical challenges associated with sampling native soft tissue for HPF (see Hohenberg *et al.*, 1996).

B. Scaffolds and Spheroids (Fig. 5)

Compact 3D cell communities of U87 glioblastoma cells were also obtained by harvesting polymer fiber scaffolds and spheroids (Suppl. Fig. 4, 5). Compared with PVT, scaffold and spheroid 3D cultures of U87 cells possessed considerably less cell surface protrusions

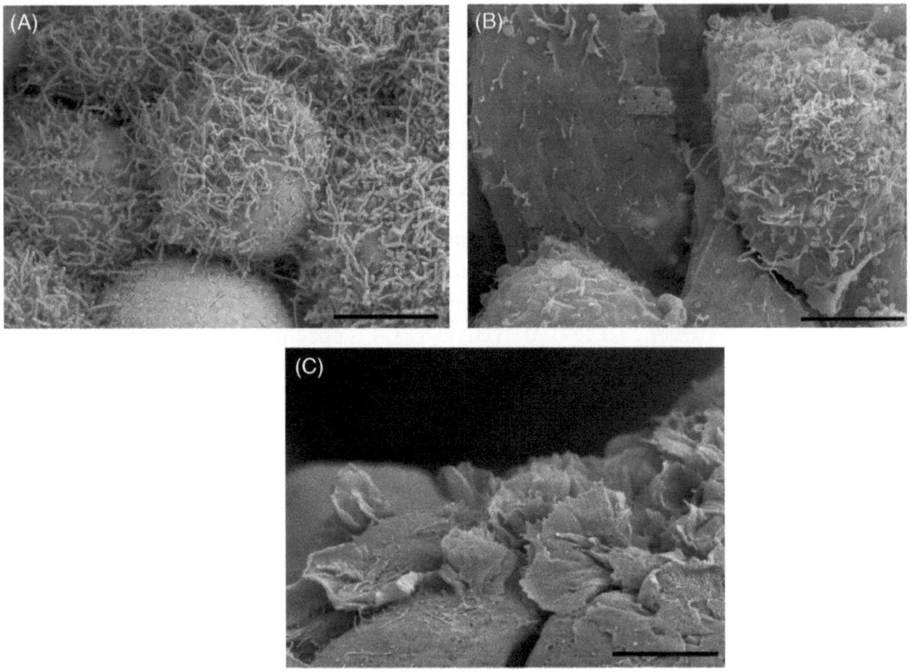

Fig. 2 SEM of U87 glioblastoma cells cultured in 3D as PVT (A) vs. fiber scaffold (B) vs. spheroid (C). The presence and shape of cell surface protrusions/microvilli depend on the cell culture conditions (see also Suppl. Fig. S7A for comparable SEM of 2D coverslip cultures). Scale bars=5 μm.

and/or microvilli (Fig. 2), possibly indicating a minor degree of cellular differentiation. As expected, the center of larger spheroids frequently showed necrotic/apoptotic areas adjacent to the bulk of virtually viable, healthy cells (Suppl. Fig. 5B), a well-known characteristic of this kind of 3D tumor cell cultures (Bell *et al.*, 2001; Kunz-Schughart *et al.*, 1998). Irrespective of cellular condition (viable, apoptotic/necrotic) the general preservation quality achieved with conventional FA/GA fixation was clearly inferior for fiber scaffold cultures and spheroids (e.g., Fig. 5A) than for PVT (Fig. 3). In contrast, cryo-processing (HPF/FS) of native, unfixed spheroids and fiber scaffolds unraveled a normally organized and well-preserved subcellular architecture of the vital cells (Fig. 5B, Suppl. Fig. 4C), even in the center of these 3D communities. The observed patterns of organelles in these HPF samples are fully consistent with previous reports on cryo-processed mammalian cells (e.g., Buser and Walther, 2008; Sawaguchi *et al.*, 2004; Wild *et al.*, 2002).

The following preliminary observations were made on CultiSpher macroporous bead scaffolds. U87 cells colonized efficiently the surface, and, to a lower extent, the interior of the sponge-like beads. However, the cells were arranged in rather loose 3D communities (Suppl. Fig. 6B–C), in contrast to other 3D cell culture systems described here. Furthermore, only the very surface, but not beads' core contained satisfactorily

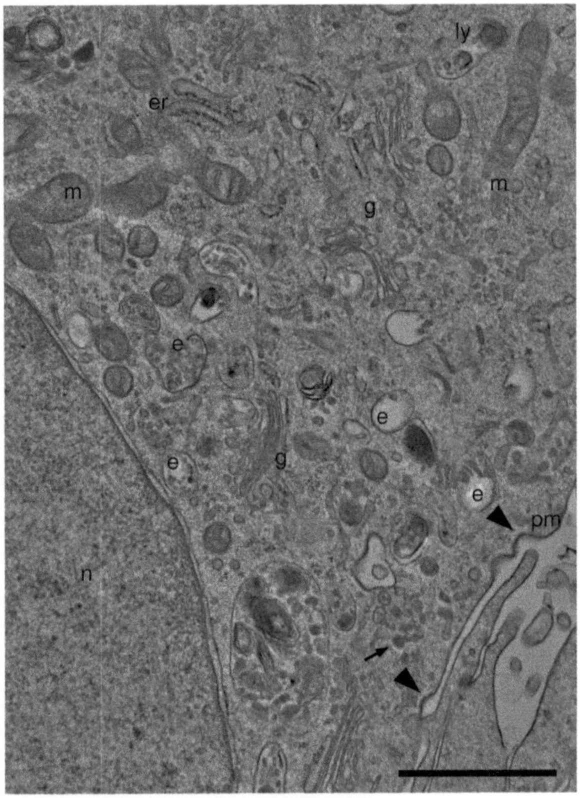

Fig. 3 PVT of U87 cells processed by conventional chemical fixation (4% FA→GA→OsO_4→UA) and Epon embedding. Note the excellent ultrastructure preservation obtained with pure FA as primary fixative. (e) endosome, (er) endoplasmic reticulum, (g) golgi stack, (ly) lysosome, (m) mitochondrion, (n) nucleus, (pm) plasma membrane. Arrow heads and arrow mark clathrin coats at the PM and at the TGN, respectively. Scale bar=2 μm.

preserved U87 cells after conventional chemical fixation (Suppl. Fig. 6D-E). HPF/FS has, to our knowledge, not been tested yet.

C. 2D Cultures (Fig. 6)

U87-MG monolayer samples derived from static 2D batch cultures using coverslips or dishes lacked cell surface protrusions/microvilli almost completely (Suppl. Fig. 7A) and appeared only modestly preserved after conventional Epon- or Tokuyasu-processing (Suppl. Fig. 7B, 3A). In contrast, HPF/FS preserved these samples excellently (Fig. 6). In addition, Fig. 6 clearly demonstrates the impact of water-containing FS media on specimen contrast. For endocytosis studies these cultures had been starved overnight in serum-free medium. As a consequence, especially the endosomal compartments suffered from inadequately low membrane contrast, when freeze-substituted with anhydrous FS media (Fig. 6A). "Watery" FS markedly improved the visibility of subcellular morphology even in these serum-starved cells (Fig. 6B); this

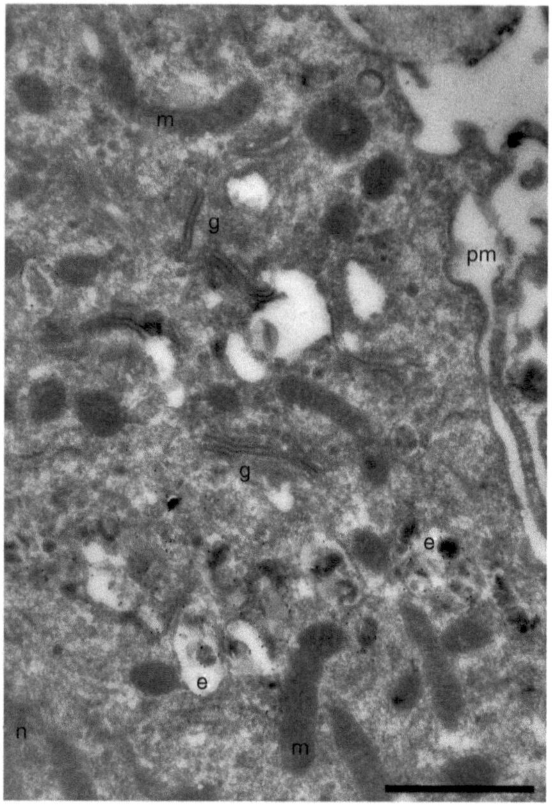

Fig. 4 Tokuyasu cryo-section from U87 PVT fixed with 4% FA and subjected to indirect immuno-gold labeling with mouse α-human LAMP1 (DSHB, Univ. Iowa) visualized by 15-nm goat α-mouse colloidal gold (British Biocell). For a high magnification view, see Suppl. Fig. S3C: Organelles lettered as in Fig. 3. Scale bar=2 μm.

was also true for other cell lines (U2OS human osteosarcoma, mouse embryonic fibroblasts (MEFs): our unpublished data).

VI. Discussion

A. Evaluation of 3D Cell Culture Systems

We observed highly varying ultrastructural patterns throughout all investigated types of 3D and 2D cell cultures of U87 glioblastoma cells, as well as a differential response of the cells to the fixation procedures applied. Notably, bulky U87 PVT cell communities did not display pathological subcellular alterations resulting from the slow action of toxic, and finally lethal fixatives, commonly termed "fixation artifacts." All other kinds of 3D and monolayer cultures studied here were not adequately preserved for EM by conventional chemical fixation, irrespective of their complexity and size. Thus,

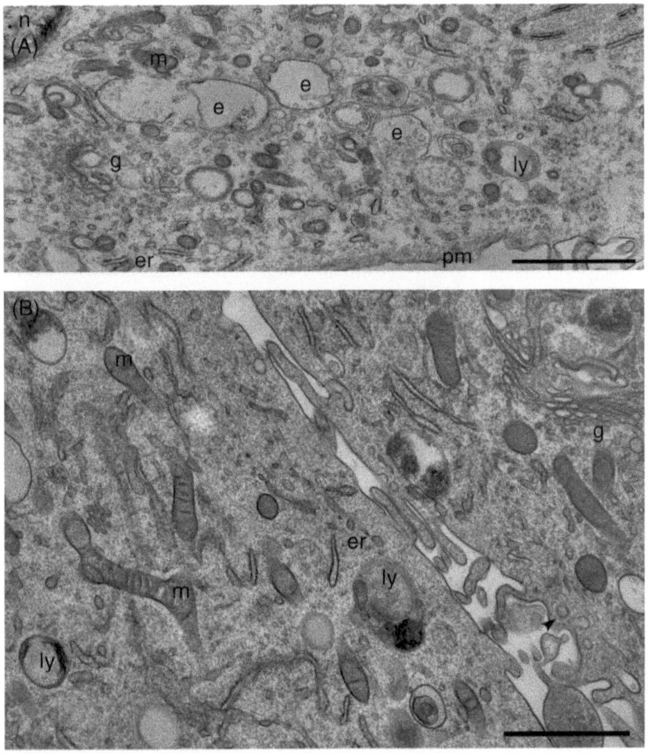

Fig. 5 Spheroid cultures of U87 processed by conventional chemical fixation (A) vs. high-pressure freezing and freeze-substitution (B). Ultrastructure preservation achieved with ambient temperature chemical fixation is in general not satisfactory for U87 spheroids, in contrast to that obtained with cryo-preparation methods (HPF/FS; FS media containing 4% water). Organelles lettered as in Figs. 3, 4. Scale bars=2 μm.

one can conclude that the size of the specimens, and, in turn, the penetration and diffusion speed of the fixatives, was not decisive for ultrastructural preservation quality in our case. More likely, the fixation patterns observed reflected the physiological conditions of the cells under investigation. These findings are fully consistent with the—widely neglected—notion by early electron microscopists that cells suffering from stress (and/or genetic modifications) are prone to fixation artifacts (Bereiter-Hahn and Voth, 1979; Trump and Ericsson, 1965); for further examples and a detailed discussion of this postulate, see (Hess, 2007) and refs. therein.

Improved vitality, pronounced cellular differentiation, and a satisfying fixation quality obtained with PVT are likely attributable to the optimized perfusion design in this type of 3D cell culture. Similarly to *in vivo* conditions, cells in PVT locate within a distance of less than 200 μm to the nearest vessel, thus providing adequate metabolite and catabolite exchange rates. In addition, the artificial capillary system is easily flooded with fixatives, analogous to perfusion fixation of laboratory animals. Recently

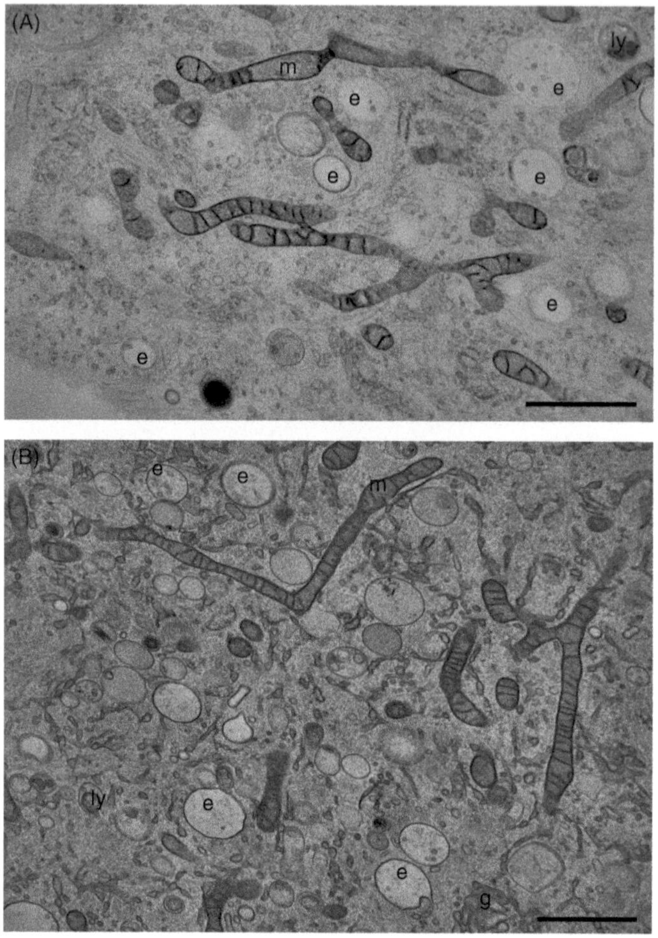

Fig. 6 HPF and FS of 2D cultures of U87 cells subjected to serum-starvation overnight. (A) Specimens processed with standard, water-free FS media suffer from low contrast, especially endosomal membranes. (B) "Watery FS" according to Walther and Ziegler (Walther and Ziegler, 2002) produces specimens with cellular membranes clearly set off the cytoplasm, including various kinds of endosomal compartments. Lettering as in Figs. 3, 4. Scale bars=1 μm.

obtained physiological and biochemical data also indicate that this type of cell culture produces stress-tolerant U87 glioblastoma cell communities, well mimicking structural and physiological patterns of certain solid tumors (our unpublished data).

B. Cryo-Processing of Cell Cultures for EM

As far as *in vitro* systems are concerned, our observations on U87 glioblastoma cells clearly showed that conventional chemical fixation is advisable only for nonstressed, "physiologically well balanced" cells under *bona fide* "optimal" cell culture conditions. Except for pseudo-vascularized 3D cultures (PVT), this requirement was not met by any 2D or 3D culture system tested here. (*In vivo*, however, this aspect appears to be less crucial. Tissues structurally and functionally embedded in the living organism seem to be more tolerant against the noxious and deleterious effects of fixatives, thus show less fixation artifacts.) Regarding ultrastructural applications, the use of cryo-preparation techniques could partly compensate for these shortcomings of cell culture systems less sophisticated than PVT. Cryo-immobilization by means of HPF, conveniently followed by FS, proved again as a highly feasible methodology for processing various 2D or bulky 3D cell cultures for TEM. Excellent preservation quality was achieved with our cryo-processed spheroid, scaffold, and coverslip cultures of U87 glioblastoma cells, but caution should be exercised when weighting the biological relevance and significance of the phenomena observed on these suboptimal *in vitro* systems. HPF/FS is now a standard preparation technique and is not only highly recommended, but also seems in most cases mandatory for the state-of-the-art cellular EM. To fully exploit its potential for immuno-EM, the combination of HPF/FS with Tokuyasu cryo-sectioning (van Donselaar *et al.*, 2007) appears promising, especially in the light of recent simplifications of this quite new approach (Ripper *et al.*, 2008; Stierhof *et al.*, 2009). FS and low-temperature acrylic resin embedding are applicable as an alternative method for the same purpose. It is less laborious and well established, though usually less sensitive than cryo-section labeling (Humbel and Müller, 1986; Schwarz and Humbel, 2007).

C. Alternative Preparation Methods Involving Chemical Fixation at Ambient Temperature

Finally, it should be mentioned that a cryo-fixation technique is not always available or feasible, or that conventional chemical fixation may be definitely preferred in certain cases. The latter pertains, for example, to highly infectious material or surgical samples. We will, therefore, briefly discuss the potential of pure FA as primary fixative, considering the good morphological and immuno-labeling results achieved here with this reagent. U87 PVT samples showed virtually no difference in preservation quality, regardless of whether pure FA or GA was the primary fixative[4]. The same is true for various mouse tissues subjected to perfusion or immersion fixation (our unpublished data). Thus, it seems reasonable to recommend pure FA as an almost "universal" primary fixative whenever cryofixation is not possible or desired. It provides adequately preserved samples equally suitable for multiple follow-up preparations such as morphology, (immuno-) cytochemistry, and *in situ* hybridization at the level of EM and light microscopy. By doing so, one need not decide how to split samples for different purposes in advance—nor during

[4] Interestingly, sequential fixation with FA and OsO_4 was occasionally used for morphology (Holt and Hicks, 1961) before the introduction of GA as primary fixative (Sabatini *et al.*, 1963). As judged from published EM micrographs (Holt and Hicks, 1961), structural preservation was acceptable, though slightly inferior to the quality obtainable with GA and OsO_4 or, as in our case, FA followed by GA and OsO_4.

specimen sampling, which is the most crucial preparation step for biological EM. This may be a considerable advantage when minute and/or structurally complex specimens are processed, e.g., lymph nodes or brain from genetically modified mice. Finally, such "formalin-cadavers" can be stored well at +4°C without losing much quality, permitting prolonged access to valuable, sometimes unique material for repeated or refined experiments.

Another modification of conventional specimen preparation involves the use of FS, instead of sample dehydration at RT. It was previously shown by others that chemically fixed tissue samples subjected to HPF with subsequent FS appeared much better preserved than those completely processed at RT (Mobius, 2009; Pfeiffer and Krupinska, 2005; Sosinsky *et al.*, 2008). However, this seems to apply mainly to tissue samples, not to cell cultures that still suffered from poor preservation, especially of endomembrane systems (our unpublished data). Moreover, it is self-evident that such aldehyde-fixed, HPF/FS samples cannot fully compete with native HPF/FS specimens, but under certain circumstances this hybrid approach might represent a pragmatic compromise (for details on this method, see the chapter by W. Möbius and colleagues (Chapter 20 of this volume)).

VII. Synopsis

Our findings can be summarized as follows (see also Suppl. Fig. 1):

- Out of the various 2D and 3D cell culture systems tested here with U87 glioblastoma cells, only pseudo-vascularized 3D cultures (PVT) produced highly differentiated and stress-tolerant 3D cell communities, mimicking *in vivo* conditions of certain tumors.
- Conventional chemical fixation of such PVT with aldehydes yielded adequate preservation, both of morphology and antigenicity, as judged from resin sections and Tokuyasu cryo-sections.
- Primary fixation with pure FA gave results equal to GA, if applied to 3D cell communities of high vitality, such as PVT or intact tissues.
- Spheroid and fiber scaffold 3D cultures, as well as 2D coverslip cultures subjected to HPF/FS, displayed excellent ultrastructure preservation, unlike after conventional chemical fixation.
- Live sampling of spheroid and scaffold 3D cultures for HPF/FS is feasible.
- Appropriate sampling of native PVT for HPF cryo-fixation is technically challenging and not optimized yet.
- Spheroids and scaffolds of less rigid consistency than polycarbonate fibers, and of a high surface-to-volume ratio, are possible candidates for HPF combined with Tokuyasu immuno-EM, but remain still to be tested.

Despite the virtues discussed here of 3D cell culture and cryo-based EM-specimen preparation, one should not forget the following aspects when interpreting the data derived from these cutting-edge methods. First, any type of cell culture remains an artificial system suffering from inherent shortcomings in comparison to tissues from living organisms. However, we have no choice than coping with these limitations for not only ethical, but also pragmatic reasons. Second, a virtually intact ultrastructure as seen in perfectly cryo-processed samples does not tell the whole truth about the actual vitality of the cells at the time of sampling. Subcellular ion distributions are more reliable in this respect. Ion shifts indicate patho-physiological processes long before any ultrastructural alteration is recognizable. These parameters should be considered, if appropriate, although the relevant EM methods are demanding (Zierold, 1997, 2000).

Acknowledgments

We acknowledge Angelia Flörl, Karin Gutleben, and Barbara Witting for their excellent technical assistance and wish to thank Eija Jokitalo and Wiebke Möbius for communicating unpublished observations, as well as Heinz Schwarz for critically commenting the chapter. This work was supported by several grants (Austrian Science Funds (FWF-P 19486-B12 to MWH), Oesterreichische Nationalbank-Jubiläumsfonds (P-11050 to MWH), Austrian Academy of Science (P-DOC-22056 to BB), Tiroler Wissenschaftsfonds (P-UNI-0404/100 to MWH), and Medizinischer Forschungsfonds Tirol (P68 to KP, TS, MWH)).
Authors' contributions: MWH, KP, and HLE performed and discussed electron microscopy; BB, DH, and TS contributed to the cell culture part of this work.

References

Bell, H. S., Whittle, I. R., Walker, M., Leaver, H. A., and Wharton, S. B. (2001). The development of necrosis and apoptosis in glioma: Experimental findings using spheroid culture systems. *Neuropathol. Appl. Neurobiol.* **27**(4), 291–304.

Bereiter-Hahn, J., and Voth, M. (1979). Metabolic state dependent preservation of cells by fixatives for electron microscopy. *Microsc. Acta* **82**(3), 239–250.

Bissell, M. J., Rizki, A., and Mian, I. S. (2003). Tissue architecture: The ultimate regulator of breast epithelial function. *Curr. Opin. Cell Biol.* **15**(6), 753–762.

Buser, C., and Walther, P. (2008). Freeze-substitution: The addition of water to polar solvents enhances the retention of structure and acts at temperatures around −60 degrees C. *J. Microsc.* **230**(Pt 2), 268–277.

Debnath, J., Muthuswamy, S. K., and Brugge, J. S. (2003). Morphogenesis and oncogenesis of MCF-10a mammary epithelial acini grown in three-dimensional basement membrane cultures. *Methods* **30**(3), 256–268.

Decker, S., Hollingshead, M., Bonomi, C. A., Carter, J. P., and Sausville, E. A. (2004). The hollow fibre model in cancer drug screening: The NCI experience. *Eur. J. Cancer* **40**(6), 821–826.

Eppenberger-Eberhardt, M., Riesinger, I., Messerli, M., Schwarb, P., Muller, M., Eppenberger, H. M., and Wallimann, T. (1991). Adult rat cardiomyocytes cultured in creatine-deficient medium display large mitochondria with paracrystalline inclusions, enriched for creatine kinase. *J. Cell Biol.* **113**(2), 289–302.

Felder, E., Jennings, P., Seppi, T., and Pfaller, W. (2002). LLC-PK(1) cells maintained in a new perfusion cell culture system exhibit an improved oxidative metabolism. *Cell Physiol. Biochem.* **12**(2–3), 153–162.

Friedrich, J., Seidel, C., Ebner, R., and Kunz-Schughart, L. A. (2009). Spheroid-based drug screen: Considerations and practical approach. *Nat. Protoc.* **4**(3), 309–324.

Funatsu, K., and Nakazawa, K. (2002). Novel hybrid artificial liver using hepatocyte organoids. *Int. J. Artif. Organs* **25**(2), 77–82.

Gingras, M., Beaulieu, M. M., Gagnon, V., Durham, H. D., and Berthod, F. (2008). *In vitro* study of axonal migration and myelination of motor neurons in a three-dimensional tissue-engineered model. *Glia* **56**(3), 354–364.

Glauert, A. M., and Lewis, P. R. (1998). "Biological Specimen Preparation for Transmission Electron Microscopy." Portland Press, London.

Griffith, L. G., and Swartz, M. A. (2006). Capturing complex 3d tissue physiology *in vitro*. *Nat. Rev. Mol. Cell Biol.* **7**(3), 211–224.

Hagen, C., and Grunewald, K. (2008). Microcarriers for high-pressure freezing and cryosectioning of adherent cells. *J. Microsc.* **230**(Pt 2), 288–296.

Haller, T., Dietl, P., Pfaller, K., Frick, M., Mair, N., Paulmichl, M., Hess, M. W., Furst, J., and Maly, K. (2001). Fusion pore expansion is a slow, discontinuous, and Ca2+-dependent process regulating secretion from alveolar type II cells. *J. Cell Biol.* **155**(2), 279–289.

Hayat, M. A. (2000). "Principles and Techniques of Electron Microscopy. Biological Applications." Cambridge University Press, Cambridge.

Hess, M. W. (1990). Spurr-resin affects the ultrastructure of freeze-fixed/freeze-substituted pollen grains. *Ultramicroscopy* **34**, 219–221.

Hess, M. W. (2007). Cryopreparation methodology for plant cell biology. *Methods Cell Biol.* **79**, 57–100.

Hess, M. W., Muller, M., Debbage, P. L., Vetterlein, M., and Pavelka, M. (2000). Cryopreparation provides new insight into the effects of brefeldin A on the structure of the HepG2 Golgi apparatus. *J. Struct. Biol.* **130**(1), 63–72.

Hohenberg, H., Tobler, M., and Muller, M. (1996). High-pressure freezing of tissue obtained by fine-needle biopsy. *J. Microsc.* **183**(Pt 2), 133–139.

Holt, S. J., and Hicks, R. M. (1961). Studies on formalin fixation for electron microscopy and cytochemical staining purposes. *J. Biophys. Biochem. Cytol.* **11**, 31–45.

Humbel, B., and Müller, M. (1986). Freeze substitution and low temperature embedding. *In* "The Science of Biological Specimen Preparation 1985" (J. J. Wolosewick, ed.) pp. 175–183. SEM Inc AMF O'Hare, Chicago.

Indovina, P., Collini, M., Chirico, G., and Santini, M. T. (2007). Three-dimensional cell organization leads to almost immediate HRE activity as demonstrated by molecular imaging of MG-63 spheroids using two-photon excitation microscopy. *FEBS Lett.* **581**(4), 719–726.

Jimenez, N., Humbel, B. M., van Donselaar, E., Verkleij, A. J., and Burger, K. N. (2006). Aclar discs: A versatile substrate for routine high-pressure freezing of mammalian cell monolayers. *J. Microsc.* **221**(Pt 3), 216–223.

Knazek, R. A., Gullino, P. M., Kohler, P. O., and Dedrick, R. L. (1972). Cell culture on artificial capillaries: An approach to tissue growth *in vitro*. *Science* **178**(56), 65–66.

Kunz-Schughart, L. A., Kreutz, M., and Knuechel, R. (1998). Multicellular spheroids: A three-dimensional *in vitro* culture system to study tumour biology. *Int. J. Exp. Pathol.* **79**(1), 1–23.

Lammermann, T., Renkawitz, J., Wu, X., Hirsch, K., Brakebusch, C., and Sixt, M. (2009). Cdc42-dependent leading edge coordination is essential for interstitial dendritic cell migration. *Blood* **113**(23), 5703–5710.

Lee, J., Cuddihy, M. J., and Kotov, N. A. (2008). Three-dimensional cell culture matrices: State of the art. *Tissue Eng. Part B Rev.* **14**(1), 61–86.

Lee, G. Y., Kenny, P. A., Lee, E. H., and Bissell, M. J. (2007). Three-dimensional culture models of normal and malignant breast epithelial cells. *Nat. Methods* **4**(4), 359–365.

Liou, W., Geuze, H. J., and Slot, J. W. (1996). Improving structural integrity of cryosections for immunogold labeling. *Histochem. Cell Biol.* **106**(1), 41–58.

Matsko, N., and Mueller, M. (2005). Epoxy resin as fixative during freeze-substitution. *J. Struct. Biol.* **152**(2), 92–103.

Maunsbach, A. B., and Afzelius, B. A. (1999). "Biomedical Electron Microscopy. Illustrated Methods and Interpretations." Academic Press, San Diego, London, Boston, New York, Sydney, Tokyo, Toronto.

McDonald, K., and Muller-Reichert, T. (2002). Cryomethods for thin section electron microscopy. *Methods Enzymol.* **351**, 96–123.

Meads, M. B., Gatenby, R. A., and Dalton, W. S. (2009). Environment-mediated drug resistance: A major contributor to minimal residual disease. *Nat. Rev. Cancer* **9**(9), 665–674.

Mobius, W. (2009). Cryopreparation of biological specimens for immunoelectron microscopy. *Ann. Anat.* **191**(3), 231–247.

Mollenhauer, H. H. (1964). Plastic Embedding Mixtures for Use in Electron Microscopy. *Stain Technol.* **39**, 111–114.

Monaghan, P., Cook, H., Hawes, P., Simpson, J., and Tomley, F. (2003). High-pressure freezing in the study of animal pathogens. *J. Microsc.* **212**(Pt 1), 62–70.

Moor, H. (1987). Theory and practice of high pressure freezing. *In* "Cryotechniques in Biological Electron Microscopy." (R. A. Steinbrecht and K. Zierold, eds.) pp. 175–191. Springer, Berlin, Heidelberg.

Moor, H., and Riehle, U. (1968). Snap-freezing under high pressure: A new fixation technique for freeze etching. *In* "Electron Microscopy 1968. Pre-Congress Abstracts of Papers Presented at the Fourth European Regional Conference. Rome, September 1–7. Vol. II." (D. A. Bocciarelli, ed.) pp. 33–34. Tipografia Poliglotta Vaticana, Rome.

Morphew, M. K., and McIntosh, J. R. (2003). The use of filter membranes for high-pressure freezing of cell monolayers. *J. Microsc.* **212**(Pt 1), 21–25.

Müller, M., and Moor, H. (1984). Cryofixation of thick specimens by high-pressure freezing. *In* "The Science of Biological Specimen Preparation." (G. H. Haggins, ed.) pp. 131–138. SEM Inc AMF O'Hare, Chicago.

Peters, P. J., and Pierson, J. (2008). Immunogold labeling of thawed cryosections. *Methods Cell Biol.* **88**, 131–149.

Pfeiffer, S., and Krupinska, K. (2005). Chloroplast ultrastructure in leaves of Urtica dioica L. analyzed after high-pressure freezing and freeze-substitution and compared with conventional fixation followed by room temperature dehydration. *Microsc. Res. Tech.* **68**(6), 368–376.

Provin, C., Takano, K., Yoshida, T., Sakai, Y., Fujii, T., and Shirakashi, R. (2009). Low O2 metabolism of HepG2 cells cultured at high density in a 3D microstructured scaffold. *Biomed. Microdevices* **11**(2), 485–494.

Rasey, J. S., Cornwell, M. M., Maurer, B. J., Boyles, D. J., Hofstrand, P., Chin, L., and Cerveny, C. (1996). Growth and radiation response of cells grown in macroporous gelatin microcarriers (CultiSpher-G). *Br. J. Cancer Suppl.* **27**, S78–S81.

Reipert, S., and Wiche, G. (2008). High-pressure freezing and low-temperature fixation of cell monolayers grown on sapphire coverslips. *Methods Cell Biol.* **88**, 165–180.

Ripper, D., Schwarz, H., and Stierhof, Y. D. (2008). Cryo-section immunolabelling of difficult to preserve specimens: Advantages of cryofixation, freeze-substitution and rehydration. *Biol. Cell* **100**(2), 109–123.

Ruemmele, F. M., Müller, T., Schiefermeier, N., Ebner, H. L., Lechner, S., Pfaller, K., Thöni, C. E., Goulet, O., Lacaille, F., Schmitz, J., Colomb, V., Sauvat, F., *et al.* (2010). Loss-of-function of MYO5B is the main cause of microvillus inclusion disease: 15 novel mutations and a CaCo2 RNAi cell model. *Hum. Mutat.* **31**, 544–551.

Sabatini, D. D., Bensch, K., and Barrnett, R. J. (1963). Cytochemistry and electron microscopy. The preservation of cellular ultrastructure and enzymatic activity by aldehyde fixation. *J. Cell Biol.* **17**, 19–58.

Sawaguchi, A., McDonald, K. L., and Forte, J. G. (2004). High-pressure freezing of isolated gastric glands provides new insight into the fine structure and subcellular localization of H+/K+-ATPase in gastric parietal cells. *J. Histochem. Cytochem.* **52**(1), 77–86.

Schwarb, P. (1990). Morphologische Grundlagen zur Zell-Zell Interaktion bei adulten Herzmuskelzellen in Kultur. PhD-Thesis. Swiss Federal Institute of Technology, Zürich.

Schwarz, H., and Humbel, B. M. (2007). Correlative light and electron microscopy using immunolabeled resin sections. *Methods Mol. Biol.* **369**, 229–256.

Shimoni, E., and Muller, M. (1998). On optimizing high-pressure freezing: From heat transfer theory to a new microbiopsy device. *J. Microsc.* **192**(Pt 3), 236–247.

Simionescu, N., and Simionescu, M. (1988). The cardiovascular system. In "Cell and Tissue Biology. A Textbook of Histology" (L. Weiss, ed.) pp. 355–400. Urban & Schwarzenberg, Baltimore, Munich.

Smalley, K. S., Lioni, M., and Herlyn, M. (2006). Life isn't flat: Taking cancer biology to the next dimension. *In Vitro Cell. Dev. Biol. Anim.* **42**(8–9), 242–247.

Sosinsky, G. E., Crum, J., Jones, Y. Z., Lanman, J., Smarr, B., Terada, M., Martone, M. E., Deerinck, T. J., Johnson, J. E., and Ellisman, M. H. (2008). The combination of chemical fixation procedures with high pressure freezing and freeze substitution preserves highly labile tissue ultrastructure for electron tomography applications. *J. Struct. Biol.* **161**(3), 359–371.

Spurr, A. R. (1969). A low-viscosity epoxy resin embedding medium for electron microscopy. *J. Ultrastruct. Res.* **26**(1), 31–43.

Stierhof, Y. D., van Donselaar, E., Schwarz, H., and Humbel, B. M. (2009). Cryo-fixation, freeze-substitution rehydration and Tokuyasu cryo-sectioning. *In* "Handbook of Cryo-Preparation Methods for Electron Microscopy." (A. Cavalier, D. Spehner, and B. M. Humbel, eds.) pp. 345–365. CRC Press, Boca Raton.

Studer, D., Michel, M., and Muller, M. (1989). High pressure freezing comes of age. *Scanning Microsc. Suppl.* **3**, 253–268, discussion 268–269

Sutherland, R. M. (1988). Cell and environment interactions in tumor microregions: The multicell spheroid model. *Science* **240**(4849), 177–184.

Tokuyasu, K. T. (1973). A technique for ultracryotomy of cell suspensions and tissues. *J. Cell Biol.* **57**(2), 551–565.

Tokuyasu, K. (1998). Immunocytochemistry on ultrathin cryosections. *In* "Cells. A Laboratory Manual. Volume 3: Subcellular Localization of Genes and Their Products." (D. L. Spector, R. D. Goldman, and L. A. Leinwand, eds.) 131.1–131.27CSHL Press, Cold Spring Harbor.

Triffo, W. J., Palsdottir, H., McDonald, K. L., Lee, J. K., Inman, J. L., Bissell, M. J., Raphael, R. M., and Auer, M. (2008). Controlled microaspiration for high-pressure freezing: A new method for ultrastructural preservation of fragile and sparse tissues for TEM and electron tomography. *J. Microsc.* **230**(Pt 2), 278–287.

Trump, B. F., and Ericsson, J. L. (1965). The Effect of the Fixative Solution on the Ultrastructure of Cells and Tissues. A Comparative Analysis with Particular Attention to the Proximal Convoluted Tubule of the Rat Kidney. *Lab. Invest.* **14**, 1245–1323.

van Donselaar, E., Posthuma, G., Zeuschner, D., Humbel, B. M., and Slot, J. W. (2007). Immunogold labeling of cryosections from high-pressure frozen cells. *Traffic* **8**(5), 471–485.

Van Harreveld, A., and Crowell, J. (1964). Electron microscopy after rapid freezing on a metal surface and substitution fixation. *Anat. Rec.* **149**, 381–385.

Vanhecke, D., Graber, W., Herrmann, G., Al-Amoudi, A., Eggli, P., and Studer, D. (2003). A rapid microbiopsy system to improve the preservation of biological samples prior to high-pressure freezing. *J. Microsc.* **212**(Pt 1), 3–12.

Vunjak-Novakovic, G., and Freed, L. E. (1998). Culture of organized cell communities. *Adv. Drug Deliv. Rev.* **33**(1–2), 15–30.

Walther, P., and Ziegler, A. (2002). Freeze substitution of high-pressure frozen samples: The visibility of biological membranes is improved when the substitution medium contains water. *J. Microsc.* **208**(Pt 1), 3–10.

Webster, P. (1999). The production of cryosections through fixed and cryoprotected biological material and their use in immunocytochemistry. *Methods Mol. Biol.* **117**, 49–76.

Weigelt, B., and Bissell, M. J. (2008). Unraveling the microenvironmental influences on the normal mammary gland and breast cancer. *Semin. Cancer Biol.* **18**(5), 311–321.

Wild, P., Schraner, E. M., Cantieni, D., Loepfe, E., Walther, P., Muller, M., and Engels, M. (2002). The significance of the Golgi complex in envelopment of bovine herpesvirus 1 (BHV-1) as revealed by cryobased electron microscopy. *Micron* **33**(4), 327–337.

Yu, S. C., Ping, Y. F., Yi, L., Zhou, Z. H., Chen, J. H., Yao, X. H., Gao, L., Wang, J. M., and Bian, X. W. (2008). Isolation and characterization of cancer stem cells from a human glioblastoma cell line U87. *Cancer Lett.* **265**(1), 124–134.

Zeuschner, D., Geerts, W. J., van Donselaar, E., Humbel, B. M., Slot, J. W., Koster, A. J., and Klumperman, J. (2006). Immuno-electron tomography of ER exit sites reveals the existence of free COPII-coated transport carriers. *Nat. Cell Biol.* **8**(4), 377–383.

Zierold, K. (1997). Effects of cadmium on electrolyte ions in cultured rat hepatocytes studied by X-ray microanalysis of cryosections. *Toxicol. Appl. Pharmacol.* **144**(1), 70–76.

Zierold, K. (2000). Heavy metal cytotoxicity studied by electron probe X-ray microanalysis of cultured rat hepatocytes. *Toxicol. In Vitro* **14**(6), 557–563.

CHAPTER 28

"Tips and Tricks" for High-Pressure Freezing of Model Systems

Kent McDonald[*], Heinz Schwarz[†], Thomas Müller-Reichert[‡], Rick Webb[§], Christopher Buser[¶], *and* Mary Morphew[#]

[*]Electron Microscope Laboratory, University of California, Berkeley, California 94720

[†]Electron Microscopy Unit, Max Planck Institute of Developmental Biology, 72076 Tübingen, Germany

[‡]Medical Theoretical Center, TU Dresden, 01307 Dresden, Germany

[§]Centre for Microscopy and Microanalysis, University of Queensland, 4072 Queensland, Australia

[¶]Department of Molecular & Cell Biology, University of California, Berkeley, California 94720

[#]Boulder Laboratory for 3D Electron Microscopy of Cells, University of Colorado, Boulder, Colorado 80309

978-0-12-381007-6
DOI: 10.1016/S0091-679X(10)96028-7

Abstract

High-pressure freezing (HPF) has been around since the mid-1980s as a cryopreparation technique for biological electron microscopy. It has taken quite some time to "catch on" but with the recent interest in cellular tomography and electron microscopy of vitreous cryosections it has been used more frequently. While HPF is relatively easy to do, there are a number of steps, such as loading the sample into the specimen carrier correctly, that are critical to the success of this method. In this chapter we discuss some of the "little" things that can make the difference between successful or unsuccessful freezing. We cover all aspects of HPF, from specimen loading to removing your sample from the carriers in polymerized resin. Our goal is to make it easier and more reliable for HPF users to get well-frozen samples for their research.

I. Introduction

In the past decade and more, there have been numerous articles about how to do high-pressure freezing—from a detailed account of using the BAL-TEC HPM 010 (McDonald, 1999) or Leica EM PACT (Studer *et al.*, 2001) to more recent works that cover much the same territory (Kaech, 2009; Vanhecke and Studer, 2009). This is in addition to many other "instructional" papers that have addressed the application of high-pressure freezing techniques for specific model system organisms such as flies (McDonald, 1994; McDonald *et al.*, 2000), yeasts (McDonald and Müller-Reichert, 2002), and worms (Müller-Reichert *et al.*, 2003; McDonald, 2007). We distinguish these papers from regular research papers by the fact that they primarily cover the "how to" aspects of HPF rather than focus on the scientific results. They are like the Methods and Materials section of regular articles, expanded to give details that most editors will not allow. We usually refer to these types of articles as "methods papers" as opposed to research articles. They are the kind of account found in books like the present publication. In this chapter, we would like to highlight some of the "little" things that one can do to make a difference between success and failure for HPF, or just make it a lot easier to do; what are often called "tips and tricks". Some of these are from our own cumulative experience of more than 70 years of high-pressure freezing, but others come from colleagues' work, both published and unpublished. This is not to say that others have not come up with the same ideas but if they are unpublished we have no way to acknowledge their contributions. There will also be items that we miss and we apologize to the authors concerned in advance.

II. Rationale

The items that follow cover all aspects of high-pressure freezing, from preparing the samples and sample carriers to removing the samples from the carriers embedded in resin. They have in common the idea that the method being discussed will make it easier or more reliable to get well-frozen cells and tissue samples.

III. Methods

A. Working with Cellulose Microdialysis Capillary Tubing

1. History

The use of cellulose capillary tubing for high-pressure freezing was the idea of Hohenberg *et al.* (1994). Since then it has found applications with samples such as plant protoplasts (Tiedemann *et al.*, 1998), single *Caenorhabditis elegans* embryos (Muller-Reichert *et al.*, 2007), 3D mouse breast cancer organoids and strips of cochlear cells (Triffo *et al.*, 2008), and correlative microscopy and tomography of worms (Müller-Reichert *et al.*, 2007) to name some.

2. Where to Buy

There are a number of sources for regenerated cellulose hollow fibers as they are called in the industry: Packs of 100 capillaries are available from Leica Microsystems (contact your local Leica Microsystems sales representative for prices and availability). In the United States, they are available from Spectrum, 23022 La Cadena Dr., Suite 100, Laguna Hills, CA 92653, USA. In the past, we ordered the Spectra/Por Hollow Fiber Bundles, 200 μm inner diameter, Cat. no. 132 290. However, this part number no longer exists at Spectrum Labs, but if you go to http://www.spectrapor.com/cell/MaxCarts.html, you will find the so-called "hollow fiber bioreactors" and some contain regenerated cellulose hollow fibers with inner diameters of 200 μm (Prod. No. 430-011) and 192 μm (Prod. No. 400-004) and molecular weight cut-off values of 30 and 20 kD, respectively. It costs around $US 500 for these "bioreactors" but it appears that they might contain thousands of fibers about 20 cm in length. The alternative at Spectrum seems to be Parts Nos. 132294 and 132295 that come 20 fibers to a package for $US 112.

In Europe, capillaries may also be purchased much cheaper from the original MICRODYN-NADIR Diaperm Dialysis Module LD OC 02, which contains 1280 cellulose hollow fibers (each with an active length of 245 mm) per cartridge. This lab module is produced and distributed under the article number 00020401 by MICRODYN-NADIR GmbH, Rheingaustrasse 190–196, D-65203 Wiesbaden; E-mail: info@microdyn-nadir.de. The price is 155 € per module. The same module is also available from Reichelt Chemietechnik GmbH, D-69126 Heidelberg (www.rct-online.de) as Thomapor MICRODYN-LD OC02, Cat. No.72571 for 390 €.

Problem 1: Cutting and sealing samples into capillary dialysis tubing requires more information than given in the original reference (Hohenberg *et al.,* 1994). In this article it was recommended that a scalpel or opthamological punch could be used to cut the capillary tubing. This is true as far as it goes, but for some samples it is also necessary to cut and crimp the tubing so that the cells inside do not leak out during subsequent processing. A new scalpel will almost always cut the tubing cleanly and it will be totally open at the ends. If you want to cut and crimp at the same time, try the following method as suggested by Andreas Kaech, University of Zurich (pers. comm.)

Solution:

1. Take a number 11 scalpel blade and cut or break off the tip so the new tip is about 1 mm across and put the blade in a scalpel handle (Fig. 1A).
2. Use a whetstone to make the tip into a chisel shape (Fig. 1B–C).
3. Test the blade by cutting dialysis tubing filled with dye under 1-hexadecene in a shallow plastic dish. Use a rocking motion perpendicular to the tubing long axis to make the cuts. You should not have to press down hard to make the cut.

Fig. 1 Details of the crimping tool for sealing microdialysis capillary tubing. (A) The tool should be mounted in a scalpel handle and labeled such that no one will throw away the blade. Depending on the frequency of use, these tools can be used for months before reshaping. (B) Side view of the tip of the crimping tool. (C) View from the top of the tip.

4. Continue shaping and testing until a single motion both cuts and seals the tubing. This has to be done for both ends. The result should look like Fig. 4A in Müller-Reichert *et al.*, 2008.
5. When the procedure is comfortable and reproducible, use on your tissues and cells of choice.

Problem 2: Capillary tubing can be used to isolate single cells, which can then be frozen and processed in the tubing. Getting cells into the tubing and sealing it can be tricky. Thomas Müller-Reichert solved this problem by using nail polish to seal some capillary tubing into a micropipette (Müller-Reichert *et al.*, 2007).

Solution:

1. Cut pieces of microdialysis tubing into 3–4 cm lengths (Fig. 2A).
2. Seal the tubing into the pipette tip with nail polish and let dry (Fig. 2B).
3. Attach the tip to a pipette and place the tubing into the liquid from which the cells will be withdrawn. The tubing will curl, and then straighten out as the walls fill with liquid.
4. Carefully draw the single cell into the tubing, transfer to a shallow plastic Petri dish with enough 1-hexadecene to cover the tubing, and cut the tubing near the pipette tip with the crimping tool (see Section above).

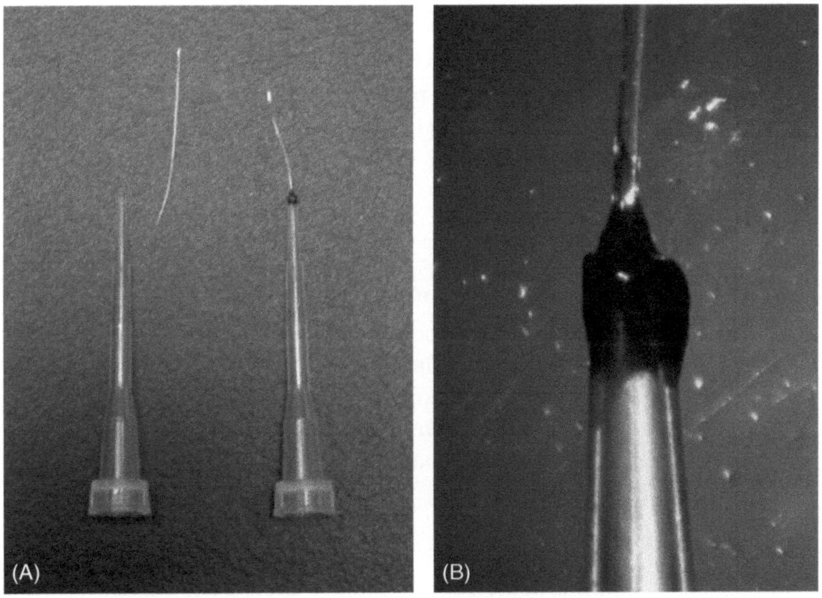

Fig. 2 Capillary tubing mounted into micropipette tips for use in controlled movement of small objects into the tubing. The tips are the small size used in 0.5–10 μl pipetmen. (A) A pipette tip and piece of capillary tubing on the left prior to assembly. On the right, a tip after sealing the tubing into the tip with nail polish. (B) Detail of the nail polish seal at the tip. It is important to make sure the polish completely seals the tubing.

5. Cut the tubing with cell into the appropriate length for the HPF specimen cup you are using. For the Wohlwend-style cups, that can be about 1.8 mm, and for Leica cups about 1.4 mm.
6. Using forceps to pick up the tubing by one of the crimped ends, place it in the cup, cover with cryoprotectant to fill the cup, and freeze.

B. Making Up 20% Bovine Serum Albumin

Problem: Making up 20% bovine serum albumin (BSA) in small quantities (less than 5 ml) can be difficult because the volume of the powder is about the same as the volume of the liquid and it does not go into solution readily. It tends to clump together and requires vortexing and heating for 5–10 min before it dissolves. If you are only using 20% BSA as a filler solution for the HPF planchettes, you may only require less than 1 ml total. If you want to use 20% BSA as a filler, try this idea from Rick Fetter of Howard Hughes Medical Institute, Janelia Farms Research Campus (pers. comm.) for making up small quantities very easily.

Solution:

1. Put 850 μl of solution (whatever you want to use for your cells/tissues) in a 2 ml Eppendorf tube. Make sure there is no solution on the side of the tube above the meniscus or the BSA will stick to it and not go into solution.
2. Weigh out 200 mg of BSA and layer it on top of the solution in the tube.
3. Spin for 15–20 s in a microfuge at 15,000 × G.

Some say that 20% BSA will extract membranes from cells, but Paul Verkade (University of Bristol, UK) reminds us that if you make up the BSA with medium containing at least 5% or more fetal calf serum, then extraction will not be a problem.

C. Working with a Very Small Volume of Cells

Problem: Sometimes you may only have a few microliters of cells and concentrating them and loading them into an HPF specimen carrier is a challenge. While the specimen carriers typically hold less than a microliter, there is often a lot of transfer loss going from a tube or filter to the carrier. Here is an idea from Alisdair McDowall of Cal Tech, Pasadena (pers. comm.) that optimizes concentrating and loading with mimimal loss.

Solution:

1. Using a flame source, such as a cigarette lighter, seal the ends of some 200 μl pipette tips (Fig. 3).
2. Concentrate the cells of interest into about 100 μl of solution in a regular small tube such as an 0.5 ml or 2 ml Eppendorf.
3. Resuspend the cells in the Eppendorf, then pipette them into the sealed pipette tip.

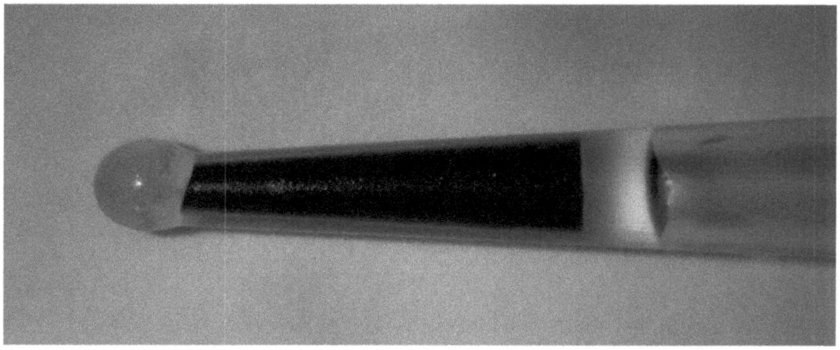

Fig. 3 A 200 μl pipette tip showing how the end looks after sealing with a flame. After centrifugation, the biological material (dark) is concentrated at the tip with a small amount of supernatant at the top to keep it from drying out.

4. Place the pipette tip plus cells into a 0.5 ml Eppendorf tube without cap, then this assemblage into a 2.0 ml Eppendorf tube without cap.
5. Create a balance tube with the same components.
6. Spin down until the cells are concentrated in the bottom of the sealed pipette tip.
7. Carefully remove most, but not all, of the supernatant above the cells with wicking paper (Fig. 3).
8. Using a very sharp blade, carefully cut off the tip of the pipette so that there is an opening for the concentrated cells to be extruded.
9. With a mouth pipette, or micropipetter, carefully extrude cells from the pipette tip directly into an HPF specimen carrier until full, then freeze immediately.
10. If there is enough material to fill more than one specimen carrier, take care not to let the very tip of the pipette with cells to dry out. Putting it up against a small piece of paper saturated with medium in the bottom of a 2 ml tube can work. Better yet, have some 1% (or less) agarose in a dish or tube and plunge the tip/cells into that immediately after extrusion for HPF.

D. Using Yeast Paste as a Filler for Difficult-to-Freeze Cell Suspensions for Vitreous Cryosectioning

Problem: Most cell suspensions (bacteria, tissue culture (TC) cells) can be concentrated and used to fill the entire volume of a specimen cup. However, some cells do not freeze well when processed this way, especially in cups 100 μm or more deep. Yeast (*Saccharomyces cerevisiae*) cells, on the other hand, freeze very well in even very deep (400 μm or more) specimen cups. Difficult cell types can be mixed with yeast to give a hybrid mix, and this will sometimes improve their quality of freezing. Mark Ladinsky of Cal Tech, Pasadena (pers. comm.) has used this strategy effectively for cutting vitreous sections of mammalian tissue culture cells.

Solution:

1. Set up cultures of yeast and the other cell type to grow in solution to early log phase.
2. Concentrate the cells by centrifugation, or another means if spinning them down could be harmful.
3. Mix the concentrated cell types in at least a 1:1 ratio. Having more yeast than the other cell type will improve the chances of good freezing.
4. Load the mixture into HPF specimen cups and freeze.

E. Using Slot Grids with the Wohlwend-Style Specimen Cups as a Variable Depth Spacer for Difficult-to-Freeze Cell Types

Problem: Some cell types do not yield 100% well-preserved tissue when frozen in the standard 100 μm deep specimen cups. With Wohlwend-style cups it is possible to buy from Wohlwend Engineering (Sennwald, Switzerland; Technotrade International, Manchester, NH, USA) specimen cups with depths of 25 and 50 μm. An alternative method to create shallower depths is to use TEM slot grids as spacers. Rick Fetter of HHMI Janelia Farms Research Campus developed this strategy to improve the yield of well-preserved *C. elegans* worms. More information on known slot grid thicknesses can be found in McDonald *et al.*, (2007).

Solution:

1. Place a piece of filter paper in a small Petri dish and saturate with 1-hexadecene.
2. Place Type B (one flat side, one 300 μm deep well on the other side) specimen cups, flat side down on the saturated filter paper. Also place the slot grid of choice on the filter paper.
3. After blotting to remove excess hexadecane, put a specimen cup (flat side up) in the tip of the HPF specimen loader (Fig. 4A).
4. Put a slot grid on top of the specimen cup in the loader (Fig. 4B).
5. Fill the cavity of the slot grid with sample.
6. Blot and place a specimen cup (flat side down) on top of the slot grid/sample (Fig. 4C). This step needs to follow the previous step quickly so the sample does not dry out.
7. Close the tip of the specimen loader and freeze the sample (Fig. 4D).

In Fig. 4, the slot grid used was very thin, about 27 μm, and this is a good size for cell suspensions such as bacteria. For thicker samples, a thicker slot grid can be used. In Fig. 5, a slot grid about 65 μm thick was used to freeze *C. elegans* worms. The worms were slightly compressed at this thickness, but by having both sides of the worm touching metal, the chances of good freezing are improved.

F. Freezing Monolayers of Tissue Culture Cells on Sapphire or Aclar® Disks

Problem: TC cells grown on a substrate are among the most widely used model systems in cell biology today. The challenge is how to grow them for high-pressure

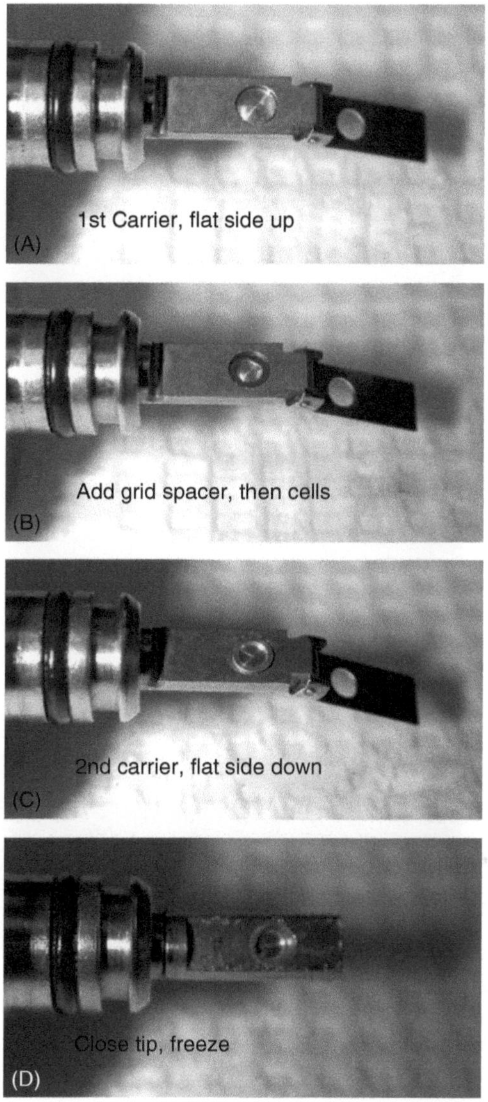

Fig. 4 A sequence of steps for using EM slot grids as spacers for creating shallow wells in ABRA HPM 010, BAL-TEC HPM 010, and Wohlwend HPF Compact 01 high-pressure freezers. (A) Place a Type B (flat on one side) specimen carrier flat side up in the specimen holder. (B) Add a slot grid as a spacer, and then fill with cells, or place a sapphire disk cell side down over the spacer. (C) Place a second Type B carrier on top of the slot grid or sapphire disk, flat side down. (D) Close the tip, secure it into the holder, and freeze.

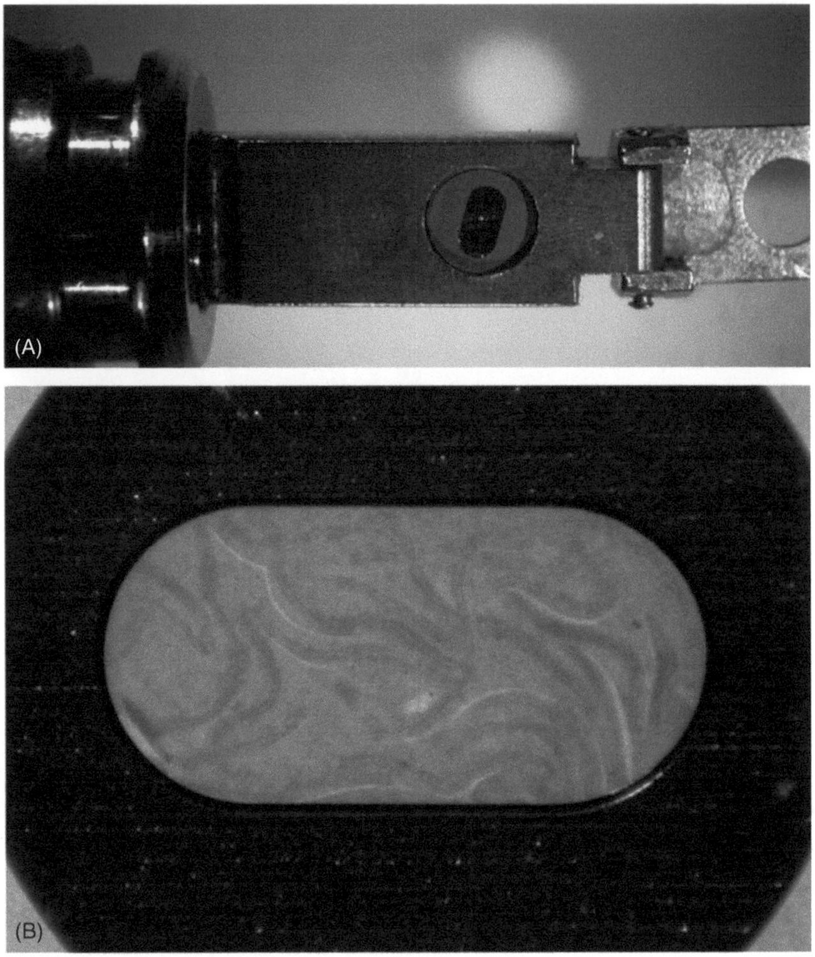

Fig. 5 Use of a slot grid thicker than the one shown in Fig. 4. (A) A slot grid in the tip prior to adding *C. elegans* worms in a paste of thick *Escherichia coli* in the slot. (B) After freezing, freeze substitution, and infiltration with resin, the worms are still trapped in the slot because the *E. coli* paste forms a solid mass after freeze substitution.

freezing and recover them for sectioning, in some cases sectioning the same cell that was observed in a light microscope. In the sections that follow we will describe the various ways in which TC cells can be prepared for high-pressure freezing and how to process them after that. We will discuss only TC cells grown on or adhered to substrates as opposed to those grown in solution and frozen in bulk. The substrates that can be used for high-pressure freezing of adherent TC cells include sapphire disks, Aclar® disks, filter membranes, HPF specimen cups, Formvar-coated gold grids, chips of Thermanox plastic Mylar® film, and microcarrier beads.

Solutions:

1. Sapphire Disks

a. History of Use Sapphire disks were first used to grow cells for high-pressure freezing by Schwarb (1990) in the Martin Mueller lab in Zurich, and later by Tse *et al.* (1997), Hess *et al.* (2000), Erlandsen *et al.* (2001), and Riepert *et al.* (2004) among others. They are preferred over glass coverslips because the coefficient of heat transfer is much greater at liquid nitrogen temperatures (Echlin, 1992). Before plating cells on them, it is best to coat with carbon or gold because cells often prefer these substrates, but more importantly, it is easy to scratch an asymmetric figure in the carbon/gold to indicate which side of the disk the cells are on.

b. Where to Buy Sapphire disks are available through Leica Microsystems at a cost of $510 per 100 for the 3 mm disks that fit the Wohlwend, Abra, and Leica HPM 100 high-pressure freezers, and $639 per 100 for the 1.4 mm disks that fit the Leica EM PACT instruments. Engineering Office M.Wohlwend GmbH (Bifig 14, CH-9466 Sennwald, Switzerland) sells the 3 mm size for either $344 per 100 (50 μm thick) or $373 per 100 (120 μm thick), and the 1.4 mm size for $442 per 100. One can also order directly from Rudolf Bruegger, S.A. in Minusio, Switzerland where ordering in bulk (500 or more) may save a little over the Wohlwend price. For example, for 500 pieces, the cost is $324 per 100.

c. Carbon or Gold Coating We use two different approaches to coating depending on whether we are planning to do correlative microscopy or not.

1. No correlative microscopy. Place the sapphire disks in a dish or other container that will fit into the coating device. Some evaporators have a metal plate with holes and slots for holding grids (Fig. 6A). We like to use the adhesive side of a Post-it® note to keep the disks separate yet put many in a small area. Evaporate enough carbon or gold onto the disk such that it will be easy to see in a stereomicroscope an asymmetric figure such as the letter "F" scratched into the coating.
2. Correlative microscopy. When using disks for this purpose it is useful to evaporate the pattern of an EM finder grid onto the surface. For the 3 mm disks used in ABRA HPM 010, Leica HPM 100, and Wohlwend Compact HPF 01 machines, this is straightforward because you just have to put an EM finder grid on top of the disk before evaporation. To keep the grid from sliding off the disk it is necessary to put them into a recess of some sort. This is where the EM grid holder plates that come with some evaporators come in handy. However, there are other easy solutions to this problem. If you have reasonable access to a machine shop, then they could make a grid holder by copying the type shown in Fig. 6A–B. Alternatively, one can use Teflon® immunostaining pads, such as Cat. No. 10526-1 from Ted Pella, Inc. (Fig. 6C). These have shallow conical depressions that will accommodate a sapphire disk and the covering finder grid nicely. There are many finder grid patterns available but we prefer those that have a relatively large "F" on the rim because it is easy to see in a stereomicroscope (e.g., EMS Cat. No. LF-135Cu; Fig. 6B and D).

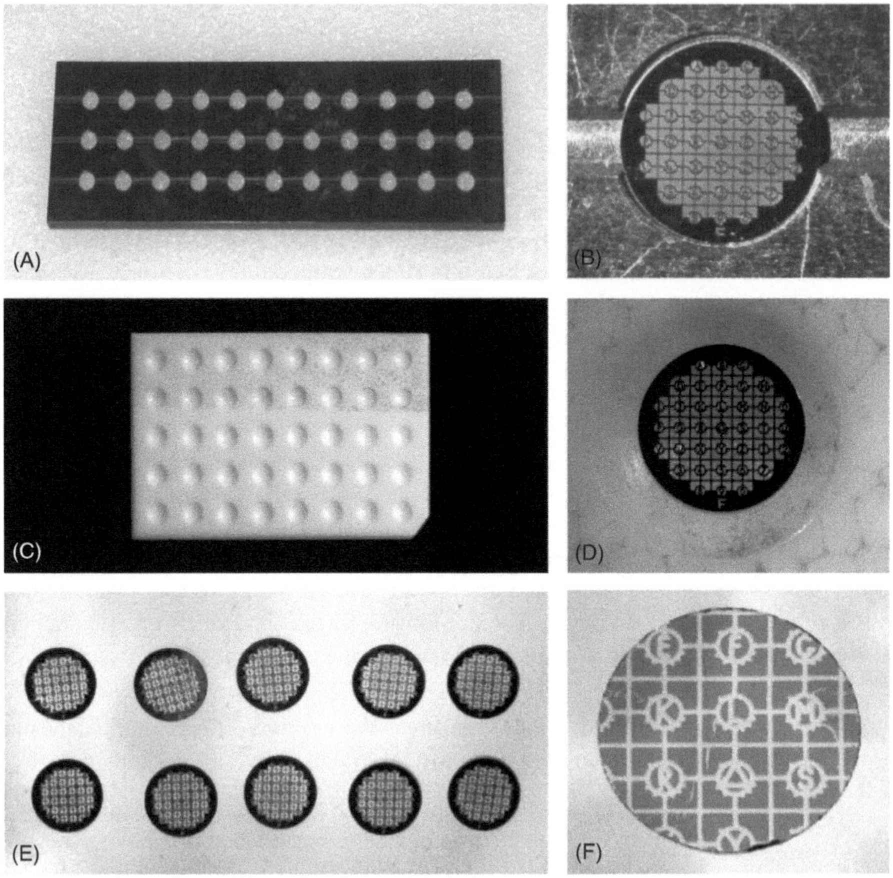

Fig. 6 Different ways to carbon coat sapphire disks with a finder grid pattern. (A) A brass grid holder with recessed holes for holding sapphire disks and finder grids. (B) Close-up of a recess shown in (A) showing the grooves on either side for picking up the sapphire disks/grids. (C) A Teflon® well plate with hemispherical depressions that can be used to hold sapphire disks and finder grids. Place a disk in the well and then put a finder grid on top. (D) Detail of (C). (E) A row of finder grids stuck down on the sticky part of a Post-it® note paper. Underneath each grid but not visible here are 1.4 mm sapphire disks used in the Leica EM PACT HPF machines. (F) The pattern on one of the 1.4 mm disks following carbon coating. Note that a heavy carbon coat is laid down, making it easier to see the pattern when handling the cells and in the resin after polymerization.

For sapphire disks in the Leica EM PACT system the approach is slightly different. Because the disks are only 1.4 mm in size they will not fit easily into one of the holders mentioned above. But here we can use the adhesive side of a Post-it® note again to hold not only the disk, but the finder grid also. First, put the sapphire disks on the adhesive about 5 mm apart. Then, place the finder grid over the disk (Fig. 6E) in a way that some asymmetric letters and/or numbers will be visible after coating. Fig. 6F shows one such pattern that is possible. With a gloved finger, press down the grid over the disk so that it sticks to the adhesive of the Post-it®.

Even if you are not going to do correlative microscopy on the same cell, using finder grids can have some advantages for processing TC cells. It can be a guide for remounting and identifying the best regions of the grid to section. This is illustrated in the Supplementary Materials file labeled SD Remount from Capsule.

d. Stabilizing the Carbon Prior to Plating Cells Before immersing the disks in medium it is recommended that the coating be stabilized so that it will not float off or curl in the liquid. For the gold, this step is not necessary. To stabilize, place the coated disks in a glass dish and bake overnight at 120°C. If the disks are to be used right away, there is no need to glow discharge them. If they are stored in a refrigerator at 4°C until use, they will remain hydrophilic for up to a month. Otherwise, glow discharge the disks before using them as cell substrates. The 3 mm size can be stored in a grid box, or on Post-it® note adhesive. For the smaller 1.4 mm size, the Post-it® note approach works, or you can store either type of disk in a Gel-Pak® box (Ted Pella, Inc., Cat. No. AD-22T-00-X8). These boxes are expensive, but they are reusable and very handy for not only sapphire disks, but also for specimen cups and other small items. The gels can be cleaned easily by stripping clean the surface with adhesive tape.

e. Culturing Cells Disks can be sterilized by ultraviolet radiation or heating in a microwave for 5 min at 800 W (Jimenez *et al.*, 2006). Make sure that after the medium is added and inoculated with cells the disks have the coated side up. Also check to ensure that there are no air bubbles collecting around the edges of the disk that might cause the disk to rise and float at the surface of the medium. This is especially critical for the small Leica EM PACT disks. If this problem recurs then consider tacking the disk down with a small amount of agarose or petroleum jelly.

f. Attaching Cell Suspensions Another strategy for getting cells onto sapphire disks is to let them settle onto the disk and attach before freezing. Some cells will do this naturally, while others may need a favorable substrate, such as fibronectin, collagen, and Matrigel®. One can also use poly-L-lysine to attach cells (Sawaguchi *et al.*, 2003).

g. High-Pressure Freezing How one goes about freezing sapphire disks depends on whether or not correlative microscopy is being done. Within the category of correlative microscopy it also depends on the time resolution required. By correlative microscopy we mean looking at a cell in the light microscope, then looking at the same cell by electron microscopy. If the cell needs to be frozen within 4–5 s of making observations, then the Leica EM PACT2 with Rapid Transfer System is the method of choice. This technique is covered in detail by Paul Verkade in his article using HPF for correlative LM–EM with about 5 s time resolution (Verkade, 2008). We would only add that there is a new rapid transfer system (RTS) loading device made by Marine Reef International that facilitates long observation times prior to freezing (Fig. 7). For those who may be interested in acquiring such a loader, they can be obtained from Marine Reef International by contacting Paul DeGeorge by email at pdegeorge@marinereef.com. The price is the same as the regular RTS loader from Leica.

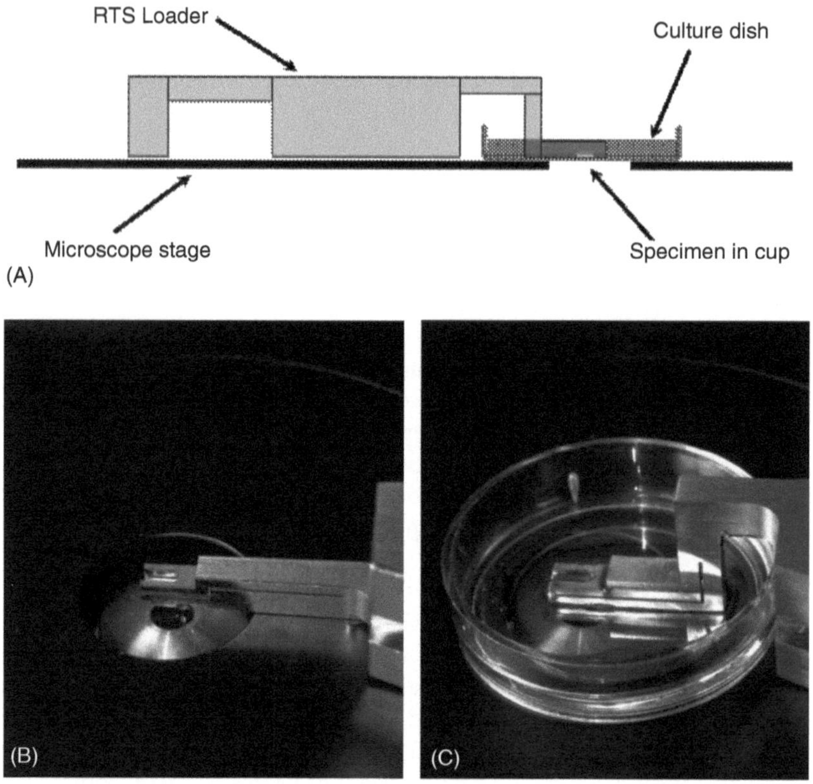

Fig. 7 A new loader for the Leica EM PACT2-RTS high-pressure freezer that allows longer incubation and observation times for cells prior to freezing. (A) Schematic representation of a loader in a culture dish on an inverted microscope stage. (B) The current RTS loader on a microscope stage. Observation times are rather short (5 min or so) because the tip is sitting on a coverslip (not shown) and the only fluid around the cells is what is trapped between the specimen carrier and the coverslip. (C) The new Marine Reef loader shown inside a culture dish on the microscope stage.

If there is no great need for speed of transfer from the light microscope to the high-pressure freezer, then it is possible to use the freezing methods described in the next sections. Except for the first method of putting disks inside the specimen cup, all the techniques described below pertain to the BAL-TEC, ABRA, or Wohlwend machines. For every method, the specimen cups, washers, and spacer grids should be coated with a release agent such as 1-hexadecene or 1% lecithin in chloroform before freezing. This will facilitate releasing the disks from the carriers under liquid nitrogen or at later stages in processing.

1. *Putting disks inside the specimen cup*. This is the method described by Reipert *et al.* (2004) for freezing 1.4 mm sapphire disks in the Leica EM PACT system. You remove the disk from the culture dish, dip it into 20% BSA solution made up with culture medium, load it into the cup, and freeze in a 100 μm deep membrane carrier.

If you like using the small sapphire disk you could also use any of the other HPF machines by loading into a 50 μm deep carrier and covering with a 25 μm deep carrier, or a flat carrier and a slot grid spacer.

2. *Sandwiching the disk between two flat carriers*. Place a carrier flat side up in the tip of the specimen loader. Remove the disk from the culture dish by picking it up from the sides with fine tweezers, and put it on top of the flat specimen carrier in the loader. Make sure the cells are facing up. You can either dip the disk in 20% BSA or use the culture medium alone. BSA will help the cells stay in place during subsequent processing because it forms a solid layer after freeze substitution. Place a Chien-style, 27 μm thick slot grid (see Fig. 4, and also McDonald *et al.*, 2007) on top of the disk, then another specimen carrier flat side down. Close the tip of the loader and freeze immediately.
3. *Sandwiching two disks separated by a slot grid between spacer rings* (Hawes *et al.*, 2007). BAL-TEC AG used to sell open washers that could be used to hold sapphire disks in the HPF specimen holder. Whether RMC or Leica Microsystems still sell these remains to be seen. To load, place a spacer ring in the tip of the HPF specimen loader, then a sapphire disk with cells up, a Chien-style slot grid, and then another sapphire disk with cell side down. Finally, place another spacer ring on top, close the tip of the loader, and freeze.

h. Freeze Substitution The thing to note about freeze substitution of TC cells on a substrate is that the substitution time can be drastically reduced compared to solid tissues. Hawes *et al.* (2007) describe a method that takes 6.5 h to reach −50°C at which point they infiltrate with Lowicryl HM20 for about 24 h, then another 4 days for complete polymerization of the resin. They believe the initial quick processing to the resin step results in improved ultrastructural preservation. Furthermore, only culture medium is used in freezing and the freeze substitution (FS) medium is acetone plus 2% or 0.2% uranyl acetate depending on whether immunolabeling is going to be performed. For more details consult the original paper by Hawes *et al.* (2007). For those preferring to use more traditional organic solvent-fixative combinations it is likely that shorter substitution times would also yield excellent results.

i. Embedding For any of the substrates used to grow TC cells it is necessary to embed them in resin and remove them after polymerization. There are different ways to do this, so in the following text we describe methods for using flat-bottomed capsules and embedding the disks in a thin layer of resin between Teflon®-coated microscope slides.

1. *Flat-bottomed capsule embedding*. Flat-bottomed capsules are available from Ted Pella, Inc. (Cat. No. 133-P) and perhaps other vendors. We prefer to use the stiff polypropylene capsules because they retain their shape when heated, especially in a microwave oven. To embed, place the capsules in a rack and put a piece of Aclar® plastic cut out with a paper punch in the bottom of each one. After labeling each capsule, fill partially with resin, then place the sapphire disk, cell side up, in the bottom of the capsule. Fill completely with resin and place in the oven for polymerization.

2. *Thin-film embedding of cells on disk substrates*. This method allows one to locate a particular cell as in the case of correlative LM–EM or to select a specific cell for its stage in mitosis. Once located, a given cell or group of cells can be cut from the thin film of resin and easily remounted in any orientation onto a blank resin block for sectioning. This technique is an adaptation of a method previously described (Reymond and Pickett-Heaps, 1983). The details of the procedure are as follows:

   a. Following freeze substitution and solvent rinses, gently wash disks (sapphire, Aclar®, or filter membranes) and remaining freezer hats out of the substitution vessel (cryovial) and into a small glass Petri dish or other glass dish. This can be done by holding the cryovial at a right angle to the glass dish and flushing the vial with acetone from a Pasteur pipette, taking care not to let the substrate with cells dry out at any time.
   b. After the disks are all transferred to a glass dish, it is best to work under a stereomicroscope. Remove the aluminum hats from the dish, leaving behind the substrates.
   c. Orient the disks so that the cells are "up" or are facing the operator. When the disks have been shadowed with a finder grid pattern prior to cell plating (see Section "Carbon or Gold Coating") this is easily done by observing the correct orientation of the letters. Otherwise, the cell side can be determined by making a small scratch on either surface to see if cells come off. When all of the disks are in the proper orientation, complete the infiltration of the samples with epoxy resin.
   d. Prepare microscope slides by coating with PTFE Release Agent Dry Lubricant (MS-143E liquid, or MS-122 spray, Miller-Stephenson, Danbury, Connecticut). Let the coating dry thoroughly (about 10 min) and then wipe the slide clean with a lab wipe.
   e. Using a forceps, remove the adhesive liner of a Secure-Seal Imaging Chamber (Electron Microscopy Sciences, Cat. No. 70327-20S, Hatfield, PA, USA) and apply the spacer, adhesive side down, onto the prepared slide and press gently. Remove the other adhesive liner.
   f. Use a forceps to transfer the disks from the resin bath to the chamber, carrying over some resin from the forceps. You may put 1–5 disks per chamber.
   g. Cover the chamber with a Thermanox® plastic coverslip, 22 mm diameter (Electron Microscopy Sciences, Cat. No. 72282, Hatfield, PA, USA), or a piece of 7.8 mil Aclar® film (Electron Microscopy Sciences, Cat. No. 50425-10, Hatfield, PA, USA) and transfer to the oven to polymerize for 48 h.

j. Remounting for Oriented Sectioning Depending on how the disks were embedded in the previous section, the remounting procedure will be different.

Disks in flat-bottomed capsules: The block is first removed from the capsule, and then resin is trimmed away from the edges of the sapphire disk. After suspending the trimmed capsule/disk in liquid nitrogen vapor for 30 s or so it is possible to separate the disk from the resin with a pair of forceps. If correlative microscopy was done and only one cell is of interest, the block can be trimmed around that cell and sectioning can begin right away.

Disks in a thin layer of resin between slides:

a. Remove the slide chambers from the oven and cool completely to room temperature. Remove the Thermanox® or Aclar® cover and use a #11 scalpel blade to release the epoxy wafer (containing the sapphire disks) from the microscope slide.
b. Invert the wafer from its original orientation so that the cells are now facing "down". Use the scalpel blade to scrape off any resin from the back of the sapphire disk and work the scalpel around the edges of the sapphire disk to expose the edges of the disk. Gently pry the sapphire off the resin wafer. Often the disk will release in one piece but it may fracture into a few pieces. If using an Aclar® substrate (see Section III.F.2; Jimenez *et al.*, 2006, 2009), simply cut away resin from the edges of the Aclar® and peel it off with forceps.
c. Once the sapphire disk/ Aclar® is removed, the cells remain in the resin at the very surface of the now-inverted wafer. If using the finder grid shadowed disks, the grid pattern will transfer to the resin wafer, as well. You may tape this wafer to a microscope slide and scan its surface to locate a cell of interest.
d. Using a #11 scalpel blade cut a small square of resin containing the relevant area and mount it onto a blank resin block using remount glue, Epoxy 907 adhesive (Miller-Stephenson, Danbury, Connecticut). Resin squares can be oriented for either cross sections or for en face sections.
e. If using filter membranes (Morphew and McIntosh, 2003) there is no need to remove them since they can be sectioned.

k. Sectioning Because TC cells are generally only a few microns thick, sectioning them en face can be challenging. However, there are ways of shaping the block face and orienting it to the knife edge that permits rather easy sampling of cells from their very bottom nearest the substrate on up to their topmost regions. Trim the block as a rectangle with an aspect ratio of about 3 to 1, with the small side about 350–400 μm in size. When aligning the block surface to the knife edge prior to cutting, tilt the top of the block slightly forward of vertical so that the knife encounters this area first when sections are beginning to cut. The tilt angle can be only a few degrees so as you section, the cells nearest the bottom of the section represent the bottom of the cell, and as you move toward the top of the section, you will be seeing areas that represent more upper regions of the cell.

2. Aclar® Disks

a. History In the mid-1960s, investigators were searching for an alternative to glass coverslips as a substrate for TC cells that would be embedded in resin for EM. While there were some earlier reports with different plastics, it was the discovery of Aclar® 33C (Honeywell, Morristown, NJ) by Masurovsky and Bunge (1968, 1989) that revealed the desirable properties of this plastic that is widely used today. They noted that it was inert to chemicals routinely used for EM specimen preparation, that it was optically clear for light microscopy, including fluorescence, and that it stripped easily from polymerized resin without losing any cells.

Although Cook (2004) used Aclar® as a substrate for high-pressure freezing algal cells, it was Jimenez *et al.* (2006) who have fully explored the potential of this plastic for high-pressure freezing of cultured animal cells. We will summarize some aspects of their work in the sections that follow, but anyone interested in using Aclar® for HPF should consult this original paper.

b. Where to Buy Aclar® 33C is sold by most EM supplies vendors and it should be noted that it comes in two thicknesses: 2 and 7.8 mil. For many EM applications, researchers prefer the 7.8 mil thickness because it is less likely to curl and is easy to handle. However, for making high-pressure freezing disks it is important to use the 2 mil (51 μm) thickness.

c. Making Disks and Coating Jimenez *et al.* (2006) used an early model of the Wohlwend HPF machine (sold by Leica as the EM HPF) for freezing so the disks were 3 mm in diameter. They were punched out with a revolving punch plier available at hardware stores. To make 1.4 mm disks for the Leica EM PACT system may require a custom-made punch of that size. Most cells will attach and grow on Aclar® without any coating but some researchers prefer to put down fibronectin, collagen, or some other material that favors cell adherence and growth. As for carbon coating it is not necessary usually and if one wants to make an asymmetric figure in the plastic to show which side the cells are on, it is easy to scratch or cut the plastic for this purpose. In a more recent study (Jimenez *et al.*, 2009), the authors have shown that, for correlative microscopy using the Leica EM PACT system, one can also prepare disks with the carbon-coated pattern of a finder grid.

d. Culturing Cells on Disks A big problem with small Aclar® disks is their tendency to float in culture medium. To prevent this, Jimenez *et al.* (2006) used a hot wire to tack the disks to the bottom of plastic culture dishes. One may also be able to use a small dab of vacuum grease, agarose, or some other inert sticky substance to affix the cells to the bottom of the dish. This group also used microwave irradiation for 5 min at 800 W for sterilization.

e. High-Pressure Freezing and Freeze Substitution HPF can be carried out as described above in Sections 1.g.1 and 1.g.2 for sapphire disks. Jimenez *et al.* (2006) use 20% dextran as a cryoprotectant, but we prefer 20% BSA because, in our experience, dextran becomes brittle after freeze substitution and is more difficult to section. Freeze substitution could also be carried out as above for sapphire disks.

f. Embedding, Remounting, and Sectioning As above, except that removing the Aclar® disk is done simply by peeling it off. Just trim around the edge with a razor blade to expose the Aclar®–resin interface, insert a forceps under the Aclar®, and lift it off.

3. Other Methods

Most of the other methods for freezing TC cells can be found in published accounts so there is little need to reprint the details here. See the following techniques and references: filter membranes (Morphew and McIntosh, 2003); HPF specimen cups (Sawaguchi *et al.*, 2003); Formvar-coated gold grids (Ladinsky *et al.*, 1999); chips of Thermanox® plastic (Marsh *et al.*, 2001); Mylar® film (McEwen *et al.*, 1998); and microcarrier beads (Hagen, C., and Grünewald, 2008).

G. Separating Frozen Tissues from HPF Carriers under Liquid Nitrogen or during Infiltration and Embedding

Problem: Some cell types and/or fillers make it difficult to separate the biological sample from the specimen cup following HPF. Attempts to "dig out" the sample with forceps or other tools while in LN_2 can break the sample into small pieces. Failing to remove the sample may mean it has to be embedded in the metal carrier and separated after resin polymerization.

Solution: Early HPF work was often done with freeze fracture in mind and some special carriers were designed to create a "dome" of frozen material that could then be fractured. Coating the top piece of the HPF sandwich with lecithin dissolved in chloroform made it possible to remove it cleanly under liquid nitrogen, leaving the biological material in the bottom piece (Craig *et al.*, 1987). For today's carriers that are like simple cups, one can coat them with either lecithin or 1-hexadecene, and that aids in the removing the material from the cup in one piece under liquid nitrogen.

1. For two-piece Wohlwend-style carriers, coat both the cup and the top with a thin layer of 1-hexadecene prior to filling and freezing. Flat-sided carriers used as a "top" piece can be put flat-side down on a piece of filter paper saturated with 1-hexadecene and blotted just before putting onto the bottom piece and freezing. Use a small paint brush or similar tool to coat the insides of cups, then blot dry with paper.
2. For Leica membrane carriers, cups, or biopsy carriers, coat with 1-hexadecene and blot dry prior to filling and freezing.
3. After HPF and while under liquid nitrogen you can try to tease out the sample by gently inserting the tip of a needle, or the tip of a number 11 scalpel blade at the edge of the cup to dislodge the sample. This works best for whole pieces of tissue such as plant leaves or disks of cartilage. For many samples it may be best to proceed with the freeze substitution and try removing the sample during the resin infiltration steps. It is not uncommon for some types of samples, such as pellets of cells, or samples in 20% BSA to come out of the cups during freeze substitution, especially if there is some agitation of the tubes.
4. If you prefer to try lecithin, you can usually find lecithin powder at health food stores. Make a 100 mg/ml solution in chloroform, then paint it onto the surfaces that you want to release from the sample, and let it dry. Fill with material and freeze. Remove sample from the carriers as above for 1-hexadecene.

H. Freezing Brain and Other Soft Materials, or Pathogenic Samples

Problem: Some tissues, brain being one of the best examples, are so flaccid that taking out a small piece for freezing is extremely difficult without creating mechanical damage. A biopsy needle may be the best approach, but not everyone has them in the small size necessary, and even with this tool there may be damage.

We are also confronted at times with users who want to high-pressure freeze pathogens of one kind or another. This poses all kinds of risks, not least the fact that any material on the outside of the HPF carrier may be aerosoled in the LN_2 exhaust.

Solution: Fix the tissue before freezing. In particular, brain and other animal organ tissues should be perfusion fixed, not by dissection and immersion methods. After fixation, the tissue will harden significantly and it will be easier to take the smaller pieces necessary to fit into HPF specimen carriers. Use of a biopsy system after fixation should also be a valid approach. A relatively recent paper by Sosinsky *et al.* (2007) provides good examples of what can be achieved with this hybrid approach.

For pathogens, especially certain bacteria, do not assume that overnight fixation in glutaraldehyde will kill the cells. Have the researcher plate out cells after fixation to see if they are still viable. One of us (KM) has had the experience that TB cells (*Mycobacterium tuberculosis*) were fixed overnight in 2% glutaraldehyde but some remained viable the next day. For additional containment of cells, it may also be a good idea to concentrate them in sealed microdialysis tubing for freezing and subsequent handling. Finally, instead of fixing them in just glutaraldehyde prior to freezing, use a cocktail containing glutaraldehyde and a small percentage of osmium as the primary fix. After 20–30 min in this cocktail, transfer to glutaraldehyde fix only for the remaining fixation time.

I. Using Agarose as a Filler

Problem: Many cell types are very difficult to freeze. One of the reasons for this can be finding a filler to surround the cells during the freezing process that does not upset the cells at all. Agarose can be used as a very effective filler in low concentrations and most cell types do not seem to be affected by being suspended in it. Low-gelling-temperature agarose (e.g., Sigma A9045) will remain perfectly liquid at 37°C and exhibit great clarity allowing easy visualization of the sample within the gel even after freeze substitution. It will also hold the cells together during subsequent processing such as freeze substitution. The agarose has the ability to bind a high percentage of the water mainly through hydrogen bonding and so changes the freezing properties of this water. It is easy to test if cells are at all affected by the agarose by growing them in media containing agarose and observing if they grow and behave as normal. Percentages between 0.7 and 1% are sufficient to be used as a filler. It can be used as a filler in the place of hexadecane or yeast paste in all situations, e.g., surrounding cellulose microdialysis capillary tubing.

Solution:

1. It is important to keep all equipment coming in contact with the agarose warm to prevent it gelling prior to freezing. A heating block and a warming platform (or a microscope heating stage) can be used for this. Micropipette tips are stored in the heating block. The end of the rapid loader for the Leica EMPACT or the loading arm for the BalTec HPM010 or Wohlwend is rested on to the platform to keep it warm and the freezing hats or carriers are stored on the platform.
2. Make up agarose in the culture media in which the cells are growing. If using cells growing on sapphire or Aclar disks, use 1%, while if using free growing cells, use 2% as there will still be some liquid remaining with the cell pellet, which will dilute this down.
3. Melt agarose and keep at 37°C in heating block.
4. If cells are growing on disks, place a drop of molten agarose in a Petri dish on the warming platform. Remove the disk from the growth media and place it into the warmed agarose dipping it a couple of times to ensure it is surrounded by agarose. Place the disk into the freezing hat and adjust the level of liquid with paper points.
5. If the cells are free then they should be spun gently into a pellet. They should then be put into the heating block to keep warm and the supernatant removed. A volume of agarose equal to the size of the pellet should be added and this mixed thoroughly. A volume of this mixture is then added to the freezing hats or carriers.

Acknowledgments

The authors want to thank their colleagues in the high-pressure freezing community for the many fruitful discussions about techniques over the years. The origins of some of these "tips and tricks" are somewhat obscure by now and we are sure that many have been developed independently in different labs. We have tried to identify the origins as best we can, but no doubt some citations have been missed and we apologize to those who may have been unintentionally left out.

References

Cook, M. E. (2004). Cytokinesis in *Coleochaete orbicularis* (Charophyceae): An ancestral mechanism inherited by plants. *Am. J. Bot.* **91**, 313–320.

Craig, S., Gilkey, F. C., and Stähelin, L. A. (1987). Improved specimen support cups and auxiliary devices for the balzers high pressure freezing apparatus. *J. Microsc. (Oxford)* **148**(1), 103–106.

Echlin, P. (1992). "Low Temperature Microscopy and Analysis." Plenum, NY. (pp. 568).

Erlandsen, S. L., Bittermann, A. G., White, J., Leith, A., and Marko, M. (2001). High-resolution cryoFESEM of individual cell adhesion molecules (CAMs) in the glycocalyx of human platelets: Detection of P-selectin (CD62P), GPI-IX complex (CD42a/CD42bα,bβ), and integren GPIIbIIIa (CD41/CD61) by immunogold labeling and stereo imaging. *J. Histochem. Cytochem.* **49**, 809–819.

Hagen, C., and Grünewald. (2008). Microcarriers for high-pressure freezing and cryosectioning of adherent cells. *J. Microsc.* **230**, 288–296.

Hawes, P., Netherton, C. L., Mueller, M., Wileman, T., and Monaghan, P. (2007). Rapid freeze-substitution preserves membranes in high-pressure frozen tissue culture cells. *J. Microsc.* **226**, 182–189.

Hess, M. W., Mueller, M., Debbage, P. L., Vetterlein, M., and Pavelka, M. (2000). Cryopreparation provides new insight into the effects of brefeldin A on the structure of the HepG2 Golgi apparatus. *J. Struct. Biol.* **130**, 63–72.

Hohenberg, H., Mannweiler, K., and Mueller, M. (1994). High-pressure freezing of cell suspensions in cellulose capillary tubes. *J. Microsc.* **175**, 34–43.

Jimenez, N., Humbel, B. M., van Donselaar, E., Verkleij, A. J., and Burger, K. N. (2006). Aclar discs: A versatile substrate for routine high-pressure freezing of mammalian cell monolayers. *J. Microsc.* **221**, 216–223.

Jimenez, N., van Donselaar, E. G., de Winter, D.A.M., Vocking, K., Verkleij, A. J., and Post, J. A. (2009). Gridded aclar: Preparation methods and use for correlative light and electron microscopy of cell monolayers, by TEM and FIB-SEM. *J. Microsc.* **237**, 208–220.

Kaech, A. (2009). BAL-TEC HPM 010 high-pressure freezing machine. *In* "Handbook of Cryo-Preparation Methods for Electron Microscopy" (Cavalier, A., Spehner, D., and Humbel, B. M., eds.) pp. 101–128. CRC Press, Boca Raton, FL.

Ladinsky, M. S., Mastronarde, D. N., McIntosh, J. R., Howell, K. E., and Staehelin, L. A. (1999). Golgi structure in three dimensions: Functional insights from the normal rat kidney cell. *J. Cell Biol.* **144**, 1135–1149.

Marsh, B. J., Mastronarde, D. N., Buttle, K. F., Howell, K. E., and McIntosh, J. R. (2001). Organellar relationships in the Golgi region of the pancreatic beta cell line, HIT-T15, visualized by high resolution electron tomography. *Proc. Natl. Acad. Sci. U.S.A.* **98**, 2399–2406.

Masurovsky, E. B., and Bunge, R. P. (1968). Fluoroplastic coverslips for long-term nerve tissue culture. *Stain Technol.* **43**, 161–165.

Masurovsky, E. B., and Bunge, R. P. (1989). Aclar film in biological electron microscopy. *J. Electron Microsc. Tech.* **12**, 172–173.

McDonald, K. (1994). Electron microscopy and EM immunocytochemistry. *In* "*Drosophila melanogaster*: Practical Uses in Cell and Molecular Biology" (L.S.B. Goldstein and E. Fyrberg, eds.). *Methods Cell Biol.* **44**, 411–444.

McDonald, K. L. (1999). High pressure freezing for preservation of high resolution fine structure and antigenicity for immunolabeling. *Methods Mol. Biol.* **117**, 77–97.

McDonald, K. L. (2007). Cryopreparation methods for electron microscopy of selected model systems. *Methods Cell Biol.* **79**, 23–55.

McDonald, K. L., Morphew, M., Verkade, P., and Müller-Reichert, T. (2007). Recent advances in high pressure freezing: Equipment and specimen loading methods. *Methods Mol. Biol.* **369**, 143–173.

McDonald, K., and Müller-Reichert, T. (2002). Cryomethods for thin section electron microscopy. *In* "Guide to Yeast Genetics and Molecular and Cell Biology, Parts B and C." (C. Guthrie, and G. Fink, eds.). *Methods Enzymol.* **351**, 96–123.

McDonald, K. L., Sharp, D. J., and Rickoll, W. (2000). Preparing thin sections of *Drosophila* for examination in the transmission electron microscope. *In* "*Drosophila*: A Laboratory Manual" (W. Sullivan, M. Ashburner, and S. Hawley (eds.) 2nd edn., pp. 245–271, CSHL Press, Cold Spring Harbor, NY.

McEwen, B. F., Hsieh, C. E., Mattheyes, A. L., and Rieder, C. L. (1998). A new look at kinetochore structure in vertebrate somatic cells using high-pressure freezing and freeze substitution. *Chromosoma* **107**, 366–375.

Morphew, M. K., and McIntosh, J. R. (2003). The use of filter membranes for high-pressure freezing of cell monolayers. *J. Microsc.* **212**, 21–25.

Müller-Reichert, T., Mantler, J., Srayko, M., and O'Toole, E. (2008). Electron microscopy of the early *Caenorhabditis elegans* embryo. *J. Microsc.* **230**, 297–307.

Müller-Reichert, T., O'Toole, E. T., Hohenberg, H., and McDonald, K. L. (2003). Cryoimmobilization and three-dimensional visualization of. *C. elegans* ultrastructure. *J. Microsc.* **212**, 71–80.

Müller-Reichert, T., Srayko, M., Hyman, A. A., O'Toole, E., and McDonald, K. (2007). Correlative light and electron microscopy of early *Caenorhabditis elegans* embryos in mitosis. *Methods Cell Biol.* **79**, 101–119.

Reipert, S., Fischer, I., and Wiche, G. (2004). High-pressure freezing of epithelial cells on sapphire coverslips. *J. Microsc.* **213**, 81–85.

Reymond, O. L., and Pickett-Heaps, J. D. (1983). A routine flat embedding method for electron microscopy of microorganisms allowing selection and precisely orientated sectioning of single cells by light microscopy. *J. Microsc.* **130**, 79–84.

Sawaguchi, A., Yao, X., Forte, J. G., and McDonald, K. L. (2003). Direct attachment of cell suspensions to high-pressure freezing planchettes. *J. Microsc.* **212**, 13–20.

Schwarb, P. (1990). Morphologische grundlagen zur zell-zell interaction bei adulten herzmuskelzellen in kultur. Ph.D. thesis, Swiss Federal Institute of Technology, Zurich.

Sosinsky, G. E., Crum, J., Jones, Y. Z., Lanman, J., Smarr, B., Terada, M., Martone, M. E., Deerinck, T. J., Johnson, J. E., and Ellisman, M. H. (2007). The combination of chemical fixation procedures with high pressure freezing and freeze substitution preserves highly labile tissue ultrastructure for electron tomography applications. *J. Struct. Biol.* **161**, 359–371.

Studer, D., Graber, W., Al-Amoudi, A., and Eggli, P. (2001). A new approach for cryofixation by high-presure freezing. *J. Microsc.* **203**, 285–294.

Tiedemann, J., Hohenberg, H., and Kollmann, R. (1998). High-pressure freezing of plant cells cultured in cellulose microcapillaries. *J. Microsc.* **189**, 163–171.

Triffo, W. J., Palsdottir, H., McDonald, K. L., Lee, J. K., Inman, J. L., Bissell, M. J., Raphael, R. M., and Auer, M. (2008). Microaspiration for high pressure freezing: A method for ultrastructural preservation of fragile and sparse tissues for TEM and electron tomography. *J. Microsc.* **230**, 278–287.

Tse, F. W., Tse, A., Hille, B., Horstmannn, H., and Almers, W. (1997). Local Ca^{++} release from internal stores controls exocytosis in pituitary gonadotrophs. *Neuron* **18**, 121–132.

Vanhecke, D., and Studer, D. (2009). High pressure freezing Leica EM PACT. *In* "Handbook of Cryo-Preparation Methods for Electron Microscopy" (Cavalier, A., Spehner, D., and Humbel, B. M., eds.) pp. 129–156. CRC Press, Boca Raton, FL.

Verkade, P. (2008). Moving EM: The rapid transfer system as a new tool for correlative light and electron microscopy and high throughput for high pressure freezing. *J. Microsc.* **230**, 317–328.

SUBJECT INDEX

K

L

M

N

O

P

R

T

Y

Z

VOLUMES IN SERIES

Founding Series Editor
DAVID M. PRESCOTT

Volume 1 (1964)
Methods in Cell Physiology
Edited by David M. Prescott

Volume 2 (1966)
Methods in Cell Physiology
Edited by David M. Prescott

Volume 3 (1968)
Methods in Cell Physiology
Edited by David M. Prescott

Volume 4 (1970)
Methods in Cell Physiology
Edited by David M. Prescott

Volume 5 (1972)
Methods in Cell Physiology
Edited by David M. Prescott

Volume 6 (1973)
Methods in Cell Physiology
Edited by David M. Prescott

Volume 7 (1973)
Methods in Cell Biology
Edited by David M. Prescott

Volume 8 (1974)
Methods in Cell Biology
Edited by David M. Prescott

Volume 9 (1975)
Methods in Cell Biology
Edited by David M. Prescott

Volume 10 (1975)
Methods in Cell Biology
Edited by David M. Prescott

Advisory Board Chairman
KEITH R. PORTER

Volume 22 (1981)
Three-Dimensional Ultrastructure in Biology
Edited by James N. Turner

Volume 23 (1981)
Basic Mechanisms of Cellular Secretion
Edited by Arthur R. Hand and Constance Oliver

Volume 24 (1982)
The Cytoskeleton, Part A: Cytoskeletal Proteins, Isolation and Characterization
Edited by Leslie Wilson

Volume 25 (1982)
The Cytoskeleton, Part B: Biological Systems and *In Vitro* Models
Edited by Leslie Wilson

Volume 26 (1982)
Prenatal Diagnosis: Cell Biological Approaches
Edited by Samuel A. Latt and Gretchen J. Darlington

Series Editor
LESLIE WILSON

Volume 27 (1986)
Echinoderm Gametes and Embryos
Edited by Thomas E. Schroeder

Volume 28 (1987)
***Dictyostelium discoideum:* Molecular Approaches to Cell Biology**
Edited by James A. Spudich

Volume 29 (1989)
Fluorescence Microscopy of Living Cells in Culture, Part A: Fluorescent Analogs, Labeling Cells, and Basic Microscopy
Edited by Yu-Li Wang and D. Lansing Taylor

Volume 30 (1989)
Fluorescence Microscopy of Living Cells in Culture, Part B: Quantitative Fluorescence Microscopy—Imaging and Spectroscopy
Edited by D. Lansing Taylor and Yu-Li Wang

Volume 31 (1989)
Vesicular Transport, Part A
Edited by Alan M. Tartakoff

Volume 32 (1989)
Vesicular Transport, Part B
Edited by Alan M. Tartakoff

Volume 33 (1990)
Flow Cytometry
Edited by Zbigniew Darzynkiewicz and Harry A. Crissman

Volume 34 (1991)
Vectorial Transport of Proteins into and across Membranes
Edited by Alan M. Tartakoff

Selected from Volumes 31, 32, and 34 (1991)
Laboratory Methods for Vesicular and Vectorial Transport
Edited by Alan M. Tartakoff

Volume 35 (1991)
Functional Organization of the Nucleus: A Laboratory Guide
Edited by Barbara A. Hamkalo and Sarah C. R. Elgin

Volume 36 (1991)
***Xenopus laevis:* Practical Uses in Cell and Molecular Biology**
Edited by Brian K. Kay and H. Benjamin Peng

Series Editors
LESLIE WILSON AND PAUL MATSUDAIRA

Volume 37 (1993)
Antibodies in Cell Biology
Edited by David J. Asai

Volume 38 (1993)
Cell Biological Applications of Confocal Microscopy
Edited by Brian Matsumoto

Volume 39 (1993)
Motility Assays for Motor Proteins
Edited by Jonathan M. Scholey

Volume 40 (1994)
A Practical Guide to the Study of Calcium in Living Cells
Edited by Richard Nuccitelli

Volume 41 (1994)
Flow Cytometry, Second Edition, Part A
Edited by Zbigniew Darzynkiewicz, J. Paul Robinson, and Harry A. Crissman

Volume 42 (1994)
Flow Cytometry, Second Edition, Part B
Edited by Zbigniew Darzynkiewicz, J. Paul Robinson, and Harry A. Crissman

Volume 43 (1994)
Protein Expression in Animal Cells
Edited by Michael G. Roth

Volume 44 (1994)
***Drosophila melanogaster:* Practical Uses in Cell and Molecular Biology**
Edited by Lawrence S. B. Goldstein and Eric A. Fyrberg

Volume 45 (1994)
Microbes as Tools for Cell Biology
Edited by David G. Russell

Volume 46 (1995)
Cell Death
Edited by Lawrence M. Schwartz and Barbara A. Osborne

Volume 47 (1995)
Cilia and Flagella
Edited by William Dentler and George Witman

Volume 48 (1995)
***Caenorhabditis elegans:* Modern Biological Analysis of an Organism**
Edited by Henry F. Epstein and Diane C. Shakes

Volume 49 (1995)
Methods in Plant Cell Biology, Part A
Edited by David W. Galbraith, Hans J. Bohnert, and Don P. Bourque

Volume 50 (1995)
Methods in Plant Cell Biology, Part B
Edited by David W. Galbraith, Don P. Bourque, and Hans J. Bohnert

Volume 51 (1996)
Methods in Avian Embryology
Edited by Marianne Bronner-Fraser

Volume 52 (1997)
Methods in Muscle Biology
Edited by Charles P. Emerson, Jr. and H. Lee Sweeney

Volume 53 (1997)
Nuclear Structure and Function
Edited by Miguel Berrios

Volume 54 (1997)
Cumulative Index

Volume 55 (1997)
Laser Tweezers in Cell Biology
Edited by Michael P. Sheetz

Volume 56 (1998)
Video Microscopy
Edited by Greenfield Sluder and David E. Wolf

Volume 57 (1998)
Animal Cell Culture Methods
Edited by Jennie P. Mather and David Barnes

Volume 58 (1998)
Green Fluorescent Protein
Edited by Kevin F. Sullivan and Steve A. Kay

Volume 59 (1998)
The Zebrafish: Biology
Edited by H. William Detrich III, Monte Westerfield, and Leonard I. Zon

Volume 60 (1998)
The Zebrafish: Genetics and Genomics
Edited by H. William Detrich III, Monte Westerfield, and Leonard I. Zon

Volume 61 (1998)
Mitosis and Meiosis
Edited by Conly L. Rieder

Volume 62 (1999)
Tetrahymena thermophila
Edited by David J. Asai and James D. Forney

Volume 63 (2000)
Cytometry, Third Edition, Part A
Edited by Zbigniew Darzynkiewicz, J. Paul Robinson, and Harry Crissman

Volume 64 (2000)
Cytometry, Third Edition, Part B
Edited by Zbigniew Darzynkiewicz, J. Paul Robinson, and Harry Crissman

Volume 65 (2001)
Mitochondria
Edited by Liza A. Pon and Eric A. Schon

Volume 66 (2001)
Apoptosis
Edited by Lawrence M. Schwartz and Jonathan D. Ashwell

Volume 67 (2001)
Centrosomes and Spindle Pole Bodies
Edited by Robert E. Palazzo and Trisha N. Davis

Volume 68 (2002)
Atomic Force Microscopy in Cell Biology
Edited by Bhanu P. Jena and J. K. Heinrich Ho¨rber

Volume 69 (2002)
Methods in Cell-Matrix Adhesion
Edited by Josephine C. Adams

Volume 70 (2002)
Cell Biological Applications of Confocal Microscopy
Edited by Brian Matsumoto

Volume 71 (2003)
Neurons: Methods and Applications for Cell Biologist
Edited by Peter J. Hollenbeck and James R. Bamburg

Volume 72 (2003)
Digital Microscopy: A Second Edition of Video Microscopy
Edited by Greenfield Sluder and David E. Wolf

Volume 73 (2003)
Cumulative Index

Volume 74 (2004)
Development of Sea Urchins, Ascidians, and Other Invertebrate Deuterostomes: Experimental Approaches
Edited by Charles A. Ettensohn, Gary M. Wessel, and Gregory A. Wray

Volume 75 (2004)
Cytometry, 4th Edition: New Developments
Edited by Zbigniew Darzynkiewicz, Mario Roederer, and Hans Tanke

Volume 76 (2004)
The Zebrafish: Cellular and Developmental Biology
Edited by H. William Detrich, III, Monte Westerfield, and Leonard I. Zon

Volume 77 (2004)
The Zebrafish: Genetics, Genomics, and Informatics
Edited by William H. Detrich, III, Monte Westerfield, and Leonard I. Zon

Volume 78 (2004)
Intermediate Filament Cytoskeleton
Edited by M. Bishr Omary and Pierre A. Coulombe

Volume 79 (2007)
Cellular Electron Microscopy
Edited by J. Richard McIntosh

Volume 80 (2007)
Mitochondria, 2nd Edition
Edited by Liza A. Pon and Eric A. Schon

Volume 81 (2007)
Digital Microscopy, 3rd Edition
Edited by Greenfield Sluder and David E. Wolf

FSC
www.fsc.org
MIX
Papier | Fördert
gute Waldnutzung
FSC® C083411

Kolibri 360 GmbH
Ferdinand-Jühlke-Straße 7
99095 Erfurt, Deutschland
produktsicherheit@kolibri360.de